U0224510

我的论文选评

求索与感悟

廉慧珍

中国建材工业出版社

图书在版编目（CIP）数据

我的论文选评 求索与感悟/廉慧珍．—北京：
中国建材工业出版社，2015.6
ISBN 978-7-5160-0991-8

Ⅰ．①求… Ⅱ．①廉… Ⅲ．①混凝土-研究
Ⅳ．①TU-528

中国版本图书馆 CIP 数据核字（2014）第 238294 号

内 容 简 介

本书是廉慧珍教授 40 余年从事建材专业工作经验的总结和智慧的结晶，其中阐述的道理和方法具有很高的参考、借鉴价值，值得各行业的人士阅读和学习。

本书纵贯廉慧珍教授的学习经历、毕业后从业的辗转、教学与科研的过程，以及退休后在工程实践中的研究和思考。全书包括四个部分：“走出初级阶段研究的过程”、“想创新，就要站在前人的肩膀上”、“对高性能混凝土的理解、推广和反思”、“思维方法与观念的转变比技术更重要”，共选择 54 篇有代表性的学术与科技论文，介绍当时发表的背景，用现在和认识和观念加以评述，指出当时认识的局限和不足及错误和需要进一步研究的问题。观点鲜明，反思坦诚。从每一篇论文都能让读者纵观建材行业半个世纪来的发展概况，对今后建材的研发和生产工作具有广泛的指导意义和深远的影响。在本书的结束语中，廉慧珍教授没有再对全书做什么结论，只使用了 20 页的文字提出了若干意义重大的研究课题，实属罕见，具有前瞻性，对当前和未来科技研究及学科的发展都有很大的启发。

我的论文选评
求索与感悟
廉慧珍

出版发行：中国建材工业出版社
地　　址：北京市海淀区三里河路 1 号
邮　　编：100044
经　　销：全国各地新华书店
印　　刷：北京雁林吉兆有限公司印刷
开　　本：787mm×1092mm　1/16
印　　张：34
字　　数：836 千字
版　　次：2015 年 6 月第 1 版
印　　次：2015 年 6 月第 1 次
定　　价：98.00 元

本社网址：www.jccbs.com.cn　　微信公众号：zgjcgycbs
本书如出现印装质量问题，由我社网络直销部负责调换。联系电话：（010）88386906

献给

热爱混凝土事业的人们

序

作者是一位在业内德高望重、年逾80且仍在工程技术领域服务的先生。这是一本前所未有的在混凝土材料科学与工程技术领域的专著。阅读时我欲罢不能，掩卷时心情激动、难以平静！

在本书中，读者会处处感受到，先生德高望重却不居高临下，而是向后辈们敞开心扉，坦诚分析自己作为一个研究者的成长历程、走过的弯路，剖析自己在追求真理、攀登知识高峰时的不足、局限甚至认知错误；先生是一位激情燃烧、永不停步的追求科学真理的先行者，在无涯的学海中，像一座高耸的灯塔，燃烧着自己，照亮着热爱混凝土事业的跟随者；先生是一位跨混凝土材料、结构与工程技术领域的大师级专家，向我们传授其近50年从事教学、科研和工程技术服务所积累的宝贵知识和经验。

我也已近花甲，在国内外从事研究30多年，却从来没读过像这样别具一格的专业著作。这是一本教科书，你可以看到，为师者，答疑解惑，诲人不倦。对一些基本概念的建立，特别是对一些业内广为流传的错误概念（例如“高性能混凝土”）的纠正，不遗余力；这是一本严肃的学术著作，你可以看到，为学者，读书而知不知，实践而知不足。在本书的最后一章，先生提出了许多目前尚未可知的重要学术和技术问题，为后人攀登知识高峰，解决工程实际问题指出了方向；这是一本教你怎样做科研的书，先生以其对沸石岩和沸腾炉燃煤固硫渣作为水泥混合材的研究为例，讲述了思维方法与观念比技术更重要的道理，分析了什么是分解论研究方法，什么是整体论研究方法、什么是系统论研究方法，以自己的深刻感悟为后人攀登高峰提供了巨人肩膀；这甚至是一本关于怎样做人的书，做事要先做人。为人者，不仅贵有自知之明，更贵有公开解剖自己的勇气。先生在自评部分毫不客气地指出自己过去研究工作的不知、不足、局限乃至错误认知。

这本书内容之丰富、时间跨度之大，一时间，作为后辈的我甚至不知该怎样写这篇序。1974年当先生从事高强混凝土研究时我还是一个懵懂无知的插队知青。这本书虽然只是选择性地介绍了先生在长达50年从事教学、科研和工程实践的一些工作，但这些方面确是反映了中国水泥混凝土材料科学与工程技术的发展历程。在这里我愿意向读者介绍我阅读本书的一些收获，与大家分享。

本书应用对立统一、量变到质变、内因与外因的关系等哲学思想，建立科学研究的思维方法和观念，强调“正确的技术决策取决于正确的观念，正确的观念取决于正确的思维方法”。这一科研方法在其对沸石岩火山灰活性的研究中得到充分的展示。先生挑战传统思维，在科学理论的指导下，通过多学科系统严谨的研究，最终纠正了错误的观点，建立了对反应机理的科学认知。

本书深入分析总结了在分解论和整体论方法指导下研究工作的利弊，指出要采用系统论方法研究混凝土这一看起来简单实际非常复杂的系统，通过科学地划分子系统，建立子系统行为的对立统一规律，在复杂纷繁的影响下把握主要矛盾和解决问题的方向与关键措施。先生以实验室研究结论和工程实际观察结果的对比，批评了许多过去似是实非的结论，这对于当代中国混凝土科技工作者是一剂清醒药，不仅有助于理解过去的研究和结论的局限性，而且对于今后的研究工作具有重要的指导意义。

本书指出，混凝土是一种用简单技术制作的高度复杂的体系。从材料到结构，从实验室到工程，所涉及的学科领域、工程结构、环境条件以及非技术性因素非常之多。显然，科研工作者必须具备多学科跨学科的知识、对技术和工程问题的通识能力，才能准确地发现问题、认识问题，制定科学合理的研究方法。本书所介绍的研究工作，从微观结构到宏观性能、从普通混凝土到高性能混凝土、从胶凝材料到混凝土结构、从结构设计与材料设计到工程施工、从混凝土学科发展的历史沿革到今天的知识体系，反映了先生深厚的跨界通识，对问题的分析不仅深入浅出，而且广征博引，不断拓展读者的眼界和深度。

为了帮助读者建立创新思维，先生明确指出要打破三种思维定势：传统思维（或习惯性思维）、从众思维和机械思维，要建立辩证思维。先生例举了思维定势问题在技术标准和规范中不能与时俱进的表现——“用老方法评价新事物”，例如矿渣的含碳量指标问题，膨胀混凝土试件的养护方式问题，实验室混凝土收缩值的测定问题，掺混合材后混凝土的强度发展特性与现行 28 天定强度的矛盾问题，等等。如果有人说在实验室检测耐久性能好的混凝土在混凝土结构中提前劣化可能是因为施工质量或别的非技术因素所致，但面对实验室检测性能差的混凝土在现场却有很好的表现时，则不能不承认是测试评价方法的问题，例如混凝土的碳化试验、抗冻融试验、氯离子侵入试验等。先生强调，试验方法必须尽量符合现场实际情况，如果试验方法与材料特性有关的话，材料发展变化了试验检测和评价方法则必须相应改变。先生不仅直言不讳，针砭时下的科研学术不端，而且专门用了一大篇章来指导创新：“想创新，就要站在前人的肩膀上”。先生告诉我们，科研工作没有捷径，不能取巧，必须老老实实学习掌握前人的工作，才能创新。

先生一再强调，“任何个人，乃至作为人类，认识世界的知识体系——科学，都不会永远正确和绝对正确”，应着重于建立科学的思维方法和分析解决问题的能力。在阅读中，读者一定会受到启发，深入思考，提高对一些问题的认识。我自己在几个方面就深受启发。

例如对于高性能混凝土的认识。先生对我国高性能混凝土的发展作了历史性的回顾，包括当初词语的翻译、概念的建立，以及在发展过程中产生的种种认识误区，如误以为高性能就一定要求高强度等。先生指出，高性能混凝土不是只要有配合比就能得到的，必须从原材料控制、拌和物生产和整个施工过程的协同努力才能实现。阅读时我感悟到，要学会站在不同的方面看问题。结构工程师、施工工程师、材料工程师所理解的高性能混凝土是有所不同的。上世纪 80 年代，高性能混凝土其实是相对于普通混凝土存在的问题而提出和开发的。虽然迄今仍未有统一的定义，但内涵逐渐厘清，简单说是相对于普通混凝土难以达到的更高性能。例如，更高的强度（如果需要的话），更容易浇筑成型，且不离析、不开裂，具有更好的耐久性，混凝土结构在各种环境中都能完好服役所要求的时间。如果再加上绿色的要求，则具有更小的环境负荷（在混凝土生产使用中的碳足迹最轻，建筑物废弃后混凝土能高值重复利用、资源化再生）。这四个方面“更高、更易、更好、更小”的要求基本上可以概括为绿色高性能混凝土的内涵。

再如对混凝土耐久性的认识也受到启发。耐久性是对结构和材料在一定的服役环境中抵御环境侵蚀，持续满足安全性和使用功能的要求。可以说，服役中的所有材料性能都是耐久性的函数，即都会随时间演变。不同的服役环境中建筑物所要求的耐久性不同。因此，笼统地一般性地说耐久性是没有意义的，必须予以明确。例如，材料强度，没有耐久性保证的强度是没有意义的。先生多次指出 28 天后强度性能的演变的重要性。对于混凝土结构来说，耐久性首先要求的是混凝土结构的整体性，即不开裂。这对材料性能设计、施工工作性设计和施工技术都提出了很高要求。

当我掩卷之时，不禁想，是什么力量让先生有这样大的勇气，为我们后人提供了这样一本独一无二的学术著作？是她对真理求索的使命感。一个受人敬重的大师，拿起实事求是这把手术刀，毫不留情地解剖自己的工作，这是什么表现？这是无私，是大公，是大德！我想，我们也许不会也不必同意先生的每一个学术观点，但是，我相信每一个读过这本书的人都会被那颗追求真理的赤诚无私之心、为先生一生为混凝土事业的奉献所感动、所激励。

为了中国的混凝土和水泥制品事业，先生近十年来花了很大的精力帮助我在协会做好服务混凝土企业和引领行业发展的工作，受惠之多，无以回报！当得知先生编写此书且很多单位要求支持此书的出版时，我要求让协会组织出版发行，除了此书的重要价值外，也希望以此作为对先生的一点感恩回报，承蒙先生同意，此书即将付梓。至于此书出版的意义，古人曰，桃李不言，下自成蹊。

中国硅酸盐学会理事长

中国混凝土与水泥制品协会会长 徐永模

2015.3.10

自　　序

虽有嘉肴，弗食不知其旨也；虽有至道，弗学不知其善也。是故学然后知不足，教然后知困。知不足，然后能自反也；知困，然后能自强也。故曰：教学相长也。　　　　——《礼记·学记》

学而不思则罔，思而不学则殆。　　　　——《论语·为政》

援引上述经典，是想说明，做为一个教师，一生中所做的事就是学而不止，教而不倦；越学越觉得所知甚少，就越想学；教的过程才知道自己有许多尚不明白之处。任何个人，乃至作为人类，认识世界的知识体系——科学，都不会永远正确和绝对正确，而是无限趋近于真理。正所谓学无止境，因而需要教学相长，经常反思，以不断补充，不断纠正。尽管这似乎人人皆知，但是重要的是实践。在此愿与大家共勉。

我是一个普通的大学教师。1953 年从华东区考入清华大学土木工程系工程测量专业。当年土木工程系有 6 个专业，招收 14 个班：工业与民用建筑结构（重点培养工程结构设计与研究人才，2 个班），工业与民用房屋建筑（重点培养房屋建筑施工技术的人才，4 个班），以及工程测量土木系、公路工程、给水与排水、暖房与通风各 2 个班。入学后第二年继续 1952 年的院系调整，把工程测量、公路工程两个专业分别合并到同济大学和测绘学院。一年级学生过去，我们二年级学生就地转到木土系其他专业和水利系水工结构专业。我被转到工业与民用房屋建筑专业，在原来 4 个班以外增设了一个“房 85 班”（按毕业年份编号。意即 58 年毕业的第 5 班。那时的“工业与民用房屋建筑专业”和现在的“工业与民用建筑”专业不同，我们比“工业与民用建筑结构”和现在的“工民建”专业多上了“施工技术”、“施工机械”、“施工组织计划”等课程）。其他同学分到其他专业的增设班。我在学生时期因做过两年的半脱产政治辅导员，而延迟于 1959 年正式毕业，工龄从 1958 年计算。这就是我的履历上为 1958 年毕业，而毕业证书是 1959 年颁发的缘故。

毕业后留校，在“十年动乱”以前，我一直从事校刊编辑工作；1972 年要求回系，并毅然选择了一个当时“冷门”的专业——建筑材料。这时，我的专业水平落后于我的同班、同级同学 14 年，而且对我来说，尽管大学期间所学的建筑材料课程达 100 多学时，而对混凝土材料专业来说，也毕竟是皮毛。从专业基础的物理化学、硅酸盐物理化学、岩石与矿物学、胶凝材料学、x 射线衍射学、现代仪器分析等，到专业知识的胶凝物质工学、水泥工艺学原理、混凝土学、混凝土制品工艺学等等，都从头学起，边做——按教学和科研需要学习和研究，边学——听课和自学，以自学为主。教学给我宽度和广度；科研给我深度。为了给学生一碗水，自己要准备一桶水。更重要的是又进入一个新的天地，使我的头脑中产生很多很多我想知道的事物、很多很多我想得到解答的问题。

曾任国家教委主任的何东昌说过，一个人的知识只有 20% 是在学校获得的，其余都是在以后得到的。（当然这里的 20% 只是一个概念，并非定量。）在学校里所学的课程，有很多后来都忘了，那么在学校期间为什么费那么大的劲去理解，去记忆？上学为了什么呢？这是很多年轻人都会困惑的问题，我也曾困惑过。在毕业后的工作中，每次遇到什么问题时，常会感到“似曾相识”，仿佛在哪一门课中学到过，例如涉及力学上的问题，是在物理课学的？还是理论力学？材料力学？还是……？只从自己大脑“储存器里”提取不出来时，就去图书馆、资料室查询、再阅读复习。因为目的明确，就比当初上课时学得更深刻些。当时的努力理解和记忆变成了以后再学习的一种能力。在学校学习时如果只是单纯记住所学的知识，那就不进校门自己看书就是了。然而其实还是不一样：在学校里，老师通过讲授基本知识来引导我们如何学习和思考。学校的教育是“打基础”。这个“基础”，一是基本概念；二是能力，而能力比知识更重要。

“能力”有丰富的内涵，其中首先就是获取知识的能力。而必须有发现问题的能力，才能有针对性地去获取知识。这就是“带着问题学”。随处都有我的老师，都有需要学习的知识。我的成长受益于我周围所有的人——我的亲人、老师、朋友、学生，……；到处都是课堂——书报、期刊、互联网、电视、会议交流……。这不是套话，有很多故事可以说明这种真实的感受。举一个例子：有一次出差，同伴是一个比我的儿子还小的年轻的朋友，他一路上都在对我说：“你们这些老先生，自己概念不清，还用一些错误的概念去误导别人。这是在害人!”我感到能这样对我说话，是真正对我好。尽管他骂的并不一定就是我，却对我有很大的启发，使我常常反省所说过的话有那些概念性的错误。有时别人不经意的一句话，都可能对我是一次启发。我越来越感到我的所知实在太少，摆在我们面前的和我们没想到而发生的问题太多。探索、研究真是一件美妙的事。知识的来源可以说是无处不在，而读书无疑是最重要的，但是读书也要会读，重要的是思考，不能囫囵吞枣。读书不思考，也有可能“中毒”。而思考才能发现问题，然后又对问题进行思考。读书之后经过实践去验证，在实践中再读书，就可能会有更深的领会，使经验达到一个高度。一知半解就会断章取义，往自己片面的经验上套。经验很宝贵，而经验主义却会害死人。

有人曾问我：“你如果当初毕业后一直从事专业而不去做行政工作，是否会比现在更有成就?”我连想都不想地回答说：“未必”。人生如旅，经常会遇到十字路口，不同的选择会得到不同的结果，总要选择一条路。不同的人有各自不同的选择理由。我们老校长蒋南翔对毕业生说过：“不要怕改行，改行是革命者的常规”。我们总说“想做什么就能做什么才能快乐。”现在想来，重要的是社会需要。人不能脱离社会而一个人活着，那么，就不能脱离社会的需要而自行其是地“想做什么就做什么”。每个人都在为自己生存而做事的同时，也在为别人的生存提供条件和服务，这个社会才能可持续地良性发展，每个人也才能得以生存。这就是社会需要。不论选择了做什么，只要用心去好好做，就会不断发现问题而促使我们去思考，去研究，总会有新的认识和成果；我们谁都并不是天生就喜欢混凝土，现在热爱混凝土，是因为用心去做后，发现混凝土太有意思了。有一位做混凝土的老板对我说，他发现做混凝土就像做人一样的有意思，他要把混凝土做得和别人的不一样。经过几年的努力，他果然做得“与众不同”。尽管为此而遭妒，受到残忍的打击，他仍然痴心不改。因为他热爱混凝土这项事业。正像有一位朋友所说：“不是因为喜欢才选择，而是因为选择才喜欢。”这句话的意义深长。相信如果当时选择了别的，进入角色后，同样也会喜欢。这正是我中学母校——上海敬业中学的校训所说的“敬业乐群”。不管做什么，都会不断地积累——经验、知识、能力的积累。不管是什么专业或者什么学科，在方法和观念上都是相通的，这些积累都是可以相互借鉴的，只是对象不同而已。其间的桥梁就是哲学。从这个意义上来说，不管曾经从事过什么工作，换个看似不相干的职业，过去的积累都有用，而且面会更宽。

除了在学校的教学外，我的科研经历大体有三段。退休前是一段，主要承担过海军后勤部委托的1000＃混凝土的研制（1974年）、国家教委项目“钢筋混凝土海上采油平台可行性研究”中的“高强流态混凝土”以及茂名石油公司委托“页岩灰用于高强流态混凝土”的研究（1984年～1986年），完成了两个科学基金和两个国家重点科技攻关专题的研究。两个科学基金中的一个是现在的国家自然科学基金的前身中科院自然科学基金（“结晶态铝硅酸盐物质活性机理的研究”1984年），另一个是国家建材行业科学技术发展基金（“水泥浆体微结构与宏观行为的关系及其定量分析”1989年）；在国家“七五”和“八五”重点科技攻关专题中，主持并负责了有关循环流化床锅炉燃煤固硫废渣建材资源化的系统研究。其中“七五”攻关的研究被课题组（专题的上游组织）负责人张绪祎评价为“完成的成果出色”，与课题形成“高效低污染循环流化床锅炉脱硫及灰渣资源化成套技术”，获国家计委、国家科委、财政部颁发的国家“七五”科技攻关重大成果表彰荣誉证书；“燃煤固硫渣制水泥”获国家教委、国家环保局、中国科学院颁发的国家“七五”科技攻关重大成果表彰荣誉证书。得以享受国务院政府津贴。

今年，我退休已整整20年，前几年如果算作是第二段的话，主要与企业合作，结合工程进行了一些技术性的研究。深感混凝土的教学与科研只有结合工程才能不断发现问题而具有生命力。在为工程服务的同时，经过工程的实践、学习和在朋友们的相助下，对混凝土有了较深的接触，才发现我们对混凝土其实很不了解，对混凝土的认识犹如对人的身体和生命的认识一样的贫乏，常有很多为之困惑的问题渴望答案。不出校门时，可以按教科书把混凝土做到符合预计的效果；进入工程后越来越发现，影响混凝土最终质量的因素实在太复杂，其中大量的并非技术问题更难解决。近10年来，可以说是第三阶段，主要是从根本上对混凝土材料、混凝土工程以及混凝土结构进行思考和讨论。20年来，混凝土和混凝土结构工程发生了很大的变化。变化是必然的，思维方法和观念的转变比技术更重要。这个观点缘于从30多年前我们教研组两位老师那里得到的启发。其中从化学教研组转来的张淑清在实验研究中经常强调“方法最重要”；另一位是从日本回来的李桂芝，当试验结果出现异常时，她总是首先从方法上找原因。当然她们所说的方法主要是试验的方法，这对于实验研究确实是最重要的。有一位有经验的工程师曾质疑说，“方法比技术重要这种说法不对，先有技术才有方法。技术比方法重要”。这是对“方法”的狭义理解。技术是解决问题的策略、工艺、步骤和手段，等同于英文中的Technology；方法有不同的层次，他所理解的“方法”恐怕就是最低层次的方法，即解决问题的技巧、技能、使用的工具和操作技艺，等同于英文里的Skill；也应包含在技术中。而广义地说的方法，是为解决问题的经常性的和系统性的、详细的、逻辑有序的计划和行为，通常是指为完成某事的途径（或者叫路线），即英文中的Methodology。思维方法属于哲学方法的范畴，是指导一切科学的最高层次的方法，又是科学研究的基础，科学研究则是技术的基础。思维方法决定观念。对于混凝土这种高度复杂的非均质多相体系，如果在观念上有偏颇甚至错误，就会误导工程技术的决策。概括地说，技术是“术”，而方法是“道”。“道”是对“术”的引领。因此在近10年来，我的论文大都涉及思维方法和观念，这也是我最重要的一点感悟。

朋友们要求我出一本论文集，我之所以迟迟未果于行动，是因为回顾和反思，在我曾发表的100多篇论文中，不乏某些错误的或偏颇的认识，实在没有什么无瑕之作能“拿得出手”。审视之余，我想，往者无需悔，来者犹可追，任何事物都不会完美无缺。探索无止境，法国著名作家左拉说过：“生命的全部意义在于无穷地探索尚未知道的东西。”不断思考才会有所感悟。每一次感悟都会有新知，从中选出一部分，用现在的认识和观念加以审评，又有感悟。这是一种乐趣。把这种乐趣奉献给大家共享，也可作为对所有帮助过我和关心我的人们的一种回报吧。

对这本集子会有各种不同的看法，我很希望能有不同看法来丰富我的思想，并再纳入我的知识中。一个人只要做事，总会有人说好话也有人说坏话，这才正常。如果只有好话而没有坏话，那其中必有假话；如果只有坏话没有好话，说明这个人恐怕真的是个坏人了。当然也有另外一种情况：按照“真理往往掌握在少数人手里”的规律，也可能这个人太超前了，大多数人不理解，而少数理解的人不敢说真话，这也是正常现象。比如1815年，23岁的俄国数学家罗巴切夫斯基大胆质疑欧氏几何的平行公理，当时只有高斯理解他，但因有忌于众非而不表态，直到罗巴切夫斯基死后12年，意大利数学家贝尔特拉米找到了适用于非欧几何的曲面，才使其他数学家也能理解。当然我绝没有这样的道行。至于学术观点的不同意见，则更是非常正常的。欢迎评论，希望通过讨论互相收获。

2014.6

收稿日期：本书稿截止2014.6.

目　　录

前言 …… 001

第一部分　走出初级阶段研究的过程 …… 005

一、对沸石凝灰岩在水泥中作用机理的研究 …… 005

论文之一　沸石岩在水泥中的作用机理探讨 …… 007

论文之二　MORPHOLOGY OF INTERFACE BETWEEN ZEOLITE-TUFF AGGREGATE AND CEMENT PASTE …… 013

论文之三　CEMENT PASTE / ZEOLITE－TUFF BOND …… 020

论文之四　EFFECT OF STRUCTURE AND COMPOSITION ON REACTIVITY OF ZEOLITE-TUFF USED AS BLENDING MATERIAL FOR PORTLAND CEMENT …… 028

论文之五　沸石岩火山灰活性的研究 …… 035

二、啃一下别人啃剩下的硬骨头 …… 044

论文之六　MECHANISM OF HYDRATION OF BLENDED CEMENT ——I. Strengthening Effect of Mineral Powder on Cement Used in Mortar or Concrete …… 047

水泥浆体微结构与宏观行为关系的研究及定量分析研究工作报告 …… 054

论文之七　硬化水泥浆体微结构的定量分析及其与抗压强度的关系“水泥浆体微结构与宏观行为关系的研究及定量分析 …… 061

第二部分　想创新，就要站在前人的肩膀上 …… 069

一、选这部分论文的理由 …… 069

二、循环流化床燃煤脱硫技术的优势及其废渣资源化的依据和攻关成果 …… 071

论文之一　因势利导，利用流化床脱硫燃煤含硫废渣 …… 081

论文之二　固硫渣中硫的形态与特性 …… 088

论文之三　沸腾炉燃煤固硫渣活性评价及其影响因素的研究 …… 097

论文之四　固硫渣中的 SO_3 对水泥的调凝作用 …… 106

论文之五　固硫渣对水泥中过量 SO_3 作用的抑制 …… 112

论文之六　流化床燃煤固硫渣用于补偿收缩混凝土的可行性 …… 117

论文之七　固硫渣双掺复合硅酸盐水泥的强度复合效应 …… 123

论文之八　对固硫渣中的碱、SO_3 和其他微量元素的评价 …… 130

论文之九　火山灰质材料活性的快速评定方法 …… 138

第三部分　对高性能混凝土的理解、推广和反思 …… 146

概述 …… 146

论文之一　高效能混凝土（HPC）…… 150
论文之二　高效能混凝土实现的可行性——国内外 HPC 材料研究现状及问题 …… 155
论文之三　高强与高性能混凝土的配制及施工质量控制和验收　Ⅰ.原材料和配合比对高强与高性能混凝土性能的影响…… 165
论文之四　无防冻剂的高强混凝土冬季施工的研究和实践…… 175
论文之五　国内外自密实高性能混凝土研究及应用现状…… 182
论文之六　自密实高性能混凝土施工性能的评价方法…… 193
论文之七　养护温度对低水灰比的掺粉煤灰外加剂砂浆强度增长的影响…… 201
论文之八　高性能胶凝材料的实验研究之一——混凝土的可持续发展与高性能胶凝材料…… 207
论文之九　高性能胶凝材料的实验研究之二——FK 系列高性能水泥的性能及其检测…… 213
论文之十　高性能胶凝材料中石膏的优化…… 220
论文之十一　高性能混凝土在深圳地铁一期工程中应用研究的工程试点裂缝控制 …… 226
论文之十二　混凝土施工中裂缝的控制…… 237
论文之十三　对“高性能混凝土”十年来推广应用的反思…… 248
论文之十四　对“高性能混凝土”的再反思…… 257

第四部分　思维方法与观念的转变比技术更重要…… 265

一、关于思维方法和观念…… 268
论文之一　水泥基复合材料科学研究中的辩证思维…… 268
论文之二　混凝土结构耐久性的系统分析…… 278
论文之三　思维方法和观念的转变比技术更重要之一 ——打破专业藩篱是时代发展的需要和必然…… 288
论文之四　思维方法和观念的转变比技术更重要之二——传统思维和从众思维对混凝土技术进步的影响…… 295
论文之五　思维方法和观念的转变比技术更重要之三——再谈从众思维…… 305
论文之六　思维方法和观念的转变比技术更重要之四——对标准、规范的认识和使用…… 312
论文之七　思维方法和观念的转变比技术更重要之五——性能检验表现的差异缘于对变化了的材料使用了不变的方法…… 322
论文之八　思维方法和观念的转变比技术更重要之六——混凝土工程中的分解论和整体论…… 332
论文之九　面对混凝土工业可持续发展的挑战…… 342
二、关于混凝土的原材料与配合比的思考…… 349
论文之十　水泥的品质和混凝土质量的关系…… 349
论文之十一　外加剂和混凝土工程质量的关系…… 357
论文之十二　砂石质量是影响混凝土质量的关键…… 367
论文之十三　矿物掺和料由谁掺好？…… 376

论文之十四 关于混凝土配合比选择方法的讨论之一——当前混凝土配合比“设计”存在的问题 …… 384

论文之十五 关于混凝土配合比选择方法的讨论之二——关于当代混凝土配合比要素的选择和配合比计算方法的建议 …… 391

论文之十六 关于混凝土配合比选择方法的讨论之三——体积法计算混凝土配合比中原材料密度的问题 …… 400

论文之十七 评《普通混凝土配合比设计规程》(JGJ 55—2011)—— 兼谈如何认识和编制及使用技术标准和规范 …… 407

三、关于回弹法和混凝土强度的关系问题 …… 420

论文之十八 质疑“回弹法检测混凝土抗压强度” …… 420

附件：文恒武、魏超琪：对“质疑‘回弹法检测混凝土抗压强度’”一文中几个问题的看法 …… 425

论文之十九 对文恒武、魏超琪二位同志《对“质疑‘回弹法检测混凝土抗压强度’”一文中几个问题的看法》的讨论 …… 429

四、主要的不是技术问题 …… 437

论文之二十 中国混凝土向何处去？——写在中国混凝土与水泥制品协会预拌混凝土分会成立之际 …… 437

论文之廿一 谁对工程质量负责？ …… 451

论文之廿二 对建材利废观念和行为的思考 …… 461

论文之廿三 解决混凝土结构工程质量的根本途径——变革混凝土工程的生产关系 …… 469

论文之廿四 从水泥和混凝土的关系对企业转型升级的思考 …… 481

结束语 我还有什么问题希望研究？ …… 489

附录 廉慧珍论文选评《求索与感悟》编者按 …… 509

科学求真 高层建瓴——写在廉先生《求索与感悟》出版之际 …… 梁文泉 509

耕耘不懈的恩师益友——回顾与廉先生在一起的日子 …… 陈恩义 511

路漫漫兮其修远——读廉老师《求索与感悟》感言 …… 江加标 513

春风化雨 润物无声——写在《求索与感悟》出版之际 …… 宋少民 515

我向您致敬——真正的行业领军人物和值得敬佩与追随的老师 …… 李玉琳 517

平实求真 厚德载物——读廉慧珍老师《求索与感悟》有感 …… 韩小华 518

思想和行动上的巨人 …… 师海霞 519

我认识的廉老师 …… 郭保林 520

科学的诚实 …… 杨雄利 521

致广大而尽精微 …… 张大康 522

前　言

在这本集子中所选论文有三类：第一类是学术性的研究；第二类是工程技术性的研究，包括工程技术的基础性研究；这两类常常是不可分的。学术性研究是工程技术研究能动性的基础，工程技术性研究是生产实践能动性的基础；反之，工程技术研究为学术性研究提供问题，生产实践又为技术性研究提供条件、数据和信息。生产实践才是科技研究的源泉和最终目的，尽管由于生产整体水平所限而使这个目的未必能在短期实现。第三类主要是关于方法和观念的论述。这正是我在研究、学习和为工程服务（合作研究、咨询、事故诊断等）中的不断求索和感悟。从总体上来看，第三类应该是更重要的，也是比较有意义的，所以选入的论文最多。最终，所选论文分成四部分：

第一部分主要是较早时期的基础性探索，欲用以说明我是怎样开始从研究的初级阶段走出的；我一开始在教学之余的科研，主要是在带领工农兵学员开门办学时，针对工程需要做一些试验，例如那时发展装配式建筑，检验预制外墙板拼接时所用胶粘剂的选择和检测方法；框架轻板体系建筑的无梁空心板、柱材料和构件的设计与检验等。在 1000＃混凝土研究和后来的高强流态混凝土的研究中，也只是做配制试验——改变配合比，检测强度。那时很投入，所想的就是如何能降低水灰比，以满足强度要求。这样一些实验对于我来说，只是一种初级的探索阶段，只是原材料选择、按包罗米公式计算水灰比，查表选择砂率和用水量，强度试验，最多加上结构检验，没有问题的分析，没有科学假设，没有技术路线，属于基本的实验，只能写出简单的实验报告。现在这部分所选论文是属于两个科学基金项目的结果成果。其中对沸石岩机理的研究，在申请科学基金之前的论文明显属于“初级阶段”而未选入；论文之一尚未完全脱离初级阶段，仍然是做对文献结论证实的试验。对机理研究，主要的还是一种推测。虽然采用现代仪器分析的手段表明了沸石岩与消石灰、石膏混合加水后确实发生了反应，确认了水化产物形貌，根据沸石的特性提出含沸石矿物的沸石岩特殊的物理作用的观点，而试验量尚不足。从之后的论文可以看出，逐渐形成科学研究方法和学术观点：分析和提出问题→针对问题提出研究结果的预想和技术路线→科学假设→用物理、化学以及物理化学方法的试验设计→有根据的结果分析和结论。这部分的另一项科学基金的研究则已形成和运用了这一科学研究方法。并且再一次证明，传统的物理方法和化学方法有时可以更直观和有效地解决问题，而先进测试的技术如果目的不明确，或掌握不得法，只能作为论文的一种点缀。技术是为目的服务的。例如对沸石岩的研究，证明在水泥浆体中沸石微孔的水气交换作用所用的试验，除了物化测试外，用超过沸石热稳定性的温度破坏沸石结构（铵离子交换量为 0，沸石晶体对 x-射线衍射的特征峰消失）的试样，进行相同条件下的强度的对比试验，就是有效的；又如，对水泥浆体微结构与宏观行为关系的研究，在用 x-射线定量分析微结构元时，所使用的斜发沸石岩无天然纯样，也无法人工合成，最后使用了铵离子交换量法可靠地检测出沸石岩中斜发沸石晶体的含量，再通过差减法将沸石晶体和凝灰岩中的无定形组分分别计入 H（hard）粒子和 L（light）粒子中，使试验得以继续进行。

第二部分是“七五”和“八五”期间国家重点攻关专题研究，试图用以作为总结我对科学研究方法的思考与实践，说明科技创新的思路和传承的意义；在这部分可见我的研究方法是基础研究和工程技术结合。两个攻关专题都是围绕着沸腾炉（后来发展成循环流化床）燃煤脱硫废渣资源化的研究，其关键就是要把固定在废渣中的硫化害为利，并高值化。因此首先分析研究渣中硫的形态和特性及其对渣活性的影响，最后有针对性地用到工程中去。为20多年后的今天重启循环流化床发电灰渣利用的研究提供借鉴。

第三部分是退休后投身于工程所做工程技术性的研究，主要是从介绍不同国家对高性能混凝土（当时译成高效能混凝土）涵义的不同理解和应用开始，发表我们对对高性能混凝土的理解和观点。在工程推广中所涉及的混凝土主要有：掺粉煤灰的高强混凝土（包括不用防冻剂的冬季施工），自密实混凝土，膨胀剂的使用，合成纤维的应用，大掺量矿物掺和料混凝土，“高性能混凝土专用水泥的开发”等等。不带上“高性能混凝土”的帽子，只从工程的需要来说，确实积累了一定的经验，并逐渐对混凝土有了更进一步的认识。经过10多年的实践，人们已经逐渐接受“高性能混凝土”的提法，并且有了自己不同的理解。作为理念，如果对“高性能混凝土”不当成一个品种而正确理解和实施，对混凝土及其结构工程的进步是会有促进作用的。但由于没有真正理解“Performance”的涵义，出现不少问题，不禁令人困惑。为此我反复学习和思考：过去我们在教学中都是讲混凝土的性质（properties）。著名英国混凝土专家A. Neville具有重要影响的著作《Properties of Concrete》，第一版印刷过三次，我国李国泮和马贞勇将其第三版译成中文，使用的书名为《混凝土的性能》。当时我们在阅读时，没有去想他们为什么这样翻译？“性能”和“性质”有什么区别？在翻译High Performance的时候盲目听取了别人的意见，把“高效能混凝土”改为大家容易接受的“高性能混凝土”，当成了一个混凝土的新品种。现在“高性能混凝土”出现的各种问题，究竟是这项技术本身有问题，还是由于理解的误区所造成的？于是两次反思，特选入与大家共享。

第四部分是近年来对我在工程实践中感悟的论述。这部分所选论文最多，按同类归成四组内容：

第一组是在工程中遇到和处理问题时所思考和感悟的总结，主要是关于与思维方法和观念的论述。思维方法中最根本的是辩证思维的方法，用联系的、变化的辩证方法进行思维，就会随着客观事物的发展而更新认识、转变观念，取得技术的进步。除了选入曾发表过的6篇系列文章——思维方法和观念的转变比技术更重要之一～之六外，另选入较早关于对水泥基材料中辩证关系认识的文章，以分析思想的来源与发展。

第二组是关于现代混凝土配合比设计的见解。现代混凝土应该是指广泛使用高效减水剂的混凝土。有人说现代混凝土的特点是使用高效减水剂和矿物掺和料，其实，是高效减水剂的使用为矿物掺和料的大量掺用创造了条件。从根本上来说是因为外加剂的广泛使用改变了混凝土的一切——改变了水泥强度和混凝土配合比与性质的关系；因为可以降低水胶比，使混凝土能掺用较多矿物掺和料而不影响强度，因而需要改变人们对水泥的理念；扩大了混凝土强度和流动性的范围，因而扩大了混凝土应用的范围；适应并促进了施工技术的发展；等等。同时，外加剂使用，为技术人员进行混凝土的配合比设计与试配，提供了发挥自己的智慧和才能的更大空间。只要目的明确和正确，遵循混凝土配合比设计的原理与原则（应当由规范规定），他们可以“八仙过海，各显其能”。事实也证明如此。文章中介绍的关于法律、法规和标准、规范、指南、手册等技术性的文件概念的科学理解，是对论文第四部分之六的补充。对以上所有的不足之处，均有自评。

第三组针对当前在工程中影响很大的“用回弹法检测混凝土强度”问题进行根本性的讨论。

第四组选入几篇不是技术而对技术有重要影响的问题，和大家讨论。

最后提出我对科学技术研究中还想研究的问题，希望能抛砖引玉。科学技术研究不能“没事找事”，不能为了完成出论文的任务而凭想象制造出一些题目来“研究”，研究必须明确意义和目的，写论文应当是有感而发，而不能“凑份子”。为了说明此意，我在每一部分都首先交代的论文的背景和来源，有选择性地用现在的认识进行一些反思和分析，真诚地希望与大家讨论。今天反思过去，明天再来反思现在，才能总是前进。

不同论文中用于从不同角度说明不同问题引用的参考图，会有一些的重复。希见谅。

未来属于两种人：思想的人和劳动的人。实际上这两种人是一种人，因为思想也是劳动。

——雨果

只有人们的社会实践，才是人们对于外界认识的真理性的标准。真理的标准只能是社会的实践。

——毛泽东

在学术上没有权威或宗师这回事——这些只是仰慕者对他们的称呼；我们不要被名气吓倒了。任何高手都可以错，所以他们的观点或理论也只能被我们考虑及衡量，不可以尽信。当然，高手的推论较为深入，值得我们特别留意。我们应该对高手之见做较详尽理解，较小心地去衡量。但我们不可以为既是高手之见，就是对的。高手与低手之分，主要就是前者深入而广泛，后者肤浅而狭窄。

——经济学家张五常新浪微博《思考的方法》

本书所选论文都保留原样，只对错别字、严重的语病、图表和文字对应等编辑上的问题做修改。在自评中，基本上只对学术和技术的观点、水平问题加以评论。

第一部分　走出初级阶段研究的过程

一、对沸石凝灰岩在水泥中作用机理的研究

1979年我校建材教研组与北京地质研究所合作，共同进行天然沸石岩用做建筑材料的实验研究，在当时的国家教委科技司立项。科研组先后由5人组成，分别以沸石岩用做水泥混合材及其应用、沸石岩载体引气混凝土、膨胀沸石岩轻骨料及其混凝土等为主进行分工合作的研究。我在参加各课题研究的同时，以沸石岩在水泥中作用机理的研究为主。1979年11月在苏州的全国新型建材及试验性建筑技术交流大会上代表科研组发言《膨胀沸石岩轻骨料的特性》，被当时同济大学老教授黄蕴元评价为“在国内第一次注意到混凝土中界面问题”。1980年应教研组要求，发表关于沸石岩作用的第一篇论文《沸石岩在水泥中的作用机理探讨》。

现在从这篇论文看来，当时的研究尚属于“初级阶段”——对别人的研究结论的证实。当然，科学研究允许证实，也允许证伪，但是证伪容易，证实难；因为“如果看见100只白天鹅，不能下结论说天鹅都是白的，而如果看见一只黑天鹅却可以说天鹅不都是白的”（英国卡尔. 波普尔 Karl Raimund Popper）。那时还没有这样的觉悟。当时在我国，沸石岩刚刚被发现，只因沸石岩在化学成分上和硅酸盐水泥的相似，就尝试将其磨细以不同掺量做硅酸盐水泥混合材的试验，发现在一定掺量下有一些增强作用。按传统观点，只是笼统地认为是沸石与水泥水化生成的$Ca(OH)_2$发生了火山灰反应。确实，在得到很有限的信息中，国内外的有关报道，都是“沸石和水泥水化生成的$Ca(OH)_2$反应”。于是在一开始时就接受了这个观点。所谓初级阶段，就是由于从众思维的指导，简单地接受别人的观点，作简单的配制和性质的试验；走出初级阶段则需要在现象的背后发现本质，发现问题，进行思考，提出自己的观点和命题，首先经得起自己的反驳，使命题成立。

在低水平上简单重复别人的研究成果难以创新，只有科学地挑战传统才能发展。这里选的几篇论文反映了当时的试验、思考和感悟的过程。这个过程概括地说是：

——初级阶段的误区：接受别人关于沸石火山灰性的观点，简单地证实；

——思考和提出问题：对意大利学者的提问“何以结晶态的沸石有比火山灰高的活性?”进行思考：①这个问题实际上是一个命题：“结晶态的沸石有比火山灰高的活性”和“相同物质的结晶态是热力学稳定的”这一规律相悖，要么热力学错了，要么这是一个伪命题；②沸石岩是什么？活性是从哪里来的?

——调研沸石的结构和特性：证明沸石岩就是意大利沸石化的凝灰岩。火山灰烬沉积成凝灰岩，其中的无定形火山灰质物质长期在水、碱金属离子、压力和温度作用下，逐渐结晶变成架状含水铝硅酸盐矿物沸石。沸石岩含有不同种类沸石。提纯的或人工合成的沸石可以做分子筛和催化剂、吸附剂、干燥剂等，还可以利用其离子交换的性质在海水中提取钾，净化水等。所有以上的用途中沸石的结构都保持不变，而且既耐酸又耐碱。

——形成观点的命题：沸石不可能具有火山灰活性，沸石岩活性来自无定性的凝灰岩。

——进一步的问题：凝灰岩的活性为什么会因沸石的存在而提高？

——第二个命题：沸石特殊结构的作用。

人的智慧掌握着三把钥匙：一把开启教学，一把开启字母，一把开启音符。知识、思想、幻想就在其中。

——雨果

我要做的只是以我微薄的绵力来为真理和正义服务。

——爱因斯坦

近代科学的目标是什么？就是探求真理。科学方法可以随时随地而改换，这科学目标，蕲求真理也就是科学的精神，是永远不改变的。

——竺可桢

论文之一

沸石岩在水泥中的作用机理探讨*

清华大学建筑材料教研组　廉慧珍

试验表明，天然沸石岩粉磨至一定的细度，按一定比例掺入硅酸盐水泥，可提高水泥强度，并改善其他性能。在提高水泥产量、节省能源、调节水泥性能方面有着广阔的前途。

本世纪初已有人将沸石岩当作火山凝灰岩用于水泥的活性混合材料[1]，但是"直到现在，在美国，对沸石矿物在活性掺料中所起的特殊反应仍未完全了解"[1]，意大利 R. Turriziant 认为沸石和 $Ca(OH)_2$ 作用形成水化硅酸钙和 C_4AH_8[2]；日本专利报道，用斜发沸石或丝光沸石同石灰混合成料浆经高压水热反应，可生成雪硅钙石和硬硅钙石[3]；南斯拉夫 B. Držaj 等人则认为沸石同石灰反应的早期，生成 C-S-H，后期 C-S-H 转变为雪硅钙石[4]；中国郭竞雄、梁春林研究证明结晶质的沸石岩与石灰反应的活性比玻璃态 SiO_2 的活性高，并提出"脱铝—解硅—吸钙多阶段化学反应的模式"[5]。从现有资料来看，人们对沸石岩的活性是肯定的，但关于沸石岩对水泥特殊作用的资料和沸石岩与石灰反应产物的显微形态研究尚未见报道。

我们在研究沸石岩水泥强度规律及其物理力学性能的同时，进行了较大量的扫描电镜观察和 x-射线衍射、红外吸收光谱、差热分析的测定，并探讨了沸石岩因其结构特性而在水泥水化时所产生的特殊物理作用。

一、沸石岩同石灰反应的产物

为了简化沸石岩在水泥复杂的水化过程中的作用，用斜发沸石岩与消石灰、石膏混合加水成型，在标准条件下养护 7 天和 28 天进行测定。

扫描电镜观察　在沸石岩同消石灰混合物水化 28 天的试样中，布满宽度为 0.1～0.2 微米、长度为 1 微米左右的蠕虫状水化物(图 1)。经微区 x-射线能谱测定，确定为水化硅酸钙相(C-S-H)。在沸石岩和消石灰、石膏混合物水化 7 天、28 天的试样中，均可见到上述 C-S-H 相和六棱柱状水化物互相穿插(图 2)。经能谱测定为水化硫铝酸钙相(Aft)见图 3，并有较多的硅进入。这可能是由于沸石岩中的高硅含量所致。

x-射线衍射分析　在隔绝 CO_2 的条件下，将上述试样做 x-射线衍射分析。在沸石岩和消石灰混合物水化 7 天的试样中(图 4)，沸石和 $Ca(OH)_2$ 峰都很明显，水化 28 天后，则 $Ca(OH)_2$ 的强峰几乎消失，沸石峰高下降，出现 C-S-H 很弱的宽峰。这是因为 $Ca(OH)_2$ 掺量很少，28 天后几乎都参加了反应。在沸石岩和消石灰、石膏的试样中也有上述现象，但由于石膏的存在，还出现 AFt 相特征峰(图 5)。

差热分析　以上两种试样水化 28 天时，在差热图上都有 880～890℃的 C-S-H 放热峰，掺有石膏的试样还出现 125℃的 AFt 相吸热峰(图 6)。

---

* 原发表于《（1980、1981 年）水泥学术会议论文选集》，中国建筑工业出版社，1981 年第一版。

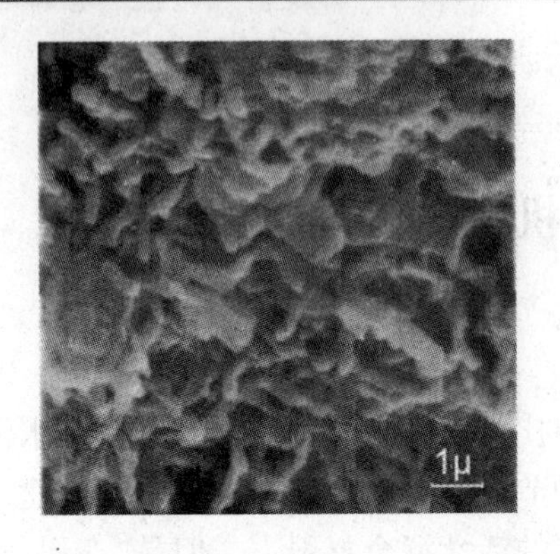

图 1　沸石岩和消石灰混合物水化 28 天扫描电子显微镜图×10000

图 2　沸石岩和消石灰、石膏混合物水化 28 天扫描电子显微镜图×5000

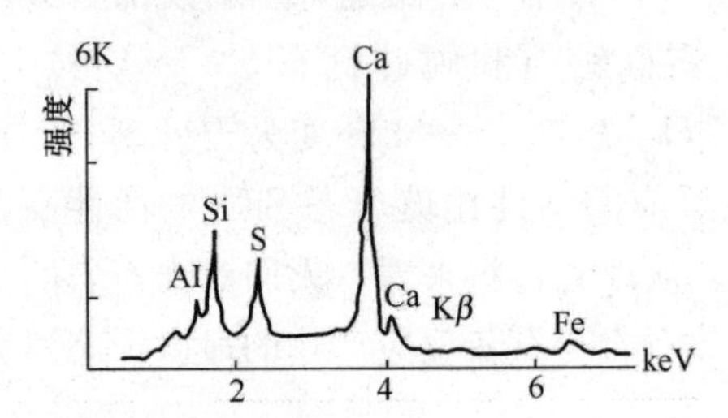

图 3　沸石岩、消石灰、石膏混合物水化 28 天能谱图

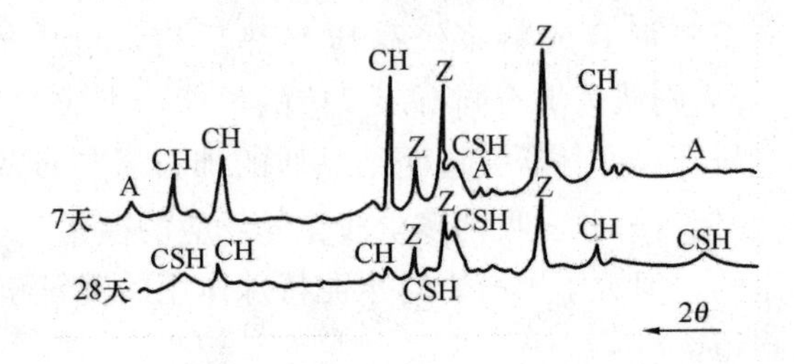

图 4　沸石岩和石灰混合物水化试样 x-射线衍射
E—AFt 相；Z—沸石；CH—$Ca(OH)_2$相；
CSH—C-S-H 相；A—含铝水化物相

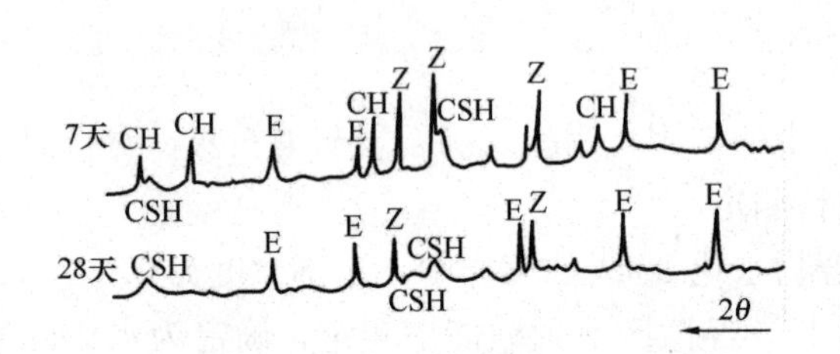

图 5　沸石岩和消石灰、石膏水化试样 x-射线衍射图　图中符号同图 4

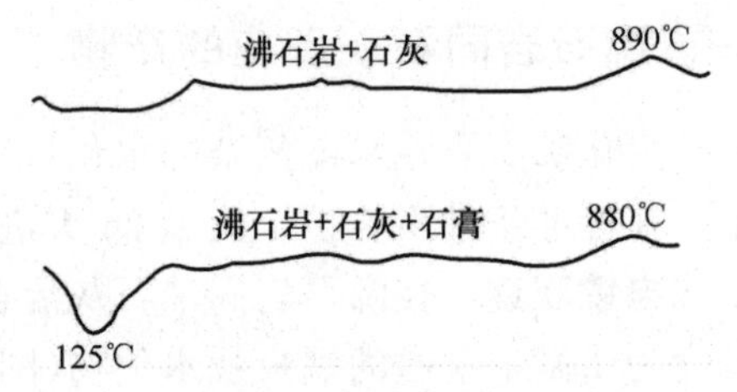

图 6　沸石岩和消石灰以及沸石和消石灰、石膏混合物水化 28 天差热分析

综合上述分析可见，沸石岩和消石灰加水混合可反应生成 C-S-H 相和含铝相，当有石膏存在时，含铝相与石膏反应生成 AFt 相。由于沸石岩中 SiO_2 含量大，使 AFt 相中有较多硅存在。在纯熟料水泥中掺入不同量的沸石岩进行差热和 x-射线衍射分析，可得到同样结论。

二、沸石岩在水泥浆中特殊的物理作用

沸石岩掺入水泥，其主要的化学反应是和水泥水化生成的 $Ca(OH)_2$ 进行二次反应，生成 C-S-H 相，以及和 $Ca(OH)_2$、石膏反应生成含硅的 AFt 相。但是化学组成通过结构而决定物质的性质[6]。如 SiO_2 活性的差别，往往是由于结构和孔隙等物理性的不同，而不是由于化学组成的差异[7]。因此对于相同组成的水泥浆来说，决定其宏观特性的，主要是其内部

结构。传统的观点认为，以无定形和玻璃态为主的 SiO_2 物质比晶态的 SiO_2 活性高。但是晶态的沸石却具有很高的活性。相反，将沸石岩经高温烧熔水淬成玻璃态与消石灰混合成型，在常温下，其水化速度极慢，标准养护 28 天几乎不具有机械强度。这说明晶态物质在一定的结构状态下可能比无定形玻璃具有更高的活性。固体物质的物理状态可以决定强度[8]。

1. 沸石结构的作用

沸石结构的特点是由 Si-O 四面体组成的具有大量多种形式的开放性孔穴和孔道的架状构造。其内表面积达近千平方米/克[9]。在常温饱和蒸气下，沸石微细空腔可吸附水约 100mg/g。这种水可自由脱、附而不影响结构，称为沸石水。

由于沸石结构的特点，使沸石岩在水泥浆中有如下的作用：

(1)“微粉效应”：将磨细的沸石岩掺入水泥浆，对水泥颗粒及其水化物起分散作用，加大水泥水化的空间，促进水泥中、后期的水化反应。其他活性混合材料如粉煤灰也有这种“微粉效应”[10]，沸石由于其巨大的比表面积，这种作用更为突出。

(2)吸附：沸石结构中一部分 Si-O 四面体由 A1-O 四面体取代而形成电性不饱和，因而有结合较弱的阳离子存在。多孔结构的沸石磨细后，由于静电力和色散力的共同存在，具有强烈的吸附作用。在水泥浆中吸附 $Ca(OH)_2$，并同其反应，降低 $Ca(OH)_2$ 的浓度，又促进水泥的继续水化。

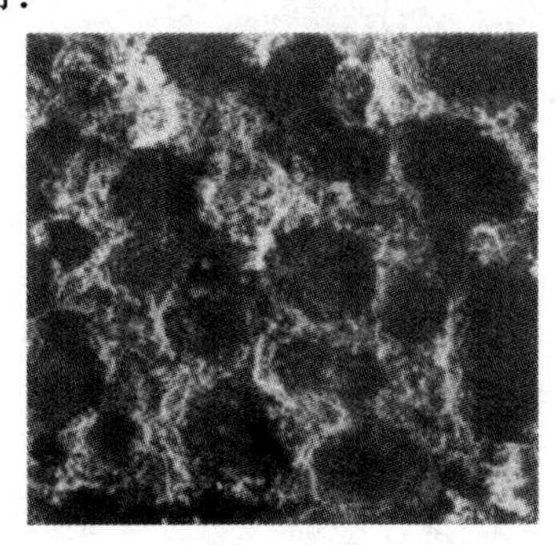

图 7　水泥浆体与沸石岩界面气孔形貌(水泥面)×50

(3) 沸石水：沸石岩掺入水泥后，其微孔很快地吸附水泥浆中的自由水，同时将微孔中原来吸附的少量空气排出(可由奥萨特空气分析仪测出)，使沸石岩颗粒与水泥浆界面薄弱。这是沸石岩水泥早期强度较低的一个原因。将沸石岩磨平、抛光，埋入水泥浆中，养护 28 天后，断开粘结面，在扫描电镜下观察，可见到在其界面所形成的气孔形貌(图 7)。但是，由于沸石岩吸收了水分，降低了浆体水灰比，却有利于水泥后期强度的发展。随着水化的进行，浆体中自由水分减少后，沸石微孔中的水又逐渐放出，对水泥水化起自养护作用，使界面不断密实加强，提高了水泥石的强度。

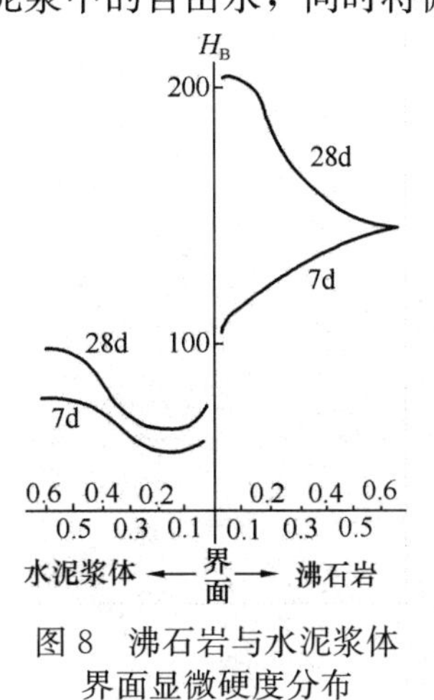

图 8　沸石岩与水泥浆体界面显微硬度分布

将沸石岩颗粒与水泥浆混合成型，养护 7 天和 28 天，表面抛光，测定沸石岩与水泥浆界面两侧显微硬度分布如图 8 所示。水化 7 天的浆体，由于界面的影响，靠近水泥和沸石界面的硬度低于沸石岩的硬度；水化 28 天，靠近界面处的硬度显著提高，超过沸石岩本身的硬度。表明界面随水化的进行而逐渐得到加强。

为了证明沸石结构及沸石水的作用，将沸石岩在 800℃高温下烘烧 30 分钟，使其结构破坏而不熔融。这时，测定其比表面积为 0.09m²/g，比未烘烧的沸石岩比表面积(19.55m²/g)①，降低 99.5%；铵离子交换量为零；用 x-射线衍方法测定，晶胞参数 c 由

① 用氮气吸附法测定 0.6～1.25mm 颗粒比表面积。所用沸石岩中沸石含量为 60%。

15.95 Å 变成 16.56Å。这些都说明这些烘烧过的不同结构沸石岩对水泥强度的影响，烘烧后的沸石岩不再具有沸石结构特性。以相同量掺入纯熟料水泥中成型，测定其不同龄期的软练强度进行对比，其结果列于表 1。

表 1　不同结构的沸石岩对水泥强度的影响

编　　号	龄　期		
	7 天	28 天	90 天
	水泥硬练抗压强度(kg/cm²)		
30%烘烧	160	250	404
30%未烧	179	386	488

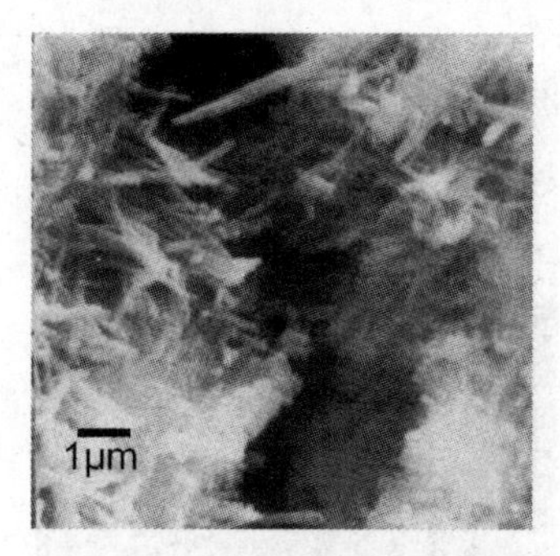

图 9　沸石岩颗粒与水泥浆界面裂缝扫描电子显微镜图像，龄期 28 天图中左侧为水泥浆体，右侧为沸石岩

由表 1 可见，烧沸石岩所配制的水泥强度比生沸石岩所配制的水泥强度低；沸石岩掺量越大，影响越大，当掺量为 20%时，烧沸石水泥 7 天强度为生沸石岩水泥强度的 87%，而掺量为 40%时，则下降为 79%。这说明沸石岩水泥强度的提高，是由于其结构和沸石水的作用。

在扫描电镜下观察沸石岩颗粒与水泥浆的界面(图 9)，可见界面裂缝被填充的情况。

2. 改善水泥浆显微颗粒级配

硬化水泥浆是由不同尺寸、形态的各种水化物、不同大小和形状的孔、孔隙中不同状态的水化以及未水化颗粒所组成的非均质多相体系。这些“颗粒”混合的随机性以及级配的情况，对硬化水泥浆的宏观性质起着决定性的作用。从宏观来看，在相同工艺和配合比的情况下，混凝土不同粒径的骨料级配好时，就可以得到较高的机械强度。水泥从亚微观和微观上类比，可看做是各种“骨料”粒径都在 10nm～几个 μm 数量级的“微集料”混凝土。加拿大 Feldman 和 Beaudoin 用“水泥浆孔隙率-颗粒类型-强度关系”说明硬化水泥浆微结构和强度的关系[11]。水泥的水化产物固然是影响水泥浆宏观性质的主要因素，但是决定水泥浆强度的因素则是水泥浆中的孔和不同形态、比重、结晶度、大小的各种粒子的级配。水泥中未水化颗粒作为最大的“微集料”参加这种级配。在硅酸盐中加入一定量的磨细沸石岩，促进水泥中、后期水化，使未水化水泥颗粒的尺寸减小、数量减少而沸石岩本身未水化颗粒则作为“微集料”代替原来未水化的水泥颗粒，参加水泥浆的颗粒级配，沸石岩同 $Ca(OH)_2$ 二次反应的产物又作为新的粒子参加反应。当沸石岩掺量适当时(如 20%左右)，作为“微集料”的未反应沸石岩颗粒和沸石岩反应产物被水泥水化物所包围，互相穿插，紧密结合，在扫描电镜下观察长龄期试样，可见结构紧密，处理试样时的打击裂缝清晰可见。当沸石岩掺量增加到 50%时，在扫描电镜下可见到明显形态的 C-S-H 和 AFt 相水化物，并极少有打击的裂缝，说明为粒子间断裂(图 10)。这种水泥的强度比 20%沸石岩水泥和纯熟料水泥的强度低。这是因为沸石岩掺量大时，沸石岩反应产物和未反应沸石岩颗

图 10　50%沸石岩水化 93 天断面×500

粒过多，使水泥浆中薄弱界面增多，“级配”不良的缘故。沸石岩掺量适当时，水泥浆可以得到较好的微颗粒级配，结构紧密，沸石岩结构的特性发挥得好，则可以得到高的机械强度和较好的其他物理力学性能。

三、沸石岩在水泥水化中的作用机理模型

为说明沸石岩在水泥水化的复杂过程中的作用，简化为图 11 的模型，表示沸石岩掺入水泥后引起的水化的特殊过程，其中只讨论主要水化物相。

1. 沸石岩(以下均指磨细粉末)在水泥水化过程中起分散作用，增加水泥水化的空间；

2. 沸石岩微孔吸附浆体的水，同时排出少量空气，使浆体界面减弱，早期强度低；同时浆体实际水灰比下降，而低于表观水灰比，有利于水泥的后期强度；

3. 沸石岩吸附水泥熟料水化生成的 $Ca(OH)_2$ 并与之反应生成 C-S-H，同时与 $CaSO_4$ 反应生成 AFt 相，浆体中 $Ca(OH)_2$ 浓度降低，促进熟料继续水化；熟料水化物包裹沸石岩反应产物，互相穿插；

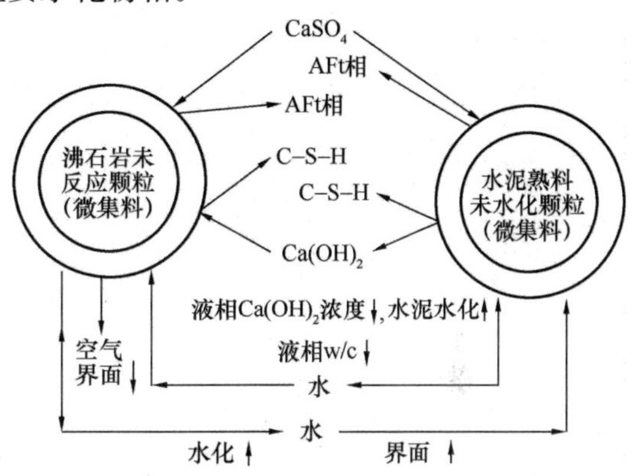

图 11 沸石岩在水泥中的作用机理模型示意图

4. 由于水化不断进行，浆体自由水逐渐减少，沸石岩逐渐释放出沸石水，对水泥水化进行自养护，使界面附近反应继续进行，反应产物不断填充界面微裂缝，界面附近浆体逐渐紧密而得到加强；

5. 沸石岩促进水泥熟料水化，使熟料未水化颗粒减少，同沸石岩未水化颗粒一起作为“微集料”改善水泥浆级配状况。

参考文献

[1] Fredeik A. Mumpton.《Commercial utilization of Natural Zeolites. Industrial Minerals and Rocks》4th Edition 1265-1266(1975).

[2] R. Turriziant, Aspects of the Chemistry of Pozzolanas.《The Chemistry of Cements》Edition by H. F. W. Taylor 80-81(1964).

[3] 日本特许厅公开特许公报 77194529，ケイ酸カルシォゥム硬化体の制法.

[4] B. Držaj S. HoČevar M. Slokan, Kinetics and Mechanism of Reaction in the Zeolitic tuff-CaO-H_2O Systems at Increased Temperature. Cem. Concr. Res. 8 711—720(1977).

[5] 郭竞雄、梁春林，沸石在水泥水化中作用机理的研究，硅酸盐学报，第八卷，第三期(1980).

[6] 唐有祺，结晶化学 126(1965).

[7] F. M. 李. 水泥和混凝土化学，38. (1960).

[8] T. C. Powers, Physical Properties of Cement Paste.《Proc. 4th Int. Symposium Chem. Cem.》V01. 2. 602(1960).

[9] 古阶祥. 沸石. 34(1980).

[10] 张令茂. 粉煤灰水泥掺料研究，建材试验研究，37，西安冶金建筑学院建材试验室(1979).

[11] H. F. W. Taylor. Discussion of the Paper "Microstructure and Strength of hydrated Cements" by R. Feldman and J. Beaudion. Cem. and Concr. Res. 7 465(1977).

任何人都不能一个人活在这世界上。所以，只有让别人生存，自己才能生存；让别人活得好，自己才活得好。

希望所有的人都活得好，甚至为了别人的生存放弃自己的利益，这是“境界”。

至少不妨碍别人的生存，不侵犯别人的利益，不破坏社会的环境，这是“底线”

其中，通过立法程序明文规定下来的，是“法律底线”；

在社会生活中约定俗成，大家都共同遵守的，是“道德底线”；

各行各业必须坚守的原则，是“行业底线”和“职业底线”。

境界不一定人人都有或要有，底线却不能旦夕缺失。因为底线是基础，是根本，是不能再退的最后一道防线。

——易中天

论文之二

MORPHOLOGY OF INTERFACE BETWEEN ZEOLITE-TUFF AGGREGATE AND CEMENT PASTE*

Lian Huizhen
Department of Civil and Environmental
Tsinghua University, Beijing. People's Republic of China

INTRODUCTION

Zeolite with framework alumino-silicate can react with lime to produce C-S-H phase, when gypsum exists in the system, AFT phase can also be produced[1~4], thus zeolite-tuff may be used as active aggregate for concrete, however as strength of zeolite-tuff is rather low, it is generally used as reactive blending material for cement, instead of aggregate for concrete. If the cement paste were regarded as a kind of micro-aggregate concrete[5], then the finely-ground zeolite-tuff might be used as reactive micro-aggregate of this"concrete"also. Morpho logy of interface between zeolite-tuff and cement paste is of great concern to the properties of this"concrete".

TEST AND RESULT

1. Materials to be used for test were clinoptilolite from Hebei Province of China, Portland cement and c. p. CaO. The water-cement ratio was ca. 0. 4. Specimens were cured at room-temperature in a sealer.

2. Zeolite-tuff particles with 2 ~ 3mm of diameter were mixed with cement, and formed. Specimens were cured at scheduled ages and examined in zeolite-tuff-cement interface by SEM. Gaps with different widths were observed on different age. Gap width decreases as the age of concrete increases. It was ca. 50 μm for 1-day curing, and 10 μm for 28-day curing. It almost disappeared over 3-months. Hydrates might be seen in the gaps(fig. 1).

3. Zeolite-tuff surface was polished, and set in cement paste, then cured for scheduled ages. On the scheduled age, the specimens were broken open along the bonding surface between zeolite-tuff and cement paste. The broken surface of each specimen piece was examined by SEM. Numerous pores all over the surface both of zeolite-tuff and cement paste were observed at lower magnification (fig. 2). In bottom of zeolite-tuff pore, some roundlets were observed (fig. 3). By higher magnification, few of the roundlets appeared almost somewhat in their longitudinal morphology (fig. 4). In order to identify their morphology, the specimen was rotated, and some of the roundlets appeared like stubs, and come from the

---

* Published in BONND IN CONCRETE. September, 1984. Scotland.

zeolite-tuff surface. The orientation was more or less perpendicular to the zeoiite surface. Parts of the particles were of ca. 0. 4 μm in diameter, and of ca. 1 μm in length. The dim hexagon section could be discerned. They were presumably of AFT phase. Some particles nestling closely to the zeolite-tuff surface showed epitaxy even more evidently. Their length was not greater than their width, and their maximum diameter was ca. 0. 2 μm (fig. 5). Examination by EDAX confirmed that they were of C-S-H phase with low calcium-silica ratio (fig. 6).

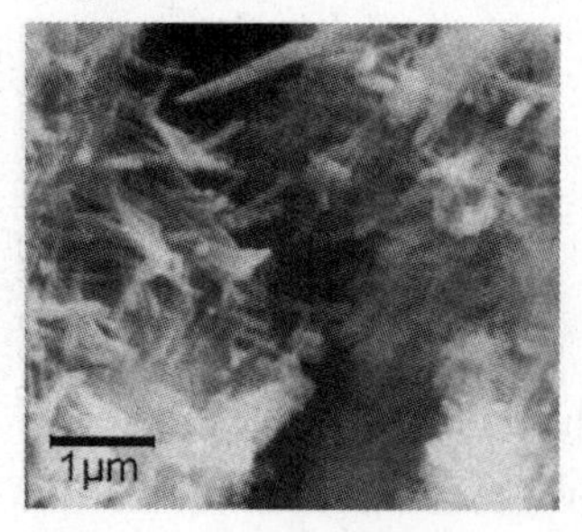

Fig. 1 Hydrates interface Gap on 28th day

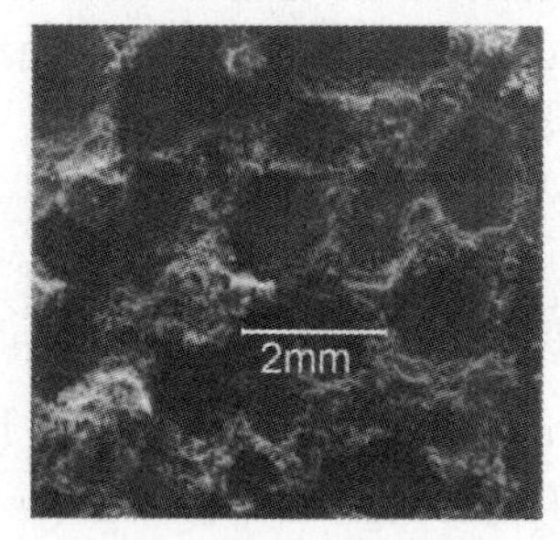

Fig. 2 Pore on bonding surface of zeolite-tuff particle

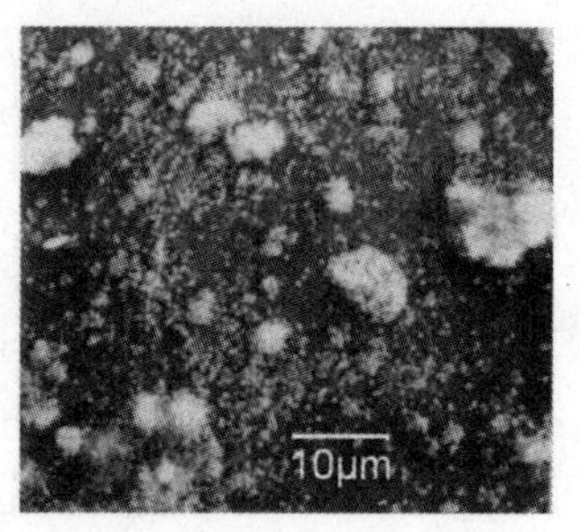

Fig. 3 Bottom of pore on surface of zeolite-tuff particle

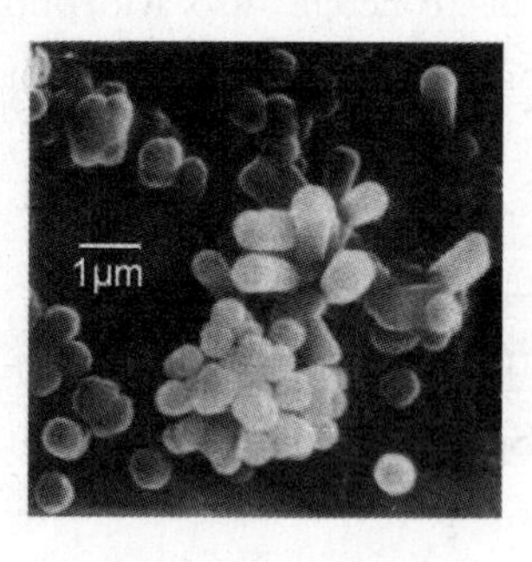

Fig. 4 Local magnification of roundlets in Fig. 3

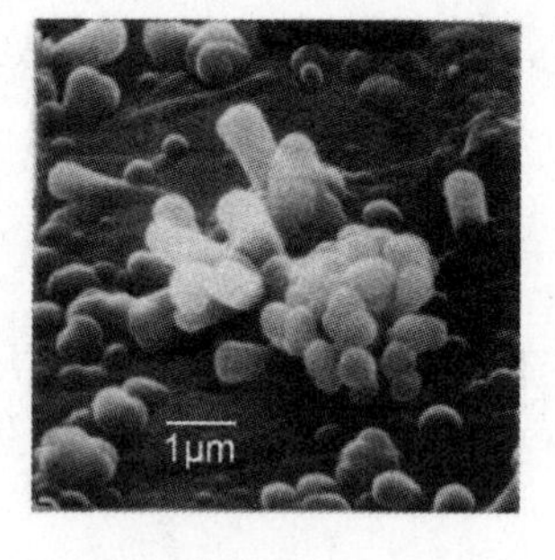

Fig. 5 After rotation by 45° of Fig. 4

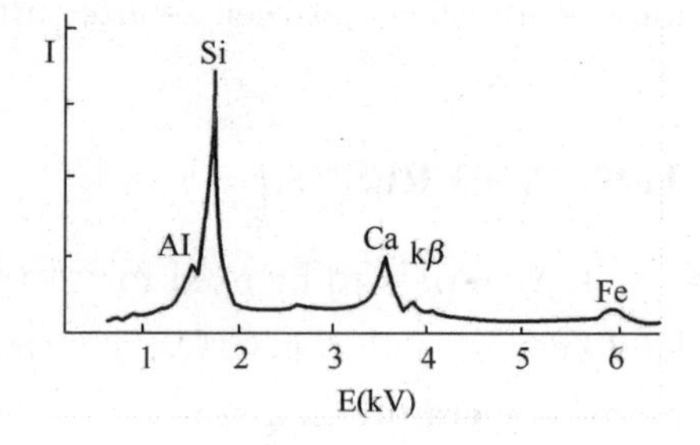

Fig. 6 EDXA on the roundlet after rotation of Fig. 5

At the thicker part of hydrates as viewed through pores on zeolite surface, spoke balls or networks formed of rod-like particles were observed (fig. 7 & Fig. 8). The diameter of a single particle was ca. 0. 4 μm and the length over 3 μm. The particles were of the shape of hexagonal prism when it is were observed at higher magnification (fig. 9). They were identified by EDXA as of AFt phase containing silicon (fig. 10). At the pore edges there was AFt phase but without apparent epitaxy (fig. 11). At the pore edges on the cement surface, there was also AFt phase without apparent epitaxial which was consistent with the case of zeolite-tuff surface. But in bottom of pore on cement surface. AFt phase had no epitaxial (fig. 12). When the specimen was segregate along water film existed between zeolite-tuff surface and pores, the pores disappeared on both the surface, but some roundlets were observed with the same morphology as those in bottom of pore on zeolite-tuff surface, but their length was smaller than their diameter. On the zeolite-tuff surface, the thickness of hydrate products in pore bottom was only several μm on early age. The thickness increases, the pores became smaller and the hydrates had no epitaxy, as the curing period increases.

Whether the examination was carried out by SEM or XRD, there was very little $Ca(OH)_2$ in water-film in the interface between cement paste/zeolite tuff.

This result is different from that on the interface between inactive aggregate and cement paste.

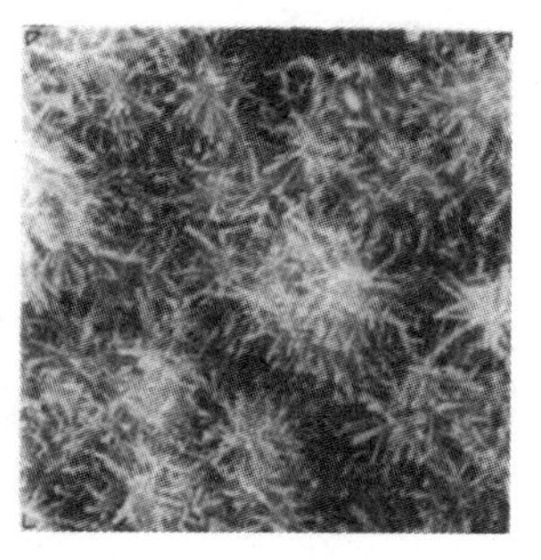

Fig. 7 Spoke-ball from hydrates in pore on zeolite-tuff surface

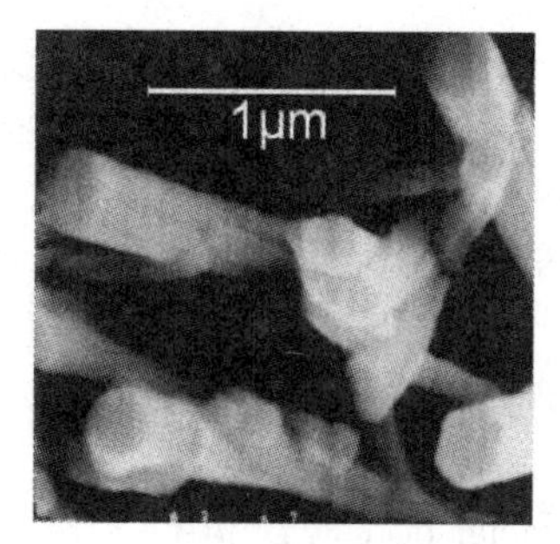

Fig. 8 Network form hydrates in pore on zeolite-tuff surface

Fig. 9 Local magnification of Fig. 8

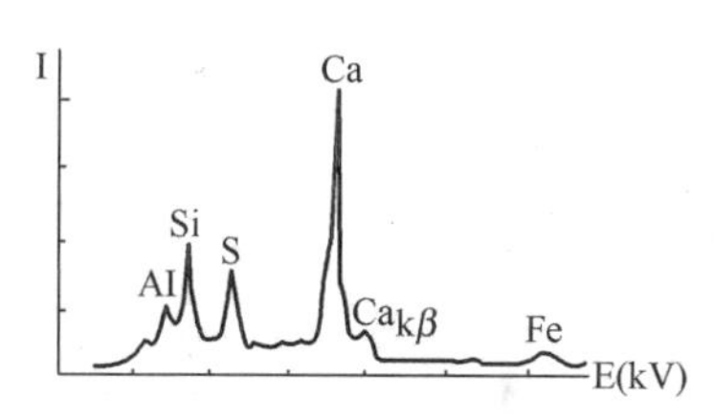

Fig. 10 EDXA of hydrates showed in Fig. 8

Fig. 11 AFt phase at the edge of pore on zeolite-tuff surface

Fig. 12 AFt phase in interior of pore on cement paste surface

4. Zeolite-tuff powder with cement fineness was mixed with c. p. CaO by 1 : 1, and formed. The specimens were cured for ages 1, 3, 7 and 28 days, and examined by XRD separatly. It had been found that C-S-H phase occurred after 7 days, but AFt phase came to appear at 1 day. This phenomenon illustrates that when zeolite-tuff reacts with gypsum and lime in cement paste, AFt phase produced firstly, then C-S-H phase (Fig. 13).

CONCLUSION AND REMARKS

1. As zeolite structure contains a great number of micro-cavities, it possesses strong absorbability for certain gases and liquids. As soon as the zeolite-tuff aggregate comes in concrete with cement paste, it could absorb the free water from paste, and releases its original absorbed air thereof, thus to formed numerous pores in the interface, as the hydration of cement proceeds continually. The free water in paste would decrease accordingly and zeolite releases its absorbed water to supply the water needed for cement hydration and the reaction between zeolite-tuff and cement paste. Therefore, the pores in concrete would become smaller with age, and the space among the pores would be increased, thus the

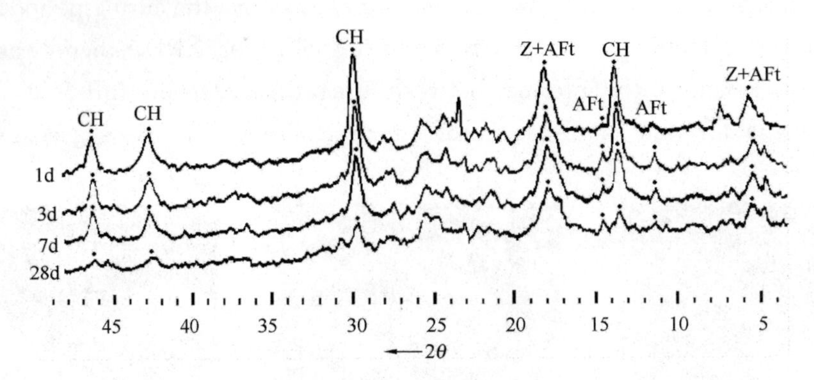

Fig. 13 XRD by the specimen with zeolite tuff-CaO-gypsum
The mark in diagram: Z—zeolite, CH—$Ca(OH)_2$, AFt—ettringite

interfacial bond would be strengthened.

2. In interface between polished-surface zeoiite-tuff and cement paste, the water-film and pores connected or not (fig. 14). But when the zeolite-tuff surface very rough, there exists mutual pore interference, pores and water-film would communicate each other, thus a wider gap is formed. This gap could be narrow gradually as the concrete ages. The bonding surface between polished-surface zeolite-tuff and cement paste could be broken easily along interface pores on an early age, as the pores became smaller when it ages, it could be broken easily along the interfacial water-film.

3. Gypsum and $Ca(OH)_2$, produced by hydration of cement and zeolite-tuff proceed secondary reaction in the zeolite-cement Interface. This process produced AFt phase containing silicon, and C-S-H phase with larger diameter particles and very low calcium-silicon ratio. Hydrates near the zeolite-tuff surface are epitaxial, as distance from the zeolite-tuff surface to the bottom and the edges of pores differ, hydrates with different morphology and degree of epitaxy are produced at pore bottom and the edges. The degree of epitaxy is decreasing from the zeolite-tuff surface toward the interior of cement paste (Fig. 15)

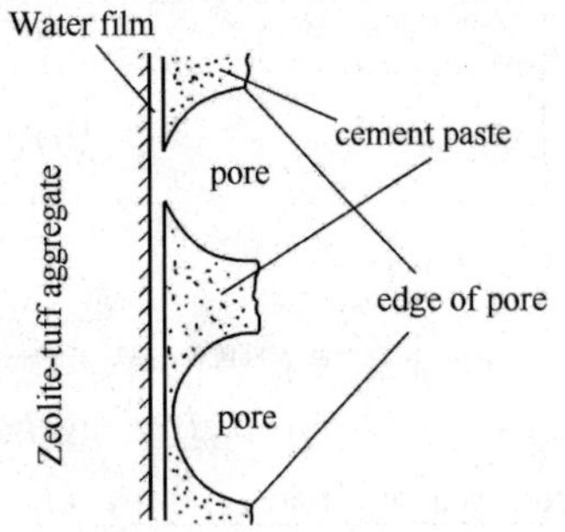

Fig. 14 Relationship between pore And water-film on polished surface of zeolite-tuff aggregate

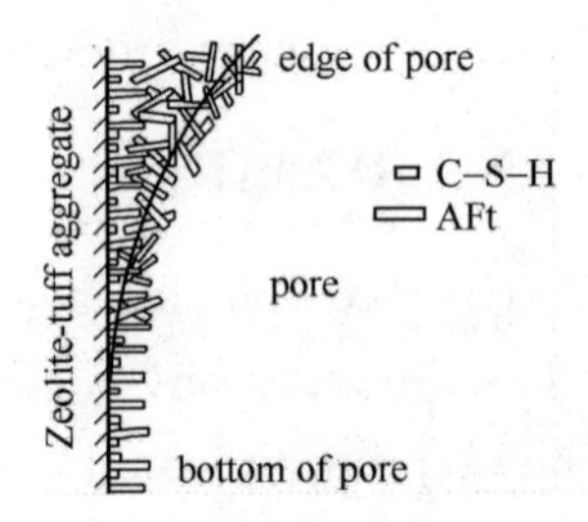

Fig. 15 Epitaxy degree of hydrates in interface between zeolite - tuff aggregate and cement paste

4. As the chemical reaction proceeds in the Interface, there is very few $Ca(OH)_2$ in the water-film, and the secondary hydrates product occur on the zeolite-tuff surface.

Consequently, such an active aggregate possesses better bonding property with cement paste. But insufficient $Ca(OH)_2$ would probably affect the CaO-SiO_2 reaction in interface and disadvantageous to the early bond apertures and micro-strength of interface. If lime and gypsum were added by proper amount, it might be probably advantageous to increase the early bond strength.

5. The cement paste blended with fine-ground zeolite-tuff might be regarded as "micro-aggregate concrete" with zeolite-tuff as the "micro-aggregate". The interface condition between this "micro-aggregate" and cement could be treated analogically as mentioned above. Besides, both the inactive micro-particles and the secondary products would served as the "micro-aggregate", and the aggregate grading in cement paste could be improved[6], Consequently, the macro-properties of cement paste and concrete could be improved. If formation of the interfacial pores could be avoided or reduced, the interfacial bond strength might be strengthened, and concrete strength would be improved, especially its early strength.

ACKNOWLEDGEMENT

This paper was given direction from prof. Wu Zhongwei, and SEM examination was conducted in Tsinghua University Analysis laboratory center.

REFERENCE

[1] R. Turriztan. "Aspect of the chemistry of Pozzolanas". The Chemistry of Cement. edited by H. F. W. Taylor Vol. 7 80~81.

[2] Patent bulletin of Japan 77194529 "The Method of Manufacturing Hardened Calcium-Silicate".

[3] B. Držai S. Ho Čar M. Slokan. "Kinetics. and Mechanism of Reaction in the Zeolite tuff-CaO-H_2O Systems at Increased Temperature". C. &C. R. Vol. 8 No. 6 1977 711~720.

[4] Guo Jingxiong& Liang Chun-lin" Research of Reaction in Zeolite and Cement[J]. Silicate Journal, P. R. China. No. 3 1980.

[5] H. F. W. Taylor. Lecture in China. Beijing, 1980.

[6] Lian Huizhen. Exploration on Reaction Mechanism of Zeolite-tuff in cement . Selections of Academic Symposiums on Cement, 1980-1981, china publishing house for construction industry , 1983.

自评

从这两篇文章可见，显然是接受当时流行的观点，即认为沸石岩活性的来源是“沸石与水泥水化生成的 $Ca(OH)_2$反应”。只是“尚未见报道沸石岩对水泥特殊作用的资料和沸石岩与石灰反应产物的显微形态研究”。也就是说，该文的创新点为该反应产物的形貌学研究和沸石岩与其他火山灰质材料相比的特殊物理作用。尽管有些新意，也仍和其他人的思维方法一样，囿于传统的和从众的思维观点。现分析如下。

1. 在“沸石岩同石灰反应的产物”一节中，主要在证实“沸石岩与水泥水化生成的 $Ca(OH)_2$反应”。这种证实比所引用文献的报道有进步之处是把“沸石”的反应换成“沸石岩”的反应。因为研究的对象是含有斜发沸石和丝光沸石的“沸石岩”，而不是结晶态矿物“沸石"。但是却从x-射线衍射结果得出结论：经过水化28天后"沸石峰高下降"，意思是结晶态的沸石参与了反应。这与热力学是相悖的。是否有可能是试验制样误差产生的？或者对发生x-射线衍射的晶面判断有误？不同龄期x-射线衍射图形强度量程是否相同？仔细分析论文之一的图4和图5，其中从7天到28天的衍射峰无论升高或降低，同样物质各峰高变化并无比例关系，甚至在图5中还有相反的变化或不变的情况。文中说"……$Ca(OH)_2$掺量很少，28天后几乎都参加了反应"，但$Ca(OH)_2$掺量是多少？28天会"都参加了反应"吗？结论未免武断。Taylor在1980年来华讲座中曾说过："……不要从很少量的x-射线资料中推论过多"。当时没有注意，现在反思起来，觉得真是"应当慎重"(Taylor语)。因为从后来的实验发现，在常温下，活性再高的火山灰质材料也不可能在28天以前耗尽硅酸盐水泥水化生成的$Ca(OH)_2$。

2. 文中差热分析只有28天试样的谱图，反应情况难以分析确定。从理论上来说，沸石在加热过程失去沸石水时会有连续吸热的表现，而且还能出现热稳定性的临界点。另外如果125℃处的吸热谷是因钙矾石脱水，则这个温度下脱去的水只是一部分，直到钙矾石脱去2个石膏，并只剩下12个结晶水时，还应当有另外的脱水谷。以上该图都没有表现，很可能是所使用的仪器精度所致，文章对此没有分析。该图在此意义不大。

3. 现在再来分析一下论文之一前言中所引文献的报道：日本专利报道"用斜发沸石或丝光沸石同石灰混合成料浆经高压水热反应，可生成雪硅钙石和硬硅钙石"是可信的，因为不仅无定形SiO_2，而且磨细的石英砂，与石灰混合在压蒸条件下都会生成雪硅钙石和硬硅钙石，只是该专利并未指明"斜发沸石或丝光沸石"是含该矿物的沸石岩还是单矿物？内行人会知道此提法有误。其他所报道的产物都归之于"沸石"的提法均需推敲。沸石岩实际上就是意大利沸石化的凝灰岩或者就叫沸石凝灰岩，其中的矿相除有不同含量和不同种类的结晶态沸石(已发现的沸石有30多种)外，还有未沸石化的无定形SiO_2和Al_2O_3。对斜发沸石，目前还没发现能从天然沸石岩中提纯或人工合成的报道，估计他们都不会采用纯的沸石矿物进行试验研究。那么，他们是怎么知道那些产物就是结晶态沸石而不是凝灰岩中无定形成分反应生成的呢？

4. 虽然提出了沸石岩特殊的物理作用。但是结论却是"晶态的沸石具有很高的活性"，发表这篇文章时距开始研究的时间只有1年多，就这样做出推翻经典热力学理论的结论，未免缺乏慎重。由于沸石的物理作用，是否就如文中所述的"沸石水"、"吸附"、"微粉应"？弄清楚这个问题对选择沸石岩的品种和品位是有意义的。

5. 论文之二是1982年在苏格兰发表的，介绍从形貌学角度对沸石凝灰岩和水泥浆体之间界面的水气交换作用进行的研究，这是真正研究的开始。选入此文是想以此为例说明扫描电镜是怎样用来做研究的。现代仪器的使用必须是根据需要，而不是用来做论文中的点缀。

后记

对矿物掺和料有正确认识才会有正确的使用，因此机理研究的科学性就很重要。研究结果对生产使用会有指导意义的，但是任何立论或命题都要经得起质疑和反驳，首先是自己不断提出问题，才能使认识不断深化。实际上，当时为了完成教研组的任务而匆忙发表的这篇文章尽管曾得到同行们的认同，其中的作用机理模型还曾得到吴中伟先生的称赞，认为是一种"运用了物理模型的科学研究方法"，但却并没有使我自己满意。该文中对沸石物理作用的解释除“沸石水”因素外，都不能算是沸石独有的特点。此后，意大利学者 Massazza F. 曾提出的问题“像沸石这样结晶物质何以具有很高的反应能力？”一直困扰着我。为了深入了解沸石的特性，我开始进行跨学科的调研，到本校化工系、石油化工研究所、大连化学研究所等单位请教过沸石结构及沸石作催化剂和分子筛的应用；向美国西北大学化学系、日本大学短期大学部应用化学系等的大学的教授请教过沸石吸附与离子交换特性的利用；和美国来中国交流的克里夫兰大学教授讨论用天然沸石做分子筛的处理工艺条件。同时，跳出建筑材料范围，尽可能多地查阅国外对天然沸石岩用途的报道。为此，于 1984 年申请了中科院自然科学基金(后改为国家自然科学基金)“结晶态铝硅酸盐胶凝物质活性机理的研究”，证明了沸石岩中的沸石是热力学稳定的矿物，沸石岩在水泥中的化学活性并非来自结晶态的沸石，而是来自沸石凝灰岩中无定形的 SiO_2 和 Al_2O_3，实验研究证明了沸石化的凝灰岩有比凝灰岩等其他火山灰质材料更高的活性，是源于沸石这种具有特殊结构的结晶矿物在水泥中的特殊物理作用。

我每天都要一百次提醒自己，我的精神生活和物质生活都要依赖于活着的和已死的其他人，我必须努力，以便按照我所接受到的同样份量来付出。

——爱因斯坦

论文之三

CEMENT PASTE / ZEOLITE-TUFF BOND*

Lian Huizhen

Tsinghua University, Beijing, China

ABSTRACT Zeolite is a kind of crystalline material with framed structure. There are a lot of micro-cavities and micro-pores in it, and zeolite-tuff, i. e. zeolitized pozzolana, has good reactivity with lime. To be blended with suitable dosage of zeolite-tuff, the highest strength of cement can be obtained, up to 15% higher than that of the plain cement without any blending. Cement paste/zeolite-tuff interface was examined by SEM, EDAX and XRD, it was showed that the reaction can occur between zeolite-tuff and $Ca(OH)_2$ which is released by hydration of cement. But by XRD and test of alkali treatment, it is showed that the reactive component is the uncrystallized portion rather than crystalline zeolite itself in zeolite-tuff. Experiments confirm that the reactivity of zeolite-tuff is caused by the special structure of zeolite, possessing numerous micro-cavities and micro-pores, which are characterized with adsorption, zeolite aqua and water-air exchange at cement paste / zeolite-tuff interface.

Keywords cement paste bond zeolite-tuff

INTRODUCTION

Zeolite is a kind of crystalline aluminosilicate mineral with framed structure. Zeolitized pozzolana has good reactivity with lime[1][2]. Hydrates of zeolite-tuff with lime or portland cement have been studied by many scientists[3~5]. But Why such crystalline material is more active than many uncrystallized volcanic seems still not clear[6][7].

If materials with special zeolite structure were blended into portland cement, the properties of cement paste could be improved. High strength and flowing concrete with compressive strength of 70 MPa and slump over 150 mm could be obtained by blending zeolite-tuff, and the strength was about 10 MPa higher than that of the comparative concrete which has same water/cement ratio and without any blending materials[8]. In order to explain what happened between zeolite-tuff and lime, the cement paste/zeolite-tuff bond is here investigated.

MATERIALS

Materials used for test were zeolite-tuff, portland cement and CaO(analytically pure);

---

* Presentation in 1988 Engineering Foundation Conference on ADVANCE OF CEMENT USE AND MANUFACTURE, July 31-August 5, 1988. Missouri, USA and Published in Proceedings of that Conference with the same title. New York, 1989.

zeolite-tuff from Hebei province comprises of clinoptilolite with about 70%; from Nenjiang, Heilongjiang province comprises of about 84% clinoptilolite; and from Zhejiang province comprises of mordenite with about 60%, etc. Moreover, the mordenite sieve sorbent was used for comparison. Specimens of zeolite-tuff were either finely-ground or in particle form for different purposes.

TEST AND RESULTS

1. Zeolite-tuff powder was blended into portland cement with various dosages from 0 to 30%, by weight of total. As showed in Fig. 1, the compressive strength of cement at optimum dosage of zeolite-tuff was not only higher than that of the comparative cement without any blending materials, but also the highest among the specimens. A great deal of data showed that zeolite-tuff has good pozzolanic reactivity.

2. In order to understand hydrates of zeolite-tuff in cement, zeolite-tuff was mixedwith CaO, and observed by SEM. The main reaction product is wormlike CSH with low C/S ratio (Fig. 2). When zeolite is mixed with lime and gypsum, the main reaction product is AFt phase besides CSH(Fig. 3)

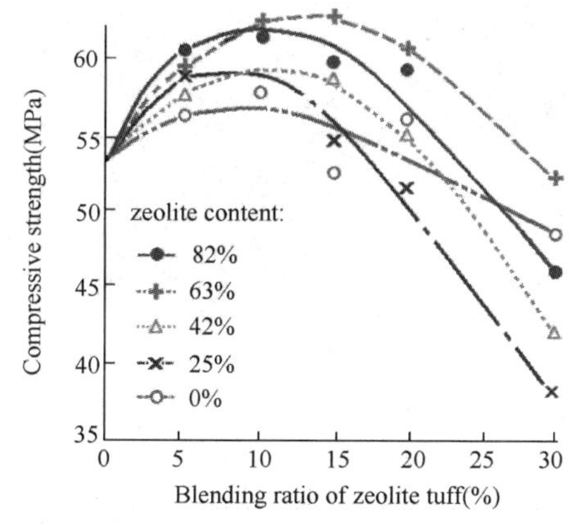

Fig. 1 Typical curve of compressive strength of cement blended with varied dosage of zeolite tuff

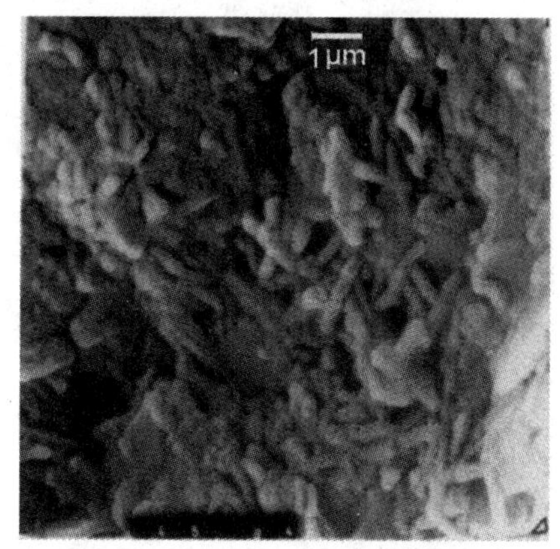

Fig. 2 Wormlike CSH of products of zeolite-tuff and lime

3. A piece of zeolite-tuff particle was set in cement paste and cured to 7-day and 28-day. At the indicated ages, surface perpendicular to interface between zeolite-tuff and cement paste was polished, and micro- hardness on both sides of interface on the polished surface was tested (Fig. 4). It shows that hardness of cement paste, near the interface, is lower than that at bulk of cement paste at 7th day due to interfacial influence. But at 28th day, hardness of both sides increases markedly. Particularly on zeolite side, the hardness near interface is higher than that at the bulk. It indicates that cement paste / zeolite-tuff bond was strengthened gradually as advance of hydration. By SEM, width of the interfacial gap is decreased as age increased (Fig. 5, 6), and the interfacial gap almost disappeared after 3 months (Fig. 7), Hydrates might be seen in the gap at 28-day (Fig. 8). Uninterrupted reaction between cement paste and zeolite-tuff was observed.

Fig. 3　AFt phase from reaction from reaction of zeolite-tuff and lime

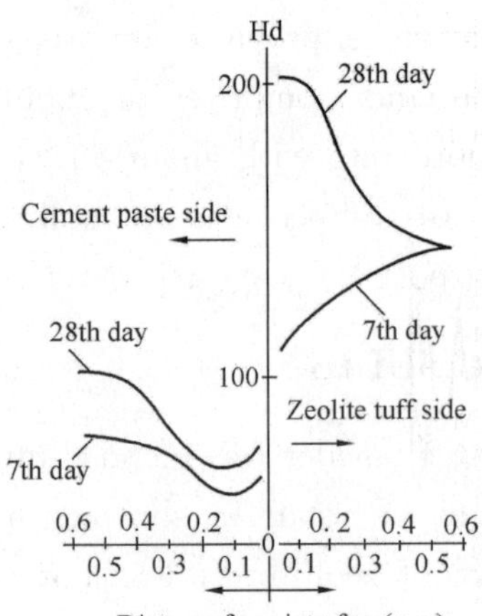

Fig. 4　Distribution of micr-hardness on both side of interface between cement past and the Zeolite-tuff respectively

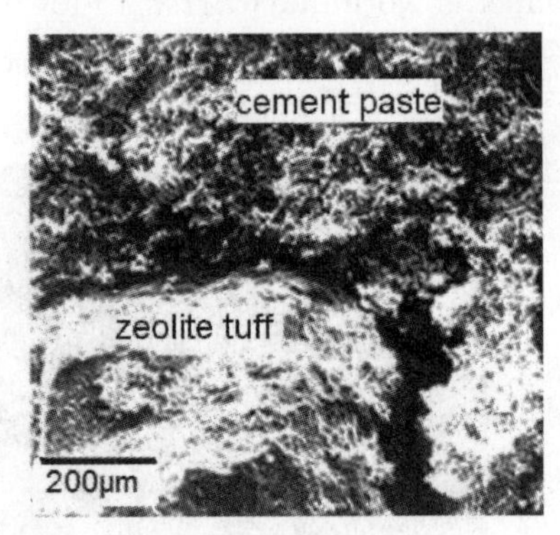

Fig. 5　Gap between cement paste/ zeolite-tuff at 1st day

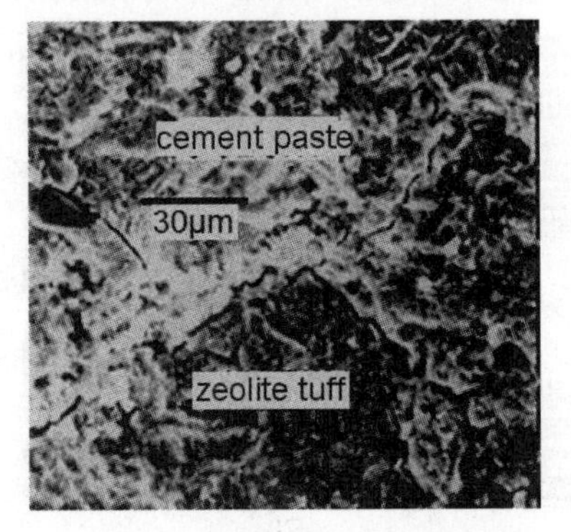

Fig. 6　Gap between cement paste zeolite-tuff at 28th day

Fig. 7　Gap between cement paste/ zeolite-tuff on 3-month

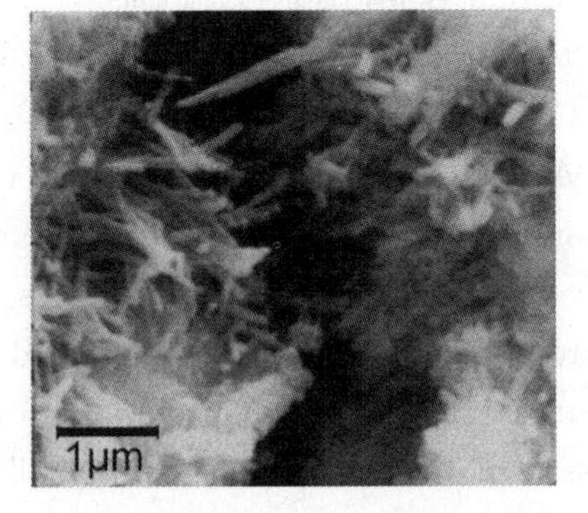

Fig. 8　Hydrate in interface between cement paste /zeolite-tuff on 28th day

4. Zeolite tuff-CaO-gypsum hydrated specimens were also examined by XRD at various ages(Fig. 9). It is showed that from 1-day to 28-day, $Ca(OH)_2$ was getting less, but the

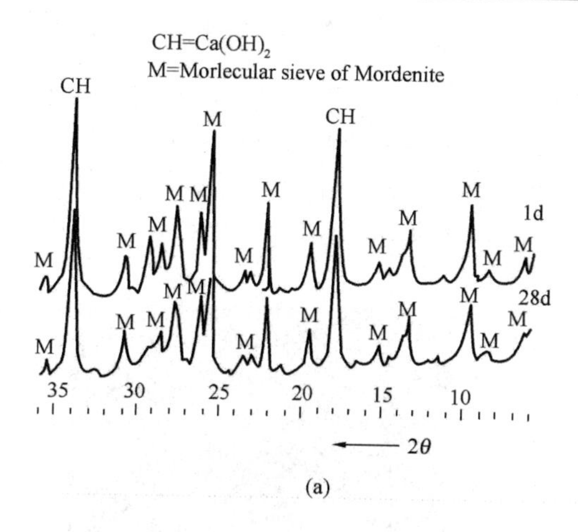

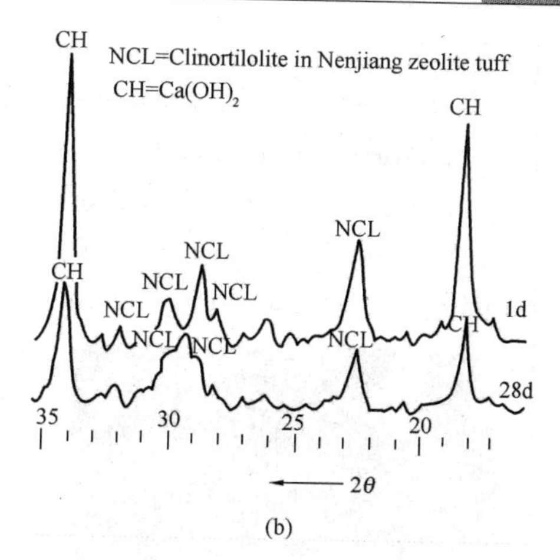

Fig. 9　XRD of zeolite tuff-CaO-gypsum

(a) Zeolite sieve sorbent with Mordenite; (b) Zeolite-tuff containing Clinoptilolite

peak of XRD on zeolite had little drop in height. It indicates that zeolite-tuff can react with CaO and gypsum. But the reactive component is not the crystalline zeolite itself mainly, at least on earlier ages. In order to confirm this opinion, alkaline stability was examined. Zeolite-tuff was treated with NaOH solution of various concentration (0.1M, 0.5M, 1M, 3M) and 0.1%, $Ca(OH)_2$ solution by boiling for 1.5 hours, then were examined by XRD (Fig. 10) and chemical analysis. It could be found that the structure of zeolite had not been changed apparently with NaOH below 0.1M, SiO_2 / Al_2O_3 ratio of that did not reduce substantially (from 4.27 to 4.21). The peak of XRD on zeolite had only little change in height, and so did that with 0.5M(SiO_2 / Al_2O_3 is 4.02), and with 1M of NaOH, the main peak of zeolite lowered and the structure of zeolite began to be destroyed. It vanished fully with 3M of NaOH. When zeolite was treated with 0.1% $Ca(OH)_2$, similar results were found as with 0.5M of NaOH. It could be estimated that the structure of zeolite could not be destroyed by alkalinity in cement paste. The strength of the cement blended with above treated zeolite-tuff was decreased as concentration of NaOH increased. When concentration of NaOH was increased to 3M, both active SiO_2 and active Al_2O_3 were increased, but the strength of cement was markedly reduced even though structure of zeolite had been fully destroyed. It confirms that the reactivity of zeolite-tuff in cement does not depend on the chemical reaction, but mainly on its structure.

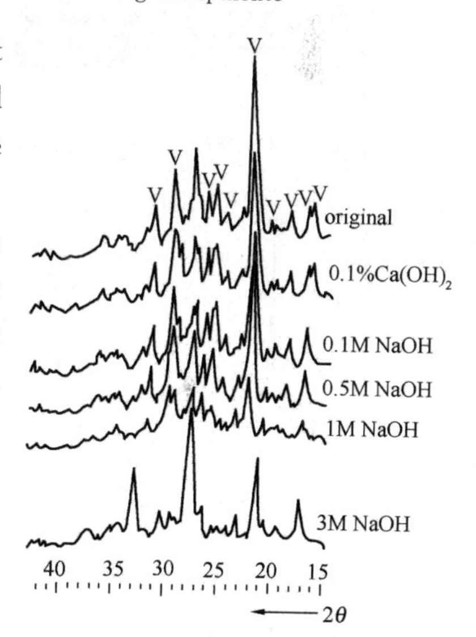

Fig. 10　XRD of zeolite-tuff treated by 0.1 % $Ca(OH)_2$ and different concentration of NaOH

5. A piece of zeolite-tuff with a polished surface was set in cement paste and cured to indicated ages, splitting was made along the cement paste/zeolite-tuff interface. Both

surfaces on cement paste and zeolite-tuff sides were examined by SEM separately. Numerous pores all over both surfaces were observed (Fig. 11). At the bottom of pores on the zeolite-tuff surface, there are stub-like hydrates growing out from and perpendicular to the surface (Fig. 12). Some of them are CSH phase with low calcium / silica ratio (Fig. 13). The others are AFt phase containing silicon (Fig. 14). Orientation of the hydrates are weakened as the position is farther off the surface of zeolite-tuff.

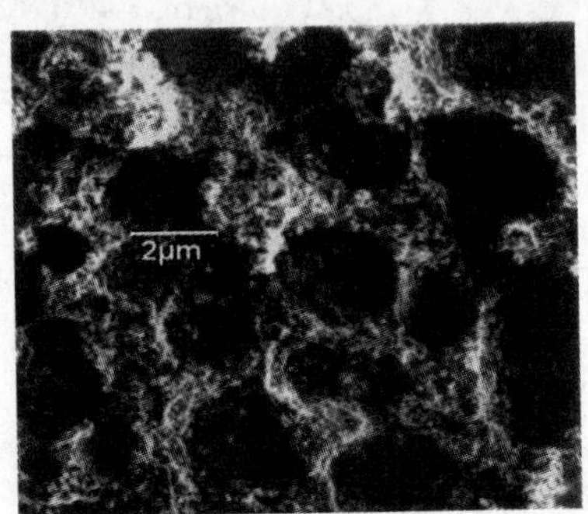

Fig. 11 The interfacial pores on surface of zeolite-tuff

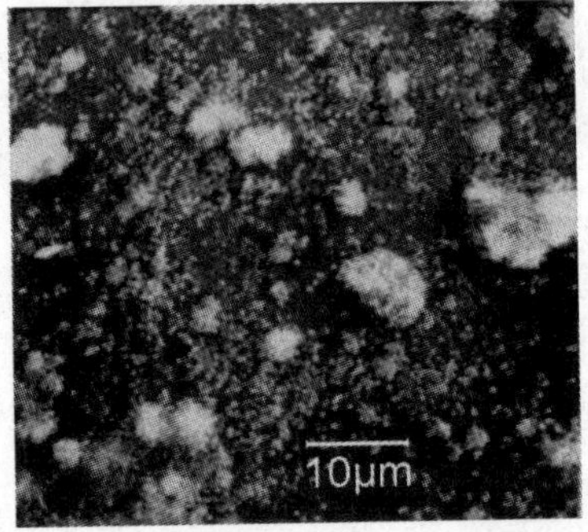

Fig. 12 Morphology at the bottom of interfacial pore

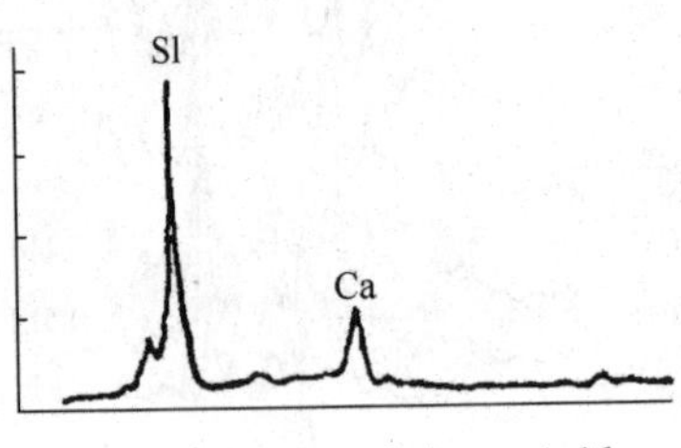

Fig. 13 EDAX of the stub-like hydrates

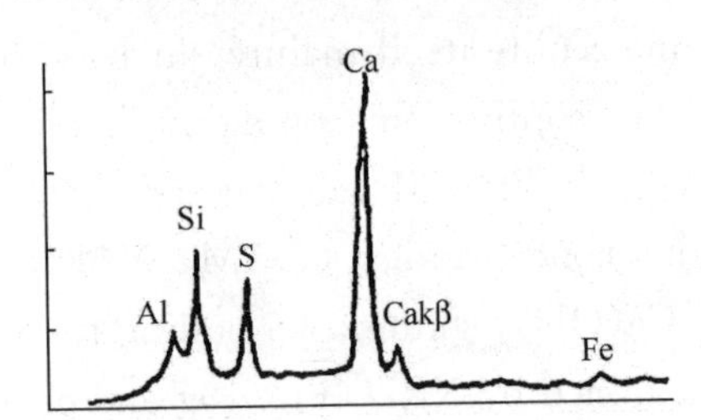

Fig. 14 EDAX of the longitudinal particles in pore on zeolite-tuff surface

On the both surfaces of cement paste and zeolite-tuff, plenty of AFt phase but no $Ca(OH)_2$ can be found at 28-day. Pitch between pores was only about several μm at earlier age, and increased with the curing period, so that the pores were becoming smaller and smaller, the hydrates had no epitaxy.

6. Zeolite tuff particles of about 2 mm size were dried at temperature of 300℃ for 3 hours, and then were cooled in dry air at room temperature. When the specimens were put into water, numerous bubbles raise. The gas was collected and examined by Ozart gas-analysis meter. It was known that gas produced was air adsorbed by zeolite during cooling stage. When zeolite tuff blended in cement paste was mixed with water, some water was absorbed by zeolite, and air adsorbed originally in zeolite was released into paste. The existence of pores around the zeolite particles may cause the relative lower early strength of cement paste blended with zeolite-tuff. But as hydration continued, the water adsorbed by zeolite originally(named zeolite aqua) would be gradually released to supply hydration water after the curing water exhausted in favor of hydration of cement and reaction between cement and zeolite-tuff. So that the cement paste / zeolite-tuff bond could be strengthened with

continuous curing.

7. Such water-air exchange effect in cement paste bond is peculiar to structure of zeolite. It can be confirmed by other way:

(1) 30% of finely ground zeolite-tuff and 30% of granulated slag were blended into cement separately. Both the two kinds of blended cement were used to make concrete and cured for 7 days and 28 days under natural condition. The same cement without any blended materials was used in comparison. The results of compressive strength test are showed in table 1. It is showed that from 7-day to 28-day, the concrete strength increase rate varied with w/c and blending materials. Blended with zeolite-tuff, concrete had the highest increase rate, as water-air exchange effect existed in cement paste / zeolite-tuff bond.

(2) Zeolite-tuff powder was burned at high temperature (according to thermostability of certain zeolite) to destroy the structure (confirmed by XRD). Both the burned and unburned zeolite-tuff was blended into cement. It is showed that strength increasing rate of cement blended with burned zeolite-tuff is lower than that of cement paste with unburned zeolite. For example, the former was 16 MPa at 7th day, 25 MPa at 28th day, while the latter was 17 MPa at 7-day, 38 MPa at 28-day, because there is no water-air exchange in burned zeolite-tuff as zeolite structure was already destroyed (see table 2).

Table 1 Comparison of strength of concrete cured under natural condition

No.	Blending material	w/c	Compressive strength of concrete (MPa)		
			7th day	28th day	Increase rate(%)
1	Zeolite-tuff	0.4	13.4	26.0	94
		0.6	7.50	20.2	169
2	Granulated slag	0.4	16.3	23.2	42
		0.6	9.9	19.9	101
3	without	0.4	18.9	28.9	52
		0.6	9.20	17.8	98

Table 2 Comparison of strength of concrete with zeolite-tuff in different structure

No.	structure of zeolite	Compressive strength of concrete (MPa)		
		7th day	28th day	Increase rate (%)
1	burned	16	25	56
2	unburned	17	38	120

(3) The concrete strength increase rates from 7-day to 28-day are also different when used cement is blended with different dosages of zeolite-tuff and cured under natural condition. With 30%, the increase rate is 1.06 and with 50%, it is 1.32 (with the same w/c). More zeolite-tuff blended shows better effect of water-air exchange (table 3).

Table 3 Comparison of increase rate of concrete strength with different dosage of zeolite-tuff

No.	dosage of zeolite-tuff	increase rate of concrete strength from 7th day to 28th day (%)
1	30	160
2	50	132

CONCLUSION

1. Chemical reaction between cement paste and zeolite-tuff can occur to produce AFt phase containing much more silicon and wormlike CSH as "secondary" hydrates in cement paste;

2. The reactive component is uncrystalline portion in zeolite-tuff rather than crystalline zeolite itself, zeolite-tuff has high activity just because of its special porous microstructure and zeolite aqua therein;

3. Pores produced by water-air exchange and inadequacy of $Ca(OH)_2$ for secondary reaction may be a reason of weak bond at earlier age, but as hydration continues the reaction between cement paste / zeolite-tuff and zeolite aqua can further strengthen the bond in latter ages;

4. Due to active bond, as zeolite-tuff which blending into cement paste behaves as active "micro-aggregate", the performance of cement paste or concrete can be markedly improved. The strength of cement paste blended with zeolite-tuff can be about 10 MPa higher than the original.

ACKNOWLEDGMENT

The research concerned in this paper was performed under the direction of Professor Wu Zhongwei and his theory about" The effect of central mass".

REFERENCE

[1] Sersale R. , Sabatelliv. Pozzolanic Activity of Zeolites. Rend. Accad. Sci. Fis. e Mat. (Soc. Maz. Sci. , Lett. Naples) 27. Ser. 4 1960.

[2] Malquori G, . Portland-pozzolanic Cement. Proc. 4th Int. Sym. on the Chemistry of cement. Washington, 1960.

[3] R. Turriziant. Aspectsof the Chemistry of pozzolanas. The Chemistry of Cements. Ed. by H. F. W, Taylor. 1964.

[4] Guo jingxong et al. Study on Mechanism of Reaction of Zeoliteduring Hydration of Cement. Journal of Chinese. Silicate, No. 3, vol. 8 1980. (in Chinese).

[5] B. Držaj et al. Kinetics and Mechanism of Reaction in the Zeolite tuff-CaO-H_2O Systems at Increased Temperature. C. & C. R. , 8, 1977.

[6] Fredeik A. Mumplon. Commercial Utilization of Natural Zeolite, Industrial Minerals and Rock, 4th edition, 1975.

[7] Massazza F. The Chemistry of Pozzolanas and Blended Cement, Proc. 6th Int. Sym. on Chemistry of Cement, Moskow. 1976.

[8] Guo Yushun et al. Production of High-strength and Flowing Concrete, Concrete and reinforce concrete, No. 4, 1986. (In Chinese).

本论文的背景

美国著名水泥化学家、Grace 化学公司水泥研究所的所长 Jan Skalny 在 1979 年发起了一个工程基金会讨论会，每三年举行一次，邀请世界顶级水泥科学家和工程师自愿拿出他们最前沿的想法和最新研究成果，公开地、无拘无束地交流、自由讨论。1988 年，在美国密苏里州圣·路易斯的度假村举办了题为《水泥制造和使用的进展》的美国工程基金会第四届会议，在全世界邀请了 40 人，我国吴中伟被邀请做关于混凝土中的界面问题的发言。吴先生说他年纪大了让我代表他去参加。论文之三就是会上的发言稿。由于被邀请人多带助手或学生，这个会实际参加的人数已达约 100 人。参加这次会的感受是非常轻松，上午开会，下午自由活动，可以参加即兴安排的专题报告或讨论（ad hoc），也可打猎、钓鱼、划船、游泳、聊天，等等，晚上是会议提供场所和其他条件的更加自由的社交时间（social hours）。参加这次会，除了学术上的收获之外，还有个收获，那就是学术会议可以这样轻松、自如地举行！自由交流可提升学术气氛，“碰出火花”。这种活动值得我们借鉴。我现在正在努力推行这种沙龙式的讨论会。希望对自由讨论能起到一点推动作用。

本篇论文和论文之二的区别是，论文之二只是对界面的形貌研究；而论文之三则是专门分析水泥浆体和沸石岩之间界面水气交换的行为，采用的是物理化学的亚微观研究和宏观实验的佐证。虽然也属于模拟的、分解论的方法，但是研究总要由此开始。而实践证明研究的结果对工程应用是有用的。

由于全球经济问题造成世界性的短期行为风行，企业纷纷紧缩科研开支，加上基本建设饱和，水泥混凝土研究的课题需求无源，Skalny 退休后，Grace 公司这方面的研究也就萎缩了。所以我们现在经常会看到国外也出现一些“没事找事”的“研究”。而我国混凝土的工程应当是正处在发展的上升时期。2013 年我国水泥产量为 24.1 亿吨，以此推算，包括预拌、现拌、预制构件、砂浆等的混凝土总量约 80 亿 m^3，占世界混凝土总量一半多。这么大量的混凝土工程为研究工作所提出的题目是很丰富的，我们理应针对一些工程中发现的问题，把现有混凝土做好。在混凝土的科学技术研究方面做出出色的成绩。企业应当有此雄心壮志和投入。

我们所能体验的最美丽的事情是神秘。它是一切真艺术和真科学的源泉。

——爱因斯坦

论文之四

EFFECT OF STRUCTURE AND COMPOSITION ON REACTIVITY OF ZEOLITE-TUFF USED AS BLENDING MATERIAL FOR PORTLAND CEMENT*

Lian Huizhen
Department of Civil Engineering, Tsinghua University, Beijing, P R CHINA

SUMMARY It is known that zeolite is a kind of crystalline with framing structure, but the pozzolanic reactivity of zeolite-tuff, i. e. zeolitizated tuff, is better than of many uncrystallized volcanic. In order to explain such a unusual behaviour, effect of zeolite content, Si/Al, positive ion and zeolite aqua in cavity of zeolite, fineness, etc. on reactivity of zeolite-tuff are tested. Among of them, the zeolite content and fineness are most sensitive rather than the chemical reactivity. It is confirmed that effect of blending materials should not be only considered according to their chemical reactivity. Compositions take effect on cement by way of their structure. What is the cause of the high reactivity of zeolite-tuff, such a kind of crystalline matter, can be explained.

Keywords: zeolite-tuff reactivity effect factor

摘要 沸石是一种具有架状结构的结晶，但沸石凝灰岩（即沸石化的凝灰岩）的活性却比非晶质的活性好。为了解释这种不正常的行为，对沸石凝灰岩中的沸石含量、Si/Al 比、孔穴中阳离子和沸石水、细度等对沸石凝灰岩活性的影响进行了试验研究。在这些因素中，沸石含量和沸石岩细度比其化学活性更敏感。这证明混合材料的作用不应只是考虑其化学活性。化学成分通过其结构起作用。沸石凝灰岩这种晶态物质高活性的原因就可以得到解释。

关键词 沸石凝灰岩 活性 影响因素

INTRODUCTION

Zeolite is a kind of crystalline with framing aluminosilicate. According to thermodynamics, stability of crystalline is much more than of amorphous. But as early as 1960, some scientists reported that reactivity of zeolite is more capable than that of glass[1][2]. Consequently, products of zeolite-tuff reacted chemically with lime have been studied continually[3~6]. Interface reaction between zeolite-tuff and cement paste has been investigated, and morphology of products thereof has been described[7][8]. Mokoto Asaka et al treated clinoptilolite-tuff by sodium hydrate solution. It is proved that structure of zeolite cannot be collapsed until above 2M of NaOH[9]. They reported that when concentration of NaOH is up to 5M, there was not zeolite any longer in tuff, and in solution of 0.1% Ca

* Published in 9th International Conference on the Chemistry of Cement and the proceedings of that conference. New Delhi, India 1992.

$(OH)_2$ (this is near the alkality of cement paste), the structure of zeolite is no changed similar to that in 0.5M of NaOH[10], Owing to such a stable characteristic, zeolitecan be used as the sieve sorbent and catalyser.

In order to understand the source of reactivity of zeolite-tuff all-sidedly, effect of varied zeolite content of tuff blended in Portland cement with varied blending amount on strength of cement paste, reactivity of zeolite-tuff, containing varied zeolite, with lime, effect of Si/Al ratio of zeolite and cation in cavity of zeolite on reactivity of zeolite-tuff, etc., are investigated and introduced in this paper.

ZEOLITE-TUFF USED FOR EXPERIMENTS

Two types of natural zeolite-tuff: Clinoptilolite-tuff from Nenjiang (in China) and mordenite - tuff from Zhejiang(in China) are used. Amount of ion exchange with NH_4^+ and corresponding zeolite content as well as composition of them is shown in table 1.

Table 1 Composition of zeolite-tuff used for experiments

Type of zeolite-tuff	Amount of NH_4^+ exchange (mg/100g)	Zeolite Content (%)	Compositions (%)							
			SiO_2	Al_2O_3	Fe_2O_3	CaO	MgO	K_2O	Na_2O	LOI
Clinoptilolite	178.92	84	61.60	14.40	2.20	3.79	1.85	3.13	0.70	12.83
Mordenite	131.57	60	68.96	10.28	1.97	0.92	0.48	0.60	3.40	13.72

Uncrystallization of natural zeolite-tuff: According to the thermostability of the two types of zeolite-tuff, 400℃ and 800℃ are used for treating above zeolite-tuff for half an hour separately. XRD indicated that structure of zeolite in mentioned tuff has been collapsed and crystallized substantially.

EXPERIMENTS AND ANALYSIS OF RESULTS

Effect of zeolite-tuff with varied zeolite content on strength of cement paste blended with zeolite-tuff by varied amount.

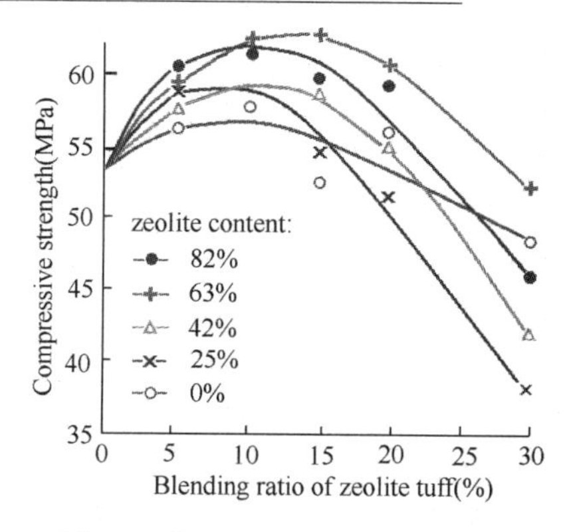

Fig. 1 Compressive strength vs. blending ratio of zeolite-tuff with varied zeolite content at 28th day

Uncrystallized and original clinoptilolite are mixed in ratio of 0 : 1, 1 : 3, 1 : 1, 3 : 1, 1 : 0 to make specimens with varied zeolite content with 82%, 63%, 42%, 25%and 0%. Those specimens are blended into Portland cement by 5%, 10%, 15%, 20%, and 30% (by weight of total binder) separately, and then to be formed in 2cm × 2cm × 2cm with 0.35 of w/c. The compressive strength is tested after curing for 28 days. Results of tests are shown in fig. 1. Every numerical datum is average from 30 groups of specimens. It is shown in fig. 1 that :

1. When the blending amount of tuff is 5%~10%

(by weight of the total binder), distinction of strength among specimens in which the zeolite content of tuff is varied seems very little. In fact, owing to "micro powder effect", if some inert material such as quartz sand is ground to be a certain fineness, strength of cement blended with it can also be elevated somewhat[11]. Distinction of strength among those containing varied amount of zeolite is increased with blending amount.

2. There are optimum blending amount of zeolite-tuff to make strength of cement higher for varied zeolite content separately. Alternatively, there is optimum zeolite content to make strength of cement highest for each blending amount of zeolite-tuff.

3. The optimum zeolite content of zeolite-tuff is over 42%, preferably about 63% and the optimum blending amount in that content is 10% ~ 20%. All the strength of uncrystalline zeolite-tuff in varied blending amount are lower than that of zeolite-tuff containing zeolite over 42%. When the blending amount is over 30%, strength of cement paste reduced markedly.

As comparison and proofing, the synthetic mordenite sieve sorbent and uncrystallized mordenite tuff are mixed by 0 : 10, 3 : 7, 5 : 5, 7 : 3 and 10 : 0 (by weight of total binder) respectively. Results of tested compressive strength are shown in table 2.

Table 2 Compressive strength of cement paste blended with 10% of zeolite-tuff containing varied amount of mordenite

Mordenite to uncrystalline thereof ratio	0 : 10	3 : 7	5 : 5	7 : 3	10 : 0
Compressive strength of cement paste (MPa)	66.2	71.7	72.9	69.2	61.9

It is shown from table 2 that varied amount of mordenite tuff has also optimum blending amount; that is, for the same blending amount, there is optimum zeolite content in tuff containing mordenite as blending material of Portland cement.

Chemical reactivity of zeolite-tuff with lime

Specimens as shown in table 2 are mixed with CaO (analytical purity) by ratio of 1 : 1 (by weight of total binder), and formed with water separately. After curing for 3, 7, 28 days, the soluble SiO_2 and the Al_2O_3 of that paste are examined by means of cold hydrochloric [12]. Above results of 28th day are shown in fig. 2.

From results of test, it can be seen that:

1. The highest strength of specimen made of zeolite-tuff mixed with CaO can be obtained by suitable zeolite content.

2. The largest amount of soluble SiO_2 occurs in specimens with mentioned ratio of about 5 : 5.

3. The least relative amount of $Ca(OH)_2$ by XRD in specimens occurs in specimens with mentioned ratio of 5 : 5 in table 2.

4. Difference strength of specimens containing varied mordenite content is much more

than difference of amount of soluble SiO_2 and Al_2O_3 as well as the relative amount of $Ca(OH)_2$. It is considered that the chemical reaction is not the only factor effecting strength of cement paste. Results as above mentioned tally with XRD of zeolite tuff-lime reaction, e. g., in 28-days, the peak values of clinoptilolite in zeolite-tuff are hardly changed, but of $Ca(OH)_2$ are changed much[13], i. e. the reaction is induced by the amorphous portion of tuff rather than the crystalline zeolite.

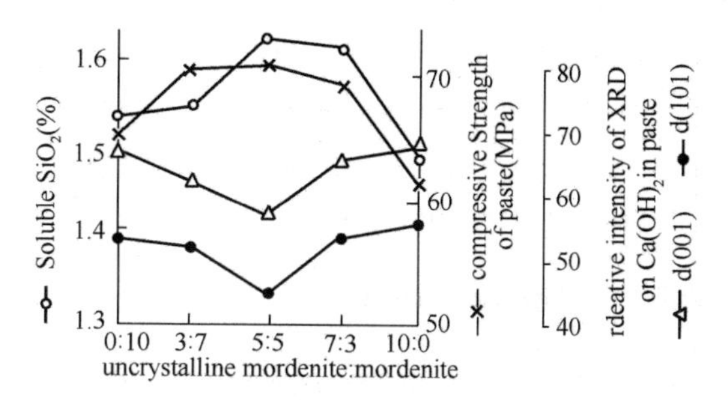

Fig. 2 Soluble SiO_2 and Al_2O_3, relative amount of $Ca(OH)_2$ by XRD and strength of specimens comprising of lime and zeolite-tuff in varied ratio at 28th day

Effect of composition on reactivity of zeolite-tuff

1. The cation in cavity of clinoptilolite is changed by ion-exchange, and it results in amount of a kind of cation to be over 50% (by weight of total cation). Mentioned zeolite-tuff, which cation have been changed and the original zeolite-tuff are mixed into Portland cement by 10% respectively. Strength of mortar made of those specimens are tested and the results are shown in table 3.

Table 3 Change of zeolite water and strength of cement blended with zeolite-tuff by 10% after ion exchange of zeolite

Type of cation	Compressive strength of cement (MPa)			Zeolite water (%)
	3rd-day	7th-day	28th-day	
Original	15.3	24.9	39.2	
NH^{4+}	14.1	24.9	44.2	9.90
Ca^{+}	16.6	27.6	44.3	7.78
Na^{+}	18.7	28.2	42.0	6.83
K^{+}	17.2	27.0	39.5	4.32

It is shown from table 3 that the strength of cement paste at 28th day, specially the strength increase rate with ages, is increased with zeolite water. This is due to induction by effect of zeolite water in interface between zeolite-tuff and cement paste[14]. The best effect of cation is Ca^{2+}, Na^{+}, (NH_4^{+} does not exist in natural zeolite-tuff).

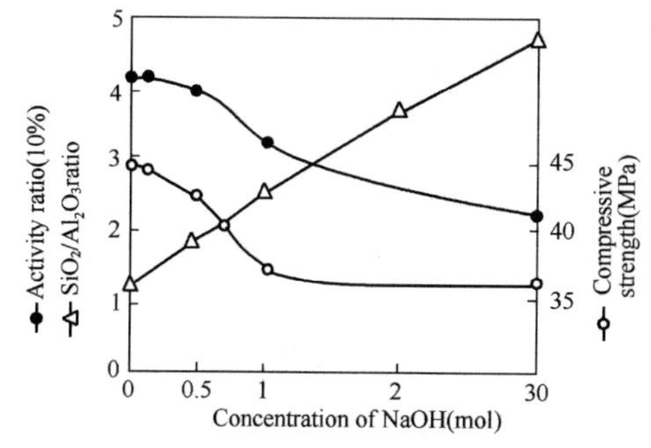

Fig. 3 Si/Al and activity ratio of zeolite tuff and strength of corresponding cement vs. normality of NaOH

2. Si/Al ratio and activity ratio of zeolite-tuff after treated by NaOH with varied normality. And strength of cements blended with 10% of above treated zeolite-tuff is tested separately. Results are shown in fig. 3. It is shown that Si/Al of zeolite is decreased with

increasing normality of NaOH, and that is identical with change of strength of corresponding cement. But the activity ratio is increased with the normality of NaOH. It is indicated that effect of structure is much more than of the chemical activity. This is different from the case of the other blending materials, strength of which is sensitive to the activity ratio[15].

Effect of fineness of tuff

Zeolite-tuffwhich was ground respectively to be 1300m²/kg, 1600m²/kg and 1900 m²/kg of special surface area are mixed into Portland cement by 10%, 20% and30% (by weight of total binder) respectively, and w/c is 0. 3. Compressive strength of those cement paste at 28th day are shown in fig. 4.

It is seen in Fig. 4 that the more amount of zeolite-tuff is blended, the more effect of fineness is. When the blending amount is about 10%, the highest strength occurs by appropriate fineness rather than by the finest, And for the blending amount over 20%, the finely-ground is better. The optimum blending amount is various for varied fineness of zeolite-tuff, and more with the fine.

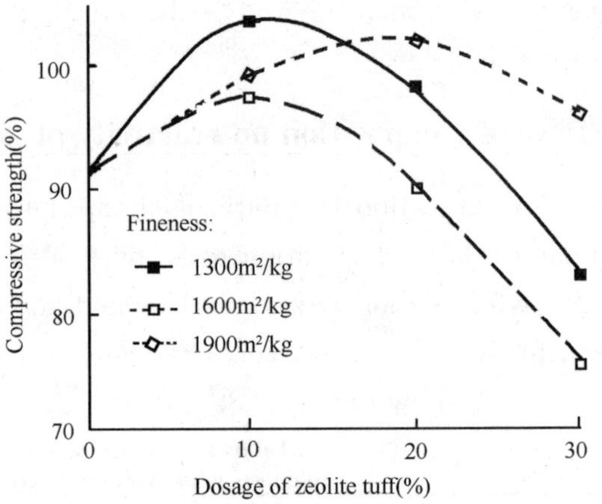

Fig. 4 Strength of cement vs. dosage of zeolite-tuff with varied fineness

CONCLUSION

1. Reactivity of zeolite-tuff is not due to collapse of crystalline structure substantially, but the reactive portion of amorphous tuff is. And that reactivity of zeolite-tuff is more than of glass is through its special structure-microcrystalline and micro cavity therein as well as zeolite water in cavity.

2. The optimum blending amount which makes the highest strength of cement zeolite-tuff is 5%~20% varied with the zeolite content. The optimum zeolite content which makes the optimum blending amount mostly is about 60% or more.

3. The suitable zeolite-tuff for blending material of cement is possessing Ca^{2+} or Na^{+} (Ca^{2+} is better to avoid harmful effect of Na^{+}) for clinoptilolite-tuff because of change of zeolite water coordinated with the positive ion in cavity.

4. The strength increase rate with special surface area increasing of cement blended with tuff by 30% is more than by 20%, but the optimum, position is varied. When blending amount is less, appropriate fineness is expected rather than the finest and when blending amount is over 20%, the finer is better.

REFERBNCES

[1] Sersale, R. et al . Pozzolanic activity of zeolites. Rend. Accad. Sci. Fis Mat.(Soc. Maz. Sci., Lett. Naples) 27, Ser. 4, 1960.

[2] Malquori, G. Portland-pozzolanic cement. 4th Int. Sym. on the Chemistry, Washington, 1960.

[3] Turriziant, R. Aspects of the chemistry of pozzolanas. The Chemistry of Cements, ed. by H. F. W. Taylor, 1964, p 69～86.

[4] Japanese pattern Bureau open report No. 194526, "Method to produce the hardened calcium silicate".

[5] Držaj, B. Hocevar, S. and Slokan, M. Kinetics and mechanism of reaction in the zeolite tuff-CaO-H_2O systems at increased temperature. C. and C. R. 8, 1977.

[6] Guo Jingxiong et al. Study on mechanism of reaction of zeolite in hydration of cement. Jornal of Chinese Silicate No. 3, Vol. 8, 1980.

[7] Lian Huizhen. Morphology of interface between zeolite-tuff and cement paste. Bond in Concrete ed. by P. Bartos, Applied Science Publisher, London, 1980, p 4～39.

[8] Lian Huizhen. Cement paste/zeolite-tuff bond. Proceedings of 1988 Engineering Foundation conference on advance of Cement Use and Manufacture, ed. by E. Gartner, 1988, 259～268.

[9] Makoto Asaka et al. Change in physics chemical properties of clinoptilolite tuff by chemical treatment (III) — Effect of Sodium Hydrate Solution Treatment. Journal of Society of Rock, Mineral and Deposit, 75, 257～265, 1980.

[10] Lian Huizhen. Cement paste/zeolite-tuff bond. Proceedings of 1988 Engineering Foundation conference on advance of Cement Use and Manufacture, ed. by E. Gartner, 1988, 259～268.

[11] Yamazaki, H. et al. Influence of pozzolana and fly ash on concrete. Concrete Technology, Nov., 1981 (in Japanese).

[12] Lea, F. M. The chemistry of cement and concrete. 3rd ed. Edward Arnold, 1970, p 530.

[13] Lian Huizhen. Cement paste/zeolite-tuff bond. Proceedings of 1988 Engineering Foundation conference on advance of Cement Use and Manufacture, ed. by E. Gartner, 1988, 259～268.

[14] Lian Huizhen, Chen Enyi. Mechanism of hydration of blended cement I. Strengthening effect of mineral powder on cement used in mortar or concrete. Proceedings of International concrete conference, 1990, IRAN.

[15] Lian Huizhen. et al. Evaluation on reactivity of FBC ash and effect factor thereof. Documents of Appraisal to a special topic 7th Five-Year-Project (in Chinese), Nov. 1990.

自评

1. 本文是 1992 年在印度召开的国际水泥化学会议的发言稿，研究证明结晶态沸石是一种热力学稳定的矿物，沸石凝灰岩在水泥中与 $Ca(OH)_2$ 的反应并非发生于结晶态的沸石，而是凝灰岩中的无定形 SiO_2 和 Al_2O_3，解释了含结晶态沸石的沸石凝灰岩比完全无定形的火山灰材料有更高的活性，是缘于含有丰富细微孔穴的架状铝硅酸盐结构的沸石特殊结构，并证明“化学成分通过结构决定性质”。1992 年在印度的第 9 届国际化学会议上，我对意大利学者 Sersale, R. 说：“你们提出的关于‘为什么结晶态沸石有比无定形火山灰更高的活性?’的问题我现在解决了。”他表示肯定。

2. 除了化学和物化方法的试验外，本研究还做了某些宏观规律的试验，其中主要是强度，但是文中未给出实验条件的控制，例如论文之三和之四中的图 1 是不同沸石含量的沸石岩不同掺量对水泥浆体强度的影响，未说明试件制备的条件，例如是否水灰比相同？还是流动性相同？如果水灰比相同，则可以比较相同掺量下晶态含量的影响；而相同晶态含量下沸石岩掺量的影响就不好比，因为浆体需水量会随沸石岩掺量的增加而增加，掺量不同时浆体流动性不同。同时，净浆的表现和混凝土的表现差别会很大，不能直接指导工程。混凝土的试验是必须的。

以上所选四篇论文中，只有第一篇是中文的，而其中的研究尚属初级阶段，并未突破传统的和从众的思维；另三篇都发表于国外。此前，1986 年 11 月，我在中国硅酸盐学会第三届水泥学术年会（柳州）上发表过“沸石岩活性的研究”的论文（未列入本选集），报道了区别“沸石”和“沸石岩”的系统实验研究，证明沸石岩的活性并非来自结晶态的沸石，而含结晶态沸石的凝灰岩活性比完全无定形的火山灰活性高，其机理却与结晶态沸石有关——结晶态沸石特殊结构的特殊物理作用。由于某种特殊原因，于 2001 年持我校土木工程系开具的知识产权证明，将该论文缩写后发表在《硅酸盐学报》第 30 卷第 4 期。作为研究的最终结论。以下是该篇论文和说明及自评。文中实验结果大都是前面几篇论文中所用过的，只是作为我对沸石岩活性研究的总结性的论文，对我的研究方法形成有代表性，故此选入。对论文之三中的评述，合并在论文之五的评述之中。

本世纪人类知识的发展突飞猛进，跨学科的知识更是如此。事实上，大部分科学的创新都是通过不同学科的融合、擦出火花来完成的。很多人都同意这个看法，但却忘记了一个重要的事情，就是有能力融合不同学科的学者，其能力和知识水平都要跟这些不同学科的专家大致相等。即使在某方面的知识跟不上，他也能理解问题的困难所在，找合适的专家求教，正如沃森找到克里克和爱因斯坦找到格罗斯曼帮忙一样。

摘自美国现代数学家、菲尔兹和沃尔夫奖得主 丘成桐 2012 年 5 月 13 日在学生科协“星火论坛”上所作的特邀报告《为学与做人》

论文之五

沸石岩火山灰活性的研究*

廉慧珍
（清华大学建筑材料研究所 北京 100084）

摘要 用化学的、物理的和物化的方法证明，沸石凝灰岩中的沸石是热力学稳定的物质，硅酸盐水泥中的碱度在常温下不足以破坏沸石的结构。沸石凝灰岩的火山灰活性来自凝灰岩中无定性的 SiO_2 和 Al_2O_3。含有结晶态沸石的凝灰岩之所以有比完全非晶态的火山灰较高的活性，是因为沸石特殊的结构所起的“水气交换”作用以及其晶体颗粒调节浆体中微结构的作用。沸石含量不同的沸石凝灰岩在水泥中的掺量以 5%～20%为宜，合适的类型为钙型和钠型。

关键词：沸石凝灰岩，火山灰活性，水气交换，自养护，微颗粒级配
中图分类号：TQ 172.4

STUDIES ON POZZOLANIC ACTIVITY OF ZEOLITE-TUFF*

Lian Huizhen
(Building Materials Research Institute, Tsinghua University, Beijing 100084)

Abstrat: By experiment of chemical, physical and physical chemistry method, it is proved that zeolite in tuff is stable in thermodynamics. Structure of zeolite cannot be destroyed by alkalinity in Portland cement paste at normal temperature, Pozzolanic activity of zeolite-tuff is from amorphous SiO_2 and Al_2O_3 therein. What activity of tuff containing crystalline zeolite is higher than that of uncrystalline materials is because of function of water-air exchange by the special structure of zeolite as well as effect of adjusting grading of micro-grain by crystalline zeolite. The suitable blended amount of zeolite-tuff in cement is 5-20% depending on zeolite content in tuff. The suitable type of zeolite in tuff used for Portland cement is that of calcium-type and sodium-type.

Keywords: zeolite-tuff, pozzolanic activity, water-air exchange, self-curing, grading of micro-grain

沸石是一种含水架状结晶的铝硅酸盐矿物，是由酸性火山熔岩玻璃或碎屑凝固沉积在地下后，与间隙水反应，并在温度和压力作用下形成的。如凝灰岩、流纹岩、珍珠岩、角砾岩、玄武岩等都可能产生这种变化。最常见的是沸石化的凝灰岩，称为沸石凝灰岩。20 年前，在我国兴起了开发利用天然沸石岩的热潮。我国地质工作者将以沸石为主要造岩矿物的

* 1984 年中国科学院科学基金资助课题（84）科基金（技）准字第 244 号，1985.2～1987.7。
收稿：2001-06-19，修改稿收到日期：2002-03-07。

岩石称做沸石岩。尤其是用作水泥混凝土的活性掺和料，沸石岩比其他火山灰质材料具有较高的活性，在一定的掺量下，可以提高强度。本文作者用沸石岩粉末和石灰混合加水后，确实有反应产物出现。在电子显微镜下观察，其反应产物的形貌与矿渣-石灰和火山灰-石灰的反应产物相同[1]。郭竞雄等人研究认为是沸石矿物的铝硅酸盐结晶在 $Ca(OH)_2$ 中的 $(OH)^-$ 作用下逐步破坏，并反应生成铝酸钙和硅酸钙等水化产物，提出"脱铝-解硅-吸钙多阶段反应"的模式[2]。实际上，早在20世纪中期，人们就发现这种"结晶态的沸石比其他完全非晶态的火山灰具有很高的反应能力"。此后，有人曾认为是沸石与石灰的反应[3]。然而，这种结论显然从热力学上无法解释。

从热力学上来说，晶态物质有比相同成分的无定形物质高得多的稳定性。沸石架状结构的硅（铝）氧四面体就具有相当高的稳定性，因而可用作分子筛和催化剂。为了提高硅铝比以提高沸石的催化活性，脱铝要经过很难的工艺处理。日本的浅贺质等人[4]用不同浓度的NaOH处理沸石含量为75.2%的斜发沸石岩，结果表明，NaOH浓度直到0.1mol时，沸石含量和沸石的硅铝比都无变化；NaOH浓度从0.25mol增加到1mol时，由于除去了不纯物，沸石含量稍有增加，硅铝比变化不大；NaOH浓度达3mol时，沸石含量和沸石的硅铝比都大大下降；NaOH浓度达5mol时，试样中已检测不到沸石的存在。该实验表明，沸石结构的破坏是很困难的，不是在常温下与水泥混合就能实现的。因此可认为，沸石岩中起化学反应的不是结晶态的沸石，而仍是非晶态的火山灰物质；沸石化的凝灰岩（即含有晶态沸石的火山灰）之所以有很高的活性，是由于沸石特殊结构所起的特殊作用。本文将介绍通过实验对以上设想的分析验证。

1. 实验用的原材料

嫩江斜发沸石岩、河北独石口斜发沸石岩和浙江缙云丝光沸石岩，磨细通过0.08mm方孔标准筛，沸石含量和成分见表1。

表1 实验所用沸石凝灰岩的成分

Tab. 1 Compositions of zeolite-tuff used for experiment

Type of zeolite-tuff	Amount of NH_4^+ exchanged (mEq/L of sample)	Zeolite content (%)	Composition (%) (by mass of total sample)								Temperature for thermal stability (℃)
			SiO_2	Al_2O_3	Fe_2O_3	CaO	MgO	K_2O	Na_2O	LOI	
Clinoptilolite for Nenjiang	179	84	61.60	14.40	2.20	3.79	1.85	3.13	0.70	12.83	400
Clinoptilolite of Hebei	138	64.7	66.52	11.37	1.55	3.26	0.98	2.25	0.75	12.80	
Mordenite of Zhejiang	137	60	68.96	10.28	1.97	0.92	0.48	0.60	3.40	13.72	800

其他有关样品的一般情况作如下介绍：

①大连化学研究所的合成丝光沸石分子筛、分析纯CaO。CaO消解成 $Ca(OH)_2$ 备用。

②水泥：柳州水泥厂五羊牌水泥，化学成分为(质量%)：SiO_2 22.07，Al_2O_3 7.01，Fe_2O_3 4.90，CaO 57.32，MgO 1.61　Na_2O 1.42，K_2O 0.5。

③不同晶态含量的沸石岩：

将10mm左右的嫩江斜发沸石岩颗粒在500℃下焙烧恒温半小时后水淬，干燥后磨细通过 0.08mm 方孔标准筛，经 x-射线衍射表明已无沸石存在，出现很宽的非晶态干涉区。以不同比例将非晶化处理的沸石岩与原岩样混合(0∶1、1∶3、1∶1、3∶1、1∶0)，制成斜发沸石含量(%)分别为 82、63、42、21 和 0；用合成丝光沸石分子筛与浙江丝光沸石岩分别按 1∶0、1∶1、0∶1 混合配成晶态含量不同的试样。

2. 实验及其结果分析

2.1 晶态沸石稳定性的验证

2.1.1 将 100 克嫩江沸石岩分别加入 4000 c.c. 不同浓度的 NaOH 溶液和 0.1%的$Ca(OH)_2$溶液，沸煮 1.5 小时后过滤、洗涤，提取干试样进行 x-射线衍射，结果如图 1 所示。

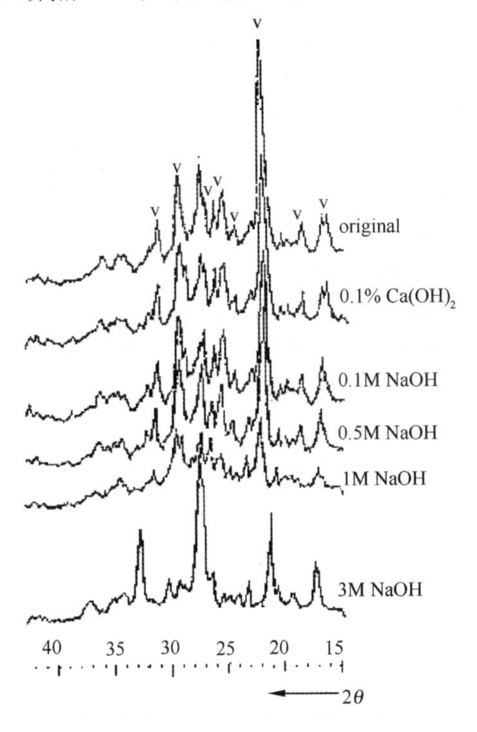

图 1　分别用不同浓度的 NaOH 和 $Ca(OH)_2$ 处理的斜发沸石凝灰岩试样 XRD

Fig. 1　XRD of Zeolite-tuff treated by various Concentration of NaOH and 0.1% $Ca(OH)_2$

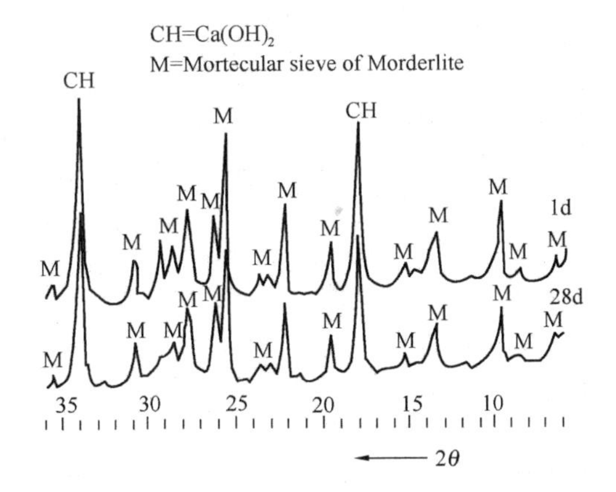

图 2　沸石分子筛与 Ca $(OH)_2$混合的试样 x-射线衍射分析

Fig. 2　XRD of sample mixed with zeolite sieve sorbent and $Ca(OH)_2$

由图 1 可见，当 NaOH 浓度低于 0.5mol 时沸石结构无明显变化；当 NaOH 浓度为 1mol 时，斜发沸石主衍射峰高度明显降低，结构开始破坏；NaOH 浓度增至 3mol 时，则斜发沸石所有晶面衍射峰全部消失，出现新的结晶，表明沸石结构完全破坏。经 0.1% $Ca(OH)_2$溶液[在 20℃时$Ca(OH)_2$饱和溶液浓度为 0.13%]处理的试样，与浓度低于 0.5mol NaOH 处理的试样一样，保持斜发沸石的结构无明显变化。由此可推断水泥浆体中的碱度不足以破坏沸石的结构。

2.1.2 合成丝光沸石、嫩江斜发沸石岩与消石灰分别按 1∶1 混合加水成型养护 1 天和 28 天，进行对 x-射线的衍射，如图 2、图 3 所示。可见，合成丝光沸石分子筛与消石灰的试样中，丝光沸石和$Ca(OH)_2$的特征峰均无明显变化，说明基本无反应；嫩江天然沸石岩与消石灰的试样，则 28 天的$Ca(OH)_2$特征峰比 1 天时的显著下降，而斜发沸石的特征峰则极少变化，说明天然沸石岩中参与反应的不是结晶态的沸石。

2.1.3 用独石口斜发沸石岩分别以10%、20%和30%等量取代水泥，测定水化7天和28天及水胶比分别为0.4和0.35的试样铵离子净交换容量，扣除相应水泥的铵离子交换容量，同时根据沸石岩中的沸石含量和沸石岩在水泥中的取代量计算试样水化前的沸石含量理论值作为对比，计算结果列于表2。如果沸石在水泥浆体中参与了反应，则沸石含量应随水化龄期的增长而减少，而由表2可见，实测试样中的沸石含量并不随水化龄期的增长（从7天到28天）而减少，与水胶比也无明显的关系，而是只与沸石岩掺量有关，并且与水化前的各试样中沸石含量计算值相符。此结果表明沸石在水泥中并不参与反应。

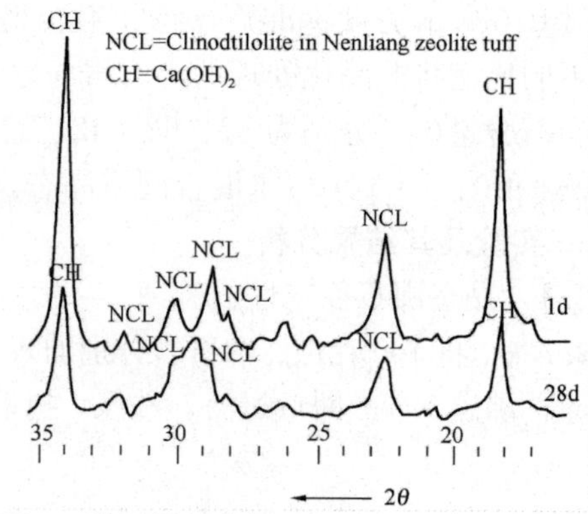

图3 沸石凝灰岩与$Ca(OH)_2$混合的试样x-射线衍射分析

Fig. 3 XRD of sample mixed with zeolite-tuff and $Ca(OH)_2$

表2 水泥浆体中沸石含量的实验值和计算值

Table 2 Results on test and calculation of zeolite content in cement paste

Name of sample	Amount of NH_4^+ exchange (M/100g of sample)	The real Amount of NH_4^+ exchange（mol/100g of sample）		Zeolite content in sample（%）	Calculated zeolite content before hydration of sample（%）
		Amount of NH_4^+ exchange in cement	Real value		
Clinoptilolite of Nenjiang	0.138	—	—	0.065	—
Pure Cement	0.049	—	—	—	—
10%Zeolite-tuff w/c=0.4，28d	0.059	0.044	0.015	0.007	0.006
20%Zeolite-tuff w/c=0.4，28d	0.068	0.039	0.030	0.014	0.013
30%Zeolite-tuff w/c=0.4，28d	0.073	0.034	0.039	0.018	0.019
30%Zeolite-tuff w/c=0.4，7d	0.078	0.034	0.041	0.020	0.019
30%Zeolite-tuff w/c=0.35，28d	0.079	0.034	0.045	0.021	0.019

2.2 沸石结构作用的研究

2.2.1 沸石结构充满大小均一的细微空穴和孔道，具有吸附、阳离子交换的性质和脱附自由的沸石水。将沸石岩颗粒置于水中，可见到从沸石岩颗粒放出气体，而水面有所下降。用奥萨特气体分析仪分析该气体为空气。这表明，沸石岩中的沸石对水的吸附性大于对空气的吸附性。干燥的沸石岩中沸石的孔在空气中吸附了空气，遇水后，沸石则吸附水而释放出空气。用斜发沸石岩的抛光面与水泥浆体粘结，待水泥浆体硬化后，将试件从界面处断开，在电子显微镜下观察界面两侧形貌。发现在沸石面和水泥浆体面上均有气孔产生，如图4所示为沸石岩表面粘结的水泥浆体层出现的气孔。这就是沸石岩中的沸石吸附了水泥浆体中的水而放出空气形成的孔。

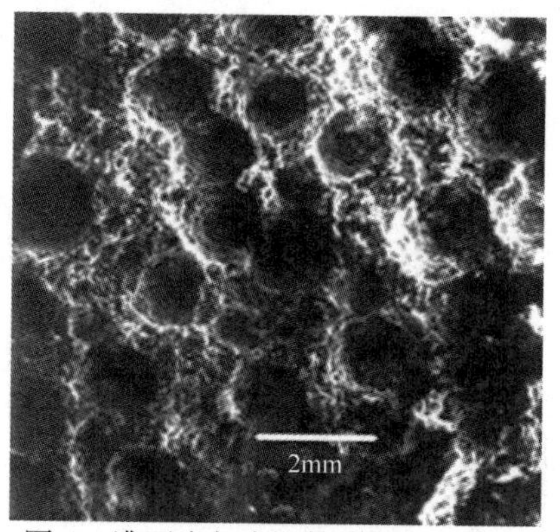

图 4 沸石岩与水泥粘接面沸石岩表面的界面孔

Fig. 4 The interface pores on surface of zeolite-tuff

2.2.2 以沸石岩颗粒为骨料，用水泥浆体包裹成型，标准养护到一定龄期，在扫描电镜下观察其界面缝的变化，如图 5 所示。

由图 5 可见，水化 1 天时，上述试样沸石岩骨料和水泥浆体之间的界面缝宽度约为 100 μm；水化 28 天时，界面缝宽约 3 μm；3 个月后，将图象放大到 3000 倍也看不到界面缝的存在。这一方面是由于沸石岩中的活性组分和水泥浆体不断反应的产物填充界面缝，另一方面也由于界面的“水气交换作用”：当浆体中的自由水随水化龄期的增长消耗而减少后，沸石孔中所吸附的水又释放出对界面处的浆体进行自养护，并提供沸石岩中活性组分与水泥反应所需要的水，使界面不断得到加强。由此可推断，沸石特有的结构导致与水泥有相同细度的沸石岩粉末颗粒在水泥浆体中的“水气交换”行为，因而表现出沸石岩的“较高活性”。

2.2.3 分别用 30%的磨细沸石凝灰岩和磨细矿渣等量取代掺入水泥，以相同配合比配制混凝土，另以无掺和料的空白试样作对比。将以上各试样自然养护至 7 天和 28 天，检测其抗压强度见表 3。

1st day

cement paste
30μm
zeolite tuff

28th day

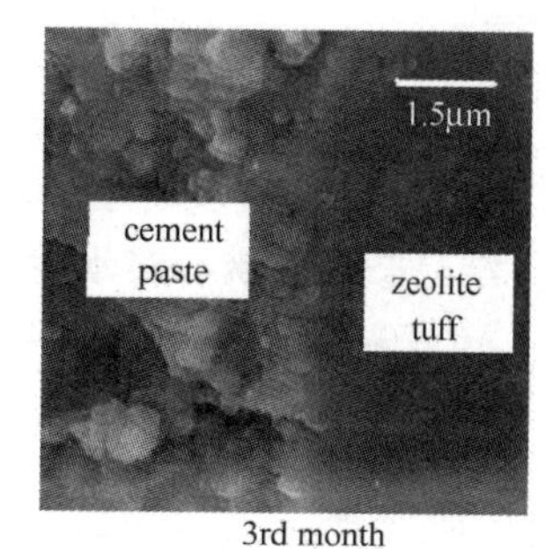

3rd month

图 5 沸石凝灰岩和水泥浆体之间的界面缝随龄期的变化

Fig. 5 Changes of interface gap between zeolite-tuff and cement past with ages

表 3 用掺入 30%不同掺和料的混凝土在自然条件下养护的试件强度对比

Table 3 Comparison of strength of concrete with various blended under natural condition

No	Blended material	w/b	Compressive strength of concrete（MPa）		
			7 day	28 day	Increase rate（%）
1	Zeolite-tuff	0.4	13.4	26.0	94
		0.6	7.5	20.2	169
2	Granulated slag	0.4	16.3	23.2	42
		0.6	9.9	19.9	101
3	Without	0.4	18.9	28.9	52
		0.6	9.2	17.8	98

由表 3 可见，用不同掺和料和不同的水胶比的混凝土，从 7 天到 28 天的强度增长率有明显的差别。掺入矿渣和无掺和料的强度增长率相差无几，而掺沸石凝灰岩的强度增长率则高得多。

2.2.4 将经高温（加热到沸石热稳定温度）非晶化处理的凝灰岩（称烧沸石岩）与原样以 30%分别掺入水泥，配制混凝土在自然条件下养护至 7 天和 28 天，检测其抗压强度如表 4 所示。

表 4 掺入不同结构沸石凝灰岩的混凝土强度发展对比

T able 4 Compressive of strength of concrete with zeolite-tuff in different structure

Blended zeolite-tuff	Compressive strength of concrete (MPa)		Increase rate (%)
	7 day	28 day	
Burned (uncrystallized)	16	25	56
Unburned (original structure)	17	38	120

由表 4 可见，沸石凝灰岩原样的混凝土，后期有很高的增长率，而经高温处理非晶化后的凝灰岩因沸石结构破坏，失去其颗粒与水泥浆体之间界面的"水气交换"作用，则其混凝土后期强度增长率大为下降。

2.2.5 分别以 30%和 50%的磨细沸石凝灰岩掺入水泥，配制混凝土，检测不同沸石岩掺量对混凝土强度增长率的影响见表 5。

表 5 掺入不同掺量斜发沸石凝灰岩的混凝土强度增长率对比

Table 5 Comparison of increase rate of concrete strength with varied dosage of clinoptilolite-tuff

Dosage of blended zeolite-tuff (%)	Increase rate of concrete strength from 7 to 28 day (%)
30	106
50	132

表 5 所示沸石凝灰岩掺量为 50%的混凝土比掺量为 30%的混凝土后期强度增长率大，也可证明沸石凝灰岩中沸石结构的作用。

3. 结晶态物质沸石对水泥浆体微结构的作用

1976 年，Taylor H. F. W 根据加拿大学者 R. F. Feldman 和 J. J. Beaudoin 的研究结果提出[7]，对于强度来说，水泥浆体组成（即不同质地的颗粒和孔）存在最佳的级配，强度并不是只取决于孔隙率。在孔隙率一定时，合适的晶胶比可使浆体获得最高的强度[7]。浆体合理的微结构可通过改变水灰比和掺和料种类、细度、掺量以及养护工艺等来得到。含有结晶态沸石的凝灰岩之所以有比非晶态火山灰较高的活性，应当也会有调节水泥浆体微结构的作用，图 6 所示为不同沸石含量的凝灰岩（见本文第一节）的掺量和浆体抗压强度的关系。试样尺寸为 2cm×2cm×2cm，水胶比相同。图中的每个点为 30 个数据的平均值。图 6 表明，晶态的沸石含量不同的凝灰岩在浆体中的最佳掺量范围是不同的。当掺量超过 5%以后，沸石含量与浆体强度的关系并非是单调增的。用丝光沸石分子筛和浙江天然丝光沸石岩按不同比例混合后，再分别与化学纯的 CaO 以 1∶1 混合加水成型，养护到一定的龄期，检

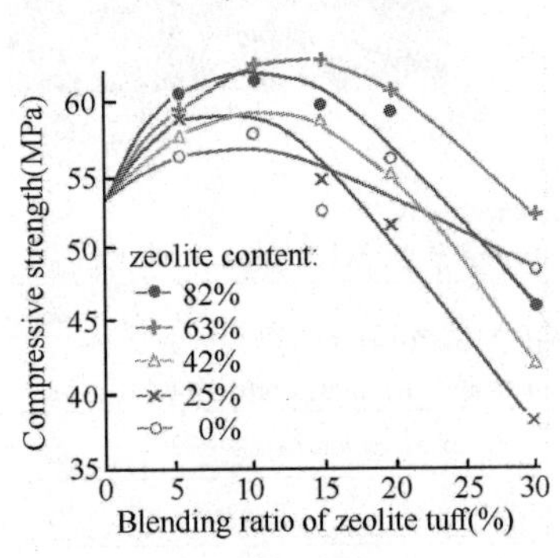

图 6 不同沸石含量的沸石凝灰岩掺量不同时混凝土 28d 抗压强度的关系

Fig. 6 Compressive strength vs. blended amount of zeolite-tuff with varied zeolite content at 28th day

测试样强度、活性 SiO_2 和 Al_2O_3 和 XRD 图谱中 $Ca(OH)_2$ 相对衍射强度，如图 7 所示。可见与图 6 所示相似的结果。

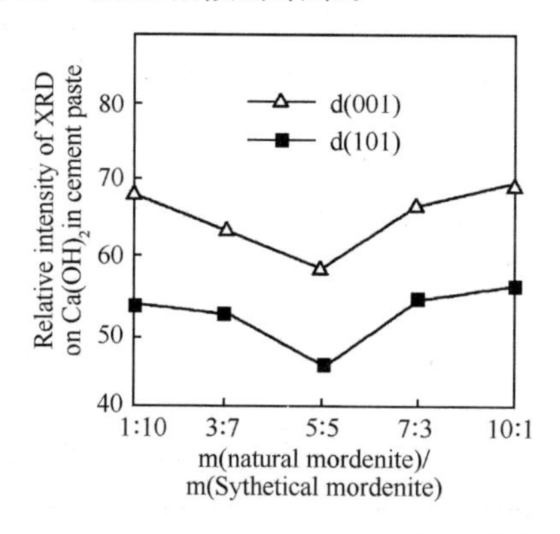

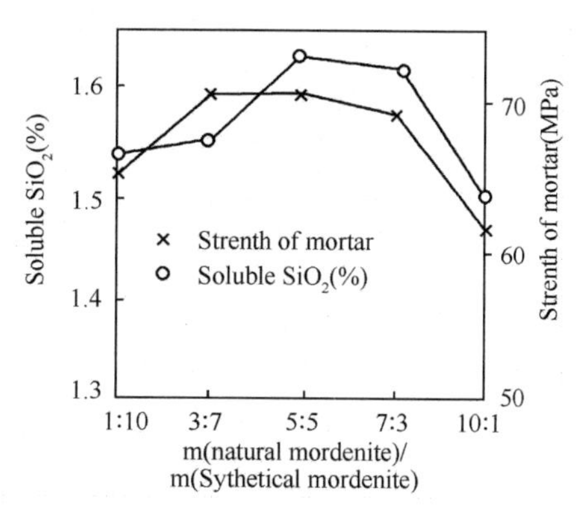

图 7　晶体含量对沸石凝灰岩活性的影响

Fig. 7　Influence of crystal content on activity of zeolite-tuff

4. 沸石中阳离子的作用

沸石孔穴中的配位阳离子是沸石结构的重要特点，而沸石水是和阳离子配位的。不同种的阳离子有不同数量的沸石水，沸石凝灰岩就会表现出活性的差别。清华大学郭玉顺用离子交换的方法改变斜发沸石孔穴中的阳离子测定沸石水的数量，并以 10%的沸石凝灰岩掺入水泥，测定试样的胶砂强度[5]，如表 6 所示。

表 6　沸石孔中阳离子交换后掺沸石中的沸石水和掺凝灰岩的水泥浆体强度的变化[5]

Table 6　Zeolite water and strength of cement paste blended with zeolite-tuff after cation exchanged in cavities of zeolite

Type of cation	Compressive strength of paste (MPa)			Amount of zeolite water (%)
	3 day	7 day	28 day	
Original (multiform)	15.3	24.9	39.2	—
NH_4^+	14.1	24.9	44.2	9.9
Ca^{2+}	16.6	27.6	44.3	7.78
Na^+	18.7	28.2	42.0	6.83
K^+	17.2	27.0	39.5	4.32

由表 6 可见，掺入沸石凝灰岩的水泥浆体 28 天强度随沸石中沸石水的增加而增加，表明沸石水参与了“水气交换”的作用。效果最好的阳离子是 Ca^{2+} 和 Na^+（自然界没有 NH_4^+ 型沸石）。

5. 结论和讨论

（1）架状铝硅酸盐结晶的沸石是热力学稳定的矿物，用做水泥掺和料时，水泥浆体的碱度不足以破坏沸石凝灰岩中沸石的结构。因此，不存在沸石结构的“脱铝-解硅-吸钙”。

（2）沸石凝灰岩的火山灰活性来源于凝灰岩中无定形的 SiO_2 和 Al_2O_3。

（3）含有晶态沸石的沸石凝灰岩之所以具有比非晶态物质高的活性，是由于沸石特殊的结构所起的物理作用，即沸石的微孔穴吸收水泥浆体中的水随水泥水化的进行而逐渐释放而起自养护的作用。用于水泥混凝土掺和料的沸石凝灰岩中晶态的沸石含量并非越多越好，纯度为100%的合成沸石不具有火山灰活性；60%左右的较好。

(4) 为得到水泥浆体中合适的微颗级配（晶胶比），沸石含量不同时，有不同的最佳掺量。但都在5%～20%的范围。

(5) 由于不同种类的沸石有不同的微结构，含有显微孔径小的沸石（如四元环的方沸石）的沸石岩不宜用做水泥混凝土掺和料，最合适的是自然界储量很大的含钠型和钙型斜发沸石（五元环）和丝光沸石（六元环和八元环）的凝灰岩。

参考文献

[1] 廉慧珍(Llian Huizhen). 沸石岩在水泥中的作用机理探讨[A]. 1980、1981年水泥学术会议论文选集[C]，北京：中国建筑工业出版社(China Construction Industry Press) 1980：228～233.

[2] 郭竞雄(Guo Jingxiong)，梁春林(Liang Chunlin)，沸石在水泥中作用机理的研究[J]. 硅酸盐学报(J Chin. Ceram. Soc.)，1980，8(3)：244～257.

[3] Držaj B，Ho evar S，Sokan M. Kinetics and Mechanism of Reaction in the Zeolite tuff-CaO-H_2O System at Increased Temperature[A]. Cem. Concr. Res. (J)，. 1977，7(4) 8：711～720.

[4] 浅贺质，崛田正己，鳥居一雄．化学处理による斜プチロルプッ石の物理化学的特性の变化(第三報)，水酸化ナトリゥム溶液处理による影響，岩石鉱物床学会志[J]．Mineral Deposit of Rock) 1980，75：257～265.

[5] 郭玉顺(GuoYushun)等．沸石岩轻骨料膨胀机理的研究[J]. 硅酸盐学报(J Chin Ceram Soc)，1985，13(1)：12～18.

自评

1. 这篇论文对我的科学研究方法有一定代表性，概述如下：

(1) 对已有现象和结论的质疑，提出问题、思考和探索：人们都认为沸石有和石灰反应的活性，但沸石是结晶态物质，从热力学上来说不通；──→通过调研和思考，认为结晶态的沸石是热力学稳定的，常温常压下没有火山灰活性，否则沸石就不能用做催化剂和分子筛；含有沸石的沸石凝灰岩的活性应当是来源于凝灰岩中的无定形部分；──→那么为什么含沸石的凝灰岩有比无定形火山灰更高的活性？──→研究沸石结构的特点：其晶体的微结构中有大量开放的孔穴和通道、能自由吸附、脱附的沸石水与可交换的阳离子。

(2) 对研究结果的科学预想：沸石岩中的结晶态沸石在水泥中是不发生化学反应的，沸石岩的化学活性来自于凝灰岩中的无定形SiO_2和Al_2O_3；沸石结构的作用主要是沸石微孔在与水泥浆体界面的水气交换作用和晶体颗粒在胶凝材料体系中的微粒级配作用，如果沸石结构被破坏，沸石岩的活性就会降低。

(3) 研究技术路线的制定：分别改变沸石岩中晶体沸石结构和晶体沸石含量，用化学的和物化的以及物理的方法从不同角度证明沸石结构的稳定性、界面水气交换作用、微颗粒级配作用。

(4) 该实验的结果对正确使用沸石岩作为掺和料的用途是：①证明了沸石岩的活性并非晶体的沸石本身的活性，而是结晶态沸石矿物的存在对火山灰质材料活性起促进作用，因此不是沸石含量越大的沸石岩活性越高；②因为沸石特殊作用的存在，适合于做水泥和混凝土掺和料的沸石岩需要有30%以上的沸石含量，所以沸石岩的质量标准中要有沸石含量的指标；③由于沸石在水泥浆体中的作用主要由于孔穴的存在，不是任何沸石都适合用于水泥混凝土，如方沸石就不适用。

（5）由于多微孔的沸石岩吸水率较大，在水泥混凝土中的掺量不宜大于30%，最佳掺量是不超过20% 。但如果改变使用方法，是否会产生另外的效果呢？

2. 对文中问题的说明：

（1）x-射线衍射试样养护天数　养护1天是因为时间很短，无论掺和料是否有活性，1天基本上不会有明显的反应。养护到28天的$Ca(OH)_2$特征峰峰高比1天时的显著下降，而斜发沸石的特征峰则极少变化，这说明在28天以前天然沸石岩中参与反应的不是结晶态的沸石。

（2）关于x-射线衍射特征峰的变化：图2中$Ca(OH)_2$只有18°处的4.90Å晶面距的衍射峰有微小下降，可能由杂质造成，其余所有特征峰的均无显著变化；图3中$Ca(OH)_2$ 2.63Å和4.90Å主强特征峰峰高均发生显著变化，表明沸石岩与石灰发生了反应；因天然沸石岩中沸石含量的波动可能造成制样误差，而使斜发沸石特征峰强有细微变化，但相应于$Ca(OH)_2$，则变化不显著，表明反应不是沸石发生的。严格地说，如果进行定量分析，只用衍射峰高(衍射强度)是很不准确的，可以用于定性分析，但是不能根据一个强峰下结论。

（3）用简单的物理方法的证明：由表2可见，沸石含量检测值和计算值的绝对误差都是±0.001个百分点，和水胶比与养护龄期无关，只和沸石岩掺量有关，因此该组数据对证明沸石在水泥中无反应更有说服力。

（4）沸石岩经过非晶化处理后沸石结构被破坏的证据：HN_4^+离子交换容量为0，对x-射线衍射花样中沸石特征峰已全部消失。

3. 尚存在需要研究的问题

（1）所有试验都是在养护到28天以前进行的，沸石属于酸性矿物，在浆体中处于硅酸盐水泥这种碱性环境中，尤其在目前水泥中碱含量普遍较大的情况下，28天以后的长期效应如何？沸石结构在水泥的碱性环境中会永远稳定吗？

（2）用x-射线衍射方法进行定性分析较直观，而定量分析需要操作严格，尤其对于成分波动的原材料；在真正需要准确定量时，往往要多做几次，以便达到稳定状态，同时最好使用计数或峰面积计算，峰高或半高宽都是很粗略的。

（3）在当时的实验室试验中，磨细沸石岩的细度一般都是粉磨时间相同，全部通过0.08mm筛，试验中未考虑胶砂浆体或混凝土的流动性，实际上掺入沸石岩后，需水量都会增加，改变掺量时未改变水胶比。这是当时未结合生产造成的缺憾。

（4）沸石结构的特性对掺入水泥或混凝土后的其他物理性质影响如何？例如开裂敏感性、因多孔、吸水性大，是否可能有抗冻性问题？

（5）以上的研究都是在常温下的实验室条件进行的，在混凝土结构实体中，则由于混凝土内部的温升，已大不同于实验室条件。已知石英砂磨细至与水泥相当的细度时，加水与石灰成型，在水热条件下会发生反应生成托勃莫来石；在压蒸条件下会生成体积更加稳定的硬硅钙石。那么，沸石岩中的结晶态沸石是否也会有这种效果呢？即使不使用水热合成的方法，在混凝土结构实体中，最高温度可达约80℃，如果有必要使用沸石岩做掺和料时，在常温下不反应，而在高温下如何？在压蒸条件下又会如何？

以下作为我的学术性研究的第二个实例

二、啃一下别人啃剩下的硬骨头

1976 年，加拿大学者 R. Feldman 和 J. Beaudoin 在 C.&C.R. Vol.6 发表了 MICROSTRUCTURE AND STRENGTH OF HYDRATED CEMENT。H. Taylor 介绍说，他们用孔隙率为横坐标，抗压强度的对数为纵坐标作图分析不同水灰比、不同养护温度下水泥浆体的孔隙率和抗压强度的关系，如图 1 所示。AB 线是室温下得出的，曲线左端水灰比很小(约可能为 0.2)，右端水灰比大（可能约为 0.8)。很明显，高水灰比的孔隙率大，从而强度低。CD 线被该作者称为高密度线，但不是指整体高密度而是指颗粒具有高的真密度。CD 线的右下端是不含 SiO_2 的水泥浆体，水灰比约为 0.8，经压蒸后所得数据，压蒸的浆体微结构由像 α-C_2S 水化物这样的晶体所构成，颗粒尺寸较大，真密度也较大，但是孔隙率很高，强度比常温养护的水泥浆体的强度低；左上端是 D. Roy 用热压力成型的浆体，水灰比约为 0.2，也是由较大的颗粒构成，其中包括未反应完的熟料和晶体水化产物。这一点说和压蒸的物质相似，但孔隙率却非常低，所得强度高于相同孔隙率的常温养护浆体的强度。以上表明，在高孔隙率下，由于颗粒之间距离大，将颗粒结合在一起是重要的问题，如欲提高强度，就需要细颗粒，而往往这是低密度的无定形颗粒，如图中的 EF 线，因掺入火山灰（H. Taylor 认为可能是粉煤灰——1982 年来华讲座），该强度曲线落在最上方；在低孔隙率的情况下，颗粒之间距离小，其间粘接力就是次要的，颗粒本身强度更重要，则需要高密度、大晶体的颗粒。Feldman 和 J. Beaudoin 认为 EF 线是达到了孔隙率与强弱颗粒的最佳配合。(H. Taylor 1980 年 4 月来华讲座，地点为国家建材院)。

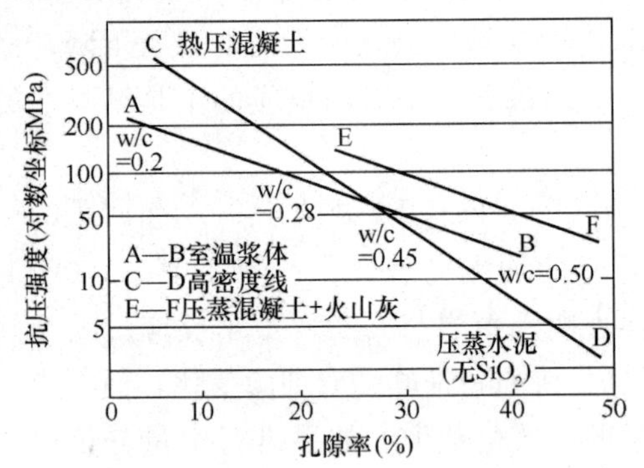

图 1 Feldman 和 Beadoin 用不同材料、不同配比和不同工艺试验得出孔隙率和抗压强度的结果

H. Taylor 根据 Feldman 和 Beaudoin 的结果用不同的方法作图（DISCUSSION OF PAPER " MICRO-STRUCTURE AND STRENGTH OF HYDRATED CEMENTMI" BY R. FELDMAN AND J. BEAUDOIN, C. & C. R., Vol. 7 pp 465～468, 1977）如图 2 所示，纵坐标是孔隙率，横坐标的参数无法给出定量的数值，Taylor 称之为粗大、致密的晶体物质含量。这只是一示意图。图中的曲线是等强度曲线。曲线上的数值单位为 MPa（括弧内数值的单位是 Psi×10^{-3}）。AB、CD、EF 各线均与图 1 中的同义。（DISCUSSION OF PAPER "MICRO STRUCTURE AND STRENGTH OF HY DRATED CEMENTMI" BY R. FELDMN AND J. BEAUDOIN,（C. & C. R., Vol. 7 pp 465～468, 1977）。为方便分析，图中的虚线是我加的，可见各曲线的

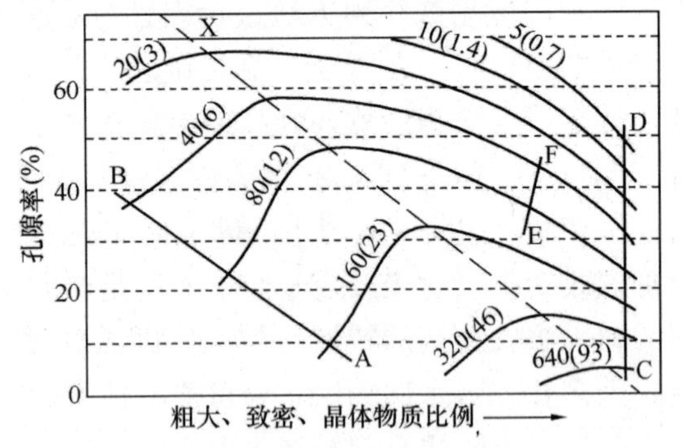

图 2 H. Taylor 根据 Feldman 和 Beaudoin 的试验结果作图

极值随孔隙率的下降而右移。

图2中的趋势和图1的趋势是一样的，即孔隙率越大，水泥水化程度越高，出现粗大、密实、结晶的颗粒越少；孔隙率越低，留下未反应完的水泥越多，则粗大、密实、结晶的颗粒越多。对于一定的孔隙率，如果引进较多的粗大、致密的结晶物质，就可以使强度提高。这就表明“物质的化学成分通过结构决定强度”（唐有祺《结晶化学》p.126，人民教育出版社1965年）。

H. Taylor这个等强度曲线的缺点是横坐标无定量，纵坐标也只是孔隙率，而未考虑孔的结构（孔结构包括孔隙率、大小孔径的级配、孔的形状、孔在空间位置的分布）。但是所提出水泥浆体的孔和颗粒的级配已是混凝土微结构研究的进步。H. Taylor在分析R. Feldman和J. Beaudoin的实验结果时，注意到了在高孔隙率的浆体中掺入细颗粒（如火山灰），可以提高强度。那么，作为矿物掺和料的其他细颗粒也应当有此作用，而不同的矿物掺和料的作用应当是不同的，水泥浆体中所需要的细颗粒不一定只是低密度的无定形颗粒。这引起了我的兴趣——水泥以外的不同的细颗粒作用结果如何？是否能变化不同养护温度、不同水胶比和不同细颗粒等宏观参数来得到不同微结构？是否能先研究将H. Taylor的等强度曲线量化？当然在当时还没有悟到这是一种分解论的研究方法。

1989年我向当时的国家建材局申请了一个建材行业科技发展基金项目《水泥浆体微结构与宏观行为的关系及其定量分析》。知道的人说，这个题目做不成，是“啃别人家啃过的硬骨头”。的确，国外已有报道，认为用x-射线衍射方法定量分析水泥浆体之所以不可能，困难之一是某些组成得不到纯样，例如钙矾石，天然矿物很难找到，即使找到岩石，也不可能满足纯度的要求，人工合成也达不到90%以上的纯度，所以在JCPDS（国际粉末衍射标准联合会）卡片中查不到钙矾石的K值。用作结晶态物质和无定形物质共存的掺和料，沸石是最典型的，但是我国储藏量大而可用的斜发沸石没有人工合成的纯样，也很难从天然沸石凝灰岩中提纯。困难之二是熟料中的C_2S、C_3S和水泥浆体中$Ca(OH)_2$碳化生成的$CaCO_3$这些重要的晶相物质对x-射线衍射的特征峰重叠。但是我想别人啃不动的我们不一定也啃不动；不去啃一下，怎知一定啃不动？哪怕啃下来一点点，也会有收获。我的做法是先想好要解决的问题：从FBT模型得到的命题是：一定的宏观参数可以得到不同的微结构；有一定的微结构就会有一定的宏观行为；结果应当是可以不问过程（即微结构）而直接用宏观参数设计与预测宏观行为。FBT的等强度曲线需要解决两个问题，一个是纵坐标的孔隙率改成孔结构会更加合理，但是这个难度太大，首先就是微结构的特征化的问题。目前涉及孔结构，大多都假设孔的形状是球状或墨水瓶状，最常用的是汞压力测孔技术，得出的数据都属于开放孔，用“最可几孔径”或“小孔和大孔的比例”作为孔结构的量化特征，且不说测控技术的可靠性和特征化参数的代表性，但就工作量来看，够几个博士生做的。另一个问题是量化横坐标。按FBT的观点来看，可以把固相颗粒简化成两类——一类是粗大的、致密的、结晶态的，另一类是细小的、松软的、无定形的，最可量化而且可测得的分界线就是晶体和非晶体。当时吴中伟先生经常参加我们的讨论，他提出，可以就用H（Hard）粒子与L（Light）粒子的比值（即H / L）做横坐标。

在这里选入三篇文章：一篇是在研究开始时现做的一次探索性试验，以了解不同掺和料的性质，作为微结构元界定的依据；第二篇是所说的基金项目鉴定文件中研究工作报告，目的是说明当时的研究的思路和研究方法：明确研究的目的（要解决的问题难点及对策）、预期结果设想、对微结构元特征化的假设和选择、技术路线和实验方案的制定。在此只选择当时所发表的主要论文。

读书有两种态度。一种人是躺在书上读书，作者怎么说，他就怎么听，完全听凭作者牵着鼻子走；另一种人是站在书上读，经常同作者进行争论，作者讲得对就听，讲得不够清楚就想办法替作者讲清楚。前一种人即使读一辈子书，充其量也不过是个书呆子，不会有大作为，另一种人不仅吸取了前人的成果，还会看到未被开垦的荒原，从而去开拓，去耕耘。”所以要“站着读书”，而不是“躺着读书”。

——理论化学家、吉林大学教授　唐敖庆

唐敖庆院士是国际学术界遐迩闻名的科学家，是中国现代理论化学的开拓者和奠基人。唐敖庆先生以渊博的知识、丰富的经验和惊人的毅力献身于国家教育与科学事业。他把自己坚实的数理基础有机地应用于理论化学的研究工作中，形成了独特的科学研究风格，被国际上誉为“中国学派”。

论文之六

MECHANISM OF HYDRATION OF BLENDED CEMENT
I. Strengthening Effect of Mineral Powder on Cement Used in Mortar or Concrete*

Lian Huizhen, Chen Enyi
Tsinghua University, Beijing, China

ABSTRACT Portland cements are blended with various mineral materials powder separately. The blending materials are silica fume, zeolite-tuff, oil shale ash, fly ash, kieselguhr and ground quartz sand. There are peak values in the curves of relationship between strength and blending amount, and there is a crest curve relationship between the peak values, that is, there is a different effect on strength of cement among various blended materials. Such effect is related to the active SiO_2 and active Al_2O_3 therein, but the chemical composition takes its effect by means of its structure. Therefore, for determining the activity of blending materials, both the structure and chemical activity are important.

Keywords: mechanism blended cement mineral powder activity strengthening effect Hydration

摘要 以硅灰、沸石凝灰岩、油母页岩、粉煤灰、硅藻土和磨细石英砂等不同的矿物掺和料分别掺入硅酸盐水泥，当所制成的胶凝材料中各掺和料掺量变化时，在其胶砂强度与掺和料掺量关系曲线上有不同的峰值，即，随掺和料掺量的增加，其强度并非单调增或减，而是具有先升后降的趋势，而且不同掺和料具有不同的强度效应。这种效果和掺和料的活性SiO_2与活性Al_2O_3有关，但是物质的化学成分通过其结构决定强度。因此，评价矿物掺和料时，其结构和化学活性都重要。

关键词 机理 混合材水泥 矿物掺和料 增强效应 水化

INTRODUCTION

A kind of quality blending materials acts not only to save energy but also to increase strength of cement even concrete in an optimum blending amount.

It is known that silica fume is considered as an excellent blending material for high strength concrete. By suitable amount of silica fume, strength of concrete can be higher than that of plain concrete with the same w/c. But by a great amount of experiments, it is found that besides silica fume, some other natural mineral rock or industrial waste powder also have such effect so long as the suitable amount is blended, such as natural zeolite rock

---

* By the aid financially of National Building Materials Science and Technique Foundation.
Published in Proceedings of INTERNATIONAL CONCRETE CONFERENCE, 1990. IRAN

(zeolitization tuff or zeolite-tuff)[1][2], oil shale ash from fluidized-bed combustion[2], burned kieselguhr, etc. A kind of strengthening agent of concrete, containing zeolite-tuff has been produced and used in practice, the strengthening rate of that is about 15%[3]. Japanese H. Yamazaki et al reported that even though quartz sand which is inert material is added to portland cement in suitable dosage, strengthening effect on cement mortar can be obtained[5].

In above mentioned example, fineness of those mineral powders is similar to that of cement used. That is much coarser than of silica fume, but cost of the former is less thanthat of the latter. Therefore, to find the regular pattern of the strengthening effect of mineral powders on cement concrete or mortar is useful for practice.

MATERIALS USED

1. Six kinds of mineral materials possessing typical structural characteristic separately are used for test, they are as follow:

Silica fume, in supper fineness, and rich in amorphous silicon, sometimes contains a little crystalline matter;

Zeolite-tuff, containing aluminosilicate crystallite of clinoptilolite, rich in micro-cavity and micro-pores;

Burned kieselguhr ,loose and porous amorphous silicious pozzolanic material;

Oil shale ash, porous and high-active as burned at optimum temperature for producing pozzolanic activity;

Fly ash, containing glassy matter produced at high- temperature of power station;

Ground quartz sand, compressing of inert crystalline silicon.

All of above mentioned materials are ground into the similar fineness except silica fume which has the inherent fineness .

2. Portland cement, from Jidong cement factory of Hebei province of china.

EXPERIMENTS

1. Test on active SiO_2 and active Al_2O_3

It is considered that the active SiO_2 and the active Al_2O_3 are what can react chemically with lime in condition which water exists. Accordingly, above various mineral powders are put into saturated solution of $Ca(OH)_2$, and to keep concentration of solution by water circuit. After boiling for 3 hours, to determine the soluble SiO_2 and Al_2O_3 therein. The result is shown in Table 1.

2. The six kinds of mineral powder are mixed with Portland cement by various dosages and formed into samples accordance with standard test for strength of cement. In order to obtain the similar consistence, suitable water-reducer was used, and the same w/c is used for comparable strength. Those samples are cured in standard condition. Relationships between average compressive strength of 28th day of mentioned above samples and blended amount are shown in Fig. 1.

Table 1 Active SiO_2 and active Al_2O_3 in mineral powders

Name of mineral powder	Principal composition (%)				
	SiO_2	Al_2O_3	Active SiO_2	Active Al_2O_3	$\frac{\text{Active}SiO_2+\text{Active }Al_2O_3}{\text{Total }SiO_2+\text{Total }Al_2O_3}$
Silica fume	98.28	1.00	47.65	0.25	47.65
Oil shale ash	61.19	23.57	15.78	11.71	32.43
Zeolite-tuff	64.60	18.27	15.88	9.52	30.48
Kieselguhr	67.37	17.97	8.21	8.02	19.02
Fly ash	65.88	14.56	7.25	2.64	12.29
Ground quartz sand	84.85	1.88	3.20	1.61	5.55

3. The six kinds of mineral powder are mixed with analytically pure CaO by 1:1 of weight separately, and w/c of 2.5, but 3.5 for silica fume, is used. Additionally, gypsum is added by 5% of above total powder separately. Then to be mixed and formed, and cured at temperature of ca. 20℃ and humidity over 95%. At required ages after curing, to stop hydration by water-free ethanol, and to test by XRD. The results are shown in Fig. 2, Fig. 3 and Fig. 4.

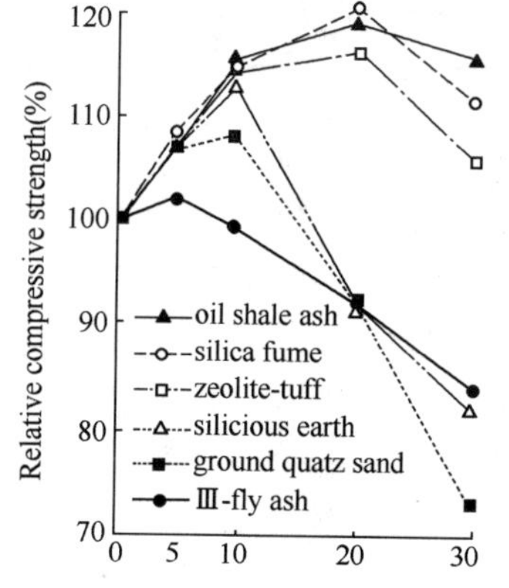

Fig. 1 Blended daosage of mineral powder(by weight of tatol powder) %

ANALYSIS OF RESULTS

1. In Fig. 1, it shows that:

(1) All of the six mineral powders appear their strengthening effect on cement mortar by given blending amount. When the blending amount is lower (e.g. 5%), the strengthening effects are the same. Difference of the strengthening effect is enhanced increased with increasing the blending amount, particular to that between the high-active matter (oil shale ash, silica fume, zeolite-tuff) and the low-active matter (kieselguhr, fly ash, burned quartz sand) .

(2)There is a peak in each curve. The peak place of high-active matter (oil shale ash, silica fume, zeolite-tuff) is in ca. 20%, and of low-active matter is in 5%~10%. The peak values of high-active matter are larger than that of low-active matter.

(3) Low-active even inert matter also has some strengthening effect by blending amount below 10% and given fineness. When the blending amount is over 10%, strength of that drops down rapidly, particular to quartz sand, the inert matter.

2. From Table 1, it shows that:

(1) The mineral powder with amount of active SiO_2 plus Al_2O_3(compared with amount of total SiO_2 plus Al_2O_3) over 30%, is considered as of high-activity, they have the similar strengthening effect on cement. By optimum blending amount, the increased rate is ca. 10%

~20%.

(2) When amount of the active SiO_2 plus Al_2O_3 (compared with amount of the total SiO_2 plus Al_2O_3 is less than 20%, the mineral powder is considered as of low-activity. They also have a certain strengthening effect on cement, but the replaceable amount for strengthening cannot be more than 10%, and it must be finely ground to a suitable fineness.

3. The hydration degree of samples was analyzed by change of $Ca(OH)_2$ and appearance of new phase in Fig. 2, Fig. 3 and Fig. 4.

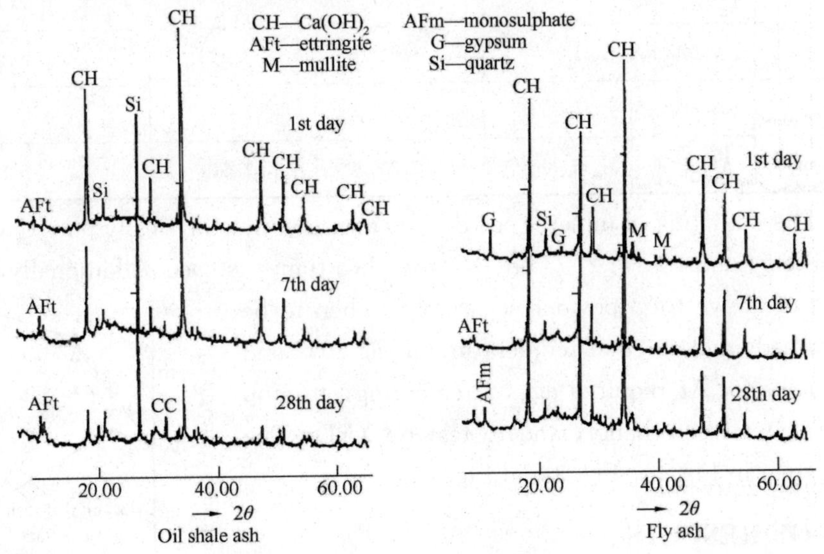

Fig. 2 XRD pattern of hydrated mixture with oil shale ash/fly ash and CaO as well as gypsum at varying age

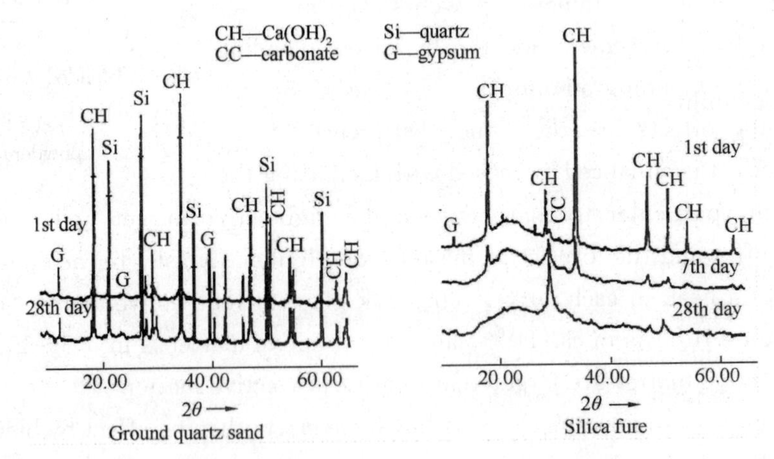

Fig. 3 XRD of hydrated mixture with ground quartz sand /silica fume and CaO as well as gypsum at varying age

(1) For high-active matter, from the 1st day to 7th day, the peak value of $Ca(OH)_2$ is reduced markedly. Of cause, reaction of peak value of $Ca(OH)_2$ is also involved with carbonation (e. g. silica fume), but the scattering zone of curve has widened and Al_2O_3

phase hydrate or CSH phase appears, i. e. gel has been increased. When active Al_2O_3 is satisfied to produce AFt phase with gypsum and CaO, AFt can be found in the curve at the 1st day, and AFt can be changed to AFm, if gypsum is exhausted (e. g. kieselguhr, fly ash, oil shale ash that have more active Al_2O_3 than the others).

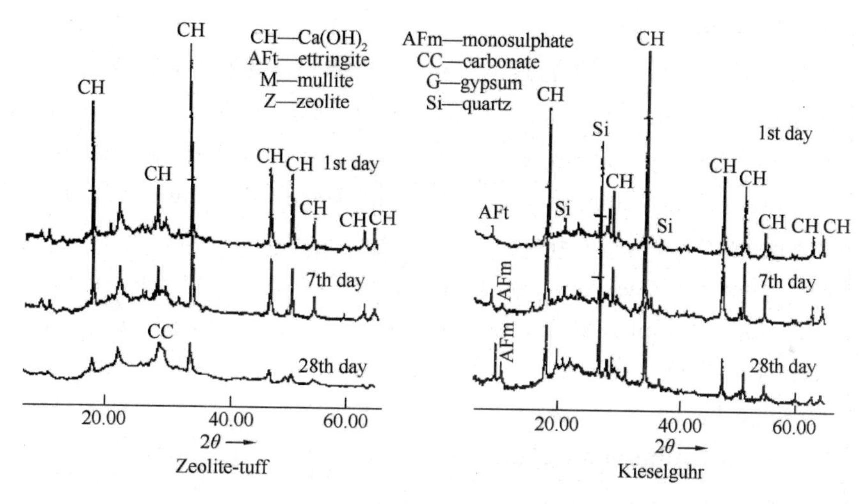

Fig. 4 XRD of hydrated mixture with zeolite tuff /kieseiguhr and CaO as well as gypsum at varying age

(2) For low-active matter, before 7th day, there is scarcely any change in peak value of $Ca(OH)_2$, although a little amount of AFt can be seen. From 7th day to 28th day, the change is marked.

(3) For ground quartz sand, such an inert matter, there is no change at any age substantially.

(4) It shows in curve of all of those samples that the crystalline inert component, quartz, zeolite, mullite, are no change at any ages (at least until 28th day). It means that the reactive component is the amorphous matter rather than the crystalline matter, even as zeolite (there is a question as "why is the activity of crystalline zeolite higher than that of uncrystalline matter?"[6] [7].

(5) It means that the strengthening effect of most blending materials on cement is not only due to the chemical reaction, but also involved with the other factors. Yamazaki considered that is due to a "micro-powder effect"[5], H. F. W. Taylor considered that the "micro-aggregate grading" is important[8]. And it is known that any substance with fineness to a certain degree can be a structure itself. Austrian scientist Čzernin demonstrated that finely ground quartz with a surface area of 20,000cm^2/g were mixed with water without any addition, and pressing formed into a cylinder sample, the sample could support more than 10kg. It is that the physical state of solid materials may be responsible for strength[9]. Therefore, study on hydration mechanism of hydration of blended cement should be processed not only by chemical method but also from physical view.

CONCLUSION

(1) Either reactive or inert mineral powder, natural or waste, has the strengthening effect on cement in different degree, so long as blending amount is suitable. But the strengthening effect and the optimum blending amount of the high-active matter is higher than that of the low-active matter, particularly than that of the inert matter. The former is 10-20%, and the latter is 5-10%。

(2) The activity of mineral materials is due to existence of active SiO_2 and active Al_2O_3 that can react with lime, but chemical composition takes its effect with the aid of structure (fineness, pore, crystallinity and crystal content), and to improve the structure of cement paste even concrete, so that to improve properties of cement paste even concrete .

(3) There is no reaction between the inert matter, such as quartz sand powder, and lime substantially, but there is a certain degree strengthening effect on cement in low blending amount, that is due to improvement of structure of cement paste by "micro-aggregate grading" and promotion of hydration of cement by "micro--powder effect". Therefore, physical effect should be conspicuous.

(4) Mechanism of hydration of blended cement is very complex and varying slightly for various mineral powders as different structure characteristic. It must be in progress to classify the mineral powder that is usable for blended cement, accordance with their structure characteristic, and to study separately. Of cause, we are interested in the high-active matter, such as silica fume, oil shale ash and zeolite-tuff or others.

REFERENCE

[1] Feng Naiqian et al. The Zeolite-tuff Cement and Concrete Thereof, Concrete and Reinforced Concrete, vol. 1, 1981, (in Chinese) .

[2] Lian Huizhen. Cement Paste/Zeolite-tuff Bond, 1989 Foundation on "Advance of Cement Manufacture and Use. August, 1988.

[3] Wang Zhimin et al. Studies on Strength and Workability of High-strength and Flowing Concrete Utilizing Low Calsium Oil Shale Ash, International Conference on Oil Ahsle Shale Oil, May , 1988.

[4] Feng Naiqian. THE STRENGTH EFFECT OF F MINERAL ADMIXTURE ON CEMENT CONCRETFE. C. & C. R. Vo. 18, 1988.

[5] H. Yamazaki et al. Influence of Pozzolana and Fly Ash on Concrete, CONCRETE TECHNOLOGY, Nov. 1981 (in Japanese).

[6] Fredeik A, Mumplon. Commercial Utilization of Natural Zeolite. Industrial Minerals and Rock, 4th edition, 1975.

[7] Massazza F. The Chemistry of Pozzolanas and Blended Cement, Proc. 6th Int. Sym. on the Chemistry of Cement, 1976.

[8] H. F. W. Taylor. Discussion of the paper Microstructure and Strength of Hydrated Cements" by R. F. Feldman and J. J. Beaudion, CCR. Vol. 7 No. 4 1977.

[9] T. C. Powers. Physical Properties of Cement Paste, Proc. 4th Int. Sym. on the Chemistry of Cement, 1960.

自评

1. 这是该基金项目启动伊始对不同活性矿物掺和料特性与在水泥中作用的探索性试验的论文。用几种典型的矿物掺和料，分别与分析纯 CaO 混合加水成型，硬化后在不同龄期检测 $Ca(OH)_2$ 衍射强度的变化。其目的是排除其他因素干扰，只研究矿物掺和料的化学活性。这是一种典型的分解论方法。结论中已提出“物理作用更加重要”。

2. 在该实验中得出的结论之一是，粉煤灰和磨细石英砂一样，常温下在 28 天以前基本上不参与化学反应，这对掺粉煤灰的水泥水化机理的研究和应用很重要，但是从工程出发，尚应考虑温度因素。也就是说，在工程中的混凝土温度都会较高，粉煤灰的火山灰反应可能会开始得早些。

3. 至今在混凝土材料与工程以及工程结构工程设计与施工中，绝大多数所采取的重要范式就是“28 天”，认为 28 天以后混凝土的性质只长不消。以 28 天为界来评价掺和料是不公平的，更重要的是忽略了混凝土区别其他任何人工材料的特性，那就是在自然环境中劣化和自愈的消长平衡。鉴于混凝土的复杂性，从基础上研究矿物掺和料的作用机理仍然任重道远。

4. 从所选英文论文来，那时的英文水平真是有欠推敲。但因本集选文主要目的在于内容，故对文字方面未做评价。现在只就本篇选文中的一段做一修改建议，供读者评阅的参考。原文见本文第三节“EXPERIMENTS”第 3 段。该段英文的主要问题是语法的错误和行文的修词不当。

以下选入该基金项目研究报告：介绍基础研究的一种方法——命题、思路、假设和技术路线。

长寿的秘诀在于保持思考的状态，但不要只是考虑自己。

——诺贝尔生物学奖得主，百岁老人丽塔．莱维-蒙塔尔奇尼

建材行业科学技术基金项目

水泥浆体微结构与宏观行为关系的研究及定量分析研究工作报告

廉慧珍（清华大学土木工程系）

一、本项目的依据，来源及研究目的

水泥混凝土开始使用以来，国内外学者在水泥及混凝土化学方面的研究对合理使用水泥混凝土、改善其性能起了很大的作用，随着生产的发展，人们认识到，支配水泥混凝土宏观行为的因素除其组份外，更重要的是这些组份在水泥混凝土内部的组合，即微结构。水泥混凝土的结构是水泥浆体、集料和孔缝及其界面的组合，其中水泥浆体的微结构是固相组份、孔缝及其界面在空间的组合。在一定范围内，例如在常用水灰比范围采用常压常温养护等条件时，固相组份与孔缝体积比的变化（如用孔隙率或水灰比或容重来表示）对其宏行为起决定性作用……。这是迄今为止用以指导水泥混凝土生产的主要依据。随着工艺和技术水平的发展及工程实际的需要，水泥混凝土生产和使用条件开始突破传统的范围，如各种外加剂的使用、特殊的成型和养护工艺，高强、超高强及流态混凝土的出现等，在固相组份和孔隙率不变的情况下，水泥浆体的宏观行为也会有明显的差异[1]。

近几十年来，人们对水泥浆体孔结构和混凝土中界面的研究产生极大的兴趣。不同尺寸孔数量的比例（即孔径分布或孔级配）、孔的形状和孔在空间位置的分布都影响水泥浆体的宏观行为。于是人们纷纷从孔结构（主要是孔径分布）的变化来研究工艺或组份的改变对水泥浆体及其混凝土宏观行为影响的机理[2]。

七十年代，加拿大学者 R. Feldmen 和 J. Beaudoin 提出“微结构与硬化水泥浆体强度的关系”[3]，认为水泥浆体的强度不仅是孔隙率的函数，同时还和浆体内部颗粒的尺寸、结晶程度、密实度有关。英国著名学者 H. F. W. Taylor 很重视他们这项研究结果，据此提出了孔隙率-固相结构-强度的关系曲线[4]。但这组曲线只是个模型，对固相结构只有定性描述而无量化，只考虑孔隙率而未考虑孔结构。这组曲线的量化是一件非常困难的工作，不仅因为水泥浆体是个极复杂的非均质多相体，而且也因为测试和特征化的困难。其中对孔结构的特征化，曾有各种尝试，例如最可几孔径、分级孔隙率等。而对固相颗粒的量化则更加困难，主要是因为固相结构几乎无法量测。

本项目研究试图利用现代仪器分析的手段探索对硬化水泥浆体微结构的固相组份进行量化，由国家建材行业科学技术基金资助。期限从 1989 年到 1991 年。已于 1990 年年底交出总结。

本项目计划研究内容为：

1. 确定水泥浆体微结构控制参数的试验研究
2. 上述控制参数与水泥浆体微结构关系的试验研究
3. 水泥浆体微结构元物理，力学模型的建立及计算机实现
4. 水泥浆体宏观行为与计算结果的比较

二、技术路线及试验方案

1. 调整水泥浆体微结构的方法

从材料科学理论上来说，有一定的微结构，就有一定的宏观行为，而与其形成方法无关。则可通过宏观参数预测其宏观行为。本项研究旨在改变宏观参数以调控水泥浆体微结构的变化，使水泥浆体表现出不同的宏观行为。传统的做法是调整宏观参数或条件（成型工艺或配比，或加入外加剂量等措施），通过试配来满足对水泥浆体宏观行为的要求。但这种做法同材料设计相去甚远，多少带有盲目性。应当是建立起来水泥浆体微结构与其宏观行为的关系后，直接通过调整宏观参数来控制其微结构，以保证满足对其宏观行为的要求。因此，核心问题是建立水泥浆体微结构与其宏观行为的关系。本项研究只采取用不同细度的不同类型的矿物粉末（即混合材料）以不同掺量加入不同水灰比的水泥中，来实现水泥浆体微结构的变化，建立力学模型的试样采用相对简单的纯水泥浆体。

2. 水泥浆体微结构体系和结构元的选择

如前所述，水泥浆体微结构元的选择取决于所选择的体系，即不同的研究尺度[5]。体系越大越接近实际越符合系统论观点[2]，但研究的难度也越大。在研究体系的选择上，有如下的发展：

①微观尺度上研究水化物中的晶体和胶体比例的影响，不考虑孔和未水化相，所选体系是水化物，即固相的一部分。

②取固相中量最大的 C-S-H 凝胶相为体系，研究 C-S-H 的 Ca/Si 比以至 C-S-H 中的 Si-O 聚合度，这是更小的体系。

③考虑体系由固相和孔组成，不考虑固相的结构，只研究孔的结构，认为扣除孔结构就是固相的结构。但孔结构目前只能近似测定不同孔径的孔在体系中的分布，并由于测试条件限制，结果的误差很大，孔的形状和孔的空间位置可通图象分析大体进行测定，但尚无法特征化。

④也取固相和孔组成的体系，不考虑孔的结构，将固相看成由不同颗粒组成，研究固相颗粒的结晶度、密度和大小以及孔隙率的变化对强度的影响。Feldman-Beodion-Taylor 曲线（简称 FBT 曲线）[4]就是这个体系，只是曲线横坐标无量化。理想的是将此曲线横坐标量化，并将孔隙率结合孔结构。

我曾选择沸石岩为混合材料，研究其含不同比例晶态与非晶态物质时，以不同细度、不同含量掺入水泥时，与相同水灰比水泥浆体强度的关系[8]，作为固相量化的探索。

我国著名混凝土科学专家吴中伟的“中心质效应”假说[2]，对在不同尺度上研究水泥混凝土的结构提出了一个符合系统论的模型见图 1。

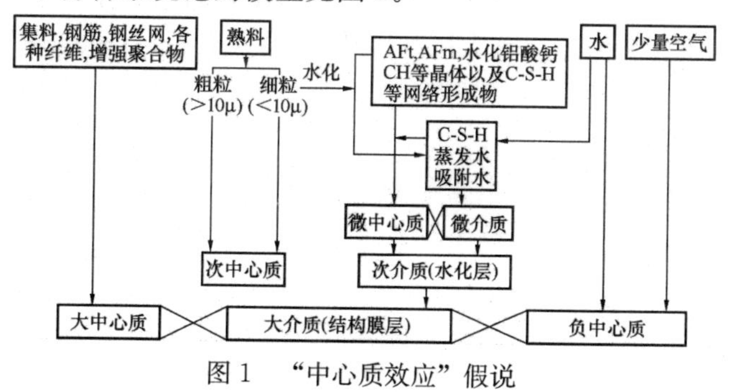

图 1 “中心质效应”假说

本研究选择亚微观粒子尺度上的体系，即研究次中心质和次介质的组合，因此考虑水泥浆体微结构见图 2。

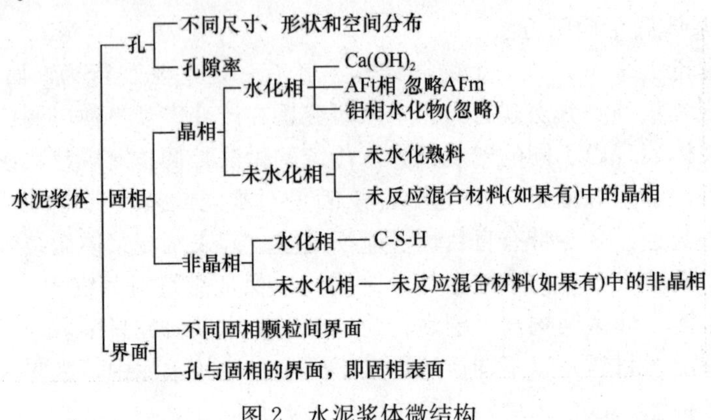

图 2　水泥浆体微结构

按系统论的观点[2]，水泥浆体基本上不单独使用，应当取混凝土或砂浆为体系，即应取包含水泥浆体，集料及其界面为体系。在集料选定的情况下，水泥浆体和界面的结构决定混凝土的宏观行为，从上述体系出发，延展到混凝土的体系，可更好地把细观力学和宏观力学结合起来。

经测试，晶相水化物以 $Ca(OH)_2$ 为主，AFt 次之，AFm 和铝相水化物在目前量测精度下可忽略不计，未水化熟料是强度高、密度大（约 3.10/cm^3）的活性微集料；当加入混合材料时，混合材料中的晶相主要是石英、$CaCO_3$ 等高强度的矿物，基本上是非活性，但因其与活性部分共同存在于材料的颗粒之中，所以可将该未水化颗粒也看成是一种微集料，其密度较小（2.5g/cm^3 左右），强度视晶态物质种类和含量而不同，考虑体系如下的结构元及假设：

①水化熟料，高活性、高强度，高密度的微集料；

②未反应混合材料，较低密度的微集料，其强度和活性视其中晶态物质种类含量而不同；

③结晶态水化相 $Ca(OH)_2$ 和 AFt ，其对浆体强度的贡献与方向有关，考虑为乱向随机分布而忽略其非晶态和其他晶态水化物；

④上述三项计为晶相颗粒；

⑤凝胶相主要为熟料水化产生的 C-S-H 和混合材料与 Ca（OH）$_2$ 反应生成的 C-S-H，此外还有很少量其他非晶相水化物，认为其内部结构一定，为均匀相；

⑥孔认为在空间均匀分布，以孔隙率（P）和最可几孔径（Dp）为结构元；

⑦不考虑界面结构的变化。

3. 试验方案及试样制备

① 改变水灰比，调整纯硅酸盐水泥浆体的微结构。

② 采用不同类型的混合材料，以不同细度，不同掺量改变不同水灰比的水泥浆体微结构。所选原材料为：

■ 硅酸盐水泥（简称纯水泥），用以制作基准浆体。

■ 河北独石口斜发沸石岩（简称沸石），即沸石化的凝灰岩。沸石是架状结构的铝硅酸盐矿物，内部有大量的细微孔穴，有巨大的比表面积，孔穴中有可交换的阳离子和与之配位的沸石水，这样的晶态的沸石不具有化学反应性，而以其特殊的结构在水泥浆体中起作用[8]，沸石岩中存在于非晶态的凝灰岩有较高的活性率。

■茂名油母页岩沸腾炉灰（简称页岩灰），其中的活性组分存在于由黏土质矿物在沸腾炉的温度（850～950℃）下分解成的非晶态物质中，有较高的活性率，晶态物质主要是石英。由于可燃部分的烧失，灰粒具有较大比表面积。

■磨细石英砂（简称石英砂）将检验水泥用的标准砂磨细使用，其成分为非活性的晶态 SiO_2。

③ 水灰比采用三个水平。所有混合材料均采用三种细度、三种掺量。按标准养护 3 天、7 天、28 天及长龄期成型试样如下：

■ 4cm×4cm×16cm 胶砂试块（每组 3 块），供测定抗压强度；

■ 3cm×3cm×3cm 净浆试块（每组 6 块），供测定抗压强度和 XRD、TG，SEM 以及测孔用；

■ 1cm×1cm×6cm 净浆试块（每组 6 块）供切片供 SAXS 分析用；

■ 3cm×3cm×10cm 净浆试块（每组 3 块），只做纯水泥试样，三个水灰比，供测定弹性模量用。

以上试样共计 2000 组。

三、测试及数据处理工作

1. 化学分析

①原材料全分析。

②原材料活性率的测定用饱和 $Ca(OH)_2$ 溶液萃取可溶 SiO_2 和 Al_2O_3，以其总量与全部 SiO_2 和 Al_2O_3 总量之比表示[9]，取得数据 6 个。

③沸石 NH_4^+ 交换量的测定：以其换算沸石含量，XRD 分析掺沸石的水泥浆体时，沸石特征峰与其他物质特征峰重迭部分难以分离，沸石含量的测定证明沸石不参与化学反应，故各掺入沸石岩的试样中后，晶态含量只与沸石岩掺量有关，不随龄期而变化，则沸石含量可通过计算取得。共测试和处理数据 20 个。

④用酸溶法测定养护到指定龄期的试样中可溶 SiO_2 和 Al_2O_3 含量，作为非晶态相中的水化物相[10]。从定量分析的非晶态相总量中扣除水化物相，即得混合材料中非晶态相的未水化相。

以上 SiO_2 用比色法测定，Al_2O_3 用络合滴定法测定。共测定处理数据 360 个。

2. 混合材料颗粒分布、细度及密度测定

①分别用透气法和 BET 法测定比表面积。

②用激光颗粒沉降法测定颗粒分布及比表面积。

③用李氏比重瓶测定密度。

以上共取得数据 40 组。

3. XRD 定量相分析

将改进的 Rulend 法和传统的 *K* 值法结合，编制了适合于水泥浆体复杂体系定量相分析的软件，用 TiO_2 为参比样，测定水泥浆体中各晶相 *K* 值进行定量相分析，用清华大学材料科学研究究所的数据处理和多重峰分离软件处理测定结果。改进软件的编制和调试由国家建材院水泥研究所物化室 x-光组完成，测试、数据处理及分析工作由清华大学完成。共计测试和处理数据 5000 个。这种相定量测定方法解决了用传统 K 值法对硅酸盐类水泥（尤其是有混合材料掺入的硅酸盐水泥）浆体定量相分析时所存在的问题。

4. SAXS（x-射线小角度散射）分析水泥浆体微孔结构，汞压力测孔仪测定浆体孔级配和总孔隙率。共测试样 45 个。

5. TG 定量测定各试样中 $Ca(OH)_2$，$CaCO_3$ 含量，作为校核。共处理取得数据 270 个。

6. 除 1cm×1cm×6cm 试件外，均在指定龄期进行抗压强度试验，共取得数据 1742 组。

7. SEM 观察

①试样在不同龄期微结构形貌及各结构元的分布，取得照片 92 帧。

②在受力破坏后的试件中分别取垂直于荷载和平行于荷载方向的试样，观察固体颗粒对浆体破坏的作用(对裂缝阻挡还是引发)，取得照片 20 帧。

四、计算机模拟及分析

1. 考虑体系由孔和固体颗粒为分散相，水泥凝胶[11]为基相而组成。水泥浆体中的孔缝实际上是不规则形状并随机分布的，固体颗粒中未水化熟料和未反应的混合材料颗粒也是不规则形状的，而有相当数量的 $Ca(OH)_2$ 虽为规则形状结晶，但其方向则是随机分布的，故需考虑最不利的情况但又能够计算的模型，在平面上假设孔为方形，固体颗粒为长条状，用有限元法分别对这样的模型进行计算。结果证明了固体颗粒在水泥浆体中所起的作用，以及孔隙率相同时孔径分布对浆体力学行为的影响，从力学的角度证明了孔和固体颗粒在水泥浆体中的作用。

2. 测定不同水灰比的纯水泥浆体的弹性模量，为有限元法计算提供数据。

3. 根据试验所得微结构元的数据和强度数据分析各结构元与强度的关系及系统影响。

4. 除选用强度为水泥浆体宏观行为研究的对象外，为了将来扩展到混凝土以致钢筋混凝土，还引入数论随机模拟方法，探索了钢筋混凝土结构或构件寿命的评价，考虑以钢筋锈蚀定作为耐久性行为，以钢筋混凝土微结构和外部条件为参变量，用计算机产生随机数，计算结果为寿命分布及其可靠性分析。

五、研究成果

1. 用 x-射线衍射 K 值法系统地、大量地对掺混合材料的硅酸盐水泥浆体进行定量相分析，提出了水泥浆体微结构模型和可计算的微结构元，并定量分析了不同宏观参数与水泥浆体强度的关系，证明宏观参数中影响水泥浆体形成敏感的因素是浆体水灰比、混合材料的颗粒组成、掺量和活性。在一定的水灰比下，水泥浆体存在使强度最高的最优晶相颗粒与凝胶比(晶/胶比)。w/c 增加时，此最优比下降，混合材料的活性越高，越能得到此最优比。

2. 用有限元法计算了水泥浆体中孔的尺寸、位置和固相颗粒对水泥浆体力学行为的影响，进一步证明，当晶相颗粒总量或体积一定时，在一定范围内，其数量越多，对提高强度有利，当总孔隙率一定时，小孔比大孔对提高强度有利，而与孔的位置关系不大。

3. 用灰色系统理论分析了混合材料颗粒组成对水泥浆体微结构和强度的影响，计算出不同颗粒组成在不同条件下对水泥浆体的不同影响。此项工作可为使用一定颗粒组成和活性的混合材料时，选择形成所需水泥浆体合适的工艺参数提供参考。

4. 根据所建立的水泥浆体微结构模型，可认为高强度水泥浆体的理想结构是：尽可能少的但足以将晶相颗粒胶结成整体的凝胶，尽可能多的具有活性的晶相颗粒，无大孔。

六、需进一步研究的问题

1. 亚微观层次的水泥浆体微结构研究是物理的、化学的、力学的、数学的和物理化学的方法的结合，每一层次研究了都应有相邻层次研究的支持，将力学和数学方法引入水泥浆体微结构的研究，还必须为其提供更多的参数，以便从力学角度进一步定量分析。

2. 孔是水泥浆体中不可少的结构元，但实验中测孔数据太少，而且离散性大，不便分析，需进一步研究可靠的测孔方法，取得大量孔结构的数据，以研究孔结构和固相结构与强度的共同关系。

3. 养护条件是形成水泥浆体微结构更敏感的因素，需进一步研究养护条件对水泥浆体微结构变化的影响规律，为水泥混凝土的应用提供更合理的工艺参数。

4. 水泥混凝土其他宏观行为参数与强度也不一定是线性关系，应用于混凝土时，还需综合考虑其他的行为如变形、断裂、损伤和耐久性。并进一步用有限元法或块体理论进行研究。并研究如何应用于混凝土及其构件。

参考文献

[1] 吴中伟．水泥和混凝土的理论研究，武汉建材学院讲课讲义，1981.

[2] 吴中伟．水泥基复合材料科学导论，清华大学研究生课程讲义，1983.

[3] R. Feldman，J. Beaudoin. MICROSTRUCTURE AND STRENGTH OF HYDRATED CEMENT Vol. 6，pp，1976：398～400.

[4] H. Taylor. DISCUSSION OF PAPER "MICROSTRUCTURE AND STRENGTH OF HYDRATED CEMENTM" BY R. FELDMN AND J. BEAUDOIN，C. &C. R.，1977，7：465～468.

[5] S. Diamond. CEMENT PASTE MICROSTRUCTURE——AN OVERVIEW AT SEVERAL LEVELS. Proceedings of Conference on Hydraulic Cement Paste，Their Structure and Properties. University of Sheffied. UK. 1976.

[6] Proceedings of 7th International Conference on Cement & Concrete. Paris，1980.

[7] 廉慧珍．沸石岩的活性，中国硅酸盐学会水泥专业委员会第三届学术年会，柳州，1986. 11.

[8] L ian Huizhen. Cement paste/Zeolite-tuff Bond，Proceedings of 1988 EFC ON ADVANCES OF CEMENT MANUFACTURE AND USE，St. Louise，USA. 30 July 1988.

[9] Lian Huizhen & Chen Enyi. Mechanism of Hydration of Blended Cement——I. Strengthening Effect of Mineral Powder on Cement Used in Mortar or Concrete，INTERNATIONAL CONCRETE CONFERENCE，IRAN，1990.

[10] F. M. Lee. Chemistry of Cement and Concrete.

[11] H. F. W. Taylor. Is the C-S-H Gel Substitute of Cement Gel ? Proceedings of 1988 EFC ON ADVANCES OF CEMENT MANUFACTURE AND USE，St. Louise，USA. 30 July，1988.

自评

1. 在第二节“技术路线及实验方案”的“1”中，“有一定的微结构，就有一定的宏观行为，而与其形成方法无关。则可通过宏观参数预测其宏观行为。”的提法可能会引起一些误会。应当改成“化学成分一定时，有一定的微结构，就有一定的宏观行为……”。“而与其形成方法无关。”如果改成“因为微结构是通过一定的成分和一定的工艺过程而形成的，因此可通过宏观参数……。”那么，就不难理解该研究的技术路线了。

2. “五、研究成果”中的第4点所述“可认为高强度水泥浆体的理想结构是：尽可能少的但足以将晶相颗粒胶结成整体的凝胶；尽可能多的具有活性的晶相颗粒，无大孔。”其中凝胶的条件是尽量低的孔隙率或优化的孔结构。孔结构指的是孔的形状、空间分布、大小级配、连通和开放的状态。细小的、圆形的封闭的孔强度高。“尽可能多的具有活性的晶相颗粒”，应改为：“尽可能多的粗大的、高密度的、结晶的活性或非活性颗粒”。但是，孔结构的研究太复杂，难度在于检测与特征化模型的建立。传统的汞压力测孔技术，由于高压力汞的压入会伤害孔的原始结构而往往测不准，CT（断层扫描）装置又价格昂贵，至今尚难以有更好的研究方法，有待于未来相邻学科的发展。

3. 从材料科学来说，化学成分通过结构决定物质的性质；相同化学成分通过不同的工艺过程可得到不同的结构。仿生学仿造宏观结构以仿造出生物行为的实例很多，例如飞行器、雷达、机械手等等。对于人工材料来说，如果说成分仿生容易的话，微结构仿生或者说过程仿生的难度则很大，原因是天然材料的“微结构”形成过程非常复杂，难以用现有工艺过程来实现。例如对蜘蛛丝的仿造，科学家们做了很大的努力提取到非常相近的蛋白质等成分，却仍然达不到与天然蜘蛛丝相同的性能，关键的就是过程仿生的困难。众所周知曾经一度引起人们关注的“凝石”，其始作俑者想对天然岩石的仿生，失败之处就在于企图用简单而快捷的过程去实现经过亿万年复杂环境作用的过程。

4. 混凝土是一种太复杂的不确定性人工体系，想要通过模拟实现对其计算预测的，并不乏其人，但是在当前的条件下几乎不可能实现。在本集中选了这部分内容的目的并非想说明混凝土现在的可计算性，只是想说明一种科学研究方法的思路。

不要屈从权威。如果人们觉得权威能够解决所有的问题，那么很可能是因为他们并没有进行足够全面的观察，各位应该从另外的角度来探索问题。

——美国国家科学院院士、加州大学圣地亚哥分校首席科学家 Ronald Graham

论文之七

硬化水泥浆体微结构的定量分析及其与抗压强度的关系*

廉慧珍 王智敏 李兆仲　　　杨人和
清华大学土木工程系　中国建材研究院水泥所

摘　要　用不同细度的不同类型矿物粉末，以不同掺量加入不同水灰比的水泥浆体中，以实现水泥浆体微结构的变化。提出水泥浆体微结构模型和可计算的微结构元，将改进的Rulend法和传统的K值法结合，编制了适合于水泥浆体复杂体系定量相分析的软件，测定水泥浆体中各晶相的K值，进行定量分析，用计算机进行了大量的数据处理。定量分析了不同宏观参数与水泥浆体强度的关系。研究结果表明，在一定的水灰比下，水泥浆体存在使其强度最高的晶相颗粒与凝胶最优比（晶胶比）。w/c增加时，此最优比下降；混合材活性越高，此最优比越小，所得到的强度越高；水灰比越低，晶胶比和强度关系越离散。在一定的水灰比下，在此最优比前，浆体强度随晶胶比增加而提高；超过此最优比后，由于凝胶组份减少到不足以胶结晶态颗粒而使浆体强度随之下降；不同类型矿物掺合料在水泥浆体中有不同的作用，与其结构特性、活性、掺量以及浆体水灰比有关。

关键词：水泥浆体 微结构 定量分析

一、前言

水泥的水化过程、水化产物及其化学性质历来是水泥研究的重要内容。水泥的水化是水泥浆体凝结硬化的基本条件，但是水化程度与浆体力学行为之间并不存在确定的关系。例如，在工艺条件与龄期相同的条件下，当混凝土的水灰比较低时，其水泥浆体的水化程度也较低，但却有较高的抗压强度；MDF和DSP的水灰比低于0.2，而强度却高达几百MPa[1]，此外，按照热力学的观点，相同成分的晶态物质比非晶态物质具有更大的化学稳定性，但是，含有50%以上晶态沸石的沸石凝灰岩却是水泥具有较高活性的混合材料[2]。以某些基本没有活性或活性很低的材料等量取代水泥，如果取代量合适，硬化水泥浆体仍能达到原有强度，甚至有所提高[3]。种种现象表明，仅从化学角度对硬化水泥浆体进行研究，不能满足材料科学对材料组成、工艺和性能关系定性与定量化发展的要求。

对硬化水泥浆体固相微结构与宏观行为关系的研究越来越受到重视，不少学者认为材料结构对性能的关系比化学组成更重要。定量分析硬化水泥浆体微结构与其宏观行为的关系，可以为实现材料设计以及对现存混凝土结构寿命进行评估提供依据。

上世纪七十年代，加拿大R. Feldman和J. Beaudoin提出“微结构与硬化水泥浆体强度的关系”[4]，认为水泥浆体的强度不仅是孔隙率的函数，同时还和浆体内部颗粒的尺寸、结晶程度、密实度有关。在一定的孔隙率下，浆体中大的、密实的、且结晶良好的组分在水泥浆体中占有最佳比率时，能使浆体获得最高的强度。在孔隙率较大的浆体中，此比率较低，而孔隙率较低时，则此比率较高。英国著名学者H. F. W.

* 建材行业科学技术发展基金项目，发表于第六届全国水泥化学及测试方法学术会议论文集1994年10月，郑州。

Taylor很重视这项研究结果，据此提出了孔隙率-固相结构-强度的关系曲线[5]，但这组曲线只是个模塑（简称FBT模型），对固相结构只有定性描述而无量化，对孔只考虑孔隙率而未考虑孔结构。这组曲线的量化是一件非常困难的工作。不仅因为水泥浆体是极复杂的非均质多相体，而且也因为测试和特征化的困难。对固相颗粒的量化则更困难，主要是因为固相结构几乎无法量测。

对硬化水泥浆体微结构的研究已有大量文献发表。但是，对各种控制水泥浆体微结构的因素及水泥浆体微结构与宏观行为关系的定量分析还缺少系统研究的报道。用QXRD定量分析水泥浆体的相组成有很多问题尚未解决，如无法得到结晶度接近100%的纯样，硬化水泥浆体的XRD花样中多重叠峰分峰困难，等等。

本项研究不考虑孔的影响，采取用不同细度的不同类型的矿物粉末（即混合材料）以不同掺量加入不同水灰比的水泥中，来实现水泥浆体固相微结构的变化。用改进的RuLand法测定“纯样”结晶度，校正各“纯样”的K值，在QXRD定量分析中采用了多重峰分峰软件和根据需要编制的软件，用计算机处理计算硬化水泥浆体微结构的固相组成[6]。探索水泥浆体固相组成与其强度的关系。

二、原材料的选择及其结构特性和活性的定量分析

1. 原材料选择

水泥：冀东水泥厂盾石牌525＃硅酸盐水泥

混合材料：

① 油母页岩沸腾炉灰，简称页岩灰，为广东茂名油母页岩在沸腾炉850℃左右的温度下燃烧后的灰渣，用以作为非晶态物质为主的混合材料代表。

② 沸石凝灰岩，简称沸石岩，选用以斜发沸石为主的河北赤城二层矿的沸石岩，作为以晶态物质为主的活性混合材料的代表。

③ 磨细石英砂，用磨细检验水泥强度用的标准砂，作为以晶态物质为主的惰性混合材料的代表。

此外，试验时还用兰州产的硅灰以及嫩江产的斜发沸石岩、浙江缙云丝光沸石岩和北京西高井粉煤灰做对比。

2. 分析的内容及方法

①用化学分析和发射光谱分析方法分析化学成分。

②检测混合材料在饱和石灰水的可溶SiO_2、Al_2O_3，实验结果用活性率表示。

③用XRD对各材料矿物组成定性分析，并用K值法定量；因无法取得斜发沸石标准的较纯试样，故用铵离子净交换量方法测定沸石岩中的沸石含量。试验结果表明晶体的沸石在28d以前不参与水泥的水化反应，故定量分析掺沸石岩的水泥浆体中的沸石含量时，也用铵离子净交换量方法，证明可行。

④分别用李氏比重瓶测定材料密度，用Blain透气法、BET法和颗粒分析仪测定磨细至不同细度的材料比表面积以及颗粒粒径分布；用水筛法测定细度。

⑤对部分材料按GB 2874—81测定胶砂强度比。

以上部分测试结果如表1所示。

由表1可见，石英砂粉和硅粉中都含有93%的SiO_2，但是硅粉中主要是非晶体，而石

英砂粉中则都是α-石英晶体；页岩灰主要成分是SiO_2和Al_2O_3，其中α-石英晶体约占三分之一，其他晶态物质很少；XRD分析表明，页岩灰和硅灰在2θ约为22°处有一个很宽的散射带，除细度差别外，二者有近似的结构和活性；XRD定量表明，沸石岩中除64.7%的斜发沸石外，其他晶态物质含量都很低。根据分析，属于斜发沸石晶体的SiO_2约为43.5%，其可溶部分主要属于非晶体，是沸石岩中具有活性的组分。

表1　混合材料主要矿相和活性

编号	名称	化学成分（%）		活性率（%）	主要结晶矿物含量（%）		胶砂强度比（%）
		SiO_2	Al_2O_3		α-石英	斜发沸石	
G	硅粉	92.85	0.75	51.18			
Y	页岩灰	59.81	20.53	34.22	22.19		107.7
D	沸石岩	65.63	14.10	31.86	1.01	64.71	87.7
F	粉煤灰	51.53	30.85	20.39	18.00		
S	石英岩	92.76	1.95	5.08	93.40		65.3

三、实验及结果分析

将所选择的各混合材料粉磨至不同细度（以粉磨时间合比表面积分析控制），分别以0、10%、20%、30%掺入水泥制成浆体，水灰比为0.30、0.35、0.40。试件尺寸依不同用途而异。标准养护至7天和28天后，分别测定其抗压强度和进行XRD定量相分析。用热重方法测定$Ca(OH)_2$含量和用酸溶法测定活性SiO_2，作为校核。测试结果及分析如下：

1. 水泥浆体固相组成分析

XRD定量分析各试样中的固相组成如下：

未水化熟料颗粒

XRD定量分析表明，各试件中都含有一定量的未水化的C_2S和C_3S，根据其含量和水化前水泥原样中硅酸钙的含量，计算水化程度。结果表明，在水灰比为0.3～0.4的纯水泥浆体中，养护28天，全部硅酸盐钙的水化程度为55%～62%，而C_3S水化程度约90%。

不同的混合材料、掺量、水灰比影响浆体水化程度（硅酸钙含量），有如下规律：

①随水灰比的增加，浆体中硅酸钙减少，说明水化程度提高。

②水灰比相同时，掺入混合材料后，浆体中硅酸钙减少，掺量越大，硅酸钙越少，说明掺入混合材料可提高水化程度。

③非活性材料比活性材料影响大；密实的晶态材料比非活性材料影响程度大；比表面积大的影响大；但像页岩灰，因需水量大，掺量为30%时，水化程度则有所降低。

氢氧化钙

XRD定量和热重分析均表明各试样中都含有一些$CaCO_3$，水泥和混合材料中都不含$CaCO_3$，考虑此部分$CaCO_3$为浆体中$Ca(OH)_2$碳化所致，故将其折算计入$Ca(OH)_2$总量中。测试结果规律如下：

① 掺石英砂粉的试样中，$Ca(OH)_2$含量随水灰比的增加而增加，在各掺量下该增加的趋势与纯水泥的平行，由此可认为石英粉基本不参与反应。但$Ca(OH)_2$含量并不随石英粉的增加等比例地减少，而是在计算值*上按石英粉掺量分别增加2%～4%。这说明非活性晶态混合材料对水泥水化有促进作用。

水泥浆体中的α-石英来自石英砂粉和页岩灰，其他试样中没有。浆体中的α-石英含量和混合材料掺量之间有很好的线性关系，并与计算值吻合得很好，证明XRD定量可信。

斜发沸石

② 在掺页岩灰的试样中，$Ca(OH)_2$含量与计算值相比，随页岩灰掺量增加按一定比例减少，说明页岩灰发生了火山灰反应，所减少的量约为4.4%；

③ 在掺沸石岩的试样中，低掺量时，$Ca(OH)_2$含量略有增加，掺量30%时则略有降低。

α-石英

测定沸石岩不同掺量和不同水灰比的水泥浆体养护7天和28天后的胺离子净交换量的结果表明，浆体中胺离子净交换量只与沸石岩掺量有关，因此可用胺离子净交换量换算试样中的沸石晶体含量。

此外各试样中均含一定量的AFt相和铝相水化物，但与其他晶相量比较，在此可忽略不计。

凝胶相固体

凝胶相固体包括C-S-H以及很少量的非晶态$Ca(OH)_2$、AFt、AFm、铝胶等。从XRD定量测出的晶相含量，用三种方法计算凝胶量，结果基本吻合。考察$Ca(OH)_2$/凝胶比和浆体水灰比的关系，结果表明，当混合材料以晶相为主时，不同掺量的该关系曲线基本平行，而掺入页岩灰时，则随掺量的增加，水灰比越大，$Ca(OH)_2$/凝胶比越小。

以上分析均表明宏观参数中影响水泥浆体形成敏感的因素是浆体水灰比、混合材料的颗粒组成、掺量和活性。

2. 微结构元的选择及微结构模型

选取亚微观粒子尺度固相和孔组成的体系，不考虑孔的结构，将固相看成由不同颗粒组成，研究固相颗粒的结晶度、密度和大小以及孔隙率的变化对强度的影响，即FBT模型。

经测试，水泥浆体中晶相水化物以$Ca(OH)_2$为主，AFt次之。AFm和铝相水化物在目前量测精度下可忽略不计；未水化熟料是强度高、密度大(约3.1g/cm^3)的活性微集料；当加入混合材料时，混合材料中的晶相主要是石英、$CaCO_3$等高强度的矿物，基本上为非活性，但因其与活性部分共同存在于材料的颗粒之中，所以可将该未水化颗粒也看成是一种微集料，但密度较小(2.65g/cm^{3}左右)，强度视晶态物质种类和含量而不同。考虑体系如下的结构元及假设：

①未水化熟料：高活性、高强度、高密度的微集料。

---

* 即按混合材料不参与反应考虑，水泥由于掺入混合材料取代熟料而减少所生成的$Ca(OH)_2$。

②未反应混合材料，较低密度的微集料，其强度和活性视其中晶态物质种类含量而不同。

③结晶态水化相 Ca（OH）$_2$和 AFt，其对浆体强度的贡献与方向有关，考虑为乱向随机分布；忽略其非晶态和其他晶态水化物。

上述三项计为晶相颗粒。

④凝胶相：主要为熟料水化生成的 C-S-H 以及混合材料与 Ca（OH）$_2$二次反应生成的 C-S-H；此外还有很少量其他非晶相水化物。认为其内部结构一定，为均匀相。

根据以上分析，提出水泥浆体模型如图 1 所示。为了可计算，模型中结构元取晶相颗粒、凝胶和孔的数量取决于凝胶量，与原始水灰比有关，不考虑混合材料填充作用引起的孔结构变化。

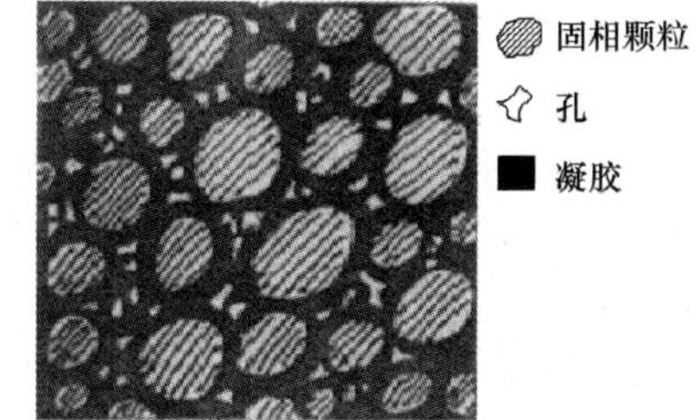

图 1　水泥浆体微结构模型

微结构以晶相颗粒与凝胶比值（以晶胶比为特征。其中晶相定义为较粗大的、结晶较好的、强度较高的、致密的晶相水化物和未水化颗粒含混合材料）；凝胶定义为细微的、结晶差的、孔隙率较大的、凝胶性强的 C-S-H 和非晶相水化物。

3. 水泥浆体微结构和抗压强度强度的关系

根据 QXRD 测定的水泥浆体组成，按所提出的模型微计算各结构元，得出微结构特征参数，将该数据和所测定的 28 天抗压强度数据输入计算机，拟合处理结果如图 2 所示（由计算机自动生成）。

图中不同符号为不同水灰比下的数据。由图 2 可见，不管原材料条件如何，水泥浆体只要在一定的孔隙率下形成一定的晶/胶比，就产生一定的强度；在一定的 w/c 下（代表一定的孔隙率）存在一个使强度最高的范围，低于此范围，起骨架作用的晶体减少而强度下降；高于此范围，则起胶凝作用的凝胶减少而强度下降；w/c（孔隙率）越大，此最佳晶/胶比范围越趋向于减小的范围。

将不同水灰比下的数据分别处理如图 3～图 5 所示。图中符号代表用不同混合材料。由图可见，水灰比越低，晶/胶比和强度关系越离散；混合材料活性越高，最佳晶/胶比越小，所产生的强度越高；混合材料活性相近，则晶/胶比和强度关系越相近。

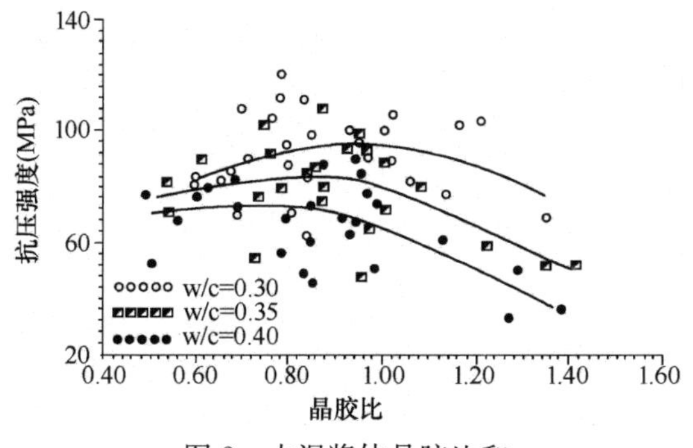

图 2　水泥浆体晶胶比和抗压强度的关系

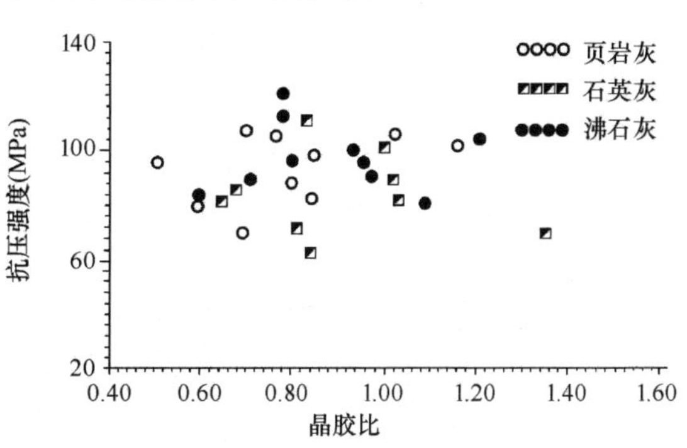

图 3　水胶比为 0.3 时浆体晶胶比与抗压强度关系

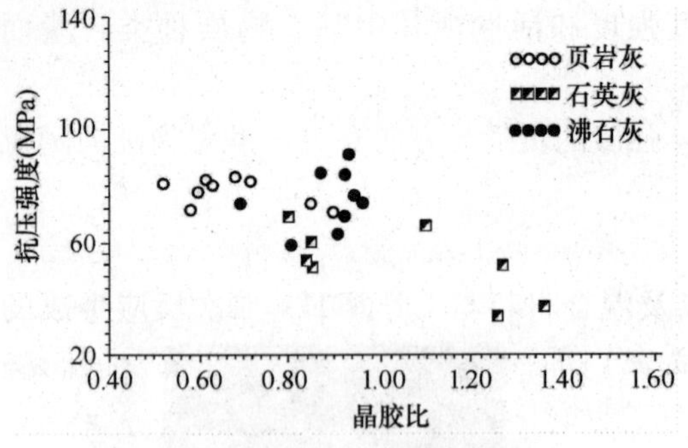

图 4　水胶比为 0.35 时浆体晶胶比与抗压强度关系

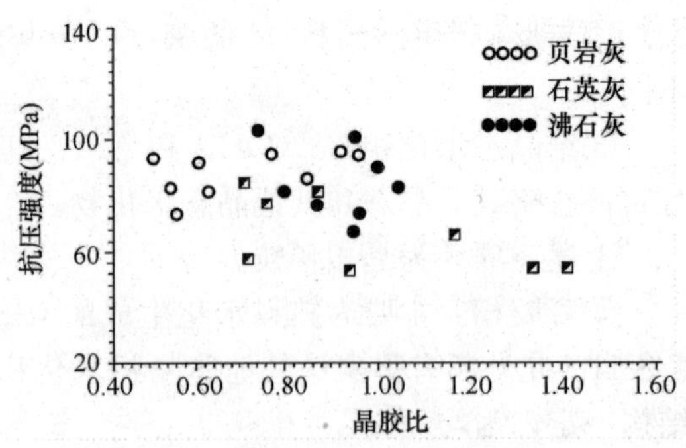

图 5　水胶比为 0.4 时浆体晶胶比与抗压强度关系

结论

1. 用改进的 Ruland 法测出水泥浆体中各相“纯样”结晶度修正 K 值后，用 XRD 定量分析硬化水泥浆体相组成时是可行的。

2. 水泥浆体微结构由孔和固相组成，固相由起到骨架作用的晶相和起胶凝作用的凝胶组成。凝胶把晶相颗粒胶结在一起。水灰比≤0.4 时，在一定范围内，强度随晶胶比的增加而提高；超过此范围时，随晶胶比的增加，凝胶数量不足以胶结晶相颗粒而使浆体强度下降。优化的水泥浆体结构有一个最佳晶态含量。

3. 除改变水灰比外，在一定的工艺条件下掺入混合材料可改变水泥浆体微结构，对固相结构，则改变晶胶比。水泥浆体最优晶胶比的位置随混合材料细度、掺量和活性的不同而变化；依水灰比大小而不同；水灰比降低时，最优晶胶比提高；混合材料活性越高，易于得到此最优比。

4. 根据所建立的水泥浆体微结构模型，可认为高强度水泥浆体的理想结构是：尽可能少的但足以将晶相颗粒胶结成整体的凝胶；尽可能多的具有活性的晶相颗粒；无大孔。

5. 水泥浆体微结构中的固相结构和孔结构都不是独立的变量，不仅有各自的形成规律和对宏观行为的影响，而且相互影响和制约。所以完整的微结构模型应当包括孔结构在内。应对孔结构-固相结构-宏观行为的三维关系进行进一步深入研究。

参考文献

[1]　吴中伟. 水泥和混凝土的理论研究，武汉建材学院讲课讲义. 1981.

[2]　廉慧珍. 沸石岩的活性，中国硅酸盐学会水泥专业委员会第三届学术年会，柳州，1986.11.

[3]　Lian Huizhen. Cement paste/Zeolite-tuff Bond，PROCEEDING OF 1988 EFC.

[4]　R. Feldman，J. Beaudoin，MICROSTRUCTURE AND STRENGTH OF HYDRATED CEMENT C. &C. R 6：398～400.

[5]　H. Taylor. DISCUSSION OF PAPER “MICROSTRUCTURE AND STRENGTH OF HYDRATED CEMENTMI” BY R. FELDMN AND J. BEAUDOIN，C. &C. R.，7. 465～468.

[6]　王智敏. 用 x 射线衍射一计算机定量分析方法（K 值法）测定硬化水泥浆体的晶态物质含量，第五届全国水泥化学与测试方法学术会议论文集，1992.

附件：硅酸盐学报（双月刊）第21卷第5期（1993年6月）报道

“水泥浆体微结构与宏观行为的关系及其定量分析”项目通过鉴定

“水泥浆体微结构与宏观行为的关系及其定量分析”是清华大学建筑材料教研组承担的国家建材局建材行业科学技术发展基金项目，于1992年12月21日在北京通过专家鉴定。该项目由清华大学建筑材料教研组教授廉慧珍负责，与国家建材院水泥所工程师杨人和、清华大学结构力学教研组教授刘西拉组成跨单位、跨学科的研究力量共同完成。

70年代，加拿大学者R. Feldmam和J. Beaudoin提出微结构与硬化水泥浆体强度的关系，认为水泥浆体的强度不仅是孔隙率的函数，同时还和浆体内部颗粒的尺寸、结晶度、密实度有关。英国著名学者H. F. W. Taylor据此提出了孔隙率-固相结构-强度的关系曲线。但这组曲线只是个模型，对固相结构只有定性描述而无量的表征；只考虑孔隙率而未考虑孔结构。这组曲线的量化是一件非常困难的工作。不仅因为水泥浆体是个极复杂的非均质多相体，而且也因为测试和特征化的困难。

本项研究旨在改变宏观参数以控制水泥浆体微结构的变化，使水泥浆体表现出不同的宏观行为。只要形成一定的微结构，就有一定的宏观行为，而与其形成方法无关。建立起水泥浆体微结构与其宏观行为的关系，就可直接通过调整宏观参数来控制其微结构，以保证满足对其宏观行为的要求。为此，本项研究只采用不同细度的不同类型矿物粉末（即混合材料以不同掺量加入不同水灰比的水泥中来实现水泥浆体微结构的变化），制成用于进行各种测试的试件2000多组。各试件养护至不同龄期。成功地解决了硅酸盐水泥由于无法制得结晶度为100%的纯样而难以用x射线衍射K值法定量的问题，系统地、大量地对各试样进行了定量相分析，并进行了热重、电镜等和化学方法的分析，用计算机处理了几万个数据，提出水泥浆体微结构物理模型和可计算的微结构元，并定量分析了不同宏观参数对水泥浆体微结构的影响，证明宏观参数中影响水泥浆体形成的敏感因素是浆体水灰比、混合材料的颗粒组成、掺量和活性。定量分析的结果很好地符合了Taylor的曲线。证明：在一定的水灰比下，水泥浆体存在使强度最高的晶相颗粒含量与凝胶的比值，即最优晶胶比。孔隙率增加时，此最优比减小；混合材料的活性越高，达到相同强度时的最优的晶胶比越小。同时建立了力学模型，用有限元法计算水泥浆体中孔的尺寸、位置和固相颗粒对水泥浆体力学行为的影响，进一步证明了此结论。其中，还用灰色系统理论分析了混合材料颗粒组成和水泥浆体微结构的关系，计算出不同颗粒组成在不同条件下对水泥浆体的不同影响。

专家们认为，该项研究“所建立的模型把很复杂的问题简化了，纯化了，而且有物理模型和力学模型，进行了定量分析，向材料设计发展了”，“做得很深、很系统”，“运用了新手段，提出了新观点”，“比Taylor有很大的发展”，“达到了国际先进水平”。

自评

附件中的报道是作为硅酸盐学会的专家而参会的《硅酸盐学报》编辑根据本项目研究工作报告和专家们的发言所写的报道，之所以列在这里，想表明我们啃了一下这个硬骨头，很难啃，却得到专家们的肯定，表明专家认可这个硬骨头被啃下来了一点：从材料科学上来看，在研究方法、技术路线和难题的解决等方面有学术上的进步，然而，当时对混凝土的复杂性并未深刻认识，所得到的数据离散程度很大，分析结果最多只是一种很粗略的趋势。抗压强度是混凝土的宏观行为中最简便易测的指标，当时只用了这项指标，而且只选择了28天这一个龄期，现在看来，不仅不能代表混凝土各种宏观行为，而且“28天”如今也说明不了什么问题了。

混凝土的研究者从来没有放弃对混凝土进行计算，不仅发达国家在基础设施的土木工程建设高潮已不再的情况下，混凝土工作者有暇研究如何通过计算设计或预测混凝土的宏观行为，而且我国在20世纪的最后十年里，就有人在尝试用“神经网络系统”预测混凝土的强度。用“灰色系统理论”、“分形几何”等计算混凝土的孔结构，等等。至今也未见什么有意义的成果。20多年前我校材料系吕允文开发了“神经网络系统”的软件，用于材料设计，在金属、陶瓷及其他特种材料方面证明都是有效的，后来他找到我们建材教研组，尝试用于混凝土，结果是毫无效果。

通过这项研究和反思，可得到这样的启发：工程问题的学术性研究和技术研究是有区别的，和纯理论的材料科学研究也不同。学术性研究主要是研究事物的本质和规律，探索已有事物或存在的现象是什么（what）和为什么（why），目的是发现；技术性研究是根据工程实际的条件运用科学原理总结的经验和技巧，解决怎样实现（how）和用哪种工艺实现（which）的问题，目的是创造还没有的东西。两类研究往往是分不开的。科学家进行学术性研究的题目必须从实践中来而用到实践中去，并经过实践的检验；工程师进行技术性研究是运用学术性研究的成果实现对世界的创造，并在实践中提出问题与科学家共同研究。工程实践是科学研究的源头，科学研究是工程实践的依据。发现和创造同等重要，互为基础和依据，没有高低之分，只是对象和目的有别。纯理论的材料科学研究主要由于对原材料有纯度要求，工艺易控，可以不问材料的用途，只研究材料本征特性和规律。对目前的混凝土来说，则完全不适用。

混凝土是一种用看似最简单的工艺制作的最复杂的人工体系，其简单是因为需要，其复杂是简单的结果。混凝土的原材料不能提纯而成型工艺又必须简单易行，否则就不能成为广泛使用的大宗建筑材料，于是原材料地方性强、成分波动、品质差异，再加上其微结构的形成和发展随时间而发展，且与环境温度、湿度关系密切，而环境又是复杂的，不确定的，这就造成混凝土微结构与宏观行为的不确知性和不确定性，成为在人工材料中最复杂的非均质多相混沌体系。然而按系统论的观点，再复杂的体系都还是有其自身规律的，例如天气的变化可谓无常，过去天气预报的不准是众所周知的，而现在的准确性已大大提高。当前世界上的科学前沿是对混沌体系“不确定性科学”的研究。混凝土早晚也应当能计算，条件是原材料相对稳定而可控，并且我们的数学水平大幅度地提升，实验数据还要足够地多。我国有人所研究的“神经系统网络”用到混凝土中不成功的原因，是他们只考虑在实验室的条件下混凝土初始参数的确定，而尚未考虑混凝土还是一个动力学系统，影响其宏观行为的还有不稳定的环境因素。有人说，这样复杂的体系可能需要分段程序。美国气象预报专家Edward N. Lorentz用计算机分析天气形势并进行预报时，编制了13个程序，不知是否因此大大提高了有效性？我们期待着未来跨学科的努力能解决混凝土的计算问题。目前还得多着眼于具体问题具体分析地对待工程实际中需要解决的问题——工程实际中有太多的问题需要解决。

第二部分　想创新，就要站在前人的肩膀上

一、选这部分论文的理由

前几年在评审期刊的送审论文时，发现有几篇研究循环流化床燃煤脱硫排放（灰）渣的文章，内容都是我们20几年前研究过并经过实践检验的，但不仅没有超过我们当时研究的水平，而且还都是处在“火山灰二次反应”的初级阶段。后来了解到还有的企业自备电厂竟然还使用了早就因脱硫效率低、热效率损失大而被否定了的氨法脱硫技术。目前采用循环流化床脱硫的不多，少数采用了的地区，也因直接使用石灰石颗粒以致钙利用率不足而存在成本问题。而且大多数人弄不清不同脱硫技术所排放灰渣特性的区别，未能发挥燃煤脱硫后排放的含硫废渣可资源化的优势。这部分的内容主要是说明我们在“七五”科技攻关专题和“八五”科技攻关专题中做了什么和怎样做的。目的是使后来者了解过去的工作，以避免在低水平上做重复的投入，而能够做出创新性的贡献。

自从清华大学热能系张绪祎开发了我国第一台沸腾炉后，发展到今天的循环流化床，已如雨后春笋般地广泛应用于中小型供热和供气的工业锅炉。在“六五”国家重点科技攻关期间，张绪祎又开发了快装循环流化床锅炉，并开发了用于脱硫的人工脱硫剂。1986年由清华大学热能系牵头承担了国家环保局“七五”重点攻关项目《大气污染防治技术研究》（编号75－58）中的课题《燃煤固硫新型沸腾炉技术开发》（75－58－03），与土木系（技术负责85－58－03－03）、自动化系和化学系合作，形成跨专业合作攻关的技术链，证明了循环流化床燃煤脱硫有很大的优势，在各种脱硫技术中，投资最少、运行成本最低，氮氧化物排放量最小，脱硫剂中钙利用率最高，使用脱硫剂所减少的热效率会因石灰石分解而补偿，更重要的是可以燃用劣质煤，而所排放的废渣又是水泥和混凝土优质的矿物掺和料。通过在我国高硫煤地区若干试点工程使用，也表明这是行之有效的、符合可持续发展要求的环保技术；更证明了跨学科的合作研究的优势：不仅是各专题之间分工合作，而且能互相借鉴、促进和渗透，例如通过我们在对固硫渣资源化的研究，所提出对锅炉工况的要求已被锅炉厂接受并采用推广至今；这得益于“上游”专题（“高效低污染循环流化床技术开发”）对我们进行的锅炉知识普及；新型高效脱硫剂的研究又得益于对建材知识的了解；……之后在“八五”期间，我们这个技术链一方面承担国家环保局“循环床燃煤脱硫技术完善化及工程配套应用技术”的攻关专题，同时在国家计委“八五”重点工程攻关项目中还继续发挥技术链作用。张绪祎承担了“220吨/小时循环流化床电站锅炉及其废渣资源化的研究”的课题，研制开发出我国第一台用于发电50MW的电站锅炉。我们土木系在其中承担了85－206－02－04专题“大型循环床电站锅炉固硫渣资源化研究”和85－912－03－03－03子专题“固硫渣膨胀剂的研制及工艺和应用研究”。从1986年到1995年的十年间，我国循环流化床的技术得到很大的进展，从“七五”期间对几吨/小时的沸腾炉改造成循环流化床，到“八五”

期间 75 吨/小时循环流化床用于工业发电的实现和 220 吨/小时循环流化床的研制、设计，我们的技术链一直在配套运行。

然而，由于若干非技术的原因，使当时的技术未能延续实现。例如环保意识普及的问题，因技术工人匮乏所致流化床运行技术的水平问题，环保政策的力度问题，缺乏整体效益分析以及其他，等等。

很多领域的发展都遵循着辩证法中“否定之否定”的规律。这个规律符合人对客观世界认识的规律。在混凝土工程技术中当然也有很多“否定之否定”的规律的实例，例如轻骨料混凝土、氯氧镁水泥、装配式建筑的预制混凝土构件等。这个规律之所以必然存在，是在一定时期，某种事物的发展受到当时客观条件和认识的局限而受阻，或者有新的事物取代；若干年后，由于其他相关技术的发展，条件有变化，又会有人“重操旧业”，但是必须也必然会在原来的基础上有进一步的更新和发展，这种发展会周而复始、螺旋式地上升。每经过一次否定所取得进展的幅度取决于现实的条件和具体情况。更重要的是对历史的传承——继承和发展的关系就像是生物的遗传和变异。如果没有遗传，物种就没有延续；如果只有遗传而没有变异，则生物就没有进化。同样，对历史批判地继承才能使人类社会不断发展。生物的遗传和变异常常是对环境自发的适应，而人类社会的继承和发展在很大程度上需要人的自觉行为。这就是学校里要学历史课的原因；从事工程与科学技术的人们要学习本行业的科学技术发展史的意义也在于此。现今多数人因被应试教育所笼罩而对历史的漠视程度是空前的。很多过去的资料因无电子存储而难于检索，缺乏传承已经成为当前科技中的一大浪费——耗费大量人力物力和财力去做一些低水平的重复性工作。此外，目前一些管理部门认为凡是近期的都是先进的，论文或申报材料引用文献的发表日期越近越好。殊不知由于发表文章主要是为应对提职称或申请学位之需，当前发表的文章可以说数量多得铺天盖地，而质量却泥沙俱下，够得上水平的太少。抄袭、代写，甚至造假数据的不乏有之。许多期刊、杂志的编辑为此苦不堪言。但也有个别期刊来者不拒，只收版面费而不送审，赢得大量稿源和广告费。相反，早年间，没有那么多功利驱使，有些值得保存的文章、资料，却因保管乏术而失落。例如有的由完成课题的单位资料室或图书馆保存的印刷本，外人无法检索，有时在库存清理时就可能被淘汰而失落。我就曾在系图书馆清理时去“淘宝”而“捡到”了我个人藏书。这些书在现在重读后，又有了许多新的收获。我的感受之一是书要反复读，尤其是当遇到问题时；二是不能忘记历史的经验，许多经典的老书（包括老的论文），都有不少对现在的研究、生产和检测有启发的内容，等等。例如在上世纪 70 年代和 80 年代，我国河南地方建筑材料研究所和上海建材研究所、硅酸盐研究所等老一代的科学家们对石膏和粉煤灰都有较高水平的研究，载有那些论文的“过刊”都正在被尘封着，很少有人问津，现在年轻人极少了解。他们多热衷于用“现代仪器”所自动给出的“数据”和信息点缀自己的论文，却解决不了当前生产中的问题。

近 10 年来，现任清华大学热能工程系教授、工程院院士岳光溪继承了他的老师张绪祎的事业，开发出了单机组发电量 300MW 的循环流化床，并实现了热、电、气三联供技术。经过“十五”和“十一五”期间的公关，他又研制开发出世界上第一台发电量 600MW 的循环流化床，并建成投入试用。现在，因可持续发展的需要，国内外对环保事业都开始加大力度，循环流化床用于发电、脱硫，势在必行。为了在循环流化床燃煤脱硫的固硫灰渣能够高值化利用方面，不再走低水平重复性实验研究之路，在这里，用我们在攻关中发表的报告和论文告诉大家我们过去已经做过哪些工作。用岳光溪的话说："想创新，就必须站在别人的肩膀上"，"创新需要一个漫长的掌握别人的东西的过程"。希望通过这一部分的介绍，能提供给后人一副尽管并不宽厚的"肩膀"。

此外，还想说明，"攻关"的目的是推动科学和生产技术的进步，必须紧密结合工程实际，同时进行知其然而又知其所以然的基础性研究，相辅相成。

我们两次攻关研究的特点是首先明确要攻的是什么关，先进行大量的调研——了解循环流化床锅炉的特性和燃煤脱硫的机理，分析国内外燃煤脱硫技术现状、不同脱硫技术的特点和利弊、不同脱硫技术产生的废渣（或灰）的特性及其资源化的难点。研究中的文献综述不是摆样子，而是使自己的研究目的、目标、技术路线具有科学性的依据。最后的工程应用的试点很重要，是从实验室走向工程的过渡。

二、循环流化床燃煤脱硫技术的优势及其废渣资源化的依据和攻关成果

1. 循环流化床燃煤脱硫技术的依据

沸腾炉又称循环流化床。我国的第一台沸腾炉的开发和应用是1960年由清华大学张绪祎承担并实现的。过去主要是用于需要使用蒸汽的工厂的中、小锅炉，适合于燃烧劣质煤。因为沸腾炉的燃烧温度低，特别有利于脱硫、脱硝（大气中的SO_2和氮氧化物）。如图1的相图所示。在燃烧过程的氧化气氛中，$CaSO_4$稳定存在的温度为1000℃以下；温度超过1000℃时，$CaSO_4$开始分解；超过约1200℃，则完全分解而排出SO_2。还原气氛下（氧气不足），则还会生成硫化钙CaS。沸腾炉的温度恰好低于1000℃。还原气氛生成的CaS与水泥混合加水后，会生成$Ca(OH)_2$，对制品有利，但是同时释放出H_2S气体又污染空气，这是不希望的。

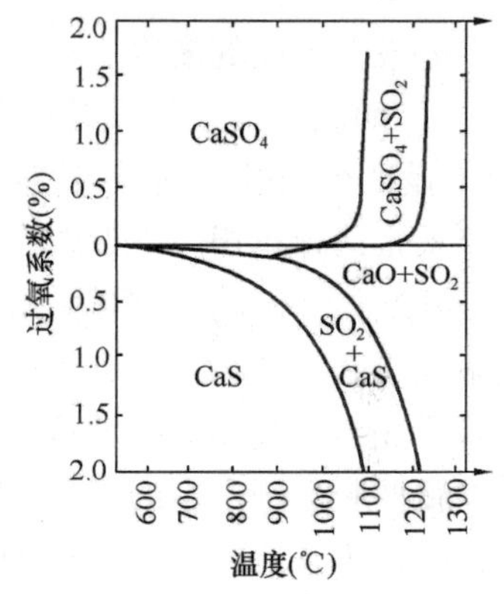

图1 Ca-O-S高温相图（《瑞典造纸》．第55期第622～631页．1952.9.4.）

沸腾炉燃煤从溢流口排出的溢流渣为2～10mm的颗粒，在沸腾段内翻腾，燃烧充分，含碳量低，约占灰渣总量的35%；其中大于10mm的颗粒因相对较重，只能在沸腾段下部翻腾，无法从溢流口排出而沉积在床层上，定期排出，称做冷渣，约占灰渣总量的20%～25%，含碳量也很低；0.5～2mm的颗粒因体轻而随气流上升至悬浮段，在省煤器和空气预热器中由于气流速度下降而沉积下来，叫做沉降灰，约占灰渣总量的25%，未燃尽的可燃物含量较高；在锅炉尾部由收尘器收集的飞灰，约占灰渣总量的15%，为小于0.5mm的颗粒受气流的带动，而在床内停留时间很短，很快进入锅炉的尾部，含碳量很高，有时可达20%。采用带分离器的新型循环式流化床锅炉时，没有溢流渣；由于燃烧完全，循环床锅炉中颗粒为200μm～2mm的对流渣含碳量很低，脱硫充分，活性最高。溢流渣和冷渣的密度为2.5～2.8g/cm³，松堆密度为900～1500kg/m³；磨细后，比表面积可为4000cm²/g。因所含不同量的赤铁矿而呈不同程度的紫褐色或灰褐色。在当时的锅炉水平上，飞灰含碳量高，固硫量却不高，所以我们在"七五"和"八五"

期间使用的就是上述沸腾炉的溢流渣、冷渣和循环床的对流渣。

燃煤废渣能用于制建材，是因为煤炭燃烧时在合适的温度下，其中的黏土质矿物分解成无定形 SiO_2、Al_2O_3而具有火山灰活性。黏土矿物中高岭石的分解温度为600～900℃。由于煤中黏土矿物并不会是单一的高岭石，如果煤中主要黏土矿物不是高龄石而是其他如蒙脱石或伊利石、云母等，则分解温度会超过1000℃，即在沸腾炉中难以分解，产生的灰渣就会活性较低。沸腾炉所用煤炭中的黏土矿物以高岭石为主时，所排灰渣会具有较高的活性。最不希望的是含蒙脱土为主，不仅分解温度高而分解不完全，灰渣活性低，而且用于混凝土会产生有害的作用。

在沸腾炉燃煤时加入脱硫剂，则在燃烧过程中可将煤中所含硫固定在渣中，防止 SO_2 向大气中排放。这种脱硫技术在环保领域中得到很大的关注，被认为是燃煤技术的一个方向。固硫渣是在煤中按一定的钙硫比加入含钙的固体脱硫剂在流化床锅炉中燃烧后排放的含硫残渣，因在燃煤过程中脱硫剂吸收的 SO_2 被固定在其中，故称其为固硫渣。固硫渣含有大量的 SO_3，即使煤中灰分很低，而由于加入25%以上的脱硫剂，也会产生相当数量的固硫渣。“七五”期间的实验研究表明，在正常工况下，固硫渣不仅具有火山灰活性，而且还含有大量以无水硫酸钙形态存在的硫和脱硫剩余的 f-CaO，其本身就具有不同程度的水硬性。因此，固硫渣的活性应包括其火山灰活性及其自身水硬性两部分。

2. “七五”攻关的成果

“七五”期间，国家环保局负责的《大气污染防治技术的研究》科技攻关项目，（编号75-58）下设四个课题：其中《燃煤固硫新型沸腾炉技术开发》课题（编号75-58-03）由清华大学热能系张绪祎承担，下设四个专题：《高效低污染新型循环床技术开发》、《耐水性高活性脱硫剂开发》、《沸腾炉燃煤固硫渣制建材的实验研究》（编号75-58-03-03）、《循环床DIC自动化系统研究》（编号75-58-03-04）。课题中的75-58-03-03专题《沸腾炉燃煤固硫渣制建材的实验研究》由我作为技术负责。使用01专题开发的6t / h快装循环床和75-58-03-02专题研制的新型脱硫剂产生的含硫废渣——固硫渣（不能叫脱硫渣，因为从字义上，脱硫渣是脱了硫的渣，不含硫）进行实验研究。针对当时国外普遍地认为固硫渣中的 SO_3 有害的观点，从研究固硫渣中 SO_3 的形态及其与固硫渣活性的关系入手，基础研究与应用研究同时进行，化害为利。研究成果已用于亚运会自行车比赛老山训练场跑道不裂面层、北京八达岭缆车索道支柱地脚锚杆锚固、宜昌市建委食堂刚性防水屋面（一年后因放水效果很好而改为接待室）、宜昌市四马路、市长路和长江江边码头（见以下各图）。

索道支柱

车站支柱

北京八达岭缆车索道支柱锚杆锚固

北京亚运会自行车老山训练场跑道不裂面层

宜昌市建委接待室刚性防水屋面

宜昌市长江码头

宜昌市 1989 年市长路

在此选入当时我提交技术鉴定会的“沸腾炉燃煤固硫渣制作水泥研究的工作报告”，以介绍研究的方法和技术路线。尽管因诸多非技术的原因而未能推广，从研究成果和工程试点应用来说，至今仍有借鉴价值。该报告如下：

75-58-03-03-03 沸腾炉燃煤固硫渣制作水泥试验研究的工作报告

清华大学土木工程系建筑材料教研组于 1986 年开始，承担国家“七五”科技攻关环保项目“大气污染防治技术”（75-58）第三项课题“燃煤固硫新型沸腾炉技术的开发”（75-58-03）第三项专题“燃煤固硫渣制作建材的研究”（75-58-03-03）第一项子专题“燃煤固硫渣制作水泥的研究”（75-58-03-03-01）。合同规定主要目标是利用沸腾炉脱硫燃煤后含 SO_3 的废渣与 525 号硅酸盐水泥熟料（GB 175—92）配合，生产 425 号及 325 号水泥，研究固硫渣中 SO_3 有益无害化及其对水泥性质的影响。

（1）研究工作的过程主要是根据理论分析制定试验研究方案，理论研究与宏观试验同时进行，并互相补充、借鉴和修正，不断实践、不断总结分析、不断完善方案。得出可靠结论后，进行中间性生产试验和应用性的试验研究。研究内容及技术路线如下图所示。

（2）在理论研究的基础上，五年来进行了大量试验工作，取得近万组数据，研究了固硫渣及其中的Ⅱ-$CaSO_4$ 对水泥的需水量、凝结时间、强度、变形等的影响规律以及固硫渣的安定性、对钢筋锈蚀的影响、对不同减水剂的适应性等。为了验证固硫渣中 SO_3 后期无害，还测定了水泥硬化浆体中的钙矾石生成量以及剩余 SO_3 量，观测了长期强度及变形等。所有对强度的检验均进行 31 组统计性平行试验，并于 1989 年 4 月和 1990 年 8 月分别进行了中间性生产试验，共生产固硫渣水泥 680 吨，已全部用于混凝土工程，使用情况良好。研究工作主要结论如下：

1）由于沸腾炉燃烧的特点，与其他燃烧方式排放的废渣（如粉煤灰）相比，沸腾炉渣

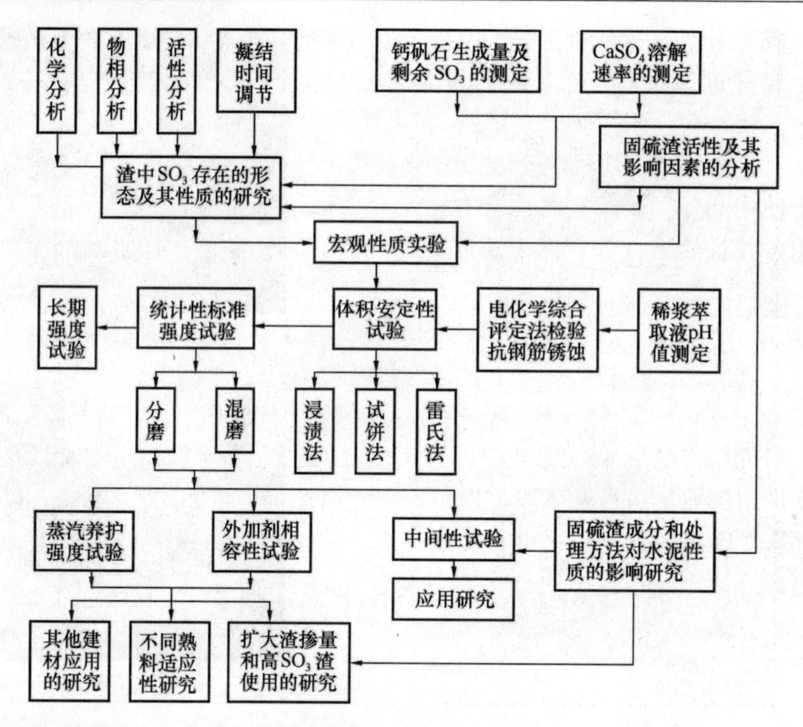

因燃烧充分而含碳量很低，具有较高的活性；脱硫燃烧后，所得固硫渣具有自身水硬性，用于生产水泥时，具有更高的活性：

①在正常燃烧工况情况下，固硫渣中 SO_3 主要以Ⅱ型无水硫酸钙（Ⅱ-$CaSO_4$，即不溶性石膏）的形态存在。Ⅱ-$CaSO_4$ 并非完全不溶，而是与二水石膏和Ⅲ-$CaSO_4$ 相比，溶解速率低但溶解度相近。因此，可根据“短时多溶”的原则，在水泥中 SO_3 总量比用二水石膏调凝时的 SO_3 总量大（但＜3.5%）的情况下，可利用Ⅱ-$CaSO_4$ 调节水泥凝结时间。由于Ⅱ-$CaSO_4$ 的溶解特性，除使水泥凝结时间合格外，与硅酸盐水泥和其他火山灰质水泥相比，初凝和终凝时间间隔还有所缩短，这正是施工所希望的。

②用 SO_3 为 6%～9%的固硫渣 30%和 40%分别与 70%和 60%525＃硅盐水泥熟料（GB 175—92）混合共同磨细至孔边长 80μm 的筛筛余＜8%，可分别得到各种性能符合现行国家水泥标准的固硫渣水泥。固硫渣中的 SO_3 在水泥硬化后期不会产生任何不良作用。

2）固硫渣水泥具有以下特点：

①当固硫渣中的Ⅱ-$CaSO_4$ 可以使水泥中的 SO_3≥2%时，生产中不必加入二水石膏。

②需水量比火山灰质水泥的低；故强度检验时按胶砂流动度 125～135mm 计算加水量。

③与其他水泥相比，在初凝时间相近或较长的情况下，初凝与终凝时间间隔较短，有利于施工。

④施工时不得使用木钙类的减水剂。

⑤有较强抗硫酸盐腐蚀的能力和抗钢筋锈蚀的能力。

⑥固硫渣具有自身水硬性，作为原材料也有储存性能问题。

由于上述特点，固硫渣水泥不宜列入火山灰质水泥，而应另外制订标准。

3）为进一步推广应用，尚有以下问题需继续研究：

①固硫渣原煤矿物组成和燃烧温度对固硫渣活性的影响规律及评定方法和标准。

②关于固硫渣对不同熟料的适应性进行了少量研究，但在全国推广时，尚需按不同地区可得的熟料、煤种进行试验，进行熟料的矿物组成对固硫渣水泥性能影响规律的系统性研究。

③含碳量对不同固硫渣活性的影响及其掺量限值的确定。

④对由于使用高 SO_3 渣或提高固硫渣掺量而使水泥中的 $SO_3>3.5\%$ 的可行性已由已进行的试验证明，并且效果良好，但为了能充分利用各种固硫渣，还应增加试验量以进一步研究。

⑤ 因具有自身水硬性，单独加水成型即可得到强度，特别适合于蒸汽养护；掺固硫渣的水泥成型试件蒸养后，在自然条件下，强度仍持续增长。

（3）除按合同完成研究外，尚进行了用高 SO_3 固硫渣制作膨胀水泥以及水泥混凝土用膨胀剂的研究。研究结果表明，固硫渣中的Ⅱ-$CaSO_4$ 作为主要膨胀组分，而用固硫渣作为强度组分，可生产出膨胀能大、膨胀稳定期短、强度协调好、生产工艺简单、成本低廉的不同膨胀性能的膨胀水泥和膨胀剂。其中一种固硫渣膨胀剂已成功地用于第 11 届亚运会北京老山自行车训练场跑道不裂面层、由法国设计和中国施工的八达岭缆车索道支柱锚杆锚固工程、宜昌市建委新建食堂 150 号刚性防水屋面。该屋面已经过雨季的考验而未漏雨。该膨胀剂尚有以下问题需进一步研究：

①对不同品种和标号水泥适应性的研究。

②对常用外加剂相容性的研究。

③进行掺固硫渣膨胀剂混凝土物理力学性质的系统研究（包括各种具有抗渗、防裂、膨胀要求的混凝土）。

④膨胀剂储存性能的试验研究。

⑤制定固硫渣膨胀剂的生产和使用规范，建立和完善固硫渣膨胀剂原材料以及产品质量检验的有关标准。

⑥研究提高固硫渣膨胀剂膨胀效能（自应力值）的方法和途径，进一步挖掘固硫渣在制作膨胀剂方面可能具有的潜力，拓宽固硫渣膨胀剂的应用范围，努力使其产品系列化。

⑦进行掺固硫渣的膨胀水泥水化和膨胀机理的研究。

（4）本次鉴定提供鉴定文件如下：

①国内外燃煤固硫渣利用现状（附查新证明）。

②沸腾炉燃煤固硫渣中硫的形态及特性。

③沸腾炉燃煤固硫渣活性评价及其影响因素的试验研究。

④用沸腾炉燃煤固硫渣制作水泥的配制及性能的试验研究。

⑤固硫渣水泥中间性试验及分析报告（附原始记录）。

⑥固硫渣水泥混凝土应用的研究（附使用证明）。

⑦用固硫渣制作水泥混凝土膨胀剂的试验研究（附使用证明）。

（5）在本研究工作中，宜昌市环保局作为专题负责单位，积极配合，大力支持，为提供试验用的固硫渣，前后共在宜昌市麻纺厂、印染厂、造纸厂等锅炉进行了六次脱硫试验，提供了十四种固硫渣和两种无硫渣，并联系、协调了二次中间性试验，保证了任务的完成。在中试中，宜昌市水泥厂在厂领导的支持、徐旭总工程师的具体组织下，克服很多困难，加班加点完成了任务。宜昌市建委对本专项研究和推广给予了很大的支持；宜昌市政二公司、宜昌市预制构件厂在将固硫渣水泥应用于工程方面给予了积极的支持，在此谨对上述单位表示衷心的感谢。

3. “八五”国家重点科技攻关

为了使循环流化床更大程度上发挥其脱硫、脱硝的优势，国家计委设置了“八五”国家科技攻关项目85－206，我们承担了其中第二个课题85－206－02《220吨/小时循环流化床电站锅炉研制与其固硫渣资源化研究》中的03专题。当时攻关的目标是与某热电站合作，将所研制的220吨/小时锅炉安装在该电厂，用以发电50MW小时，并使用所研制的脱硫剂进行脱硫试验。脱硫所产生的固硫渣用于研究固硫渣扩大资源化，加上“七五”期间制作水泥及其硅酸盐制品的技术，有望使该电厂达到零排放目标，并从固硫渣资源化得到经济效益。然而，于1994年，该厂突然变卦，决定从国外进口一台发电用的循环流化床锅炉而终止了合同。已经做的工作由国家教委组织验收而结束。整个课题使用了上千万元的资金，却没能取得预期的成果，这是一件很遗憾的事。但是从科技攻关的经历来说，证明了我国完全有能力自主开发发电用的循环流化床，并进一步证明了循环流化床燃煤脱硫技术的优势，不仅能燃用高灰分的劣质煤，也不仅能有效地脱硫，而且还能避免或大为减少氮氧化物的排放，同时固硫渣是水泥和混凝土的优质资源，对实现节能降耗减排具有重要的意义。为了避免今后一旦需要大规模使用循环流化床燃煤脱硫并利用其排放的灰渣时，不至于还从头开始。现选入当时提交验收的工作报告供今后参考：

85-206-02-03大型循环流化床电站锅炉固硫渣资源化研究工作报告

攻关目的

将固硫渣进一步资源化，使其既能扩大消耗量，又能产生较高价值，提高固硫渣的环境和经济效益。

合同中专题分解和主要内容如下：

1. 商品固硫渣的研究

（1）商品化固硫渣品质评定标准及生产要求。

（2）固硫渣对不同水泥的适应性。

（3）与外加剂的相容性。

（4）固硫渣用于普通混凝土和高强混凝土的掺量。

2. 固硫渣轻质多孔混凝土的研究

（1）固硫渣轻质多孔混凝土的配制和性能。

（2）发泡剂的选择及配比、掺量对制品性能的影响。

（3）研究引气及成型工艺，以降低制品容重，提高强度，稳定体积。

成果简介

1. 商品固硫渣用于普通和高强混凝土的实验研究

（1）水灰比为0.3～0.45，固硫渣掺量30%～40%，混凝土28天配制强度为60～80MPa，坍落度60～200cm。

（2）水灰比为0.45～0.6，固硫渣掺量＜40%时，C30～C40混凝土28天配制强度为40～50MPa，坍落度100～200cm。

（3）水灰比≤0.35时，掺固硫渣5%～40%对混凝土强度与和易性都几乎没有什么影响，这说明固硫渣最适合用于高强混凝土。

（4）在高强混凝土中掺入固硫渣，可降低混凝土绝热温升、节省水泥，并使混凝土后期强度有较大的持续增长率。

（5）当固硫渣中 SO_3 为 6％～10％时，控制混凝土中 SO_3 占水泥和固硫渣总量为≤4％时，混凝土强度随 SO_3 量的增加而提高。故混凝土中使用固硫渣时，水泥中 SO_3 可放宽至 4％。

2. 商品固硫渣用于高性能混凝土

（1）用 30％体积（25％重量）的固硫渣取代水泥，混凝土强度≥60MPa。

（2）用 50％体积（44％重量）的固硫渣取代水泥，混凝土强度≥50MPa。

（3）用 60％体积（44％重量）的固硫渣取代水泥，混凝土强度≥40MPa。

混凝土坍落度均≥220mm，和易性良好。

（4）固硫渣所含 SO_3 可以补偿混凝土在空气中的收缩，使混凝土的干燥收缩小于开裂极限。SO_3 从 2.7％增加到 4％时，混凝土在空气中的收缩减小约 55％。因此固硫渣是高性能混凝土的优质的掺和料。

（5）固硫渣对混凝土中过量 SO_3 产生的膨胀有抑制作用。

3. 固硫渣轻质多孔混凝土的研究

使用固硫渣 60％的制品性质指标：单位质量为 700～800kg/m^3，抗压强度达 4～5MPa，取代价格昂贵的铝粉为自制的泡沫剂发气，改高压蒸养为常压蒸养。

4. 固硫渣复合双掺水泥的试验研究

（1）组合效应：较低活性的混合材料如Ⅱ级以下的粉煤灰，水泥强度与固硫渣和粉煤灰的比例成正比。

（2）超叠效应：活性较高的混合材料（如矿渣和天然沸石岩）、较低活性的混合材料 90 天复合水泥强度超过单掺的水泥强度，甚至超过无掺和料水泥的强度。

（3）固硫渣复合水泥的长期性能和耐久性可靠，特别适合于蒸汽养护。

5. 固硫渣品质评定标准的研究和标准的制定

（1）影响固硫渣活性的因素及其评价方法研究

①研究固硫渣中 SO_3 含量和固硫渣粉磨细度对活性率影响的规律与对抗压强度的影响规律一致；说明用活性率评价固硫渣品质，不仅可反映其化学活性，也可反映物理活性。

②分析影响活性率测定结果的因素。确定了对测定结果干扰最小的操作方法。

③进行活性率测定的误差分析及验证。结果表明该试验方法本身闭合性好，使用可靠。

（2）研究固硫渣中 SO_3 对水泥性质影响的研究

①确定可用的固硫渣中 SO_3 含量的范围

②确定代替或部分代替二水石膏的规则。

（3）固硫渣中微量有害物质的分析

①固硫渣对混凝土中的碱-骨料反应有明显的抑制作用。

②固硫渣中所含的 Na_2O、K_2O 无害性。

（4）提交固硫渣品质标准建议稿及其编制说明。

提交验收的研究报告

之一：商品固硫渣用于配制混凝土的实验研究

之二：固硫渣双掺复合水泥的研究

之三：固硫渣多孔混凝土的研制

之四：制定固硫渣品质评定标准的研究

附件一：国家建材局标准——用于水泥的固硫渣（建议稿）

附件二：用于水泥的固硫渣（建议稿）编制说明

在上述“八五”期间计委主持的攻关项目中攻关专题研究的同时，我们还承担了国家环保局关于大气污染防治技术攻关项目 8503912 中的专题《循环床燃煤脱硫技术完善化及工程配套应用技术》下的子专题《固硫渣膨胀剂的研制、工艺和应用研究》。在贵阳与贵阳水泥厂合作建立了一条钙基高效人工脱硫剂和固硫渣膨胀剂共用的生产线。但是当时的环保意识无法和实际利益抗衡，在政策上，为了鼓励燃煤脱硫所制定的排污费力度不足，使企业宁可交排污费也不愿脱硫。已建成的上述生产线被迫废弃。相信今后环保问题必然会日益尖锐，我们的技术也总会有一天经过否定之否定后被启用。尽管该项成果属于一个子专题，而对了解我们已经做的工作以供将来再启用和发展时的研究，会有一些参考价值。以下选入该子专题的研究工作报告，以供对该项成果的了解。

“八五”国家环保攻关项目子专题 85-912-02-03-03
固硫渣膨胀剂的研制及工艺和应用的工作报告

本子专题在“七五”攻关研究的基础上，根据固硫渣的特性建立模型，设计了适用于含不同成分、具有缓慢发展微膨胀性能的膨胀剂和用于补偿收缩的膨胀剂。在实际生产应用中，用不同固硫渣配制膨胀剂时，进行了各组分及其配合比对产品性能影响的实验研究，对优选配方的固硫渣膨胀剂进行了生产工艺和混凝土性能的实验研究。

研究主要内容和结果

1. 粉磨方式：在分磨和混磨两种粉磨工艺中，分磨的比混磨的膨胀率较大，但工艺较麻烦，能源消耗较多。综合考虑，建议采用混磨方式。

2. 细度：太细时由于钙矾石生成速度快，膨胀能较多消耗在塑性阶段，有效膨胀能就小，对补偿收缩不利，但膨胀稳定期缩短；当固硫渣膨胀剂较粗时，钙矾石形成速度慢，限制膨胀率会受到一定的影响，且膨胀稳定性差，对混凝土后期强度不利。膨胀性能较好的膨胀剂比表面积在 3330cm^2/g 左右。固硫渣膨胀剂的比表面积宜控制在 2800～3500 cm^2/g 的范围。

3. 对蒸汽养护的适应性：蒸汽养护制度为：

$$20℃ \xrightarrow{\text{升温 1h}} 40℃\text{（恒温 1.5h）} \xrightarrow{\text{升温 1.5h}} 80℃\text{（恒温 7.5h）} \xrightarrow{\text{降温 4h}} 20℃$$

无约束试件在蒸汽养护后有效膨胀率有所损失，但继续水养护后，膨胀仍继续发展；强度的规律也如此；如用于自应力压力管，由于是带模养护，受到较大限制，可提高制品的密实度，并产生较大自应力，但自应力压力管要求的自应力要大得多，需另行配制高膨胀能的膨胀剂。

4. 固硫渣膨胀剂对水泥的适应性：使用了 14 种不同品种、标号和厂家的水泥进行对比试验，表明水泥不同时，掺膨胀剂混凝土的强度、膨胀性能都有差别，但除个别外，影响不大。而其中掺较多混合材的低标号水泥并不因其中 C_3A 含量低而降低膨胀剂的作用。因此并无水泥的禁忌。

5. 膨胀剂掺量的影响：当固硫渣膨胀剂掺量为6%～18%时，随着固硫渣膨胀剂掺量的增加，标准稠度用水量稍有增加，但变化并不大；凝结时间有所缩短，但都符合JC 476—92的要求，而限制膨胀率却随固硫渣膨胀剂掺量的增加而明显增大。当固硫渣膨胀剂掺量超过12%时，水中养护14天时的限制膨胀率超过0.02%，达到JC 476—92的要求。因此固硫渣膨胀剂掺量应超过12%。当掺量超过15%以后，限制膨胀率增长率较大，同时强度下降，7天时强度损失明显。虽然28天时得到恢复，强度指标符合规范要求，但从满足补偿收缩、强度和造价综合考虑，固硫渣膨胀剂掺量以12%～15%为宜（取代水泥重量）。

6. 固硫渣膨胀剂和其他外加剂的相容性：试验表明，固硫渣膨胀剂和其他外加剂无不相容性问题，但对于不同品种外加剂，影响规律有所不同，实际使用时应根据试配选定。固硫渣中的$CaSO_4$和木钙减水剂不相容性的问题在此并未表现的原因是，固硫渣膨胀剂在混凝土中的掺量比在水泥中的小，无水石膏比例在混凝土中比在水泥中的更小，其影响也就不大。但实际使用时还是应当慎重。

7. 储存性能的试验：将优选配方和其他来源固硫渣按相同配方配制的固硫渣膨胀剂封入防潮塑料编织袋中存放，定期取出，以15%的掺量测试砂浆限制膨胀率。结果表明，各配方的固硫渣膨胀剂储存6个月后各项性能变化不大，预计按本试验所用储存方法储存时间至少可达半年，符合现行有关膨胀剂储存有效期的规定。

8. 长期稳定性：用优化的配方进行掺固硫渣的水泥砂浆长期的强度和限制膨胀率试验，并与当前市售膨胀剂作对比。试验结果表明，掺固硫渣膨胀剂的砂浆90天之前强度一直在增长；在水中养护14天的胀剂率是90天的80%以上，7天后膨胀率增长很少。而其他膨胀剂在水中养护14天的胀剂率是90天的60%以下，42天到90天仍在增长。故固硫渣膨胀剂的长期稳定性是可靠的。

9. 固硫渣膨胀剂用于基础锚固的研究：在普通水泥砂浆中分别掺不同比例UEA膨胀剂和固硫渣膨胀剂，用于锚杆的锚固，进行抗拔力的比较试验，分析对锚固性能的影响，以得到其在锚固用砂浆中的最优掺量；测定钢筋应变的变化，以求得其表面的粘结剪切应力分布规律，探讨锚杆传力的锚固机理，为推广固硫渣膨胀剂在锚固工程中的应用提供理论依据和基础参数。研究结果表明，掺有膨胀剂的水泥砂浆能明显提高锚固强度，锚固强度随膨胀剂的掺量的增加而提高，但掺入过多时反而使锚固强度下降，固硫渣膨胀剂的适宜掺量为25%左右；水泥砂浆中掺入固硫渣膨胀剂后锚固效果与掺UEA的差不多，但由于固硫渣膨胀剂工艺简单、成本低廉，可利用工业废渣，综合效果优于UEA。因此，固硫渣膨胀剂在地下工程的应用领域有广阔的前景。

提交的技术鉴定文件

1. 固硫渣膨胀剂的研制、工艺和应用研究报告
2. 固硫渣膨胀剂生产性实验研究（附生产工艺设计）
3. 固硫渣膨胀剂锚固性能的试验研究

后记

参加以上工作的先后还有硕士研究生7人，博士研究生1人，讲师3人，工程师2人，实验员2人，另有若干名本科生。参加工作的教师除我之外，有郭玉顺、王志敏、李桂芝、

崔京浩、刘西拉等。每个研究生入校后第一次和科研组见面，首先明确他们是与我和其他教师共同承担国家攻关专题研究的科研合作者。第一学期，他们就开始介入专题研究，除上课外，要求他们一面通过协助上届研究生和老师做些辅助性的工作，了解专题研究目的、技术路线、工作进展，尽快进入状态；另一方面做自己开题的准备，告诉他们不能把一年级研究生当成继续做（或六年级）大学生。所以他们的工作都很踏实有效。只要是在校，除参加科研组的学术活动外，各人的实验研究工作随时通气、交流、讨论，以便及时发现问题调整计划。他们是（按入学先后）：刘燕芳、梁文泉、陈恩义（1989 年取得硕士学位后保留博士生学籍留校的教师，1992 年入学攻读博士学位）、江加标、刘爱明、王迎华、王丹松、李英，本科生李海燕、王珂珂、陆文胜等，他们都为“七五”或“八五”攻关专题做出过出色的贡献。

自评和思考

1. 从上述报告可见，攻关研究要有很明确的针对性和实际工程应用的目的性，有基础研究、应用研究，有实验室研究和 31 批以上批量小试、工厂中试生产、工程应用等。符合工程技术研究的要求。

2. 85-206-02-03 研究工作报告中介绍的成果所列“商品固硫渣用于高性能混凝土”和“商品固硫渣用于普通和高强混凝土的实验研究”，显然是把“高性能混凝土”看成了是混凝土的一个品种，误认为大坍落度是高性能混凝土的指标。这是当时对“高性能混凝土”认识的误区。其实应该都合在一起作为混凝土的研究。但从内容和成果来看，还是针对了固硫渣的特点，可扩大用途，化害为利。

3. 陈恩义在博士论文工作期间，研制利用固硫渣制作膨胀剂，进行了用于钢筋混凝土结构构件本构关系的初步研究，并用不同膨胀源的膨胀剂进行试验，提出膨胀剂材料设计模型，这应当是固硫渣特性利用具有的创新性的研究。然而和混凝土其他原材料一样，固硫渣源于煤炭的脱硫燃烧，也具有天然资源成分和组成波动的特性。通过本研究也是提供一种思路，随着煤炭来源、锅炉工况、脱硫技术、灰渣排放方式等的不同，必须具体问题具体分析，有针对性地进行重演性试验。任何时候都不可能有一成不变的技术。

以下是本部分所选论文。

生意人有钱就赚；商人有所为，有所不为；企业家要承担企业、社会、国家发展的义务！

——马云

论文之一

因势利导，利用流化床脱硫燃煤含硫废渣*

北京清华大学土木工程系　廉慧珍　陈恩义

国内外已有的研究表明，由于燃烧温度合适，流化床燃烧的炉渣（称沸腾炉渣）具有比粉煤灰高的活性[1]。因此已被作为一种人工火山灰材料用于水泥的掺和料和制作蒸汽养护的硅酸盐砌块或砖[2][3]。但当燃烧高硫煤时，为了防治SO_2对大气的污染而加入脱硫剂后，燃煤释放的SO_2就被固定在所排放的废渣中。对这种含硫的废渣，许多国家都认为是有害的，而在将其用于建筑材料时，要经过预处理，使渣中的SO_3预先反应掉[4]。

对水泥混合材料，各个国家都有对其中SO_3含量限制的标准。主要原因是，当水中SO_3超过该标准时，会引起水泥强度的急剧下降和产生体积膨胀，引起不良后果[5]。在上述含硫废渣（以下称固硫渣）中，依煤中含硫量和灰分的不同，含SO_3量最高可达20%左右。我国目前燃烧的大量含硫劣质煤固硫渣中SO_3约为3%～12%（<3%时可按普通沸腾炉渣处理）。将固硫渣作为火山灰质混合材料加入水泥中时，就会使水泥中SO_3超过国家标准。如能因势利导地将SO_3变害为利，则制作建材就成为固硫渣最广阔的出路。

一、固硫渣中SO_3存在的形态和特性

1. 脱硫剂的主要成分是CaO，在流化床中，当燃煤排出SO_2气体时，SO_2即与CaO结合。对固硫渣进行x-射线衍射分析表明，这种结合的产物为$CaSO_4$。

$CaSO_4$是石膏的主要成分。天然存在的石膏有二水石膏（$CaSO_4 \cdot 2H_2O$）和无水石膏；用作胶凝材料的是半水石膏（$CaSO_4 \cdot 0.5H_2O$）或其他类型脱水石膏（α型或β型脱水半水膏）。石膏胶凝材料加水凝结后又变成与天然二水石膏不尽相同的二水石膏。半水石膏继续脱水后，即可生成无水石膏（$CaSO_4$）。在不同温度下，无水石膏又具有不同形态：Ⅲ型$CaSO_4$（又有α型和β型）和Ⅱ型$CaSO_4$或称不溶性硬石膏。在更高的温度下（1200℃左右）可由于部分$CaSO_4$分解而生成高温煅烧石膏。经x-射线衍射和以天然硬石膏为对比样进行物化和化学分析证明，固硫渣中的SO_3属于Ⅱ-$CaSO_4$（图1），即不溶性无水石膏，

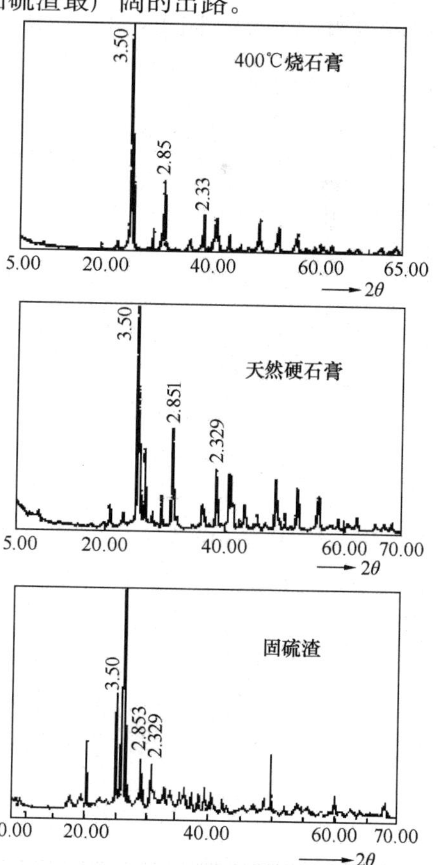

图1　无水石膏的XRD

* 原载于《硅酸盐建筑制品》1995年1月。

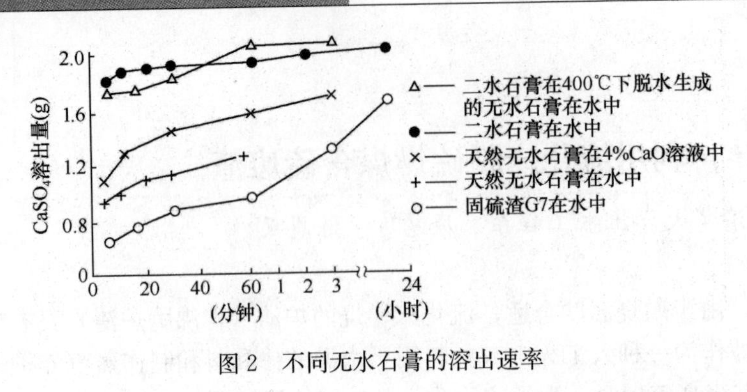

图 2　不同无水石膏的溶出速率

具有与天然硬石膏相似的结构和性质。结晶紧密而稳定，难溶于水，其溶解速率低于二水石膏的溶解速率，但溶解度稍大，凝结极慢，甚至完全不凝结。即无水硬活性。图 2 所示为不同石膏的溶出速率。

据文献报道，通过加入碱性激发剂（如石灰、煅烧白云石、硅酸盐水泥等）或某些盐（如硫酸钠、硫酸钾）、半水石膏或二水石膏等激发剂，可激发无水石膏的活性[7]。图 3 为溶液中 CaO 不同浓度对无水石膏溶解度的影响。该图表明，随溶液 CaO 浓度的增加，无水硫酸钙的溶解度提高，但当 CaO 浓度超过 4%时，无水硫酸钙溶解度又开始下降。

当有活性的 Al_2O_3 存在时，受激发的 Ⅱ-$CaSO_4$ 生成钙矾石（AFt 相，$3CaO \cdot Al_2O_3 \cdot 3CaSO_4 \cdot 2H_2O$），成为强度组分或膨胀组分。固硫渣中恰好有较多的活性 Al_2O_3，在不同浓度的 CaO 溶液中，测定 $CaSO_4$ 溶出量，同时将其干燥后的滤渣进行 x-射线衍射分析，表明有不同量的 AFt 相生成。AFt 多时，则 $CaSO_4$ 溶出量少，说明 $CaSO_4$ 与 CaO 和渣中的活性 Al_2O_3 发生了反应（图 4）。

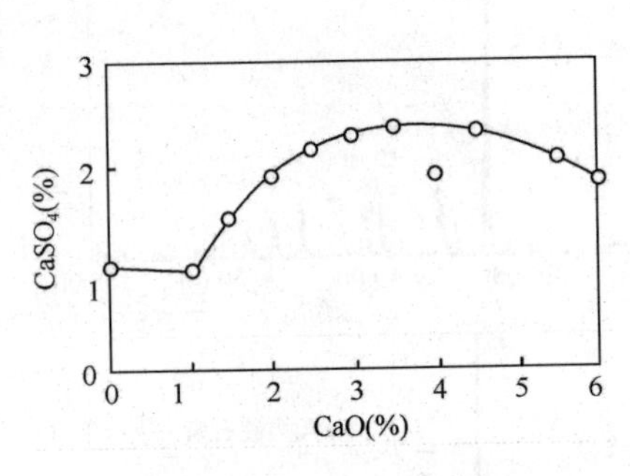

图 3　CaO 不同浓度的溶液中 $CaSO_4$ 的溶解度

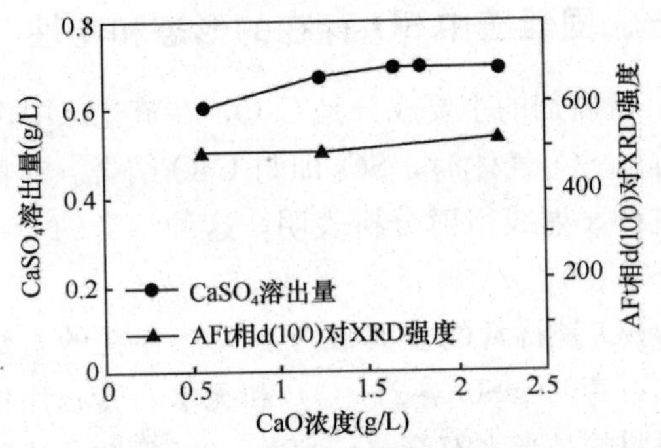

图 4　固硫渣在不同 CaO 溶液中的 $CaSO_4$ 溶出量及 AFt 的 XRD 强度

2. 固硫渣中 SO_3 的无害性：虽然二水石膏的溶解速率远大于硬石膏的，但后期硬石膏的饱和溶解度却稍大于二水石膏的，因此掺固硫渣的硬化水泥浆体中剩余 SO_3 含量的测定，对固硫渣水泥也是十分重要的。如果后期的硬化浆体中 SO_3 含量较大，在适当的条件下就还会形成钙矾石相使试体膨胀而影响性能。

通过 5%甘油饱和石灰水-石膏萃取剂萃取的方法，分别用掺固硫渣 G6 以 10%、15%、20%、30%、45%和熟料配制水泥，并用烧制的无水石膏调节水泥的 SO_3 总量，使其大体相同。对养护 28 天的硬化水泥浆体中 SO_3 进行测定，各试样均萃取 24 小时，结果见表 1 所示。萃取 30 小时时，结果基本不变。由此得出，28 天硬化水泥浆体中这样微小的 SO_3 含量对其长期性能并无不利的影响。

表 1　掺固硫渣的水泥 28 天剩余 SO_3

固硫渣掺量（%）	水泥中 SO_3 量（%）	28 天剩余 SO_3（%）
15	3.6	0.36
20	4.0	0.40
30	3.8	0.52
45	3.8	0.42

二、固硫渣的活性

1. 燃煤脱硫后，随钙硫比和钙利用率的不同，在固硫渣中还含有一定量剩余的游离 CaO（称为游离 CaO，写作 *f*-CaO）。这种 *f*-CaO 不同于立窑水泥熟料中有害的过烧 *f*-CaO，而是在低于 1000℃的温度下存在的，具有很高的水溶性。

将固硫渣加水后，其中的 *f*-CaO 就成为 $CaSO_4$ 的激发剂。所以含有 *f*-CaO 的固硫渣本身具有自身水硬性。即无任何外加物质，只加水成型经养护后，可产生强度。最高的 28 天抗压强度可高于 10MPa（100 号）（表 2）。

表 2　固硫渣单独水硬性及受石灰的激发

渣号	SO_3（%）	*f*-CaO（%）	抗压强度（MPa）		加 5%石灰后的强度(MPa)		备　注
			28 天	蒸养	28 天	蒸养	
G4	3.35	3.32	3.8	6.3	8.8	12.4	含碳量 10.8%
G5	4.21	2.40	1.6	1.6	8.5	8.9	球状脱硫剂
G6	7.31	4.97	10.0	7.9	15.5	18.8	片状脱硫剂
G7	11.81	2.36	13.7	9.4	15.2	16.4	片状脱硫剂
G8	6.69	0.97	2.5	2.6	8.4	8.5	未排出，随炉温冷却
FK	2.62	0.32	—	—	11.7	11.9	无脱硫剂，无法成型

由表 2 可见，固硫渣自身水硬性随 *f*-CaO 量的增加而提高，当 *f*-CaO<0.35%时，固硫渣不具有自身水硬性。加入 5%石灰时，各固硫渣强度可进一步提高。

2. 固硫渣的活性可用活性率 K_a 表示：

$$K_a = \frac{\text{活性 } SiO_2 + \text{活性 } Al_2O_3}{\text{全 } SiO_2 + \text{全 } Al_2O_3} \times 100\%$$

图 5 为固硫渣不同活性率 K_a 和加入 5%CaO 的固硫渣试件蒸养抗压强度的关系，图 6 为 K_a 与 30%固硫渣和 70%的熟料配制的水泥强度发展的关系。

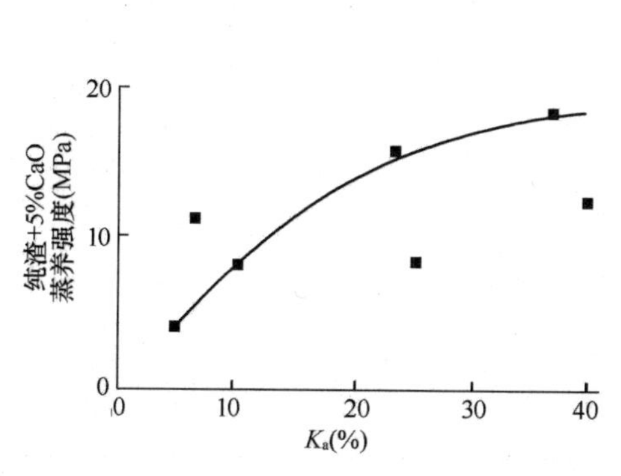

图 5　活性率 K_a 与加入 5%CaO 的固硫渣试样蒸养抗压强度的关系

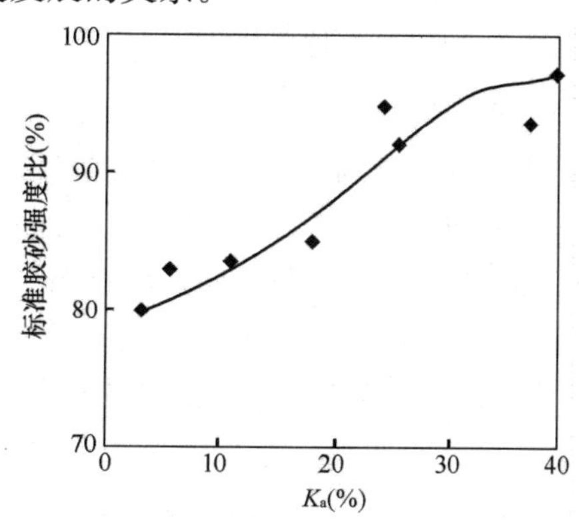

图 6　活性率 K_a 与加入 30%固硫渣的水泥抗压强度的关系

三、固硫渣中SO_3的利用

1. 部分或全部代替生产水泥用的二水石膏，调节水泥凝结时间：

由于施工的要求，国家标准规定水泥加水后初凝（即开始失去流动性）时间不能少于45分钟。在水泥生产时加入适量二水石膏（简称石膏），即可调节水泥凝结时间达到要求。已有人研究，硬石膏可以部分代替二水石膏用于生产水泥。据报道，硬石膏和二水石膏各半可得到最好的水泥调凝效果[8]。目前市售水泥一般凝结时间都在2小时左右。加入固硫渣后，调整SO_3含量，测定水泥的凝结时间。结果表明，不仅凝结时间合格，而且在初凝时间相当的情况下，掺固硫渣的水泥终凝时间较短（表3）。

表3　掺固硫渣的水泥凝结时间

固硫渣编号	掺量（%）	SO_3含量（%）	水泥初凝时间（h：min）	水泥终凝时间（h：min）
G8	30	2.61	3：28	6：23
	40	3.48	3：33	6：59
G7	30	3.5	3：30	5：57
	40	4.3	4：20	6：28
相同熟料的纯熟料水泥		1.67	2：35	6：30

2. 作为膨胀组分制作水泥混凝土用的膨胀剂或膨胀水泥：

利用固硫渣中的$CaSO_4$和f-CaO作为产生膨胀的组分，渣中活性Al_2O_3和SiO_2为强度组分，可制作膨胀水泥和混凝土用膨胀剂。

普通的水泥硬化和使用时温湿度变化都会发生体积的收缩。因水泥在使用时大都是受限制的情况（如体积、面积、钢筋、模板、新老混凝土结合等），这种收缩必然引起开裂。因此水泥混凝土几乎不可能不透水。但如果在水泥水化过程中同时还能产生膨胀能则可在限制的条件下补偿水泥的收缩；如其膨胀能超过补偿的需要量，则可积蓄起来而产生自应力，延长钢筋混凝土构件受力后裂缝的开展。膨胀水泥和膨胀剂可用于刚性防水、接缝、灌浆、锚固、修补及自应力管道等工程。据专家预测，未来的水泥有可能被膨胀水泥代替，膨胀剂会像现在的减水剂、早强剂一样被广泛使用。

固硫渣制作的膨胀剂和膨胀水泥性能好、强度匹配好、成本低廉、工艺简单。已用于中法合资八达岭缆车索道基础锚杆锚固、亚运会自行车训练场跑道不裂面层、宜昌市建委宿舍接待室屋顶刚性防水等工程。

3. 制作无熟料或少熟料的硅酸盐制品：

①无任何外加材料时，纯固硫渣加水成型蒸汽养护强度可达5MPa（即50号）以上。

②掺入5%石灰，蒸汽养护强度高于10MPa（最高18MPa）。

③掺入80%熟料和某种外加剂，蒸汽养护强度可达10MPa左右（100号）。

④掺入50%熟料，蒸汽养护强度可超过100号。

随熟料量的增加，制品强度也提高。当熟料达80%时，可得到最高的强度。并且高于熟料强度（图7）。为了尽量消耗固硫渣，生产硅酸盐制品时，熟料用量最好在20%以下。

在上述制品成型时加入一种特制的发泡剂以代替骨料或改用各种轻质、特轻质材料（如废弃的加气混凝土碎块，废弃的聚苯乙烯泡沫塑料包装板）为骨料，采用常压蒸汽养护，即可制成不同容重的轻型保温、隔热板、块。其性能与高压蒸汽养护的铝粉加气混凝土性能相当。如容重＞700kg/m³ 时，抗压强度＞5MPa；容重＞800kg/m³ 时，抗压强度＞7MPa；容重＞600kg/m³ 时，抗压强度＞3MPa。有轻质骨料的，代替普通黏土砖用于墙体时，厚度可减少1/3～1/2。对非承重墙，其20厘米厚可达三七砖墙的保温效果。这种制品中固硫渣消耗量占所用胶凝材料的约60%以上，成本低、投资少，工艺不复杂。由于使用固硫渣，制品比其他多孔蒸气养护制品的收缩小。

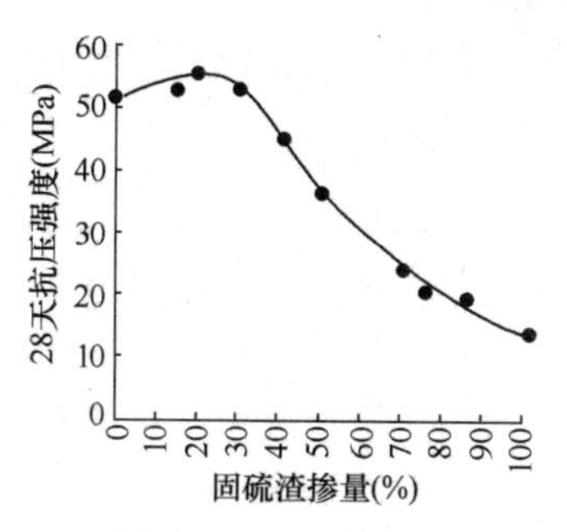

图7 固硫渣掺量对水泥强度的影响

4. 用固硫渣作为高性能混凝土的掺和料可补偿混凝土的收缩，提高混凝土的体积稳定性。

实验研究表明，固硫渣具有较高活性，在一定掺量下，可使水泥强度不低于熟料的强度；与掺用其他火山灰材料相比，掺量较高而掺量相同时，掺固硫渣的混凝土强度较高；同时固硫渣还对水泥中过量 SO_3 引起的膨胀有调节作用[10][11]。因此固硫渣作为辅助胶凝材料，可配制出无收缩的高性能混凝土。固硫渣体积掺量为60%时，坍落度为200mm左右的混凝土，强度可大于50MPa；固硫渣体积掺量为60%时，上述混凝土强度可大于60MPa。

全部使用普通水泥和混凝土在水中养护14天后，再在空气中养护28天，其收缩率超过0.04%；用占水泥重量40%的粉煤灰和15%的UEA膨胀剂等量取代水泥时，在上述相同条件下，混凝土的收缩率为0.034%；而用含 SO_3 为7.12%的固硫渣以30%等量取代水泥作为辅助胶凝材料配制高性能混凝土时，使水泥中的 SO_3 总量达4%，在上述相同条件下，混凝土14天在水中的膨胀率与上述粉煤灰加UEA的相同（约0.02%），在空气中28天的收缩率则小于0.02%，低于混凝土的开裂极限。

四、结论

1. 固硫渣中所含 SO_3 以Ⅱ型无水石膏的形态存在，与天然无水石膏和天然二水石膏在400℃下脱水而成的无水石膏有相似的性质；比二水石膏溶解速率慢，但溶解度相当，在激发剂的作用下，可提高其溶解度，能代替二水石膏用作水泥的调凝材料而后期无害。

2. 由于固硫渣中还有在800～950℃反应残留的游离CaO，使固硫渣具有自身水硬性，是建筑材料优质的资源。

3. 固硫渣中所含 SO_3 可作为建材的强度组分、膨胀组分、调凝组分而化害为利。与熟料配制成的水泥，具有早期强度高、需水量较小、初凝和终凝时间间隔短等特点。

4. 利用固硫渣为强度组分、膨胀组分，可制成补偿收缩混凝土用的膨胀剂，其膨胀性能与市售其他膨胀剂的性能相同，而耐久性好、成本低、生产工艺简单、效益高。

5. 在固硫渣一定的掺量下，可使水泥强度不低于熟料强度，可降低熟料消耗而生产高标号水泥；在较大掺量下，标号相同时，比用其他火山灰节省熟料10个百分点。因此使用固硫渣生产水泥不仅有环境效益，而且有利于节能。

6. 由于固硫渣有较高活性，可利用其开发更多的建材制品，扩大固硫渣消耗量。

7. 由固硫渣制建材所得经济效益补偿脱硫所增加的费用后，还可获得利润。

参考文献

[1] 同济大学等. 胶凝材料学[M]. 北京：中国建筑工业出版社，1980.

[2] 李万有. 利用沸腾炉渣做混合材生产火山灰质硅酸盐水泥.

[3] 胡春芝. 沸腾炉渣活性及其与蒸压硅酸盐制品强度和水化物的关系，中国硅酸盐学建材专业委员会1984年学术年会.

[4] GB 175—77.

[5] 国家建材院，南京化工学院. 修改三氧化硫指标的研究.

[6] J. Lotze, Kenndaten und Verwertungsmöglichkeiten von Aschen aus einer Feuerungslage mit zirkulierender wirbdschicht. Teil Ⅰ：Verfahrenstechnik und Charakteristik, Z-K-G, S239－243, 5. 1985.

[7] I. Odler et al. Hydraulic properties of fluidized bed combustion. ashes, MATERIALS RESEARCH SOCIETY SYMPOSIUM PROCEEDINGS Vol. 178, edited by Robert L. Day et al.

[8] П. П. 布德尼柯夫. 石膏的研究与应用[M]. 中国建筑工业出版社，1963.

[9] 林春玉，太连奎. 硬石膏作水泥调凝剂的研究，全国第三届水泥学术会议.

[10] Irassar, E. F. A Discussion of the paper "Internal and Extra Sources of Sulphate Ions in Portland Cement Mortar" by C. Ouyang et al, C &C. R. 19, 1989.

[11] Lian Huizhen. Inhibition of Excessive Sulphate's Action in Cement by FBC Ash, Proceedings of the 3rd International Symposium on Cement and Concrete, 1993, 2.

自评

本篇论文是应邀对“七五”和“八五”两次攻关中固硫渣利用研究成果的总结性论文。现在看来，有些问题还不够成熟，分析如下：

1. 从图1所见，虽然各图横坐标量程不同，但还是可以判断，固硫渣中的石膏、用天然二水石膏在温度400℃下灼烧的硬石膏和天然硬石膏，三者有一致的相组成。但在图2中，几种硬石膏却有着不同的溶解速率，其机理何在？且在图1和图2中所用的固硫渣也不是同一来源。不同来源固硫渣中硬石膏的溶解特性是否会有差别？

2. 图5中的数据只有7个点，显然有很大的波动性，却用了其中的四个点作出曲线，缺乏说服力。倒不如做成自评图1的表达方式，说明总体上有正相关的趋势，但是在直觉上感到有些欠缺，似乎原图的对数趋势比较合理。但因数据太少，难以确认。有这样大的离散性，原因何在？这表明该项试验量不足，数据太少，只能定性地说明固硫渣可用于制作不用水泥的硅酸盐制品，如免烧砖、多孔砌块、纤维板材等，但是需要对配制规律及其机理进行系统研究。

3. 同理，图6的8个数据点也太少，能处理成直线是否和坐标的量程有关。试比较两个图，这两个图中的数据是同一组，只是量程不同。自评图2左图的 K_a 处多了一个点（空心点标记），自评图2右图（来自本部分论文九）中约为 $K_a=21\%$ 处多了一个点（空心点标记）。

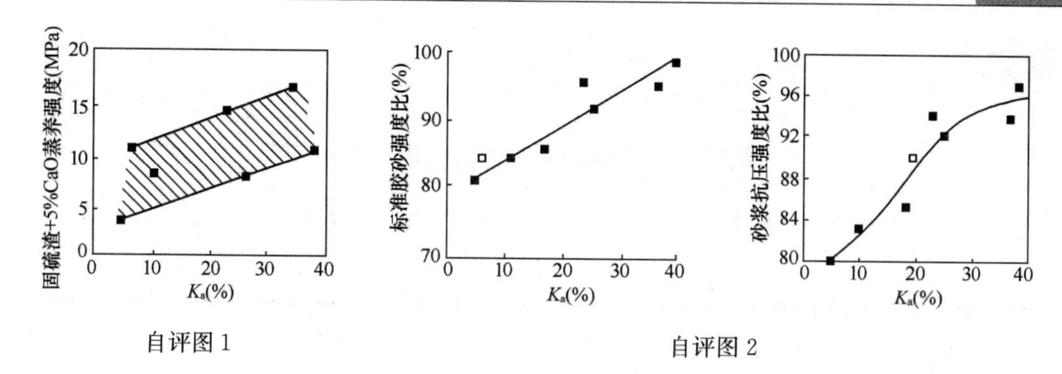

自评图 1　　　　自评图 2

似乎左图做成直线关系显得比较合适，这是因为两个图坐标量程不同造成的视觉差异，而从逻辑上看，右图似更合理：K_a 在 0.25 以前与砂浆强度大体呈现线性相关，超过 0.25 时，影响趋势减弱。恐怕需要更多的实验数据回归会更趋于合理。

4. 第三节第 3 段“制作无熟料或少数料的硅酸盐水泥”的叙述不够确切。根据表 2，纯固硫渣蒸养强度并非都达 5MPa 以上，六个数据中有两个未达到；③、④两点未见有数据支持。

人要在诱惑面前学会拒绝，欲望要有底线，贪婪一定要付出代价，甚至是自己的生命。

——马云

凡是较有成就的科学工作者，毫无例外地都是利用时间的能手，也都是决心在大量时间中投入大量劳动的人。

——华罗庚

论文之二

固硫渣中硫的形态与特性*

清华大学土木工程系　廉慧珍　李桂芝

摘　要　分别用XRD、SEM及化学分析等方法对天然硬石膏、400℃烧石膏、700℃烧石膏及几种不同的固硫渣中的无水硫酸钙进行分析测试。结果表明，渣中无水硫酸钙与天然硬石膏及400℃烧石膏具有相同的结构，属于Ⅱ-$CaSO_4$；由对渣中Ⅱ-$CaSO_4$各种特性的研究，证实了硫的可利用性，同时发现天然硬石膏、400℃烧石膏及渣中无水硫酸钙在不同的溶剂中溶出速率的差别，以及出现差别可能的原因。从理论上分析硫酸钙在水泥中作用机理。

关键词　沸腾炉（流化床）　固硫渣　硫形态　特性

一、前言

当煤炭中的黏土矿物以高岭石为主时，在沸腾炉（俗称流化床）的温度下分解成无定形Al_2O_3和SiO_2。在含硫的煤炭中掺入含氧化钙成分的脱硫剂时，SO_3与CaO反应而被固定在废渣中。被固定下来的SO_3的形态与炉内工况有关[1]。保证固硫效率的工况应保证温度<1000℃的氧化气氛。

据有关资料表明[2][3]，天然硬石膏能部分或全部代替二水石膏在水泥中起缓凝作用，那么固硫渣中的无水硫酸钙是否也具有天然硬石膏的特性？因此，研究固硫渣中硫的形态及特性，对于利用固硫渣制作水泥和其他产品以及用渣中无水硫酸钙部分或全部代替二水石膏和控制水泥中SO_3含量的可行性，可为脱硫燃煤的残渣变害为利提供可靠的依据。

二、试样及实验方法

1. 实验样品与溶剂

实验中所用试样全部通过4900孔/cm^2筛。使用石膏的化学成分如表1；固硫渣的化学成分如表2；溶剂种类如表3所示。

表1　实验所用石膏的化学成分

化学成分（%）								备　注
样品名称	SiO_2	Al_2O_3	CaO	Fe_2O_3	MgO	SO_3	LOI	—
南京天然硬石膏	18.65	3.95	37.73	1.35	2.06	35.61	—	—
400℃烧石膏	7.99	2.40	48.24	0.58	1.76	43.64	—	用天然二水石膏灼烧4小时
700℃烧石膏	8.21	1.61	47.90	0.58	2.06	44.19	—	用天然二水石膏灼烧4小时

* 原发表于国家“七五”科技攻关环境保护项目论文集《大气污染防治技术研究》（1993．环境工程出版社）。

表 2　固硫渣的化学成分

样品名称	化学成分（%）									
	LOI	SiO_2	Al_2O_3	CaO	Fe_2O_3	MgO	SO_3	*f*-CaO	K_2O	Na_2O
G4	10.81	36.50	18.58	14.82	0.90	—	5.35	3.32	—	—
G5	4.85	48.82	18.26	11.18	8.02	1.36	4.21	2.40	3.35	0.48
G6	5.80	41.92	15.05	18.38	6.89	2.10	7.13	4.98	0.60	2.58
G7	1.48	50.33	13.51	14.66	6.12	0.75	11.81	2.36	0.77	0.04
G8	0.48	54.94	17.94	7.85	8.08	1.53	6.69	0.97	1.23	0.19

表中各渣来源及特征见文献[4]《沸腾炉燃煤固硫渣活性评价及其影响因素的研究》。

表 3　溶剂的种类

溶剂名称	CaO 含量（g/L）
蒸馏水	—
CaO 溶液	0.7
CaO 溶液	1.2
CaO 溶液	40*

*　过饱和，使用清液

2. 实验方法

准确称取通过 4900 孔/cm^2 筛的天然硬石膏、烧石膏 0.3g（用固硫渣时折合含 $CaSO_4$ 0.3g）于 250mL 锥形瓶中，加入 100mL 溶剂（分别为蒸馏水和 CaO 溶液），温度为 25℃±2℃，用橡皮塞盖紧，在电磁搅拌器上连续搅拌至各龄期，过滤后，吸取滤液 50mL 于 250mL 烧杯中，加入 100mL 蒸馏水，加热至微沸，再加入 5 克“732”苯乙烯型强酸性阳离子交换树脂，立即置于电磁搅拌器上搅拌 5 分钟后，以定性滤纸过滤，用热水洗涤 7～8 次，向滤液中加入 7～8 滴酚酞指示剂，用 0.05mol/L NaOH 标准溶液滴定至红色。根据单位体积 NaOH 标准溶液相当于 SO_3 的重量和滴定时消耗 NaOH 的体积，计算出溶解液中 $CaSO_4$ 浓度。

选择无水硫酸钙与溶剂的比例为 0.3g∶100mL，这是根据在液相达到接近于石膏饱和溶度（2.05g/L）时仍有少量的硬石膏存在的条件而定的，这样可明显地测定出各种硬石膏在溶液中溶解速率的变化[5]。

x-射线衍射仪为日本理学 P/max-IIIA，铜靶，Ni 滤光片。

三、实验结果及分析

1. 固硫渣中硫的形态的探讨

硫酸钙以两种稳定的形态存在于自然界中，一种是未水化的天然无水石膏（$CaSO_4$），另一种是水化程度最高的二水石膏（$CaSO_4\cdot 2H_2O$），介于这两种稳定形态之间的工业硫酸钙是二水石膏焙烧后的产物，根据燃烧的条件不同，可以有几种不同的变体。一般工业上常见的二水石

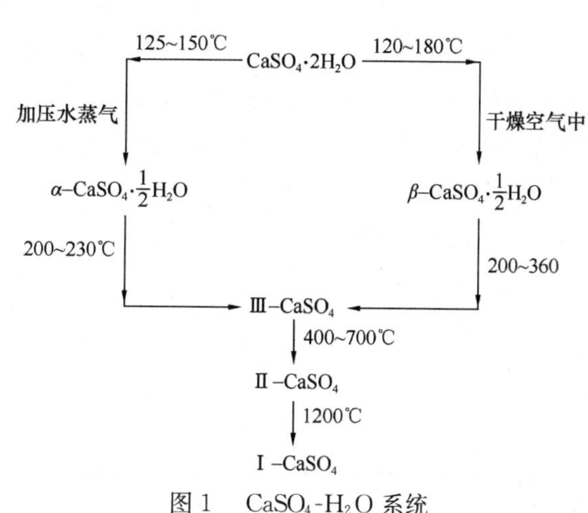

图 1　$CaSO_4$-H_2O 系统

膏的脱水转变温度如图1所示。

据资料表明[6]，从结晶学和动力学角度看，无水石膏Ⅱ型$CaSO_4$是一个稳定相（斜方晶系）。大自然中存在的天然无水石膏也是这个结晶相，其结晶形态与Ⅱ型完全相同。

由于沸腾炉燃煤固硫渣是在800～950℃下烧成的，为了考察固硫渣中无水硫酸钙的形态，分别对硬石膏、400℃烧石膏、700℃烧石膏（将天然二水石膏分别在400℃、700℃下恒温4小时烧成）进行了XRD分析，结果如图3所示。天然硬石膏的特征峰d值（取三强峰）分别为：3.50Å、2.85Å、2.33Å。400℃烧石膏与天然硬石膏几乎完全相同，d值为：3.498Å、2.851Å、2.328Å，而700℃烧石膏与前两者不同，特征峰d值为3.373Å、2.939Å、2.767Å。由$CaSO_4 \cdot H_2O$系统相图可知，400℃以及700℃的烧石膏都属于Ⅱ-$CaSO_4$，两者结晶形态相同，但是700℃ 烧石膏比400℃烧石膏结晶体更加致密（图2，晶面距减小）。

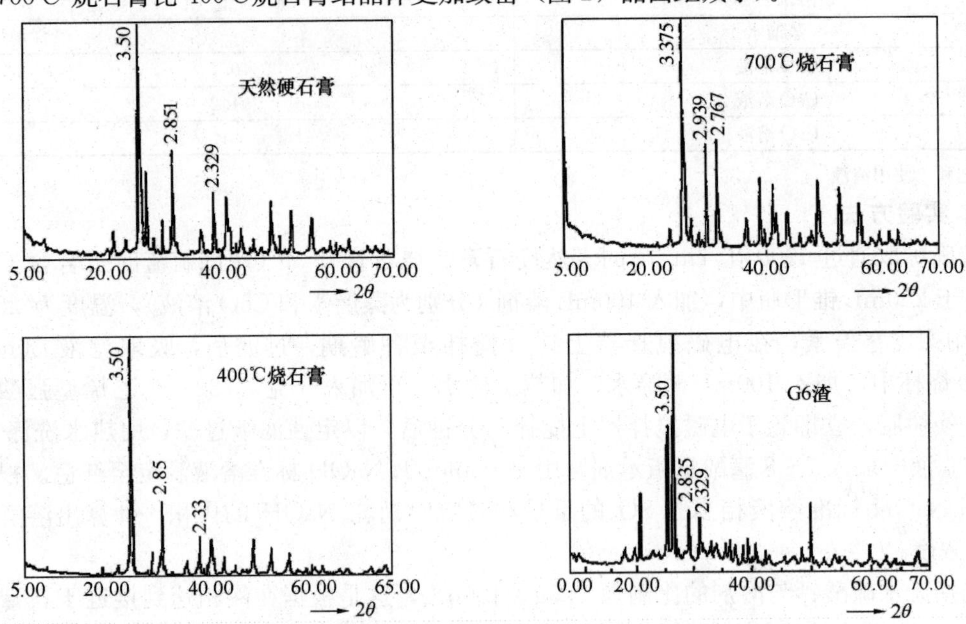

图2 700℃和400℃烧石膏、天然硬石膏和固硫渣G6的对比

分别对其他不同来源的渣进行XRD分析，如图3所示。

几种渣中无水硫酸钙的特征峰基本相同，三强峰分别为3.50Å、2.853Å、2.328Å。如图3中所标出黑点的三强峰。以G6为例（图2），其特征峰正好与400℃烧石膏存在相同的位置。所以，从结晶学上可判断：固硫渣中无水硫酸钙与天然硬石膏及400℃烧石膏有相同的结构，属于Ⅱ-$CaSO_4$。图中的FK2为无硫渣，故无任何硫酸钙的特征峰。

2. 固硫渣中Ⅱ-$CaSO_4$特性

2.1 固硫渣中Ⅱ-$CaSO_4$在水中的溶解速率

为了考察固硫渣中Ⅱ-$CaSO_4$是否能部分或全部代替二水石膏在水泥中起缓凝作用，分别测定了G6、G7、G8渣在水中的溶出量，其结果如图4所示。

由图4可以看出，二水石膏在纯水中5分钟内就达到饱和溶解度（2.05g $CaSO_4$/L）的90%以上，1小时基本上达到饱和溶解度，以后变化甚微。固硫渣中Ⅱ-$CaSO_4$的溶出速率随着时间的增加而增大。G6、G7渣搅拌3天的溶出量分别为1.30g $CaSO_4$/L、1.59g $CaSO_4$/L，以后不再变化，而G8渣的溶出量则一直随搅拌时间而增加，搅拌14天时为2.05g $CaSO_4$/

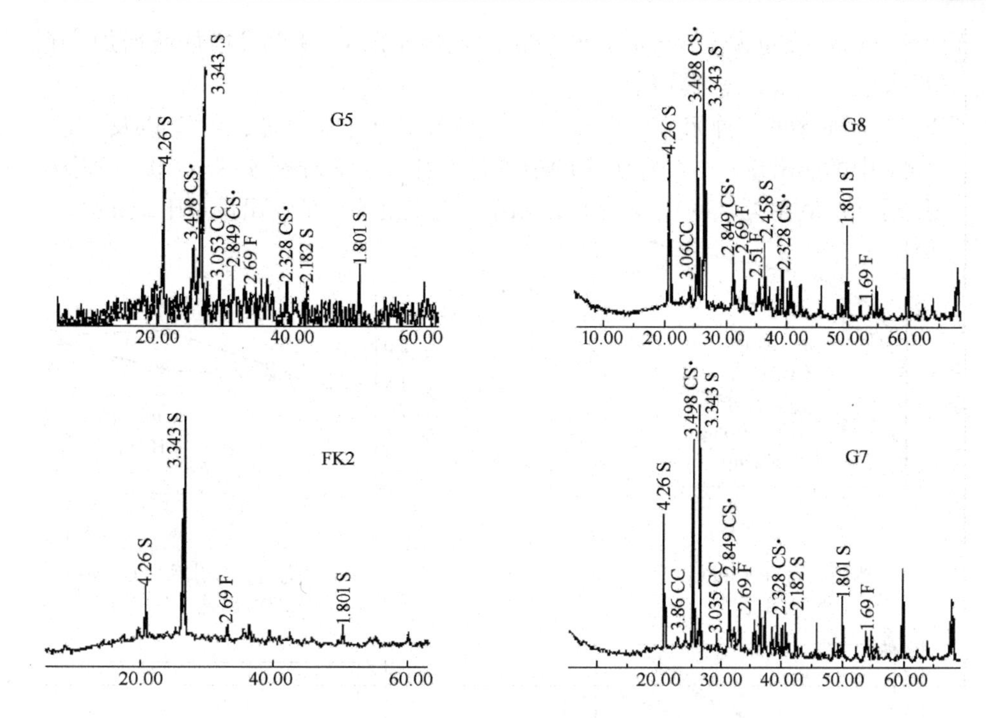

图 3　不同来源固硫渣的 XRD

（图中 CS 为Ⅱ-$CaSO_4$，S 为石英，CC 为 $CaCO_3$，F 为 F_2O_3）

L，以后不再变化。由渣中Ⅱ-$CaSO_4$的溶出速度可知：一定量的渣经过足够的时间，或者较大量的渣在较短的时间内可有较多的无水硫酸钙溶出。因此，可根据需要以不同的掺量加入水泥中使其部分或全部代替二水石膏起缓凝作用。当渣掺量较少时，由于渣中的Ⅱ-$CaSO_4$有早期溶出速率低的特点，可加入少量的二水石膏使其早期溶解以提供钙矾石所需要的硫酸根离子，而待渣中有足够的Ⅱ-$CaSO_4$溶出后，即可代替二水石膏。如渣掺量较大，就会“短时多溶”，有可能全部代替二水石膏。这样既可节省天然石膏，又可变废为利。

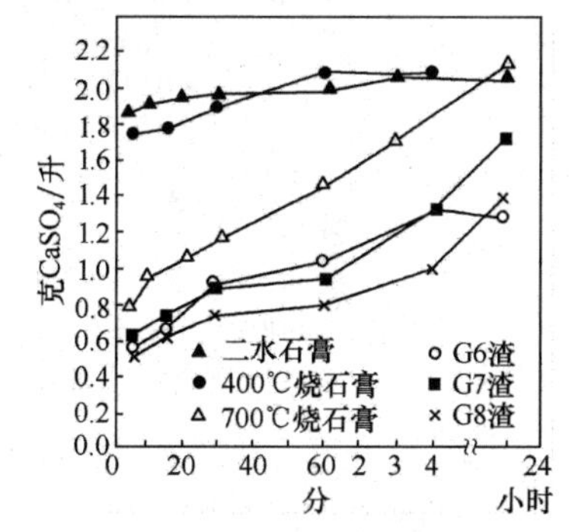

图 4　二水石膏、700℃烧石膏、G6、G7、G8 渣在水中的溶出速率

2.2　渣中Ⅱ-$CaSO_4$在 CaO 溶液中的溶出速率

关于使不溶性硬石膏活化的研究早在 1919 年就开始了，当时，苏联的布德尼科夫等人和格拉绥那帕经过系统研究，搞清了硬石膏的特性，同时证实了在凝结过程中起决定作用的是硫酸钙在高温分解时所生成的 CaO，因此考虑用加 CaO 粉末的方法来活化不溶性硬石膏，获得成功。同时通过进一步的研究表明，有少量其他碱、酸以及中性盐和酸性盐类存在时，也能促进不溶性硬石膏的水化。后来的很多研究也都证实布德尼可夫等人的结论是正确的，其中 CaO 对不溶性硬石膏水化的影响如图 5 所示[7]，CaO 含量在 0.6%（占 2 克硬石膏重量的百分数，如以水表示时，将其扩大 1.33 倍）范围内时，硬石膏水化不因 CaO

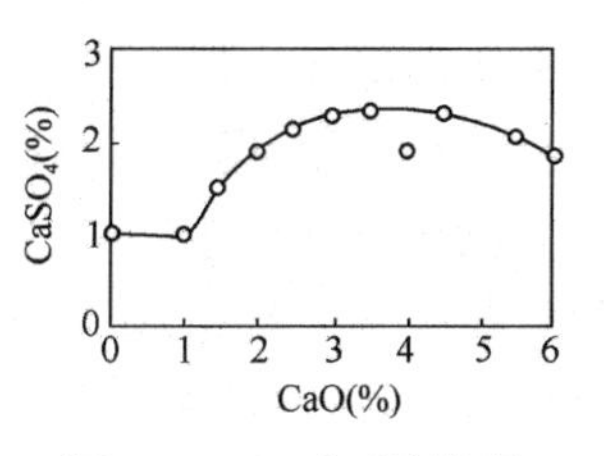

图 5　CaO 对不溶性硬石膏水化的影响[7]

的存在而变化，CaO 含量大于 1%时，硬石膏水化速率随 CaO 含量增加而增加，到 3%～4%时达到最大值，其后又逐渐降低。

为了研究相同形态的几种无水 $CaSO_4$ 的性质，测试了天然硬石膏、400℃以及 700℃烧石膏在 4%CaO 溶液中的溶出量，与其在水中的溶出量对比。结果分别如图 6、图 7 和图 8 所示。进一步又测试了 G7 渣在 4%CaO 溶液中的溶出量，与其在水中的溶出量对比（图 9）。

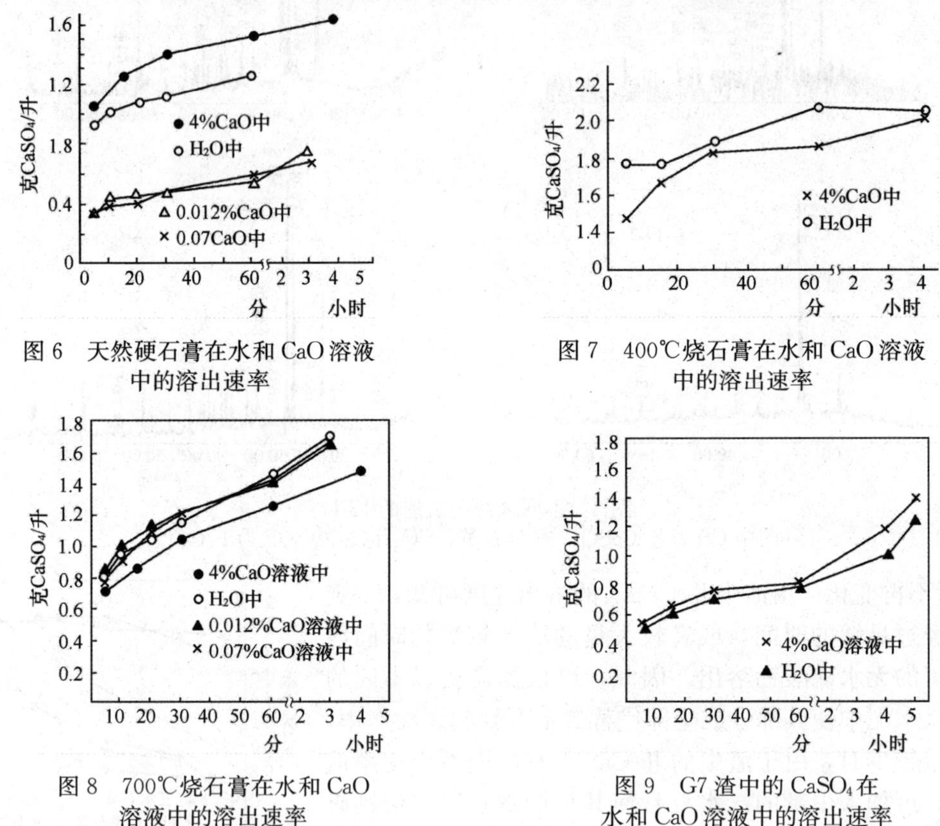

图 6　天然硬石膏在水和 CaO 溶液中的溶出速率

图 7　400℃烧石膏在水和 CaO 溶液中的溶出速率

图 8　700℃烧石膏在水和 CaO 溶液中的溶出速率

图 9　G7 渣中的 $CaSO_4$ 在水和 CaO 溶液中的溶出速率

由图 6 可见，天然硬石膏在 0.012%CaO 及 0.07%CaO 溶液中溶出速率几乎相同，而此浓度范围内正好是图 5 中 0.6%CaO 以内的水平线段，即该实验结果正好与图 5 相符。而在 4%CaO 溶液中溶出速率远大于在纯水中的，故用不同实验方法同样可以得出结论：天然硬石膏在某浓度范围内的 CaO 存在时，会受到激发而活化。400℃烧石膏（图 7）、700℃烧石膏(图 8)在纯水与 CaO 不同浓度溶液中溶出速率趋势与天然硬石膏的不同。400℃烧石膏在 4%CaO 溶液中溶出速率略低于在纯水中的溶出速率，而 700℃烧石膏在任何浓度（0.07%、0.012%）CaO 溶液中的溶出速率都与在纯水中相似，而不因 CaO 的存在使石膏溶出速率降低，但在高浓度的 CaO 溶液（4%）中，CaO 的存在对上述二种石膏的溶出均有抑制作用，使其溶出量有所降低。

比较 400℃与 700℃烧石膏在纯水中的溶出速率，400℃烧石膏在纯水中 5 分钟就达到饱和溶解度（实验条件下为 2.09g/L）的 80%以上，以后变化不大，1 小时即达饱和；而 700℃烧石膏 5 分钟只达饱和溶解度的 70%，1 小时才近 90%，4 小时达到饱和。因此，由其溶出速率也可看出，700℃烧石膏在结构上比 400℃烧石膏更加密实，更加稳定。

天然硬石膏与 400℃烧石膏虽然有着相同的结晶形态，但其水化速度不同。并且，在不同溶剂中的变化趋势也不同。两者相比较，400 ℃烧石膏在纯水中的溶出速率较天然硬石膏的大，但 400℃烧石膏与天然硬石膏分别在不同溶剂中相比较时，400℃ 烧石膏在纯水中的溶出速率略大于其在 4%CaO 溶液中的溶出速率，而天然硬石膏则相反，实验结果是天然硬石膏在 4%CaO 溶液中的溶出速率大于其在纯水中的溶出速率，而且在 4%CaO 溶液中比在纯水的溶出速度增长幅度很明显。700℃烧石膏与 400℃烧石膏情况相似。也就是说，天然硬石膏在 CaO 存在（一定范围内）的情况下可受激发而活化，而对烧石膏，当有过量 CaO 存在时，则会抑制其水化。由图 9 可见，固硫渣中无水 $CaSO_4$ 在水中和在 4%CaO 溶液中的溶出性质相差不多。

将一定量 G7 渣加入不同浓度 CaO 溶液中，在振荡器上振荡 1 小时，测定不同浓度 CaO 溶液中渣中 $CaSO_4$ 的溶出量，结果如图 10 所示。将振荡 1 小时后的滤饼干燥后进行 XRD 分析表明有少量 AFt 相生成。图中右侧纵坐标表示于在 AFt 相最强峰（d=9.6）的强度。由图 10 可见：

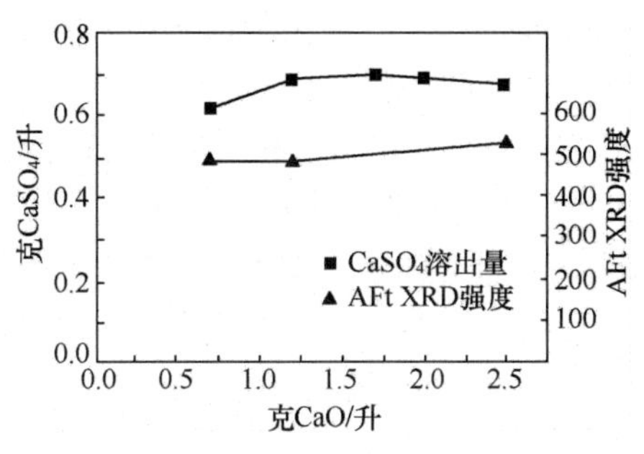

图 10　G7 渣在不同浓度 CaO 溶液中的溶出速率及 AFt 相 XRD 强度

（1）当 CaO 溶液浓度在 0.7～1.2g/L 的范围内时，渣中 $CaSO_4$ 的溶出量随 CaO 含量的增大而增加（与天然硬石膏不尽相同）；CaO 浓度在 1.2～2.0g/L的范围内时，渣中 $CaSO_4$ 的溶出量基本不变；而当 CaO 浓度大于 2.0g/L 以后，溶液中 $CaSO_4$ 的浓度呈下降趋势。

（2）当溶液中 $CaSO_4$ 浓度偏低时，AFt 相量相对增加。由此可以说明，天然硬石膏在 4%CaO 溶液中会受到激发而活化，但固硫渣中的Ⅱ-$CaSO_4$ 受激发的 CaO 浓度为1.2～2.0g CaO/L。这与固硫渣水泥的宏观性能（如凝结时间、前期抗压强度等）是吻合的。

2.3　渣中Ⅱ-$CaSO_4$ 在水泥中可部分或全部代替二水石膏起缓凝作用

有资料表明，天然硬石膏能部分或全部代替二水石膏在水泥中起缓凝作用[2][3]。XRD 及化学分析均表明，固硫渣中无水硫酸钙与天然硬石膏有相同的结构形态以及对水有相同的反应性能，那么，这种无水石膏是否也像天然硬石膏那样可部分或全部代替二水石膏在水泥中起缓凝作用呢？为此，分别测定了加石膏无渣、加固硫渣而无石膏及加部分固硫渣和部分石膏时水泥的凝结时间，其配比及结果如表 4 所示（试验按国标 GB 1346—77 规定的有关内容进行）。结果见表 4。

表 4　不同配比水泥的凝结时间

固硫渣名称	编号	水泥配比（%）			水泥中 SO_3（%）	标准稠度用水量（%）	凝结时间（h：min）	
		熟料	石膏	固硫渣			初凝	终凝
	A	96	4	0	1.78	26.5	2：07	4：21
FK	B	66	4	30	2.08	29.0	3：28	5：38
	C	56	4	40	2.18	30.0	4：02	6：15
G5	D	68	2	30	2.15	28.5	2：00	4：31
	E	58	2	40	2.57	30.0	2：17	4：57
G6	F	70	0	30	2.61	28.5	3：28	6：23
	G	60	0	40	3.48	28.5	3：33	6：49

由表 4 测定的结果可以看出，硅酸盐水泥 A、分别加入 30%和 40%无硫渣配制的水泥 B 和 C、用固硫渣 G5 中无水硫酸钙来代替部分二水石膏的水泥 D 和 E、用固硫渣 G6 中无

水硫酸钙来全部代替二水石膏的水泥 F 和 G，其凝结时间均满足国标 GB 1344—85 的要求。结果表明，当固硫渣的掺量适当时，完全可由渣中的Ⅱ-$CaSO_4$全部或部分代替二水石膏起调凝作用。

2.4 硬化水泥浆体中剩余 SO_3的无害性

虽然二水石膏的溶解速率远大于硬石膏的，但后期硬石膏的饱和溶解度却稍大于二水石膏的，因此掺固硫渣的硬化水泥浆体中剩余 SO_3 含量的测定，对固硫渣水泥也是十分重要的。如果后期的硬化浆体中 SO_3 含量较大时，在适当的条件下就还会形成钙矾石相，应当会使试体膨胀而影响性能。

通过 5%甘油饱和石灰水-石膏萃取剂萃取的方法，分别用掺 G6 渣 10%、15%、20%、30%、45% 和熟料配制水泥，并用宜昌烧石膏调节水泥的 SO_3 总量使其大体相同。对养护 28 天的硬化水泥试体中 SO_3 含量进行测定，各试样均萃取 24 小时，其结果是：掺渣量为 15%（水泥中 SO_3 总量为 3.6%）时，剩余 SO_3 为 0.36%；掺渣量为 20%时（水泥中 SO_3 总量为 4.0%），剩余 SO_3 为 0.4%；掺渣量为 30%（水泥中 SO_3 总量为 3.8%）时，剩余 SO_3 为 0.52%；掺渣量为 45%（ SO_3 总量为 3.8%）时，剩余 SO_3 为 0.42%；将各试样萃取 30 小时，其结果基本不变。由此得出，28 天硬化水泥浆体中这样微小的 SO_3 含量对其长期性能不会产生什么不利的影响。

四、固硫渣中Ⅱ-$CaSO_4$在水泥浆体作用机理的探讨

为了解释固硫渣中Ⅱ-$CaSO_4$在水泥中的作用机理，首先模拟在水泥矿物成分的环境中使其反应，用 XRD 分析产物，具体方法是：在一定量的 G7 渣中加入一定量的 C_3A 及适量的 CaO（0.17～1.18 g/L）以纯水为溶剂在磁力搅拌器上搅拌 5 小时后过滤。待滤饼干燥后，进行 XRD 分析，结果如图 11 所示。由图可见有 AFt 相生成(图中 d=9.562 Å、5.541Å)。表明固硫渣中的Ⅱ-$CaSO_4$在水溶液中不断溶出，且与 C_3A 相反应，生成不溶的钙矾石。

对不掺二水石膏只掺 30%G6 渣及掺 40%、水化 3 天的硬化水泥浆体形貌观察表明（图 12），浆体中有大量的针状钙矾石相生成，这也说明固硫渣中的无水硫酸钙与水泥中的 C_3A 或渣中的游离 CaO 及活性 Al_2O_3反应生成了钙矾石。由于钙矾石难溶于水，固硫渣中溶出的Ⅱ-$CaSO_4$与 C_3A 反应形成的钙矾石起初在 C_3A 周围形成一层保护膜，可阻碍 C_3A 的水化，延缓水泥的凝结时间，起到与二水石膏一样的缓凝作用。而且由于针状钙矾石相的生成，也有利于浆体强度的发展。

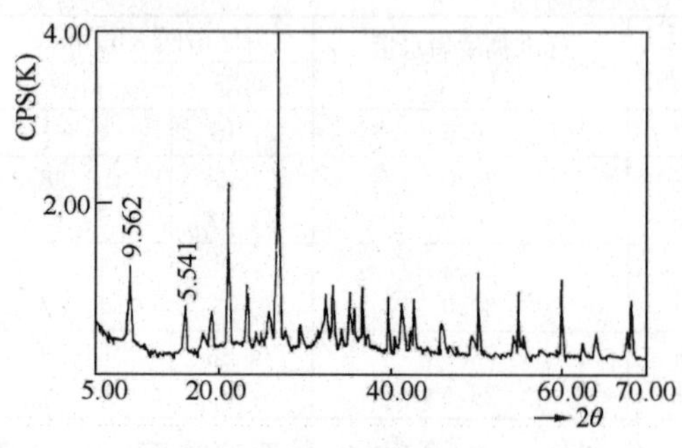

图 11 （G7 渣＋C_3A＋CaO）水化后 XRD 图

图 12 无外加二水石膏时固硫渣水泥水化三天试样的 SEM 图

五、结论

1. 固硫渣中的无水硫酸钙属于Ⅱ-$CaSO_4$，其结晶形态和天然硬石膏及用二水石膏在400℃下灼烧成硬石膏的相同。

2. 在4%CaO溶液中天然硬石膏的活性会受激发，而对烧石膏及渣中无水石膏在此浓度时的$CaSO_4$溶出量则有抑制作用。

3. 400℃烧石膏与天然硬石膏及渣中的无水硫酸钙比较，其结晶颗粒大小及密实程度有所差别；故其溶解性质有所不同。

4. 掺量合适时固硫渣中的无水硫酸钙可以部分或全部代替二水石膏在水泥中起缓凝作用。

5. 在较高掺量的固硫渣硬化水泥浆体中，剩余SO_3对水泥石的长期性能不会产生不利影响。

参考文献

[1] Slake. Sulphur Dioxide Removal from Waste Gases Noyes Date Corperation Noyes Building USA. 1971.

[2] 林春玉. 硬石膏作水泥缓凝剂的研究. 中国硅酸盐学会水泥专业委员会第三届水泥年会，1986.

[3] 龚关田. 硬石膏作水泥缓凝剂的试验研究[J]. 水泥，1983.

[4] 廉慧珍，江加标. 沸腾炉燃煤固硫渣活性评价及其影响因素的研究. 国家“七五”科技攻关85030301沸腾炉燃煤固硫渣制作水泥的研究鉴定文件，1990.

[5] 成希弼. 水泥中所用石膏种类与其溶解速度的关系[J]. 硅酸盐学报. Vol. 15，No. 2，1987. 4.

[6] 法国石膏工业协会著，杨得山译. 石膏.

[7] П. 布德尼柯夫著. 石膏的研究与应用[M]. 北京：中国建筑工业出版社，1963.

自评

1. 在硅酸盐水泥的生产中，最终是以熟料和适量石膏共同粉磨至不同细度而得到水硬性产物。所用石膏，是根据熟料中C_3A的含量和溶解速率而选择的，过去因煅烧工艺和工矿的差别，西方国家熟料的C_3A溶解速率快，多为二水石膏和半水石膏各半掺和使用，我国则需要与天然硬石膏掺用。但是，在我国的大多数水泥厂长期以来并不问C_3A的特性，因硬石膏溶解速率慢而只用二水石膏，石膏的用量只根据水泥的凝结时间确定，而且在原材料和生产较稳定的情况下，一般很少经常调节，大多控制在SO_3=1.7%～1.8%。1990年以后，因提高强度和节能的需要，发展到使用干法窑，熟料中C_3A含量增大，SO_3随之增大到2.2%～2.5%。尽管标准中对SO_3的限值是3.5%（对矿渣硅酸盐水泥是4%），为了不增加成本，几乎没有用到该限制的。现代混凝土拌和物因水泥与外加剂的相容性引起的流变性能问题、硬化混凝土的体积稳定性问题等，其影响因素之一就是石膏的形态和含量。在循环流化床燃煤的温度下，所掺入脱硫剂中分解出的CaO捕获煤炭燃烧排出的SO_2而生成$CaSO_4$。由于石膏有不同的变种，其溶解性质不同，研究$CaSO_4$的形态对正确使用石膏具有有重要意义。

2. 这篇文章所述的研究是有成效的，但是天然硬石膏、400℃烧石膏和不同固硫渣中的硬石膏，同为一种结构的石膏为什么溶解特性会有差异？对此问题仍不清楚。是否与天然原料中所含杂质有关？例如纯的硬石膏SO_3含量应为58.8%，由本文表2所示，所用天然硬

石膏 SO_3 为 35.61%；400℃烧石膏 SO_3 为 43.64%；700℃烧石膏为 44.19%，则其纯度分别为 61%、74.2%和 75.2%。由此分析图 6～图 9，可以理解为：纯度较高是 400℃烧石膏比天然硬石膏溶出速率快的原因，700℃烧石膏虽然 SO_3 含量和 400℃烧石膏的相当，但是其中的 $CaSO_4$ 结构致密（晶面距减小），因此一开始时的溶出量比 400℃烧石膏的小，却和天然硬石膏的相当；400℃烧石膏 4 小时即达饱和量，而 700℃烧石膏达到饱和量的时间要到超过 20 小时。

3. 对固硫渣中石膏的分析，只用了 SO_3 含量较大的 G7 渣（SO_3 为 11.81%），表明其中的 SO_3 属于和 400℃烧石膏结构相同的Ⅱ-$CaSO_4$，却没有对不同来源的固硫渣进一步的分析，可能是当时的工作量和原材料供应量的问题，但是既然 SO_3 含量不同，可能还有其他组分的差异，是否会影响其中Ⅱ-$CaSO_4$ 的性质，还是做一些试验为好，这是希望今后继续研究者能做的事。

4. 图 1 对固硫渣中 SO_3 在水泥中的作用机理是个有力的证明。AFt 对 x-射线衍射的特征峰 $d_{(100)}=9.73$Å，$d_{(110)}=5.61$Å；当有其他离子取代时，晶胞尺寸会有变化。本部分论文之三中的图 10 和图 11 所用试样是排渣时浇水降温后的固硫渣，表明未加任何其他物质的原状固硫渣因浇水而生成 $d=9.73$ 和 5.61 的钙矾石。可旁证固硫渣的自身水硬性。而本文图 1 中的 9.562 和 5.541，是否这种情况？或者 d 值的计算有误？不过图 12 可以是又一个佐证。

静安先生第一境写的是预期。第二境写的是勤奋。第三境写的是成功。其中没有写天资和机遇。我不敢说这是他的疏漏，因为写的角度不同。但是，我认为，补上天资与机遇，似更为全面。我希望，大家都能拿出“衣带渐宽终不悔”的精神来从事做学问或干事业，这是成功的必由之路。

——季羡林

若是一个人的思想不能比飞鸟上升得更高，那就是一种卑微不足道的思想。

——莎士比亚

论文之三

沸腾炉燃煤固硫渣活性评价及其影响因素的研究*

清华大学土木工程系　廉慧珍　江加标

一、前言

在沸腾炉的燃烧温度下排出的炉渣与粉煤灰相比，有较高的活性[1]，可作为硅酸盐水泥优质的混合材。国内在这方面有很多研究[2~5]。

在沸腾炉燃煤时加入脱硫剂，则在燃烧过程中可将煤中所含的硫固定在渣中，防止二氧化硫向大气中排放。这种脱硫技术在环保领域中得到很大的关注，被认为是燃煤技术的一个方向。但脱硫燃煤所产生的渣就含有大量的硫。即使煤中灰分很低，由于加入25%以上的脱硫剂，也会产生相当数量的含硫的废渣——固硫渣。实验表明，在正常工况下，固硫渣不仅具有火山灰活性，而且还含有大量以无水硫酸钙形态存在的硫和脱硫剩余的f-CaO，其本身就具有不同程度的水硬性。因此，固硫渣的活性应包括其火山灰活性及其自身水硬性两部分。研究影响固硫渣活性的因素，可有效地利用固硫渣，使这种为治理大气污染而将形成二次污染的废渣得到最大量的消耗和有效的利用。

据报道，沸腾炉渣的火山灰活性与燃烧温度有关。温度在800～950℃，原煤的灰分中的矿物以高岭石为主时，渣的活性最高[6]。本文讨论在该温度下排放出的固硫渣中化学成分和含碳量以及结构对固硫渣活性的影响。同时用山东沂蒙化肥厂流化床1100℃左右的固硫渣做对比试验。

二、原材料

1. 固硫渣

1.1　固硫渣的物理性质

沸腾炉燃煤固硫渣是在煤中按一定的钙硫比加入含钙的固体脱硫剂在流化床锅炉中燃烧后排放的含硫残渣。其中从溢流口排出的溢流渣，为2～10mm的颗粒，在沸腾段内翻腾，燃烧充分，含碳量低，约占灰渣总量的35%；其中大于10mm的颗粒因相对较重，只能在沸腾段下部翻腾，无法从溢流口排出而沉积在床层上，定期排出，称作冷渣，约占灰渣总量的20%～25%，含碳量也很低；0.5～2mm的颗粒因体轻而随气流上升至悬浮段，在省煤器和空气预热器中由于气流速度下降而沉积下来，叫做沉降灰，约占灰渣总量的25%，未燃尽的可燃物含量较高；在锅炉尾部由收尘器收集的飞灰，约占灰渣总量的15%，小于0.5mm的颗粒受气流的带动而在床内停留时间很短，很快进入锅炉的尾部，含碳量很高，有时可达20%。采用带分离器的新型循环式流化床锅炉时，没有溢流渣；由于燃烧完全，循环床锅炉中颗粒为200μm～2mm的对流渣含碳量很低，脱硫充分，活性最高。本研究在

* 原发表于国家“七五”科技攻关环境保护项目论文集《大气污染防治技术研究》(1993. 环境工程出版社)。

试验室研究阶段所用固硫渣为溢流渣和冷渣。其密度为 2.5～2.8g/cm^3，松堆密度为 900～1500kg/m^3；磨细后，比表面积为 4000cm^2/g。因所含不同量的赤铁矿而呈不同程度的紫褐色或灰褐色。

1.2 固流渣来源及编号（表1）

表1 固硫渣来源及编号（其中二号脱硫剂为耐水型高强脱硫剂）

编号	G4	G5	G6	G7	G8	G9	G10	FK2	SG
脱硫剂	华宜牌	华宜牌	二号	二号	二号	二号	二号	无	华宜牌
来源	宜昌市麻纺厂流化床	宜昌市印染厂流化床	宜昌市印染厂流化床	宜昌市造纸厂快装循环流化床	同G7，在炉内同时降温	宜昌市造纸厂快装循环流化床水冷渣	同G9储存于水泥厂大料堆，处于水热条件	同G7	山东沂蒙化肥厂流化床，燃烧到1000℃

1.3 固硫渣化学成分

固硫渣的化学成分主要取决于原煤的成分。原煤中除可燃部分外，还有一些黏土质矿物如高岭石、蒙脱石、绿泥石、云母等。含硫的煤主要还含有硫化铁和有机硫。燃烧后的残渣中就主要是 SiO_2、Al_2O_3、Fe_2O_3、CaO、MgO 等成分。加入脱硫剂后，原煤中的硫被固定下来，则固硫渣成分中还有 SO_3。为提高脱硫效率，脱硫剂和原煤混合时，钙硫比必须高于反应的理论值，则固硫渣中还存在脱硫剂所残留的游离状态的 CaO。这种 *f*-CaO 不同于水泥熟料中过烧的游离 CaO，而是在 960℃下存在的活性的游离 CaO。固硫渣中的含碳量用950℃的烧失量表示。

试验研究所用固硫渣化学成分（%）列于表2。

表2 固硫渣化学成分

No.	SiO_2	Al_2O_3	Fe_2O_3	CaO	MgO	SO_3	K_2O	Na_2O	LOI	*f*-CaO
G4	36.50	18.58	6.90	14.82	5.88	5.35	—	—	10.81	3.32
G5	48.82	18.26	8.02	11.18	1.36	4.21	3.35	0.48	4.58	2.40
G6	39.92	15.21	5.99	19.47	1.45	8.70	2.79	0.46	4.74	4.97
G7	50.33	13.51	6.12	14.66	0.75	11.81	0.77	0.04	1.48	2.36
G8	54.94	17.94	8.08	7.85	1.53	6.69	1.23	0.19	0.48	0.97
G9	42.70	11.80	12.34	12.38	2.66	16.00	—	—	2.79	1.64
G10	42.70	14.16	9.81	8.71	3.91	12.59	—	—	6.39	0.27
FK2	69.17	16.26	8.14	1.22	1.00	2.62	1.31	0.16	1.22	0.34
SG	46.64	23.48	10.74	9.79	1.02	4.41	1.84	0.34	1.30	0.51

1.4 固硫渣矿物组成

沸腾炉燃煤后，可燃物挥发，黏土质矿物分解成无定形的 SiO_2、Al_2O_3 和 Fe_2O_3；而原煤中的石英在 800～950℃下结构不破坏，仍以*α*-石英存在于渣中；脱硫燃烧后，在氧化气氛形成的固硫渣中，还有无水石膏以及脱硫剂残留的游离 CaO 及其碳化成的碳酸钙。从固硫渣对 XRD 图形中可见，所有固硫渣中主要都有*α*-石英（4.26、3.343、2.458 Å）、无水硫酸钙（3.498、2.849、2.328Å）、碳酸钙（3.035、3.86、2.812Å）、*α*-Fe_2O_3（3.69、2.51、

1.69 Å)。高岭石、绿泥石等黏土质矿物的衍射峰都已消失。说明这些黏土质矿物已分解为无定形相。图 1 为无硫渣 FK2 及固硫渣 G5、G7 和 G8 的 x-射线衍射图。

采用基体冲洗法[7]测定固硫渣中的矿相如表 3 所示。

表 3　固硫渣的矿物组成

编号	矿物组成（%）								
	α-石英	无水硫酸钙	α-F_2O_3	碳酸钙	*f*-CaO	LOI	无定形相	其他	不溶性晶相含量
G4	7.4	10.9	5.9	9.8	3.3	10.3	55.1	5.9	35.1
G5	10.1	7.0	4.7	4.6	2.4	4.6	57.0	5.9	31.4
G6	11.8	14.1	6.9	8.2	4.7	5.0	43.4	5.2	46.3
G7	22.2	20.1	7.8	4.7	2.4	1.5	41.7	1.5	48.9
G8	21.0	13.0	8.0	1.9	1.0	0.5	51.9	3.0	41.8
G9	23.8	22.6	11.7	2.0	1.7	1.2	53.3	2.7	64.9
G10	29.9	21.3	14.7	7.1	0.3	4.8	30.4	3.9	73.0
FK	24.1	痕量	9.5	0.0	0.3	1.2	35.4	2.6	61.2

1.5　固硫渣火山灰活性

按 ISO《火山灰活性试验》方法检验，所有使用的固硫渣火山灰活性均合格。

2. 其他

试验中所用石灰中有效 CaO 含量为 82%。熟料为宜昌市水泥厂生产的硅酸盐水泥熟料。用该熟料配制对比用的硅酸盐水泥，所加的石膏为宜昌水泥厂使用的二水石膏，SO_3 含量为 33.53%。

三、试验及结果分析

1. 活性硅、活性铝分析

固硫渣中含有大量 SiO_2、$A1_2O_3$，其中能与石灰反应的部分称为活性硅、活性铝。用回流的方法在饱和石灰水中沸煮三小时后，测定其中与石灰反应的硅、铝，即为活性硅、活性铝。所用固硫渣的活性硅、铝见表 4。

表 4　试验用固硫渣中的活性率

编号	活性 SiO_2（%）	全 SiO_2（%）	活性SiO_2/全SiO_2（%）	活性 Al_2O_3（%）	全 Al_2O_3（%）	活性Al_2O_3/全Al_2O_3（%）	K_a ❷（%）	A_c ❸	备注
G4	12.15	36.50	33.21	9.55	18.58	51.40	39.40	0.12	—
G5	8.50	48.82	17.41	8.53	18.26	45.73	25.12	0.11	—
G6	13.58	39.92	34.00	6.80	15.21	44.70	36.97	0.17	—
G7	7.84	50.30	15.59	7.25	13.51	53.66	23.65	0.09	—
G8	4.97	54.94	9.05	2.69	17.94	15.00	10.51	0.07	慢冷渣
G9	微量❶	42.72	0.00	3.46	11.80	29.23	6.34	0.08	水冷渣

续表

编号	活性 SiO_2 (%)	全 SiO_2 (%)	$\frac{活性SiO_2}{全SiO_2}$ (%)	活性 Al_2O_3 (%)	全 Al_2O_3 (%)	$\frac{活性Al_2O_3}{全Al_2O_3}$ (%)	K_a ❷ (%)	A_c ❸	备注
G10	0.55	46.10	1.60	2.44	14.26	17.25	4.96	0.02	水冷渣
FK2	7.29	69.17	10.554	4.10	16.26	25.22	13.33	0.02	无硫渣❹
SG	9.41	46.64	13.60	3.12	23.38	13.28	17.90	0.03	山东高温渣❺

❶已扣除水化反应部分

❷固硫渣活性率 $K_a=\frac{活性\ SiO_2+活性\ Al_2O_3}{全\ SiO_2+全\ Al_2O_3}\times 100\%$

❸固硫渣自硬性指数 $A_c=\frac{f\text{-}CaO}{SO_3+活性\ SiO_2+活性\ Al_2O_3}$

❹$SO_3<3\%$

❺炉温 1000℃

由表 4 可见，在 850～950℃下正常排放的沸腾炉燃煤固硫渣中，活性 SiO_2 和活性 $A1_2O_3$ 总量占全部 SiO_2 和 $A1_2O_3$ 总量的比例都比较高，即含有较多的活性组分；而无硫渣和在炉内缓慢冷却的 G8 渣、出渣浇水冷却的 G9、G10 渣、温度高于 950℃烧成的 SG 渣，活性组分比例都较低。固硫渣中的活性组分可用活性率 K_a 表示。

2. 强度

以 30%固硫渣与 70%硅酸盐熟料混合，以 w/c=0.44 成型（GB 175—1982），标养至 7 天和 28 天；全部使用固硫渣而无任何其他材料混合时，用水泥标准检验方法成型（w/c=0.40），48 小时后拆模。一组在 90℃下蒸养，另一组在 R.H.>90%的室温空气中养护 7 天后，再置于 20℃±3℃的水中养护至 28 天，分别测其强度；固硫渣与石灰（有效 CaO 82%）混合加水成型，养护条件与纯固硫渣试件相同；以 70%固硫渣与 30%熟料混合，按 w/c=0.4 成型，标养至 7 天和 28 天，另一组在 90℃下蒸养，分别测其强度。

全部试验均用标准砂，灰砂比为 1∶2.5。

试验结果见表 5。由表 5 可见：

①正常排放的固硫渣均具有不同程度的自身水硬性，不加任何其他物质，单独加水成型，不仅经蒸汽养护后具有强度，而且在标准条件下养护 28 天后也具有强度。

②不同来源固硫渣的实际活性有较大差别。

③>1000℃燃烧排放的 SG 渣和浇水排放并焖置的 G10 渣实际活性低于其他固硫渣的活性。

表 5　固硫渣制品抗压强度（MPa）

编号	K_a (%)	30%熟料+70%固硫渣（MPa）			$\frac{70\%熟料+30\%固硫渣}{硅酸盐水泥}$ (%)			100%固硫渣（MPa）		5%石灰+95%固硫渣（MPa）	
		7 天	28 天	蒸养	7 天	28 天	蒸养	28 天	蒸养	28 天	蒸养
水泥	—	33.5	43.4	—	—	—	—	—	—	—	—
G4	39.40	20.2	38.0	30.0	74	96	98	3.8	6.3	8.8	12.4
G5	25.12	10.0	26.6	21.2	67	78	89	1.6	1.6	8.5	8.9

续表

编号	K_a (%)	30%熟料+70%固硫渣（MPa）			70%熟料+30%固硫渣 / 硅酸盐水泥 (%)			100%固硫渣 (MPa)		5%石灰+95%固硫渣（MPa）	
		7天	28天	蒸养	7天	28天	蒸养	28天	蒸养	28天	蒸养
G6	36.97	8.9	28.7	23.8	87	98	95	10.0	7.9	15.5	18.8
G7	23.65	8.4	24.5	83	95	96	13.7	13.7	9.4	15.2	16.4
G8	10.51	7.0	21.2	21.5	81	80	84	2.5	2.6	8.4	8.5
G9	6.34	5.4	13.4	18.9	68	76	84	4.3	4.7	11.5	12.1
G10	4.96	3.8	8.8	16.6	47	54	67	*	*	7.2	4.0
FK	13.33	7.3	17.6	18.8	75	78	86	*	*	11.7	11.9
SG	17.90	—	—	—	62	65	66	—	—	—	—

* 无胶凝性，无法成型。

3. 固硫渣活性影响因素分析

3.1 固硫渣的自身水硬性

固硫渣本身存在水硬性的原因是脱硫后残余的 f-CaO 和活性 SiO_2 及活性 Al_2O_3 反应生成硅酸盐和铝酸盐的水化物，同时由于硫的存在而生成钙矾石。固硫渣的自身水硬性随渣中 f-CaO 量的增加而提高，其关系的趋势见图 2 所示；考虑到 f-CaO 的影响与渣中的 SO_3、活性 SiO_2、Al_2O_3 有关，还可采用下式作为固硫渣的自硬活性指数：

$$A_c = \frac{f\text{-CaO}}{SO_3 + \text{活性 } SiO_2 + \text{活性 } Al_2O_3}$$

固硫渣自身水硬性与自硬活性指数 A_c 和 f-CaO 的关系趋势见图 2；固硫渣标准胶砂强度和 A_c 的关系见图 3。

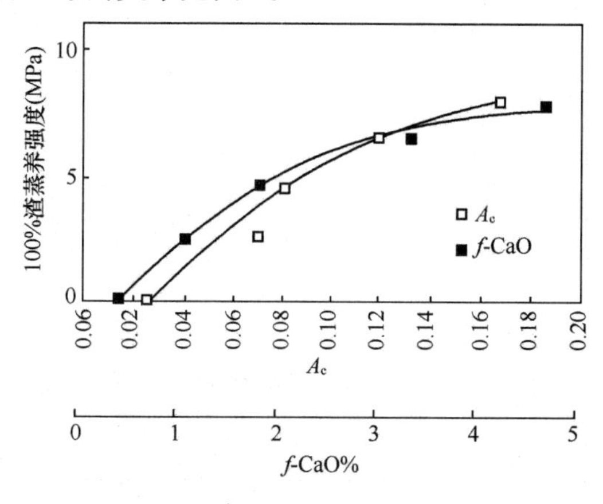

图 2　100%固硫渣蒸养强度与 A_c 和 K_a 的关系

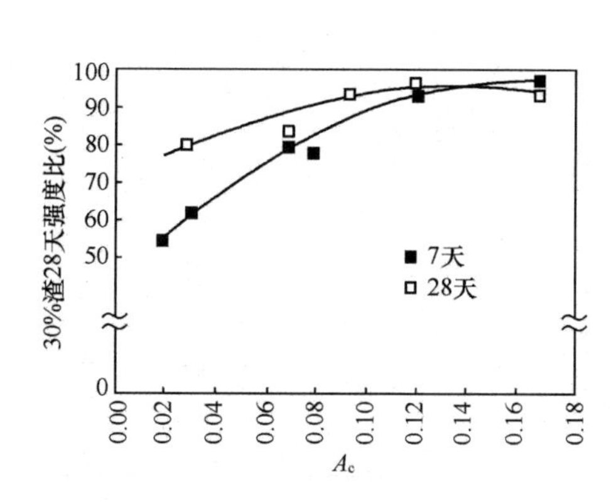

图 3　标准胶砂强度和 A_c 的关系

由表 3、表 4 和图 2、图 3 可见：

①纯固硫渣蒸养强度和 A_c、f-CaO 的关系规律基本一致，说明固硫渣具有自身水化的能力。当渣中含有一定量活性硅、铝时，随 f-CaO 的增多，其单独水化能力增大，所产生的强度也提高；A_c<0.02 或 f-CaO<0.35%时，固硫渣不具有自身水硬性；当 f-CaO>3%时，f-CaO 对固硫渣活性的影响减小。

②当渣掺量为30%、熟料为70%时，随 A_c 和 f-CaO 的增加，试件与相同熟料硅酸盐水泥试件抗压强度比值增加，而7d的比28d的显著。说明活性石灰有提高早期强度的作用；而当渣掺量为70%时，因熟料量小，A_c 影响明显；但当 A_c 大于0.1或 f-CaO>3%时，以上影响均减小；用30%的固硫渣配制水泥，加入不同量化学纯的CaO，也证明活性石灰对提高早期强度的作用而当CaO量过大时，则影响减小，甚至不利（表6、图4）。

表6　CaO对掺30%固硫渣的水泥抗压强度的影响

CaO（%）	抗压强度（MPa）			早期强度比	
	3天	7天	28天	R_3/R_{28}	R_7/R_{28}
0	19.7	30.1	49.9	39.0	60.2
2	23.5	33.1	52.5	45.0	63.0
5	18.1	34.7	50.8	36.0	68.5

3.2　固硫渣活性的评价方法

当固硫渣和硅酸盐水泥熟料掺配时渣中活性的 SiO_2、Al_2O_3 可与熟料水化生成的 $Ca(OH)_2$ 反应而生成具有水硬性的水化产物。故活性 SiO_2、Al_2O_3 明显影响用于制水泥的混合材料的活性。混合材料中活性的 SiO_2、Al_2O_3 数量可用活性率 K_a 表示：

$$K_a=\frac{活性\ SiO_2+活性\ Al_2O_3}{全\ SiO_2+全\ Al_2O_3}\times 100\%$$

实验表明，该活性率也适用于其他火山灰质混合材料[8]。

分别用不同活性率 K_a 的固硫渣加入5%CaO，经蒸汽养护后测定其抗压强度，结果对 K_a 作图（图5）；分别用活性率 K_a 不同的固硫渣30%和70%的熟料配制水泥，测定其抗压强度和 K_a 的关系如图5所示。由图4、图5可见：

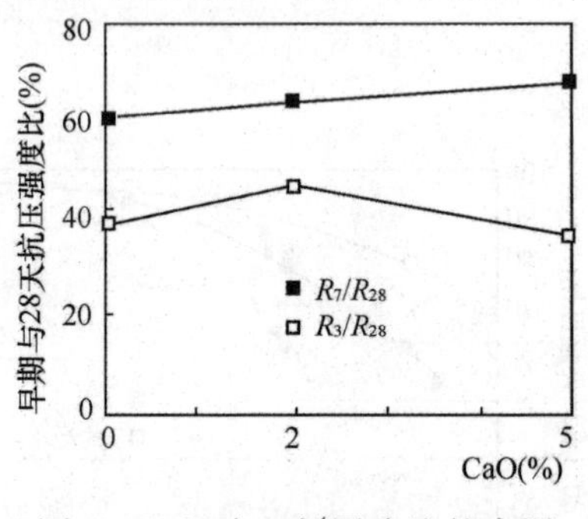

图4　CaO对30%固硫渣的水泥强度发展的影响

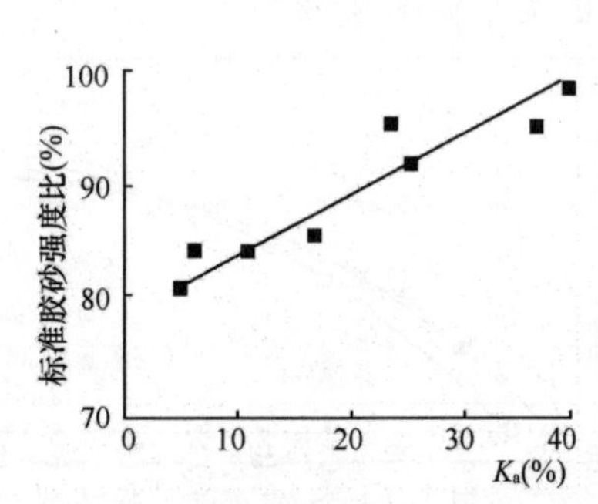

图5　K_a 和固硫渣标准胶砂强度比的关系

①当活性的CaO量较大时，固硫渣中活性的 SiO_2、Al_2O_3 对固硫渣的活性起明显的作用。例如在渣中掺入石灰后，原来没有单独水化能力的G10、FK2渣也能产生强度；但由于活性CaO总量比渣中的 f-CaO大得多，不同试件的强度和 A_c 没有明显的关系（表3与表4对比），而与渣的活性率 K_a 关系明显（图5）；因外掺CaO量一定，当 K_a 超过一定值后（在图5中为25%），CaO量相对不足，K_a 的影响减小。

② 对30%固硫渣+70%熟料的试样，因熟料水化生成的 $Ca(OH)_2$ 比5%CaO的数量还要多，K_a 对水泥强度的影响大体为线性关系（图5）。

3.3　晶态物质含量对固硫渣活性的影响

固体物质的化学成分通过结构起作用[9]。固体物质的物理状态决定其强度[10]。水泥水化后所形成亚微观水平上的微颗粒级配比水化物数量还要重要[11]。故除活性组分的数量外，混合材料中所含晶态物质含量不同，会在水泥浆体中造成微颗粒级配的变化，而影响水泥浆体的强度[12]。图6为30%固硫渣与70%熟料配制的水泥强度和纯熟料水泥强度的比值随渣中不溶性晶态物质（即除 $CaSO_4$ 外的晶态物质）含量的变化趋势。由图6可见，当晶态物质含量适当时，可以使水泥浆体得到最好的颗粒级配而获得最高的强度。图中晶态物质最佳含量约为33%～50%。随水化龄期的增长，该值有所下降。

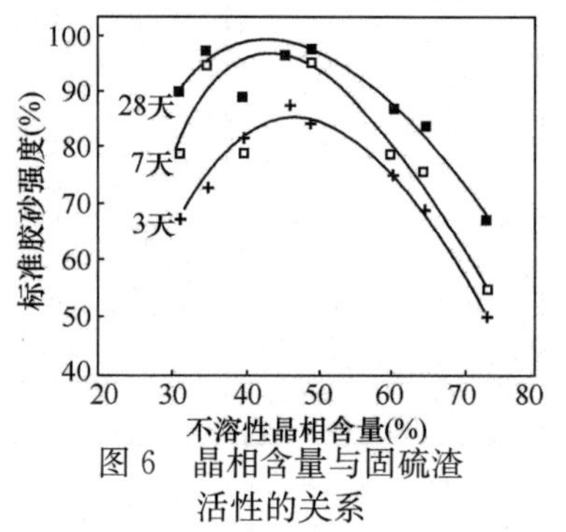

图6 晶相含量与固硫渣活性的关系

3.4 固硫渣中的 f-CaO

由于燃烧温度较低，固硫渣中的游离CaO不同于煅烧水泥时所形成的游离CaO，熟料中的 f-CaO 是在煅烧水泥的高温下形成的过烧石灰，而固硫渣中的 f-CaO 是CaO在沸腾炉或循环流化床温度下吸收 SO_2 剩余的正烧石灰，具有很高的反应活性；固硫渣向水泥所提供的 SO_3 即使小于3.5%，也比其他混合材料水泥中的 SO_3 多，可以在早期生成更多的钙矾石，有利于水泥的强度，尤其是在固硫渣中存在一定 f-CaO 的情况下。因此固硫渣的活性一般高于普通沸腾炉渣的活性。

4. 含碳量对固硫渣活性的影响

用含碳量较大的G4渣（烧失量为10.8%）在700℃下烧灼至不同的时间，得到不同含碳量的G4渣，再将其各以30%与70%的熟料配成水泥，测定其强度列于表7。

表7 固硫渣含碳量对水泥强度的影响

编号	掺渣量（%）	标准稠度（%）	含碳量（%）	抗折强度（MPa）				抗压强度（MPa）			
				3天	7天	28天	%	3天	7天	28天	%
PC	0	26.0	—	6.6	7.4	8.7	100	35.3	48.8	58.0	100
AG4-1	30	29.0	10.8	5.3	6.5	9.1	104	26.3	39.6	56.7	98
AG4-2	30	28.3	7.1	5.3	6.5	9.1	104	26.5	40.1	61.3	106
AG4-3	30	27.8	5.1	5.7	6.6	9.1	104	27.9	39.9	61.3	106
AG4-4	30	27.3	3.0	5.7	7.1	10.3	118	28.9	40.8	65.1	112
AG4-5	30	27.0	0	5.3	6.4	10.0	115	27.1	47.1	68.5	118

由表7可见，以含碳量0为基数，当含碳量从0增加到10.8%时，水泥强度下降17.2%，需水量增加7.4%。

5. 反应产物分析

将纯固硫渣试件蒸养后进行XRD析（图7），发现有钙矾石生成，原有的无水硫酸钙减少，有二水硫酸钙析出，还有少量的CSH凝胶生成。SEM析也表明了这一点（图8）。在没有固硫的沸腾炉渣FK2中 f-CaO 和无水硫酸钙都很少，固硫渣G10中的 f-CaO 很少而无水硫酸钙较多，FK2、G10两者都没有单独水化能力，这就充分说明固硫渣自身水化的能力来源于渣中所存在的 f-CaO。固硫渣G10的XRD（图9）和SEM（图10）分析表明由于出渣后泡水，其本身已经水化生成了AFt相（9.73，5.61，4.69Å），并有一部分无水硫酸钙转变成二水硫酸钙。

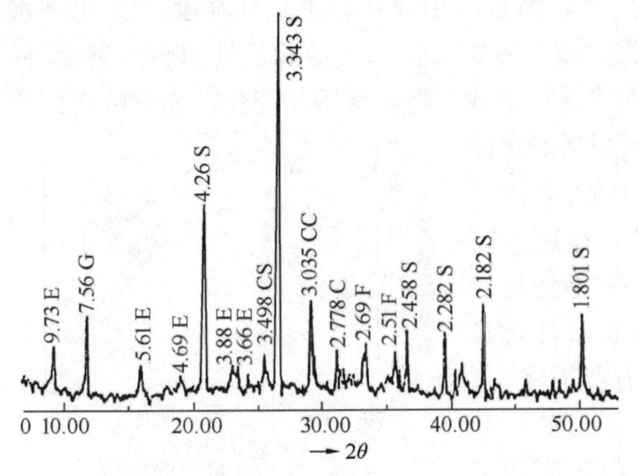

图 7　固硫渣 G7 加水成型蒸养后 XRD
XRD 图中符号：E—钙矾石；G—二水硫酸钙；S—α-石英；CC—碳酸钙；CS—无水硫酸钙；F—赤铁矿

图 8　图 7 所用试件 SEM 形貌

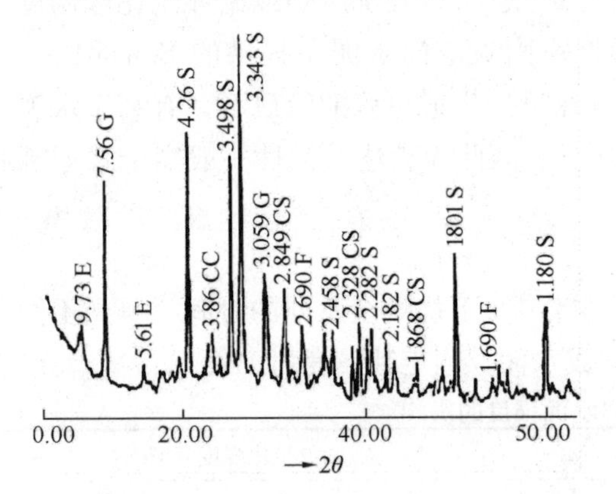

图 9　水冷渣 G10 料堆中取样的 XRD
XRD 图中符号：E—钙矾石；G—二水硫酸钙；S—α-石英；CC—碳酸钙；CS—无水硫酸钙；F—赤铁矿

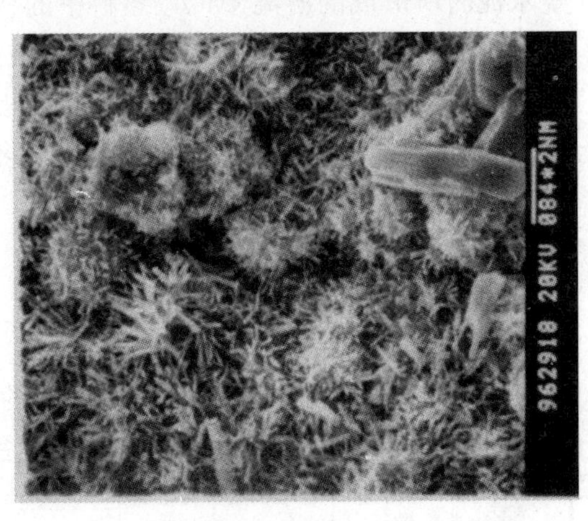

图 10　图 9 所用试样 SEM 形貌

6. G9、Gl0 渣特点分析

G9、G10 渣是燃煤固硫后用水冷却形成的残渣。从 XRD（图 6）和 SEM（图 7）可见，G10 渣表面上有钙矾石和 C-S-H 凝胶生成，这是因为固硫渣经水冷后堆放时，其内部仍有很高的温度，经一定时间后，燃煤脱硫后在渣中剩余的 f-CaO 与渣中的活性硅、活性铝发生水热反应，在有硫酸钙的情况下反应还生成钙矾石，降低或丧失了原有的水硬性和活性。为了防止这种情况发生，固硫渣不能用水冷却，而应改用空气自然冷却或者加置空气冷却机来强制冷却。固硫渣储存时应当避免受潮，储存时间不宜过长。

四、结论

1. 在 850～950℃下烧成的固硫渣活性最高，为了充分利用硫渣的活性，应控制沸腾炉的温度不超 950℃。活性率 K_a 达 20%以上的固硫渣强度比（按规范加入 30%混合材料的水泥 28 天抗压强度与同标号熟料的硅酸盐水泥 28 天抗压强度的比值）可在 85%以上。活性率 K_a（活性 SiO_2＋活性 Al_2O_3）（全 SiO_2＋全 Al_2O_3），可用以作为评定固硫渣活性的一个指标。

2. 沸腾炉燃煤固硫渣具有自身水硬性，其蒸养制品抗压强度随渣中的 f-CaO 和自硬活

性影响指数 A_c 的增大而提高。A_c 大于 0.1 或 f-CaO 大于 3%时，影响下降；A_c 小于 0.02 或 f-CaO 小于 0.35%时，固硫渣不能单独水化；自硬活性影响指数 $A_c = f$-CaO/（$CaSO_4$ + 活性 SiO_2 + 活性 Al_2O_3）可作为评定固硫渣活性的另一个指标。

3. 当 f-CaO 量相同时，活性率 K_a 大的固硫渣活性高，但和 f-CaO 有匹配的关系。当 f-CaO 量大到一定程度时（例如熟料量 70%以上），渣的强度比和 K_a 的关系大体呈线性。

4. 固硫渣中不溶性晶态物质含量影响制品的强度，在不同龄期存在一最佳值。由于渣中不溶性晶态含量与原煤煤种和所加脱硫剂成分及数量有关，相同活性的固硫渣有可能会对水泥浆体有不同的作用。

5. 固硫渣的活性随其含碳量的增加而下降。因含碳量影响试验所用的是活性最高的固硫渣，虽然含碳量达 10%以上，用其所配制的水泥强度仍然很高。考虑到一般情况，最好控制含碳量不超过 5%。

6. 由于固硫渣的自身水硬性，遇水后会降低其活性，甚至失去单独水化能力；固硫渣单独水化能力越强，活性降低得越多。故排渣时不能浇水，储存时应避免与水接触。

参考文献

[1] 高校统编教材，胶凝材料学．1980.

[2] 胡春芝等．沸腾炉渣活性及其与蒸养硅酸盐制品强度和水化产物的关系中国硅酸盐[C]．学会建材委员会第二届学术年会论文，1984.

[3] 李万有．利用沸腾炉渣做混合材生产火山灰硅酸盐水泥，“劣质流化床燃烧灰渣综合利用”课题鉴定文件，1985.

[4] 蒋继久等．沸腾炉炉渣做水泥混合材试验总结[J]．吉林建材，1981，3.

[5] 湖南省煤炭研究所．劣质煤流化床燃烧灰渣综合利用，“劣质流化床燃烧灰渣综合利用”课题鉴定文件，1985.

[6] 武汉工业大学．流化床灰渣在胶凝材料中应用的基础研究，“劣质流化床燃烧灰渣综合利用”课题鉴定文件，1985.

[7] Hark Von Rode. XRD Measurement of Glass Content in Fly Ashes and Slag. C. C. R. V01. 17 1987.

[8] Lian H. Z.，Chen E. Y.. Mechanism on Hydration of Blended Cement：I. Strengthening Effect of Mineral Powder on Cement Used for Mortar or Concrete. Proceeding of The International Conference on Concrete Structure and Materials，Tehran，1990.

[9] 唐有祺．结晶化学．1965.

[10] T. C. Powers. Physical properties of Cement Paste，Pro. 4th Int. Sym. on the Chemistry of Cement，1960.

[11] H. F. W. Taylor. Discussion of the paper"Microstructure and Strength of Hydrated Cement"by R. F. Feidman & J. J . Beaudion. C. C. R. Vo1. 7 1977.

[12] 廉慧珍等．沸石岩的活性，第三届全国水泥学术年会．1986.

[13] 廉慧珍．沸石岩在水泥中的作用机理[C]. 1980.

自评

对本文和论文之九一起自评，此处提一个关于表达方法的问题：文中表 5 主要表述用不同配比和工艺的固硫渣制品抗压强度，期间欲与水泥强度对比，把水泥强度列入第一行却处于第三大列“30%熟料+70%固硫渣”。

论文之四

固硫渣中的SO_3对水泥的调凝作用*

廉慧珍　江加标

摘　要　循环流化床燃煤固硫渣可直接与水泥熟料配制成合格的水泥，渣中SO_3可代替或部分代替（当渣中SO_3不足时）二水石膏在水泥生产对起调凝作用。当无二水石膏掺入，固定固硫渣量时，水泥中SO_3总量≤3.5%，当SO_3量增加，凝结时间延长；而当水泥中SO_3≥4%以后，含固硫渣水泥凝结时间缩短。此规律和用二水石膏调凝时相同。当固硫渣SO_3不足时，需另加适量二水石膏加以补充；但所需SO_3总量随渣所提供的SO_3量增加而增加，直到渣中SO_3足以单独起调凝作用，此时该最低量为1.85%，最好是2.2%（占水泥总量）。当SO_3总量高到一定值后，SO_3的形态不同时对水泥的调凝作用已无区别。还对固硫渣中SO_3的溶解性能做了试验，以验证其“短时多溶”原则。对其与减水剂的相容性进行研究表明，固硫渣与木质素类减水剂的相容性极差。

关键词　固硫渣　三氧化硫　水泥调凝

引言

国家标准规定硅酸盐类水泥中SO_3不得超过3.5%，否则，会引起强度急剧下降以及发生体积膨胀而使制品劣化甚至破坏（国家建材院情报标准所，1977）。

为防治燃煤造成的大气SO_3污染，在流化床锅炉中加入脱硫剂，将煤中的硫固定在所排放的废渣中。但用这种含硫废渣做水泥混合材料时，就可能使水泥中SO_3总量超过标准。例如将含SO_3为8%的固硫渣以30%加入水泥中，水泥中SO_3总量可达3.4%～5.4%。然而，做水泥掺和料的大量是废渣，虽然是防止燃煤脱硫造成二次污染的最简便的途径，但并不是百利无一害。若将含SO_3 6%～9%的固硫渣30%或40%分别与525号硅酸盐水泥熟料70%或60%混合，则可生产425号或325号的合格水泥。当固硫渣所提供的SO_3不足时，需另外加入适量二水石膏予以补充。为指导生产，就需要给出另加二水石膏量同固硫渣所提供SO_3量之间的关系。

国外在流化床固硫渣利用方面的研究也是刚刚起步。其所利用的是流化床排放的细灰，灰分少，含SO_3量很大，多用于道路、砌块或预处理后制成轻骨料，并且尚未见生产性应用的报道。用于水泥的研究，只有德国曾报道过用含SO_3 5%～6.5%的循环流化床废渣以5%掺入水泥的报道（Lotze，1985）。以上均未见有对SO_3作用研究的报道，而且许多国家都视该种废渣中的SO_3为有害。因此只有研究固硫渣中SO_3的可利用性及其利用的规律，才能使固硫渣中的SO_3无害—有利—高值化。

1　试验用材料

（1）固硫渣　不同脱硫剂脱硫燃烧排放的灰渣，其化学成分见表1。

* 原发表于国家“七五”科技攻关环境保护项目论文集《大气污染防治技术研究》（1993. 环境工程出版社）。

表 1　固硫渣化学成分（%）

编号	SiO_2	Al_2O_3	Fe_2O_3	CaO	MgO	SO_3	K_2O	Na_2O	LOI	f-CaO*
G5	48.82	18.26	8.02	11018	1.36	4.21	3.35	0.48	4.58	2.40
G6	39.92	15.21	5.99	19.47	1.54	8.70	2.79	0.46	4.74	4.79
G7	50.33	13.51	0.12	14.66	0.75	11.81	0.77	0.04	1.48	2.36

* 该 f-CaO 为在低于 1000℃的温度下生成，为脱硫后所剩余，具有很高的反应活性，不同于水泥煅烧生成的过烧 f-CaO。

（2）沸腾炉无硫渣　为不加脱硫剂的沸腾炉燃煤废渣，有时含少量 SO_3（<3%），其化学成分（%）：

SiO_2=58.92，Al_2O_3=21.85，Fe_2O_3=8.7，CaO=1.88，MgO=1.45，SO_3=1.00，LOI=0.92，f-CaO=0.17 及其他。

（3）二水石膏和无水石膏　所用无水石膏为天然二水石膏在 400℃下烧灼而成。经 XRD 分析，证明该无水石膏的结构与天然硬石膏的相同。其化学成分见表 2。

表 2　二水石膏及无水石膏化学成分（%）

试样	SiO_2	Al_2O_3	Fe_2O_3	CaO	MgO	SO_3	K_2O	Na_2O	LOI	f-CaO*
二水石膏 1	1.87	0.55	0.02	31.84	1.85	44.48	0.15	0.04	4.44	17.7
无水石膏	19.70	4.02	1.56	26.00	1.90	39.62	0.76	0.76	4.70	—
二水石膏 2	16.97	3.49	1.35	22.35	1.65	33.53	0.65	4.00	4.00	14.20

* 同表 1。

（4）熟料　所用熟料为湖北省宜昌市水泥厂生产，其化学成分（%）为：SiO_2 20.15，Al_2O_3 5.33，Fe_2O_3 4.37，CaO 68.40，MgO 1.36，K_2O 0.57，Na_2O 0.29，SO 0.36，LOI 2.72，f-CaO 2.92。

2　试验及结果分析

试验 1　测定不同比例固硫渣水泥的凝结时间。结果列于表 3。

表 3　不同比例固硫渣水泥的凝结时间（h：min）

固硫渣编号	渣量（%）	SO_3（%）	石膏（%）	总 SO_3（%）	凝结时间（h：min）	
					初凝	终凝
0	0	0	4	1.78	2：07	4：21
FK1	30	0.3	2	1.19	闪凝	—
	40	0.4	2	1.21	闪凝	—
	30	0.3	4	2.08	3：28	5：38
	40	0.4	4	2.18	2：06	4：55
G5	30	1.26	0	1.26	闪凝	—
	40	1.68	0	1.68	快凝	—
	30	1.26	2	2.15	1：42	3：45
	40	1.68	2	2.57	2：06	3：55
	45	1.89	0	1.89	1：16	3：55
	20	1.74	0	1.74	快凝	—

续表

固硫渣编号	渣量（%）	SO_3（%）	石膏（%）	总 SO_3（%）	凝结时间（h：min）	
					初凝	终凝
G5	30	2.61	0	2.61	3：28	6：23
	40	3.48	0	3.48	3：33	6：49
G7	20	2.27	0	2.27	0：53	2：48
	30	3.50	0	3.50	3：00	5：57
	40	4.32	0	4.32	4：29	6：28

注：闪凝指初凝时间＜20min，快凝指初凝时间＜45min。

由表3可见，固硫渣中的SO_3可以用于调节水泥凝结时间。对硅酸盐水泥来说，用二水石膏调凝，SO_3为1.78％时，初凝时间可达2h，固硫渣水泥不加二水石膏，SO_3超过1.89％时，初凝时间才能超过1 h。

掺无硫渣的水泥中加入二水石膏，SO_3超过2％，初凝时间超过3 h。当渣中SO_3少于5％时，如不另加二水石膏，即使固硫渣量达40％，水泥仍快凝。尽管G6渣中SO_3达8.7％，但20％的渣所提供的SO_3（1.74％）仍不足以调凝。这说明，固硫渣提供的SO_3越多，调凝所需SO_3总量也越大，直到固硫渣能提供的SO_3达到1.8％以上时，可不再另加二水石膏。

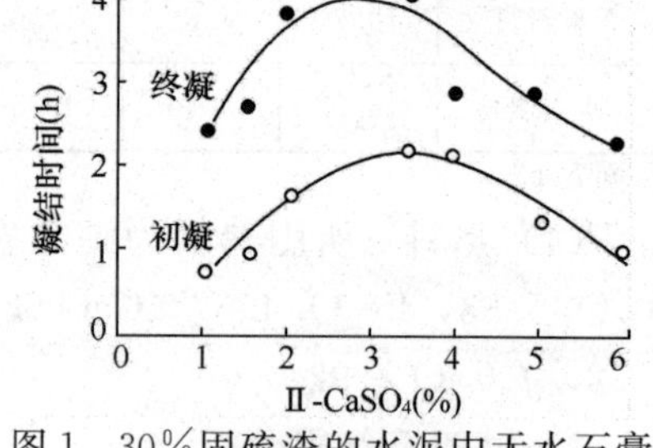

图1　30％固硫渣的水泥中无水石膏对水泥凝结时间的影响

试验2　用无水石膏调节掺无硫渣30％水泥的SO_3量，测定水泥凝结时间。结果见图1（以ⅡSO_3％表示无水石膏的SO_3量）。

由图1可知：

（1）水泥凝结时间随ⅡSO_3量的增加而延长，但当ⅡSO_3超过3.5％后，初凝时间开始变短。此规律与二水石膏的影响相似。

（2）固硫渣为30％时，当其中SO_3超过2％，只用渣中无水硫酸钙即可使水泥凝结时间合格。可以认为，用固硫渣中SO_3调凝的最小量为2％，而ⅡSO_3超过2％以后，SO_3量的增加对水泥凝结时间的影响减弱。

试验3　控制水泥中ⅡSO_3为2.2％，改变固硫渣掺量，测定水泥凝结时间。结果表明，当SO_3量一定时，水泥凝结时间随固硫渣掺量的增加而增加（图2）。这表明，随着固硫渣量的增加，不另加二水石膏而只利用渣中SO_3调凝的最小SO_3量可相应减少。

试验4　以3∶7混合无硫渣FK2和熟料，用无水石膏调节渣中的ⅡSO_3量。水泥中SO_3总量由另加二水石膏来控制。则得到在不同初凝时间水平上渣所提供ⅡSO_3量不同时，水泥所需另加二水石膏（以二水SO_3％表示）量或SO_3总量(图3)。

由图3可见，当由于固硫渣中SO_3较少或掺量较低时，固硫渣所提供的ⅡSO_3不足，需另加二水石膏补充。ⅡSO_3越多，所需SO_3总量越大，当ⅡSO_3超过1.7％时，ⅡSO_3已可单独使水泥凝结时间正常，不需再另加石膏。即当固硫渣中SO_3为6％～9％时，掺固硫渣30％和40％配制的水泥可不另加二水石膏，固硫渣中SO_3可代替全部二水石膏。

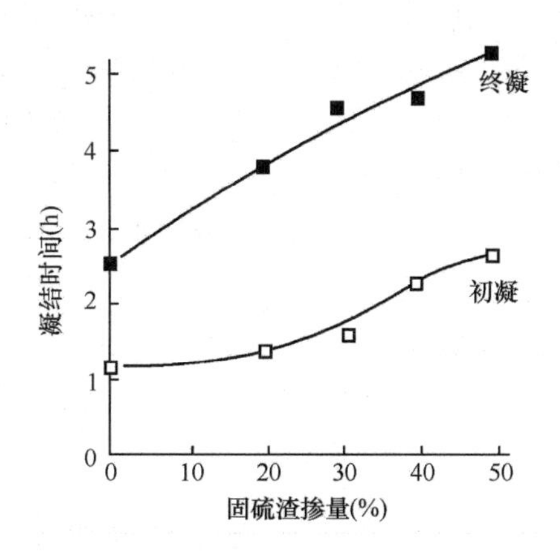

图 2　SO_3 总量一定（SO_3 为 2.2%）时固硫渣掺量对水泥凝结时间的影响

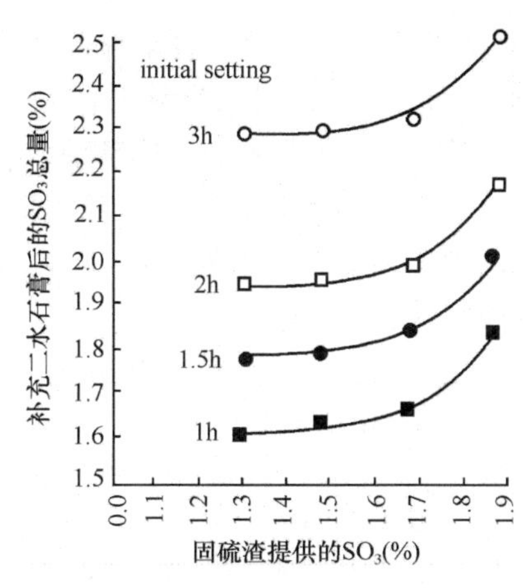

图 3　凝结时间一定时 SO_3 总量与无水 $CaSO_4$ 提供的 SO_3 量的关系

试验 5　将固硫渣 G6 和 G5 以不同固水比与水拌和，在不同时间测定渣中无水硫酸钙的溶解量。结果如图 4 和图 5 所示。

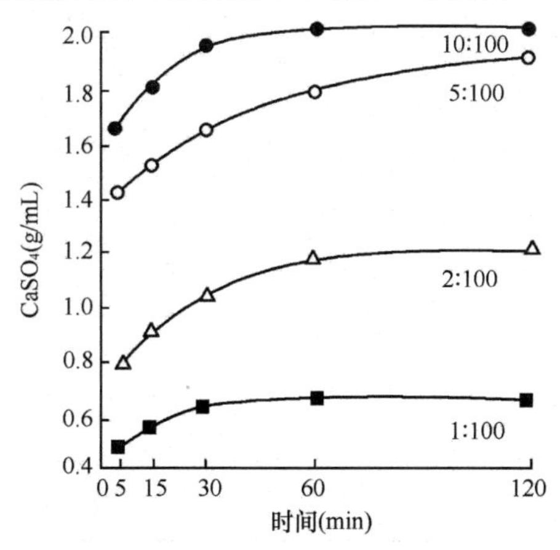

图 4　G6 渣中 $CaSO_4$ 在不同固水比下的溶出量

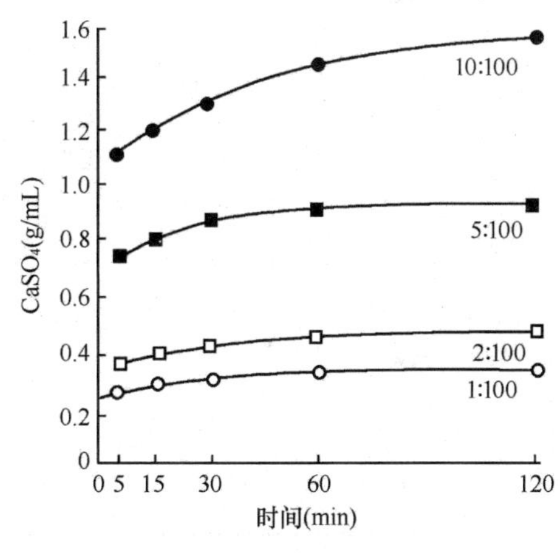

图 5　G5 渣中 $CaSO_4$ 在不同固水比下的溶出量

由图 4 和图 5 可见，在固水比相同的情况下，G6 渣中 $CaSO_4$ 溶出量几乎是 G5 渣的 2 倍，这是因为 G6 渣中 SO_3 量（8.70%）约为 G5 渣 SO_3 量（4.21%）的 2 倍。在前 30min，固水比越大，溶出量增加得越快，此后曲线平缓。这说明，溶解速率低的无水硫酸钙，可通过多量来提高其溶解速率。这就是“短时多溶”的方法。固硫渣中 SO_3 能对水泥起调凝作用，但 SO_3 总量要大些，就是这个原因。

试验 6　改变 G5 渣与熟料的比例，用二水石膏调节 SO_3 总量，测定水泥凝结时间。结果见图 6。

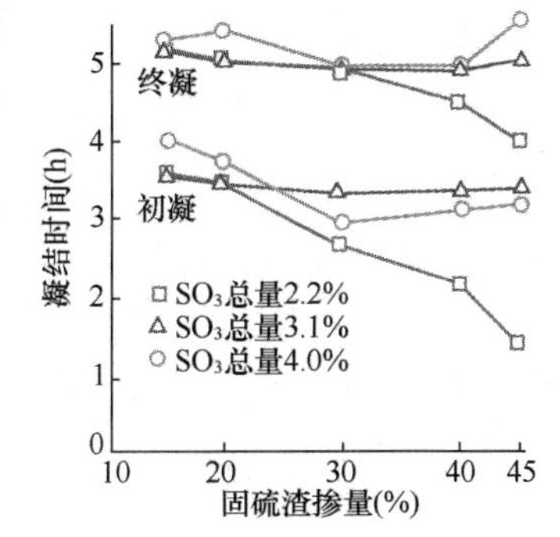

图 6　SO_3 总量不同时 G5 渣对水泥凝结时间的影响

由图 6 可见，在 SO_3 总量较低（例如 2.2%）时，固硫渣量越大，SO_3 总量中无水硫酸钙越多，凝结时间越短。但当SO_3总量超过3%后，固硫渣掺量（即其所提供的无水硫酸钙）不再影响凝结时间；当固硫渣掺量超过40%时，凝结时间反而因渣量增加而延长。可以认为，当SO_3超过一定量（如3%）后，SO_3的形态对固硫渣水泥凝结时间的影响不再有差别，即这时不论二水石膏还是无水石膏，都能提供足够的SO_3使水泥凝结时间得到调节。

试验7　测定不同减水剂对固硫渣水泥的适应性。结果列于表4，所用减水剂为常用木钙类和萘系。

表4　减水剂对固硫渣水泥凝结时间的影响（h：min）

外加剂	凝结时间 h:m ＼ 配料	纯熟料水泥	30%G5[a]	40%G5[a]	30%G6	40%G6
0	初凝	2：07	1：42	2：06	3：28	3：35
	终凝	4：21	3：45	3：55	6：23	6：49
木钙 0.2%	初凝	3：04	0：24[b]	0：55[b]	0：15[b]	0：25[b]
	终凝	6：06	6：04	8：32	0：35[b]	1：17
糖蜜钙 0.3%	初凝	1：37	—	0：11[b]	0：20[b]	0.44[b]
	终凝	9：14	—	5：12	1：03	2：44
J-1 0.5%	初凝	2：17	—	—	2：38	2：58
	终凝	6：31	—	—	6：14	5：48
NF 0.5%	初凝	2：43	1：37	2：51	2：20	1：07
	终凝	4：57	7：17	7：49	4：15	2：47
FDN 0.5%	初凝	0：25	—	—	2：21	1：51
	终凝	6：22	—	—	6：03	5：11

a. G5水泥中均另加2%二水石膏。
b. 说明使水泥初凝时间太快。

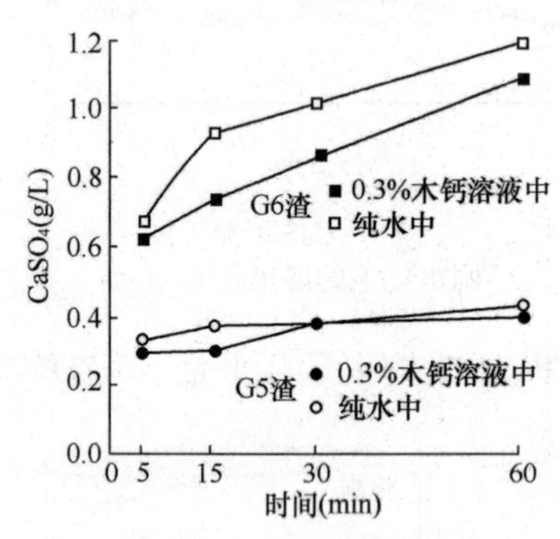

图7　G5渣和G6渣中$CaSO_4$的溶解速率（灰水比为2：100）

由表4可见，木钙类减水剂（糖蜜）属缓凝型，可使水泥缓凝，但却使固硫渣水泥速凝，而萘系减水剂则无影响。这种现象可通过固硫渣中$CaSO_4$在木钙溶液和在水中的不同溶解性质来解释。图7为固硫渣G5和G6中$CaSO_4$的溶解速率测定结果。

由图7可见，在前30分钟，$CaSO_4$在木钙溶液中比在纯水中的溶出量小得多。对于G5，60分钟以后$CaSO_4$溶出量在木钙中开始大于在纯水中的，故早期由于$CaSO_4$溶出量小而使水泥快凝（初凝阶段），而到后期（60分钟后），无水$CaSO_4$溶出量增加，与二水石膏共同作用而调凝，使水泥终凝时间延长。

G6渣中$CaSO_4$在木钙溶液中的溶出量始终比在纯水中的低得多，故初凝、终凝都很快。

3　结论

（1）固硫渣中SO_3为6%～9%时，30%或40%固硫渣所提供的SO_3可完全代替二水石膏对水泥调凝，而水泥中SO_3总量≤3.5%。

（2）当固硫渣中SO_3不足或固硫渣掺量较低时，需另加适量二水石膏来补充，所需SO_3

总量与固硫渣中提供的 SO_3 和二水石膏提供的 SO_3 比例有关，该比值越大，所需 SO_3 总量越大；直至Ⅱ SO_3 足以单独调凝，该值最好是2%。

（3）当 SO_3 总量超过3%时，不同形态的 SO_3 对水泥调凝作用不再有差别。

（4）木钙类减水剂用于无水硫酸钙是会发生闪凝，不能用于固硫渣水泥。

参考文献

[1] 国家建材院情报标准所. 水泥标准的修订[M]. 北京：中国建筑工业出版社，1977.

[2] Lotze, J., Kenndaten und Verwertungs möglichkeiten von Aschen aus einer Feuerungslage mit zirkulierender Wirbelschicht, Teil2, Eignung und Anwendung inder Bauwirtschatt, Z-K-G, 7.1985：374-378.

自评

1. 石膏在水泥中的作用很重要，除了调节凝结时间外，对混凝土拌和物的流变性能（需水量、与外加剂相容性等）、硬化混凝土的非荷载变形和强度都有重要影响。其中对水泥凝结时间的影响是水泥生产者最为关心的。对固硫渣中 SO_3 的利用问题，除了需要研究其中石膏的形态和性质外，还必须研究其用法。在本研究中，通过对凝结时间的试验，提出固硫渣中无水石膏和外加二水石膏匹配的规律和“短时多溶”的原理，在固硫渣水泥中试生产中证明是可行的。

2. 研究证明了无水硫酸钙和木钙类的外加剂高度不相容。在当前二水石膏资源紧缺的情况下，一些水泥厂正在使用各种来源的石膏，而木钙（或糖蜜）又是混凝土最方便的缓凝剂，水泥和外加剂相容性的问题应引起注意。此外，过去系统研究都在使用萘磺酸盐类减水剂的条件下进行，如今已有30多年大量工程使用的经验，当前正在推广聚羧酸类的高效减水剂，而聚羧酸减水剂更加复杂，使用的经验尚不足。当前人们往往着重于调外加剂组分去适应水泥，其实水泥才是和外加剂相容性的矛盾主导方面，外加剂生产商更需要了解水泥的因素，双方合作进行研究。

3. 固硫渣中的 $CaSO_4$ 在木钙溶液中的溶出速率低是木钙和无水硫酸钙不相容的机理，但是 $CaSO_3$ 在木钙中溶出速率低的机理又何在？

4. 结论中的第二条最后的“该值最好是2%”指的是什么？意思不清楚。

一位法国心理学家说：“应当教育孩子敬重老师，但更要教育孩子敬重真理。怀疑并不是缺点，总是没完没了的怀疑才是缺点。只有敢于怀疑，才能减少盲从。有怀疑的地方才有真理，真理是怀疑的影子。”

论文之五

固硫渣对水泥中过量SO_3作用的抑制*

廉慧珍　李海燕　江加标

摘　要　固硫渣中SO_3以Ⅱ-$CaSO_4$的形态存在，与天然硬石膏的形态相同。当用固硫渣与熟料配制水泥时，可用渣中的Ⅱ-$CaSO_4$代替天然二水石膏调节水泥凝结时间。而当固硫渣中SO_3较高时，就会使水泥中的SO_3总量超过国家规定的标准（3.5%）。通过对不同SO_3量和不同掺渣量水泥的凝结时间、强度以及线膨胀率规律的研究表明，当SO_3较高而同时渣量也大的水泥，过量SO_3所引起的凝结时间缩短、强度下降和体积膨胀就会受到抑制。当固硫渣掺量为40%时，水泥中SO_3可达4.7%而无害。

关键词　固硫渣　水泥调凝　三氧化硫

引言

国内外都有对水泥中SO_3限量的规定[1]，主要原因是过量的SO_3可能使水泥的凝结加快，并在水泥浆体硬化后产生有害的膨胀和强度的下降[2]。利用治理SO_2对大气污染的燃煤固硫渣制作水泥，为了使水泥中SO_3不超过国家标准的规定，用30%固硫渣生产425号水泥时，固硫渣中SO_3含量不得大于11%，用40%固硫渣生产325号水泥时，固硫渣中SO_3含量不得大于8.5%。当固硫渣中SO_3较高时，固硫渣的利用就受到了限制。例如研究中所用固硫渣G7，为宜昌市造纸厂快装循环流化床锅炉燃煤脱硫的废渣，含SO_3 11.8%。用其生产水泥时，掺量就不能达到30%。实际上SO_3引起水泥浆体的有害膨胀不仅取决于SO_3的含量，而且还与其他因素有关，如细度、熟料的相组成、硫酸钙的形态等。因此，优化各种因素，在水泥中增加SO_3的添加量是有可能的。德国学者Odler（1988）研究用硬石膏代替二水石膏的结果表明，即使SO_3达到6%，细度为3000 cm^2/g的水泥砂浆标准养护三年后，其膨胀量只有5.27mm/m，当比表面积提高到4500 cm^2/g时，掺硬石膏的试件三年后膨胀率只有0.78mm/m；而掺含同量SO_3的二水石膏时，无论细度如何，试件均开裂[3]。这说明硬石膏有比二水石膏更明显的无害性，而细度对膨胀的影响很大。Ouyang等认为在水泥中加入粉煤灰是抑制内部硫酸盐侵蚀的一种有效的方法[4]。阿根廷Irassar E. F. 的试验表明，粉煤灰对水泥砂浆受外部硫酸盐作用而产生的膨胀只起延缓的作用，而对内部的硫酸盐引起的膨胀来说，只是增加了速率，却可减小膨胀率，三个月以后膨胀即停止，而未加粉煤灰的水泥砂浆膨胀则一直在发展[5]。这原因主要是一种对C_3A的“稀释效应”（P. K. Mehta，1986）[6]。

固硫渣中的SO_3以Ⅱ-$CaSO_4$的形态存在，是与天然硬石膏结构和形态相同而性质相近的无水石膏[7]。研究固硫渣对SO_3作用的抑制，有利于扩大固硫渣应用的范围，具有环境效益和经济效益。

---

* 原发表于国家“七五”科技攻关环境保护项目论文集《大气污染防治技术研究》（环境工程出版社，1993年）。

1 试验材料

（1）固硫渣：用未经脱硫燃烧的无硫渣 FK2 和无水石膏配制成不同 SO_3 含量的固硫渣，以及宜昌市造纸厂快装循环床锅炉用清华大学研制的高活性人工钙基脱硫剂脱硫燃煤的固硫渣 G7，和宜昌市印染厂沸腾炉用清华大学研制的华宜牌脱硫剂脱硫燃煤的固硫渣 G5。其化学成分如表 1 所示。

表 1 固硫渣和无硫渣化学成分（%）

编号	SiO_2	Al_2O_3	Fe_2O_3	CaO	MgO	SO_3	K_2O	Na_2O	LOI	f-CaO
FK2	69.17	16.26	8.14	1.22	1.00	2.62	1.31	0.16	1.22	0.34
G7	50.53	13.51	0.12	14.66	0.75	11.81	0.77	0.04	1.48	2.36
G5	48.82	18.26	8.02	11.18	1.36	4.21	3.35	0.48	5.58	2.40

（2）无水石膏：用天然二水石膏经 400℃焙烧 4 小时而得，经 XRD 检验，其成分为Ⅱ-$CaSO_4$，含 SO_3 47.31%；此外还用南京天然硬石膏，SO_3 为 51%。

（3）熟料：用宜昌市水泥厂和柳州市第二水泥厂的硅酸盐水泥熟料 Ys 和 Ls。其化学成分见表 2。

表 2 试验用熟料化学成分（%）

编号	SiO_2	Al_2O_3	Fe_2O_3	CaO	SO_3	LOI	f-CaO
Ys	19.97	5.38	4.75	62.71	1.67	1.84	2.63
Ls	20.81	5.67	5.37	63.96	0.12	2.48	—

2 实验及结果分析

（1）用含 SO_3 为 4.21%的固硫渣 G5 以不同掺量与熟料 Ys 配制成水泥，并用二水石膏调节水泥中 SO_3 总量，测定水泥凝结时间。结果如图 1 所示。

在图 1 中，随固硫渣量的增加，SO_3 总量中无水石膏增多。当 SO_3 总量较低时（如 2.2%），固硫渣掺量增大，则无水石膏比例增大，凝结时间缩短，而当 SO_3 总量超过 3%以后，SO_3 总量的增加，不同形态 SO_3 的比例都不再对水泥凝结时间有明显的影响。这是因为 SO_3 量较大时，无水硫酸钙由于“短时多溶”而可提供足够的 SO_3 调节水泥的凝结时间。与二水石膏不同的是，因其溶解速率较低，当 SO_3 超过 3.5%以后反而会使凝结时间缩短。

（2）用无水石膏调节 SO_3 总量，测定在不同固硫渣掺量下，标准养护的水泥 28 天抗压强度随 SO_3 总量的变化。结果如图 2 所示。

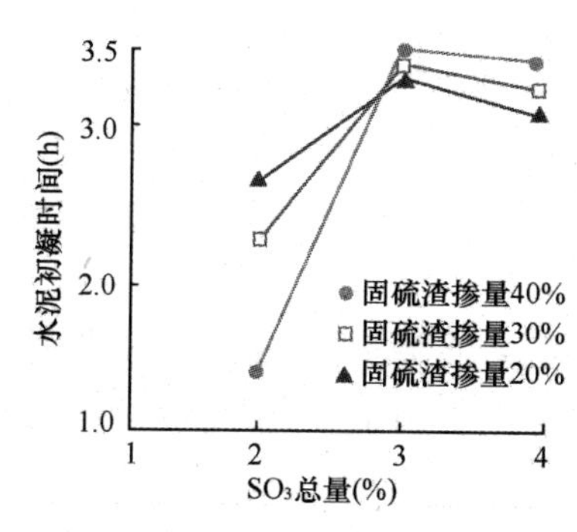

图 1 SO_3 总量对不同掺渣量水泥凝结时间的影响

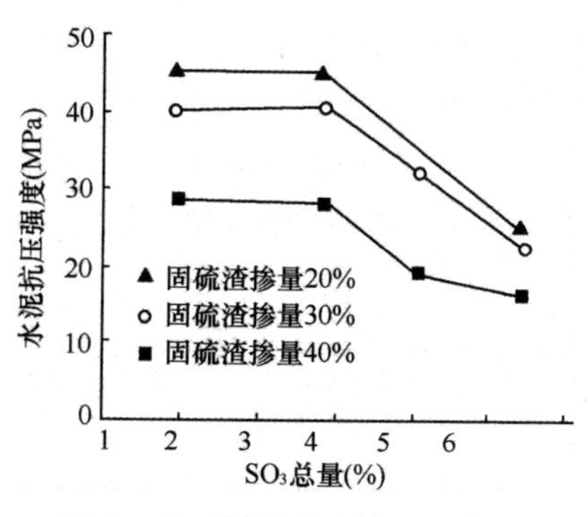

图 2 在不同固硫渣掺量下水泥抗压强度随 SO_3 总量的变化

由图 2 可见，在各固硫渣掺量下，当 SO_3 超过 3.5％以后，水泥强度随 SO_3 的增加而下降，但随渣量的增加，这种下降的幅度减小。当固硫渣量达 40％时，SO_3 超过 5％以后，SO_3 再增加就对水泥强度几乎不再有影响，当 SO_3 总量达 6％时，固硫渣掺量对水泥强度的影响也减小。

（3）分别测定 SO_3 为 2.62％、4％和 6％时固硫渣掺量对水泥净浆线膨胀率的影响。图 3 中 SO_3 为 2.62％，图 4 中 SO_3 为 4％；图 5 中 SO_3 为 6％。

由图 3～图 5 可见，在各 SO_3 量的情况下，掺固硫渣后水泥膨胀率均下降，龄期越长，下降越明显；高 SO_3（4％和 6％）比低 SO_3（2.62％）更为显著，在高 SO_3 情况下，掺渣量的影响并不明显。

（4）固定固硫渣掺量为 40％时，改变 SO_3 量，测定水泥净浆线膨胀率随不同龄期发展的变化。结果如图 6 所示。

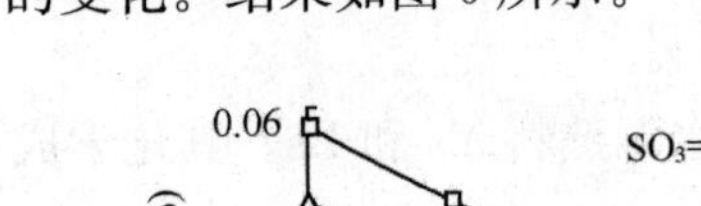

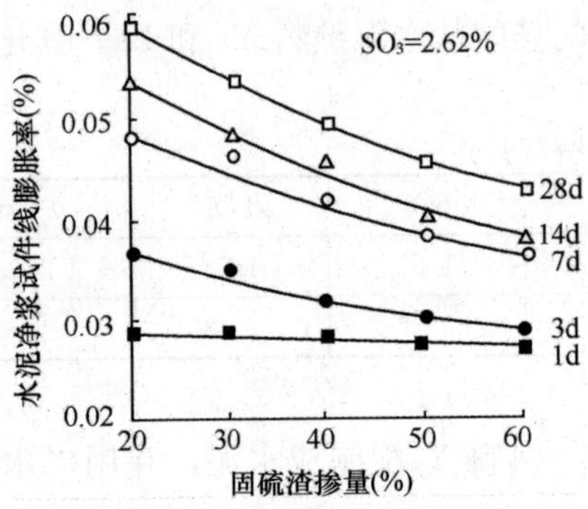

图 3　SO_3 为 2.62％时固硫渣掺量与水泥线膨胀率的关系

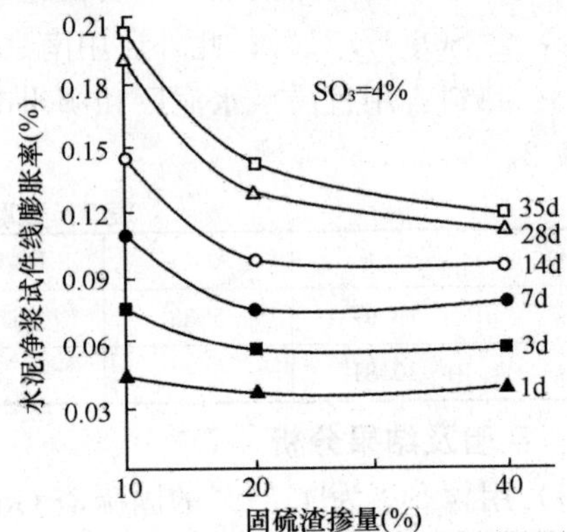

图 4　SO_3 为 4％时固硫渣掺量与水泥线膨胀率的关系

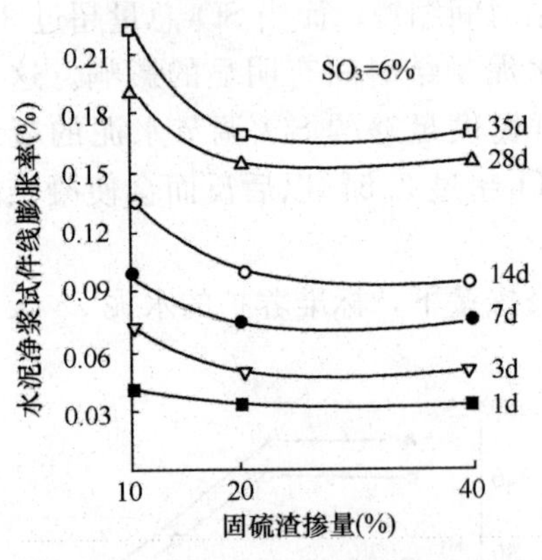

图 5　SO_3 为 6％时各龄期水泥净浆线膨胀率随固硫渣掺量的变化

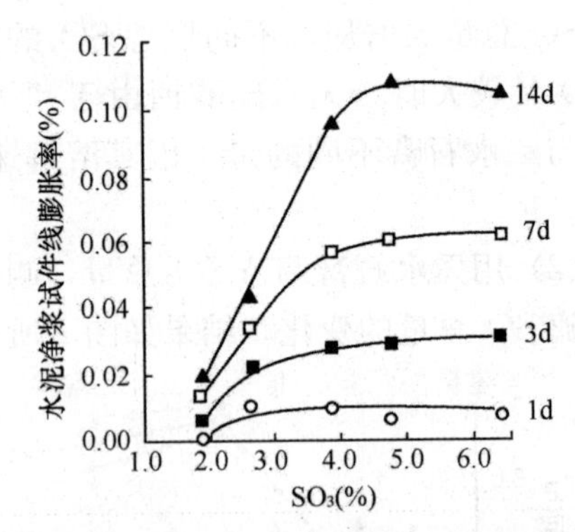

图 6　固硫渣 40％时各龄期水泥净浆线膨胀率与 SO_3 含量的关系

由图 6 可见，掺入固硫渣 40％以后，在 SO_3 低于 4％时，水泥膨胀率随 SO_3 的增加而增加得很快，并随龄期增长而加快。但当 $SO_3>4\%$ 以后，各龄期膨胀几乎不再随 SO_3 的增加而增加，14 天时 SO_3 为 6％的水泥膨胀率甚至似有下降。这可能是由于固硫渣对水泥中 C_3A 的“稀释效应”所引起的。

（5）将固硫渣 G7 以不同掺量与熟料配合，制成水泥，测定当 G7 渣量变化时，G7 渣所提供的SO_3量的变化对水泥膨胀的影响，结果如图7所示。

由图7可见，当G7渣增加时，SO_3也增加，此时水泥膨胀的变化是由这两个因素共同作用所造成的。G7渣掺量从20%增加到40%时，SO_3从2.36%增加到4.7%，但由于固硫渣的抑制作用，此时14天和28天膨胀只增加了一倍，如果固硫渣为20%不变，水泥中SO_3从2.36%增加到4.7%时，水泥的28天膨胀率则增加1.3倍（图8）。而且在图7中G7渣40%时膨胀率最大，也不超过0.5%；当G7渣为50%时，膨胀率反而下降。

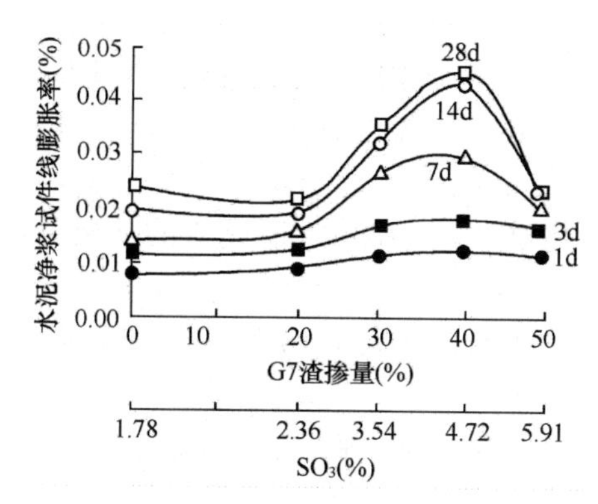

图7　G7渣掺量对水泥净浆膨胀率影响

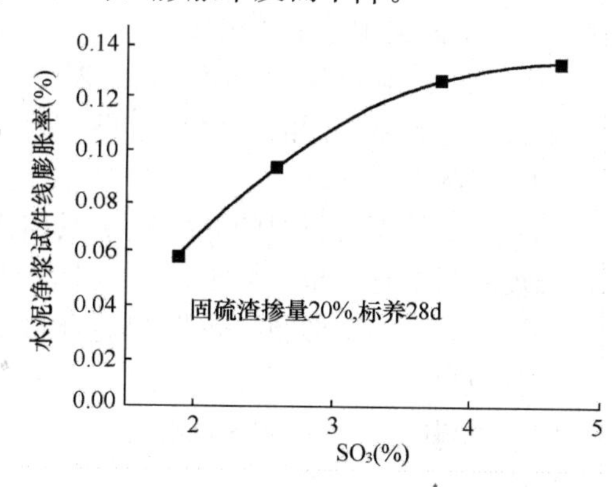

图8　固硫渣20%时水泥膨胀率随SO_3增加的变化

3　结论

（1）用固硫渣和水泥熟料配制成水泥时，可用渣中固定的SO_3代替天然石膏调节水泥凝结时间。当SO_3超过3%时，SO_3量对凝结时间的影响不大；

（2）SO_3>3.5%以后，水泥强度随SO_3含量的增加而降低，但当固硫渣掺入量为40%时，这种降低幅度减小；而且当SO_3为6%时，由于固硫渣的增加而降低强度的幅度也减小；

（3）水泥中SO_3过量时，其膨胀率随之增大，并随龄期明显增长；掺固硫渣后，这种现象得到缓解。当固硫渣达40%时，SO_3从5%增加到6%，则水泥膨胀无明显增加；

（4）用SO_3为11.81%的固硫渣40%与60%熟料配制成水泥，水泥中SO_3达4.72%，但28天净浆膨胀率也不超过0.5%，因而无害。

参考文献

[1] 杨斌. 国内外水泥标准选编[M]. 北京：中国标准出版社，1987.

[2] 国家建材院情报标准所. 水泥标准的修订[M]. 北京：中国建筑工业出版社，1977.

[3] Odler, I.，The Possibility of Increasing SO_3 Content in cement，Engineering Foundation Conference in Advance on Manufacture and use of Cement and Concrete，Missouri，USA. July，1988.

[4] Chengsheng Ouyang，Antonio Nanni and Wen F. Chang. Internal and External Sources of Sulfate Ions in Portland Cement Mortar：Two Type of Chemical Attack，Cement and Concrete Research，Vol. 16，1988.

[5] Irassar，E. F.. A discussion of the Paper"Internal and External Sources of Sulfate Ions in Portland CementMortar：Two Type of Chemical Attack"by C. Ouyang，A. Nanni and W. F. Chang，Chemical Attack，Cement and Cocrete Research，Vol. 19，1989：662～664.

[6] Li，G. et al. Form and Characteristics of SO_3 in Sulphur-fixed Boiler Ash，In Proceedings of Shanghai 1991 Ash Utilization Conference，1991，1：100～101.

[7] Mehta, P. K.. ACI Journal, 1986. 83: 994～1000.

自评

1. 在固硫渣对过量SO_3抑制作用的研究中尚用粉煤灰和沸石凝灰岩做过试验，在第9届国际水泥化学会议（1992年，印度，新德里）发表了论文《Inhibition of Blending Materials on Effect of Excess SO_3 Content in Cement》（混合材料对水泥中过量SO_3的抑制作用）。现将该文中结论译出如下：

（1）水泥因含过量SO_3而导致的强度下降，可通过混合材料的掺入而得到抑制，但不同种类的混合材料有不同的作用和效果，因此用"C_3A稀释效应"无法解释上述混合材料对过量SO_3抑制的机理；

（2）与不掺混合材料的硅酸盐水泥相比，即使SO_3超过6%，在掺入固硫渣的情况下，水泥强度降低的程度也可减小，膨胀率的减小程度也更低；

（3）对过量SO_3抑制的效果和混合材料的活性有关，固硫渣和沸石岩的活性率都大于20%，其效果比所使用活性率不到12%的粉煤灰的要明显；

（4）可以认为，在硅酸盐水泥中掺入高活性混合材料时，可将SO_3总量控制到4%～5%而无害。

该文所述试验结果还表明，用硬石膏替代二水石膏时，所述抑制作用更为明显。

2. 上世纪末至今对膨胀剂使用的褒贬虽然主要是使用技术问题，但与机理不清楚不无关系。上述对膨胀抑制作用的机理同样有待于更深入的研究。机理研究的目的是能动地掌握和运用客观规律。

3. 固硫渣所含SO_3既可被利用为膨胀源，固硫渣又可用作矿物掺和料，对膨胀有抑制作用，因此对固硫渣的应用存在对立统一规律的运用，推想现在因矿物掺和料的使用而抑制膨胀剂效率的问题是否也可运用此规律予以解决？

古希腊时，有个病人满脸愁容地问哲学家安提丰："人为什么活着啊？"安提丰回答："这正是我还没解决的问题，所以我要活着。"

论文之六

流化床燃煤固硫渣用于补偿收缩混凝土的可行性*

清华大学土木工程系　廉慧珍　陈恩义　阮庆革

摘　要　对流化床燃煤固硫渣实验研究表明，固硫渣不仅具有较高活性，是水泥的优质掺和料，而且还可利用固硫渣中的 SO_3 作为水泥膨胀组分补偿混凝土的收缩，后期变形稳定，在较高 SO_3 含量下也是安全的，并使混凝土具有较高的后期强度增长率，并大大降低混凝土的成本。

关键词　流化床燃烧固硫渣（FBCSA）　强度　变形　混凝土

1　前言

流化床锅炉（fluidized bed boiler）的燃烧温度低，最适合于煤炭灰分中黏土矿物分解成无定形物质，所排放的废渣具有较高火山灰活性，同时最适合燃烧过程中脱硫。在流化床锅炉燃用含硫的煤炭时，加入一种含钙质的脱硫剂，将燃煤时释放出的 SO_2 固定在废渣中，称为流化床燃烧固硫渣（fluidized bed combustion sulphur-fixed ash，以下简称固硫渣 FBCSA）[1]。清华大学在承担我国“七五”国家重点科技攻关专题研究中，对固硫渣做了系统的基础性研究和应用研究，证明由于燃烧温度合适，固硫渣具有较高火山灰活性[2]。固硫渣所含 SO_3 以 Ⅱ-$CaSO_4$ 存在，并含有少量脱硫反应剩余的高活性石灰，还具有自身水硬性。刘燕芳[3]、陈恩义[4]分别对固硫渣用于制作膨胀水泥和膨胀剂进行了系统试验研究。在日本，有学者提出用加入膨胀剂的方法来减少混凝土的收缩，并将膨胀剂用于高性能混凝土的研究[5]，但未见有工程实施的报道。

以固硫渣为掺和料，利用其中所含的 SO_3 为水泥膨胀组分，配制高性能混凝土，以补偿混凝土的收缩，可大大降低混凝土的成本，并使其具有较高的后期强度增长率。固硫渣用于配制补偿收缩混凝土将为固硫渣开辟更加广阔的工程应用领域[4]。

2　实验用原材料

2.1　固硫渣

1990 年初在广西柳州化工厂 10 吨/时沸腾炉进行工业脱硫（脱硫剂为宜昌产“华宜”脱硫剂）试验所排废渣，编号为 L4。其主要化学成分如表 1 所示。

表 1　L4 渣化学成分（%）

样品	SiO_2	Al_2O_3	Fe_2O_3	MgO	CaO	*f*-CaO	Na_2O	K_2O	SO_3	LOI
L4 渣	46.32	24.6	5.71	1.17	11.1	0.54	0.35	1.50	3.95	4.04
水泥	20.66	5.55	3.02	1.26	62.5	0.75	0.51	0.86	2.20	—

* “八五”期间发表的论文很少，这篇文章是在计算机中留底的副本，其中所述为“八五”攻关时所做工作的综述，较有代表性，故选入。

2.2 水泥

冀东水泥厂生产的“盾石”牌525号硅酸盐水泥，细度为（0.08mm筛筛余）3.2%，其化学成分如表1所示，熟料矿物组成见表2，强度检验结果见表3。

2.3 石膏

分别用山西太原产的二水石膏和北京燕山水泥厂生产用的二水石膏，磨细至全部通过0.15mm筛。在400℃下加热4小时，x-射线衍射证明其为Ⅱ-$CaSO_4$，与固硫渣中的SO_3有相同的形貌和性质[4]。其化学成分见表4。

表2 盾石牌硅酸盐水泥熟料矿物组成（%）

成分	C_3S	C_2S	C_3A	C_4AF	$CaSO_4 \cdot 2H_2O$
含量	38.89	29.89	9.58	9.18	5.09

表3 “盾石”525号硅酸盐水泥强度检验结果

龄期	3天		7天		28天	
强度（MPa）	抗压	抗折	抗压	抗折	抗压	抗折
	38.24	6.75	54.04	8.06	58.18	9.75

表4 两种烧石膏化学分析结果（%）

样品名称	SiO_2	Al_2O_3	Fe_2O_3	MgO	CaO	SO_3	LOI
太原石膏	9.75	2.00	0.80	0.93	36.4	42.7	6.72
燕山石膏	—	—	—	—	39.7	45.4	—

2.4 其他原材料

砂石、水：试验用砂为昌平龙凤山产河砂，细度模数3.0，属中砂偏粗，含泥量为1.8%；石子为昌平龙凤山产碎石，粒径5～20mm；水为普通自来水。

减水剂：采用广东湛江外加剂厂生产的FDN高效减水剂。

粉煤灰：北京市售Ⅱ级粉煤灰，0.08mm筛筛余4.6%，烧失量2.3%。

U型混凝土膨胀剂：北京建材院研制的U-Ⅱ型混凝土膨胀剂。

3 实验及结果分析

3.1 强度试验

3.1.1 控制混凝土坍落度相同，水泥中SO_3量相同，不同水胶比的条件下，混凝土的强度与固硫渣掺量的关系如图1所示。

由图1可见：

①w/c≤0.35时，固硫渣掺量对混凝土后期抗压强度影响不大，强度值在80MPa线上波动，而当w/c＞0.35以后，混凝土强度随渣掺量增加而下降，当固硫渣掺量从5%到40%，强度均下降10MPa；由此可见固硫渣用于高强混凝土时更为有利；

②当固硫渣掺量≤5%时，混凝土强度变化不大；即在相同强度等级要求下，加入固硫渣可节省水泥5%；

③w/c＞0.35时，在各固硫渣掺量下，混凝土强度和c/w基本为线性关系；而当w/c≤0.35以后，w/c对强度的影响敏感性下降。

3.1.2 相同坍落度、水胶比相近的情况下，固硫渣掺量和强度的关系如图2所示。

实验表明，固硫渣掺量小于25%时，混凝土28天强度不低于60MPa；固硫渣掺量为44%时，混凝土的28天强度＞50MPa；用60%体积（54%重量）的固硫渣取代水泥时，混凝土的 28 天强度≥40MPa；即对 28 天抗压强度 40MPa 以上的泵送混凝土（坍落度均大于 220mm），固硫渣掺量可达 50%以上。该结果和图 1 中在不同水灰比条件下固硫渣掺量对混凝土强度的影响具有相同的规律，表明这一规律具有较好的重演性。

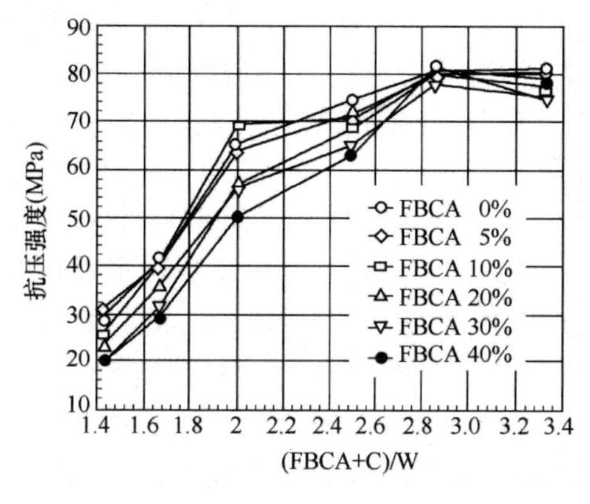

图 1　不同 w/c 混凝土固硫渣掺量与强度的关系

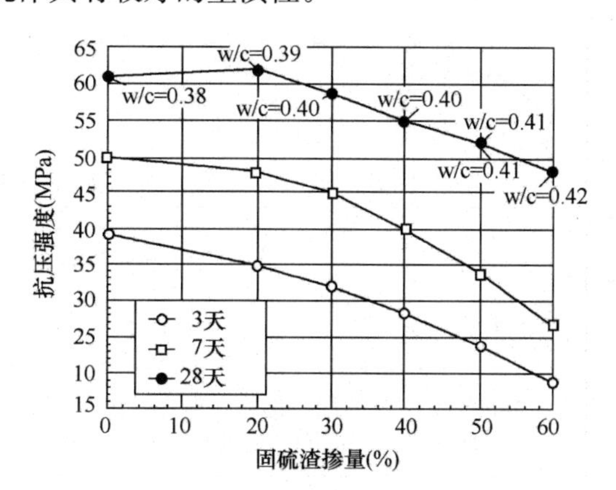

图 2　同坍落度混凝土固硫渣掺量与强度的关系

3.2　水泥中 SO_3 含量对混凝土变形行为影响的试验

3.2.1　改变水泥中石膏的含量，测定水泥中 SO_3 含量对混凝土强度和变形的影响，结果如图 3 所示。

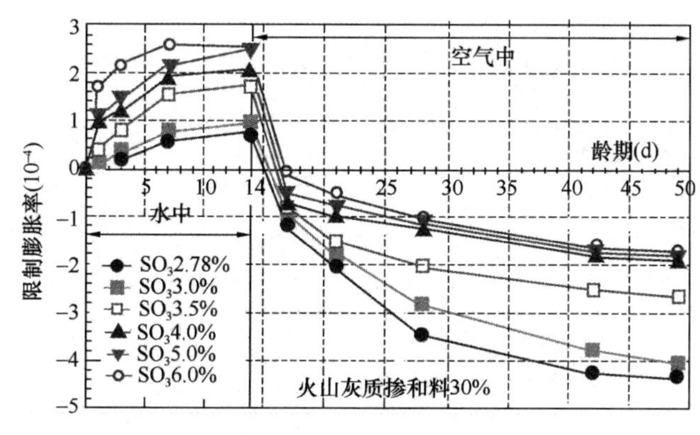

图 3　水泥中 SO_3 含量对混凝土变形的影响

图 3 表明，水泥中 SO_3 含量小于 4.0%时，随着水泥中 SO_3 含量的增加，混凝土在水中的膨胀率增加，而在干空气中收缩减少；当 SO_3 达到 4.0%以后，SO_3 对混凝土的变形的影响就不再显著。在水中膨胀后到空气中收缩，期间变形的差值称作“落差”。对于膨胀剂来说，这个“落差”比水养护产生的膨胀值更重要，更能反映补偿收缩的能力。当 SO_3 为 2.8%时，混凝土在水中养护 4 天的膨胀值不到 0.01%，而在转入空气中 35 天（总龄期为 49 天）时，收缩值为 0.043%，落差约为 0.05 个百分点；当 SO_3 增加到 4%时，在水中膨胀值约为 0.022%，收缩值为 0.019%，小于混凝土极限收缩率 0.02%，落差约 0.04 个百分点。

当水泥中的 SO_3 超过 4%时，对混凝土的水养护后转入空气中的干缩不再有明显影响，即补偿收缩的作用不再随 SO_3 增加而增大。可见，欲改善现行硅酸盐水泥干缩的性能，将 SO_3 量增加到 4%似有必要。

3.2.2 固硫渣对混凝土体积稳定性的作用

用固硫渣体积掺量30%制作的水泥配制混凝土，另用体积掺量为40%的粉煤灰和质量占15%的UEA膨胀剂掺入混凝土，分别检测二者补偿收缩效果进行对比，如图4所示。由图4可见，在水中养护14天，二者有相同的膨胀率。随后转入空气中存放5周，掺固硫渣试样的干缩值为0.019%，小于混凝土极限收缩率；UEA试样的干缩值为0.031%；前者比后者，落差约小0.012个百分点。这表明，虽然粉煤灰因含有硬度较大的玻璃微珠而能减少混凝土的收缩，但固硫渣因含有溶解速率慢的无水石膏石，既可保证后期不致因水泥中SO_3耗尽而引起早期生成钙矾石向单硫盐的转化，又可因后期提供SO_3而继续生成钙矾石，加强补偿收缩的功效。

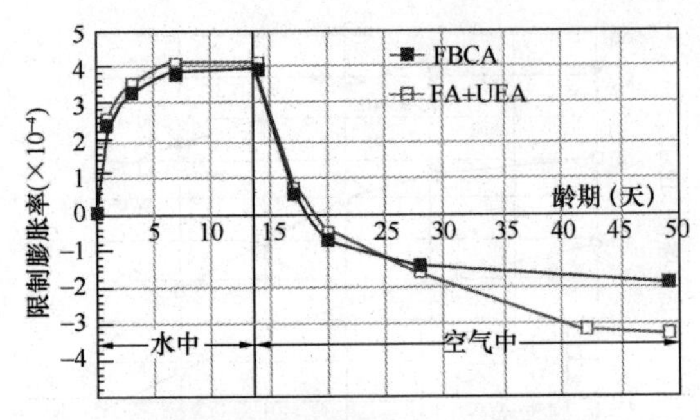

图4 用固硫渣配制的混凝土与用粉煤灰加U型膨胀剂配制的混凝土的变形对比

当固硫渣作为掺和料掺入混凝土时，其干缩值比不掺膨胀剂的约小40%，固硫渣中的SO_3可以作为水泥中SO_3的补充。如图5所示，当水泥中SO_3总量较大（如5.0%）时，随着固硫渣掺量的增加，混凝土在水中养护时的膨胀率减小，但随后放入干空气中养护40天左右时，不同混凝土试件却具有几乎相同的补偿收缩效果；而当胶凝材料中SO_3相同时，随着固硫渣掺量的不同，对混凝土的变形影响甚微。

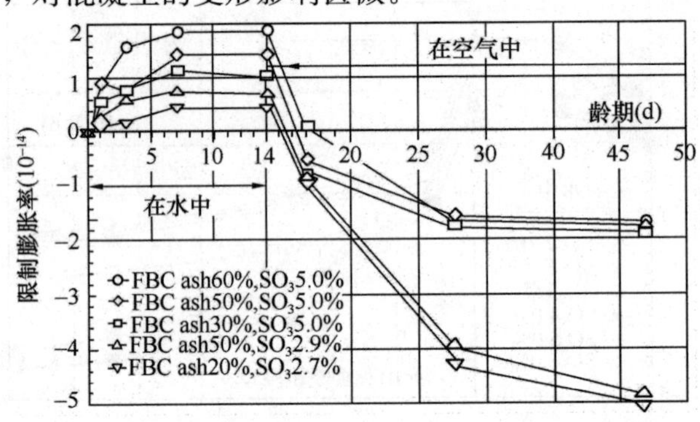

图5 相同SO_3下固硫渣掺量对混凝土的变形影响

4 结论与展望

（1）采用525号硅酸盐水泥，当水灰比为0.30～0.45，固硫渣掺量为10%～40%时，可以配制出C50～C70（配制强度为60～80MPa）的高强混凝土，此时混凝土拌和物坍落度为60～200mm；当水灰比为0.45～0.60，掺固硫渣量小于40%时，可以配制C30～C40（配制强度为40～50MPa）的混凝土，混凝土拌和物坍落度为100～200mm。

（2）当水灰比≥0.40时，混凝土强度随固硫渣掺量的增加而降低；而当水灰比≤0.35时，掺固硫渣5%～40%对混凝土后期强度和和易性几乎无影响。因此，固硫渣较适合于配制高强混凝土。在高强混凝土中掺入固硫渣可降低水化温升，节约水泥，并使混凝土具有较高的后期强度增长率。

(3) 采用525号硅酸盐水泥，以30%（体积比）取代水泥，可配制28天抗压强度高于60MPa的混凝土；以50%（体积比）取代水泥可配制28天抗压强度高于50MPa的高性能混凝土；以60%（体积比）取代水泥则可配制28天抗压强度高于40MPa的高性能混凝土。以上混凝土拌和物均具有能满足施工性能的要求，并能保证混凝土后强度的持续增长。

(4) 可有效地利用固硫渣中的SO_3使混凝土在水化早期产生适量的体积膨胀，以补偿混凝土后期的干收缩，降低混凝土开裂敏感性，提高混凝土耐久性；但水泥中SO_3总量宜不超过4.5%。本实验变形的检测采用砂浆试件，而当水胶比相同时。砂浆比混凝土的干缩值大，可以预见，用于工程中的混凝土时，干缩值可望控制在混凝土收缩极限以内。

(5) 当混凝土所用水泥中SO_3含量相对较高时，掺入固硫渣可以有效地抑制由于超量SO_3引起的体积膨胀，提高混凝土的体积稳定性。

(6) 现行水泥中的SO_3普遍约为2%～2.4%，由本实验推断，用固硫渣的SO_3作为水泥中SO_3的补充应不失为一种经济而可行的方法。

(7) 在混凝土中掺入矿物掺和料时，原有水泥中的SO_3会被稀释，影响混凝土拌和物硬化后的性质，而固硫渣由于其特有的品质，用作混凝土掺和料时可以补充所稀释的SO_3。

(8) 由于煤种矿相、成分以及燃煤工况的差异，具体应用时应对原材料做认真的了解和检测。

参考文献

[1] 廉慧珍. 因势利导利用流化床脱硫燃煤含硫废渣[J]. 硅酸盐建筑制品，1995.

[2] 廉慧珍，江加标. 沸腾炉固流渣活性评价及其影响因素的实验研究. “七五”国家重点科技攻关专题“沸腾炉燃煤固流渣制建材的研究”鉴定文件之三，1990.

[3] 刘燕芳. 用沸腾炉燃煤固硫渣制作膨胀水泥的研究[D]. 清华大学工学硕士论文，1988.

[4] 陈恩义. 沸腾炉燃煤固流渣制作混凝土膨胀剂的实验研究. “七五”国家重点科技攻关专题“沸腾炉燃煤固流渣制建材的研究”鉴定文件之七，1990.

[5] 寺村悟，坂井悦郎. 高强度化のための混合材の開発[J]. コンクリート，No.546，Aug.，1992.

自评

1. 固硫渣含有与天然硬石膏相同的结构与性质，用于补偿收缩混凝土时，既可在水养护中产生必要的膨胀率，还能在其后的空气中收缩值小于混凝土收缩极限，从膨胀到收缩的落差小，这是工程应用对膨胀剂所要求的重要品质。之所以会有这样的效果，是因为固硫渣所具有的特性使其既能利用所含SO_3产生膨胀，又能对过量SO_3有抑制作用。用固硫渣制作膨胀剂的安全性也在于此，这是对固硫渣利用中辩证法的又一实例。

2. 粉煤灰有抑制膨胀的作用，可按薛君玕对钙矾石膨胀机理的解释：生成钙矾石的膨胀与否和钙矾石形貌有关，钙矾石形貌和在环境中$Ca(OH)_2$的浓度有关，掺入粉煤灰会稀释浆体中$Ca(OH)_2$的浓度，待粉煤灰开始反应后，更加稀释。但其机理仍需继续研究。

3. 因固硫渣中的无水硫酸钙溶解速率慢，可以在浆体中维持较长时期钙矾石的生成，而且还因固硫渣含有活性很高的CaO，以维持浆体中$Ca(OH)_2$浓度，促进钙矾石膨胀。这可能是固硫渣膨胀剂的膨胀后期落差小的原因。由此可推断，如果在U型膨胀剂中引进的石膏换成硬石膏，应当也会减小其后期的膨胀落差。”这在实际工程中已得到证明。

4. 因天然资源成分的波动，在具体使用时必须具体分析和对待。

什么是科学?

日前，德国生物学家汉斯·思高勒尔在汕头大学作了一次演讲，台下坐的都是高二的学生。

“在座的各位当中，有多少人将来想要成为科学家?”演讲正式开始之前，思高勒尔首先抛出这样一个问题。80%的学生不假思索地举起了手。“有多少人已经开始思考如何成为一名科学家?”这一次，学生们略显迟疑，最终举起手的人数少了一大半。

“在我像你们这么大的时候，甚至从没想过将来要当科学家。”学生时代的思高勒尔喜欢在放学后，到家附近的树林里游荡。他会长时间蹲在植物旁边，目不转睛地观察瓢虫如何捕食蚜虫；或者，把咖啡粉撒在地上，观察蚂蚁沿着粉末一路迁徙；有的时候，他还会把甲虫塞进口袋，直接导致他的母亲每次洗衣服的时候都会抓狂。“我对生物学的兴趣，或许就是在那些如痴如醉观察虫子的讨程中慢慢产生的。”思高勒尔总结道，“兴趣，对于从事科学研究至关重要。”

与思高勒尔相似，女科学家海伦·玛格丽特·布劳童年的时候也喜欢在附近的黑森林里游荡。她喜欢搜集动物的骨头，然后钻到海德堡大学的图书馆里，从字里行间，寻找骨头的来历。布劳博士认为，科学就是一个循环往复的提出疑问、假设，再寻求解答的过程，永无止境。就像她12岁时寻找动物头骨来历的过程一样。她曾假设它是一只羊的头骨，但查阅书籍之后，推翻了这种想法；通过观察头骨的牙齿部分，她又假设它是一头牛，调查比对之后，这种假设也被推翻了。经过多次失败的假设，布劳最终发现，它属于一只雌鹿。布劳认为，科学家应该像侦探福尔摩斯那样，循着蛛丝马迹，一路追查下去。

（2015.5.18《报刊文摘》摘自2月11日《中国青年报》作者张莹）

论文之七

固硫渣双掺复合硅酸盐水泥的强度复合效应*

清华大学土木工程系　廉慧珍　刘爱明

摘　要　以活性较高的固硫渣（FBCSA）分别与粉煤灰、矿渣、天然沸石等矿物掺和料复合，加入适量石膏与硅酸盐水泥熟料混合，制成复合硅酸盐水泥。当与较低活性混合材复合时，即固硫渣对水泥强度的贡献与其含量成正比，这称为“组合效应”，可在水泥熟料量一定的情况下，优化复合比例而提高水泥强度，尽量消耗较多的低活性掺和料；当与较高活性混合材复合时，在优化的复合比例和掺量下，水泥可得到高于同掺量下各组分单独掺入时的强度，甚至高于纯熟料水泥的强度，这称为“超叠效应”。两种复合效应都有利于提高工业废料利用效率，改善水泥性能。不同混合材、不同熟料、不同龄期对复合效应有不同的影响。

关键词　固硫渣　双掺复合水泥　强度效应　组合效应　超叠效应

前言

我国国家标准规定可用的水泥混合材料主要有粒化高炉矿渣和火山灰质材料（含粉煤灰和天然沸石岩等）两大类。其他各种废渣从结构和性质来说，很多不属于上述范围。我国有的单位从 80 年代初就有在水泥生产时掺两种以上工业废渣（简称双掺或多掺水泥）的尝试，取得了很好的效果[1][2]。工业废渣资源化，我国在 1992 年颁布了复合硅酸盐水泥（简称复合水泥）的国家标准。允许在水泥中掺入总量大于 15%的两种或两种以上经过启用鉴定的工业废渣。据统计，有 13 个在 1983 年以前就有允许在水泥中掺两种或两种以上混合材料的水泥标准，而且还用以配制成高强及超高强混凝土。

固硫渣是沸腾炉或循环流化床燃煤脱硫排放的含硫废渣。试验研究表明固硫渣活性较高，但还具有与其他火山灰质材料不尽相同的性质。利用其较高的活性，可在复合水泥中起积极的作用。

复合水泥的复合效应包括宏观复合效应和微观复合效应；微观复合效应是产生宏观效应的机理。在宏观效应中，人们首先关注的是强度效应，尤其是抗压强度的效应。此外还有其他的宏观效应如水化热、耐久性等的复合效应。

本文中所述固硫渣为一种组分，分别与矿渣、粉煤灰和天然沸石岩进行双掺，对其抗压强度复合效应的试验研究，以作为固硫渣双掺复合水泥研究的突破。文中所用水泥标号均按 GB 175—1977 检验。

一、试验用原材料

1. 水泥熟料

选用首都水泥厂 525 号硅酸盐水泥熟料、柳州水泥厂 525 号硅酸盐水泥熟料和北京燕山

---

* 原载《第六届水泥化学及测试方法学术会议》论文集，1994.10，郑州。

水泥厂机立窑525号硅酸盐水泥熟料。其化学成分相近，只有 f-CaO 和矿物组成的差别如表1所示。

表1 水泥熟料矿物组成和 f-CaO 的含量（%）

编号	来源	C_3S	C_2S	C_3A	C_4AF	f-CaO
CK-Y	燕山	28.94	42.46	8.76	10.85	1.69
CK-LL	首都	33.92	37.91	4.80	5.96	0.37
CK-LZ	柳州	49.39	21.51	5.62	15.47	1.10

2. 混合材料

选用粉煤灰、矿渣、沸石岩为代表与固硫渣进行复合。其结构特点和活性指标示于表2。

表2 所用混合材料特性

编号	材料名称	来源	特 性	活性率（%）	胶砂强度比
L4	固硫渣	柳州	含 SO_3，有自身水硬性，活性高	14.82	85.7
FA-N	粉煤灰	南宁	玻璃体为主，Ⅱ级，活性较低	—	82.1
FA-G	粉煤灰	北京		—	77.6
SL-D	矿渣	大连	有自身水硬性，高活性	—	1.43*
SL-S	矿渣	北京		—	1.28*
ZE-L	沸石岩	河北	以结晶态为主，较高活性	20.56	78.0

* 用矿渣水泥强度/纯熟料水泥强度×（1－矿渣掺量）表示。

二、试验及结果分析

采用固硫渣-粉煤灰、固硫渣-矿渣、固硫渣-沸石岩三种双掺复合方案，复合比例从0：10到10：0变化，以15%、20%、30%、40%与水泥熟料和适量石膏混合后共同磨细。其中熟料用三种对比，粉煤灰和矿渣各用两种对比。

以胶砂流动度不小于116mm确定水灰比，测定上述各水泥试样不同龄期标准抗压强度。试验结果分析如下。

1. 强度复合效应方式

从图1～图4可见不同龄期有不同的复合效应：

① 3天和7天时，无论与何种材料复合，单掺固硫渣的强度都是最高的，随另一种混合材料比例的增加，强度单调下降；但在相同掺量下，以任何比例与固硫渣双掺的水泥都比另一种混合材料单掺时的强度高。两组分对强度的贡献与其所占比例成正比。这称为"组合效应"；粉煤灰28天以前，在任何双掺比例下都没有明显的峰位，主要是"组合效应"。

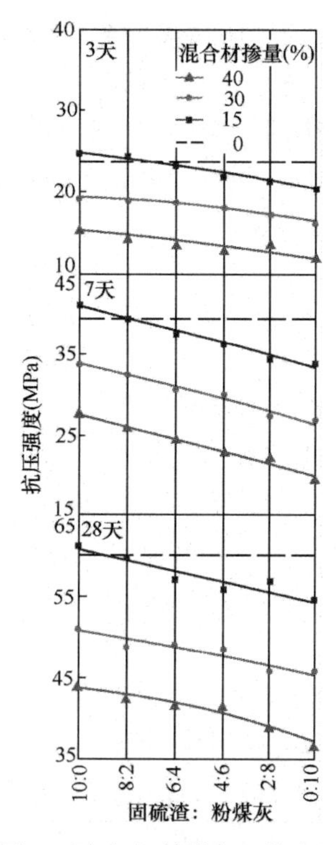

图 1　固硫渣-粉煤灰双掺水泥强度复合效应（熟料 LL）

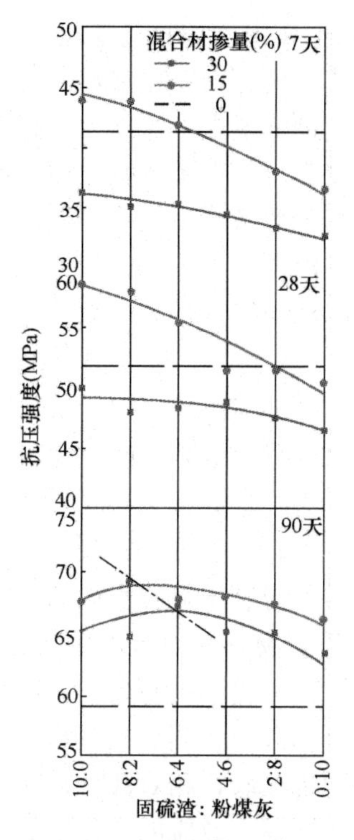

图 2　固硫渣-粉煤灰双掺水泥强度复合效应（熟料 LZ）

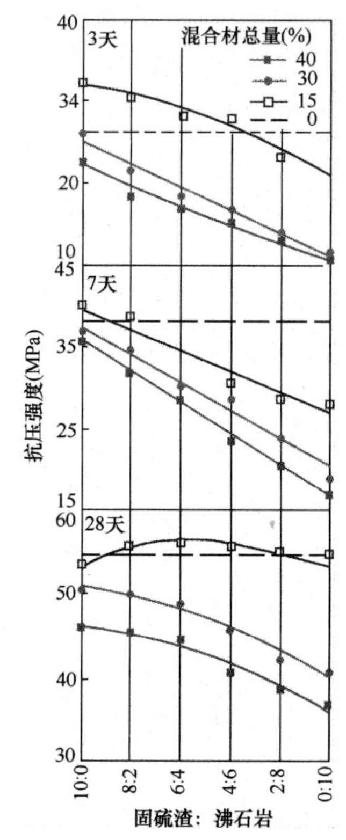

图 3　固硫渣-沸石岩双掺水泥强度复合效应（熟料 LL）

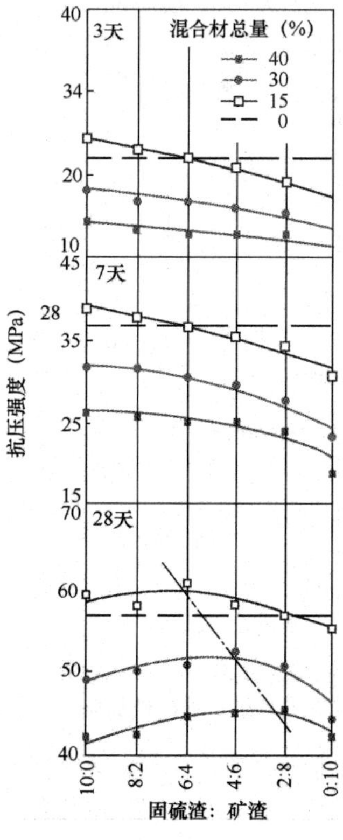

图 4　固硫渣-矿渣粉双掺水泥强度复合效应（熟料 LL）

② 固硫渣-矿渣双掺总量超过30%而达到28d时，双掺比例与强度关系曲线出现峰值，即在合适双掺比例下，双掺水泥的强度高于各自单掺时的强度。在固硫渣-矿渣任何双掺比例下都出现上述峰值点。这称为“超叠效应”，即以适当比例双掺时，在适当总掺量下，复合水泥的强度高于各自单掺时水泥的强度。这种“超叠效应”的结果，甚至可使该峰值点的强度超过熟料的强度，如图4所示。但对固硫渣-沸石岩来说，只有在低掺量下才有这种“超叠效应”（图3）。

③“超叠效应”的峰值位置随总掺量的增加而向右移动，即向固硫渣比例小的方向移动。

④ 各自单掺时强度差别不大的混合材料双掺才会出现“超叠效应”，而各自单掺时的强度差别大的混合材料双掺只能出现组合效应。如图3，与沸石岩单掺量大时强度损失较大的规律一致。

⑤ 28天只有“组合效应”的固硫渣-粉煤灰试样，90天后也出现“超叠效应”。可见，水化龄期长时，有利于复合水泥的“超叠效应”。

2. 不同混合材料和熟料对强度复合效应的影响

图1和图4为用不同粉煤灰、回转窑熟料实验的结果，图5用不同粉煤灰、立窑熟料实验的结果。可见前者无明显“超叠效应”，而用立窑熟料时，则出现“超叠效应”。图5中各曲线由于非同批试验，相对数据有些差异，但各曲线基本平行，而且峰位置基本无变化，说明不同粉煤灰对复合效应的变化影响不大。从表1可见，三种熟料虽然矿物组成有差别，但28天强度差别不大，故可不考虑矿物组成的影响；所不同的是燕山熟料为立窑生产，*f*-CaO较多，说明含*f*-CaO较多的立窑水泥更适合用于固硫渣-煤灰双掺复合。

与矿渣双掺时，熟料和矿渣的差别对强度复合效应的影响都不明显，只要各自单掺时强度差别不大，都可出现“超叠效应”，并且都有随总掺量增加而峰位向右移动的规律（图6）。

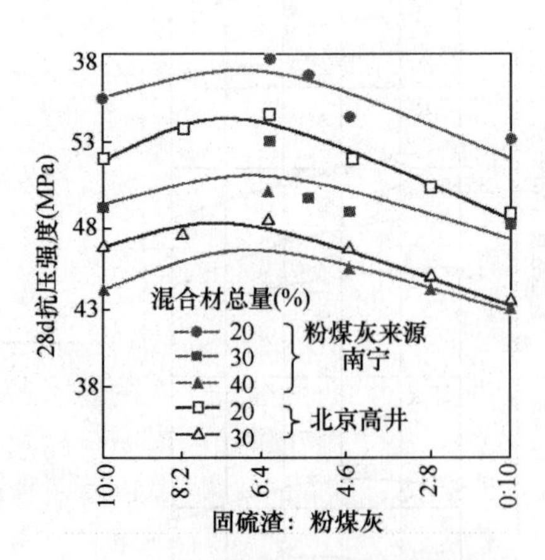

图5 不同粉煤灰对固硫渣-粉煤灰双掺水泥28d抗压强度的影响

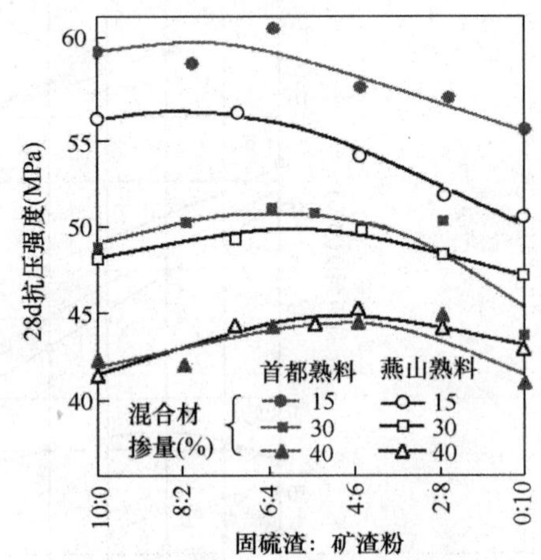

图6 不同熟料对固硫渣-矿渣双掺水泥28d抗压强度的影响

3. 双掺复合水泥强度发展的效应

根据正交设计优化，用柳州日用化工厂固硫渣和北京高井电站粉煤灰分别单掺和双掺，与熟料配制成复合水泥，检测其3天、7天、28天、90天和180天抗压强度，列于图7。结果表明，固硫渣掺入量为15%时，单掺与双掺都可提高水泥早期和后期强度；水泥强度随龄期而增长，随总掺量的增加而降低；28天以后，所有掺入混合材的水泥强度增长率都高于纯硅酸盐水泥；混合材总掺量相同时，双掺的增进率又高于单掺的。表明固硫渣与粉煤灰复合水泥的超叠合效应随龄期的增长而更加显著。

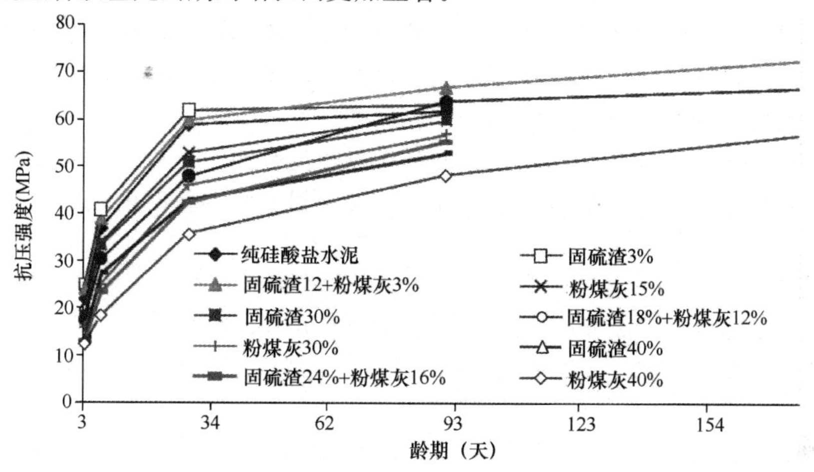

图7　固硫渣和粉煤灰单掺和双掺的复合水泥与纯熟料水泥强度发展对比

4. 固硫渣复合水泥复合效应的机理

分别测定固硫渣和其他混合材的活性率。活性率可反映火山灰质材料的化学反应活性，其测定方法是用饱和石灰水沸煮后，用冷盐酸中和溶液，测定所溶出的SiO_2和Al_2O_3量，计算其合量与全分析中SiO_2和Al_2O_3合量的比值：

$$活性率=\frac{活性\ SiO_2+活性Al_2O_3}{全SiO_2+全Al_2O_3}$$

使用加权法计算其按不同比例混合后的活性率，与实测值进行对比，如表3所示。

表3　固硫渣复合水泥活性率的实测值与计算值比较

组分		固硫渣	固硫渣＋粉煤灰	固硫渣＋粉煤灰	固硫渣＋粉煤灰	粉煤灰	固硫渣＋沸石岩	固硫渣＋沸石岩	固硫渣＋沸石岩	沸石岩
固硫渣比例（%）		100	80	60	20	0	80	40	20	0
活性率（%）	实测	14.82	15.55	15.27	15.48	12.80	16.81	20.25	21.99	20.56
	计算	—	14.42	14.04	13.20	—	15.97	18.26	19.44	—

在表3中，复合组分各自活性率按照组合比例加权计算的活性率值都小于实测值，这表明，两种混合材复合使用时，其反应活性并非二者简单加合，而是相互作用后而产生“超叠效应”。同济大学张雄在对工业废渣进行复合效应的研究中认为，高钙类的矿渣与高铝类的粉煤灰共同水化时，可相互诱导激发对方的水化活性，从而产生化学的复合作用[4]。实际上矿物掺和料的作用及其不同种类之间的复合效应要复杂得多，不仅表现于强度，其他性能如凝结、变形等行为有时更重要；不仅是化学性的，物理作用对复合形成的微结构也更重要。

结论

1. 固硫渣与其他活性混合材料双掺制成复合硅酸盐水泥时，不仅产生“组合效应”，而且在一定条件下可产生“超叠效应”。

2. 固硫渣双掺复合水泥的复合效应随龄期而不同。和粉煤灰复合而用回转窑的熟料时，28 天以前只有“组合效应”，90 天才出现“超叠效应”，而用立窑熟料则 28 天就出现“超叠效应”。这可能和熟料中所含游离石灰含量不同有关。

3. 固硫渣双掺复合水泥的复合效应与混合材料种类有关。当各自单掺时强度差别不大时，双掺后可产生“超叠效应”，即复合后水泥强度超过各自单掺时的强度，甚至可超过熟料的强度。

4. 用 525 号硅酸盐水泥熟料，当固硫渣和粉煤灰或矿渣双掺时，总掺量为 15%、30%、40%时，相应固硫渣占混合材料分别为 80%、60%、40%，可分别生产出 325 号、425 号、525 号双掺复合硅酸盐水泥。对相同标号水泥，比单掺粉煤灰时可节省熟料 10 个百分点。

参考文献

[1] 李岳庆. 高强复合硅酸盐水泥的研制与应用[J]. 湖南建材，1991，2.

[2] 马瑞庆等. 复合硅酸盐水泥的研究与生产效果[J]. 山东建材，1991，4.

[3] 王幼云等. 大力发展一种新型通用水泥——复合硅酸盐水泥的论述[J]. 水泥，1989，10.

[4] 张雄. 工业废渣复合胶凝材料体系效应及其数据库网络模型[D]. 上海：同济大学，1991.

自评

1. 文中只注重固硫渣与其他混合材复合用于生产硅酸盐复合水泥时，可产生不同的强度复合效应，实际上由于不同混合材具有不同特性，各有其利弊，其复合效应不一定只表现在强度上。可能在强度上只有“组合作用”，而在其他性能上会产生“超叠效应”。例如设想一下，固硫渣中 SO_3 合适时能使混凝土减小收缩或产生膨胀，粉煤灰因含玻璃微珠而会提高混凝土的抗裂性，那么是否有可能复合使用后会更提高混凝土的抗裂性？更有利于混凝土的耐久性？

2. 关于对混合材活性的评定，在原材料一节中，表 1 中水泥矿物组成似乎令人生疑：三种熟料都是当时的标号 52.5＃，相当于现行标准的强度等级 42.5。现在看看其矿物组成，其中的 CKLL 和 CKLZ 都能生产出需水量低、开裂敏感性低、长期性能发展好的水泥，但是四种矿物总量似不足，不知是否计算错误。因原始化学成分已无法得知，故不知是哪一项的错误。虽然是立窑的燕山熟料也会有不错的性质。遗憾的是不知早期效果如何？这是面对现在的熟料我们所关心的。

3. 表 2 中矿渣的“胶砂强度比”注中的计算公式怀疑有误，应注明来源。

4. “在对图 1～图 4 所示不同复合方式的复合效应实验结果分析中，第②条实际上只讨论的是 28 天的情况，是不全面的。应在一开始加一句：“固硫渣和矿渣复合，在早龄期没有任何复合效应，无论总掺量和二者比例如何，在早龄期总是固硫渣∶矿渣比值越小，抗压强度越低，但在龄期 28 天时，则在各种掺量下都在合适的双掺比例下出现峰值，掺量超过 30%后更加明显。这表明所用固硫渣比所用矿渣有更高的活性。

5. 本文所述只是所使用的原材料情况下有限试验量得出的结论，尤其是现在熟料的变化很大以后，不知会有什么样的影响。

6. 对复合效应机理的研究有助于按照工程的需要进行设计，但是混合材的复合效应机理涉及所用混合材的特性。在传统上，将混合材分为火山灰质材料和有潜在水硬性的材料，实际上，同为火山灰质材料，其性质也会有明显的差别，他们的主要作用也会不同，本文中的数据也表明如是的事实。相信复合效应的产生也符合系统论的理论：化学成分通过结构决定强度（唐有祺：《结构化学》）。当前许多对机理的研究只是或者热衷于用某些仪器测试一下，加上几张扫描电镜照片，作一些主观想象的说明，或者按传统说法想当然地附和一下，作为一种点缀。以至于至今对矿物掺和料的作用尤其是复合效应机理仍然不清楚，在生产中的使用多具有盲目性。鉴于混凝土体系的复杂性，不可能有立竿见影的研究结果，需要精心实验和潜心思考，需要传承和积累。

7. 顺便一说，x-射线衍射仪法作为相组成分析是快捷和有效的，现在已不用再去查JCPDS（the joint commitee on powder diffraction standards）的卡片，而由电脑自动判断给出分析结果。其结果是任何人都能用来点缀自己的论文了，但是却忽视了“电脑不如人脑”的道理，近年来杂志上发表的论文中由此而出现的常识性的错误判断并非鲜见。用浮躁的心态是无法进行科学研究的。

张大康评

1. 图 7 显示固硫渣与粉煤灰复合效应的长期效果不完整，少了 180d 的数据。线条不易区分，没显示“超叠效应”，看不出纯硅酸盐水泥是哪条线。纵坐标使用对数标度是否会好些？

2. 如果固硫渣中石膏的溶解速率慢的结论可靠，则在结论中应提及这一特性及其作用。

3. 在本文没有注意到各种混合材料的物理作用是一个遗憾。不同混合材料的粒度分布不同，掺入水泥后对复合水泥的粒度分布影响不同，对水泥强度会产生不同的影响。对于活性较低的混合材料，其粒度分布的影响甚至可以超过其化学活性的影响，特别是对较早龄期的强度。同时，胶凝材料的粒度分布还影响混凝土的密实度、孔结构，从而影响其渗透性。

我们常常会遇到一些重要的科学问题或关键的技术问题，单靠我们自己的努力难以解决，而合作往往会简化难题，最终解决难题。因此你们要学会合作，善于合作。

摘自清华大学化学系教授张希院士 2014 年 9 月 3 日为清华 2014 级研究生新生所作的“开学第一课”

论文之八

对固硫渣中的碱、SO_3和其他微量元素的评价*

王英华　廉慧珍

清华大学土木工程系

摘　要　固硫渣用于生产水泥或商品化用作混凝土掺和料时，需制订标准。本文是“八五”攻关专题852060204专题研究验收报告“制订固硫渣品质评定标准的试验研究”中的一部分。正交试验表明，固硫渣中的碱和SO_3所引起的膨胀相互之间没有交互作用，可考虑为简单叠加，而碱含量和固硫渣掺量之间、SO_3和固硫渣掺量之间却存在交互作用，前者大于后者。在有碱-骨料反应条件存在时，掺入≥20%的固硫渣可抑制因碱-骨料反应产生的膨胀；当SO_3含量为4%时，掺固硫渣40%即可完全抑制；而当SO_3含量为8%时，掺固硫渣40%可使该膨胀率下降约60%。固硫渣中所含的碱可溶性的只占全分析量的十分之一，不会引起碱-骨料反应。由于固硫渣对膨胀的抑制作用，掺固硫渣的水泥中SO_3总量可放宽至4%。对微量元素的分析认为，固硫渣用于混凝土后对渣中有害元素可起到固化、稀释和减少淋溶的作用，应无害。燃煤也需要避免煤炭中有害物质的排放污染环境，这方面的要求和其他燃煤排放物的要求一样。

关键词　固硫渣的无毒无害性　碱-骨料反应　对膨胀的抑制　微量元素

前言

国外对含SO_3的循环床排放灰渣，多进行预处理，使其中的SO_3和CaO尽量消耗掉。一般用于制作压蒸的人造轻骨料、道路垫层、小型砌块等。用于水泥时，德国有允许在硅酸盐水泥中掺入不超过30%固硫渣的专利。我们在“七五”期间的试验研究和试点应用证明，固硫渣中的SO_3是可以利用得很好的。“八五”期间研究将固硫渣商品化用作混凝土的矿物掺和料，这就需要制定产品标准，对固硫渣中SO_3与碱的作用及其对混凝土体积稳定性等提出评价。此外同其他工业废料资源化一样，必须进行其他微量元素的毒害分析和评价。

碱-骨料反应（AAR）是水泥中的碱和骨料中的活性硅酸、硅酸盐或碳酸盐在有水存在的情况下反应，生成破坏性产物。目前美国、加拿大、欧洲、日本等地都有混凝土由碱-骨料破坏的事例，其修补和重建的费用十分昂贵。混凝土工程发生碱-骨料反应导致建筑物膨胀开裂毁坏，已成为一个全球的灾害性的问题。我国目前按国际公认规定水泥中以Na_2O当量计碱含量不大于0.6%，混凝土中碱含量不大于$3kg/m^3$。我国许多地区混凝土用的骨料都含有能与碱发生反应的活性物质（称为碱活性）。如北京大部分石子都有碱活性。有些固硫渣碱含量很低，而有的固硫渣如G5（含碱的Na_2O当量为2.684%）、G6（含碱的Na_2O当量为2.296%）；当掺量为20%、30%时，水泥中的碱含量很有可能超过0.6%。但国内

* 节选自852060204验收文件。

外对硅灰、粉煤灰、矿渣等混合材料作为掺和料对碱-骨料反应的抑制作用已有报道[1]，因此，有必要对固硫渣这种活性较高的混合材料在水泥中对碱-骨料反应的作用进行研究。为商品化固硫渣品质标准的制定提供依据。

一、渣掺量、总 SO_3 含量、总含碱量对膨胀的交互作用研究

目前在我国出现的碱-骨料反应一般属于碱-硅酸反应，碱-碳酸盐反应（ACR）较少。因此，本文仅着重讨论碱-硅酸反应的情况，以下所指的碱-骨料反应均指的是碱-硅酸反应。

以下所说的碱含量均指 Na_2O 当量 $R_2O(Na_2O+0.658K_2O)$。

1. 试验用原材料和方法

（1）碱活性骨料的选择和鉴定

为了提高试验的可靠性，参照目前有关资料[2]，用高纯度的石英玻璃作为高碱活性的人工骨料，其中 SiO_2 含量大于 99.99%，分别用 SD 105—82 化学法和砂浆棒法检验其潜在的碱活性。测定结果证明该骨料是一种有效的活性骨料，可用以进行碱-骨料反应的研究。

将 20%的这种级配的碱活性骨料与 80%的标准石英砂混合，作为本试验的骨料。

（2）水泥

从实验的可比性考虑，本试验所用的水泥全部采用北京首都水泥厂生产的水泥熟料与大红门天然二水石膏按比例混合而成。

（3）采用化学纯 NaOH 作为外加碱，调整水泥的含碱量。在试验时，先将所需的 NaOH 溶于要加入水泥浆中的水中，待水溶液冷却至室温后和物料共同搅拌。

（4）固硫渣

用 G5、G6、G7、L4 渣以及在 L4 渣中掺 400℃大红门烧石膏的 L′4 渣，其 SO_3 含量见表 1。

表 1　试验用固硫渣的碱和 SO_3 含量

固硫渣名称	G5	G6	G7	L4	L′4
来源	宜昌印染厂	宜昌印染厂	宜昌造纸厂	柳州日化厂	同左，外掺 SO_3
SO_3 含量（%）	4.21	7.13	11.81	3.95	8%
Na_2O 当量（%）	2.68	2.30	1.32	2.84	4%

（5）试验方法

试验依据水利电力部《水工混凝土试验规程》SD 105—82 进行，同时参考美国材料试验协会 ASTM C221、ASTM C227 和 ASTM C289 标准。

水泥∶砂浆的质量=1∶2.50，每组三个试件共用水泥 400g、砂 900g；砂浆用水量按规程中水泥浆流动度方法选定。固硫渣的掺加方法采用内掺法。

按上述标准制作的试件标准养护后测量其各龄期的线性的自由膨胀率。

2. 试验方案的设计及结果分析

由于涉及 SO_3 的作用、碱的作用及渣的作用，研究起来比较复杂，现采取化学模拟的方法，定性地研究固硫渣在这一体系中的作用。

由于SO_3也会引起一部分膨胀，所测定的膨胀既包括SO_3引起的膨胀，又包括由碱-骨料反应引起的膨胀，即用总膨胀来研究碱-骨料反应带来的影响是不合适的。因此首先要研究这两种作用的关系。

用有交互作用的正交试验法，采用正交表中L8（2^7）方案进行试验研究，见表2。

表2　因素水平表

因素 / 水平	总SO_3量 A	总含碱量 B	固硫渣掺量 C
1	2.188	1.34	15
2	3.162	1.60	30

由直观分析和方差分析可知，无论哪个龄期，固硫渣掺量的作用都是非常显著的，而且无论哪个龄期，SO_3总含量×总含碱量对膨胀率都无影响。即SO_3与碱引起的膨胀作用之间无交互作用，可以说是互不干扰的，因此总膨胀率可看成这两个作用之和。

在大多数龄期，影响膨胀率的顺序为：固硫渣掺量（以下简称掺量）＞总SO_3含量＞总含碱量×掺量＞其他。由此可知，固硫渣掺量对膨胀有非常显著的抑制作用；总SO_3含量对前期、后期都有显著的膨胀作用；总含碱量×掺量对膨胀亦有一定的作用，而总SO_3含量×掺量只有龄期为90天时作用显著，即碱含量与掺量的交互作用及SO_3含量与掺量的交互作用都对膨胀有一定的作用。

180天时，总含碱量×掺量的作用远大于总SO_3含量×掺量的作用，即碱含量与掺量的交互作用大于总SO_3含量与掺量的交互作用，说明长龄期时，掺量对由碱-骨料反应引起膨胀的抑制作用要大于对由SO_3引起膨胀的抑制作用。

3. 固硫渣掺量及渣中的SO_3对碱-骨料反应的影响

考虑掺量及渣中不同SO_3含量的影响，试验方案如表3所示，实验结果见图1和图2。

表3　固硫渣对碱-骨料反应影响规律的实验方案

固硫渣名称	固硫渣掺量（%）	活性骨料用量（%）	总碱量（%）	石膏掺量（%）
L4（SO_3=4%）	0，10，20，30，40	20	2.2	4
L4（SO_3=4%）	10，20，30，40	0	2.2	4
L′4（SO_3=8%）	0，10，20，30，40	20	2.2	4
L′4（SO_3=8%）	10，20，30，40	0	2.2	4

从图1和图2可看出，在各龄期都出现相同的规律，固硫渣掺量≥20%时，随着渣掺量的增加，总膨胀率明显下降。渣中SO_3含量小的渣（L渣，SO_2=4%）对总膨胀的抑制作用更加显著。渣中含SO_3为4%的L渣掺量为40%时，可以完全抑制总膨胀；而当渣中含SO_3为8%时（图2），在掺量为40%时仍有一部分膨胀。这是由于随着掺量的增加总SO_3含量也增加较大，这时固硫渣已经不能完全抑制由SO_3与碱引起的总膨胀。因此当固硫渣中SO_3含量较大时，在有碱-骨料发生的必要而充分条件时，应控制固硫渣的掺量。

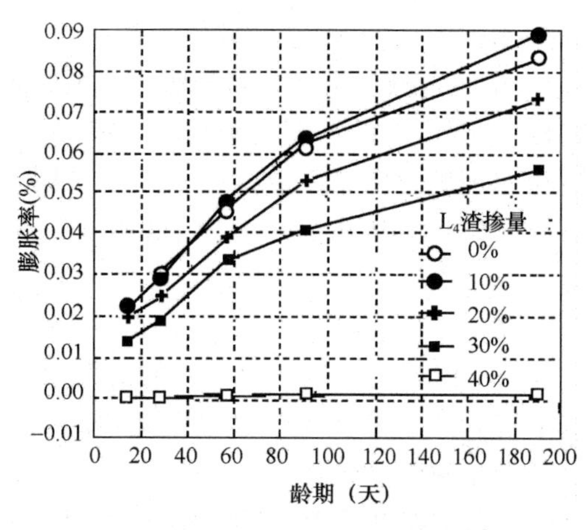

图 1　L4 渣不同龄期渣掺量与总膨胀率关系

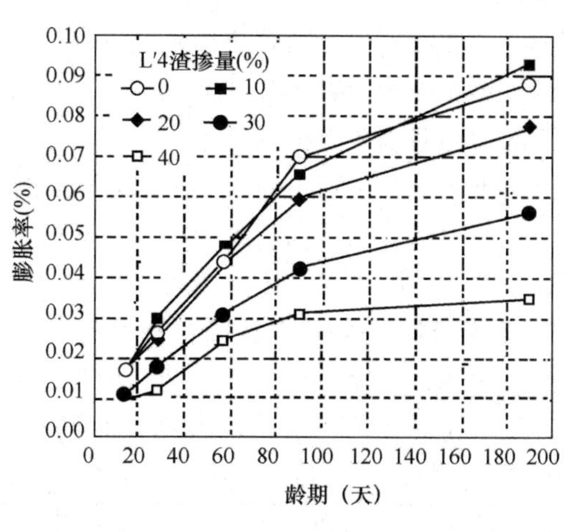

图 2　L′4 渣不同龄期渣掺量与总膨胀率关系

4. 固硫渣对碱-骨料反应膨胀性的抑制作用

采用含碱量较高的固硫渣来验证其在碱-骨料反应中的无害作用，试验方案见表 4，试验结果如图 3、图 4。

由于如上所述，SO_3和碱引起的膨胀没有交互作用，在试验结果中，可以对由SO_3引起的膨胀与碱引起的膨胀分开来考虑，则从总膨胀率中扣除由SO_3引起的膨胀率，即可得出由碱作用引起的膨胀率。图 3、图 4 是从总膨胀率中扣除由渣中SO_3引起的膨胀率而得到的。结果表明，渣掺量在大于 20%时，固硫渣掺量的增加可以抑制碱-骨料反应。图 3 表明，当渣中含SO_3为 4%时，在渣掺量为 10%时出现一个最大值，即这时对碱-骨料反应的抑制效果不佳。从总的情况来看，固硫渣可抑制碱-骨料反应，并且掺量越大抑制作用越明显。

表 4　验证固硫渣对碱-骨料反作用的试验方案

固硫渣名称	固硫渣掺量（%）	活性骨料含量（%）	总碱量（%）
G5	0，30，40	20	1.32，1.75，1.89
G6	30，40	20	1.66，1.75
G7	30，40	20	1.14，1.05

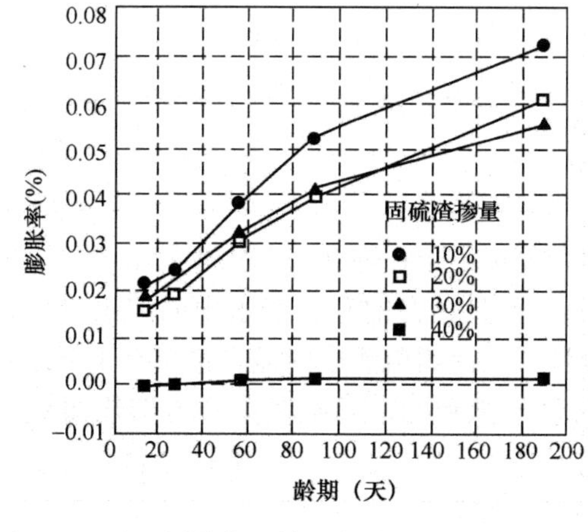

图 3　L4 渣不同龄期渣掺量与碱引起的膨胀率关系

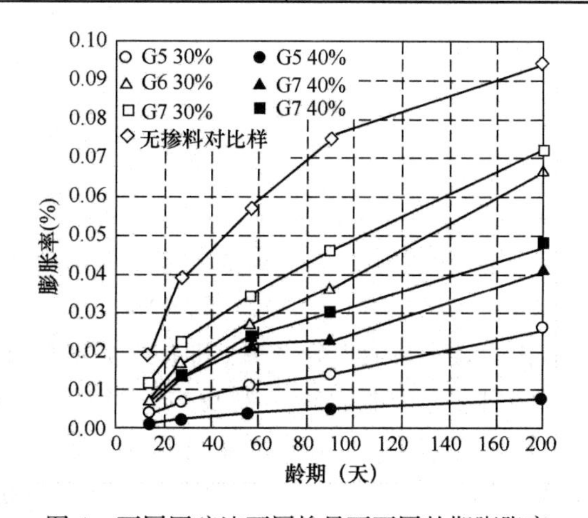

图 4　不同固硫渣不同掺量下不同龄期膨胀率

由图4亦可看出，随着渣掺量的增加，G5渣（水泥中总SO_3量为2.2%，掺量为30%、40%时，水泥中的含碱量分别为1.75%、1.89%），虽然含碱量增加，但膨胀也降低，说明渣掺量的增加对膨胀有抑制作用。在相同的掺量下，掺G6渣的水泥含碱量要比掺G5渣的低，但由于SO_3含量高于掺G5渣水泥的，掺G6渣水泥的总膨胀率要高于G5的；从掺G7渣的水泥中也能看出同样的规律。可见固硫渣引起的膨胀作用主要来自于固硫渣中的SO_3，而不是由碱-骨料反应引起的。

5. 固硫渣抑制碱-骨料反应的机理分析

如前所述，固硫渣是一种特殊的活性混合材料，不仅可与水泥水化生成的$Ca(OH)_2$反应，生成水化硅酸钙和水化铝酸钙，而且由于含有无水石膏，还可生成水化硫铝酸钙，因此，对碱-骨料反应有抑制作用，并且掺量越大这种抑制作用越显著。

固硫渣中的碱含量虽然不少，但在碱-骨料反应中应该是可溶出的部分起作用，因此有必要研究渣中的碱形态，测定渣中可溶碱含量。试验方案如表5所示，结果如表6所示。

表5　固硫渣中碱含量测定方案

固硫渣名称	溶液种类	浸泡时间
G5	蒸馏水	20天
	饱和石灰水	20天
G6	蒸馏水	20天
	饱和石灰水	20天
G7	蒸馏水	20天
	饱和石灰水	20天

表6　固硫渣中Na_2O、K_2O的溶出量

固硫渣名称		G5	G6	G7
全分析的含量（%）	K_2O	3.35	2.79	0.77
	Na_2O	0.48	0.46	0.04
饱和$Ca(OH)_2$溶液中的溶出量（%）	K_2O	0.389	0.351	0.201
	Na_2O	0.36	0.031	0.012
水溶液中的溶出量（%）	K_2O	0.100	0.060	0.060
	Na_2O	0.004	0.002	0.001

由结果可以看出，固硫渣中的碱在水中的溶出量要远小于在饱和石灰水中的溶出量，这可能是石灰水中的Ca^{2+}可以置换Na^+和K^+的缘故。而且，在饱和石灰水中的溶出量只是全分析中碱当量的十分之一。即在水泥水化反应过程中，使固硫渣中的碱含量只有十分之一可以溶出参加可能的碱-骨料反应。当$Ca(OH)_2$含量不饱和时，固硫渣中的碱溶出量就更少。因此可以认为，在水泥熟料碱含量适当的情况下，掺入固硫渣不会引起碱含量超标及导致碱-骨料反应而破坏。而且当水泥中碱含量超标时，固硫渣还可抑制碱-骨料的反应所产生的膨胀。

二、固硫渣中 SO_3 对水泥性能的影响

1. 缓凝作用验证

试验采用鞍山无硫渣加入 400℃大红门烧石膏，配成含 SO_3 分别为 0、4%、8%、12%的固硫渣 A0、A1、A3、A5。熟料为琉璃河硅酸盐水泥熟料，采用试样配比及凝结时间测定结果如表 7 所示。试验结果表明，凝结时间均合格，验证了无水石膏掺入水泥中（以 SO_3 计）为 2.2%左右时，可以代替二水石膏起到很好的缓凝作用。由于二水石膏溶解速率比无水石膏快，当无水石膏中 SO_3<2.2%时，可加入二水石膏增加溶液中的硫酸根离子的量，起到缓凝作用。因此，固硫渣中的 SO_3 在 7%～12%而掺量为 30%时，可起到缓凝作用；固硫渣中的 SO_3 在 3%～7%时，可以加入适量的二水石膏调节凝结时间。

表 7　凝结时间的测定

序号	编号	试样配合比（%）			渣中 SO_3 含量（%）	标准稠度用水量（%）	凝结时间（h：min）	
		熟料	石膏*	固硫渣			初凝	终凝
1	K	95	5	0	0	28.0	0：45	1：26
2	A0	66	4	30	0.3	28.5	2：25	3：41
3	A1	68	—	30	4.0	27.0	2：55	3：38
4	A3	70	—	30	8.0	26.5	3：08	3：35
5	A5	70	—	30	12.0	29.0	1：08	2：50

* 石膏中 SO_3 含量为 43%。

2. 体积稳定性

采用固硫渣 L4（SO_3 含量 4%）、L′4（SO_3 含量 8%），在不同掺量下，测定不同龄期水泥砂浆自由线膨胀率，结果见图 5。

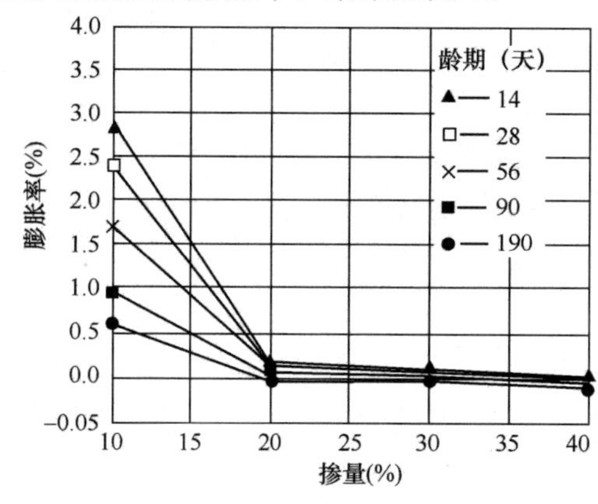

图 5　固硫渣掺量对水泥砂浆在不同龄期的膨胀率的影响

图 5 表明，低硫固硫渣，随着掺量的增加，可明显抑制膨胀，即使掺量增加带来 SO_3 含量增加，膨胀率仍在下降，而且 190 天膨胀率在掺量大于 20%时，与 14 天的相比增加并不大。可见，SO_3 在限定的指标内，固硫渣掺量增加对膨胀有明显抑制作用。因此，掺有固硫渣的水泥，在固硫渣掺量较大的情况下，水泥中的 SO_3 量可放宽至 4%。

三、对固硫渣中微量有害物质的分析

与其他燃煤方式所排副产品如粉煤灰一样，固硫渣也是由燃煤产生，有可能含有与粉煤灰中相似的微量有害物质，大致分为四类：①放射性物质；②重金属微量元素和非金属有毒元素；③毒性有机物质；④导致矽肺病的物质。

1. 关于放射性物质

燃煤废渣（或灰）中放射性物质来自原煤。根据清华大学核能技术研究所[3]，国家建筑材料测试中心，辽宁省环境监测中心站等单位发表的资料表明，我国粉煤灰中天然放射性水

平接近标准限值。由于在燃煤过程中，加入15%～25%以氧化钙及碳酸钙为主的脱硫剂，使灰渣排放量比原来增加约20%，这样，如果原煤中含有放射性物质，则对固硫渣中的放射性核素的浓度起一个稀释的作用。从而与粉煤灰相比，相同煤种的固硫渣中天然放射性水平要低得多。即使在100%使用固硫渣的情况下，其放射性也符合标准。实际上，固硫渣在水泥中的掺加量最多为40%，则在混凝土中只有2%～5%，因此，一般来说是安全的。但为安全起见，当出现煤中含有的放射性物质超标时，对其固硫渣，在利用前应进行放射性检测。实际上，对其他燃煤方式所排废渣（例如煤粉炉的粉煤灰）的利用也如此。

2. 其他微量有害物质

（1）重金属微量元素和非金属有毒元素

固硫渣中除含有硅、铝、铁、钙、镁、硫等主要元素以外，炉底灰中存在的微量元素有硼、铈、镧、铷、钐、钍、钡、钴、铪、钪、锶、钒、钛、铕、锰、钽、铬、镍、钇、镱、锆、铀等。这些微量元素由于煤种、矿藏等不同，含量变化很大。

人们曾对粉煤灰有害微量元素进行研究，指出有害微量元素基本与土壤本底含量接近，认为是在安全范围以内。但用于混凝土时，即使含有有毒元素，由于混凝土的固化作用使其淋溶性降低，对环境保护也没有害处。同理，由于固硫渣制水泥必将用于混凝土中，利用混凝土的固化作用，使有毒微量元素淋溶性降低。而且固硫渣中的有害微量元素亦应与土壤本底含量相近，甚至低一些，对环境保护也无害处，相反从利用固硫渣的角度来说，对环境保护还有益处。

（2）毒性有机物质

煤炭中的有机物质一般在燃烧过程中都会挥发掉，但同时也会生成另一些有机物，如3、4苯并芘等毒性物质。这类物质在粉煤灰中含量极微，且3、4苯并芘在水中的溶解度约0.0025%，固硫渣分散于混凝土中，对这类物质也能起固化和降低淋溶性的作用。这样不使固硫渣堆积起来，而加以分散固化，实际上是对这些微量毒性物质最有效的处理方法。

（3）导致矽肺病的物质

荷兰所进行的粉煤灰对人体健康影响的研究中，提到粉煤灰是否导致矽肺病的结论是：即使粉煤灰中存中一些α石英，也不会导致矽肺病。由于固硫渣颗粒较大、粉尘很小，也不会导致矽肺病。但在应用固硫渣的工厂中，仍严格执行除尘要求[3]。

参考文献

[1] T. E. Stamtpm. Expansion of Concrete Through Reaction Between Cement and Aggregate, Proc. ASCE，1940，1：66.

[2] 郑红卫. 抑制碱-集料反应矿物质掺合料的试验研究，清华大学硕士论文，1993.

[3] 清华大学核能技术研究所. 我国61个电厂燃煤及其灰分中的放射性水平和环境辐射影响的评价，1986.

自评

1. 固硫渣用于生产水泥的混合材或商品化用作混凝土掺和料时，需制定产品标准。产品标准的功能是出厂时进行合格性检验，其功能有三：①检验是否达到所承诺的指标；②用该指标检验产品的匀质性；③是否无毒无害。因此开发任何一种产品时，都要这样做，不能只满足强度要求。因此该项实验研究是有意义的，但是对工程普适性的应用需谨慎对待，还需有重演性试验。

2. 碱-骨料反应的试验是强制性的快速方法，实际工程中不存在这样的条件，实验的结果只说明所用骨料是否具有碱活性，或者证明在使用活性骨料时，在有充足的水供应的条件下，所用水泥（或胶凝材料）是否有发生碱-骨料反应的可能性。这里所说的充足的水供应指的是供应到胶凝材料浆体与骨料的界面区。标准试验所用固定 0.5 的水胶比条件和实际工程也有区别。对水胶比很低的高强混凝土来说，硬化后，即使环境多水，在正常情况下，外部水也无法进入，而内部水又少得会产生自收缩，也就无法提供碱-骨料反应所需要的水。现在很多人对碱-骨料反应谈虎色变是没有必要的。当然，对中低强度的混凝土，在碱、活性骨料、水三者具备的必要而充分的条件下，仍不可掉以轻心。

3. 研究表明，固硫渣中的Ⅱ型无水硫酸钙，在用于生产混合材硅酸盐水泥时，可以减小水泥的干燥收缩，用于制作膨胀剂或膨胀水泥时，可以减小后期膨胀能的落差。在本文中又证明了固硫渣中的 SO_3 的无害性。这表明固硫渣不同的使用方法可以发挥其不同的功效，使用得当时有利，使用不当时有害。要具体问题具体分析，且不可一概而论地简单化对待。循环流化床对燃用高硫煤发电具有节能、降耗、减排的优势，今后大有发展前途。必须运用辩证法的思维方法指导提高燃煤脱硫及其所排灰渣利用的技术。

4. 用正交试验研究 SO_3、碱和固硫渣交互作用的试验只提供出正交设计的因素水平表，而缺少实验方案和实验结果，使读者无法分析。

5. 在对固硫渣中 SO_3 对水泥性能影响的研究中，为了做对比，使用了鞍山的无硫渣。当时因考虑将固硫渣用作水泥混合材时的最大掺量为 40%，故定义当渣中 SO_3 含量不大于 2% 时按无硫渣使用。当时的鞍山渣 SO_3 含量 1%，配制水泥渣掺量 30%时，计算所掺 30% 固硫渣中的 SO_3 含量就成为 0.3%，是可忽略的。此外，对纯硅酸盐水泥的调凝作用，SO_3 含量和凝结时间的关系是有极值的，当 SO_3 含量超过 3.5%以后，凝结时间就会较快缩短，而对掺混合材的水泥则可放宽。现行标准允许对矿渣水泥放宽到 4%。从本文表 7 来看，编号 A3 水泥的 SO_3 为 2.4%，而 A5 的 SO_3 为 3.6%，凝结时间比 A3 快的原因可能在于此。这里面还有个石膏品种问题。水泥中 SO_3 总量和固硫渣中的无水石膏和外掺的二水石膏的比例有匹配的问题。无水石膏含量越大，调到相同凝结时间所需 SO_3 的总量越大（非正比关系）。以上是 29 年前研究的结果，而今各种原材料，尤其是石膏比那时要复杂得多，现在固硫渣的应用还需要进一步做很多的实验研究。

张大康评

当混凝土工程与食品、食品原料和饮用水接触时，应检测并慎重限定重金属及有毒物质含量和溶出量。

论文之九

火山灰质材料活性的快速评定方法*

廉慧珍[1] 张志龄[2] 王英华[3]

摘　要　用加热回流的方法测定不同火山灰材料的活性率 K_a（在饱和石灰水中反应的 SiO_2 和 Al_2O_3 总量占该材料初始全 SiO_2 和 Al_2O_3 总量的百分比），表明 K_a 与标准胶砂强度比较好的相关关系。对于相同材料，当细度变化时，K_a 率的变化和胶砂强度比的变化有相同的规律；改变含硫的循环流化床废渣中 SO_3 含量，对 K_a 和胶砂强度比的影响规律一致。验证试验表明活性率 K_a 能反映其他因素对材料活性的影响。经闭合实验及对影响测试结果因素的正交实验，表明活性率测定方法可靠，测定时间只需几个小时。

关键词　火山灰材料　活性率　快速评定　胶砂强度比

前言

火山灰质掺和料是指主要含有无定形 SiO_2（如硅灰、硅藻土）或同时含有无定形 SiO_2 和 Al_2O_3（如火山碎屑、凝灰岩、粉煤灰等），而不含水硬性矿物组成，但在常温下加水能与石灰反应生成水硬性产物的物质。2000 多年前古罗马用火山灰和石灰混合制作的混凝土建筑物（如罗马万神殿）至今仍完好地存在。经验和理论均证明火山灰-石灰-水的硬化体具有很好的耐久性，火山灰质掺和料能改善混凝土的性质，提高混凝土的耐久性。但直到 20 世纪 20 年代人们才开始寻找评价火山灰材料的简单、快速、定量的方法。所研究、试用或采用的方法有数十种之多。其中有的是根据火山灰反应的机理的化学方法，如维卡法（石灰吸收值法）、火山灰性实验法（检测掺用火山灰的水泥混浊液中石灰和碱量的变化）、酸碱溶出法等；有的是测定火山灰在强酸中溶解时的放热速率或电导率变化的物理方法。实际上火山灰质材料在水泥中的作用不止是化学的，还与火山灰质材料的结构形态有关。因此上述各种方法都不能很好地反映火山灰在水泥中的行为。传统的被认为可靠而应用最多的方法是胶砂强度试验法，但是试验周期需要 28 天。袁润章提出 WYZ 法，即以特定条件下单位比表面积上可溶硅的含量评价粉煤灰的质量[1]，沈旦申借用灰色系统理论，以细度、密度、色度三个本征参量，建立“脉搏系统”以评定和控制粉煤灰的质量[2]，谷章昭、乐美龙等将掺粉煤灰水泥的强度与其各个影响因素之间的关系建立起回归方程，以快速评定粉煤灰的活性[3]。这些方法都在一定程度上考虑了化学成分以外的因素，但是检测方法烦琐、费时，并且难以定量。埃及 Kaissar M. Hanna 和 Aly Afify 将 1∶1 的火山灰-石灰混合物和 61%熟料、4%石膏、35%火山灰材料制作的砂浆在 25℃±1℃下养护 90 天，测定前者在冷盐酸中的 SiO_2、Al_2O_3、Fe_2O_3 溶出量与后者的抗压强度，发现有以下线性关系[4]：

---

* 原载上海同济大学《建筑材料学报》2001.3。

1　廉慧珍 女 教授　清华大学土木工程系 100084
2　张志龄 男 工程师 清华大学土木工程系 100084
3　王英华 女 硕士生 清华大学土木工程系 100084

$$S = 56.0 \times \frac{溶解的(SiO_2 + Al_2O_3 + Fe_2O_3)}{初始(SiO_2 + Al_2O_3 + Fe_2O_3)} + 29.0$$

式中 S 为掺 35%火山灰的水泥砂浆 90 天抗压强度（MPa），分子为火山灰-石灰混合物中 90 天的可溶氧化物质量百分数，分母为火山灰中初始的相同氧化物质量百分数。火山灰中 SiO_2、Al_2O_3、Fe_2O 在一定养护龄期内与石灰反应的那部分与未反应部分的比例，实际上可表征火山灰的结构形态，该比例与砂浆强度的线性相关性很好，但实验周期太长，仍无法快速评定火山灰的质量。

清华大学张淑清根据相似的原理，采取回流沸煮法，加速火山灰材料中活性组分在饱和石灰水中的反应。测定反应后可溶于稀盐酸的 SiO_2 和 Al_2O_3，分别称作可溶 SiO_2 和可溶 Al_2O_3（该方法未公开发表）。但在本研究的大量实验中，无论是可溶 SiO_2、可溶 Al_2O_3，还是二者之和，都与所测火山灰材料的胶砂强度没有相关性。本研究在张淑清方法的基础上，提出火山灰材料活性率 K_a 的概念：

$$K_a = \frac{可溶(SiO_2 + Al_2O_3)}{全(SiO_2 + Al_2O_3)} \times 100\%$$

实验表明，K_a 不仅能反映火山灰质材料的化学反应性，而且与材料的其他组分和细度、结晶物质含量等物理状态有关，用活性率 K_a 可在 8 小时之内快速评价火山灰材料的质量。

1　实验原理

（1）火山灰材料的化学反应性来源于无定形的硅、铝、铁等氧化物。其中活性的 SiO_2、Al_2O_3、Fe_2O_3 与 CaO 反应，生成水化硅酸盐、铝酸盐、铁酸盐。水化物生成量和环境温度有关，和 CaO 含量有关。采用在饱和石灰水中沸煮（回流冷凝以保证溶液的浓度），既加速反应，又可使反应充分。

（2）水化硅酸盐、铝酸盐、铁酸盐可溶于稀盐酸，而未反应的部分是不溶的[5]。故可用稀盐酸将反应产物(可溶)和未反应组分(不溶)分离，并分别检测，则可测得可溶的 SiO_2、Al_2O_3、Fe_2O_3 的数量。亦即活性的 SiO_2、Al_2O_3、Fe_2O_3。

（3）化学组成通过结构决定物质的性质[6]。对于火山灰材料的质量，活性 SiO_2、Al_2O_3、Fe_2O_3 个值是没有意义的，比如活性 SiO_2、Al_2O_3、Fe_2O_3 高的，可能非活性（不溶）部分更高，其间比例合适时才会在水泥中发挥最好的作用[7]。

（4）一般 Fe_2O_3 含量和活性 Fe_2O_3 都很少，计算材料活性率时可忽略。

（5）材料中的 SiO_2 和 Al_2O_3 活性（反应，溶于稀盐酸）部分与非活性（未反应，不溶于稀盐酸）部分数量之和应等于其全量（原样初始成分全分析中的 SiO_2、Al_2O_3 全量）。

2　实验方法

2.1　活性率的测定

将磨细至与水泥等同细度的试样在 105～110℃下烘干后，称取 0.5g，置于 250mL 三角瓶中，注入饱和石灰水溶液 200mL，用回流冷凝的方法沸煮 2 小时后，加入 8mL 浓盐酸，中和剩余 CaO 后，形成稀盐酸，用于溶解反应的 SiO_2、Al_2O_3。用蒸馏水洗净回流内壁，再沸煮 5 分钟，冷却后，过滤、定容到 250mL 的容量瓶中，即待测溶液。溶液中的 SiO_2 用重量法测定，Al_2O_3 用 EDTA 容量法测定。测定结果，所得可溶 SiO_2、Al_2O_3 即活性的 SiO_2、Al_2O_3；测定原试样中初始成分，得到 SiO_2、Al_2O_3 全量。则试样的活性率为：

$$K_a = \frac{可溶(SiO_2 + Al_2O_3)}{全(SiO_2 + Al_2O_3)} \times 100\%$$

2.2 闭合实验

用取自宜昌的流化床锅炉燃煤脱硫的固硫渣 G5、GY 和取自鞍山的普通沸腾炉渣 A，进行全分析，得到全 SiO_2 和 Al_2O_3，按上述方法测定其中活性 SiO_2 和 Al_2O_3，按试样全分析方法，测定上述溶液在滤纸上残留物中的 SiO_2 和 Al_2O_3，即未反应（非活性）的 SiO_2 和 Al_2O_3。则应有：

活性（$SiO_2+Al_2O_3$）＋非活性（$SiO_2+Al_2O_3$）＝全（$SiO_2+Al_2O_3$）

测试结果如表 1 所示，实验计算与实测全量比较见表 2。

表 1　Results of test on closed loop（闭合实验测试结果）

No.	Active portion（%）		Inactive portion（%）		Total（%）	
	SiO_2	Al_2O_3	SiO_2	Al_2O_3	SiO_2	Al_2O_3
G5	9.85	4.66	39.10	13.71	48.85	18.37
A	3.48	2.31	56.43	20.75	60.10	23.11
GY	3.39	3.39	37.77	14.85	48.11	18.26

表 2　Computation of SiO_2＋Al_2O_3 from fig. 1 for closed loop（闭合实验计算校核结果）

No.	Active portion（%）	Inactive portion（%）	Total from calculation（%）	Total from test（%）	Difference（%）
G5	14.51	52.81	67.32	67.22	＋0.10
A	5.79	77.18	82.97	83.21	－0.24
GY	13.83	52.62	66.45	66.37	＋0.08

由表 1、表 2 可见，所测结果闭合得很好，该方法是可靠的。

2.3 实验条件中影响测试结果的因素分析

由实验操作、试剂和蒸馏水的纯度以及仪器等造成的系统误差是可以校正的，在这里不予讨论。需要了解的是：在条件相同的情况下，待测液制备过程中的加热时间、饱和石灰水用量（相对于一定量的试样）、浓盐酸加入量等因素对检测结果是否有影响。以该三个因素按 L_9（3^4）正交设计进行实验，检验其对可溶 Al_2O_3 测定值影响程度。正交实验因素水平如表 3 所示。

表 3　正交设计试验条件因素和水平

factors / level	Time of heat（h） A	Amt of thick hydrochloric acid.（mL） B	Amt of saturated $Ca(OH)_2$（mL） C
1	1.5	8	150
2	2.0	10	175
3	2.5	12	200

实验过程及直观分析、方差分析部分从略，这里只介绍分析结果。分析结果表明，加热时间不同时，可溶 Al_2O_3 测定值变化的均方差（0.18）接近临界误差值（0.185）；饱和石灰水用量（均方差为 0.06）和浓盐酸加入量（均方差为 0.31）的影响则不显著（其临界误差

值分别为 0.08 和 0.985）。根据计算，在本实验条件下，浓盐酸加入量 8mL 相对于剩余石灰来说已大大过量，中和剩余 CaO 后，形成稀盐酸，已可基本上溶解反应的 SiO_2、Al_2O_3。故实验条件取饱和石灰水 200mL、浓盐酸加入量 8mL、加热时间 2 小时是合适的。

3 用活性率对火山灰材料质量的评定

3.1 不同火山灰材料在水泥中的掺量与其胶砂强度的关系

使用北京琉璃河水泥厂普通硅酸盐水泥，其熟料化学成分（%）为：SiO_2 19.29，Al_2O_3 5.40，Fe_2O_3 4.83，CaO 58.92，SO_3 1.33，LOI 6.43。分别将 5 种火山灰材料和磨细石英砂，以不同掺量等量取代水泥，测定其活性率和 28 天标准胶砂强度比。除硅灰和磨细石英砂外，所有掺和料都磨细到与水泥同等的细度。用震动磨磨细石英砂比表面积约 10000cm²/g。实验结果如图 1 所示。

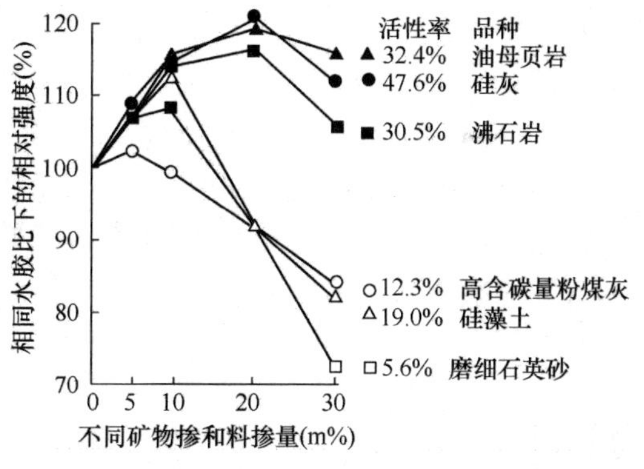

图 1 不同活性率火山灰质矿物掺和料的掺量和砂浆相对强度的关系

由图 1 可见，各种掺和料都有使水泥强度最高的掺量，该强度值与掺和料的活性率有关。将上述掺量为 30%的 28 天水泥胶砂强度与掺和料活性率的关系作图，如图 2，可见火山灰质材料活性率与标准胶砂强度之间的相关关系。

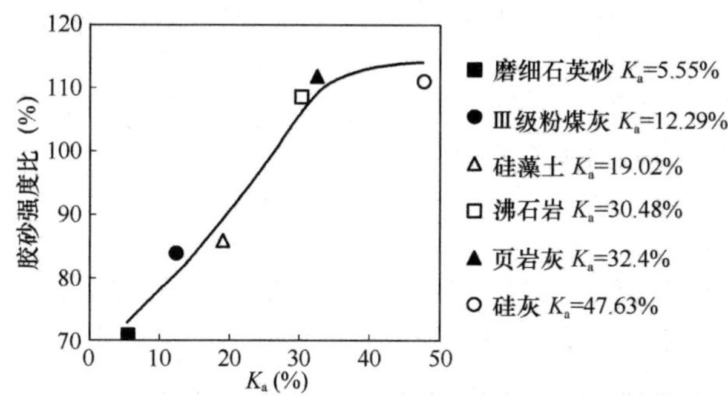

图 2 火山灰质材料活性率和标准胶砂强度比的关系

由图 2 可见，当活性率 K_a＜30%时，标准胶砂强度与活性率的关系近似于线性，活性率 K_a 接近 30%时，曲线出现拐点，活性率 K_a＞30%以后，标准胶砂强度比随活性率的增大而提高缓慢。这可能是因为当材料中的活性组分达到一定数量后，水泥水化所提供的 CaO 相对量变少的缘故。K_a＞20%时，胶砂强度比＞90%，故可认为，火山灰质材料 K_a＞20%的为高活性；当 K_a＞10%时，胶砂强度比＜80%，故可认为，K_a＝10%～20%的，火山灰质材料为中等活性；而 K_a＜10%的，胶砂强度比＜75%，简单取代用做水泥掺和料时，胶砂强度下降过多，为低活性。

3.2 固硫渣活性的评定

将钙质脱硫剂掺入含硫煤中，在流化床的温度（850～950℃）和氧化气氛下燃烧，燃煤释放的 SO_2 被脱硫剂吸收，生成 $CaSO_4$，被固定在废渣中。同时，煤中的黏土质矿物分解成无定形 SiO_2、Al_2O_3、Fe_2O_3，具有较高火山灰活性。这种含硫废渣称为固硫渣。在对固硫渣的利用进行基础研究中，对不同来源固硫渣的活性及其评价方法进行了实验研究。实验结果表明，固硫渣的活性率与固硫渣的标准胶砂强度比之间有与图 2 极相似的相关关系，如图 3所示。

3.3 活性率与材料成分和结构的关系

3.3.1 其他成分对活性率的影响

以固硫渣为例，不同来源的固硫渣中含有不同数量的$CaSO_4$，对固硫渣的活性有影响，反映在对胶砂强度比的影响[9]。活性率作为评定火山灰质材料的指标，也应能反映这种影响。用无水硫酸钙调节取自柳州的固硫渣中SO_3的含量，检测SO_3含量不同的固硫渣胶砂强度比和活性率。结果示于图4。

由图4可见，活性率-胶砂强度比的关系与$K_a<25\%$时K_a与胶砂强度比的关系（图2、图3）一致，而SO_3-胶砂强度比和SO_3-活性率的关系有极相似的规律。因此认为活性率可反映SO_3的含量对固硫渣活性的影响。

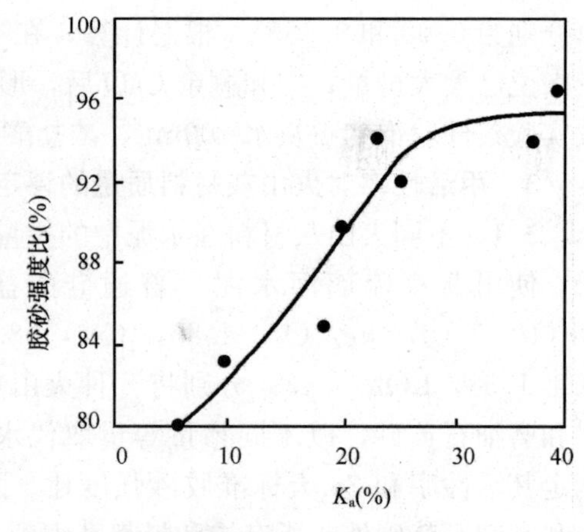

图3 固硫渣的活性率与标准胶砂强度比的关系

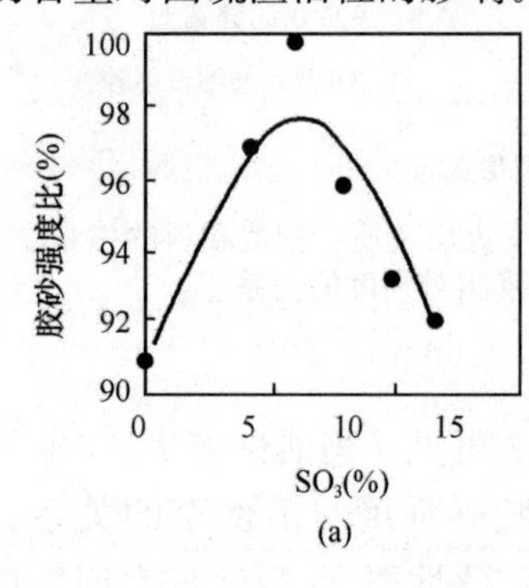

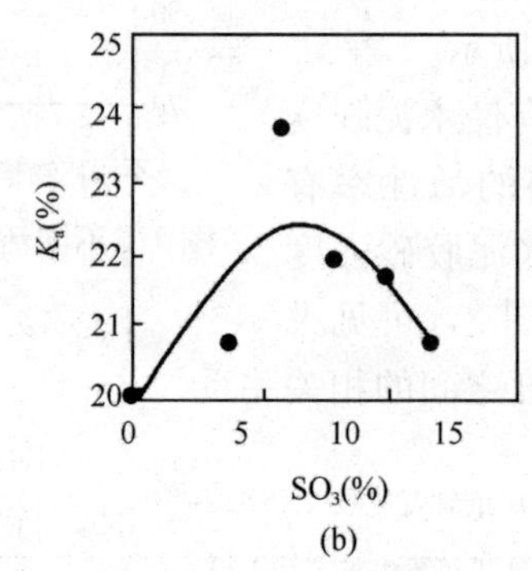

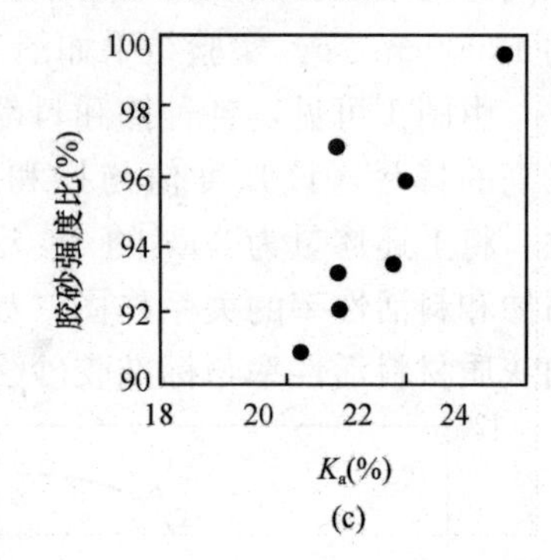

图4 固硫渣中的SO_3对活性率的影响

3.3.2 细度与活性率的关系

以固硫渣为例，考查细度的影响。将柳州含相同数量SO_3的固硫渣按三种细度分别磨细，检测渣的胶砂强度比和活性率。结果表明，比表面积对胶砂强度比和活性率的影响是一致的[图5的(a)、(b)]，活性率-胶砂强度比的关系符合$K_a<25\%$时K_a与胶砂强度比的关系。因此可认为活性率可以反映活性随细度的变化[图5的(c)]。

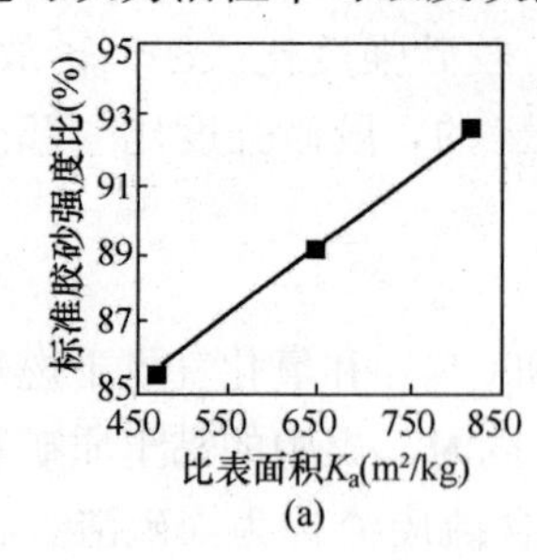

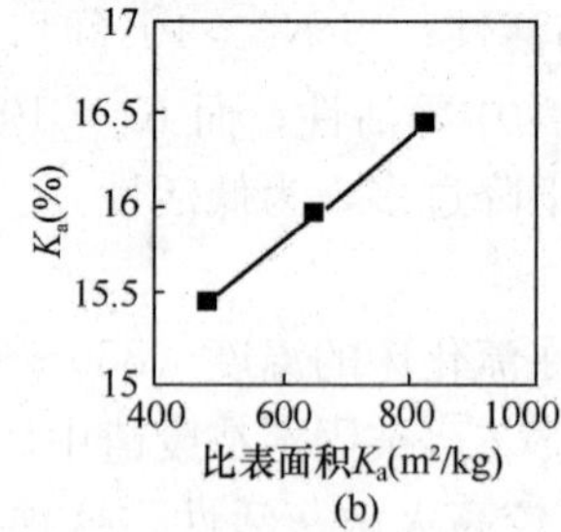

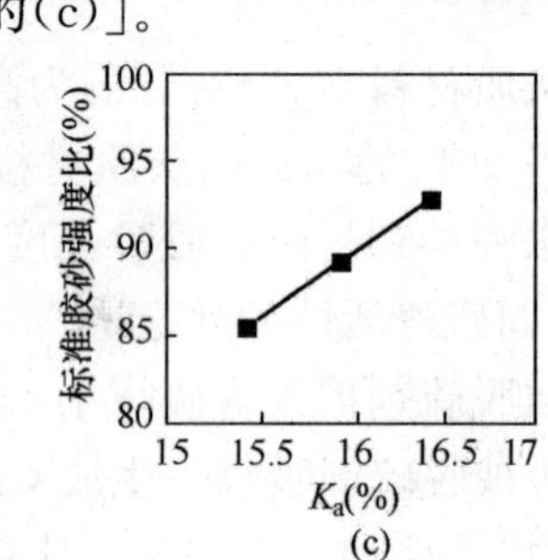

图5 固硫渣的细度对K_a的影响

结论与说明

1. 火山灰材料的活性率K_a与火山灰材料的标准胶砂强度比有较稳定的相关关系，可用以快速评价火山灰材料的质量：

$$K_a = \frac{可溶(SiO_2 + Al_2O_3)}{全(SiO_2 + Al_2O_3)} \times 100\%$$

$K_a>20\%$的为高活性，$K_a=10\%\sim20\%$的为中等活性，$K_a<10\%$的为低活性。

2. 火山灰材料的活性率$K_a<30\%$时，活性率与标准胶砂强度比呈线性相关；$K_a>30\%$时，胶砂强度比随活性率的增加而增大的幅度减小。

3. 影响火山灰材料质量的其他因素（如SO_3含量、细度等）对胶砂强度比的影响与对活性率的影响有一致的规律，因此活性率可以反映化学组分与结构对火山灰材料质量的影响。

4. 本实验中检测胶砂强度比所用水泥为北京琉璃河水泥厂525号普通硅酸盐水泥。当所用水泥不同时，由于所提供CaO量不同，K_a-胶砂强度比关系曲线上的拐点位置可能有所变化。因此当在用活性率比较数种火山灰材料质量时，应当使用相同的水泥。标准K_a-胶砂强度比关系曲线应当使用标准水泥进行标定。

参考文献

[1] 袁润章，朱颉安，章丽云．评定粉煤灰的火山灰活性方法的研究[J]．武汉建材学院学报，1982，2：169～176.

[2] 沈旦申，瞿秋云．粉煤灰物理序参量的系统化[J]．硅酸盐学报，Vol. 20，No. 4，1992，8：302～308.

[3] 谷章昭，乐美龙，伍劲夫等．粉煤灰活性的研究[J]．硅酸盐学报，Vol. 10，No. 2，1982，6：151～160.

[4] Kaissar M. Hanna & Aly Afify. Evaluation of Activity of Pozzolanic Materials，J. appl. Chem. Biotechnol. 1974，24：751～757.

[5] Lee，F. M.，唐明述，杨南如，胡道和等译，水泥和混凝土化学(第三版)[M]．北京：中国建筑工业出版社，1980.

[6] 唐有棋．结晶化学[M]．北京：人民教育出版社，1964.

[7] 廉慧珍．沸石岩活性的研究(中科院科学基金资助项目：结晶态铝硅酸盐胶凝物质活性机理的研究)．中国硅酸盐学会水泥专业委员会第三届年会，1986，11.

[8] 廉慧珍，江加标．燃煤固硫渣活性的研究[J]．四川建材，1990. 3：26～30.

对论文之三和论文之九的自评

这两篇都是研究活性的论文，现在看来的不足之处和问题需要讨论如下：

1. 在论文之九的图2和图3的数据中加另一组检测嫩江斜发沸石岩、非晶化的嫩江斜发沸石岩、浙江沸石岩、茂名油母页岩灰、吉林天然火山渣、北京高井Ⅱ级粉煤灰、北京沸腾炉灰等活性的数据，再加上“八五”期间制定固硫渣标准的研究中检测徐州固硫渣和柳州固硫渣活性的数据共28个数据混合作图，如自评图1所示。该图数据较多可以回归，虽然像混凝土这样高度非均质的复杂体系，按数理统计的要求要有超过31组的数据，但是该图已显示出总体上呈大体对数曲线的规律。

2. 用论文之三中的表4数据作图，如自评图2所示。其中有三个点应属非正常点：空心圆点的试样是经过浇水冷却的固硫渣，故其强度比应当不会比同为浇水冷却固硫渣的高出两倍。空心三角形点的试样是G5固硫渣，在配制水泥的试验时，掺量30%的31组试件平均抗压强度为 49.9MPa（标准差 3MPa），胶砂强度比为 91%，说明活性较高，因此该图此点强度检测似有误；方形空心点试样为固硫渣 G4，在胶砂强度比的实验中，测得其胶砂强度比平均为 99.81%，与 5%石灰混合加水成型蒸养强度却比 G5 和 G6 的低，怀疑有误。

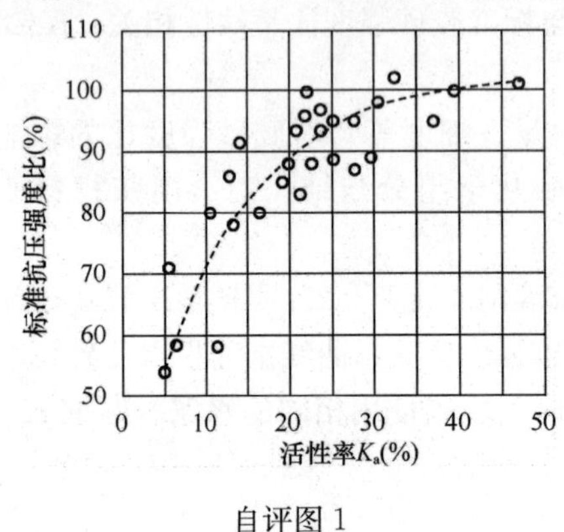

自评图 1

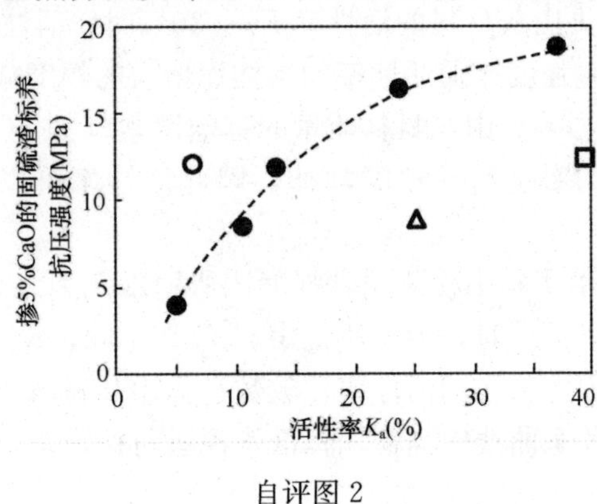

自评图 2

3. 如果换一个思路来看，上述实验是基于一种分解论的方法，亦即只考查固硫渣中的活性硅、铝和 CaO 的反应，但实际上在活性率变化的同时固硫渣其他的组分也是变化的，比如 SO_3，文章中也介绍了 SO_3 和细度的影响，但具有不同相组成和结构特点的不同掺和料微结构是否一定都会反映到强度上？都掺一样多的 CaO，未考虑固硫渣本身也还含有不同量的 CaO，对实验结果是否也有影响？不同掺和料会有不同作用，则用同一种掺量、相同水胶比、在同一龄期的胶砂强度比，是否能反映实际工程中不同掺和料在不同掺量下对混凝土的影响？是否可以怀疑强调矿物掺和料的活性有什么意义？

4. 从整体论来看，任何矿物质材料，只要颗粒尺寸和水泥的相当，即使磨的石英砂，也会因其可增大硅酸盐水泥颗粒水化的空间而促进水泥水化，并参与浆体微结构的形成，在一定掺量下，硬化后的强度不会下降，甚至会提高（本文的图 1）。本文图 1 是在相同水胶比下的结果，未考虑因需水量不同而产生的流动性变化。如今普遍使用高效减水剂，只要控制水胶比降到检测水泥的水胶比以下合适的量值，再用外加剂调整拌和物的流动性，则活性可以并不那么重要了。这就是外加剂的普遍使用所产生的区别于传统混凝土的现代混凝土。从这个观点出发，100 多年不变的水泥似乎也应当进行一次范式转换了吧？水泥只能用于混凝土（砂浆也是没有粗骨料的混凝土），因此这种转换的标志就是按混凝土的需要生产，按混凝土特点检测。

5. 研究工作最重要的是实验条件的控制。贾耀东在校攻读博士学位时给我提过两个问题，一个是：我们检测活性率时使用的是饱和 $Ca(OH)_2$ 溶液，但是混凝土中还有碱的存在，是否会有不同效果？另一个是，用不同掺和料和 $Ca(OH)_2$ 混合加水成型，检测不同龄期矿物掺和料反应程度，是在标准条件下养护的，而在实际工程的早期混凝土中温度是高于标准养护温度的，结果是否也有不同？这两个问题都提的非常好。我建议他把前者 $Ca(OH)_2$ 溶液按混凝土孔溶液组成加入适量的碱，所测出的活性率有提高；对检测反应程度的试样采用 40℃养护。这样做出来的结果当然就不同了。以后再做时还可以采用模拟变温的方法。

6. 该文投稿时无英文注释，表中英文名称均为杂志社所为。现将其错误改正如下：

位置	原文	修改
表 1 的名称	Results of test on Closed loop	The closed loop testing results
表 1 的表头	Active portion，Inactive portion	The active part；The non active part
表 2 的名称	Computation of SiO_2 + Al_2O_3 from fig. 1 for Closed loop	Results to check the closed loop testing by calculated Value
表 2 的表头	Active portion，Inactive portion	The active part；The non active part
	Total from calculation（%）	Total calculated
	Total from test	Total tested
表 3 的表头	Time of heat	Heating duration
	Amt of Thick Hydrochloric acid	Concentrated hydrochloric acid
	Amt of saturated $Ca(OH)_2$	Saturated $Ca(OH)_2$

有嫉妒心的人自己不能完成伟大事业，乃尽量去低估他人的伟大，贬低他人的伟大使之与他本人相齐。

——黑格尔

第三部分　对高性能混凝土的理解、推广和反思

概　述

1993年，留校任教的陈恩义随土木系教师代表团去日本交流，回国后带来日本的ハィ—ポルフォルマンス コンクリ—ト（High Performance Concrete 的音译，缩写 HPC）的概念和信息。我们按照日本的配合比试验时，没有做成。这说明我国原材料和日本原材料存在差距。当时通过国际联机检索查到全世界有关 PHC 的论文共 48 篇，其中最早的是 1984 年法国的 1 篇。此外，法国 HPC 工程研究项目的负责人 Y. Malier 教授 1992 年编辑出版的《HIGH PERFORMANCE CONCRETE——From material to structure》一书，其中收集了 35 篇有关 HPC 的评论及研究报告。此后，一方面对所得到的资料进行研究，了解到不同国家对 HPC 涵义有不同的理解，把这些资料翻译出来编印成册，供国内业界内部参考，并进行宣传，另一方面继续选择原材料，根据我国原材料进一步试配。当时对 HPC 的认识主要是强调混凝土的耐久性。为了耐久，必须密实成型，必须尽量降低混凝土内部的温升，以减小开裂的敏感性。所采取的技术路线主要是掺用矿物掺和料并降低水胶比（本部分所选论文使用的术语是水灰比，但已将矿物掺和料计入水泥中）。本部分所选入论文，反映当时对“高性能混凝土”的认识水平，以及推广应用后对工程所起的积极的和负面的作用。

一开始，我们把 HPC 翻译成高效能混凝土。在此选入相关论文两篇，以说明当时对 HPC 初级的认识。论文之一是对当时国外情况概括介绍的综述；论文之二结合我们在实验中的体会介绍混凝土配制的特点。从中可见当时对 HPC 的认识着重于对其原材料与配合比。当时，为了与国人共享，我们翻译、编辑了“高强和高效能混凝土译文集”提供业内同仁内部参考。第一集～第三集的题目是《高强与高效能混凝土译文集》，其中大部分译文附有译者的点评。后来，在一次全国性的讨论会上我发言后，有一位设计人员当场提出意见说：“你说的高效能混凝土，我们不习惯，觉得很别扭，先约定俗成就叫高性能混凝土吧！以后有更合适的名称的时候再改。”当时我接受了这个意见，从第四册开始，译文集就改成《高强与高性能混凝土译文集》。从此开始了在工程中对“高性能混凝土”的推广。

1993 年末，我应解放军总后营房部邀请，合作进行 C45～C80 系列混凝土的试配，又于 1994 年应深圳第五建筑公司第四工程处邀请合作试配 C60 混凝土，用于深圳莲花北住宅小区高层建筑 7 层以下的柱子。当时深圳的状况是，全市只有三家预拌混凝土，使用的是立轴强制式搅拌机，并且没有人使用矿物掺和料。与莲花北工程同期进行的还有原贤成大厦 C60 混凝土。由于是第一次使用，贤成大厦工程承接者非常谨慎地自己出车到广东粤秀水泥厂拉回刚出厂的 525＃硅酸盐水泥（相当于现在的 42.5 强度等级，时价 600 元/吨，），未使用任何掺和料；选用深圳稀有的乌石谷石灰石（时价约 70 元/吨）；当时我们在莲花北工程的技术路线是，选用价格较低（时价 460 元/吨）的其他厂 525＃硅酸盐水泥；使用当时深圳一

般产品的深康风化粗粒花岗岩石子（价格约低 30 元/m³），另掺入粉煤灰 20%（购自广东沙角，时价约 130 元/吨）。开始工作之前，设计单位安慰我们说："我们的设计是按 50MPa 的强度做的，你们只要做到 60MPa 就行。"主持工程的深圳五建四处处长说，为了技术的进步，必须达到 C60（配制强度 69MPa）的要求。工程处密切配合，认真对待我们所提的技术和质量要求。最后核算混凝土材料成本约降低了 140 元/m³，工程验收合格。期间，曾有过一个插曲：在使用钻芯取样进行结构验收时，强度多数未达到设计要求。对此，我专门去检查了芯样和操作情况。发现所使用的是以往用于普通混凝土的钻头，操作工没有经验，钻取和加工的芯样极不规范：侧面不平整而呈锯齿状，说明在进钻时钻机抖动；有的侧面出现显然是进钻时搓出的横向裂纹，说明钻头刚度不够；侧面和顶面不垂直，说明钻杆与墙面垂直度不够，顶面补平时操作不精细；更严重的是压力试验机压头尺寸远大于试件尺寸，造成试件偏心受压，……后来请中国建筑科学研究院重新钻芯检测，结果都合格了。这说明质量检测也有质量的问题。应注重对操作工的再教育与定期考核。

该工程只用了 20%的粉煤灰，却是经过多次辩论和不断抗争才争取来的。质量监督负责人激烈地反对，因为在他的经验里（尽管他比我年轻得多），粉煤灰都是黑乎乎的，掺进混凝土就影响强度，抗冻性很差。他不知道在十年里，电厂的制粉、锅炉工况、收尘等技术有了根本性的变化。现代电厂的粉煤灰颜色和水泥很难分辨；有了高效减水剂以后，可以用水灰比调节强度，不存在以往简单等量取代时的"强度下降"问题。

那时深圳因为没有人用粉煤灰，本地新建的妈湾电厂虽然有四对电场收尘设备，但所排粉煤灰却支付 25 元/车请人拉到海边填海造地，而且为防止粉尘飞扬而在装车前喷洒了海水。我去要了一些回来做实验，发现其中烧失量不到 1%，质量很好；为此我给电厂第三产业办公室写了一份建议书，建议他们改造出灰口，取消喷洒海水，以袋装出售给水泥厂或混凝土搅拌站。并告诉他们澳门水泥厂购进柳州水泥厂生产鱼峰牌水泥的熟料，掺粉煤灰混磨后，生产鲤鱼牌水泥，出口到东南亚，这是个销路。一年多以后该厂三产办公室的张新亭打电话告诉我，他们按我的意见做的结果是以 40 美元/吨的价格出售到澳门水泥厂。如今妈湾电厂的粉煤灰已供不应求。可见，思路决定出路。

1994 年，我们又与山西省第三建筑公司合作，为广东惠阳大亚湾地区成龙花园工程试配 C60 和 C35 混凝土。其中用于基础底板的 C35 混凝土中掺粉煤灰 33%，现场泵送。在试配合格后连续数日浇筑顺利。过了约半个月后，我们在现场协助施工的研究生江加标突然来电话说，拌和物流动性变差，无法浇筑。当时的混凝土是在现场拌和的，这种现象不会是坍落度损失。当时我判断是原材料有变化。去现场后发现果然施工单位更换了砂子。更换后的砂子明显细颗粒增多。当把砂子换回后，拌和物和易性又恢复正常了。后来在 1995 年与北京市二建公司合作开发自密实混凝土的过程中，也遇到过类似情况：石子的针、片状颗粒只增加 2 个百分点，原配合比的拌和物钢筋间隙通过率就大为减小（见论文之六）而堵塞。混凝土拌和物性质对原材料的细小变化很敏感，这岂非混凝土这种混沌体系的蝴蝶效应？

1995 年和北京住总集团研发中心合作，进行北京航华大厦柱子 C60 混凝土的冬季施工的研究，大量实验表明了用防冻剂的弊大于利（见论文之四）。从 9 月 25 日左右到 1996 年 3 月，取消防冻剂进行了连续施工，现场埋置传感器测温证明混凝土从未早期受冻。这是一种冬季施工观念的转变。据说现在冬季施工不用防冻剂在北京已经比较普遍了。

日本是最早开发和使用自密实混凝土的国家。1996 年与北京二建公司合作在研究日本经验的基础上，使用我国本地原材料，开发自密实混凝土并实施浇筑。无论是实验室的实验研究还是工程中的应用中，都有经验，有教训和体会，对混凝土的认识也有深化，当时发表论文三篇，现选入其中两篇。

为了适应高性能混凝土的需要，受前苏联开发BNB水泥、意大利的超塑化水泥和瑞典强力改性水泥的启发，从1994年开始与住总集团研发中心合作研制开发“高性能胶凝材料”，经过中试和工程试点，推广应用步履维艰。在之后所参与的工程中，逐渐对这项技术有了在更高层次上的认识。变革是发展生产和社会进步的需要，能否实现变革不是技术问题，而是利益问题，也需要转变思维方法和观念；而能否转变思维方法和观念则是决策者水平的问题。这里所选论文是抛砖引玉，也是立此存照。

1999年深圳地铁一期工程招标，其中有一项指标谁都做不到，那就是因地下水中含腐蚀性的硫酸盐和氯盐，设计要求混凝土耐蚀系数不低于0.85。深圳港创建筑材料有限公司的总经理谢敬华邀请我做深圳地下水腐蚀性介质对混凝土的影响和对策的讲座。我在分析深圳地下水情况和混凝土腐蚀劣化的关系后，介绍了“高性能混凝土”在严酷环境中抗劣化的机理，提出使用高性能混凝土的建议。随后和他们一起与深圳地铁公司合作立项研究“高性能混凝土深圳地铁一期工程中应用。”港创实验室根据我们的要求，按强度、变形、抗渗、氯离子扩散系数，硫酸盐环境中硫酸盐耐蚀系数等指标做了1000多组试配，于1999年年末设计和浇筑了一段地铁区间足尺断面模型，总长度为14.5m，并模拟由咬合桩形成的连续墙，与二次衬砌形成组合墙。长向中间设施工缝形成四爿墙，墙厚600mm，各长6m，用四个配合比：墙①525＃普通硅酸盐水泥（相当于现行标准的42.5等级）180kg/m^3、为提高抗硫酸盐和氯盐侵蚀并提高抗裂性而掺Ⅱ级粉煤灰180kg/m^3，为提高早期强度掺磨细矿渣80 kg/m^3，要求磨细矿渣比表面积不大于400 m^2/kg（不要太细的目的是减小自收缩），称作“高性能混凝土”；墙②深圳市当时常用C30混凝土称作普通混凝土，使用52.5＃普通水泥278 kg/m^3，粉煤灰掺量24.5%；墙③配合比同墙②但掺入0.9kg聚丙烯纤维以作对比；墙④使用525＃普通水泥150 kg/m^3，Ⅱ级粉煤灰140kg/m^3，磨细矿渣80 kg/m^3，UEA膨胀剂50kg/m^3；底板厚900mm、顶板厚700mm，宽度6.06m，混凝土配合比同墙①；顶板使用的混凝土，在施工缝两侧，各用同墙①和同墙③的混凝土。模型试验的目的是：检测混凝土拌和物和硬化后混凝土的匀质性；预埋温度传感器，监测、记录混凝土内部温升及其对开裂的影响；按所测温度变换养护水温度，检测构件混凝土内部实际强度；预埋应变计，检测混凝土内部温度、体积变化产生的应力变化；观察混凝土表面开裂情况（数量、长度、宽度、出现时间等）；进行养护试验，以确定合适的养护方法；根据混凝土的实际制备和施工情况以及检测结果，为地铁设计和实际施工提供参考资料。为了使施工人员了解和适应“高性能混凝土”拌和物黏聚性和触变性的特性，港创公司改造了搅拌机的下料口，还邀请施工单位的人观摩试验中拌和物的表现。该模型后来用作该公司的油库，同时做长期观测。2006年12月（模型建成7周年）钻芯取样得到了不同配合比混凝土强度、碳化深度、氯离子扩散系数等的发展、裂缝的变化、钢筋锈蚀等很宝贵的对比数据。

该研究项目于2000年通过专家鉴定，建议补充抗疲劳试验后，进行工程试点应用。但是由于一些非技术性的原因，当时深圳地铁的董事长发话：“凡是别人没有用过的技术，我们都不用。”没有人敢于突破传统规范“水泥用量不得小于320kg/m^3”的规定。但是在深圳市建设局总工和有关技术人员的努力争取下，为了对工程负责，决定改为“水泥用量不得大于 280kg/m^3”，还是突破了已有规范。从此改变了深圳混凝土界的观念。2002 年在深圳地铁一期工程车辆段项目经理杨骏的支持下，研究成果进行了工程试点试验。实验目的主要是研究影响混凝土墙开裂的因素和对裂缝的控制的技术。我当时的想法是先改变已有工程施工缝间距 14m 为 20m，浇筑两段各 20m，只改变硅酸盐水泥用量（240kg/m^3、280 kg/m^3）和一次浇筑高度，预埋传感器检测温度和应力的发展。使用施工单位原来常用的养护条件。在这种情况下，预期至少会有一段墙出现裂缝；如果都不出现裂缝，就会无法分析并继续进行下去。14 天后拆除模板，结果正如所料：拆模后就发现两段墙都有裂缝：水泥用量 280kg/m^{3}的墙有 1 条；水泥用量 240kg/m^{3}的墙有 2 条，此后在 3 天之内共分别出现 6 条和 7 条，水泥用量变化 20kg/m^{3}并没有显著的差别。经过分析，总结出开裂的原因，对接下来的施工提出了具体的措施，此后所浇筑的 70m 均未再发生任何裂缝，一直保持到现在。这一试验为之后至今深圳地铁的混凝土及其施工（严格地说是拌和物成型工艺）提供了重要经验，使港创公司在后来工程中所供混凝土结合售后技术服务，实现了“100m 长度范围内可见裂缝不多于 3 条”的承诺。

2007 年在和天津城市建设投资公司合作进行“以耐久性为目标的天津站交通枢纽混凝土结构设计与施工优化”的研究中，针对天津地下水潜在的中等腐蚀性和控制逆作法施工的大体积顶板混凝土温升需要，采用大掺量矿物掺和料（掺和料 50%以上）混凝土，但是还要考虑背水面混凝土碳化问题。为解决这一对矛盾，在交底时给施工提供三个方案，由他们选择。此后，任一个方案都被证明是有效的。大掺量矿物掺和料混凝土碳化及其对策是个观念问题。该项研究尚未发表论文。

从 1993 年到 1994 年我们在不同场合共发表了三篇介绍高效能混凝土的论文，并多次应邀进行讲座。此处所选论文之一是对国外文献的介绍，论文之二是结合自己的工作的进一步综述。

希望你们年轻的一代，也能像蜡烛为人照明那样，有一分热，发一分光，忠诚而踏实地为人类伟大的事业贡献自己的力量。

——法拉第

论文之一

高效能混凝土（HPC）*

陈恩义　廉慧珍
清华大学土木工程系

近年来，许多工业发达国家，如日本、美国、法国、加拿大、德国、挪威等，面临着一些基础设施老化的问题。这些设施多是混凝土结构，需要进行修补或更新。这项工程需要投入巨额资金。目前国际上已广泛认识到：采用高效能的混凝土更新原有已陈旧的混凝土结构物，具有显著经济性。这导致了上述国家关于新型水泥混凝土材料及技术的研究。

HPC（high performance concrete）是基于混凝土结构耐久性设计的概念提出来的。它不同于传统的混凝土配合比设计思路，而是把混凝土结构的耐久性作为一个重要的技术指标，以此为目标，根据对混凝土在不同阶段的性能要求也有所不同的特点，设计配制出能满足各阶段性能要求的特殊混凝土。设计 HPC 的目的，在于通过对混凝土材料若干性能的改善，提高混凝土结构的耐久性和可靠性。

因此，上述各国目前已把 HPC 作为跨世纪的新材料，投入大量人力、物力进行研究与开发。部分国家已开始将 HPC 用于一些重要工程，如法国已将 HPC 用于核反应堆防护外壳；日本目前正在修建的 20 世纪世界最长的悬索大桥——明石海峡大桥桥墩及基础部分全部用 HPC 制成。

关于 HPC 的中文表述：

关于 HPC 的研究国内尚未见文献报道。国内最早提到 HPC 的是吴中伟教授（1992 年 6 月），当时他把这种混凝土称为高功能混凝土。也有人译成高性能混凝土。对于 HPC（high performance concrete）中 Performance 一词译成性能并不确切，它实际上含有操作和运行的表现和效果之意，是一种动态的行为和现象。HPC 译成高效能混凝土较好，其涵义是混凝土在施工、硬化及在所处环境条件下使用过程等各阶段都有满足技术要求的效果和性能，即同时满足施工及强度要求，并能长期保持其强度、高韧性、体积稳定、高抗渗性、高耐久性等性能。

1　HPC 的涵义

对于 HPC 的确切涵义，各国学者有着不同的认识。

（1）法国（Prof. Y. Malier，1992 年）：

High performance means more than high strength（高效能不仅是指高强度）。HPC 的特别适用于桥梁、海港建筑、核反应堆以及高速公路等重要混凝土结构。（R_{28}＝60～80MPa 试件为 ϕ15×30cm）。

法国官方相信：HPC 应该是有长期耐久性要求的各种工程的必用材料，即使在结构上

* 刊载《混凝土工程新技术》，现代建筑技术实用丛书，中国建材工业出版社，1994 年。

不要求使用高强混凝土。

（2）美国（Prof. P. K. Mehta，1990 年）：

有一种认识已逐渐被接受，即 HPC 的高效能与高耐久性能具有相同的意义，而不是高强度。除强度外，能使（混凝土）具有高耐久性（抗化学腐蚀性）的其他重要性能是高体积稳定性（高弹模、低干缩、低徐变以及低热应变）、高抗渗性和高施工性（$R_{28}=60\sim120$MPa，试件为 $\phi15\times30$cm）。

ACI（1990 年）讨论会：

具有所要求的性质和匀质性的混凝土，靠采用传统的组分、普通的拌和、浇筑与养护方法是不可能制备出这种混凝土的。这些性能包括：易浇筑、捣实而不离析；长期保持力学性能；高早期强度；高韧性；体积稳定性；在严酷环境下的使用寿命长。

（3）加拿大［Prof. Pierre-Claude，和 Adam Neville（英），1993 年］：

HPC 除比普通混凝土抗压强度高以外，还具有高弹性模量、高密实性能、低渗透性以及能抵御多种形式侵蚀的性能；特别适用于高层建筑、桥梁以及一些暴露在恶劣环境中的结构（$R_{28}=65\sim119$MPa 试件为 $\phi15\times30$cm 圆柱体）。

Prof. S. L. Sarkar，1992 年：

HPC 属于具有较高的力学性能（如抗压、抗折和抗拉强度），高耐久性（如高抗冻融性、抗碳化性和抗化学腐蚀性），低抗渗性，具有很低水胶比的高技术混凝土家族（$R_{28}=60\sim116$ 试件为 $\phi15\times30$cm 圆柱体）。

（4）日本（Dr. Ozawa and Prof. Okamura，1992 年）：

HPC 应具有高施工性（高流动、高黏聚性、高可浇筑性）、低水化热、低干缩、高抗渗性以及足够高的强度（$R_{28}=42\sim45$MPa 试件为 $\phi10\times20$cm 圆柱体）。

虽然各国学者对 HPC 的涵义有着不同的理解，但 HPC 应具备高施工性、高体积稳定性、高抗渗性、高耐久性以及足够高的力学强度——这是统一认识。

2 各国关于 HPC 的研究与应用

各国关于 HPC 的研究与应用情况如表 1 所示。

表 1 各国关于 HPC 的研究内容与应用成就

国家 强度等级	HPC 各种性能获得的技术路线					关于 HPC 的 研究内容	HPC 应用范围
	施工性	强度	抗裂	低干缩	高抗渗		
法国 60～80MPa	流化剂、消泡剂、混合材	低水灰比（0.3～0.4）掺硅灰	加纤维（钢纤维、玻璃纤维等）	低水灰比	低水灰比加硅灰 3%～15%	（1）配制、工艺、材性标准研究 （2）应用研究多种工程应用	桥梁、海港工程、核反应堆、高层建筑、高速公路、抗爆
美国 60～120MPa	超塑化剂或超塑化剂与减水剂并用	低水灰比（0.2～0.35）掺硅灰	—	低水灰比	低水灰比加硅灰 5%～10%	（1）配制、材性研究、超早强碱激活水泥 （2）应用研究 HPC 在高速公路及高层建筑中的应用	高速公路、桥梁、高层建筑、广场、高塔等

续表

国家 强度等级	HPC 各种性能获得的技术路线					关于 HPC 的 研究内容	HPC 应用范围
	施工性	强度	抗裂	低干缩	高抗渗		
加拿大 60～119MPa	超塑化剂、混合材料	低水灰比（0.3～0.42）掺硅灰	—	低水灰比	低水灰比、加硅灰 5%～10%	（1）配制、材性研究 （2）应用研究高层建筑、高塔	广场、高塔（电视塔）等
日本 40～45MPa	AE 减水剂、增稠剂、混合材料	低水灰比（0.3～0.32）掺混合材料	膨胀剂	低水灰比、掺膨胀剂	低水灰比加磨细矿渣、粉煤灰、石灰石粉	（1）配制、材性、工艺、流变特性等 （2）应用研究高速公路、桥梁、桥墩、基础等	公路、桥梁

3　日本关于 HPC 的研究

日本土木工程师协会混凝土委员会前分会主席、东京大学土木系岡村甫（Okamura Hajime）教授及其助手于 1988 年开始进行研究，目前已进入实用阶段，取得了很好的技术经济效果。

3.1　HPC 在水化硬化的不同阶段性能要求

（1）塑性阶段：新拌混凝土可自动填充模板而无需任何振捣，这是 HPC 最重要的性能之一，可通过掺加高效能减水剂、增稠剂以及混合材料等实现。

（2）凝结硬化早期：混凝土内部不存在由于水化热的产生、水化硬化或干燥收缩等引起的初始裂缝，这些裂缝不仅会对混凝土结构的外观造成破坏，而且还会引起混凝土内部钢筋的锈蚀。可通过降低水泥用量、掺加膨胀剂、混合材料等实现。

（3）硬化后期：在具有足够的力学强度的同时，还应具有很高的抗渗性能，能够阻止许多可能引起混凝土结构破坏的潜在因素的渗透，如氧气、氯离子、水分等，尤其是在混凝土的表层。可通过低水灰比、掺加活性混合材料、膨胀剂等实现。

3.2　关于 HPC 研究的技术路线图解（图 1）

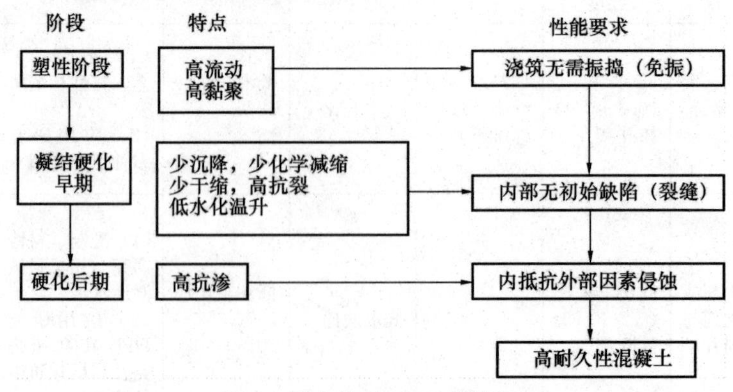

图 1　日本关于 HPC 的图解

3.3 HPC用于当代最长的悬索桥——明石海峡大桥

明石海峡大桥是一座三跨双铰接加劲桁架悬索桥，它连同大鸣门桥一起把本州、淡路岛与四国连成一体，大桥全长3910m，中央跨度1990m；大桥桥面设计为6条汽车通道，速度为100km/h，水上部分结构耗用钢材200000t，水下部分结构耗用1420000m³混凝土。明石海峡大桥于1988年5月动工兴建，整个工程预计1998年完成。这将是本世纪全世界最长的悬索桥，其中央跨度比英国的Humber大桥的长50m。

针对这一跨世纪的巨大工程，施工部门——日本本州四国联络桥公团从多方面采取措施，以保证结构的耐久性和可靠性，桥墩（基础）全部采用现浇HPC（高效能混凝土）制成。

日本明石海峡大桥桥墩及基础部分所用现浇HPC配合比及强度如表2所示。

表2 用于明石海峡大桥的HPC配合比

材料用量（kg/m³）						强度（MPa）		
水泥（低热）	水	砂子	卵石	混合材	外加剂（复合）	设计（91d）	实测	
							7d	91d
260	145	615	1137	150	C×3%	24	22	42

作为以混凝土结构耐久性设计为基础而发展起来的高效能混凝土，配合比设计思路比传统的混凝土有了一个新的突破，人们将会以一种全新的观念来认识它。同时，HPC的开发与应用，也使得水泥混凝土这一世界最大宗建筑材料上了一个新的高度，给建筑行业来非常明显的效益，首先，在建筑工程施工中应用HPC，即使无严格的技术保证措施，也能够获得高密实性混凝土，结构的可靠性可进一步增强；其次，由于HPC良好的自密实性，施工人员的劳动度大大减轻，人员减少，施工能耗降低；再次，由于无需任何振捣设备，施工人员免受噪声的危害，劳动环境明显改善，施工效率提高；最后，应用HPC可免除振捣工序，使整个工程施工体系更加有序合理，工程质量进一步得到保证和提高。

HPC在日本明石海峡大桥工程中的成功应用，就是一个很好的例证。

由于HPC的研究与应用在世界范围内才刚起步．研究工作主要限于试验室内，各国学者对它的认识也不尽相同。在从试验室到实际工程的转化中，HPC本身仍有许多问题还未得到完全解决。例如，HPC的定义和内涵是什么？传统配合比设计方法已不适用于HPC，如何建立科学的设计方法？HPC的施工性如何测定与量化？HPC与高流态混凝土的区别是什么？如何评定HPC的耐久性？为降低水化热，HPC需加人大量混合材料，这使得其早期强度偏低，不能适应现代施工高效率的要求，如何提高HPC的早期强度？除常用的组成材料（水泥、砂、石、混合材料和水）外，HPC还需加入高效减水剂、膨胀剂、增稠剂等外加剂，它们对混凝土的耐久性有何影响？且其生产成本相对较高，为HPC的普及和推广带来困难，能否在保证HPC性能特点的前提下从组成材料上降低其成本？等等。

HPC是一种很有发展前途的、性能好而节能的新型混凝土，更符合混凝土和钢筋混凝土工程的施工和使用要求，适用范围广，尤其是适用于现代化工程（如大跨度桥梁、隧道、高速公路、高层及塔式建筑、海上工程、水利工程、核反应堆以及有毒废物处理工程等）。在国外都是由国家列项并投入大量经费进行研究的。为此，我国应尽早投入该项研究。清华大学于1993年开始进行有关HPC的研究。首先是将HPC在我国原材料、工艺、设备和环境条件下予以实现，并研究其各项性能的测试和评价的方法。研究工作量很大，并有相当的难度，需要有关单位联合和跨行业、跨学科的合作研究。尤其希望得到国家和有关部门的支持。

时间是一切财富中最宝贵的财富。

——德奥弗拉斯多

我们做科研的有一个共同的秘密——乐趣，也就是必须学会去享受你现在所做的事情。古人常说的悬梁刺股，这个模式今天也许并不适用。在当今世界，不应过多关注知识的“记忆”能力，关键是要有创造新知识的能力，要有创造力。养成这种能力的唯一方式是要非常自由，参与其中，享受其中。理查德·费曼发表诺贝尔获奖感言时说，其实他在探索的过程中就已经感到获得奖励了。在持续学习的过程中你可以找到自己的方向，这对于应对明天很重要。我不在乎你们在考试中得到多少分，如果你觉得考试很痛苦，那你就失败了。当年我每次参加考试都很激动，因为我把它看作一个解决问题的过程。

——“图灵奖”得主、清华大学教授姚期智

论文之二

高效能混凝土实现的可行性——国内外 HPC 材料研究现状及问题*

清华大学　廉慧珍　陈恩义　覃维祖

前言

高效能混凝土（high performance concrete，缩写 HPC）是近年来一些发达国家提出的新概念的混凝土。HPC 的出现是科学技术和生产发展的需要。如跨海大桥、海底隧道、海上采油平台、核反应堆、有毒及有害废物处置与处理等各种在严酷环境下工作的重大混凝土结构物，施工难度大；使用时一旦出现事故，后果严重；而且修补耗资巨大。同时，不少工业发达国家面临一些基础设施老化的问题。这些设施有许多是混凝土结构，需要修补或更新。HPC 和普通混凝土的差别在于把混凝土耐久性作为重要的技术指标进行设计，而不仅仅考虑强度。高强不一定耐久。单纯追求高强度，不仅限制其应用，而且在某些情况下还会与耐久性有矛盾而不利于结构的寿命。例如为了达到高强，往往使用高标号水泥，增加水泥用量，由此增加混凝土温升，引起温度应力，而且增加硬化混凝土的收缩，不利于混凝土的耐久性。国际上已广泛认识到：高效能的混凝土用于高层、大跨、大体积结构、高耸结构以及上述特殊结构和更新原有已陈旧的混凝土结构物，具有显著的经济性[1]。美国、日本、法国、加拿大、挪威等国把 HPC 作为跨世纪的新材料，投入了大量人力、物力进行研究和开发，部分国家已开始用于一些重要工程。如法国已将 HPC 用于核反应堆防护外壳；日本目前正在修建的二十世纪世界最长的悬索大桥——明石海峡大桥桥墩及基础部分全部用 HPC 制成[2]。上述国家已有多项关于这种新型水泥混凝土材料及技术研究计划的制订和实施[3]。

挪威皇家科学与工业研究院（1986）：有关 HSC（高强混凝土）的研究与应用项目；

日本建设厅（1988）：利用高强混凝土与增强措施开发先进混凝土技术（简称为“新 RC”）的五年计划；

法国政府：“混凝土新法——含 HPC 应用研究课题”的国家项目；

加拿大政府(1989)：开发新型建筑材料、高强混凝土结构设计和开发新的制品与施工技术；

美国：以 NSF（国家科学基金会）和历时五年的 SHRP（战略公路研究计划）为主，重点包括路面与桥梁水泥混凝土的耐久性问题；NIST（国家标准与技术局）（1990）：一项关于 HPC 的国家计划。

各国学者对 HPC 的涵义有不同的见解，但 HPC 应具有高施工性、高体积稳定性、高耐久性以及足够的力学强度，这是一致的看法。其中日本的 Dr. Ozawa 和 Prof. Okamura 甚至认为，HPC 的强度只要足够（ϕ10×20cm 圆柱体试件 28 天抗压强度≥40MPa），但应具有高施工性（高流动、高黏聚性、高可浇筑性）、低水化热、低干缩、高抗渗性。设计 HPC 的目的，在于通过对混凝土材料若干性能的改善，提高混凝土结构的耐久性和可靠性。

* 刊载全国高强与高效能混凝土研究课题讨论会，1994 年 4 月 11～16 日，黄山。

由于不同于普通的混凝土，HPC 并没有万能的现成配比，而是和原材料等条件有关。在对原材料要求、配合比和配合比设计方法、试验和检验方法，以及施工控制等方面，都有与普通混凝土不同之处。现就已掌握的信息和我们的工作，介绍 HPC 的特点和对上述有关问题的观点。

一、HPC 的效能

如前所述，HPC 是基于混凝土结构耐久性设计的概念提出来的。它不同于传统的混凝土配合比设计思路，而是把混凝土结构的耐久性作为一个重要的技术指标（而不仅是强度）。以此为目标，根据对混凝土在不同阶段的性能要求也有所不同的特点，设计配制出能满足各阶段性能要求的一种新型的混凝土。设计 HPC 的目的，在于通过对混凝土材料若干性能的改善，提高混凝土结构的耐久性和可靠性。其涵义是混凝土在施工、硬化及在所处环境条件下使用过程等各阶段都有满足技术要求的效果和性能，即同时满足易施工、强度并能长期保持其强度、高韧性、体积稳定、高抗渗性、高耐久性等性能。例如日本对 HPC 效能的观点和做法如下：

阶 段	性 能	表 现	措 施
塑性阶段	高流变，体积稳定，无泌水和离析	新拌混凝土可自动填充模板无需任何振捣，或稍加振捣，这是 HPC 重要的效能之一	掺加高效能减水剂、增稠剂及混合材料等
凝结硬化早期	低水化热、低沉降、低干缩	混凝土内部无初始裂缝	降低水泥用量、掺加膨胀剂、混合材料等
硬化后期	足够的强度，很高的抗渗性，高耐久性	能阻止可能引起混凝土结构破坏的潜在因素如氧气、氯离子、水，以及其他能引起水泥腐蚀的介质的侵入，使混凝土在恶劣的工作环境中能长久保持其强度	低水灰比、掺活性混合材料、膨胀剂等

HPC 在塑性阶段和凝结硬化早期的性能和表现是为了保证硬化后期的效能。因此，HPC 的设计以耐久性为目标，以控制新拌混凝土流变特性与硬化混凝土的强度和体积稳定性为手段。

二、HPC 的原材料

1. 胶凝材料

水泥：因国外水泥品质较稳定，在文献中很少特别讨论水泥的问题．但是高强度的水泥是高强度混凝土的必要条件，这是无疑义的。但水泥强度的提高是很难的。应当研究用现有标号水泥能配制出的混凝土最高强度。日本在 1988～1993 年的“新 RC 计划”中研究了用现有标号的水泥和目前所能达到的手段配制的混凝土能达到多高的强度[4]，在我国还查不到这方面的数据，但是用目前国内的现有水泥配制出 C80 以上可施工的混凝土是完全可能的。实际上，目前水泥的问题更重要的是与超塑化剂的相容性。因为目前国内多采取提高 C_3A 的含量和水泥细度来提高水泥的早期强度，用于普通混凝土时，由于水灰比较高，流变性的问题不突出；当水灰比很低时，拌和水中 SO_3浓度大大减少，由于水泥用量相对增加，需要与之反应的 C_3A 却大大增多，超塑化剂的效果就大大下降。所以应选用低 C_3A 和不太细的水泥，最好生产 HPC 专用的水泥。俄罗斯的 VNV（俄文 BHB 的音译）系列水泥就是为高强度混凝土生产的专用水泥。在该水泥中，熟料可被 50%以上的矿物粉取代，需水量为 0.16～0.18，所配制的混凝土 28 天抗压强度可达 110MPa[6]。

混合材料：过去，尽管由于取代了一定量的熟料，掺入混合材料的水泥比硅酸盐水泥在某些性能上具有优势，但强度要受到影响。尤其是早期强度。因此掺入混合材料的主要目的是调节水泥标号、节约水泥，这在发达国家并不重要。因而对混合材料的研究只是近十几年出于节省能源的考虑而开始重视。在高强度混凝土的研究中，人们发现了硅灰对混凝土神奇的增强作用。而其他如矿渣、粉煤灰和沸石凝灰岩等矿物粉，当适当处理且掺量合适时，则不仅有利于提高新拌混凝土的和易性，而且还可在不同方面改善硬化混凝土的其他物理力学性质[7]。

粉煤灰：出于环境保护的需要，许多国家对粉煤灰在混凝土中利用的研究也多了起来。我国是粉煤灰排放的大国，粉煤灰的利用率在国际也是最高的。但是，至今不少人对在混凝土中掺用粉煤灰仍怀疑或否定。尤其怀疑在高强混凝土中是否能掺用粉煤灰。因此粉煤灰的利用大都集中在较低强度的混凝土和大体积混凝土。但对高效能混凝土的研究表明，合适的混合材料是高效能混凝土必不可少的组分。甚至可采用较大的掺量，以降低水泥的水化热、改善新拌混凝土的和易性和提高硬化混凝土的抗渗透性。

现在掺粉煤灰的效果明显比几十年前提高的原因是粉煤灰质量的提高：煤粉炉燃烧温度比过去提高，因而含碳量大大下降；制粉技术提高、使用电收尘的方法收尘，使煤粉比过去细得多；由于粉煤灰的主要组分是空心微珠，加粉煤灰成为提高混凝土和易性的必要措施，目前我国已制订了粉煤灰的标准，限定 I 级粉煤灰在边长为 45μm 筛取余量≤12%，烧失量≤5%。这样的粉煤灰使混凝土的强度损失就较小。而且 60 天以后，掺用粉煤灰的混凝土强度比不用粉煤灰的有较大幅度增长；粉煤灰用量越大，后期强度增长越多。

实验表明，当水灰比低到一定程度时（例如<0.40），粉煤灰用量在一定范围内变化对混凝土强度影响不大，如表 1 所示。

表 1　粉煤灰对混凝土强度的影响

编号	水泥用量（kg/m³）	粉煤灰（体积%）	抗压强度（MPa）			
			3d	7d	28d	90d
H513	424	10%	47.5	54.0	63.4	70.5
H515	369	20%	44.7	54.3	65.3	—
H516	342	25%	41.1	49.7	62.9	68.9

注：①为保证要同流动度，粉煤以体积差取代同体积水泥，则重量水灰比稍有变化。
②所用粉煤灰为 II 级灰，水泥为大同水泥厂 GB 175—77 525＃散装硅酸盐水泥。
③坍落度均为 190～200mm。

表 2 为按质量比掺入入粉煤灰时（质量水灰比相同）对混凝土强度的影响。

表 2　粉煤灰掺量对混凝土强度的影响

编号	水泥用量（kg/m³）	粉煤灰（kg/m³）	抗压强度（MPa）			
			3 天	7 天	28 天	90 天
T1	500	50（9%）	34.6	59.6	64.0	68.6
T2	470	60（11.3%）	39.0	58.5	63.1	67.5
T3	450	80（15%）	37.0	51.6	62.5	70.1

注：①所用为 II 级粉煤灰。
②因使用冀东水泥厂生产的盾石牌 525＃硅酸盐水泥，与减水剂相容性很差，故坍落度均为 30～50mm。

由表1和表2可见，当水灰比低于0.40时，粉煤灰掺量从10%～25%变化后，混凝土28天抗压强度均可达60MPa。用大同水泥时，混凝土坍落度在200mm左右。

硅灰：又译成硅粉或硅尘，为冶炼铁合金所排放而收集到的气态SiO_2凝聚而成的SiO_2微细颗粒尘。在国外，硅灰成为高强度混凝土不可缺少的组分。但掺硅灰后，混凝土需水量增加，尤其是在很低水灰比时。在掺硅灰时，同时掺入矿渣或粉煤灰，则可取得复合叠加的效果。

例如在加拿大的一座混凝土预拌厂生产的一种混凝土，w/c=0.25，总用灰量为542 kg/m^3，其中矿渣50%，硅灰10%，水泥只有30%，混凝土坍落度为210mm，$\phi10\times20mm^3$的试件，28天抗压强度为114MPa，一年抗压强度为137MPa。

矿渣：矿渣硬度大，与熟料混磨时不易磨细，因此矿渣水泥需水量较小，保水性差。若将矿渣单独磨细掺入混凝土中使用，则可使磨细的矿渣发挥其较高活性的作用，由于矿渣结构以玻璃体为主，当掺入量较大时，需水量也不会增加。例如我们在混凝土中掺入采用超细磨技术磨细的矿渣，与不用掺和料的混凝土相比，抗压强度发展见表3。

表3　超细矿渣的混凝土流动性和强度

编号	水灰比	矿渣掺量（%）	坍落度（mm）	强度（MPa）		
				3天	7天	28天
K101	0.264	0	210	81.0	94.9	90.3
K301	0.248	30	210	69.6	81.3	102.7

注：①总用灰量相同。
②试件尺寸为10cm×10cm×10cm。
③水泥为柳州水泥厂熟料+4%二水石膏配制而成。

日本柳田佳宽在执行“新RC计划”中研究表明，配制目标强度为800kgf/cm^2、坍落度为210mm的混凝土，用矿渣超细粉（比表面积为6000～8000cm^2/g）时，最佳掺量为30%～50%：用硅粉时，最佳掺量为10%～20%。

其他混合材料：其他活性较高的火山灰质混合材料也可用于配制HPC，沸石凝灰岩被证明是其中的一种，但因其需水量大而掺量受到限制，将沸石岩粉同其他有效组分复合制成增强剂，也被证明可用于配制高强混凝土，这部分内容已见冯乃谦教授及其助手的有关报道。

提高抗裂性能的添加剂或混合材料：在我们进行HPC的试验研究中，掺入适量膨胀剂，得到补偿收缩的效果，而当掺入流化床锅炉燃煤脱硫的固硫渣时，其补偿收缩的效果比掺有粉煤灰和膨胀剂的效果更好，如图1所示。

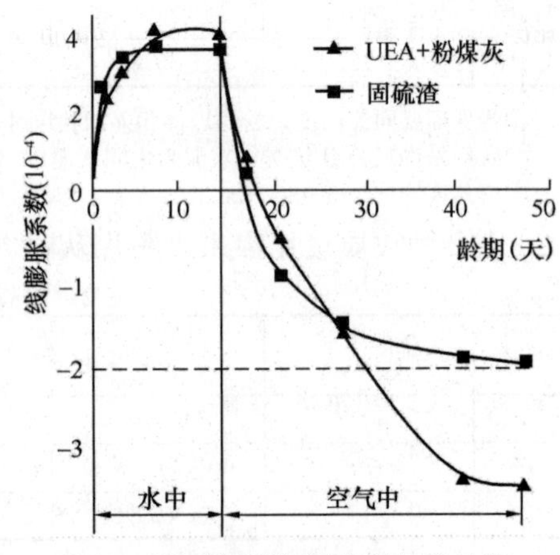

图1　用不同添加剂配制的高效能混凝土能补偿收缩效果

图1中的虚线为混凝土收缩极限（0.02%），收缩超过此限，将可能引起开裂。不加膨胀剂的混凝土通常收缩值>0.04%。

在高强混凝土中掺入一定量的膨胀剂，除补偿收缩提高抗裂性外，还会提高混凝土结构的强度。日本文献报道，使用钙矾石系列的混合材料可提高混凝土抗裂性能，抗压强度提高150kgf/m^2[9]，其道理相似。

2. 减水剂

在我国目前生产的减水剂中，宜采用高浓度的萘磺酸盐甲醛缩合物类的产品，当配制C60以上的高效能混凝土时，掺量宜≥1%。实际上用占总用灰量1.5%的FDN，用水量为145kg/m^{3}时，在没有混合材料的情况下，也达到了210mm的坍落度。在高浓度萘系减水剂掺量≥1.5%后，坍落度损失也可减小（图2），此外，还可调节水泥中的石膏掺量来调节混凝土坍落度和坍落度损失。例如冀东水泥标准稠度用水量为28.6%，加入FDN后为24.5%，再加入1%石膏后，则降至21.2%。

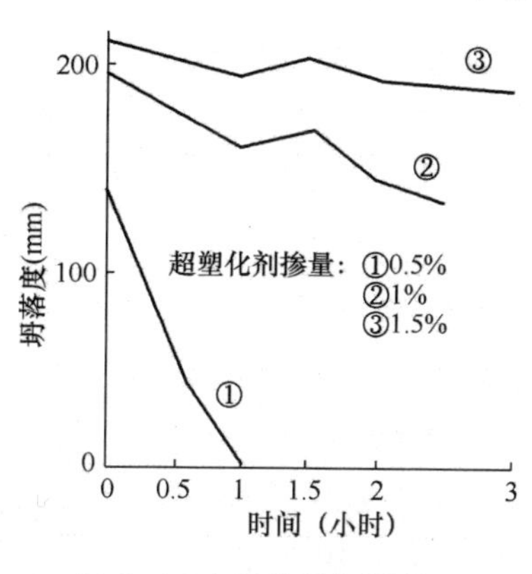

图2　减水剂掺量与混凝土坍落损失的关系

3. 粗骨料

在普通混凝土中，石子强度一般高于水泥浆体强度，普通石子强度都足够。即使是强烈风化的低强度花岗岩，其岩石抗压强度也达80～100MPa；粗粒花岗岩岩石强度也达120～140MPa；非常坚固的粗粒花岗岩岩石强度可达180～200MPa[10]。因此对C50～C80的混凝土来说，石子强度影响并不大。对于C80以上的混凝土，就要求石子的岩石强度不能低于混凝土的强度。重要的是要对石子的粒形、级配以及软弱颗粒、石粉含量的限制特别注意。这些既会影响混凝土强度，又对新拌混凝土和易性有较大影响。石子对混凝土弹性模量的影响要更重要。C80以上应选用弹性模量大些的石子。

目前还没有很好的方法评价石子强度，压碎指标可作为参考，这对评价其粒形是有用的，目前主要以所配制的混凝土强度试验来评定。由于高强混凝土骨料和水泥浆体的弹性模量相近，C50以上混凝土试件破坏时也都从骨料中间断裂。并不说明骨料强度不够。所以从混凝土破坏方式不能判断骨料的强度。

在工程中喜欢用较大粒径的石子，如水工混凝土石子粒径可达150mm，甚至采用毛石。但水工混凝土标号低，一般不超过C20，是从节省水泥的观点出发的。在建筑工程中限制最大粒径是根据结构尺寸、钢筋间距等要求。但基于增强界面的要求和碎石匀质性的考虑，高强混凝土石子最大粒径应当小。如加拿大Aïtsin和英国Neville建议，混凝土强度为60～100MPa时，石子最大粒径可以≥20mm，但强度超过100MPa时，石子最大粒径不能超过10～12mm[11]，日本建议超高强混凝土石子最大粒径在10mm以下[12]，我国高强混凝土协会制订的《高强混凝土施工指南》规定，C60～C80的混凝土所用石子最大为25mm，是为了适应目前石子生产的国情[13]。

三、HPC配合比

1. 水灰比和水泥用量

由于水泥完全水化的理论需水量是0.227，国外有文献认为0.22为混凝土水灰比的极限。实际上在w/c远大于此值时，水泥就已经不能完全水化了。w/c的极限应当从目前的技术所能得到的混凝土的和易性来确定。对普通混凝土，在一定工艺和水泥标号下，水灰比是影响强度的主要因素。对高强混凝土说，水灰比仍然是影响强度的主要因素。只是影响程度下降（图3）。如果不能达到一定的和易性，w/c太低会影响密实成型，强度也提不高。所以不能单纯从w/c考虑。由于用灰量太大会影响弹性模量和体积稳定性（徐变、收缩），应尽量限制总用灰量。建议研究在不同的强度范围规定最大用灰量。

2. 骨料用量、砂率

在一定范围内，相同工艺下的混凝土强度主要受水灰比的影响。研究表明，高强混凝土的规律也如此，只是强度与灰水比关系的曲线斜率变小。中国建筑科学研究院发表的实验数据也证明砂率的变化不影响强度[14]。基于和易性的要求，高效能混凝土砂率较大。当骨料为连续级配时，泵送混凝土砂率应提高4%～5%[15]，而当石子级配不好时，更必须提高砂率。据日本资料，平均坍落度每提高20mm，砂率应增加1%，而强度并无明显变化[16]。GB 50204—92第4.2.8条也规定，泵送混凝土砂率宜控制在40%～50%[17]。砂率对强度无明显影响的原因是，由于骨料的存在，使骨料－水泥浆体界面形成薄弱的过渡区。石子的表面积远大于砂粒的表面积，其与浆体界面过渡区相对地又长又厚，对混凝土影响很大。因此，相同水灰比的水泥净浆强度＞砂浆强度＞混凝土强度。

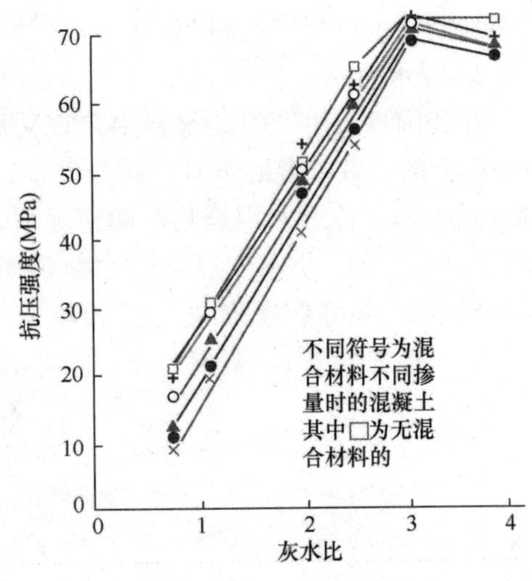

图3　混合材料掺量不同时混凝土强度和灰水比关系

但因砂率过大时会影响混凝土的弹性模量，还应对最低石子用量加以限制，日本规定C60泵送混凝土的石子用量不能少于1090kg/m³。美国加拿大实用的混凝土中石子用量有的在1000kg/m³左右[18]。需要系统地进行试验研究，很好地确定石子用量的限制。

为保证硬化混凝土足够的弹性模量，骨料总量不能太少。Mehta认为，对高强HPC来说，水泥浆体占35%（骨料总量占65%）时，似乎可以使强度、和易性和尺寸稳定性这几个相互矛盾的需要达到最佳的均衡[19]。

四、HPC的质量控制和检验

HPC除保证足够的强度外，由于主要目标是耐久性，以耐久性设计的混凝土具有很高的密实性和体积稳定性，对施工过程的控制就要更加严格，尤其是养护。使用膨胀剂的混凝土尤其对养护条件敏感。掺用膨胀剂后，如不加强养护，同样达不到预期的膨胀能，甚至照样产生不希望有的收缩。日本研究认为，HPC由于坍落度达200～220mm，可以自流平，不必振捣而自动充满模板[20]。当HPC用于泵送（坍落度为200mm左右）时，由于黏性较大，仍需机械振捣。表4为不同振捣方式对混凝土强度值的影响（表4）。

表4　振捣方式对混凝土强度的影响

编号	w/c	水泥用量(kg/m³)	混合材(kg/m³)	表观密度(kg/m³)		插捣强度（MPa）			机械振捣强度（MPa）		
				插捣	机振	3天	7天	28天	3天	7天	28天
GH-1	0.38	368	79	2429	2450	36.9	48.5	58.6	43.9	54.9	61.0
GH-4	0.38	230	197	2416	2431	24.7	33.3	46.0	26.4	35.0	50.3

在表 4 中，机械振捣的混凝土强度比手工插捣的同配合比混凝土强度提高 4%～19%。对流动性很好的混凝土稍加振捣，可提高混凝土密实度，强度也会提高。

此外，要严格控制原材料的品质和质量。特别要控制加水量的准确和砂子含水量的测定与扣除。

对 HPC 的验收，目前仍参照对普通混凝土验收的办法[21]，即按在浇筑时留取的试件标标养后期（28 天或 90 天）强度关系以及影响此关系的因素的试验研究，以便及早发现问题。准养护的抗压强度进行统计性或非统计性的验收[22]。应进行早期（甚至极早期，如 1 天）和还应当进行坍落度损失和环境温度与湿度关系的试验，以便得到控制坍落度损失的依据。坍落度损失还受水泥品质的影响，不要使用早强型的、C_3A 含量大的和很细的水泥。

对 HPC 的试验，除抗压强度外，还应有弹性模量的数据，必要时还应提供抗折强度的数据。但在施工中对 HPC 的检验仍可以标准养护试件的抗压强度验收。结构强度以现场水养护的试件抗压强度评定[23]。由于目前大多数施工部门所使用的压力机额定荷载低于 2000kN，标准试件用 10cm×10cm×10cm 的试件代替 15cm×15cm×15cm 是否可能？当对试件有疑问时，目前用于定评定普通混凝土结构强度的钻芯法是否适用于 HPC，则需要研究。国内外在高强混凝土的研究中，都已发现用钻芯法在现场检验结构强度，显著低于标准养护试件的强度（即使是圆柱体强度）。其原因可能是由于高强混凝土硬度高、脆性大，而使钻芯时对试件的损害也比普通混凝土的大。

因此有必要进行系统的试验研究，来了解相同混凝土芯样和标准养护试件间的差别，找到其间关系的规律，做出不同结构物芯样强度和设计强度间允许差异的规定。

由于进钻阻力，芯样胎体截面同心程度较难掌握，要特别仔细加工，严格按规范要求制作芯样。由于强度和刚度的差异，建议不用对普通混凝土所用的硫磺砂浆补平端面；最好用磨平的方法处理。

五、结论

1. HPC 与普通混凝土有许多不同之处，也不同于单纯高强或高强流态混凝土，国外做出来的，国内不一定能做出来；北方能做出来的，南方也不一定做得出来，必须根据所用原材料进行系统研究，提出特定的配合比设计规律。

2. HPC 是能够获得高密实性和结构的可靠性的混凝土，在工程中使用 HPC，可大大减轻施工人员的劳动强度和噪声污染，提高施工效率，降低施工能耗，工程质量可进一步得到保证和提高。HPC 以耐久性为目标，HPC 之所以能实现，又基于工业和科学技术的发展（如新型超塑化剂的开发和生产、优质混合材料的开发、搅拌与运输设备的改进等）以及对混凝土技术认识的提高。

3. HPC 要求水泥和超塑化剂具有较好的相容性，较低的用水量、相对较大的砂率；石子粒径较小、级配良好；以及必须重视混合材料的使用。但目前水泥生产的控制、混凝土的配合比设计、试验方法、结构设计规范以及检验验收都是针对长期以来传统使用的混凝土，要实现 HPC，还需要进行大量的试验研究工作。

4. HPC 的重要指标是和易性、耐久性和强度，传统的用于评定普通混凝土上述性能的方法对 HPC 进行评定已不能反映 HPC 真正的行为，需要系统试验来建立合适的标准。如 HPC 施工性的测定和量化，HPC 耐久性的评估和量化等；对于其他改性材料（如超量的高效减水剂、膨胀剂、增稠剂等外加剂）对混凝土的耐久性有何影响？如何从我国国情出发，从原材料选择、配制以及施工等方面进一步降低 HPC 的成本？等等。

总之，HPC 是一种很有发展前途的性能好而省能的新型混凝土，更符合现代混凝土和钢筋混凝土工程的施工和使用要求，适用范围广，尤其是适用于现代化工程如大跨桥梁、隧道、高速公路、高层及塔式建筑、海上工程、水利工程、核反应堆以及有毒废物处理工程等。需要有比单纯高强或高强流态混凝土更多的研究。在国外都是由国家列项，投入大量经费进行研究的。研究工作量很大，并有相当的难度，需要有关部门、单位联合和跨行业、跨学科的合作研究。

参考文献

[1] 陈恩义，廉慧珍. 高效能混凝土及其研究现状[J]. 水泥与混凝土制品，1993.
[2] 同[1].
[3] 同[1].
[4] 大滨嘉彦.《セメント. コンクリート》N0546.《特集》. 1993.
[5] A Tagnit－Hamou et al.《Proceediig of International Cogress on Chemistry of Cement》.
[6] S. A. Podmasova et al.《Proceeding of High-Strength Concrete 1993》Vol. 2.
[7] 寺村悟等.《セメント. コンクリート》N0546.《特集》. 1993.
[8] F. C. Aïtcin，《Concrete International》No1. 1993.
[9] 同[8].
[10] 同济大学等.《混凝土学》. 1982.
[11] 同[8].
[12] 同[4].
[13] 土木工程学会高强混凝土委员会.《高强混凝土结构施工指南》. 1994.
[14] 田桂茹等. 中国建筑科学研究院科技交流资料：关于 JGJ 52—92 和 JGJ 53—92 制订的背景材料. 1992.
[15] CB 50204—92.
[16] 同[14].
[17] 同[14].
[18] 同[8].
[19] P. K. Mehta et al.《Cement，Concrete and Aggregate》. 1990.
[20] 同[1].
[21] 同[13].
[22] 同[15].
[23] Nicholas J. et al.《Concrete International》Aug. 1991.
[24] 柳田加宽.《セメント. コンクリート》N0546.《特集》. 1993.

论文 1 和论文 2 的自评

1. 论文之一着重介绍国外提出高性能混凝土概念的原因和不同国家不同的理解。在国外，高性能混凝土主要是针对严酷环境下重大基础设施混凝土结构耐久性而提出的。尽管在 1990 年由美国国家标准与技术研究所（NIST）和美国混凝土协会（ACI）主办的第一届高性能混凝土的讨论会上，提出定义高性能混凝土为："靠传统的组分和普通的拌和、浇筑、养护的方法不可能制备出的、具有所要求的各种性质和匀质性的混凝土"；而对什么是高性能混凝土，如何保证耐久性的实现，不同国家和不同的人仍然有各自的理解。最早提出高性能混凝土的是西方国家，大部分认为耐久性首先要密实，高强混凝土因采用低水灰比而密实，所以强调的是高强度（美国规定高强混凝土抗压强度的低限是6000Psi，$\phi15\times30$ 圆柱体强度磅/平方英寸，约合公制 41MPa，折合为 15cm×15cm×15cm 立方体强度约为50MPa。见美国混凝土学会 363 委员会：State-of-the arts report on high strength concrete)；著名英国混凝土专家 A. M. Neville 在他的著作《Properties of Concrete》第四版（2011 年中译本译为《混凝土的性能》由刘淑华、冷发光、李新宇、陈霞等翻译，中国建筑工业出版社出版）中说："'高性能混凝土'一词带有明显的广告色彩，原来称为'高强混凝土'。但在很多情况下要求的是高耐久性，……在一些应用中，寻求的性能是高弹性模量。"

2. 日本最早是因国情的需要而开发了自密实混凝土，在西方提出高性能混凝土的概念时，他们就将自密实混凝土理解为 HPC（日本没有把 HPC 译成日文，而是音译或直接叫HPC)，对强度的界定则是 40～45MPa。1999 年，美国 ACI 的技术活动委员会（TAC，technical activities committee）提出作为 ACI 对高性能混凝土的定义建议为："高性能混凝土是符合特定工程性能组合和匀质性要求的混凝土，采用传统的原材料和一般的拌和、浇筑与养护方法，未必总能生产出这种混凝土"。当前我国某些企图以此表明自己业绩的官员不满意这个定义，因为不具体。实际上高性能混凝土只是一种理念，一种在当前工程建设发展的形势下对混凝土工程质量的需要和由此而从原材料、拌和直到成型工艺对混凝土的要求，而不是混凝土的一个什么新的品种，当然就不可能具体。不是配出来的，也不是在搅拌站能生产出来的。企图用某些具体的指标来定义高性能混凝土，是不可能的。

3. 在论文之一中明确提出"HPC 是一种很有发展前途的、性能好而节能的新型混凝土"，在论文之二中，也有"HPC 是能够获得高密实性和结构的可靠性的混凝土"的结论。显然是错误地把高性能混凝土看成了混凝土的一个品种。论文之二结合自己的经验和体会，着重介绍实施高性能混凝土的技术措施。实际上，除了强度以外，我在潜意识中对高性能混凝土的理解就是高流动性和矿物掺和料的使用，而且介绍得如此"具体"。一方面给人们提供了可操作性的指导；另一方面则无疑是对高性能混凝土的理解起了一个误导的作用。从1994 年开始，我在全国做过十余次"高性能混凝土"的讲座，对现在存在的误区，应当是有责任的。但是，和对任何事物的认识一样，这是一个过程：接受"耐久性更重要"的理念后，就有个如何操作的问题。当时是需要有些指导的。不足之处在于没有把"具体问题具体分析"放在首要的位置来强调。这和我对混凝土认识的过程有关。

4. 当时发表的文章对 HPC 的翻译是"高效能混凝土"，说明已感到 performance 和property 的词义有区别：我国业内人士都熟悉被译成中文的 Neville 名著《Properties of Concrete》中文译名是《混凝土的性能》，而"高性能混凝土"的"性能"在英文中并不是"properties"。后来在人们的要求下改成了把"高效能混凝土"改成了"高性能混凝土"，有权宜之计的意思。本来，HPC 的理念是好的，即从过去只注重实验室检测的材料性质指标，转向以那些只有在处于一定环境的工程中才能表现出的行为和效果——performance 为目标，却因为未能真正理解该词用于混凝土的涵义而只从表面上作了解释。究其原因，一方面是东西方文化差异所致，另一方面是个思维方法和观念的问题。因为过去在教学中，我们对混凝土性质的认识就是实验室检测的那些指标，直到 2004 年我在商品混凝土协会成立会上讲了一次"方法比技术更重要"，才触及这个问题。此后接连十几年在各地多次宣讲，其中“实验室检测出的指标并不代表结构中混凝土真实的行为”这一观念已被业内许多人所接受，但是多数人还是拗不过来，总认为实验室检测抗冻融循环次数等指标就是结构中混凝土的耐久性，或者有个什么相关的关系。现在很多人已认识到，那不过是在标准条件下在实验室中快速检测的指标，而标准条件和实际工程中的条件有根本上的不同，甚至不能建立其间的相关关系。

5. 高效减水剂问世已有几十年的历史，在 20 世纪 80 年代中期我国已有了自己的产品，但是我们的学生 1989 年到施工单位调研高效减水剂使用情况时，技术人员竟然大都不知道什么是减水剂。我们把高性能混凝土介绍到国内，对高效减水剂的广泛使用确实起过推动作用，而高效减水剂的使用改变了混凝土的一切，促进了我国进入现代混凝土时代。然而在“高效能混凝土”（后来改为高性能混凝土）的旗号下，这样详尽地叙述 HPC 的“技术措施”(论文之二)，却又实际上对理解高性能混凝土起了误导的作用。

6. 文章提出 HPC 仍有许多问题还未得到完全解决。例如“HPC 的定义和内涵”，“配合比科学的设计方法，施工性如何测定和量化？HPC 与高流态混凝土的差别……，如何评定 HPC 的耐久性?”等等，这些问题至今依然存在。这实际上就是因为现在混凝土的一切正常的变化和问题主要就是因为高效减水剂的广泛应用。

7. 论文之二的第 4 条结论所述“HPC 的重要指标是和易性、耐久性和强度”这三项指标，至今仍然基本上在沿用着传统混凝土性质检测的方法：和易性不论用坍落度还是坍落扩展度、倒筒流空时间……，检测的都是静态的流动度，对经过泵送后的表现无法检测或评价；用实验室的标准条件检测的抗冻性、抗渗性、碳化、抗硫酸盐腐蚀性等指标只能作为优化原材料与配合比的指标，仍无法预测混凝土用于处在一定环境中的钢筋混凝土结构服役寿命究竟有多长。对结构中混凝土强度的验收仍采用实验室的标准养护试件，最多使用与实际施工并不相同的工艺所制作的“同条件”养护的试件，却并不真的同条件，只是“同位置”而已。

8. 如果能真正正确理解高性能混凝土，则高性能混凝土主要是针对那些造价高、资源和能源消耗量大的、处于严酷环境(频繁干湿和冷热交替、冻融循环和能引起钢筋锈蚀的高湿、高温……)中的结构物(如有毒有害废物处理、辐射的屏蔽、大跨桥梁、隧道、地铁、近海或离岸、含有很严重化学腐蚀性介质的地下结构等)和大体量的重要建筑物(如高层建筑、纪念性建筑等)。提高其服役寿命关系到人类社会的可持续发展。而对于并不那么严酷的“一般环境”(即 GB/T 50476 混凝土结构耐久性设计规范中关于环境分类中的 I 类 A 级环境，没有化学侵蚀性和冻融循环作用的环境)中的建筑物，只要不具备钢筋锈蚀的条件(钢筋与氧气和水接触)，耐久性的问题就应当不会那么严重和复杂。

9. 论文之二的表 2 所列混凝土，是在相同质量水胶比下的粉煤灰掺量对混凝土强度的影响。其中粉煤灰掺量都未超过 15%（占胶凝材料总量)，且水胶比很低，故对混凝土强度没有什么影响。如果粉煤灰掺量超过 20%，则必须降低水胶比，否则混凝土不可能达到一致的强度。

论文之三

高强与高性能混凝土的配制及施工质量控制和验收 Ⅰ.原材料和配合比对高强与高性能混凝土性能的影响*

廉慧珍[1]　陈恩义[2]　清华大学土木工程系

摘　要　从流变性能的角度对低水灰比的混凝土中影响水泥和外加剂相容性的因素、辅助胶凝材料的特性和作用以及骨料对混凝土质量的影响进行分析，结果表明，使用超塑化剂的高强及高性能混凝土不宜使用R型水泥，即C_3A含量和比表面积大的水泥。所用水泥的熟料塑化程度SD应大于70%；加入优质的粉煤灰和磨细矿渣不仅有利于提高混凝土的流变性能，而且有利于较大体积(构件最小的断面尺寸大于1m)混凝土的强度发展；粗骨料的粒径和原材料选定和水灰比一定的情况下，砂率的大小影响砂浆的数量，因而影响混凝土的和易性；高性能混凝土要求较大砂率，而砂率增大，混凝土强度有所提高，弹性模量则有所下降，但砂率变化于34%～50%时，弹性模量下降幅度不大。

关键词　高强混凝土　高性能混凝土　原材料　配合比　性能　影响

前言

高性能混凝土是按耐久性设计的混凝土，首先要求混凝土具有高密实度（即使不要求很高的强度）和体积稳定性(无开裂)。为此，要求拌和物具有高流动性和体积稳定性而便于施工；当用于大体积工程时，内部不产生高温升；其配制的特点是使用超塑化剂和活性混合材料以及低水灰比。由于高层建筑的发展，大体积(断面最小尺寸>1m)混凝土的基础和柱子增多，需要配制的不同强度等级(多数为C60、C60、C40、C35等)、低温升、低渗透性混凝土，就属于高性能混凝土。近两年来，我国高强混凝土发展很快，已有许多高层建筑至少在底层柱子采用C50、C60的混凝土。许多实验室争相提高混凝土的强度。一般说来，和普通混凝土相比，C50、C60的泵送混凝土并没有什么特殊的施工工艺；有些单位也能在实验室配制出抗压强度在80MPa以上的混凝土，但就我国水泥、砂、石以及超塑化剂的质量现状来说，无论是在实验室还是在实际工程中，能稳定地、较经济地生产C35～C60的高性能混凝土(强度45～70MPa以上、坍落度200mm左右，体积稳定)的，还是很少。往往用相同的配合比在不同施工的场合会得到不同的效果。

从1994年开始，我们先后在深圳莲花北7＃ 地高层住宅、大亚湾开发区成龙花园、北京航华科贸中心大厦等工程中稳定地浇筑了约10000m^3C60泵送混凝土，实现了较高粉煤灰掺量的C35泵送混凝土的施工，均经验收合格，并在航华科贸中心大厦实现了不掺防冻剂

---

* 原发表于《混凝土》1996.10。

1　廉慧珍，清华大学土木工程系，教授

2　陈恩义，清华大学土木工程系，博士

的C60泵送混凝土柱的冬季施工。

配制试验研究和施工的实践表明，原材料、配合比、施工技术和管理水平等对高强混凝土质量的影响要比对普通混凝土的影响更大。现分两次介绍在施工中的体会：Ⅰ．原材料和配合比对高性能和高强混凝土质量的影响；Ⅱ．高强与高性能混凝土的施工质量控制和验收。本文为第一部分：原材料和配合比对高性能和高强混凝土质量的影响。施工情况如表1所示。

表1 高性能混凝土施工简况

工程及强度等级	所用525#❶硅酸盐水泥	配合比（kg/m³）	28天抗压强度(MPa)	坍落度(mm)
深圳莲花北7#地高层住宅七层以下柱及剪力墙，C60	吉林鼎鹿、大连华日、东方龙、湖南韶峰等	w/c=0.31，水泥470，粉煤灰80，FDN掺量1.1%	66～79.8	170～210
深圳莲花北7#地高层住宅厚度为1.5m的筏式基础、梁、楼板等，C35	同上	w/c=0.38，水泥285，粉煤灰30%，NF-2掺量1.4%	>50	170～210
大亚湾成龙花园58层商住楼地下柱，C60	湖南雪峰	w/c=0.30，水泥470，粉煤灰80，NF-2掺量1.4%	67～68.0	190～210
大亚湾成龙花园58层商住楼厚度为2.6m的基础筏片，C35	同上	w/c=0.37，水泥300kg/m³粉煤灰33%，NF-2掺量1.0%	35.1～58.2	145～160
北京航华科贸中心大厦8层以下柱，C60，包括不掺防冻剂的冬季施工	冀东盾牌	水泥450，粉煤灰100 NF-2-6掺量1.5%	60.5～80.1	190～220

❶ GB 175—82的标号，相当于现行水泥标准的强度等级42.5。

原材料

一、水泥

为保证水泥质量的稳定，在没有专用水泥供应时，高强混凝土不得使用立窑水泥。一般，对普通混凝土来说，使用旋窑水泥时，普通混凝土只需考虑水泥的强度；而对水灰比很低的高强混凝土，则还必须考虑水泥的流变性能。影响水泥流变性能的因素主要是水泥的需水量和水泥与超塑化剂的相容性。水泥需水量大时，不能得到有良好施工性能的高强混凝土。目前水泥标准稠度用水量的测定方法是按照不加超塑化剂的普通混凝土的需要而制订的，不能反应加入超塑化剂后水泥的需水量。高强混凝土所用水泥的流变性质，只有在使用超塑化剂时才能反映出来。这就是水泥与超塑化剂的相容性问题。水泥与超塑化剂的相容性不好时，不仅影响超塑化剂的减水率，更重要的是造成混凝土严重的坍落度损失。对水泥来说，影响其与超塑化剂的相容性的主要因素是SO_3含量和水泥中C_3A、细度和碱含量的匹配。

早在1968年，乔奇等人就提出，如果水泥中石膏含量根据水泥中的C_3A、细度和碱含量调整到最佳水平，就可避免这些因素的有害影响。石膏在水泥中的作用不仅是调节凝结时间，而且对水泥的强度、流变性能和收缩都有影响。如图1所示为在不同温度下某熟料的收

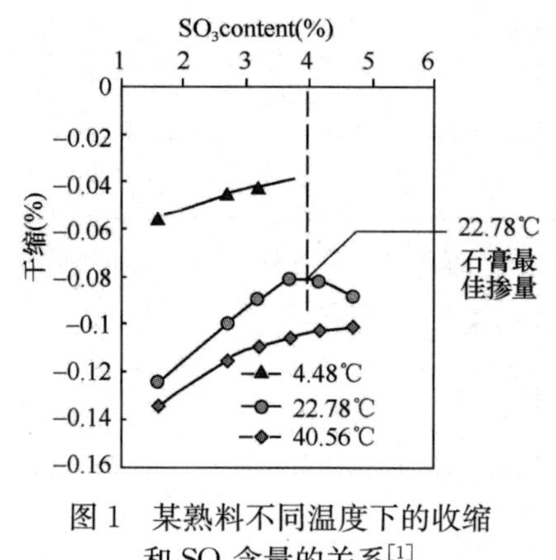

图 1 某熟料不同温度下的收缩和 SO_3 含量的关系[1]

缩和 SO_3 含量的关系，表明在常温下，该水泥的 SO_3 最佳含量为 4%[1]。水泥中的 C_3A 和碱含量越多，水泥越细，SO_3 最佳含量越大。而实际上，按 ISO 标准，在水泥生产中，石膏按符合水灰比为 0.5 左右的混凝土所需凝结时间的最小掺量加入，此后几乎是不变的，和熟料化学组成与细度无关。

1. C_3A 含量

由于超塑化剂的使用，使水灰比可降低到<0.4，则现有的 SO_3 量就不能满足混凝土流变性能的要求[2][3]。这是因为水泥中水很少，SO_3 在水泥浆体中的溶出量减少，尤其有些厂家为了提高水泥早期强度和降低烧成温度而增加熟料中的 C_3A 含量（如 R 型水泥），则溶解度比二水石膏溶解度大的 C_3A 会和石膏争夺水，SO_3 更加不足，使水泥与超塑化剂的相容性问题更为突出。

用 C_3A 含量不同的水泥配制高强混凝土，其他条件相同，达到坍落度相同时，所需水灰比不同，因而强度不同，见表 2 所示。

表 2 C_3A 含量不同的水泥配制高强混凝土的实例

水 泥	熟料中 C_3A 含量（%）	水泥强度（MPa）	磨细矿渣掺量（%）	w/c	坍落度（mm）	混凝土强度（MPa）
用柳州熟料配制	5.62	59.8	30	0.248	210	92.1
冀东 525R 硅酸盐	8.90	56.2	30	0.330	191	71.3

由表 2 可见，在水泥用量相同的条件下，用熟料中 C_3A 含量高的水泥时，达到相同坍落度的混凝土水灰比不可能很低。因此也就不能配制出更高强度的泵送混凝土。

2. 熟料中的碱含量

当熟料中含有碱时，SO_3 量要和熟料中的含碱量相匹配。加拿大 Sherbrooke 大学 Aïtcin 等人用不同的水泥以相同水灰比（0.45）和超塑化剂掺量（0.6%）进行流变性试验，用通过 Marsh 锥形筒的流动时间表示[4]，如表 3 所示。

表 3 不同水泥试样的流变性测定

No.	勃氏比表面积（m^2/kg）	流 动 时 间（秒）		熟料硫酸盐化程度 SD*（%）
		搅拌 5 分钟	搅拌 60 分钟	
4	377	53	63	71
5	372	53	63	69
6	383	54	61	103
7	386	59	77	71
8	371	53	99	68
9	353	59	139	66

* $SD=\frac{SO_3}{1.292Na_2O+0.85K_2O}$

由表 3 可见，各试样细度相近、初始坍落度相近，但 1 小时后流动性损失明显不同，而与 SD 成反比。流动性损失最小的是 6# 试样，其 SD 最大，为 103%；坍落度损失最大的是 9#试样，SD 最小，为 66 %，1 小时后流动时间延长了一倍多。

以我国首都水泥厂熟料为例，其化学成分如表 4 所示。

表 4　首都水泥厂水泥熟料化学成分

成分	SiO_2	Al_2O_3	Fe_2O_3	CaO	MgO	SO_3	K_2O	Na_2O	TiO_2	P_2O_5	LOI	f-CaO
%	22.15	5.17	5.25	61.23	2.32	1.2	1.5	0.4	0.3	0.15	1.56	0.37

用该水泥进行混凝土坍落度试验，w/c＝0.305，加入缓凝型超塑化剂 1.5 %。坍落度初始为 20cm，半小时后为 16cm，1 小时后为 7.5cm，损失达 60%以上。

其熟料硫酸盐化程度为：

$$SD=\frac{1.2}{1.292\times 0.4+0.85\times 1.5}=67\%$$

3. 颗粒组成

水泥如果粗细颗粒恰当级配，可得到良好的流变性能。好的级配应控制 5～30μm 的颗粒约占 90%，而＜10μm 的颗粒＜10%。我国水泥的生产则基本上只考虑细度，甚至用增加比表面积来提高水泥的强度。结果是水泥颗粒越细，细颗粒越多，需水量也越大，而且也加剧混凝土的坍落度损失。

4. 其他

如由于生产工艺控制的差别，使 C_3A 和 SO_3 的形态不同，而不同形态的 C_3A 和 SO_3 各自有不同的溶解速率，其间的匹配也影响水泥的流变性能。目前我国水泥也存在这方面问题。随着高强混凝土的发展，应当重视这方面的研究。

二、辅助胶凝材料

在配制混凝土时加入较大量水泥的活性混合材料，可降低温升，改善和易性，增进后期强度，提高抗腐蚀的能力。因此，在国外将这种混合材料称为辅助胶凝材料，是高性能（不仅高强）混凝土不可缺少的组分[5]。从环保和经济的角度出发，建议扩大使用粉煤灰。有些辅助胶凝材料如硅灰、沸石凝灰岩、油页岩灰等，在 w/c 不变的情况下掺入一定量时，可对混凝土有增强作用，但因需水量大，故允许的掺量有限。而好的粉煤灰则由于需水量小，可以大量掺入。磨细的矿渣需水量小且有很好的活性，但要额外消耗能源；粉煤灰对混凝土强度的贡献随龄期而增加，随水胶比的提高而增加。图 2 所示为粉煤灰体积掺量为 60 %时粉煤灰和水泥分别的贡献[6]。

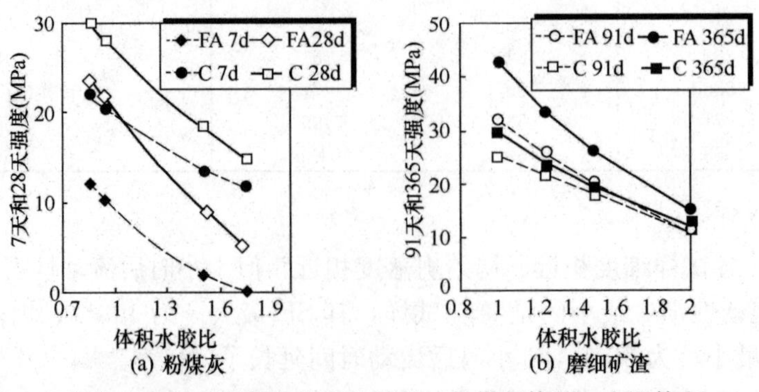

图 2　粉煤灰和水泥对混凝土强度的贡献（粉煤灰体积：水泥体积＝1：1）

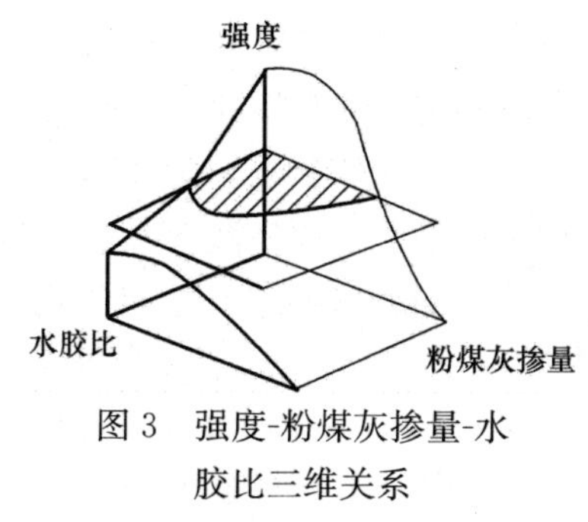

图 3　强度-粉煤灰掺量-水胶比三维关系

由图 2 可见，28 天以后，粉煤灰与水泥对混凝土强度的贡献的差距缩小，并随水灰比的降低而显著；90 天以后二者相近，一年后，则粉煤灰的贡献超过水泥的，水灰比越低，粉煤灰的贡献越大。故加入粉煤灰后，应当相应降低“水胶比”，即 w/(c+F)。以保持早期强度不降低。Dunstan 推荐的粉煤灰-水胶比-强度三维关系模型[6]（图 3）很好地说明了这个问题。但这个模型的定量，依原材料的变化如粉煤灰品质、水泥的组分和强度、砂石质量、超塑化剂种类等而不同。

由图 3 可见，等水胶比平面显示当水胶比不变时R28－F/(c+F)的关系，即粉煤灰掺量增加时，混凝土强度下降；等强度平面显示达到相同强度时 w/(c+F) － F/(c+F)的关系（图中阴影部分所示），即增加粉煤灰掺量时，要相应降低水胶比，以保持混凝土强度不变。但是，当水胶比很低时，粉煤灰掺量在一定范围内，上述影响并不显著；超过此范围时，则上述影响随粉煤灰掺量的增加而增加得很快。用北京西高井Ⅱ级粉煤灰以 8%～24%不同掺量配制混凝土，控制坍落度相同，实验结果也表明有这样的规律（表 5）。

表 5　粉煤灰掺量对混凝土强度和配合比的影响

试验编号	水泥用量（kg/m³）	粉煤灰（%）	w/c	坍落度（mm）	抗压强度（MPa）		
					3 天	7 天	28 天
H513	424	8	0.355	190	47.5	54.0	63.4
H514	396	12	0.359	190	46.5	54.1	64.9
H515	369	18	0.360	190	47.7	54.3	65.3
H516	342	24	0.365	190	41.1	49.0	62.9

硅酸盐水泥和普通硅酸盐水泥在超过 75℃的条件下蒸养时，强度会随温度的升高而降低，而当混凝土构件体积较大（最小断面尺寸＞1m）时，内部温度会超过此蒸汽养护的温度，对混凝土的强度不利；掺用粉煤灰和矿渣等辅助胶凝材料的混凝土强度则随养护温度的升高而提高，掺量越大提高的幅度越大。

三、骨料

天然石子有很高的强度，即使是强烈风化的低强度花岗岩，其岩石矿压强度也达 80～100MPa；粗粒（伟晶）花岗岩岩石强度达 20～140 MPa；非常坚硬的粗粒花岗岩强度可达 180～200 MPa[7]。因此对 C50～C80 的混凝土来说，石子强度的影响并不大。石子的粒径、粒形、表面状况、级配以及软弱颗粒和石粉含量等更应引起注意。这些既会影响混凝土的强度，又影响新拌混凝土的和易性。

高强混凝土比普通混凝土的强度对界面敏感。和大粒径的石子相比，用小粒径石子时，水泥浆体和单个石子界面的过渡层周长和厚度都小，难以形成大的缺陷，有利于界面强度的提高。同时，粒径越小，石子本身缺陷的几率越小。因此，基于增强界面的要求和碎石匀质性的考虑，高强混凝土应使用最大粒径小的石子。如加拿大 Aïtcin 和英国 Neville 建议，混凝土强度为 60～100MPa 时，石子最大粒径可以≯20mm，但强度超过 100MPa 时，石子最大粒径不能超过 10～12mm[8]；日本建议超高强混凝土石子最大粒径在 10mm 以下[9]；我国高强混凝土委员会制订的《高强混凝土结构设计和施工指南》规定C60～C80的混凝土所用石子最大粒径≯25mm，是为了适应目前石子生产的国情[10]，但这种国情要尽快改变，才能适应科技和生产的发展。高强或有抗渗要求的混凝土，最好不用卵石或卵碎石，如不得不用时，则粒径应进一步减小。对有些骨料，当水泥用量增加或水灰比降低到一定程度后，强度不再提高，“这显然不是混凝土的强度已完全开发”，也不是石子的强度不够，“而是达到了水泥浆体与这种石子界面粘结强度的极限”[11]。对高强泵送混凝土，粒形比岩石强度更重要，宁可选用颗粒强度稍差但粒形好的石子。如在深圳，用相同配合比，分别以深康粗粒风化花岗岩和乌石谷致密石灰岩石子配制混凝土，结果如表6所示。表6中结果的差别主要是因为乌石谷石子虽然强度高，但针、片状颗粒明显多；而粗粒花岗岩虽然强度低些，但石子却大多为等径状。

表6　石子对混凝土性质的影响

所用石子	混凝土28抗压强度（MPa）	坍落度（mm）
深康粗粒花岗岩	71.3	195
乌石谷致密石灰岩	68.8	148

砂子的粗细和级配比石子的级配对混凝土流变性能的影响更大；高强混凝土宜用中粗砂，尤其当石子级配很差时，砂子以偏粗为好。应严格控制砂中粉细颗粒的含量和石子的含泥量。砂子的粗细不能只看细度模量，有时，细度模量较大，但5mm以上和0.315mm以下的都多，级配很差。最好是0.63mm筛的累计筛余大于70%，0.315mm筛的累计筛余为90%左右，而0.15mm筛的累计筛余>98%。

配合比

一、水灰比和水泥用量

由于水泥完全水化的理论需水量是0.227，国外有文献认为0.22为混凝土水灰比的极限。实际上在w/c远大于此值时，水泥就已经不能完全水化了。w/c的极限应当从目前的技术所能得到混凝土的和易性来确定。对普通混凝土，在一定工艺和水泥强度下，水灰比是影响强度的主要因素，对高强混凝土来说，水灰比仍然是影响强度的主要因素，只是影响程度下降。假设使用实际强度为56MPa（1.13×52.5）的硅酸盐水泥和碎石，根据混凝土强度公式$R_{28}=mc/w-b$计算，其$m=25.8$，$b=13.8$；根据大量试配和施工经验统计，当水灰比低于0.4时，上述强度公式大体为$m=17.8$，$b=5.3$。表明高强混凝土强度与灰水比关系的曲线斜率变小。

为了降低混凝土构件内部的温升，减少干缩，水泥用量和胶凝材料总量应尽量小；浆骨比大时，和易性好、强度较高，而弹性模量较低；浆骨比小时则相反。P. K. Mehta认为高强混凝土浆骨体积比为0.35∶0.65时，可解决以上矛盾[12]。因此对于C80以下的高强混凝土，胶凝材料总量不宜大于550kg/m³。

二、骨料用量、砂率

在水泥浆量一定的情况下，砂率在混凝土中主要影响和易性。高性能混凝土由于用水量很低，砂浆量要由增加砂率来补充[13]，砂率宜较大。当骨料为连续级配时，泵送混凝土砂率应提高4%～5%[14]，而当石子级配不好时，更必须提高砂率。据日本资料，平均坍落度每提高 20mm，砂率应增加 1%，而强度并无明显变化[15]。GB 50204—92 第 4.2.8 条也规定，泵送混凝土砂率宜控制在 40%～50%[16]。中国建筑科学研究院发表的实验数据证明水灰比为 0.55～0.53 时，砂率从 34%～46%变化对强度无明显影响[17]。其原因是，由于骨料的存在，使骨料-水泥浆体界面形成薄弱的过渡区。单个石子的表面积远大于砂粒的表面积，其与浆体界面过渡区相对又长又厚，对混凝土影响很大。因此，相同水灰比的水泥净浆强度>砂浆强度>混凝土强度。为了解低水灰比时的情况，以低水灰比（0.26、0.3、0.4）配制混凝土，改变砂率从 34%～50%，测定混凝土的坍落度、强度和弹性模量，如表 7 所示。

表 7　砂率对混凝土性能的影响

试验编号	水灰比（w/c）	砂率（%）	坍落度（mm）	28 天抗压强度（MPa）	棱柱体抗压强度*（MPa）	弹性模量*（GPa）	备　注
S3-1	0.30	34	205	60.3	45.2	43.2	稍泌水
S3-2	0.30	38	205	62.1	54.3	42.9	
S3-3	0.30	42	215	67.0	58.1	41.7	
S3-4	0.30	46	240	68.6	61.8	42.4	
S3-5	0.30	50	215	72.0	61.8	40.7	黏性大
S2-1	0.26	34	1.5	73.4	—	—	
S2-2	0.26	38	6.0	72.6	—	—	
S2-3	0.26	42	4.5	72.4	—	—	
S2-4	0.26	46	4.5	75.9	—	—	
S2-5	0.26	50	3.0	75.2	—	—	
S4-1	0.4	34	155	50.7	—	—	离析、泌水
S4-2	0.4	38	180	57.3	—	—	稍离析
S4-3	0.4	42	200	58.4	—	—	
S4-4	0.4	46	190	55.3	—	—	稍黏
S4-5	0.4	50	140	61.9	—	—	黏性大

* 使用 Instron 公司 IX 系列自动材料试验系统 1.20 测试；由于试验条件所限，试件标准养护 28 天，取出在空气中存放 14 天后测试。

在表 7 中，所用石子级配很差，空隙率为 48%；所用砂的细度模量为 3.1。试验结果表明，从和易性来看，当水灰比不同时，最优砂率有所不同，但都在 38%～46%之间；超过此范围后，坍落度减小。随混凝土砂率增加，强度呈增长的趋势，而弹性模量呈下降趋势。因砂率过大时会影响混凝土的弹性模量，故应对石子最低用量加以限制。砂率的大小与砂的粗细、级配和石子的粒径、级配有关。当砂子的细度模量大而石子最大粒径小时，可减少石子用量。例如当砂子细度模量为 3.00，而石子最大粒径为 10mm 时，绝干捣实石子体积用量可到约 0.44；砂细度模量为 2.80，石子最大粒径为 20mm 时，上述状态石子体积用量可提高到约 0.62[14]。石子级配越差，该值越低。掺入密度小的辅助胶凝材料时，砂率可减小。砂率的选择可取砂浆富裕系数 K=1.6～1.9 计算。

结束语

1. HPC以耐久性为目标，与普通混凝土有许多不同之处，也不同于单纯高强或高强流态混凝土，是混凝土科学技术的一种进步。HPC的配合比受原材料的影响很大，国外能做出来的指标，在我国不一定能做出来；北方能做出来的，南方也不一定做得出来。任何工程都必须根据所用原材料认真进行试配。HPC之所以能实现，又基于工业和科学技术的发展（如新型超塑化剂的开发和生产、优质混合材料的开发、搅拌与运输设备和试验条件的改进等）以及对混凝土技术认识的提高。

2. HPC要求水泥和超塑化剂具有较好的相容性、较低的用水量、相对较大的砂率；石子粒径较小、级配良好；以及必须重视混合材料的使用。但目前水泥生产的控制、骨料的生产、混凝土的配合比设计、试验方法、结构设计规范以及检验验收都是针对长期以来使用的水灰比在0.4～0.5的混凝土，要实现HPC，还需要各部门配合进行大量的工作。

参考文献

[1] Proceeding of Fifth International Syposium on The Chemistryof Cement. Vol. Ⅲ，1～3.

[2] P. C. A. Aitcin. CONCRETE INTERNATIONAL，1993. 1.

[3] A. Tagnit-Hamou et al. PROCEEDINGS OF 9TH INTERNATIONAL CONGRESS ON CHEMISTRY OF CEMENT，1992. 11.

[4] 同[3].

[5] M. R. H. Dunstan，ACI SP 91-7.

[6] 同[5].

[7] 同[5].

[8] 同[2].

[9] 同济大学等. 混凝土学，1982.

[10] 大滨嘉彦，セメント・コンクリート，N0. 546，《特集》，1993.

[11] 土木工程学会高强混凝土委员会. 高强混凝土结构设计和施工指南[M]. 北京：中国建筑工业出版社，1995.

[12] 美国ACI 363委员会. STATE-OF-THE-ART REPORT ON HIGH-STRENGTH CONCRET，1993.

[13] 田桂茹等. 中国建筑科学研究院科技资料交流部. 关于JGJ 52—92和JGJ 53—92制订的背景材料，1992.

[14] GB 50204—92.

[15] 同[13].

[16] 同[12].

[17] 同[13].

自评

1. 本论文是在1994年与总后营房部合作进行C45～C80混凝土系列配制的研究，和在广东的深圳、惠阳大亚湾房屋建筑工程技术服务实践的总结，至今仍可供参考。文中对高性能混凝土的理解显然是将其看作是一个混凝土的品种，只是强调了耐久性，而并未理解如何判断和保证耐久性；并未有意识地针对不同地区的环境特点考虑不同部位结构的耐久性，在表1所列工程中，深圳、大亚湾显然属于年平均气温和相对湿度较高的滨海地区，和北京的环境有很大的差别。地下和地上的结构也不同，但文中均未见分析。掺粉煤灰能提高混凝土抗化学腐蚀的能力，但是像北京这样的温差较大、干燥气候较多地区的工程，除了冬季除冰盐的作用外，大多数地上和室内的构件处于一般环境（即没有化学腐蚀和冻融循环作用），那么应如何考虑其耐久性？矿物掺和料的意义是什么？

2. 这里对大体积混凝土的理解是按日本的定义，最早我国冶金建筑部门也据此而定义，后来又随日本将1m改为0.8m。由于进入21世纪以来水泥强度的提高和粉磨过细，在混凝土强度也不断提高度的情况下，即使构件断面最小尺寸只有30cm左右，也需要在施工中控制温度（S. Bernander，“Practical Measures to Avoiding Early Age Thermal Cracking in Concrete Structures ”in“Avoidance of Thermal Cracking in Concrete at Early Ages”edited by Rupert Springenschmid，Munich，March 1998.）因而文中所引大体积混凝土的界定已过时；用构件尺寸来定义大体积似显简单化。我赞成美国ACI的观念，凡需要控制由温度变化产生的应力而出现早期开裂的混凝土，都可视为“大体积混凝土”来对待。只要水泥高强度、高比表面积的现状不改变，现代混凝土结构在混凝土拌和物施工中控制温度将具有普适的必要性。

3. 配合比中对w/c的选择采用Lyse的直线方程具有普适性，因为我国长期使用的Bolomey公式原始的形式也是回归成的直线方程：

$$S = K\left(\frac{c}{V+A}\right) - K'$$

式中 S——混凝土28天抗压强度；c——水泥体积用量；V 和 A——分别为水和空气含量；K 和 K'——直线的回归系数，分别为直线的斜率和截距。

对于混凝土这种高度非均质的复杂多相体，文中所给出强度公式的回归系数只能为作者自己所用，不可能具有普适性。如果每个搅拌站有自己的回归系数，有可能会更适用。

4. 在“骨料用量、砂率”一节第一段中，所引用日本的规定，只是从拌和物流动性的角度出发；所引中国建筑科学研究院发表的实验数据只是在水灰比0.53～0.55的很窄范围内的规律。当因耐久性的需要而考虑浆骨比时，就要在一定水灰比（水胶比）下和相同施工性的条件下进行比较和优化。对现代混凝土来说，水胶比0.5和0.3可能分别是配合比各参数对拌和物及其硬化后性质影响的两个临界点。

5. 对表7数据的分析：

①每一个水灰比下骨料总量一定，只改变砂率，但是却未保持坍落度不变，水灰比为0.26时，坍落度为3～6cm，其他则为流态，这对工程使用的实际来说，不具可比性；

②水灰比0.3的，只有28天抗压强度数据，“随混凝土砂率增加，强度呈增长的趋势，而弹性模量呈下降趋势”的结论是否适用于不同水灰比的混凝土？

③水灰比0.3和0.4时对抗压强度的影响比水灰比0.26时的大，也就是说，似乎水灰比在0.3以下时影响就不大了，那么在0.26和0.3之间会怎样呢？

6. 顺便一说：关于混凝土配合比的参数选择对混凝土性质影响的辩证关系，文中所引用的“P. K. Mehta认为浆骨体积比为0.35∶0.65可解决以上矛盾”针对的是高强混凝土。因为从施工性来说，高强混凝土的浆骨比往往会较大，为了体积稳定性，这是个下限。现在不少人对任何混凝土都用这个比例是一种误解。对于耐久性来说，浆骨（体积）比应当尽量小，以提高混凝土的体积稳定性，但对低强度混凝土，从抗渗性来说，也不宜过小。

当你全心全力去做学问的时候，实在找不出时间去做不应当做的事。我看见某些朋友、学生做学问的态度，不禁慨叹权力欲望愚弄人如此。高尚的情操需要培养和坚持，良师益友、先哲懿范、文化修养都是培养这种情操不可缺少的。我希望青年人能克制私欲，以真挚的感情来欣赏和理解大自然的奥秘。我们每个人在年轻时都怀着一颗赤子之心，关爱家人、朋友，也爱慕异性，对事物充满好奇。我们何不继续保持这份赤子之心，培养孟子的“浩然之气”，昂昂然做一个顶天立地的大丈夫？我们何必受到外界的影响，要富且贵才觉得舒适？学者有了独立的精神、自由的意志，方能创出不朽。

摘自美国现代数学家、菲尔兹和沃尔夫奖得主 丘成桐2012年5月13日丘在学生科协“星火论坛”上所作的特邀报告《为学与做人》

论文之四

无防冻剂的高强混凝土冬季施工的研究和实践*

廉慧珍　　　　　　　　李玉琳　钱选青
（清华大学土木工程系）［北京住总（集团）技术开发中心］

摘　要　高强混凝土由于水泥用量大，水灰比低，当用于较大体积（最小的断面尺寸＞1m）构件时，混凝土内部有较高的温升，早期强度较高。通过在构件中预埋温度传感器实测混凝土温度，跟踪构件中的温度养护试件。实测表明，只要采取相应的蓄热措施，使混凝土温度尽量保持较长时间，在混凝土内部温度下降到负温以前，混凝土已产生足以抵抗冰冻的强度。故可不用防冻剂进行冬季施工。

关键词　高强混凝土，泵送，无防冻剂，同温度养护，冬季施工

前言

北京地区每年 11 月 15 日左右或最低气温低于 5℃时，开始进入冬季施工，则混凝土必须掺用防冻剂。为了补偿掺入防冻剂后混凝土中强度的损失，目前市售防冻剂中都含有减水剂，以降低混凝土中原配合比的水灰比。此外，大多数市售防冻剂还掺有矿物掺和料作为载体，以分散其有效组分。对于普通混凝土，这是一项有效的措施。但是对于水灰比很低的高强混凝土来说，已经掺用的高效减水剂几乎是饱和量，防冻剂中的的减水剂不再起作用，在原配比中不可能再进一步降低水灰比。也就是说，掺用防冻剂后混凝土的强度损失是不可避免的。因此有必要对高强混凝土的冬季施工进行研究。

北京航华科贸中心为超高层建筑，地下两层和地面以上 6 层钢筋混凝土框架圆柱直径为 1.6m，方柱边长为 1.6m。混凝土设计强度等级为 C60。在该工程中使用了低水泥用量、适用于大体积构件施工的 C60 泵送混凝土。1995 年 9 月末开始浇筑，施工期约 8 个月，几乎有半年时间处于冬季施工季节。该工程未使用防冻剂，利用混凝土自身的内部温升，并采取相应蓄热措施，顺利地进行了冬季施工。通过对构件混凝土内部温度变化的监测，对预留试件进行与构件混凝土同温度养护，试验结果表明，高强混凝土的水泥用量较大，用于较大体积（最小断面尺寸＞1m）时，其内部有较高的温升，如同时采取相应的蓄热措施，在北京地区的冬季施工中，是可以不用防冻剂的。而且，高强混凝土早期强度发展快，在混凝土内部温度下降至负温以前，已足够抵抗冰冻的影响。

一、原材料

水泥使用冀东 525＃（Ⅱ）硅酸盐水泥；粉煤灰为内蒙古赤峰元宝山电厂Ⅰ级粉煤灰，需水比为 90％，28 天胶砂强度比为 96％；骨料用河北省三河黄土庄镁质石灰石 5～25mm 连续级配碎石；昌平砂细度模数为 3.0；外加剂为由华迪合成材料公司提供的 NF-2-6 高效复合减水剂。

---

* 原载中国硅酸盐学会混凝土与水泥制品分会、国家建材局科教委新型建材与混凝土水泥制品组举办的《混凝土与水泥制品 1997 年学术年会》并收入论文集，1997.5。

二、混凝土配合比、和易性及强度

控制坍落度为180～200mm，坍落30秒钟的扩展度＞500mm；

施工标准差按6MPa考虑，配制强度为：

$$60+1.645\times 6 = 69.9\text{MPa}，7\text{天强度}>50\text{MPa}。$$

试配结果如表1所示；表2为现场施工混凝土拌和物性质；表3为现场施工强度统计结果。

表1　试配混凝土配合比及强度

胶凝材料组成（kg/m³）		坍落度（mm）	坍落铺展度（mm）	90分钟坍落度损失（%）	抗压强度（MPa）		
水泥	粉煤灰				7天	14天	28天
450	100	195	500	9.1	62.7	72.7	80.1

表2　混凝土拌和物性能

压力泌水比（%）	含气量（%）	坍落度（mm）		坍落流动铺展度（mm）	凝结时间差（分钟）
		出机	泵前（约30分钟）		
68	2.4	220	200	＞500	95

表3　现场混凝土强度统计结果

龄期（天）	3		7		28		60	
	实测（MPa）	达设计（%）	实测（MPa）	达设计（%）	实测（MPa）	达设计（%）	实测（MPa）	达设计（%）
平均值	46.8	81	55.4	92	71.1	118	74.4	124
标准差	4.5	—	6.7	—	4.6	—	5.1	—

三、混凝土的温度及实际强度评价

传统的评价施工混凝土强度的方法是采用现场留样标准养护和“同条件”养护，实际上因试件体积比构件体积小得多，养护环境不同，故都不是实际的强度。为了解不同环境养护对混凝土强度的影响，现场留样按三种方式养护：一组标准养护；一组在现场用麻袋片覆盖并与构件同时浇水；一组在现场露天放置。测定其各龄期强度如表4所示。

表4　不同养护条件对混凝土强度的影响

强度（MPa）　龄期（天） 养护方式	3	7	28	90
标准养护	47.4	57	70.0	72.8
现场覆盖	43.9	55.3	69.7	70.4
现场露天	29.4	45.6	47.2	—

由表4可见，标准养护的试件比现场自然养护试件的早期强度高，但28天以后接近，都达到C60。露天放置的试件无养护条件保证，强度最低；7天以后，因不仅其内部水分消耗而无补充，而且还有水分蒸发，强度提高很少。因此高强混凝土尽早水养护十分重要。

实测该混凝土绝热温升如图1所示。

由图1可见，该混凝土在22小时前升温很快，22小时以后，升温速率大大下降，32

小时升至最高，并开始稳定。混凝土绝热温升为45.88℃。当混凝土体积较大时，如浇筑温度不低于10℃，则柱子中心实际最高温度可不低于此值。本工程为高层建筑的底层柱，最小断面超过1m，应属于大体积。构件混凝土内部温度应当高于标准养和的温度。为了研究混凝土结构内部温度对强度发展的影响，用三种方式养护预留的混凝土试件：一组标准养护；一组与构件“同温度”养护；一组置于现场构件所在位置的自然环境同时养护。

其中与构件“同温度”养护的方法为：在柱内预埋温度传感器，根据所测柱中心温度，调整试件养护水的温度，测定上述三组混凝土试件的强度发展。测点布置如图2所示。各测点温度发展见图3所示。上述不同养护条件下混凝土强度发展测定结果见表5和图4。

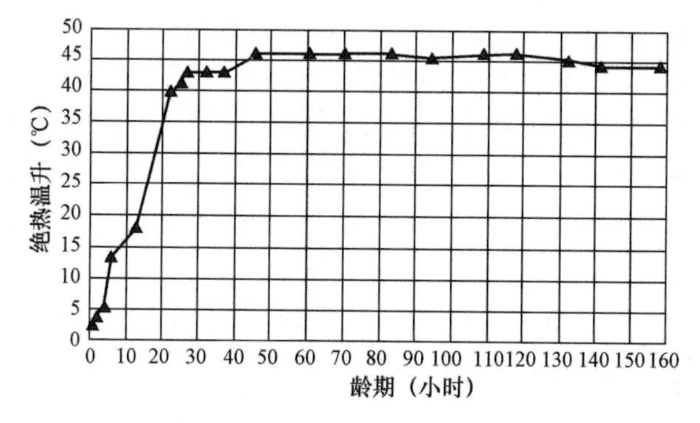

图1　柱C60泵送混凝土绝热温升

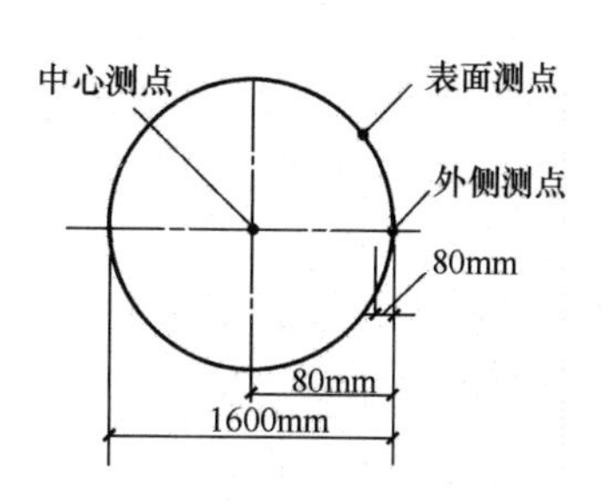

图2　柱混凝土温度测点的布置

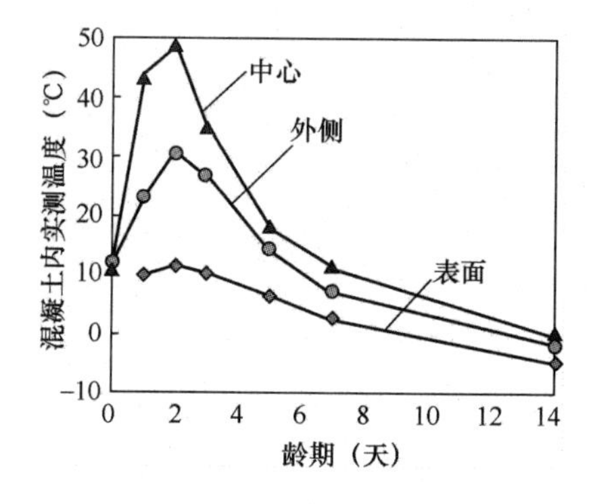

图3　各测点温度发展

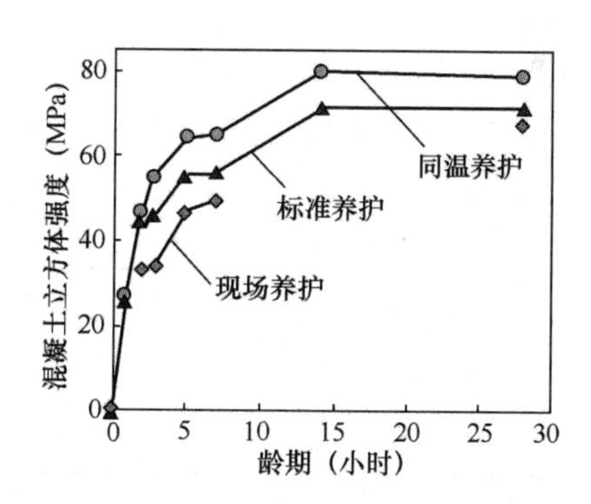

图4　不同养护条件下混凝土强度的发展

混凝土浇筑时气温为1.8～2.2℃，混凝土出机温度为14℃，成型试块温度为12℃。

表5　不同养护条件下混凝土强度的发展

龄期（天）	温度（℃）			试件抗压强度（MPa）			备　注
	柱中心	柱外侧*	大气	标准养护	同温水养护	现场自然养护	
1	44	23	1.2	26.2	26.9	—	
2	49	31.2	0.7	45.6	48.5	33.2	
3	35	27	1.0	46.0	55.0	34.2	
5	18	14	−2.9	55.1	64.5	46.5	

续表

龄期（天）	温度（℃）			试件抗压强度（MPa）			备 注
	柱中心	柱外侧*	大气	标准养护	同温水养护	现场自然养护	
7	11	−1	−2.9	56.0	65.0	49.4	
14	0	−1.5	−4	71.2	79.8**	—	
28				71.3	79.0**	67.5	

* 柱外侧指与柱的外表面水平距离 80 mm 处（见图 2）。

** “同温度养护”试件只养护到 7 天，7 天后，该试件试件取出置于现场自然环境，故 14 天强度已不再是“同温度养护”的结果，而是“同温度养护”7 天后持续发展的结果。

由图 3、图 4 和表 5 结果可见，在低温下浇筑的混凝土内部温升也很快，也有较高温度，即使是混凝土表面，也可维持混凝土不至结冰。由表 5 结果可见，与柱中混凝土“同温度”养护的混凝土强度最高。这表明，柱内混凝土温度有利于结构混凝土强度的发展。不同养护条件下强度的差异主要是因为粉煤灰对温度很敏感，尤其是早期。现场养护的试件因体积比柱的小得多，温度很快下降，故其强度比柱内混凝土的强度低的较多，不能代表结构强度。

四、C60 混凝土的冬季施工

高强混凝土水灰比低，水泥用量大，早期强度发展快，较大体积的混凝土内温升也较大。根据上述测定的柱内混凝土的温度和强度发展可知，在一定的负温下，可以不掺用防冻剂。

1. 防冻剂对混凝土在负温下强度的影响

为了确定施工方案，对掺与不掺防冻剂的上述配合比混凝土进行负温养护和负温-标准联合养护下的强度对比试验。联合养护的制度为：成型后在−10℃下养护 3 天后继续标准养护 28 天，记作“−3＋28”；依此类推为“−7＋28”、“−10＋28”。试验结果如表 6 所示。所用防冻剂经按标准进行复验合格。

表 6　在负温下养护掺与不掺防冻剂的高强混凝土强度对比

试样	养护条件	1 天	2 天	3 天	28 天	−3＋28	−7＋28	−10＋28
掺防冻剂	成型后 0℃	10.1	13.5	20.7				
不掺防冻剂	成型后 0℃	10.7	13.5	20.0				
掺防冻剂	标准养护				71.5			
不掺防冻剂	标准养护				57.3			
掺防冻剂	成型后−10℃						56.3	
不掺防冻剂	成型后−10℃					74.6	65.5	67.4

由表 6 可见，由于 C60 混凝土的强度发展快，即使置于负温下养护，强度照样增长；早期处于低温时，只要恢复养护，强度就会正常增长。无论标准养护还是联合养护，不加防冻剂的混凝土都比加防冻剂的混凝土强度高。在 0℃下养护时，不加防冻剂的混凝土和加防冻剂的混凝土强度无明显差别，说明该 C60 混凝土具有良好的抵抗负温影响的性能；即使在−10℃下养护 10 天，再进入标准养护 28 天后，强度仍然达到 67MPa 以上。有鉴于此，

决定不使用防冻剂，按原配合比进行－10℃以上的冬季施工。同时采取相应的蓄热措施，防止混凝土内温度下降过快。

2. 冬季施工的实施

进入1996年元月，气温骤降，在浇筑C60混凝土柱头时，正值最低气温为－9℃，最高气温－1℃，为了保证施工质量，首先要求搅拌站采用热搅拌，控制水温大于65℃，砂石不受冻结块；混凝土罐车有保温措施。因柱头体积较小，温升也较低，故要求控制浇筑时混凝土坍落度不大于180mm，以进一步提高混凝土强度，增强混凝土抵抗低温的能力。要求24小时拆除模板后，立即用浸水麻袋片紧裹混凝土表面，外面再包以塑料布，防止混凝土内水分损失。施工时混凝土温度如表7所示。

表7 低温下施工的混凝土温度（℃）

气温	出机混凝土温度	浇筑混凝土温度	浇筑后4小时柱中心	浇筑后4小时柱外侧
－7.5	19.0	11.9	8～9	6～7

现场试验如下：

(1) 按最不利情况，将所预留的试件成型后带模置于－8℃以下的现场大气中，不加任何覆盖，分别自然养护至30小时、75小时、96小时，取出在室温下放置一小时，测定其强度如表8所示。

表8 在实际的负温下养护的小试件混凝土强度

养护制度	养护温度和历时 / 养护历时总计 / 养护顺序	30小时	75小时	96小时
养护制度	1	－8～－3℃ 15小时	－4～－6℃ 20小时	－1～－3℃ 10小时
	2	3～6.5℃14小时	3～5℃ 24小时	3～6℃ 10小时
	3	23℃ 1小时	23℃ 1小时	23℃ 1小时
抗压强度（MPa）		3.3	9.5	13

各试件破型后无受冻现象，也无酥散的现象。由前述表4中的结果可知，这样的混凝土只要转入正温下养护，强度即可恢复增长至正常。但是实际工程的混凝土并不是处于这样的情况，而是由于内部的温升，使混凝土在7天前一直处于正温下。故实际情况要好得多。

(2) 在不同尺寸的试件中预埋温度传感器，成型后带模置于现场大气中，不加任何覆盖，分别自然养护至不同龄期，测定试件内部温度，然后在正温下放置1小时，测定其抗压强度如表9所示。

表9 不同尺寸的试件内部温度和抗压强度的发展

混凝土龄期（小时）	试件尺寸（cm^3）	混凝土内部温度（℃）	恢复环境温度（℃）	抗压强度（MPa）
23	10×10×10	－1	30	0.95
	15×15×15	0.8	25	0.89
36	10×10×10	－1.6	32	5.7
	15×15×15	1.2	32	4.9

续表

混凝土龄期（小时）	试件尺寸（cm^3）	混凝土内部温度（℃）	恢复环境温度（℃）	抗压强度（MPa）
40	10×10×10	0.1	36	7.6
	15×15×15	0.1	36	8.9
51	10×10×10	−0.4	32	11.4
	15×15×15	0.2	32	12.9
72	10×10×10	−3	33	25.6
28天	10×10×10	4	—	43.7

表9中混凝土内部温度的波动是由气温波动引起的，由于混凝土中的温升和昼夜温度变动的影响，试件中温度3天以后才开始下降较多，而混凝土在一定的负温下强度仍然增长。负温下7天强度达到负温下28天强度的58.5%。

（3）在平均气温低于0℃时浇筑混凝土，浇筑时间为21：00左右，气温为−1.2℃，混凝土进场时温度为16℃，泵送时为14℃，成型时为10℃。2小时后开始量测混凝土温度。除按图1测点预埋温度传感器外，另在表面留一个直径约1cm、深15 cm的洞，插入温度计。各测点所测温度、测温龄期、时间以及相应强度见表10所示。

表10　负温下浇筑的C60混凝土温度和不同养护条件下的强度

龄期（小时）	测温时间（时：分）	实测气温（℃）	混凝土温度（℃）				混凝土强度（MPa）		
			中心	外侧	表面内15cm	表面	标准	同温	现场
2	23：00	−1.2							
5	2：00	−3.0	17.1	7.3					
8	5：00	−3.0	22	8	11				
11	8：00	−2.2	25	8.5	12	11.3			
17	14：00	−3.2	39	11.5	17.5				
21	18：00	1.0	44	13.3	12				
27	24：00		45	13	12	9	25.2	31.4	16.2
35	8：00	2.9	47.5	23.1	31.5	12.7			
43	17：00		45	23.6	28	12.9	47.5	49.4	22.8
58	8：00		39	14.3	23.4		55.1	57	34.2
82	8：00		29	4.0	12		58	59.9	35.6
106	8：00		19	4.0	5.0				
115	17：00		6.8	0.3	1.8	−0.6	58.9	60.8	47.8

从表10可见，气温下降时，混凝土内部温升也随之降低，温度下降得也较快，但在混凝土内部的强度达到设计强度以前，混凝土一直处于正温下，即使在现场自然条件下养护的混凝土在不到5天的龄期时，强度也达到了设计强度的70%以上。

（4）采取不同的养护措施，考察对混凝土中温升和降温速率的影响如表11所示。

表11　养护状况对混凝土降温速率的影响

降温速率 \ 降温历时 最高温度	<50小时	<100小时	<150小时	14天	养护状况
52℃	1℃/小时	0.32℃/小时	0.26℃/小时	↓87%	蓄热养护稍好
49℃	0.52℃/小时	0.32℃/小时	0.26℃/小时	↓98%	蓄热养护稍差

由表 11 可见，柱内混凝土温度相同，而蓄热养护稍差时，降温速率就较快，14 天时混凝土温升下降 98%，基本与环境温度平衡。这表明加强养护对高强混凝土质量控制有至关重要的意义。除保证使强度增长的温度和湿度外，应控制混凝土温升不要太低而且尽量持久，以保证混凝土较长的正温时间，并且不产生较大的内外温差。只要混凝土处于正温，就应不断地补充养护水。

该工程的高强混凝土已成功地完成了约 1000m^{3}的冬季施工，节省防冻剂费用 9 万多元。而且由于现场少加一种组分，给施工质量控制也带来便利。

五、结论

1. 高强混凝土由于水灰比低，早期强度高；水泥用量较大，水泥水化提供的热量大。用于大体积构件的高强混凝土在气温不低于－10℃时进行冬季施工，可以不加防冻剂而充分利用其内部的温度。但需采取冬季施工的一切措施，保证浇筑温度不低于冬季施工规范的规定，并对成型的混凝土采取蓄热保温的措施，使混凝土温度尽量保持较长时间。

2. 在冬季施工期间，顺便监测了梁的 C30 混凝土跨中中心温度，浇筑后 4 天为 2.9℃，而同期柱子的 C60 混凝土中心温度为 19℃。实际上，当水泥用量相同时，水灰比大的比水灰比小的混凝土中温升高。如果梁也采用 C60 混凝土，可能混凝土内部温升要低得多，但由于至少有 4～5 天时间处于正温下，强度可发展到足以抵抗冰冻的压力（负温下 3 天强度可达 25MPa 以上）。预计也有望在冬季施工时取消防冻剂。

自评

1. 防冻剂作用机理是降低混凝土中水的冰点和提高混凝土早期强度。传统溶解度大的的氯盐和钠盐效果最好，但对硬化混凝土后期性能有不利影响。取消氯盐并改成钙盐后，效果变差，我国无氯盐和钠盐防冻剂，为了不降低混凝土强度，多掺入减水剂。在美国的工程中没有防冻剂，而是使用低水胶比、低坍落度和早强水泥或促凝剂。更重视养护和加热、保温、蓄热等措施（见 Kosmatka S. H..，Korkhoff B.，Panaress W. C.《Design and Control of Concrete Mixture》，14th edition，published by PCA，Skokie，Illnois，USA2003.）。取消防冻剂而利用高强混凝土内部的温升进行冬季施工是应当提倡的。如能进一步做中等强度混凝土的试验，可能会使冬季施工的观念有更多的改变。

2. 北京地区冬季气温一般最低不会低于－20℃，利用混凝土浇筑后内部温升抵抗冰冻是可行的，但是除了实验室按现行标准方法的试验外，本文的试验的数据都是在不低于－9℃的气温下取得的，多数实验室的气温都在－5℃以上，似欠代表性。

3. 混凝土工程无论在何种季节，施工的养护都包括保湿和控制温度，而不是机械地浇水。

论文之五

国内外自密实高性能混凝土研究及应用现状*

廉慧珍[1]　张　青[2]　张耀凯[3]

在传统上，混凝土的密实成型依靠对具有一定和易性的混凝土的振捣。超塑化剂的出现，使同样原材料和配合比的混凝土坍落度从过去的5～7cm变成18cm以上，可以泵送施工，易于振捣，大大提高了施工效率。80年代后半期，日本东京大学教授岡村甫开发了"不振捣的高耐久性混凝土"，称之为高性能混凝土（High Performance Concrete）[1]。1996年在美国泰克萨斯大学讲学和在1997年发表在《混凝土国际（Concrete International)》的论文中，岡村甫称该混凝土为自密实高性能混凝土（以下简称自密实混凝土 self compacting concrete）[2]。之所以称为高性能，是因为具有很高的施工性能，而能保证混凝土在不利的浇筑条件下也能密实成型，同时因使用大量矿物掺和料而降低混凝土的温升，并提高其抗劣化的能力，而可提高混凝土的耐久性。在国内外发表的论文、专利中，这种混凝土还有许多其他名称，如高流动混凝土（high flowing，high fluidity）、高施工性混凝土（high workability）、自流平混凝土（self-leveling）、自填充混凝土（self-filling）、免振捣混凝土（vibration free）等。

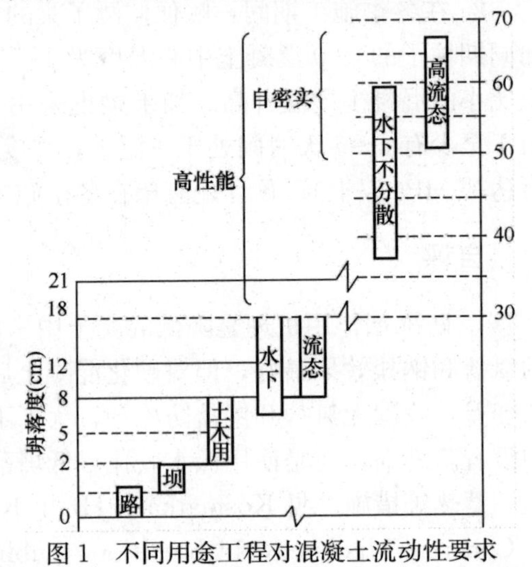

图1　不同用途工程对混凝土流动性要求

自密实混凝土即拌和物具有很高的流动性而不离析、不泌水，能不经振捣或少振捣而自动流平并充满模型和包裹钢筋的混凝土。这种混凝土虽然比相同强度等级的普通混凝土材料费用略高，但由于节省动力和劳力、加快工程进度，并解决扰民问题，其综合效益是显著的。特别是用于难以浇筑甚至无法浇筑的部位，可避免出现由于振捣不足而造成的空洞、蜂窝、麻面等质量缺陷。自密实混凝土配制的关键是满足良好的流变性能要求。图1为日本对土木工程所用不同混凝土流动性的要求[2]。自密实混凝土属于其中的高流动性混凝土的一部分。

一、国内外自密实混凝土的应用概况

至1994年底，日本已有28个建筑公司掌握了自密实混凝土的技术。可自密实的高流动

* 原载中国土木工程学会高强混凝土委员会第三届学术讨论会议（1998.10. 济南）论文集，《施工技术》1999年第5期。

1. 廉慧珍，教授，清华大学土木工程系，100084

2. 张青，副总工程师，北京建筑工程（集团）第二建筑工程公司

3. 张耀凯，高工，中国冶金建筑科学研究院

性混凝土在日本得以发展的背景是[1]：

① 施工现场作业人员高龄化，熟练作业人员的缺乏；

② 浇筑、振捣作业的节省劳动量的要求；

③ 混凝土浇筑作业机械化、自动化程度的提高；

④ 混凝土使用范围的扩大；

⑤ 混凝土结构物的大规模化、高层化；

⑥ 高密度钢筋结构物的增加；

⑦ 具有浇筑困难部位的结构物的增加；

⑧ 对混凝土高质量、高耐久性的要求；

⑨ 低水胶比、高强度、高耐久性混凝土的使用。

由表 1 可见，自密实高性能混凝土特别适合于浇筑量大、浇筑高度大、钢筋密集、有特殊形状等的工程。

如日本明石海峡大桥使用了 100 多万 m^3 不振捣的高流动性混凝土[3][4]，其中缆索锚固基础采用现浇地下混凝土连续墙露天开挖的方法施工。地下混凝土连续墙为环形平面，外浇筑混凝土。该工程混凝土浇筑量大，浇筑深度深，无法振捣，要求不间断地连续浇筑。混凝土温升要求低于 30℃。普通的混凝土无法满足施工要求。

表 1　1992～1993 年日本高流态（自密实）混凝土工程应用实例[6]

工程部位	工程概况	原材料	输送	浇筑	质量管理
木场公园大桥斜拉桥主塔	高度为 64m，浇筑量 $650m^3$，f_c=400kg/m^2	B-矿渣水泥，粉煤灰，抗分离剂	$5m^3$ 的搅拌车，约 30 分钟	$1m^3$ 的吊斗＋软管套。与普通混凝土相比每循环作业人数和需要的时间减半	
PC LNG 贮罐的钢筋混凝土防波堤开口部大门（反打）	15×4.5×0.9 m，6×3.5×0.9m，浇筑量 $80m^3$，f_c=400kg/m^2	低热水泥，B－矿渣，石灰石粉，CSA 膨胀剂、Al 粉	$4.5m^3$ 的搅拌车，约 30 分钟	泵车＋5in 软管套压送，压送速度约 $30m^3$/h	通过 50mm 筛网检查，管道冷却，监测温度应力
明石大桥锚固	P2 ϕ80 × 70m*，P3 ϕ78×67m*，浇筑量约 50 万 m^3**，设计强度 24 MPa	两组分或三组分的低热水泥，石灰石粉	直接泵送，8 英寸泵管，6 系统	6 台固定泵＋料浆入口自动启闭装置层装浇筑，最大泵送速度为 $288m^3$/h，日浇筑量约 $1900m^3$	骨料表面含水量每 30～60 分钟检测一次，混凝土每 $150m^3$ 试验一次，管冷却，温度应力计监测
过密钢筋密封层的二次覆土	长 12m，内径 3m，厚度 225mm，浇筑量 $48.4m^3$，f_c=300kg/m^2	低热水泥（3 组分）	$5m^3$ 的搅拌车，约 32～55 分钟	泵车＋5in 配管压送，压送速度约 20～$30m^3$/h	每 9 车中 4 车测坍落流动度、空气含量和漏斗流下试验；厂测 2 次，入泵前 1 次
过密钢筋密封层的二次覆土	9×3m 的块，浇筑量约 $45m^3$，f_c= 240kg/m^2	普通水泥、增稠剂	搅拌车	泵车＋5in 配管压送，喷射式浇筑	

续表

工程部位	工程概况	原材料	输送	浇筑	质量管理
奥美浓发电厂水轮机罩外壳	机罩高 2.15 浇筑量 $280m^3$，最大配筋密度为 $250kg/m^3$	普通水泥、石灰石、增稠剂	$5.2m^3$ 的搅拌车，约 60 分钟	泵车＋5in 配管压送，用每 5m 设一阀门均匀浇筑	每 5 车测 1 次坍落流动度和含气量，每 10 车测 1 次强度，每 $30m^3$ 测 1 次砂表面含水率
LPG 贮罐钢筋混凝土外槽	2.95×3m，高架，厚 1 ～ 0.55m，浇筑量 $1370m^3$	普通水泥、粉煤灰、高密矿渣、抗分离剂	$5.5m^3$ 的搅拌车，约 30 分钟	装有吊杆的泵车 4 台，其中一台有分叉配管，分叉配管大体水平完成浇筑	
铁路高架桥	D32 异形钢筋，间隔 10cm，2 段配筋，柱接头高密度配筋，浇筑量为 $100m^3$	B－矿渣水泥，粉煤灰，抗分离剂	$4.5m^3$ 的搅拌车，约 60 分钟	泵车＋5in 配管压送	

* 原文为 80m×63m×71 m，有误。该两个水下桥墩为圆形，根据文献[7][8]修改如是；
** 原文为 24 万 m^3，系指一个桥墩。两个桥墩浇筑量约 50 万 m^3。

在西方也有不振捣的混凝土的应用，如美国西雅图 62 层的双联广场钢管混凝土柱，钢管直径分别为 3.0m、2.25m、0.75m，无配筋，用钢管起纵向钢筋和箍筋的作用。按承载要求混凝土设计强度为 79MPa。但由于柱子的刚度的要求，混凝土的弹性模量成为主要的因素。设计弹性模量为 50GPa，为此 28 天抗压强度需达 115MPa。为保证混凝土的密实性，混凝土从底层逐层泵送，无振捣[5][6]。这种高强的不振捣混凝土是用于无配筋的钢管混凝土的。在美国为了保证混凝土的浇筑质量以保证钢筋和混凝土的整体性，在密筋的钢筋混凝土和几何形状复杂的结构中，也使用高坍落度而能自流平的混凝土，但强调仍需要适当的振捣以确保混凝土的足够密实[7]。近年来由于在日本不断有采用自密实混凝土成功的工程实例，美国也开始注意该项技术。日本自密实混凝土的创始人岡村甫教授 1996 年应邀在美国德克萨斯大学讲学，介绍了自密实混凝土的概念、性能及其检测和工程实例[8]。

日本在 1992 年到 1993 年各学会、技术刊物等发表的自密实的高性能混凝土在土木工程中应用实例如表 1[9]所述。

近年又不断有发展，例如用于水电站泄洪坝，墙、柱、楼板等预制构件的联结，轻质高流动混凝土，钢丝网模板体系中的浇筑，等等。

在我国，由于原材料和施工条件的差别，不能照搬日本的配合比。1995 年，清华大学陈恩义用目前市售的原材料进行试验，在北京清河 602 住宅小区工地成功地浇筑了 C25 流态混凝土，用于墙体的施工，使混凝土振捣布点减少一半，每点振捣时间减少三分之二，浇筑高度从 2m 增加到 4 m，大大减轻劳动强度，方便了施工，提高了施工效率，缓和了噪声扰民问题，并且解决了普通混凝土由于胀模而返工的问题，受到工地施工人员的欢迎。该项成果为在北京实现自密实混凝土提供了可行性的依据。

1996 年 9 月，北京城建集团总公司构件厂搅拌站进行了自密实混凝土技术的鉴定。其中 C30 混凝土用于实际工程，浇筑量达 $3000m^3$。根据鉴定文件所附的工地留样检测报告，28 天平均抗压强度达 37MPa。特点为：①由搅拌站集中生产，因而质量均匀；配合现场泵送，解决了地下暗挖施工混凝土浇筑和困难并无法振捣的问题；②使用该厂独家生产的苯酚系高效减水剂 DFS，减水率可达 25%～30%；③设计了一个带隔板的混凝土流动箱，与配筋模型试验结合评价混凝土流变性能。[10]

北京建工集团二建公司从 1994 年开始进行高流动自密实混凝土的试验研究，1996 年和 1997 年分别在三个工程试用于柱、梁、楼板和基础后，通过了技术鉴定。其特点为：①采用两种不同原材料来源的萘系高效减水进行复配，解决了拌和物离析问题；②在工地现场搅拌，用吊斗施工；③改进了现有坍落度测定装置，并参考 L-流动仪，设计和制作了“L800 型高流动自密实混凝土施工性综合测定仪”，用于现场拌和物的质量控制；④建立了不同于普通混凝土的高流动自密实混凝土配合比设计方法。[11]

由于自密实混凝土的实现适应建设事业发展的需要，在需要和条件合适的工程中应用是有利的，其他施工部门也开始研究和使用，例如北京住总集团五公司今年在光华木材厂宿舍楼浇筑了 C35 的内外墙体和楼板。

二、自密实混凝土的性能

1. 自密实混凝土拌和物的性质

自密实混凝土的拌和物除高流动性外，还必须具有良好的抗材料分离性（抗离析性）、间隙通过性（通过较密钢筋间隙和狭窄通道的能力）和抗堵塞性（填充能力）。国外大多用拌和物的坍落流动度，即坍落后拌和物铺展的直径，作为高流动性混凝土流变性能的量度。日本报道，坍落流动度一般为 50～70mm。超过 70mm 时，拌和物易产生离析；不到 50mm 时，则可能发生充填障碍。拌和物抗离析性可用坍落流动速率来评定。坍落流动速率用拌和物坍落后铺展到直径为 50cm 的时间表示。坍落流动速率快时，流动性好，但过快时容易产生离析[2]（该文献未给出定量值）。也有人在一种 L 形流动性测定装置的转角处装置传感器测定拌和物流动初始的速率，来判断拌和物的抗离析性。对于泌水量，按日本标准 JASS A 1123 的方法检测时，JASS 5 规定对普通混凝土要求$<0.5cm^3/cm^2$，而对高耐久性的混凝土则要求$<0.3cm^3/cm^2$[2]。抗离析性直接影响混凝土拌和物浇筑后的均匀性。必要时可检测水平流动至不同部位或垂直浇筑到不同高度的拌和物中粗骨料的含量，作为拌和物匀质性的评定。

一般，自密实混凝土的凝结时间较长，可达 10 小时左右，尤其是在冷天施工时。初、终凝时间间隔短，一旦凝结，强度很快就会增长；如果使用低浓度的高效减水剂，由于 Na_2SO_4 含量较高，会使混凝土凝结时间较短，甚至在夏季还需添加适量缓凝剂。

2. 硬化混凝土的性质

强度 自密实混凝土属于高性能混凝土，可有很宽的强度范围，从 C25 到 C60 以上。我国目前大量使用的是 C30～C40。为了保证及时拆模，成型后在标准条件下 24 小时抗压强度应≥5MPa[2]。对施工计划允许的、着重长期强度的、使用低热水泥等的情况，可放宽上述要求。

弹性模量 由于粗骨料用量较少，自密实混凝土比使用同一品种骨料的普通混凝土弹性模量稍低些，根据 JIS 的方法试验，标准养护 28 天时，降低值小于 10%[2]。根据北京二建的测试，因采用低水胶比，尽管有所降低，但仍能满足结构设计规范的要求[11]。

收缩 通常，由于粗骨料用量小，粉体材料用量大，自密实混凝土的干燥收缩会大些，容易产生有害裂缝。可根据结构形式、构件尺寸、施工条件、工程性质等的不同，确定不同的目标。日本的资料是在用标准条件下养护7天的试件在(20±3)℃相对湿度(60±5)%的条件下6个月的干缩为8×10^{-4}以下，比同种骨料的普通混凝土收缩增加量<10%[2]。掺用粉煤灰和少量膨胀剂有利于减小收缩[11]。

抗碳化性 普通混凝土的碳化速率和水灰比近似于线性关系。掺入矿物掺和料后，在相同水灰比下，碳化速率增加。降低混凝土的水胶比，则可达到相近的碳化速率。混凝土掺用大量混合材料后，碱度大大降低，会加速碳化而不利于对钢筋的保护，但自密实混凝土因水胶比很低，混凝土密实度高，抵抗碳化的能力强。单纯从材料来说，应当可以不怀疑其对钢筋保护的作用。而实际上，对不同的构件应作不同的考虑：对主要受压的构件如基础、墩柱以及长期处于水下的结构，可不考虑碳化问题；对受弯构件如梁、板，则因在荷载作用下产生裂缝是不可避免的，设计时允许受力后受拉区产生宽度不大于0.2mm（对预应力钢筋混凝土是0.1mm）的裂缝，则碳化问题就应考虑。在这样的情况下，抗碳化的性能和掺和料的品种、掺量有关。例如掺粉煤灰30%而水胶比为0.35时，碳化速率约和普通混凝土水灰比为0.5时相当；同样效果的矿渣掺量可达70%；水胶比为0.4、矿渣掺量达50%时，碳化速率同普通混凝土的相差无几（图2）[12]。

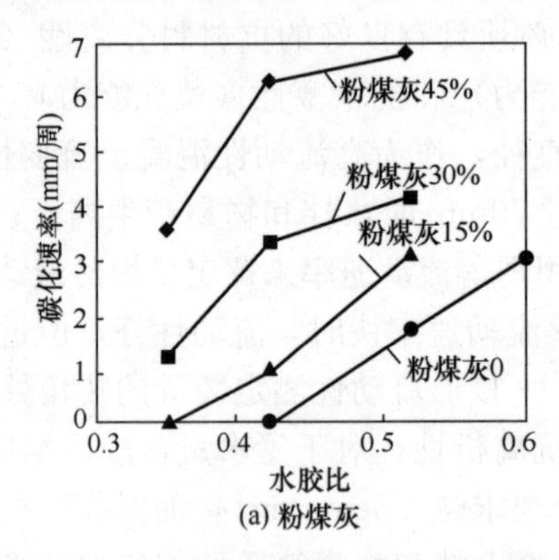

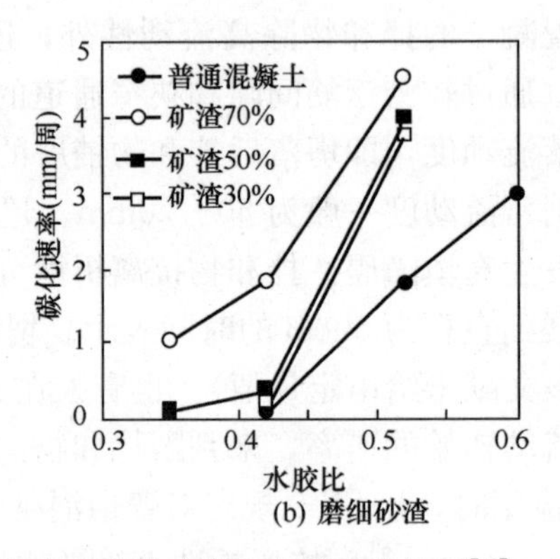

图2 粉煤灰和磨细矿渣不同掺量下混凝土碳化速率和水胶比的关系[12]

因此，对用于不同部位的自密实混凝土，可通过配合比的调整来保证其抗碳化的性能。有些矿物掺和料中往往含有一定量的碱，对保持混凝土中的pH值是否起作用，需要通过试验来证明。对自密实混凝土的抗碳化性能也需要和构件的裂缝情况结合起来进行实验研究。

其他 掺用一定量的引气剂，是低抗冻融作用特别是除冰用的有效措施。日本规范规定，经冻融循环作用后，动弹性模量必须保持80%以上，循环次数最低为200次，在冻融循环作用频繁的环境下，要求300次；含气量一般要求为3%～6%，在冻融循环作用频繁的环境下，为4%～7%。在日本多使用引气型减水剂（AE减水剂）。由于掺入较大量矿物掺和料，自密实混凝土有很好的抗化学侵蚀和抗碱骨料反应的能力。矿物掺和料抗碱骨料反应的有效掺量粉煤灰为30%，矿渣是40%[13]。

三、自密实混凝土的原材料和配合比

1. 自密实混凝土的原材料

胶凝材料 除要求温升很低的大体积工程自密实混凝土需要选用中热或低热水泥外，硅酸盐水泥、普通硅酸盐水泥和矿渣硅酸盐水泥都可选用，按目前我国标准标号应不低于425#，具有较低的需水性，还应考虑与所用高效减水剂的相容性。掺用矿物掺和料的目的是调节混凝土的施工性能、提高混凝土的耐久性，降低混凝土的温升。因此活性混合材料应具有低的需水量、高的活性。为了保证混凝土的耐久性可利用不同掺和料的复合效应。例如粉煤灰和磨细矿渣相比，矿渣比粉煤灰活性高，而抗离析性差；粉煤灰比矿渣抗碳化性能差，但收缩小。按适当比例同时掺用粉煤灰和矿渣，则可取长补短。由于采用低水灰比，强度要求较低时，可再掺用适量填充性混合材料，如石英砂粉、石灰石粉等，来保证足够的浆量。因此，日本的高流动性混凝土普遍采用水泥、矿渣、粉煤灰三组分胶凝材料，有时加上石粉，成为四组分。据日本资料，从活性来说，磨细矿渣最优的比面积为6000～8000cm²/g[14]，但从减小自生收缩来说，最好不超过4000cm²/g[15]。

骨料 骨料的粒形、尺寸和级配对自密实混凝土拌和物的施工性，尤其是对拌和物的间隙通过性影响很大。和泉意登志报道，用密实体积率为62%的日本青梅碎石，混凝土拌和物的间隙通过量是用密实体积率为67%的日本鬼怒川卵石时的一半，因此粗骨料的密实体积率大的好[16]。粗骨料的最大粒径，当使用卵石时为25mm，使用碎石时为20mm，间隙狭小的部位用15mm[16]。由于砂率大，砂子宜选用中粗砂，以偏粗为好。应严格控制砂中粉细颗粒的含量和石子的含泥量。同时要保证0.63mm筛的累计筛余大于70%，0.315mm筛的累计筛余为90%左右，而0.15mm筛的累计筛余>98%[17]。

外加剂 即使设计强度等级不高，也要用高效减水剂。日本杉本贡报道，对高流动混凝土外加剂性能的要求为：对拌和物，要有优质的流化性能、保持流动性的性能、合适的凝结时间与泌水率、良好的泵送性；对硬化混凝土：对力学性质无坏影响、对干缩和徐变无坏影响、耐久性（抗冻、抗渗、抗碳化、抗盐浸）好。为此多采用高性能引气型（AE）减水剂[18]。同时，由于自密实混凝土拌和物往往有离析的倾向，在日本多采取掺用抗离析剂或增稠剂来解决。日本的抗离析剂有纤维素水溶性高分子、丙烯酸类水溶性高分子、葡萄糖或蔗糖等生物高聚物等[19]。其中纤维素醚和甲基纤维素用得最多。但是添加抗离析剂时，对混凝土的强度有些影响。北京建工集团二建公司使用两种不同原料来源的萘系高效减水剂复合使用，对自密实混凝土的抗离析性有显著效果。

2. 自密实混凝土的配合比

自密实混凝土的配合比应满足拌和物高施工性能的要求，因此，与相同强度等级的普通混凝土相比，有较大的浆骨比，即较小的骨料用量，胶凝材料总量一般要超过500kg/m³[14]；砂率较大，即粗骨料用量较小。砂率最大可达50%左右；使用高效减水剂；由于胶凝材料用量大，必须掺用大量矿物掺和料。掺和料总掺量一般大于胶凝材料总量的30%。为了保证耐久性，水胶比一般不宜大于0.4。

自密实混凝土配合比的确定是以上各参数和混凝土强度、耐久性、施工性、体积稳定性（硬化前的抗离析性，硬化后的弹性模量、收缩徐变）等诸性质之间矛盾的统一。例如流动性和抗离析性要求粗骨料用量小，但粗骨料用量小时硬化混凝土的弹性模量低，收缩、徐变大；砂率大有利于施工性和强度而不利于弹性模量；水胶比大有利于流动性，而不利于强度和耐久性，等等。因此与普通混凝土配合比设计不同的是，首先根据上述矛盾的统一确定粗骨料的最合适用量、砂子在砂浆中的含量。图3为小泽和岡村甫建议的自密实混凝土配合比设计体系实例[20]。是作为砂浆和混凝土两个层次的体系考虑的。石子最大粒径为20mm使用中热水泥和增稠剂。

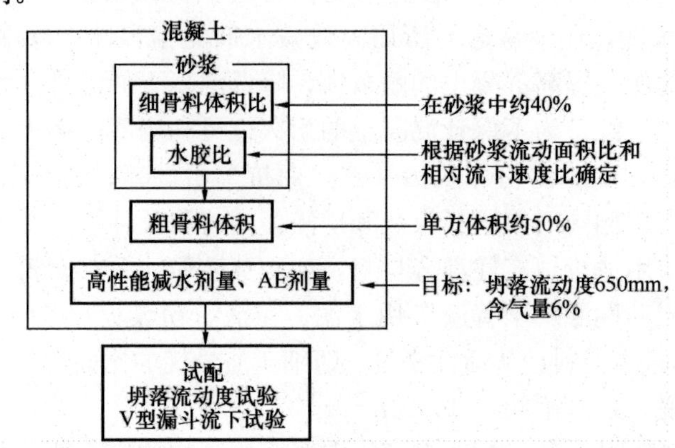

图 3　小泽和岡村甫建议的自密实混凝土配合比设计体系实例[20]

北京建工集团二建公司建议的是按混凝土、砂浆、水泥净浆、胶凝材料四层次体系设计，如图 4 所示。

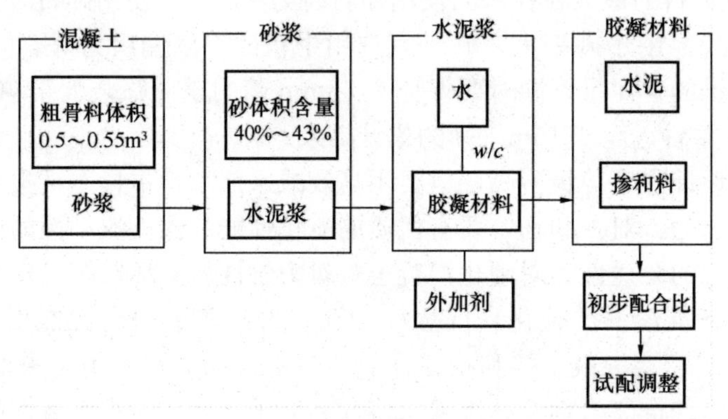

图 4　北京建工集团二建公司建议的自密实混凝土配合比设计体系

以上步骤均采用绝对体积方法计算。由此可见自密实混凝土的浆骨比变化的范围是很小的。混凝土的性质主要受水泥浆浓度和性质的支配。

表 2 所示为自密实混凝土配合比的实例。

表 2　自密实混凝土配合比实例

工程名称	现场实际强度（MPa）	单方混凝土中原材料用量（kg/m³）					
		水	水泥	掺和料	砂	石	外加剂
日本明石大桥 A1 基础[21]	91 天 62.3	142	中热 172	矿渣 172 粉煤灰 86	771	965	高效引气型减水剂＋超塑化剂＋引气剂。
北京恒基中心地下通道方厅墙、顶板、柱[10]	28 天 37.5	200	P.O.525# 280	粉煤灰 175 UEA 33	830	830	DFS　0.75%
北京凯旋大厦梁、板、柱[11]	28 天 53.3	200.8	P.S.425# 381	粉煤灰 148 UEA 20	796	760	SN　1.8%

为了便于参考，在表 3 中介绍日本文献中高流动混凝土配合比的实例。

表 3　日本高流动混凝土配合比的实例

来源	① w/p (%)	s/m (%)	单方混凝土原材料用量（kg / m³）										
			水 w	粉体材料					骨料		外加剂		
				水泥		混合材		粉体总量	砂	石	高性能 AE 减水剂	AE 减水剂	抗离析剂
				②OPC	BB	③BS	FA						
岡村[22]	30.1	44.7	159	155	—	171*	202	528	760	874	4.62	—	纤维素 0.02
小门[23]	32.0	45.1	197	178	—	382**	—	560	786	822	14.56	—	—
松岡[24]	34.0	44.4	170	—	400	—	100	500	717	907	11.0	0.9	葡萄糖类 0.08

注：① w/p 为水和粉体的重量比；s/m 为砂浆中砂子体积浓度；
② OPC 为普通的硅酸盐水泥（相当于我国的硅酸盐水泥）；BB 为 B 种高炉矿渣水泥；
③ BS 为磨细矿渣，* 比表面积 5400cm²/g，** 比表面积 5920cm²/g；　FA 为粉煤灰。

四、自密实混凝土施工的特点

自密实混凝土的质量对原材料的变动很敏感，制作和施工中各环节的控制要求严格，因此对操作工的要求低了，而对技术人员和管理人员的要求高了。由于组成材料多，必须注意搅拌均匀，目前多采用双卧轴强制式搅拌机，搅拌时间比普通混凝土的长 1～2 倍[25]，60～180 分钟甚至更长是必要的[26]。搅拌不足的拌和物不仅因不均匀而影响硬化后的性质，而且在泵送出管后流动性进一步增大，会发生离析的现象。投料顺序最好是先搅拌砂浆，最后投入粗骨料[25]。

一般来说，自密实混凝土最适合于泵送浇筑。浇筑墙或柱时浇筑高度可在 4 m 左右[27]。浇筑顺序可参考和泉意登志在文献中给出的泵管移动顺序实例[28]，如图 5 和图 6 所示。

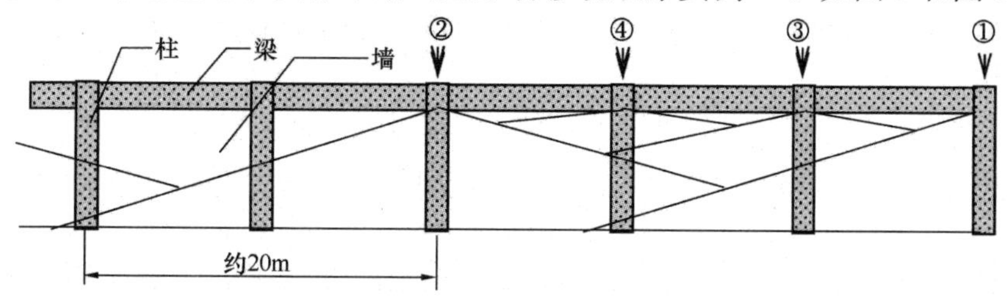

图 5　自密实混凝土浇筑顺序举例示意
▼为浇筑位置　　①～④为浇筑顺序

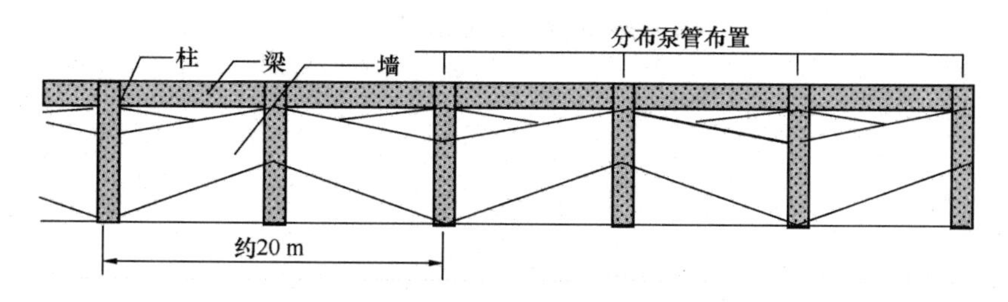

图 6　分布管浇筑自密实混凝土的情况

用吊斗浇筑时产生离析的可能性大得多，对配合比要求更严格，难度较大。在必须用吊斗浇筑时，应使出料口和模板入口的距离尽量小，必要时可加串筒。

对柱子和墙，浇筑前要严格检查和钢筋间距和钢筋与模板间的距离，最好准备一根长钎，以便必要时进行适当的插捣，排除可能截留的空气。

自密实混凝的质量对原材料和配合比的变动以及施工工艺都很敏感，因此对施工管理水平要求较高。每项工程实施前要有严格的施工规程和班前交底，尤其在交接班时，要有具体的措施来避免差错。

五、自密实混凝土应用展望及需继续研究的问题

1. 任何材料和技术都是一分为二的，缺点和优点总是同时存在，都有其最适用的时间和空间。自密实混凝土由于优异的施工性能，大大加快可施工速度，减小劳动强度，并可避免由于可能振捣不足而引起混凝土严重的质量事故；低水胶比、低温升和大量矿物掺和料，可保证混凝土的耐久性。但与相同强度的普通混凝土相比，弹性模量稍低，收缩和徐变稍大。此问题可通过提高配制强度、掺用膨胀组分和收缩小的掺和料如优质粉煤灰等措施来解决。但因掺用大量掺和料，混凝土碱度较低，宜用于主要受压的构件，特别适用于较大体积的基础底板和桩。用于受弯构件时，在目前尚无构件实验的情况下，掺和料掺量以不超过30%为宜。为此需要系统研究用自密实混凝土的主要受弯构件在荷载作用下的裂缝和混凝土性质的关系碳化及其对钢筋锈蚀的影响。

2. 本人目前只检索到Dhir等人1984年发表的《超塑化的流态混凝土：强度和变形性质》一文，其中报道了对超塑化的高施工性混凝土的抗压强度、干缩、热膨胀、弹性变形、徐变、徐变恢复等性质的系统研究[29]，此外尚未查到自密实混凝土用于预应力钢筋混凝土的报道。在这方面需要进行有针对性的系统研究。

3. 由于一般凝结时间较长，早期强度较低，冬季施工时最好不用。但在大体积混凝土中混凝土的温升对掺用掺和料的混凝土强度发展有利，自密实混凝土可在采取冬季施工的措施的前提使用。

4. 对都可达到自密实的高流动性混凝土并不是所有文献上都称之为“自密实”或其他如“免振”等，如明石大桥的混凝土无论水下还是水上，都是不振捣的；美国西雅图双联广场钢管柱混凝土也是不振捣的，但都没有叫做“自密实”或“免振”等，因为在“拥挤的部位”“尽管可得到高坍落度而自流平，但仍需稍加振捣以保证混凝土的足够密实”[9]，也就是说，并非任何工程浇筑的混凝土都能够完全不振捣，需要在施工规程中加以区别，否则反而会引起质量问题。

5. 自密实混凝土由于质量对原材料和配合比很敏感，要求严格的施工管理制度，实践表明，进行过自密实混凝土施工后，可有力地推动施工管理水平的提高，但在管理水平低、缺少技术人员的工程中不宜采用。

参考文献

[1] 友澤史纪，高流動コンクリートの现状と展望，建築技術，1996.04，特集：高流動コンクリートの基本と実際.

[2] 桝田佳宽，今，なぜ高流動コンクリートなのか，建築技術，1996.04，特集：高流動コンクリートの基本と実際，總論.

[3] 岡田凌太、版本光重、井保武寿、中川良隆，明石海峡大桥主塔基礎的の水中コンクリートの施工，コンクリート工学，Vol.30，No.12，1992.12.

[4] S. Kahima, M. Sakamoto, S. Okada, T. Iho and Nakagawa, Application of High Slag and Fly Ash, Low-Heat Cement to Antiwashout Underwater Concrete, Proceeding of Fourth International Conference on Fly Ash, Silica Fume, Slag and Natural Pozzolana in Concrete . Istanbul, Turkey, May 1992. SP-132-86.

[5] ACI 363 R-92, State of the arts Report on High Strength Concrete, 1994.

[6] FIP and CEB, State-of-the-arts Report on High Strength Concrete. 1990.

[7] Charles Nmai and Brad Violetta, The Use od Flowing Concrete in Congested Areas, Concrete International, Septemper 1996.

[8] Hajime Okamura, Self-compacting High-Peformance Concrete, Concrete International, July. 1997.

[9] 有马勇・樱井重英，高流動コンクリートの施工，コンクリート工学，Vol. 32，No. 7，特集：コンクリートの高性能・高機械化/3. 7，1994. 7.

[10] 北京市城建集团构件厂，自密实免振混凝土鉴定材料，1996 年 5 月.

[11] 北京工集团二建公司，高流動自密实混凝土的试验研究与应用鉴定材料，1979 年 9 月.

[12] 野萱胜九，高流動コンクリートの性质——硬化コンクリート，建築技術，特集：高流動コンクリートの基本と建築技术，特集：高流動コンクリートの基本と実际，1996. 04.

[13] 刘崇熙，关于碱-集料反应的几个理论问题，长江水利科学研究院学报，1990 年 12 月.

[14] 桝田佳宽，Fc=600kg/cm²が高强度でなくなるとき——总プロ/New RC が目指するもの，セメント・コンクリート，No. 545，《特集》コンクリートの强度/界限への挑战，Aug. , 1992.

[15] E. Tazawa and S, Autogenous Shrinkage by Self Desiccation in Cementitious Material, 9th international Conference on Chemisry of Cement, New Delhi, Sept. 1992.

[16] 和泉意登志，高流動コンクリートの材料・调合，建築技術，特集：高流動コンクリートの基本と実際，材料，1996. 04.

[17] 廉慧珍，陈恩义. 原材料和配合比对高强与高性能混凝土性能的影响. 混凝土，No. 596，1996 年 10 月.

[18] 杉本贡，高流動化のためにの材料——化学混合剂——，コンクリート工学，Vol. 32，No. 7，1994. 7，特集：コンクリートの高性能・高機械化/3. 3 .

[19] 早川和泉，高流動化のためにの材料——分離抵抗剂——，コンクリート工学，Vol. 32，No. 7，1994. 7，特集：コンクリートの高性能・高機械化/3. 4 .

[20] 小澤一雅・岡村甫，自己充填コンクリートの配合の考ぇ方，コンクリート工学，Vol. 32，No. 7，1994. 7，特集：コンクリートの高性能・高機械化/3. 1 .

[21] S. Kashima, N. Furuya and R. Yamaoka, High-Strength Concrete for Wall Foundation Using Ternary Blended Cement with Intermixture of Blast-Furnace Slag and Fly Ash. Proceeding of Fourth International Conference on Fly Ash, Silica Fume, Slag and Natural Pozzolana in Concrete, Istanbul, Turkey, May 1992. SP-132-78.

[22] 岡村甫，新しぃコンクリート材料への期待，セメント・コンクリート，No. 475，Sept. 1986.

[23] 小门武・三宅正人，スラグ系缔固め不要コンクリートの基础特性と充填性実験，コンクリート工学年次論文報告集，Vol. 13，No. 1，1991.

[24] 松岡康训・新藤竹文・赤冢一司，超高流動コンクリートの构造実物への适用，コンクリート工学，Vol. 30，No. 5，1992.

[25] 佐藤孝一，高流動コンクリートの製造，コンクリート工学，Vol. 32，No. 7，1994. 7，特集：コンクリートの高性能・高機械化/3. 5 .

[26] 阿部保彦，高流動コンクリートの制造・运搬，建築技術，特集：高流動コンクリートの基本と実際，材料，1996. 04.

[27] 早川光敬，高流動コンクリートの型枠施工，建築技術，特集：高流動コンクリートの基本と実際，施工，1996.04.

[28] 和泉意登志ほか，高流動コンクリートの実大施工実験(その1～13)，日本建築学会大会学術講演梗概集，1995.

[29] Dhir, Ravindra K.；Yap，Andrew W. F.，Superplasticized Flowing Concrete：Strength and Defomation Properties，Magazine of Concrete Reserch Vol. 36，No. 129，Dec. 1984.

只有经过长时间完成其发展的艰苦工作，并长期埋头沉没于其中的任务，方可有所成就。

—— 黑格尔

我只有在工作得很久而还不停歇的时候，才觉得自己的精神轻快，也觉得自己找到了活着的理由。

—— 契诃夫

论文之六

自密实高性能混凝土施工性能的评价方法*

张青[1]　廉慧珍[2]　王平[3]

一、前言

自密实高性能混凝土（以下简称自密实混凝土）即拌和物不离析而流动性很高、在不振捣或少振捣的情况下能密实地充满模型、不产生蜂窝、空洞等质量缺陷而且耐久的混凝土。这种混凝土虽然比相同强度等级的普通混凝土材料费用略高，但由于节省动力和劳力并解决扰民问题，其综合效益是显著的。自密实混凝土的配制关键是满足良好的流变性能要求。国外目前大多用拌和物的坍落流动度作为高流动性混凝土流变性能的量度。日本标准JASS 5—1996提出划分土木工程用的混凝土流动性如图1[1]所示。

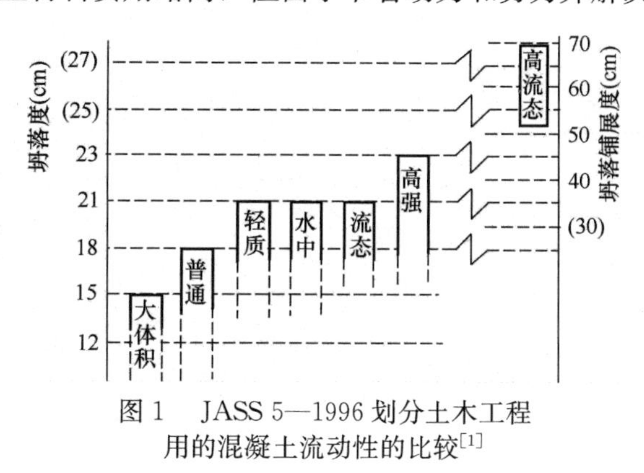

图1　JASS 5—1996划分土木工程用的混凝土流动性的比较[1]

然而，实际上，只达到上述坍落流动度指标的混凝土拌和物在实际工程施工时，尤其是用非泵送浇筑（如吊斗）时，却不一定能很好地密实成型。因此有必要对用于不同浇筑工艺的自密实混凝土施工性能评价方法进行研究。

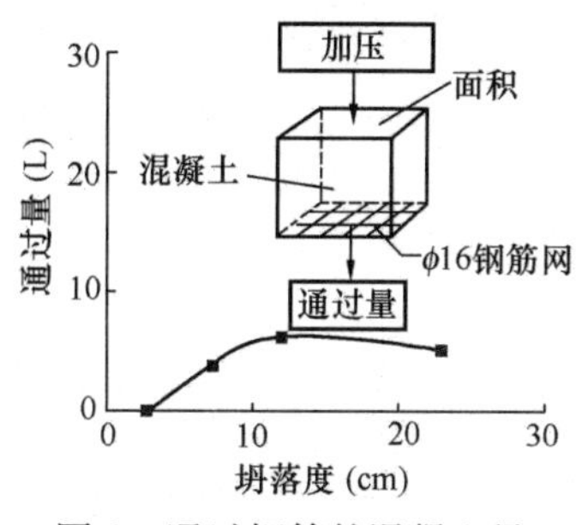

图2　通过钢筋的混凝土量和坍落度的关系[3]

二、自密实高性能混凝土流变性能特征

为了保证混凝土成型密实，要求混凝土拌和物不仅有高流动性，而且应能顺利通过钢筋间隙和狭小模板空间，填充到模板的各个角落。具有高的抗堵塞能力和充填性。如在图2中，以通过钢筋的混凝土量表示混凝土充填性[3]。混凝土水灰比为0.254，粗骨料最大粒径为15mm，含量为900kg/m^3。坍落度以用水量调整。混凝土用强制式搅拌机搅拌3分钟。试验结果表明，流动性和充填性有一定的联系。在配筋密集、模板形状复杂的情况下，流动性不足的混凝土充填性差；由于流动性主要

---

* 发表于中国土木工程学会高强混凝土委员会第三届学术讨论会议（济南），收入论文集。
1. 张青，高工，副总工程师，北京建工（集团）二建公司，100044
2. 廉慧珍，教授，清华大学土木工程系，100084，本文执笔
3. 王平，工程师，北京建工（集团）二建公司，100044

受用水量的控制，流动性随用水量的增加而增加，充填性也随之增加。但只由增加用水量而提高的流动性增加到一定的程度后，由于粗骨料在如钢筋等障碍物处的堵塞，充填性不再提高，甚至下降，如图 2 所示。混凝土拌和物的充填性受流动性、抗离析性和间隙通过性的影响；间隙通过性是混凝土拌和物抗堵塞的能力，受流动性、抗离析性、钢筋密度、模板尺寸等因素影响。对具有足够流动性的拌和物，影响充填性的主要因素是抗离析性。因此，根据上述自密实高性能混凝土的定义和和拌和物流变特性，自密实混凝土施工性能需要由流动性、抗离析性、间隙通过性和充填性四个项目综合评价。

三、自密实的高性能混凝土流变性能评价方法的选择

国外用以检验自密实混凝土流变性能的方法以日本的报道居多，如表 1 所示。

表 1　混凝土流变性能试验方法及测定值范围[5]

试验方法	测定值	符号	单位	简便性	测定值范围
坍落度试验	坍落度值	Sl	cm	◎	26～28
	坍落流动值	Sf	cm	◎	50～80
	流动 50cm 的时间	St_{50}	s	○	2～25
L-流动试验	L-流动值	Lf	cm	◎	40～80
	L-流动速度	Lv_{30}	cm/s	○	2～20
	L-流动初速度	$Lv_{5\text{-}10}$	cm/s	△	3～30
V 形筒流下试验	流下时间	Vt	s	○	4～50
环贯入试验	沉下速度	Rv_1	mm/s	◎	8～130
	沉下速度比	Rv_1/Rv_4		○	0.1～1.0
圆筒贯入试验	流入砂浆值	Fm	mm	◎	1～55
充填性试验	充填流动值	Cf	cm	○	18～60
	充填流动速度	Cv_{20}	cm/s	△	0.1～1.0
	充填性评价	Cfg/Cfm		△	0.55～1.0
钢筋通过性试验	流出比	Pr		△	0.07～0.47
附着试验	附着韧性	Ta	Pa-m	×	1～18
粗骨料冲洗试验	内外粗骨料比	Ai/Ao		△	0.88～1.2
砂浆流变试验	屈服值	τ_y	Pa	×	2.2～22
	塑性黏度	η	Pa-s	×	1.5～12

注：表中符号代表的简便性顺序为：◎>○>△>×

1. 传统坍落度法

在表 1 介绍的方法中，坍落度和坍落流动度由于操作简便，仍适合用于现场施工质量控制。在所浇筑混凝土的断面较大、钢筋不过密、泵送距离不长的情况下，目标坍落流动度为 550mm 即可施工；当构件断面狭小、钢筋密集、泵送距离较长的情况下，目标坍落流动度应为 600mm 和 650mm[4]；对于非泵送的浇筑，坍落流动度应为 650mm 到 700mm。坍落流动度 500mm 以下时，流动性不足，无法充满模型；坍落流动度超过 700mm 时，则可能发生离析[4]。但是，施工前必须使用施工时所用的原材料进行试配。对于抗离析性目前大多是

在进行坍落度试验时凭肉眼观察其抗离析性。这种观察是不准确的。往往在实验室凭肉眼观察无离析和泌水的现象，而在工地浇筑时仍会发生严重的离析和泌水，混凝土不能密实地填充。尤其是使用吊斗浇筑，而浇筑高度又较高时。

2. 流动性混凝土在管道内流动性能的测定

为了探索自密实混凝土流变性能评价的方法，首先用英国 PeterJ. M. Bartos 介绍的 Orimet法[4]测定混凝土配合比和混凝土流动性的关系。试验用原材料为：冀东 525# 普通硅酸盐水泥，北京卢凤山砂，细度模量 3.1，含>5mm 颗粒 7.0%；北京卢沟桥碎卵石，粒径 5～25mm；内蒙古元宝山Ⅰ级粉煤灰；山东来芜汶河外加剂厂 FDN。所用装置如图 3 所示。

图 3 中，混凝土拌和物从立筒中流出时需克服内摩擦力和与筒在壁间的摩擦力，故流出速率反应拌和物的黏度。如拌和物有离析的倾向，流经筒出口的斜面处时就会被堵塞而降低流出速率。

改变水胶比、胶凝材料总量、粉煤灰掺量、砂率和减水剂掺量，测定拌和物从出料口流下的时间，计算出流出速率。试验结果如图 4、5、6、7 所示。

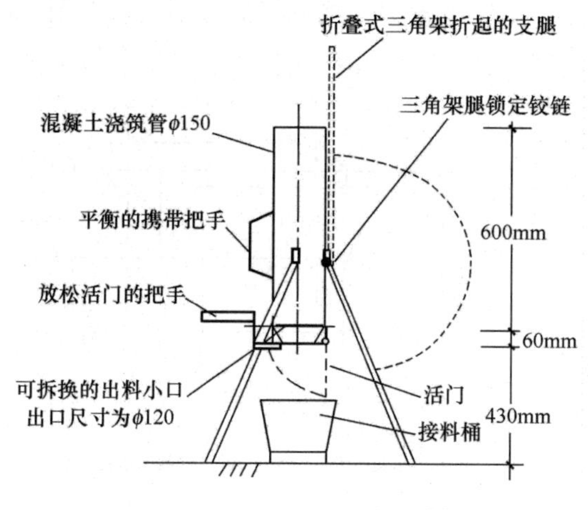

图 3　Orimet 法试验装置[4]

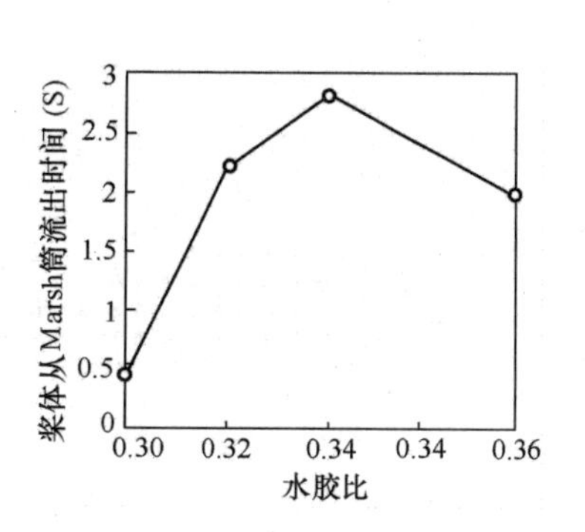

图 4　水胶比与流出时间的关系

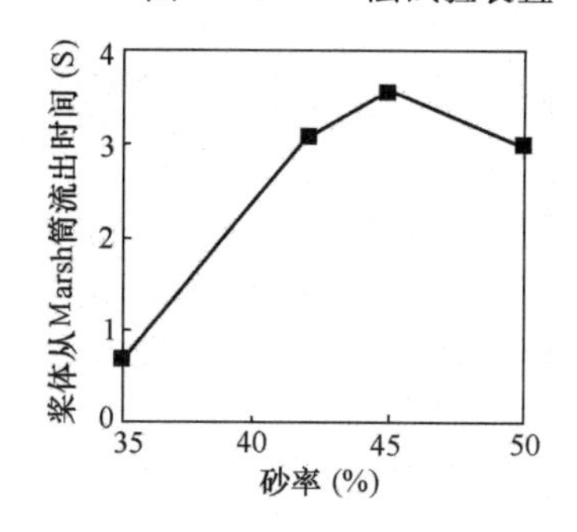

图 5　砂率与流出时间的关系

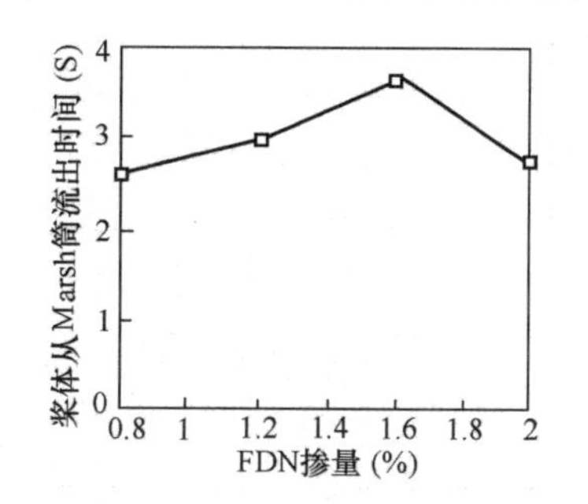

图 6　FDN 掺量与流出时间的关系

图 4 表明，在本试验的配合比下，混凝土拌和物流出速率随水胶比的提高而增大，但水胶比超过一定值时，流出速率就因拌和物离析而降低；图 5 和图 6 中砂率和减水剂掺量的影响也有类似的规律；掺入需水比小的粉煤灰不仅可增加拌和物流动性，而且可改善拌和物的黏聚性，故流出速率随粉煤灰掺量的增加而增加（图 7）。

但将与上述试验所用配合比相同的混凝土拌和物对应进行坍落度检测时，各组拌和物之

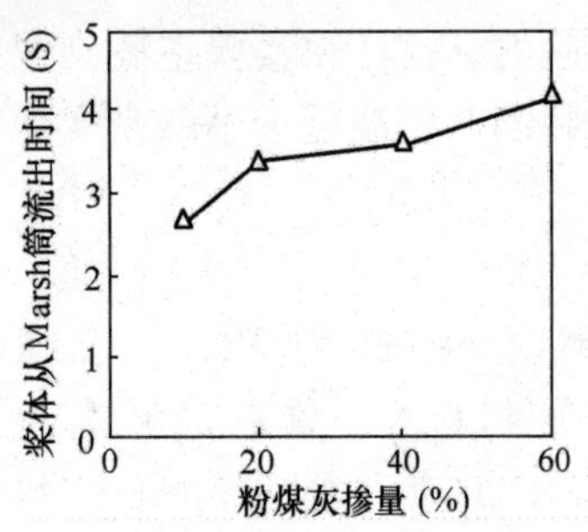

图 7 粉煤灰掺量与流出时间的关系

间却几乎没有差别，说明坍落度法对流动度很大的混凝土拌和物敏感性差，用 orimet 方法检测混凝土优于坍落度法。但是对坍落度大于 250mm 的自密实混凝土，用 orimet 方法检测时，试验结果就不能很好地反应在实际工程中流变行为。

3. 充填性试验

除表 1 所介绍的配筋 L-流动性测定装置外，充填性实验装置有多种，如平底或圆底 U 形流动充填性实验装置[5][6]。配筋 L 形流动性测定装置也有多种，图 8 是其中的一种。该装置的是垂直于混凝土流动方向的水平筋，可观察混凝土拌和物通过钢筋填充模型的程度。

钢筋通过实验装置如图 9 所示。将混凝土拌和物从上口装满后，打开侧面的活门，待拌和物停止流动后，测定流出的混凝土拌和物与原装入的全部拌和物重量的比值，为钢筋通过率。实验表明，能在如图 8 所示靴形 L—流动试验装置中充填良好的混凝土，在图 9 的装置中不一定能很好地通过钢筋。

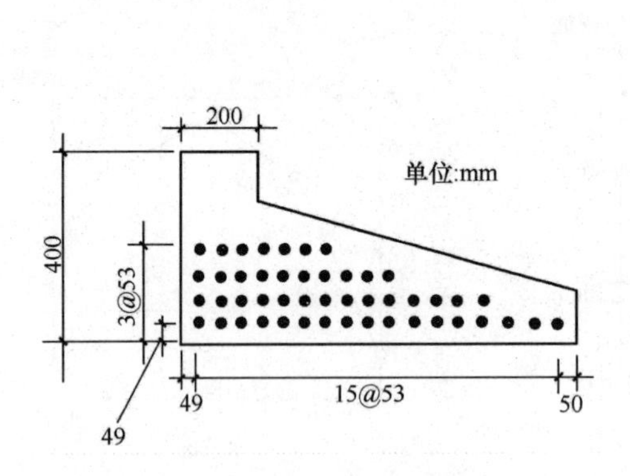

图 8 靴形 L-流动试验装置[5]

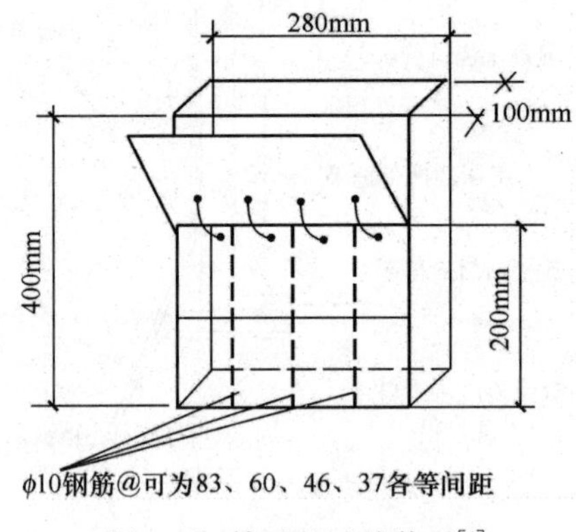

图 9 钢筋通过试验装置[4]

由以上实验和分析，吸取上述流动性、充填性和钢筋通过率测定的各自特点，设计、改装了一个综合的装置，如图 10 所示。

图 10 的装置左侧为一立筒，右侧为一水平槽。在立筒和水平槽的交接处的立同侧面设一个开口，并装有一活门。模拟工程中的密筋情况，在口部垂直装有 4ϕ12 螺纹钢筋。将混凝土拌和物装满立筒后，提起活门，使混凝土拌和物通过钢筋流出。通过该装置可测定三个指标：混凝土拌和物通过钢筋流出的速率 v、混凝土在水平槽内流动的距离 L、混凝土停止流动后立筒内混凝土坍落的高度 T。试配时，选择坍落度、坍落流动度与强度都合适而且目测无离析、泌水的配合比，进行此项实验，即可综合评定自密实混凝土的流变性能。

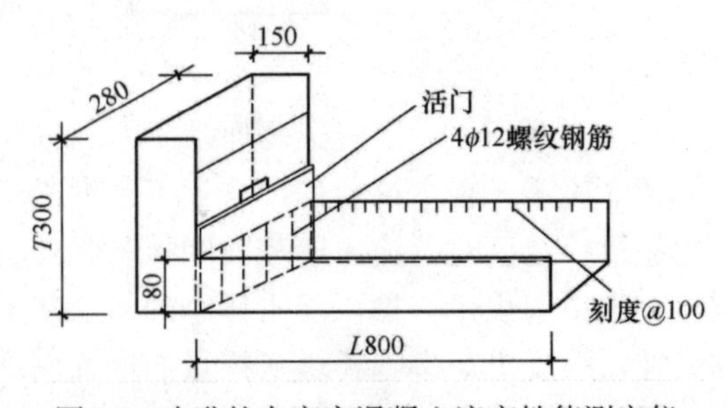

图 10 改进的自密实混凝土流变性能测定仪

为了使各项实验之间有较好的可比性，特别加工了坍落度和坍落流动度测定的台钣。该台钣底部四角装有调节螺丝，保证实验时台钣的水平；围绕台钣中心在直径 300mm 和 500mm 处做两个同心圆标记，记录提起坍落度筒后混凝土拌和物流动到该标记时的时间，计算混凝土拌和物流动速率。实验完毕，从 300mm 圆内、300～500mm 之间和 500mm 以外合取拌和物 3kg，测定该三部分混凝土拌和物中石子的含量，以评定混凝土拌和物的匀质性。

四、实验结果和讨论

按自密实混凝土原材料要求和配合比原则设计拌合物，进行流动性和强度试验，经优选后，用图 4 和图 5 两种装置进行填充性实验。实验如表 2 所示。由表 2 可见：

1. 各组混凝土的流动性极其相近，但是充填性差别很大。凡是不能充满配筋 L-模型的（如 K59），在改进的自密实混凝土流变性能测定仪实验中钢筋通过情况均不好，L＜400mm（因此凡是不能充满配筋 L-模型的，就不再进行改进的自密实混凝土流变性能测定仪的实验）；而能充满配筋 L-模型的，在改进的自密实混凝土流变性能测定仪实验中钢筋通过情况也不一定好（如 K48）。

表 2　自密实混凝土流变性能的测定

	流动性		流动时间(秒)		配筋L-流动充填性	充填性实验：通过钢筋流出一定距离时时间（s）										
	Sl mm	*D* mm	流过50cm	流过65cm		150 mm	200 mm	250 mm	300 mm	400 mm	500 mm	600 mm	700 mm	800 mm	*T* mm	*L* mm
K41	255	615			×										＜130	＜400
K43	260	645	7	18	×										＜130	＜400
K44	255	653	4	15	×										＜130	＜400
K47	255	600	5		×										＜130	＜400
K48	270	695	7	16	√	3.5	5	8	16	24	84					离析
K49	270	673	6	11	√					6	12	18	26	47	250	＞800
K50	255	695	6	14	√	2	3	4	10		14	20	26	47	254	＞800
K51	280	680	6	11	√	2		6		9	13	18	28	47	255	＞800
K53	260	695	6	17	√	2		4	6	7	15	22		44	260	＞800
k54	260	665	10	22	√										100	＜400
K55	270	665	7	12	√										90	＜400
K57	270	695			√	3	5	8	10	14	19	29	40	62	260	800
K59	265	643			×	3	6	8	14	40					100	400
K60	260	690	5	9	×											
K61	265	665	6	16	×											
K62	260	660	7	18	×											
K67	265	670	4	8	√	2	4	6	8	12	18	30	42	53	260	＞800
K68	265	700	5	9	√	2	3	4	6	8	10	14	18	34	265	＞800
K69	275	680	5	10	√	1	2	4	5	8	10	14	18	26	265	＞800
K70	280	700	5	12	√	1	2	3	5	6	7	16	20	30	255	＞800
K78	245	660			√	2	4	5	7	11	15	24	32	55	245	＞800
K81	275	670	5	10	√		2	4	6	10	16	24	30	46	250	＞800
K82	265	600	7	28	×		1	2	12	17	28	47		134	205	800
K82		685	5	12	√		2	3	8	12	18	30	40	60	240	＞800

注：①表中符号：*Sl*—坍落度，mm；
D—坍落流动扩展度，mm；
T—充填装置立筒中混凝土拌和物下降高度，mm；
L—混凝土拌合物通过钢筋流出底距离，mm；
√是 L-流动填充性实验时填充良好；
×为不能充满模型。
②所有编号的混凝土所用石子粒径均为 5～20mm，但不是同一来源。

2. 编号67～70与编号41～44和59的混凝土配合比相同，只是石子针片状颗粒仅相差2个百分点，其充填性能却大不相同。

3. 由于用坍落度测定的混凝土拌和物流动时间（速率）终点判断不准确，因此结果不可靠。

4. 充填性好的混凝土拌和物L值和坍落度Sl值的相关性很好，L值都大于250mm；通过钢筋流出至400mm处时间均小于15s。出现这种情况的原因是凡充填性不好的试料都是较大粒径或针片状颗粒的石子被堵塞在钢筋排前。充填性不好的试料所用石子中针片状颗粒含量为9.3%，而改换石子后就得到充填性很好的拌和物（针片状颗粒含量<5%）。

五、结论

1. 自密实混凝土在试配时，流变性能评价的指标不能只用坍落度和坍落流动度，必须进行充填性的检验；但可先用坍落度和坍落流动度的比值约0.4左右初选配合比。

2. 用改进的自密实混凝土流变性能测定装置检验自密实混凝土流变性能，用通过钢筋的流动能力综合评价自密实混凝土流变性能更接近于工程实际。

3. 本实验是针对非泵送自密实混凝土而进行的，当采用泵送施工时，流变性能各项指标均可放宽。

4. 目前生产条件下难以严格按规范使用最大粒径小于钢筋最小净距三分之一的石子，本实验装置所配制的钢筋净距只有45mm，石子最大粒径为20mm，只有钢筋净距的约二分之一，尤其要求自密实混凝土有足够携带粗骨料的黏聚性；由于自密实混凝土有足够多量的砂浆，粗骨料的粒形和粒径要比其级配重要得多。

参考文献

[1] 桝田佳寛，今，なぜ高流動コンクリートをのか，建築技術特集《高流動コンクリートの基本と実際》，1996.6.

[2] Peter J. M. Bartos, Assessment of Workability of Fresh Special Concrete Mixes, Proceedings of The 3rd Bejing International Symposium on Cement and Concrete, Vol. 2. Oct., 1993.

[3] 岡村甫，小澤一雅，缔固め不要コンクリートの可能性と課題，コンクリート工学，No. 2，1992.2.

[4] 铃木忠彦，高流動コンクリート試験方法，建築技術特集《高流動コンクリートの基本と実際》，1996.4.

[5] 和泉意登志，高流動コンクリートの材料・调和，建築技術，1996.04.

[7] 岡村甫，小澤一雅，缔固め不要コンクリートの可能性と課題，コンクリート工学，No. 2，1992.2.

[8] 和泉意登志など，各种高流動コンクリートの特性評价及び実大模型打設実験（その5 実大模型打設実験のフレッシュコンクリートの品质），建築大会，1993.9.

[9] K. Sakata etal, Study on the Durability of Low Heat Highly Flowable Concrete Incorperating Urea, Fourth International Symposium on the Utilization of High Strength / High Performance Concrete, Paris, France 1996.

论文之五和之六自评

1. 日本是个多地震国家，结构物修补加固的需要量较大，又因较多使用外包混凝土的钢结构，或劲型混凝土（即钢骨混凝土），修补施工操作空间比较狭小，加上劳动力素质下降，存在施工质量问题。因适应其国情而成为在世界上最早开发应用自密实混凝土的国家。但是从介绍来看，尽管日本在流变性能的检测方面做了不少研究，当时自密实混凝土还存在一些问题，主要是胶凝材料总量太大，以及浆骨比太大，砂石总量甚至低到 1600kg/m³。因此在强度等级相同的条件下，弹性模量较低，干缩较大。从论文可见，因工程发展的需要，实际上西方国家也已有类似技术，如西雅图双联广场钢管高强混凝土。近年来自密实混凝土在西方已经盛行起来，在技术上也取得进展。这两篇文章的信息量已远远不够。自密实混凝土的核心问题是流变性能。对中低强度混凝土来说，因水胶比较大，胶凝材料用量较少，浆体浓度低，拌和物泌水、离析是主要问题；对高强度混凝土来说，水胶比低，胶凝材料量大，浆体浓度大，拌和物黏度大，触变性突出，而且在一定的浓度范围，还可能发生剪胀性。触变性通常不影响泵送，剪胀性可通过调节浆体、砂浆或混凝土拌和物的浓度来解决，例如可掺用需水量小的低密度矿物掺和料（例如粉煤灰、石灰石粉、硅灰），在适当减小水胶比的条件下增加胶凝材料体积而不影响强度；有时调节砂率；合适的粗骨料粒径和级配对自密实混凝土尤为重要。调节的目的是使拌和物达到在自重下流动时屈服应力和塑性应变的协调。自密实混凝土还必须按照不同的用途进行配制。这两篇文章的信息量已远远不够。

2. 1996 年，我和北京市二建公司合作研制自密实混凝土，尽管采取了多种手段评价混凝土的施工性，在实验室试配结果令人满意，而在实际工程试验中，却和实验室的效果大不相同。这更加说明施工性能确实是一种无法在实验室量化并检测出指标的 Performance。在我们认为实验室试验已经满意了的基础上，二建公司提供他们员工自用的住宅建筑工程做工程试验。为了避免拌和物无限流淌到别处，在浇注前先做好隔堵措施。未曾想，开始浇筑后，预先做的隔堵根本挡不住，拌和物流动无法在预定的部位停住，而且浇完后很快就在顶部泌出一层很厚的稀浆。第二天拆模后，发现外墙的窗下因无法排气而留下空洞。这次试浇筑告诉我们，施工性不仅是流动性，在实验室无法模拟施工现场的浇筑高度，因此难以评价拌和物实际浇注过程的体积稳定性。此后，我们又用了半年的时间，研究提高拌和物裹携骨料的能力，才成功地用于工程。这说明，实验室对拌和物的检测如何符合工程实际，对自密实混凝土来说更加重要。

3. 现在在我国，自密实混凝土似乎被认为是很容易做到的，因为大多用于钢管内浇筑，对拌和物在钢管内的行为，实际上是不确知的。很多人担心混凝土收缩会和钢管脱离。设想一下，钢管里面隔绝了空气，干缩还存在吗？自收缩最大值能有多大？钢管和混凝土一起受压，互相能脱离吗？重要的倒是拌和物是否能通过一切障碍。例如为缓解在钢管中的加劲肋与钢管壁连接处截留空气而采取的排气措施效果如何？又如果钢筋密实得其间距只有几毫米（当前实际工程不乏此例），则只能通过砂浆了。因此自密实混凝土的实现也需要设计与施工的配合。

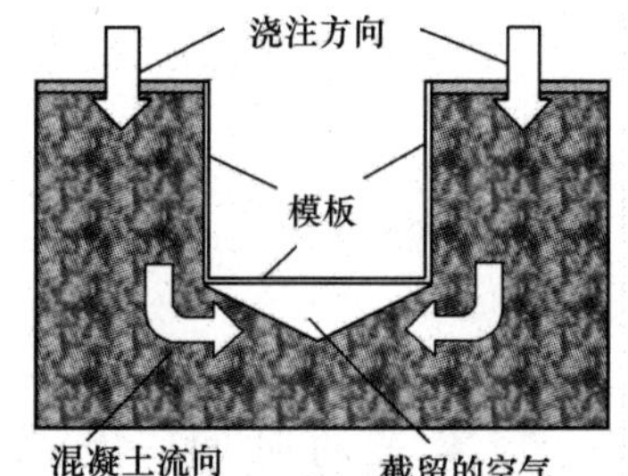

窗洞下面截留空气示意

4. 自密实混凝土的流动性很大，骨料最大粒径较小，为保证混凝土体积稳定性和拌和物的匀质性，应尽量减小浆骨比，因此需要对砂石的粒形和级配有更高的要求。应当由砂石场供应粒形好的单粒级石子，供用户自行级配，分级上料，以得到理想的级配。而且，还不只是石子，连砂子也需要精细地级配，再加上胶凝材料的合理组成，才能得到浆骨比最小而泵送性能最好的混凝土拌和物。对于自密实混凝土，流变性能是第一重要的。西方国家现在使用自密实混凝土，除流动性和抗堵塞性能外，都有塑性黏度和屈服应力的流变性能参数要求。

5. 任何材料都有其最合适的用途，主张把自密实混凝土作为混凝土发展方向并不合适，其原因是任何材料面对不同的使用对象何条件时，都有各自的优点和缺点，自密实混凝土也有其自身的缺点，如与相同强度等级的振捣混凝土相比弹性模量较低，开裂敏感性较大。但对那些难以振捣的构件来说，不得不用时，可以用其他措施作为补偿的代价。最适合的用途是钢管混凝土和狭窄空间与密集配筋的修补加固用混凝土。

零星的时间，如果能敏捷地加以利用，可成为完整的时间。所谓“积土成山”是也，失去一日甚易，欲得回已无途。

——卡耐基

时间像奔腾澎湃的急湍，它一去无还，毫不留恋。

——塞万提斯

论文之七

养护温度对低水灰比的掺粉煤灰外加剂砂浆强度增长的影响*

廉慧珍　覃维祖　徐　珩（清华大学土木系　100084）

摘　要　抗压强度为1、3、28天和60天的低水灰比掺粉煤灰外加剂的砂浆试件，在不同温度养护条件下的试验结果表明，当养护温度低于40℃时，波特兰水泥砂浆的早期强度随着温度上升而提高大，但在第7天后，混凝土强度增长缓慢。当温度高于85℃时，后期强度下降。掺粉煤灰的砂浆，强度随着温度升高而平稳增长。60天之后的粉煤灰砂浆的强度，高于不掺粉煤灰砂浆的。由于结构混凝土内温度较高，在标准条件20℃养护的混凝土强度不能反映结构中混凝土的强度。

关键词　低水灰比混凝土　粉煤灰　养护温度　强度增长

近年来科研和工程技术的发展使人们认识到，矿物细掺料在砂浆中不再是一种单纯的水泥替代物，而是高性能混凝土的必要组分之一。“由于矿物细掺料中控制其火山灰活性和胶凝性能的是其矿物组成，而不是化学组成，则强调其化学组成的规范和标准阻碍了矿物细掺料在水泥和砂浆工业中的使用。能够评价含有矿物细掺料的硬化水泥浆体微观结构性能的规范、标准和快速实验方法是必须的。……”；因此有必要对如何评价掺用大量矿物细掺料的高性能混凝土强度的问题进行研究。现行标准规定以20℃作为标准养护温度，来评价砂浆的强度。实际上由于温升而造成混凝土内部很高的温度，在强度发展最快的早期（7天以前），混凝土内部所处的温度远不止20℃。例如用550kg冀东525#硅酸盐水泥、w/c＝0.3配制的高强混凝土最高绝热温升为52℃，普通混凝土的水灰比大，但水泥用量小，最高绝热温升也高于50℃。在夏季，混凝土内部的温度可达到90℃。对于使用硅酸盐水泥或普通硅酸盐水泥的混凝土，养护温度越高，早期强度越高，而后期增长越少；当养护温度超过65℃以后，混凝土的强度随温度的升高而下降[2]。所不同的是，掺入矿物细掺料后，养护温度对混凝土强度的影响规律与此大不相同。例如Raymundo Rivera v. 对水胶比为0.6的混凝土在分别为15℃、23℃和38℃的条件下养护，对比掺30％粉煤灰和纯水泥混凝土的强度发展，结果是，纯水泥的混凝土强度随养护温度的升高而降低，掺粉煤灰的混凝土强度则随养护温度的升高而升高[3]。也就是说，现行标准养护的试件强度并不能代表结构混凝土的实际强度。

北京住宅总集团在C60混凝土柱冬季施工时，在混凝土中预埋温度传感器，然后按照混凝土实际温度养护试件，检测其强度发展，并与标准强度和现场自然条件下养护的试件强度对比，如表1所示。混凝土中粉煤灰掺量为18％，水胶比为0.30，无防冻剂，浇筑时气温为1.8℃～2.2℃，混凝土出机温度为14℃，成型试块温度为12％。可见即使在寒冷气候里，水胶比很低且掺有粉煤灰的混凝土，三天内内部温度也超过30℃，混凝土实际强度显

* 原载于《低温建筑技术》1999年第6期。

然高于标准养护的强度，现场预留试件尺寸太小，在自然条件下不能称做“同条件养护”，也不能代表结构中混凝土的强度。

表1 不同养护条件下高性能混凝土柱强度发展

龄期（天）	温度℃			试件抗压强度（MPa）		
	中心	外侧*	大气	标养	同温养护	现场养护**
1	44	23	1.2	26.2	26.9	—
2	49	31.2	0.7	45.6	48.5	33.2
3	35	27	1.0	46.0	55.0	34.2
5	18	14	−2.9	55.1	64.5	46.5
7	11	−1	−2.9	56.0	65.0	49.4
14	0	−1.5	−4.0	71.2	79.8***	—
28	71.3	—	—	67.5	—	—

* 柱外侧指与柱的外表面水平距离80mm处；

** 即通常所谓的“同条件养护”；

*** “同温养护”试件只养护到7天后，取出置于现场自然环境负温下，故14天强度已不再是“同温度养护”的结果，而是“同温养护”7天后持续发展的结果。

为了科学地评价掺粉煤灰的低水胶比混凝土强度，对在不同养护温度下不同水胶比、不同种类粉煤灰的掺量对砂浆强度的影响进行了实验研究。由于实验量很大，粉煤灰种类不同时的影响规律有所不同，为了节省篇幅，现只就其中两种粉煤灰的实验结果进行讨论。并用元宝山级粉煤灰（I—FA）进行平行对比实验。

1 实验用原材料和实验方案

主要原材料：

水泥：冀东525#R硅酸盐水泥；粉煤灰：北京西高井电厂Ⅱ级粉煤灰、元宝山Ⅰ级粉煤灰；高效减水剂：山东莱芜汶河化工厂FDN。

1.1 实验方案

固定胶凝材料总量为500kg/m³，改变粉煤灰掺量占胶凝材料总量的0%、30%、40%、50%和60%；水胶比分别为0.3、0.35和0.4；为减少实验工作量，实验用试件尺寸采用3cm×3cm×3cm；测试龄期分别为1、3、7、28、60天，养护条件见表2。

表2 试件养护条件

养护温度（℃）	湿度条件
0	冰、水混合物
20	水中
40	相对湿度>90%
85	相对湿度>95%

2 实验结果与分析

2.1 水泥用量为500kg/m³、不掺粉煤灰、不同水灰比的纯水泥砂浆在表2的条件下养护至规定龄期，测定其抗压强度，养护温度对强度发展的影响见图1中（a）、（b）所示。

由图1可见，对于不掺粉煤灰的砂浆，在水灰比一定的情况下，当养护温度不超过

40℃时，随温度的升高，早期强度增高，7 天以后则强度发展平缓，到 85℃时则强度随龄期而下降；与较低温度下的情况相比，85℃下养护的试件 28 天以后强度最低。在 0℃下养护，强度照样随龄期而增长。

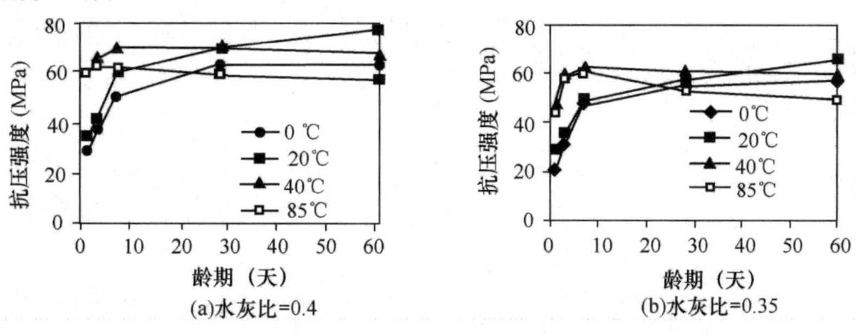

图 1　养护温度对不同水灰比硅酸盐水泥砂浆强度发展的影响

2.2　保持胶凝材料总量不变，养护温度对不同水胶比、不同掺量的西高井Ⅱ级粉煤灰砂浆强度发展的影响见图 2 各图所示。

由图 2 可见：

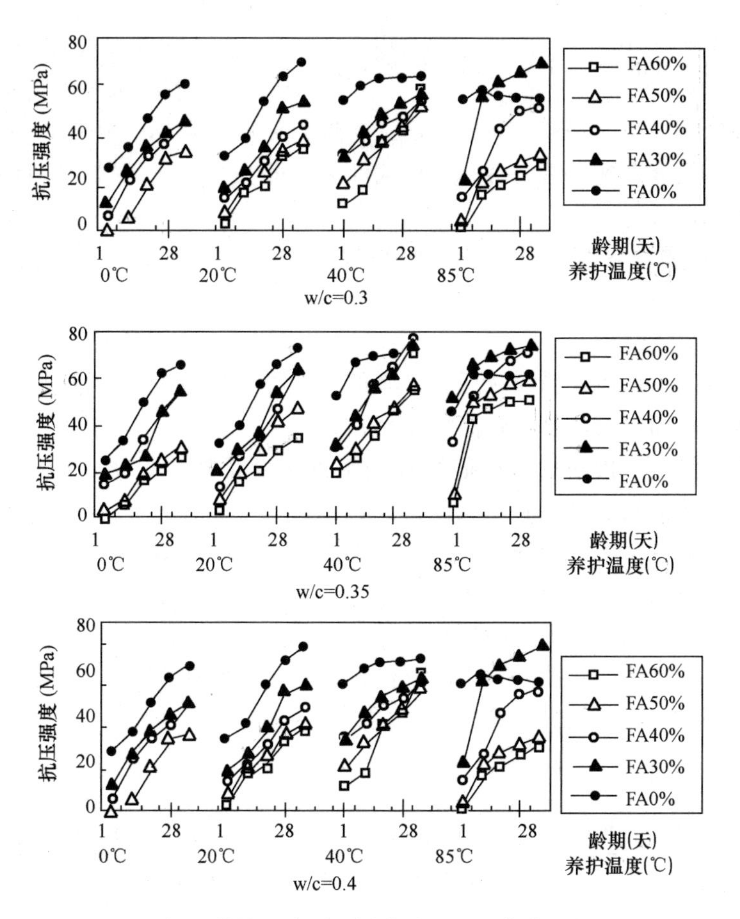

图 2　养护温度对不同水胶比、西高井Ⅱ级粉煤灰不同掺量的砂浆强度发展的影响

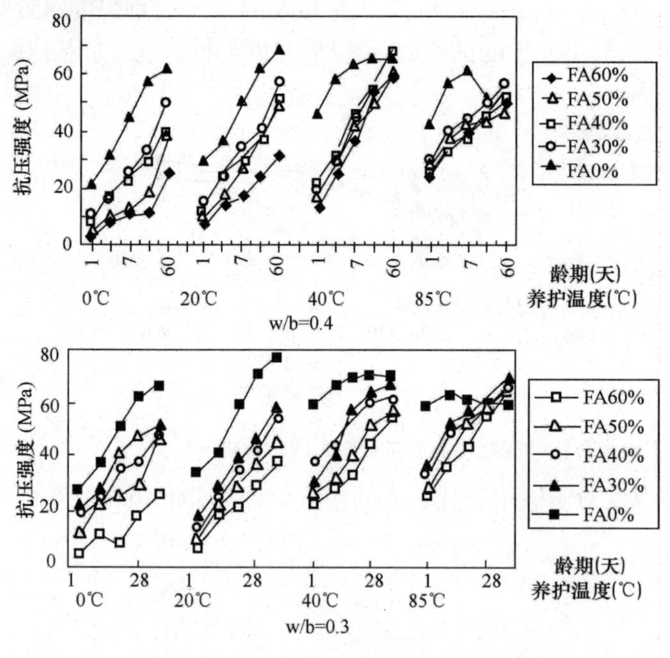

图 3　养护温度对不同水胶比、元宝山Ⅰ级粉煤灰不同掺量的砂浆强度发展的影响

（1）在任何温度下，掺入粉煤灰后，砂浆的早期强度都随粉煤灰掺量的增加而降低，但强度都随龄期的增长而增长；

（2）在任何温度和任何水胶比下，掺量为 30%和 40%的 60 天强度相近；除纯水泥外，在 40℃下，各掺量的 60 天强度相近，且随龄期进一步增长的后期，有强度超过纯水泥砂浆强度的趋势；

（3）除纯水泥外，在任何掺量下，在 40℃以下，砂浆强度都随温度的升高而提高；85℃时，掺量大的（50%和 60%）强度下降，但仍随龄期的增长而增长；掺量为 30%和 40%的 60 天强度超过纯水泥砂浆同龄期的强度；

（4）因试件尺寸太小（3cm×3cm×3cm），西高井粉煤灰活性较低，掺量太大（60%）而水胶比很低时，强度发展过慢，在 0℃时试件 1 天即冻坏，故无数据，而用活性较高的Ⅰ级粉煤灰时则无此现象。

2.3　将西高井Ⅱ级粉煤灰换成元宝山Ⅰ级粉煤灰，进行实验 2 的平行实验，结果如图 3 所示。

由图 3 可见：

（1）在任何温度下，掺入粉煤灰后，砂浆的早期强度都随粉煤灰掺量的增加而降低，但强度都随龄期的增长而增长，且有较高增长率；

（2）在任何掺量下，掺粉煤灰的砂浆任何龄期强度都随养护温度的提高而提高，达到 85℃时，则早期强度提高得多而后期提高得少；

（3）40℃下，60 天掺粉煤灰的与纯水泥的同龄期强度相近，达 85℃时，则掺粉煤灰的 60 天强度均超过纯水泥的同龄期强度；

（4）在任何温度和水胶比下，粉煤灰掺量 30%和 40%的强度及其发展相近；

（5）由于具有较高活性，掺元宝山Ⅰ级粉煤灰时，即使在 0℃下养护，强度仍然随龄期的增长而发展。

3　讨论

3.1　养护温度高时，对无掺和料的纯水泥砂浆（混凝土）的强度发展不利，甚至随龄期而倒缩；对掺粉煤灰砂浆来说，则温度高对强度发展尤其对早期强度有利；

3.2　在实际工程结构中，混凝土浇筑后内部温度较高，所以 20℃标准养护温度下的强度不能代表结构混凝土的真实强度，用以评价混凝土的强度，对纯水泥混凝土来说是不安全的，对掺粉煤灰的混凝土来说是保守的；

3.3 当使用活性较高的粉煤灰时，即使用小尺寸试件在持续0℃下养护，强度仍随龄期的增长而发展，由于混凝土内部产生温升，则在一定的负温下浇筑混凝土不用防冻剂是可行的；

3.4 粉煤灰的来源和品质不同时，温度影响的规律大体相似，具体有差别，说明目前对粉煤灰的基础研究仍待深入；

3.5 本文所述只是探索性实验，目的是企图了解结构中纯水泥的和掺有不同量粉煤灰的混凝土的试件的强度，以及评价混凝土强度的合理养护温度。

（1）各实验均只有一组数据，缺乏统计性；

（2）需要更多的实验数据验证；

（3）温度区间划分不细，例如其中缺少40℃和85℃之间的数据；

（4）试验均采用恒温养护，而实际构件中的温度则是变化的，而且是波动的，应当继续进行更接近实际的模拟实验。

参考文献

[1] P. K. Mehta，Pozzolanic and Cementitious Byproducts as Mineral Admixture for Concrete——A Critical Review，Proceedings of the First International C0nference on the Use of fly Ash Silica Fume，Slag and Natural Pozzolana in Concmte，Detroit，USA，1983.

[2] 沙凤丹．水泥强度快速测定[S]．北京硅酸盐协会水泥制品专业委员会，1962年学术年会论文报告汇编.

[3] Raymundo Pdvera V.，Effect of Temperature On the Properties of Mortars and superplasticized Concrete Containing Low Calcium fly Ash，Proceeding of the second International Conference on the Use d FLy Ash，Silica Fume，Slag and Natural Pozzolana in Concrete，1986.

自评

1. 硅酸盐水泥和掺混合材的硅酸盐水泥在不同温度下强度发展不同，这是国内外几十年前就有定论的。但工程界只是在近十几年来才确认构件内部混凝土的温度高于外界大气温度，对是否掺用矿物掺和料的混凝土强度发展有不同的影响。在处于大气温度下的实际工程中，混凝土内部温度总的趋势是升温达温峰后降温，并随着大气的温度而波动，这种过程在实验室难以模拟。本文所述试验是用变化养护水的温度以模拟混凝土内部温度，而使用的是恒温条件，这只是作为一种研究的方法，了解硅酸盐水泥的混凝土和掺有不同掺量粉煤灰的混凝土在不同温度下的行为，以改变混凝土工程验收一律采用标准养护试件的传统。现在有关规范已注意到标准养护试件不能代表构件中混凝土，提出以“同条件养护”试件验收的要求。但是因使用小试件，温度不同，仍然不是真正“同条件”。

2. 在现场采用跟踪养护的方法，会得到混凝土内部升温、降温以及随环境温度波动的结果，可得到更接近于实际构件中强度的发展。但是，由于环境温度的波动与混凝土原材料和配合比的差异，跟踪养护的方法只能在具体工程和具体条件下指导施工和验收，不可能有普适性的定量规律。

3. 混凝土构件距表面不同深度的温度是不同的，对于梁、板、柱等的构件，混凝土强度重要的并不一定在构件的中心，使用跟踪养护的方法时，传感器部位需根据需要来埋设。

最聪明的人是最不愿浪费时间的人。

——但丁

聪明的资质、内在的干劲、勤奋的工作态度和坚忍不拔的精神，这些都是科学研究成功所需要的其他条件。

——贝弗里奇

论文之八

高性能胶凝材料的实验研究之一

——混凝土的可持续发展与高性能胶凝材料*

廉慧珍　吴中伟　（清华大学土木系）

1. 混凝土的可持续发展与高性能胶凝材料的技术路线

1994年3月25日，我国国务院第16次常务会议讨论通过了《中国21世纪议程——中国21世纪人口、环境与发展白皮书》，指出："本世纪以来，随着科技进步和社会生产力的极大提高，人类创造了前所未有的物质财富加速推进了文明发展的进程。与此同时，人口剧增、资源过度消耗、环境污染、生态破坏和南北差距扩大等日益突出，成为全球性的重大问题，严重地阻碍着经济的发展和人民生活质量的提高，继而威胁着全人类的未来和生存的发展。在这种严峻形势下"，"必须寻求一条人口、经济、社会、环境和资源相互协调的，既能满足当代人的需求而又不对满足后代人需求的能力构成危害的可持续发展的道路"[1]。1993年6月世界建筑师代表大会的《为可持续的未来而相互依存宣言》提出"……可持续性设计能够在改善生活质量和经济福利的同时，大大减少人们对自然环境的有害冲击"。混凝土作为需用量猛增、发展迅速的大宗材料，必须对其资源（除原材料外，还包括土地、水、劳力等）和能源的消耗，环境付出的代价与经济、社会效益进行了认真的综合评价和协调，才能具有可持续性[2]。美国高层建筑与城市环境协会组织世界各国32位专家编著的《高层建筑设计》一书中特别论及建筑的可持续性问题。指出"可持续发展本质上是努力应用科学的、技术的和经济的知识，去修正由于无节制的技术激增所造成的负面后果。可持续发展的主要方面是：通过保护和减少浪费来更有效地利用能源和材料更大的循环利用，……"[3]。混凝土可持续发展的出路就是应用现代混凝土的科学技术增加混凝土的使用寿命，尽量减少造成修补或拆除的浪费和建筑垃圾。大量利用优质的工业副产品和废弃物，尽量减少自然资源和能源的消耗，减少对环境的污染。

混凝土的主要原材料是水泥。近年来，我国水泥产量以惊人的速度增长，1997年产量已达5.1亿吨，约占世界产量的三分之一。按照这样的速度发展，2010年可能超过8亿吨，将是该时世界产量的二分之一。在目前生产的水泥中，立窑生产的水泥超过总量的80%，生产水平低，生产要素单耗大，产品质量低，效益差。小水泥中有大量的325＃水泥，其使用价值比525＃水泥约低2/3～1/2（在很多工程中是不能替代的），但生产代价并不低。而且，不仅生产过程污染环境严重，还会给使用部门增加能耗、造价和不便。这种现状违背可持续发展的战略。如不改变，必遭淘汰。

水泥厂历来是一种污染源。现代水泥厂粉尘排放最已大幅度减少，但对有害气体如NO_x、CO_2、SO_2和微尘等。尽管目前采用高烟囱排放，减少了对地面空气的污染，却滞留

* 原载于《混凝土》1998年6月。

于地球的大气层中。其中CO_2的排放不仅来自燃料，而且还伴随着熟料的烧成产生更多的CO_2。我国由于工艺落后，生产1吨熟料的燃料生成的CO_2约400kg；生产1吨熟料平均由$CaCO_3$分解生成620kg CaO，同时产生490kg CO_2，比燃料产生的CO_2还多。再加上电耗，则生产1吨熟料所排放的CO_2总量约为1吨。1997年我国生产的5.1亿吨水泥中从低估计熟料约3.5亿吨，即已增了3亿多吨CO_2的排放。按上述初步规划，2010年水泥产量将为8亿吨，则届时地球大气层将因我国的水泥生产而增加CO_2积累量接近75亿吨之多！不仅我国的能源负担和环境难以承受，而且将为整个地球环境增加不可想象的负担。从可持续发展的角度来说，必须立即采取有效措施，减少水泥熟料的产量。我国水泥熟料的年产量保持在3亿吨上下的水平比较合理。在本世纪末出现高性能混凝土，适应了人类更大规模改善和保护环境的需要。高性能混凝土科学地大量使用矿物细掺料，既是提高混凝土性能的需要，又可减少对增加水泥产量的需求，既可减少煅烧熟料时CO_2的排放，又因大量利用粉煤灰、矿渣及其他工业废料而有利于环境保护。同时，以耐久性为设计目标，符合可持续发展战略。

目前按现行标准生产的混合材水泥，如火山灰质硅酸盐水泥、矿渣硅酸盐水泥、粉煤灰硅酸盐水泥以及复合硅酸盐水泥，多出于节省熟料和调节标号的考虑，而却以降低性能为代价，如降低强度、矿渣水泥的泌水、火山灰水泥的高需水量等。而高性能混凝土一般使用无掺和料的硅酸盐水泥或掺和料很少的普通硅酸盐水泥。根据工程需要掺用矿物细掺料和外加剂，则要求施工部门同时采购多种原材料，在配制时需多次计量和加料。造成工艺复杂，成本提高，为避免加料和计量的差错，对施工人员素质要求也较高。此外，目前国内市售的水泥质量差异很大。存在水泥和高效减水剂相容性不稳定的问题，不能适应低水灰比的高性能混凝土的需要，给高性能混凝土施工质量控制造成困难。因此人们正在寻求简化高性能混凝土的配制。目前国内这方面的研究主要有两种技术路线：在工厂将矿物细掺料（如矿渣）与高效减水剂共同粉磨至一定的细度，作为除水泥、水和骨料外的第四组分用于高性能混凝土的配制，可成为高性能掺和料。这种做法的优点是可通过细磨提高矿物细掺料的活性，同时，由于和熟料的硬度不同而造成颗粒组成和级配的不合理。例如通用的矿渣水泥，一般因矿渣的硬度较大，不易磨细，与熟料混磨后，当熟料细度合适时，矿渣颗粒太粗，活性难以发挥，并由于保水性差而造成混凝土泌水；如果使矿渣磨细，则熟料就会过细，影响水泥的性能。如可能早期强度高而后期增长很少。高性能掺和料单独粉磨至比表面积4000～8000cm^2/g，则可避免上述问题的出现。但因矿渣不易磨细，用现行球磨机粉磨时，能源消耗较大，需对粉磨设备进行改造或优选。

另一种技术路线是在生产水泥时，大量减少熟料用量，从流变性能的需要进行石膏、掺和料和外加剂等各组分的选择和配比优化，用于混凝土时，可在达到相同施工性时，得到最低的水胶比；在得到高流动性的同时，拌和物不离析、不泌水，有良好的可泵性和填充性，可称做高性能胶凝材料（实质上是高性能水泥）。这种高性能胶凝材料，从生产来说，可减少煅烧熟料的能耗和CO_2的排放，大量利用工业废料；从使用来说，具有高强度、低需求量、低水化热、低收缩、高抗化学侵蚀性等优点，有利于提高混凝土的耐久性，因此是可持续发展的。

高性能胶凝材料并不是高性能混凝土所用的胶凝材料简单的预先混合，而是通过熟料与外加剂的共同粉磨，不同矿物细掺料的组合与大量掺用，按流变性能优化石膏品种与掺量等主要措施实现其上述的高性能。

现就以下有关问题进行讨论：

①在生产高性能胶凝材料加入的外加剂(主要是高效减水剂)时比生产混凝土时所加入的外加剂效率高，有利于混凝土的流变性能；

②大量掺入矿物细掺料及其不同品种的复合是还没出现现代意义水泥的古罗马时代使用火山灰-石灰胶凝材料的否定之否定；

③从低水灰比的高性能混凝土流变性能的需要来看，现行通用水泥中的石膏掺量是不足的。掺用大量矿物细掺料后，虽然熟料减少了，但掺和料中的活性物质需要石膏与之反应以提供强度组分。因此生产高性能胶凝材料必须重新优化石膏掺量。

2. 水泥的塑化——对上述问题①的证明及国外情况

V. Alunno Rossetti 等试验研究在意大利一家水泥厂投产了一种特种超塑化水泥 SPC (Special Superplasticized Cement)，该水泥是在意大利 525 型硅酸盐水泥生产时掺入超塑化剂而制成的[3]，出发点是：使用水泥时，超塑化剂在加水之前已经先吸附在水泥颗粒的表面，以提高流化的效果；同时可避免超塑化剂通常掺在混凝土中时被骨料吸附而降低效率的问题。Rossetti 将超塑化剂用三种方式掺入水泥，测定溶液中超塑化剂溶出量，并用微型坍落度进行坍落度的经时变化：在工二厂中试验生产的 SPC（即在生产水泥时加入超塑化剂）；在使用水泥时加入超塑化剂（称 SpAD 试样）；将超塑化剂溶于水中掺入（称 AD 试样）。试验结果如图 1 所示[3]。

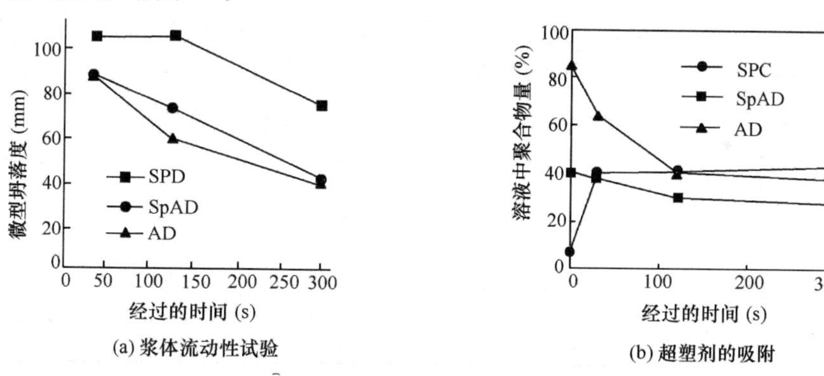

图 1　水泥对超塑化剂的吸附和浆体微型坍落度试验结果[3]

在图 1（a）中，对于相同掺量的超塑化剂，AD 试样和传统 SpAD 试样的初始流动性相同，但 SpAD 试样的流动性损失稍慢，而在水泥生产时加入超塑化剂的 SPC 试样效果最好；在图 1（b）中，SPC 试样效果最好；在图 1（b）中，SPC 由于对超塑化剂的吸附，在溶液中缓慢释放而保持时间最长。可见在生产水泥时加入超塑化剂可以提高超塑化剂的效率，减少流动度的损失。

瑞典用中热水泥和硅灰生产出一种强力改性水泥 EMC（Energetically Modified Cement)，是一种用于高强和超高强混凝土的低需水量专用水泥。改进后的水泥比基准水泥的强度提高 60%以上。可以用 0.19 的水灰比配制出 170MPa 超高强混凝土[4]，比 SPC 更进一步的是在水泥生产时除掺超塑化剂外，还掺入了硅灰。

S. A. Podmasova 等人报道了俄罗斯一种注册商标为 BHB（该符号为俄文，音译成英文时为 VNV）的低需水量水泥。已有数家水泥厂生产[5] 相对于普通水泥标准稠度用水量为

25%～30%，BHB-IDO～BHB-40 的标准稠度用水量为 16%～20%。BHB 的后缀数字代表该水泥中熟料的用量。其余组分为矿渣、粉煤灰、石英砂中的两种或三种。BHB 水泥减少熟料用量达 50%～70%，但所配制的混凝土强度可达 80～100MPa；该水泥中熟料取代量最多可达 70%，强度却比基准水泥的强度高。例如其中 BHB-50 水泥用量只有 350kg/m^3 时，混凝土强度可达 55.9MPa。这种水泥又比前两种更前进了一步，无疑是可持续发展的。

因此，在粉磨水泥时掺入高效减水剂以实现水泥的塑化，是生产高性能水泥的重要措施，和生产混凝土时掺入高效减水剂的作用不尽相同。

3. 大掺量矿物细掺料对混凝土性能的作用——对上述述问题②的思考和讨论

罗马有丰富的火山灰资源，2000 年前，罗马人用石灰、火山灰混合物建造了很多大型的建筑物，如著名的万神殿，20 英尺（约 6.1m）厚的墙，表面砌有一薄层砖，墙身用的是凝灰岩和火山灰制成的混凝土；跨度为 142 英尺 6 英寸（约 43.4m）的圆屋顶是完全用含浮石的火山灰混凝土浇筑的。公元 12～14 年罗马 Caligula 帝国时期用石灰和火山灰以 1∶2 配合，成功地用于建造了那不勒斯海港。现场观察，至今虽然被海浪磨光了表面，长满青苔，但混凝土却完好无损，数百米长的墙几乎无一裂缝。建于公元 70～82 年的古罗马竞技场，外径约 187m×126m，占地 204 公顷，容纳 5 万座位，作为斗兽和斗剑的场所，有时还需放水淹没。公元四世纪初，罗马城市有 1000 家公共浴室，最大的面积达 110000m^2。能容纳 3200 人沐浴。引水渡槽、桥梁、围堰至今还遗存 200 处。其中引水渡槽工程十分浩大。从 92 公里外引水进罗马城，共 11 座。建筑工程 310B.C. 到 226B.C.，这些都是用石灰加磨细凝灰岩与沸石凝灰岩以 1∶3.5～1∶45 建成。上述石灰火山灰砂浆与混凝土建成的桥梁和渡槽历时 2000 多年，仍然完好能使用。18 世纪中期，英国的 J. Smeaton 被委派建造位于 Cornish 海湾外的 Eddystone 灯塔。该灯塔受海浪冲刷，环境恶劣，Smeaton 用石灰-火山灰砂浆成功地建成了耐海水的灯塔。这些案例说明这种石灰-火山灰胶凝材料不仅有水硬性，而且有极好的耐久性。

石灰-火火山灰混合物之所以后来被硅酸盐水泥所代替，是因为石灰-火山灰胶凝材料凝结缓慢。早期强度很低，并由于天然材料品质的差异而不易控制其质量，不能适应资本主义兴起后大工业、交通运输和军事设施发展的需要。硅酸盐水泥正是应运而生，100 多年来成为不可替代的水硬性胶凝材料而被广泛使用。

由于使用部门不断提出提高强度尤其是早期强度的要求，水泥标号不断提高。尤其是近 50 年来，片面提高强度而忽视其他性能的倾向，造成水泥生产向大幅度提高比表面积和硅酸三钙、铝酸三钙含量发展。提高混凝土强度的办法除采用高标号水泥外，更多的是增加单方水泥用量、降低水灰比与单方加水量。使混凝土的流动性随之下降不得不采用高频振捣以期保证密实性和均匀性，增加劳动强度和能耗。与此同时，一方面建设速度的加快，一方面操作人员素质下降，混凝土工程质量得不到保证。80 年代前后，混凝土耐久性问题愈来愈尖锐。因混凝土材质劣化和环境等因素的侵蚀作用，出现混凝土建筑物破坏失效甚至崩塌等事故，造成巨大损失。加上施工能耗、劳动与环境保护，尤其是对工程安全型极端重要的均匀性要求，对混凝土施工性提出了越来越高的要求。改变传统的单一高强化的指导思想，逐渐形成高性能混凝土的概念，将耐久性放在第一位，作为混凝土配制的指标。从便于施工保证密实成型的角度，拌和物优异的施工性是高性能混凝土的重要特征。目前生产的水泥不能适应高性能混凝土的要求。优质矿物掺和料的大量掺用，又是一次应运而生。目前在美国预拌混凝土中粉煤灰掺量已达37%；英国已将粉煤灰体积用量达60%～80%的混凝土用于水坝、路面、机场停机坪等工程，在油罐、高架桥、后张法预应力预制块、给水塔等工程中．粉煤灰体积掺量为40%～60%日本新建世界上最长的悬索跨海大桥——明石大桥采用了不振道的高性能混凝土，28天强度51.9MPa的缆索锚固基础混凝土矿渣和粉煤灰掺量为60%；28天平均强度为24MPa的主桥墩混凝土掺和料用量为80%。

有2000年前古罗马使用石灰-火山灰胶凝材料的成功先例。混凝土大量掺用优质矿物细掺料后的耐久性是不容质疑的。由于高效减水剂、膨胀剂等外加剂的出现，以及其他技术措施的采用，现在使用大掺量优质矿物掺和料生产的高性能胶凝材料绝不是古代石灰-火山灰胶凝材料的简单重复，而是具有高强度、高抗化学侵蚀性、低需水量、低水化热、低收缩等高性能的新型胶凝材料；古代石灰-火山灰混凝土是在还没有钢筋混凝土，更没有预应力钢筋混凝土的情况下使用的，而现代混凝土主要用于钢筋混凝土和预应力钢筋混凝土。除了考虑凝结和强度的发展以外，还要考虑对钢筋的保护。因此，从混凝土的流变性能、温升和抗化学侵蚀性、后期强度发展等方面考虑，矿物细掺料根据其不同活性可用到极限量；而从保持混凝土一定的碱度来说，应区别构件性质而选用不同矿物细掺料不同用量的产品。有人认为高性能混凝土的密实度很高，可不考虑碳化问题，但应注意到，受弯构件在荷载作用下，由于受拉区钢筋的保护层混凝土裂缝的出现和扩展，碳化问题不能不考虑。在没有充分试验结果的支持以前，对用于钢筋混凝土受弯构件，混凝土的掺和料用量宜偏于保守。但是对柱、基础等受压构件和外包钢管混凝土则应大大放开。

4. 按流变性能的需要优化石膏掺量——对上述问题③的强调

按传统观念，在实际生产中，石膏在水泥中只起调节凝结时间的作用。一般，只按符合水胶比为0.5左右的混凝土所需凝结时间的最小掺量加入，此后，几乎是不变的。基本不管熟料化学组成与细度。由于水泥中SO_3含量达到3.5%以前，随SO_3含量的增加，凝结时间延长；强度提高，超过3.5%以后，结果则相反。因此，我国现行水泥标准规定水泥中SO_3含量≤3.5%（对矿渣水泥为≤4%），欧洲水泥试行标雉ENVl97-1-1992中则规定425＃以上任何品种水泥的SO_3含量上限均为4.0%，较我国的标准稍高。但是实际生产的水泥中，几乎没有超过2.5%的，甚至不超过2%。对水灰比较大的普通混凝土来说，没有明显的影响。而对掺用高效减水剂、水灰比很低的高性能混凝土就有显著的影响。图2所示的实例为某水泥在掺入不同水平的高效减水剂时改变石膏掺量对水泥标准稠度用水量的影响[6]。由图2可见，掺入高效减水剂后，水泥的需水量随石膏掺量的增加而明显降低。当高效减水剂掺量足够大时（如>1.5%），高效减水剂超过饱和点，石膏掺量的影响就不再显著。

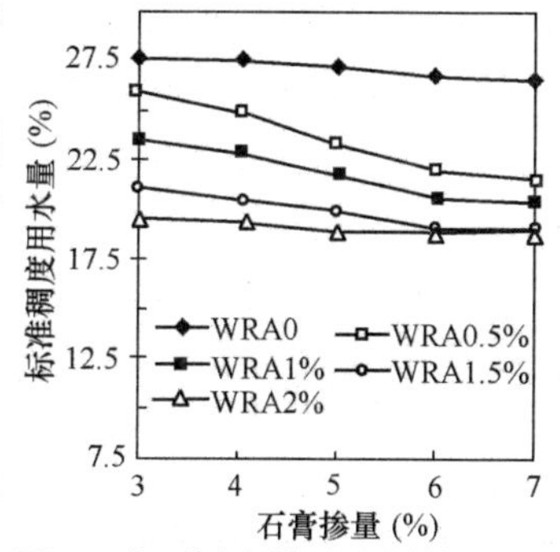

图2　在不同高效减水剂掺量水平下石膏掺量对水泥需水量的影响

实际上，水泥中SO_3含量应与C_3A和碱含量、细度等因素相匹配，否则对水泥及其制品性质会有不利影响。但当用于较大水灰比的传统混凝土时，通常不考虑这种影响。高性能混凝土采用很低的水胶比，水泥中的水很少，水化速率最快的C_3A和石膏争夺水，溶解速率和溶解度比C_3A低得多的石膏在液相中溶出的SO_3不足，会引起混凝土流变性能不良，流动性差，流动性损失快。因此用传统水泥配制高性能混凝土时，为得到高施工性，对原材料的要求较苛刻。水泥用量较大，在生产高性能胶凝材料时按流变性能的需要对石膏掺量进行优化，则可保证混凝土拌和物的良好的流变性能。

实验表明，尽管高性能胶凝材料中掺入大量矿物细掺料，对熟料中的C_3A与C_4AF有稀释作用，使得水泥的C_3A与C_4AF浓度降低，使水泥与高效减水剂的相容性得到改善。石膏掺量的变化对水泥需水量的影响不如对纯熟料水泥的那么显著，石膏掺量的优化仍能将水泥需求量降低约1个百分点[6]。而且，由于外加剂和矿物细掺料之间也存在相容性问题，在最优石膏掺量下，使用不同的矿物细掺料时，相同高效减水剂会有不同的饱和点。

石膏掺量不仅影响水泥的需水量、流动性损失的快慢，而且影响硬化水泥浆体的强度和变形性质；优化石膏掺量可提高强度，减小收缩。这部分实验结果将在其他文章中介绍。

5. 结论

水泥混凝土因其无可代替的优点，成为世界上用量最大的建筑材料，同时又由于传统的水泥还存在着固有的缺点，随着人类活动的发展，水泥混凝土材料就面临着是否可持续发展的严重问题。应运而生的高性能混凝土用高性能掺和料大量取代水泥，或用高性能胶凝材料取代传统的水泥，保证耐久性。大量利用工业废料，节省生产熟料的资源和能源，减少对环境的污染，是水泥混凝土可持续发展的出路。

参考文献

[1] 中国21世纪议程——中国21世纪人口、环境与发展白皮书[M]. 北京：中国环境工程出版社，1994.

[2] 美国高层建筑与城市环境协会原著. 罗福午，英若聪，张似赞，石永久译. 高层建筑设计[M]. 北京：中国建筑工业出版社. 1997.

[3] V Alunno Rossetti. F. Curclo and Cussino. Production Performance and Utilization of a Special Superplasticized Cement，Proceedings of 9th International Conference on the Chemistry of Cement，New Delhi，1992.

[4] J. E. Jonasson，V. Ronin. Energetically Modified Cement（EMC），Proceedings of International Conference on High-Strength Concrete，Norway. 1993.

[5] S. A. Podmasova，SH T. Rabev，YU. S. Volkov，New low wate rDemand Binder for High－Strength Concrete. Proceedings of International Conference on High-Strength Concrete，Norway. 1993.

[6] 阮庆革. 用于HPC的高性能佳凝材料实验研究[C]. 北京：清华大学硕士论文. 1996.

论文之九

高性能胶凝材料的实验研究之二
——FK 系列高性能水泥的性能及其检测*

廉慧珍　阮庆革　（清华大学土木系）
李玉琳　　　　　（北京住总集团）

1　适应高性能混凝土发展需要的高性能水泥

由于以耐久性为目标进行设计，在施工上具有施工简便、效率高而且经济等优点。近年来，高性能混凝土在我国发展很快，不仅用于高强度混凝土，而且也在 C50 以下的混凝土中应用，其中自密实高性能混凝土已由北京城建集团构件厂、北京建工集团、北京住总集团在北京应用于多个工程，浇筑了 40000 多立方米。高性能混凝土的组成材料为普通混凝土的原材料加入一定的无机和有机添加剂，主要特性为高施工性（高流动性、高抗离析性：坍落度≥220mm、可泵送，自流平）、较高强度（配制强度≥40MPa）、低收缩（可补偿水泥硬化后的收缩而不开裂或减少开裂）、低水化热、低温度应力和不透水性，尤其适用于高层、大跨、大体积、长跨桥梁、海底隧道、高速公路及严酷环境中结构物，如核反应堆、海上结构和处于有腐蚀性介质环境的结构等的建造和修补。

在高性能混凝土中，由于加入大量的矿物掺和料，既可以明显降低混凝土的温升，消除其温度裂缝，又可以改善新拌混凝土的和易性（流动性和抗离折性），使施工简便。掺用矿物掺和料的混凝土，其密实度得到提高，因此在 28 天以后强度有较大的增长率，其抗化学侵蚀能力也得到改善。由于不同的矿物掺和料对混凝土有不同的功能，往往需要复合掺用。例如粉煤灰使混凝土具有低收缩性，但抗碳化性较差；磨细矿渣抗碳化性好，但自收缩较大；粉煤灰和矿渣早期强度较低，后期强度高；硅灰早期强度高，但需水量大，自收缩大等等。复合比例恰当时，就可以取长补短，优化混凝土的性能。此外要保证足够高的强度和施工性，还必须添加超塑化剂，有时还要加入适当的增稠剂、膨胀剂等。如法国专利 FR2640962A1（1990 年 6 月 29 日公开）提出用硅酸盐水泥、流化剂、消泡剂、硅灰、石英砂粉、钢纤维等与普通砂石配制高性能混凝土；日本用于高性能混凝土的掺和料，一般从矿渣、硅粉、粉煤灰、石英砂粉、石灰石粉等矿物掺和料中选择 1～3 种，有时还需再加入增稠剂、膨胀剂、高效减水剂等有机或无机添加剂。

上述配制高性能混凝土的方法要求施工部门同时采购多种原材料，在施工现场或混凝土搅拌站多次计量和加料，造成工艺复杂，成本提高，为避免加料和计量的差错，对施工人员素质要求较高。此外，目前国内市售的水泥质量差异很大，一个工程在施工一个月内，就可能更换四五个牌号的水泥，水泥和高效减水剂相容性很不稳定，不能适应低水灰比的高性能混凝土的需要，给高性能混凝土施工质量控制造成很大困难。

---

* 原载《混凝土》1999 年第一期。

预先在工厂将混凝土所需要的各种掺和料和外加剂按一定比例混合，以合适的参数共同磨细，制成专用掺和料，再选择合适的水泥熟料，调节其他辅助材料，制成适用于不同强度等级高性能混凝土的胶凝材料（以下称高性能水泥），则可大大简化施工过程，稳定混凝土质量。对混凝土搅拌站，则可得到稳定的高质量水泥来源，生产出质量稳定的商品混凝土，并提高产量。在高性能水泥中，熟料用量只占50%左右，配制混凝土时，不需再使用任何添加剂，即可得到坍落度为16～21cm、配制强度为40～80MPa的高性能混凝土。B. Д. Сизов于1991年报道俄罗斯研制出BHB（俄文，音译成英文为VNV）系列低需水量水泥。其中BHB-100为纯硅酸盐水泥，粉磨时掺入超塑化剂C-3，BHB-50和BHB-30分别为熟料30%和50%，掺入矿物掺和料和超塑化剂C-3。所用矿物掺和料有：矿渣、粉煤灰、磨细石英砂粉等。水泥比表面积为4000～6000cm^2/g。S. A. Podmasova等人报道该注册商标为BHB的水泥还有BHB-80、BHB-60、BHB-40等型号。已有数家水泥厂生产，并用于实际工程。如耐久性要求较高的整体性地面。相对于普通水泥标准稠度用水量为25%～30%，BHB水泥的标准稠度用水量为16%～20%，可配制最高强度达100～150MPa的混凝土，可节省钢筋，带来很高的综合经济效益。本文所述高性能水泥与这种以低需水量为主要特征的水泥吻合。

高性能水泥并不是简单地使用通用水泥与掺和料、外加剂等混合，其主要技术特点是：按照流变性能的要求优化石膏、掺和料和不同外加剂的品种和掺量，并加入适当的助磨剂以合适的粉磨制度控制粉磨细度。在介绍优化实验研究之前，本文首先介绍经优化的高性能水泥的性能及其检测的方法。

2 实验用原材料

根据不同的原材料条件，高性能水泥可有多种优化方案，本文实验研究所用主要原材料如下。

2.1 熟料

北京首都水泥厂硅酸盐水泥熟料、北京燕山水泥厂硅酸盐水泥熟料，其化学成分和矿物组成如表1所示，强度见表2。

表1 熟料化学成分及矿物组成（%）

熟料名称	化学成分									矿物组成			
	SiO_2	Al_2O_3	Fe_2O_3	CaO	MgO	K_2O	Na_2O	SO_3	LOI	C_3S	C_2S	C_3A	C_4AF
首都	19.29	5.40	4.83	58.92	3.10	0.35	0.32	1.33	6.43	46.27	20.41	6.12	14.68
燕山	21.38	5.22	3.42	59.76	3.16	0.47	0.22	0.91	4.86	38.20	32.49	8.03	10.40
江南	20.02	5.05	4.61	60.92	1.08	0.48	0.40	0.87	4.36	50.02	17.54	5.56	14.01
房山	21.03	6.85	3.84	61.18	0.96	0.26	0.30	0.46	4.81	39.96	32.43	11.63	11.67

表2 熟料的强度

熟料名称	抗折强度（MPa）			抗压强度（MPa）		
	3天	7天	28天	3天	7天	28天
首都	6.58	7.63	7.82	37.1	41.3	54.7
燕山	7.5	8.22	9.1	42.9	47.2	57.8

2.2 矿物掺和料

使用首钢矿渣、元宝山Ⅰ级粉煤灰和北京东郊热电厂Ⅱ级粉煤灰，其化学成分见表3，表4为粉煤灰技术指标。

表3 掺和料化学成分（%）

种类	SiO_2	Al_2O_3	Fe_2O_3	CaO	MgO	K_2O	Na_2O	TiO_2	MnO	LOI
元宝山粉煤灰	58.64	19.78	9.56	4.42	2.08	2.64	0.87	0.91		0.89
北京东郊粉煤灰	51.96	32.61	5.61	2.61	0.63	0.78	0.17	1.21	0.06	3.46
首钢矿渣	44.12	20.69	21.21	4.43	1.10	1.09	0.44	2.30	0.06	3.55

表4 粉煤灰技术指标

技术指标	元宝山灰	北京东郊灰	Ⅰ级标准	Ⅱ级标准
细度（45μ方孔筛筛余）	6.0	18.0	≯12	≯20
需水量比（%）	94	1.03	≯95	≯105
烧失量（%）	0.89	3.46	≯5	≯8
含水量（%）	0.70	1.00		
SO_3 含量（%）	0.68	0.96		
28天胶砂强度比（%）	96.0	78.2	≯75	≯62

2.3 石膏

所用石膏分别为北京大红门石膏板厂和北京燕山水泥厂所用的天然生石膏，其化学成分见表5所示。

表5 石膏化学成分

来源＼成分	SiO_2	Al_2O_3	Fe_2O_3	MgO	CaO	SO_3	结晶水	烧失量
大红门	0.80	0.21	0.15	2.19	31.25	44.78	15.02	20.34
燕山	4.82	1.34	0.54	2.91	34.51	39.60	9.09	15.86

2.4 高性能水泥优化方案

采用正交实验优化高性能水泥的组分和粉磨工艺，编号为FK系列。

3 FK高性能水泥的性能

用现有原材料进行FK型高性能水泥配合比及粉磨工艺的优化（见本研究论文之三、之四）。选用其中FK-Ⅰ型和FK-Ⅱ型，检测其物理力学性能。其中FK-Ⅰ型用于低于C50的中、低强度等级混凝土，FK-Ⅱ型用于C50～C70的混凝土。

3.1 需水量和凝结时间

优化的FK型高性能水泥标准稠度用水量和凝结时间检测结果见表6所示。

表6表明，FK型高性能水泥的需水量远低于同样熟料的普通水泥的标准稠度用水量，也低于在水泥中掺入同量外加剂和掺和料时的标准稠度用水量。

此外，送由国家建材院和冶金建筑科学院检测编号为DF4K1（属于FK-II型）的标准稠度用水量分别为18%和17.4%。

表6 FK型高性能水泥的标准稠度用水量和凝结时间

性质＼编号	YFK-Ⅰ		YFK-Ⅱ			对比样		
	1	2	2	3	5	首都525#	首普+FDN	首普+FDN+FA+K^+
比表面积（m^2/g）	390	396	430	415	420	—	—	—
需水量（%）	17.5	18.5	19.0	20.0	16.5	27.0	23.0	23.0
初凝时间（h：m）	4：20	5：00	4：50	5：00	4：30	3：00	2：20	3：50
终凝时间（h：m）	6：30	7：00	6：30	7：30	7：00	6：20	4：30	6：50

3.2 强度

检测表7中所列试样的强度示于表8。

由表7可见：

①高性能水泥比用同样首都水泥厂熟料生产的普通水泥的强度高；

②与将相同掺料以相同掺量掺入水泥相比，在粉磨时掺入优化掺量的掺料（石膏、矿物掺和料、外加剂）可得到更好的效果。

为了取得统计性数据，用YFK-Ⅱ-2、YFK-Ⅱ-5进行批量性实验，实验进行15批，并以燕山水泥厂硅酸盐水泥进行平行对比实验。结果如表8所示。

用相同配比试样分别送国家建材院和冶金建筑研究院检验，结果示于表9。

表7 FK型高性能水泥强度实验结果（MPa）

强度值＼编号		YFK-Ⅰ		YFK-Ⅱ		首普		首普+FDN+FA	首硅*	冀东	冀东+FDN+FA
		1	2	2	3	5	425#		525#	525#	
抗折	3天	7.28	7.05	9.20	7.75	7.35	5.81	5.02	6.58	5.48	5.26
	28天	9.48	9.35	11.2	11.1	11.3	8.62	8.05	8.82	8.70	8.54
抗压	3天	32.8	30.6	42.5	41.2	40.6	25.9	23.6	37.1	30.6	26.5
	28天	56.5	55.6	67.2	65.5	68.7	47.7	48.9	54.7	53.6	51.5

表8 FK型高性能水泥强度批量实验结果（MPa）

强度＼龄期＼编号	YFK-Ⅱ-2				YFK-Ⅱ-5				燕山水泥			
	抗折强度		抗压强度		抗折强度		抗压强度		抗折强度		抗压强度	
	3天	28天	3天	28天	3天	28天	3天	28天	3天	28天	3天	28天
最小值	8.26	10.05	39.8	66.6	6.65	9.90	39.8	67.4	6.96	8.65	36.8	53.0
最大值	10.02	12.10	45.6	72.4	8.60	11.96	44.9	74.0	7.85	10.04	42.6	58.5
平均值	9.27	11.09	43.1	70.0	7.73	10.91	42.6	71.4	7.4	9.3	40.1	55.5
标准差	0.52	0.59	1.78	1.72	0.57	0.59	1.64	1.84	0.32	0.45	1.85	1.61

表 9 FK 型高性能水泥强度复验检测结果

编号	检测单位	抗折强度（MPa）			抗压强度（MPa）		
		3 天	7 天	28 天	3 天	7 天	28 天
YFK-Ⅱ-2	建材院	5.3	7.0	9.2	30.5	47.0	71.3
YFK-Ⅱ-5	冶建院	4.9	6.2	8.2	28.9	40.2	64.5
DFK-Ⅱ-2	建材院	4.7	6.8	8.9	9.8	44.8	69.9

3.3 收缩

因实际使用的混凝土绝大多数都处于限制条件下，故用限制膨胀测定方法来检验高性能中泥的收缩并与普通水泥比较。即在 20℃ 水中养护 14 天后，取出置于相对湿度（50±5%）、20℃的空气中养护 28 天。由北京建工院第一检测所检测结果如表 10 所示。

表 10 水泥变形性质检测（变形单位为%）

种类	FK-Ⅱ高性能水泥		对比普通水泥		标准水泥	
龄期	14 天水中	28 天空气中	14 天水中	28 天空气中	14 天水中	28 天空气中
变形	+0.014	−0.014	+0.029	−0.050	+0.014	−0.022

由表 10 可见，高性能水泥的收缩值，明显低于普通水泥，而且低于混凝土的开裂极限。在水中养护膨胀到 14 天后，再在空气中干燥到 28 天，高性能水泥比普通水泥和标准水泥的变形落差都较小。因此使用 FK 高性能水泥时，无需加膨胀剂即可实现低收缩。

3.4 水化热

分别用溶解热法和直接法测定水泥水化热结果见表 11。

表 11 高性能水泥水化热（J/g）

编 号	检测方法	3 天	7 天
空白	熔解热法	263	276
F4K0-2	熔解热法	212	259
F0K5-2	熔解热法	183	235
F0K4-1	直接法	167	188
F0K4-2	直接法	184	205
F0K5-2	直接法	146	184
FK-Ⅱ-2	直接法	111	176

由表 11 可见，高性能水泥水化热明显低于通用水泥的水化热，但由于混磨水泥（如 FOK4-2 细度 0.8%；FOK5-2 细度 1.2%）过细，比表面积可达约 5000cm²/g，水化热降低不十分显著；比表面积较小的 FK-Ⅱ-2（0.8mm 筛筛余 1.8%，比表面积约为 4000cm²/g）则水化热明显降低，相当于标准中对低热矿渣水泥水化热的规定。

3.5 储存性能

因高性能水泥粉磨细度较细，需了解存放过程中其强度的变化。将 FK-Ⅱ-2 水泥和 FK-Ⅱ-5 分别存放于用塑料袋密封（简称“密封”）和装袋装存放于相对湿度 30%的空气中（简称“防潮”）。分别在 1 个月、3 个月和 5 个月或 2 个月、3 个月和 4 个月检验其 28 天强度。结果表明，高性能水泥在密封或防潮的条件下储存 3 个月后强度损失不大于 10%；密封存放 5 个月后强度损失小于 15%，防潮存放 5 个月后强度损失小于 20%；夏季露天袋存放 3

个月，强度损失小于20%。因此高性能水泥应采用内套塑料袋封装出厂，储存时间一般可到5个月。

4 关于高性能水泥检测方法

4.1 关于标准稠度用水量

目前水泥标准稠度用水量的测定方法符合不加超塑化剂的普通混凝土的需要，而不能反映加入超塑化剂后水泥的需水性。适合用于高性能混凝土的高性能水泥的流变性质，只有在使用超塑化剂时才能反映出来。因此需做适当调整：①由于需水量的大大降低，需用调整用水量的方法；②实验表明对掺超塑化剂的净浆，现行标准搅拌时间不足，以反转时延长15秒为宜；

4.2 关于强度

水泥、混凝土的强度应在达到相同流动性的条件下进行对比，高性能水泥的需水量很低，不能按现行水泥标准中规定的加水量0.44或0.46进行强度检验。参考复合硅酸盐水泥标准中的规定："当流动度小于116mm时，须……将水灰比调整至胶砂流动度达到不小于116mm。"和微集料火山灰水泥、微集料粉煤灰水泥标准中的规定："当使用需水量较大的活性混合材料时，应按胶砂流动度在120～130mm时的水灰比加水。"以及硫铝酸盐、铁铝酸盐水泥标准中有关规定，考虑到高性能水泥用于流动度大的混凝土，建议强度检验时按胶砂流动度在125～135mm时的水灰比加水。

4.3 关于收缩

混凝土结构工程施工及验收规程中规定，"混凝土浇水养护的时间，对采用硅酸盐水泥、普通硅酸盐水泥或矿渣硅酸盐水泥拌制的混凝土，不得少于7天……"，因此掺有膨胀剂的混凝土或使用具有微膨胀性质的水泥，应当先在水中养护14天后，再存放在空气中检测其干缩值；同时因一般混凝土都处于受限制的条件下，故检测这类水泥的变形性能时，宜采用限制的条件。因此，建议按混凝土膨胀剂限制膨胀率的方法检验高性能水泥收缩的性质。

5 结论

(1) FK型高性能水泥需水量很低，其标准稠度用水量为16%～21%；

(2) FK型高性能水泥比通用水泥的水比热3天低20%～30%，7天低10%～15%；

(3) FK型高性能水泥在水中养护14天后继续在空气中养护28天，收缩率不到万分之一，远低于普通水泥的，而且远低于混凝土的收缩限。

(4) FK型高性能水泥强度可达50～70MPa；

(5) FK型高性能水泥凝结时间，初凝4～7小时，终凝6～10小时。

(6) 高性能水泥性能检测方法需做适当调整，另行制订标准。

参考文献

[1] В. Д. Сизов Прочность Бентона ВНВ，Бетон и Железобетон，1991.

[2] S. A Podmasova，SH. T. Babev，YU. S. Volkov，New Low Water Demand Binders for High-Strength Concretes. Proceedings of International Conference on High-Strength Concrete，Norway，1993.

[3] В. И. Соломатв Бетоны на ВНВ, Модифицированные Ацетоно Фермалъ-Дечидной Смолой, Бетон и Железобетон, 1990. 10.
[4] Н. Х. Наназасивили, Структурообрзование древесно-Цеметны х кемпозитов на оснрве ВНВ, Бетон и Железобетон, 1992. 12.
[5] В. Н. Рябошапко, Перспективы Развития Строителъства, Москвы, Бетон и Железобетон, 1994. 10.
[6] C. Jolicoeur, M. A. Simard and P. C. Aïtcin, Cement - Superplasticizer Compatibility in High Perfomance Concretes: The Role of Sulphates, Selected Papers on Superplasticizer Prepared by P. C. Aïtcin, Hundv Chemical Ltd. CA. 1995.

我并没有什么方法，只是对于一件事情很长时间很热心地去考虑罢了。

——牛顿

时间，就像海绵里的水，只要愿挤，总还是有的。

——鲁迅

论文之十

高性能胶凝材料中石膏的优化*

廉慧珍[1]，阮庆革[1]，李玉琳[2]
（1. 清华大学土木工程系，北京 100084；2. 北京住总集团，北京 100026）

摘　要　高性能胶凝材料主要用于高性能混凝土，具有强度高、需水量小、水化热低和收缩小等特点。按掺高效减水剂的胶凝材料的流变性能来优化石膏的掺量是高性能胶凝材料的特有技术之一。试验研究了石膏的掺量对高性能胶凝材料的需水量、凝结时间、强度、收缩性的影响，以此作为优化石膏掺量的依据。

关键词　石膏　高性能胶凝材料　流变性

大力推广应用高性能混凝土是混凝土可持续发展的举措[1]。适应高性能混凝土需要的高性能胶凝材料，即以合适的熟料在工厂预先与高性能混凝土所需的各种无机和有机添加剂按一定的比例混合，以调节水泥的颗粒级配，并优化石膏的掺量，以合适的参数和工艺磨细至一定的细度而制成，可用于不同强度等级的高性能混凝土。该胶凝材料性能的特点为低需水量、低水化热、低收缩、高强度[2]。按流变性能优化石膏掺量是该胶凝材料的技术特点之一。

1　优化水泥中石膏掺量的必要性

水泥是用熟料加入适量石膏共同粉磨至一定的细度而生产出来的。加入石膏主要是为了控制熟料中 C_3A 的水化从而调节水泥的凝结时间。实际上，石膏在水泥中的作用不仅是调节凝结时间，而且对水泥的强度、流变性能和收缩都有影响。图 1 为在不同温度下某熟料的收缩和 SO_3 含量的关系。表明该水泥的 SO_3 在常温下的最佳含量为 4%[3]。水泥的 SO_3 最佳含量与水泥中的 C_3A 和 R_2O 含量及水泥细度有关[4]，如表 1 所示。

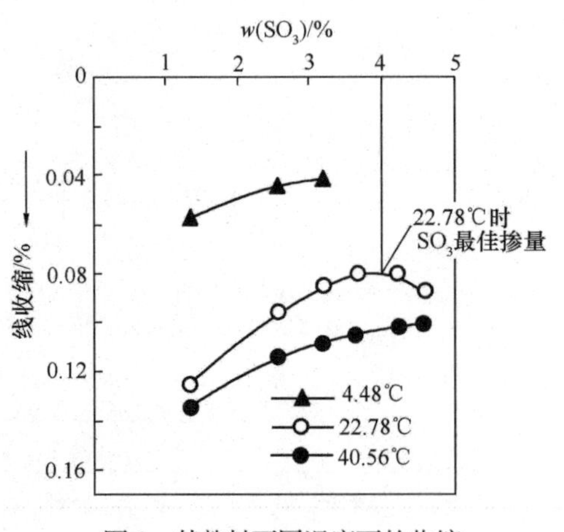

图 1　某熟料不同温度下的收缩和 SO_3 含量的关系[3]

水泥的 SO_3 最佳含量还和使用温度有关。C_3A 和碱含量越大，水泥细度越细，温度越高，SO_3 最佳含量越大。在实际工程中，由于水泥水化放热，构件内早期水化温度较高，混凝土收缩最小而强度最高的最佳 SO_3，含量更高些。S. M. Khalil 使用 C_3A 含量为 9.22%的Ⅰ型水泥和萘磺酸盐超塑化剂进行了 SO_3 含量对混凝

* 原载《山东建材学院学报》Vol. 3. No. 2. 1999. 6。

土的温升和坍落度损失的试验研究。结果表明，SO_3含量对混凝土保持坍落度和提高强度有一个最优范围，并且这一最佳范围随养护温度而不同。养护温度为25℃时，最佳SO_3含量为3.15%（对比样相同水泥的商品混凝土中SO_3含量为2.15%）；养护温度为40℃时，最佳SO_3含量为5.65%（一般混凝土的温升使构件内部混凝土实际养护温度>40℃）。在SO_3最佳含量下，浇筑后30～70min的坍落度等于或大于对比混凝土浇筑后20min的坍落度，28d强度提高12%[5]

表1　SO_3含量和C_3A、R_2O的关系[4]

$w(C_3A)$ /%	$w(R_2O)$ /%	$w(SO_3)$ /%
<6	0.5	2
<6	>1.0	3～4
>10	0.5	2.5～3
>10	>1.0	3.5～4

目前ISO标准是在水胶比为0.5而不掺外加剂的砂浆的条件下制定的。在这样的拌和物中，所存在的大量拌和水可使水泥颗粒分散，并使离子在溶液饱和前自由地进入溶液，对这种砂浆的流变性能起主导作用。由于超塑化剂的使用，使水胶比可降低到<0.4，甚至<0.3。水泥颗粒间距减小，能进入溶液的离子数量减少，水泥开始水化的动力学与水胶比为0.5时的情况大不相同[6]。因为水泥中的水很少时，SO_3在水泥浆体中的溶出量很少。尤其当水泥中C_3A含量和比表面积较大时（如为了提高水泥早期强度和降低烧成温度而生产的R型水泥），水泥水化加快，其中水化速度极快的C_3A要和石膏争夺水，溶解速率和溶解度比C_3A低得多的石膏在液相中溶出的SO_3更加不足，因此，市售水泥中的SO_3量就不能满足高性能混凝土流变性能的要求。图2所示的实例为某水泥在掺入不同水平的高效减水剂时改变石膏掺量对水泥需水量的影响。

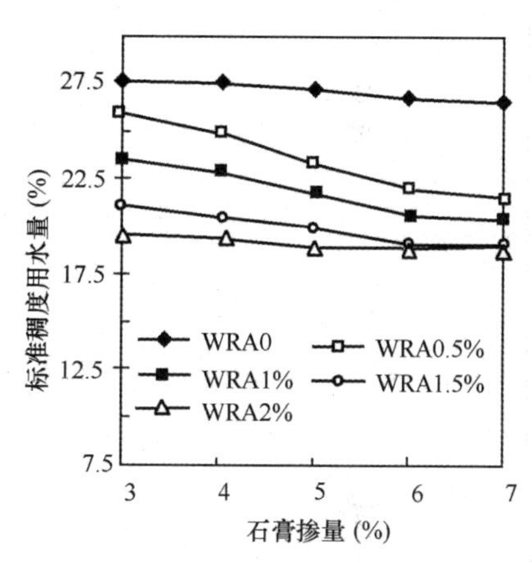

图2　减水剂不同掺量下石膏掺量对水泥需水量的影响

由图2可见，掺入高效减水剂后，水泥的需水量随石膏掺量的增加而明显降低。当高效减水剂掺量足够大时（例如>1.5%），超过饱和点，石膏掺量的影响就不显著了。

2　实验用主要原材料

石膏：所用石膏分别为北京大红门石膏板厂和北京燕山水泥厂所用的天然生石膏，其化学成分见表2所示。

表2　石膏化学成分（%）

来源	SiO_2	Al_2O_3	Fe_2O_3	MgO	CaO	SO_3	结晶水	烧失量
大红门	0.80	0.21	0.15	2.19	31.25	44.78	15.02	20.34
燕山	4.82	1.34	0.54	2.91	34.51	39.60	9.09	15.86

熟料：主要有首都水泥厂旋窑熟料、燕山水泥厂旋窑熟料、江南水泥厂旋窑熟料、房山

周口店水泥厂立窑熟料等。各熟料化学成分见表 3 所示。各熟料粉磨细度见表 4。

表 3　熟料化学成分及矿物组成（%）

熟料名称	化学成分									矿物组成			
	SiO_2	Al_2O_3	Fe_2O_3	CaO	MgO	K_2O	Na_2O	SO_3	LOI	C_3S	C_2S	C_3A	C_4AF
首都	19.29	5.40	4.83	58.98	3.10	0.35	0.31	1.33	6.43	46.27	20.41	6.12	14.68
燕山	21.38	5.22	3.42	59.76	3.16	0.47	0.22	0.91	4.86	38.20	32.49	8.03	10.40
江南	20.02	5.05	4.61	60.92	1.08	0.48	0.40	0.87	4.36	50.02	17.54	5.56	14.01
房山	21.03	6.85	3.84	61.18	0.69	0.26	0.30	0.48	4.81	36.98	32.43	11.63	11.67

表 4　熟料粉磨细度

熟料名称	江南	首都	燕山	房山
水筛法筛余（%）	2.4	3.0	2.0	3.2
比表面积（m^2/kg）	400	404	419	395

高效减水剂：山东莱芜汶河化工厂 FDN。

矿渣：首钢水淬矿渣，质量见表 5 所示。

表 5　试验用磨细水淬矿渣

化学成分（%）										筛析法筛余（%）	比表面积（cm^2/g）
SiO_2	Al_2O_3	Fe_2O_3	CaO	MgO	K_2O	Na_2O	TiO	MnO	LOI		
33.56	11.40	0.33	40.39	11.2	0.57	0.57	1.34	0.09	0.07	0.8	443

3　实验与结果分析

3.1　石膏掺量对水泥需水量的影响

分别用 4 种熟料，掺入高效减水剂，改变石膏掺量制成水泥，测定水泥的需水量和凝结时间，结果如图 3 所示。

由图 3 可见，水泥需水量随石膏掺量的增加而降低，熟料中 C_3A 含量越高，降低得越多。如房山熟料中 C_3A 比其他熟料明显要高，石膏掺量的影响显著，其他熟料中 C_3A 含量相差不多，影响趋势也一致。

以上述熟料掺入 50%磨细矿渣，测定在高效减水剂作用下石膏掺量对水泥需水量的影响，结果示于图 4。

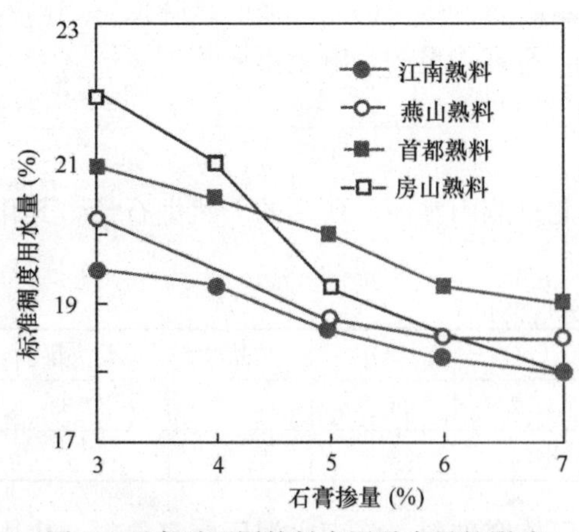

图 3　石膏对不同熟料水泥需水量的影响

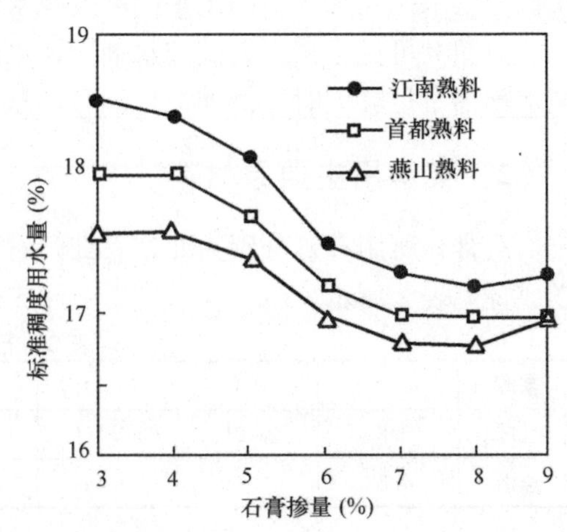

图 4　石膏掺量对高性能胶凝材料需水量影响

由图4可见，与纯熟料水泥相比，掺入50%磨细矿渣后，由于矿渣的需水量比较小，水泥需水量降低，但仍随石膏掺量的增加而进一步降低，只是降低的幅度减小。3种熟料的C_3A含量相差不大，石膏掺量的影响趋势一致，石膏的饱和掺量为6%～7%，水泥中SO_3含量为3.5%～4%。当SO_3含量超过4%（相应于石膏的饱和掺量）以后，再增加石膏掺量时，水泥需水量不再降低。

3.2　石膏掺量对水泥凝结时间的影响

上述试样凝结时间见表6。掺入矿渣后，水泥的凝结时间有所延长，石膏掺量对凝结时间影响不大。

表6　石膏掺量对水泥凝结时间的影响

组成	石膏掺量(%) / 凝结时间(h：min)	3		4		5		6		7	
		初凝	终凝	初凝	终凝	初凝	终凝	初凝	终凝	初凝	终凝
无掺料	江南熟料	3：00	4：20	3：20	4：30	3：40	4：50	3：40	4：50	3：50	5：00
	首都熟料	2：40	3：40	2：50	4：00	2：50	3：50	3：10	4：10	3：00	4：20
	燕山熟料	2：50	4：00	3：00	4：20	3：10	4：30	3：20	4：30	3：00	4：30
掺矿渣	江南熟料	3：00	4：30	3：00	5：10	3：00	5：20	3：00	5：20	3：40	6：00
	首都熟料	3：00	6：00	3：20	6：10	3：30	6：10	3：50	6：40	3：40	6：30
	燕山熟料	5：00	6：50	5：20	7：00	5：30	7：00	5：30	6：50	5：40	7：00

3.3　石膏品种影响的考虑

不同品种的石膏溶解速率和溶解度差别较大，对水泥性质的影响不同。例如生石膏（天然二水石膏）和天然硬石膏的溶解度相近，但溶解速率差别很大，半水石膏则不仅溶解速率很快，溶解度也大得多。严格地说，不仅应当优化石膏的掺量，而且应当根据熟料中C_3A的不同活性优化石膏的品种。按我国的国情，一般不宜使用半水石膏。相反，根据熟料情况掺适量的天然硬石膏倒是有利的。本实验用两种石膏测定石膏掺量对水泥需水量的影响，结果如表7所示。从表7可以看出，产地不同的石膏及不同的掺量对水泥需水量均有影响。究其原因，主要是化学成分的差异。

表7　不同来源石膏对水泥需水量的影响

石膏掺量（%）	3	4	5	6	7
大红门	23.6	22.6	21.6	20.5	20.5
燕山	22	21.5	21.2	20.8	20.7

大红门石膏中的$CaSO_4$含量较大，其结晶水比纯二水石膏的结晶水少7%；而燕山石膏中的$CaSO_4$含量较少，其结晶水比纯二水石膏的结晶水少8.71%。石膏掺量越大，二者差别越小。因此在本实验中暂不计其差别。如果品种差异更大，则不可忽视，需做更进一步的实验。

3.4　石膏掺量对水泥强度的影响

石膏不仅影响C_3A的水化，而且也会影响C_3S的水化，SO_4^{2-}还可能进入C-S-H的层间，对水泥强度产生影响[7]。加之水泥的需水量随石膏掺量的增加而降低，这也会影响到强度。因此，石膏掺量的优化，还要考虑强度的因素。

用首都水泥厂和燕山水泥厂熟料，加入 FDN 和大红门石膏。改变石膏掺量，测定水泥强度，结果示于图 5。

由图 5 可见，在 SO_3 含量为 2%～4.5%的范围内，水泥的抗压强度随石膏掺量的增加而提高，强度值与熟料强度有关。

3.5 石膏掺量和水泥收缩的关系

石膏在水泥中会生成体积膨胀的反应物，SO_3 含量合适时，在限制条件下，可补偿水泥因干燥而引起的收缩。因此，石膏掺量的优化应考虑石膏掺量对水泥收缩的影响。

用燕山熟料掺入 FDN，改变石膏掺量，按 JC 476—92 规定的方法测定水泥砂浆在限制条件下的变形。养护条件和测定结果见图 6 所示。其中水养护温度为(20±3)℃，空气养护的条件为温度(20±3)℃、相对湿度(55±5)%。所用试件砂浆用钢筋骨架限制其变形。

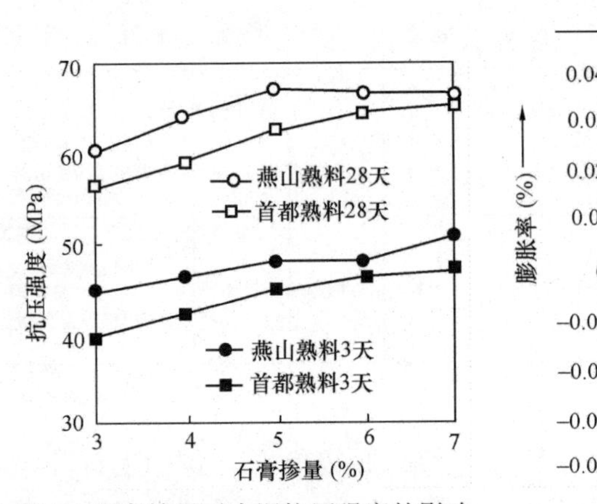

图 5 石膏掺量对水泥抗压强度的影响

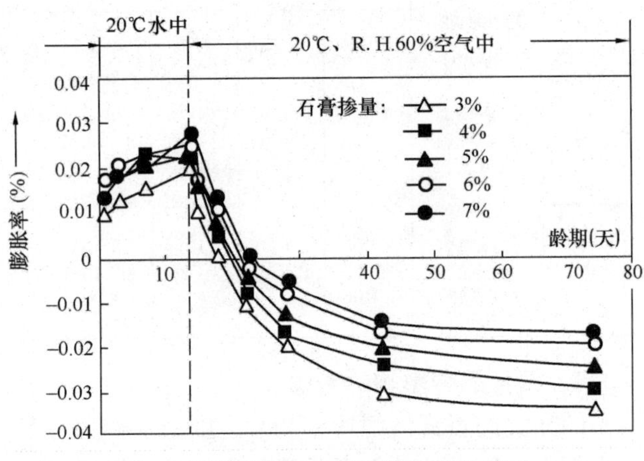

图 6 石膏掺量和水泥砂浆变形的关系

由图 6 可见，砂浆的收缩随石膏掺量的增加而减小。当石膏掺量超过 5%（SO_3 含量为 3.57%）以后，该试件在水中养护 14d 后，再在空气中养护 28d 的限制收缩率不大于 2%，符合 JC 476—92 规定的掺膨胀剂补偿收缩的要求。而当石膏掺量达到 6%［w（SO_3）%＝4.02］时，空气中养护 60d 的限制收缩率也不超过 2%。石膏掺量超过 6%时，再增加掺量时，体积收缩的变化就不大了。

4 结论

（1）石膏在水泥中的作用不仅是调节凝结时间，还影响水泥的流变性能、强度和变形性质。这一点对使用高效减水剂且水胶比很低的高性能混凝土来说尤其重要；

（2）用于高性能混凝土的高性能水泥并非混凝土胶凝材料各组分的简单混合，优化石膏掺量是获得水泥高性能的一项重要技术措施；

（3）高性能水泥中石膏掺量优化的目标是使水泥的流变性能好、强度高、体积稳定性好、凝结时间合适；

（4）在本实验所用原材料的条件下，SO_3 的优化掺量为 3.5%～4%。

参考文献

[1] 廉慧珍，吴中伟. 混凝土的可持续发展与高性能胶凝材料[J]. 混凝土，1998.（6）.

[2] 廉慧珍，阮庆革，李玉琳. 高性能水泥的性能及其检测[J]. 混凝土，1999.（1).

[3] Verbeck J J，Helmuth R A. Structures and physical properties of cement paste[C]. Proeedings of Fifth International Symposium on the Chemistry of Cement，Tokyo. 1968.

[4] Khail S M 著，陈冀宇译. 水泥中硫酸盐含量对掺超塑化剂混凝土放热和坍落度损失影响的研究[J]。武汉建材学院译丛，1982.（4）

[5] Tagnit A，Baalbaki M，Aïtcin P C. Calcium—sulphate optimization in low water/cement ratio concretes for rheological purposes [C]. Proceedings of 9th International Conference on Chemistry of Cement. New Dehli. 1992.

[6] Jolicoeur C Simard M A，Aitcin P C. Cement—superplasticizer compatibility in high performance concretes：The role of sulphates. Selected Papers on Superplastieizer [M]. Hundy ChemicaI Ltd. CA. 1995.

[7] Aïtcin P C，Joficoeur C，Mac Gregor J G. Superplasticizers：how they work and why they occationaly don't[J]. Concrete International，1994.（5).

论文之八～论文之十自评

1. 在 1997 年我国水泥总产量为 5.1 亿吨时，吴中伟就曾设想过，未来硅酸盐水泥熟料维持当时的产量，再掺入 50％的矿物掺和料以提高水泥产量，即可满足建设的需要，又能达到节能、降耗、减排的目的。他提出《绿色胶凝材料》、《环保型胶凝材料》就是这个意思。以上三篇文章是此类研究中最主要的内容：论文之八阐述国内外同类研究的现状、必要性与可行性；论文之九介绍该项研究的技术路线、产品性能及其检测方法；论文之十是该产品关键的技术之一。

2. 该项技术看似为把混凝土拌和物中的水泥和矿物掺和料简单地预混均化，实际上是对搅拌机中拌和物胶凝材料组成的优化，是对传统水泥（hydraulic cement，水硬性胶凝材料）观念的转变。

3. 从文中可见，当原材料来源不同时，各试验规律趋势相近，但是量值有较大的变化。尤其是当今的水泥由于细度过细，其强度发展的规律大不同于过去的水泥；配料和烧成工艺的变化使现在的 C_3A 结构和性质也有变化，则对石膏的优化不仅有量的也有品性的问题，等等。因此论文所提供的是一种思路，解释一种可行性，而实际用于生产时，尚需针对具体情况进行优化。

4. 在粉磨时将高效减水剂一并加入，其目的是使其均匀分散并吸附到矿物颗粒上提高其利用效率，同时起助磨作用。论文中所介绍的都是使用萘磺酸盐类的产品，对其他品种的需要进行工艺上和性能优化的研究。

5. 检测方法和实时的标准有关，对检测方法及其指标如何用于指导混凝土的配合比设计需要进一步的研究。

论文之十一

高性能混凝土在深圳地铁一期工程中应用研究的工程试点裂缝控制*

廉慧珍　阎培渝

一、试验目的

根据深圳港创建材有限公司、清华大学土木工程系与深圳市地铁有限公司订立的《深圳地铁一期工程竹子林车辆段挡土墙高性能混凝土应用试验（合同编号：DT10-KY08/2002)》研究项目合同，及在2000年地铁公司科研项目“高性能混凝土在深圳地铁工程中应用的试验研究”所取得的成果基础上，通过试验段的实际施工，进一步验证高性能混凝土在地铁工程中使用的可行性，为在地铁工程建设中大规模推广高性能混凝土积累经验。按照ACI和我国混凝土技术前辈吴中伟对高性能混凝土的定义，高性能混凝土是能满足给定工程的各项性能和匀质性要求的混凝土，并符合可持续发展的战略要求。高性能混凝土必须由设计、材料和施工各部分协作共同努力来实现。本次试验除在混凝土的制备上采用低水泥用量、大掺量矿物掺和料、低水胶比以降低混凝土内部温升、减小温度应力和提高混凝土对地下土壤中盐类腐蚀的抗力外，对施工缝间距、浇筑顺序、模板及拆模时间、养护、温度和早期裂缝的控制进行试验，并对设计提出相应建议。

二、工程概况

计划在深圳地铁一期工程竹子林车辆段挡土墙浇筑高性能混凝土的试验段，位于深南大道红树林段南侧，北侧毗邻车辆段机车维修车间，试验段长度原定为80m，实际完成110m。该挡土墙横断面及配筋情况如图1所示。

如图1所示，挡土墙高度从趾板和踵板顶部到墙顶为6m，墙厚底部为815mm，顶部为500mm，从顶部到底部坡度为5%，上下左右每隔2m设置一个泄水孔。

挡土墙的配筋：在4.55m高度范围内，内侧竖筋为Φ20@100；4.55m以上内侧

竖筋为Φ20@200，而外侧竖筋全部为Φ14@200。水平筋为Φ12@200。竖筋之间的拉筋为Φ8@400。

三、试验方案和内容

试验方案：

与深圳地铁公司、深圳港创建材公司合作进行的《高性能混凝土在深圳地铁一期工程中应用的实验研究》，浇筑了地铁区间足尺模型，成果鉴定专家们建议进行工程试点试验。试验于2002年在深圳地铁一期工程车辆段进行。试验方案如下：

* 该论文未曾发表。

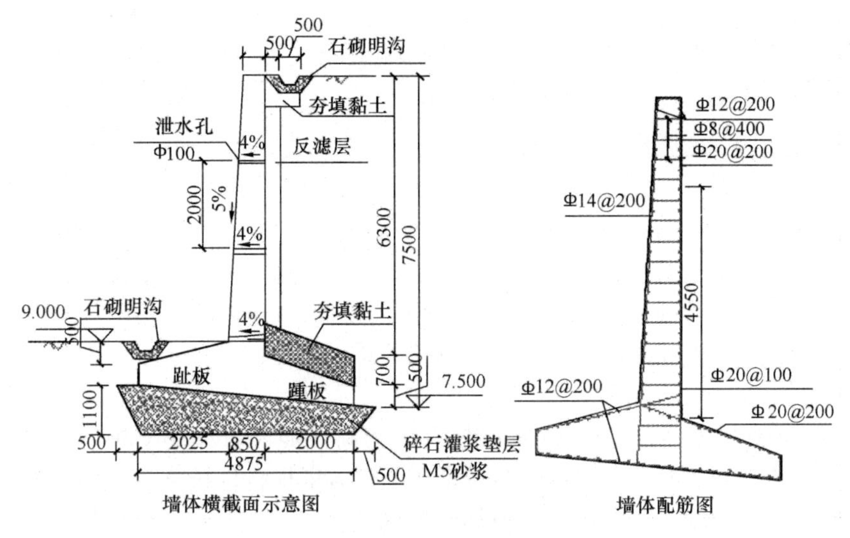

图 1　试验段挡土墙断面及配筋图

1. 试验采用的混凝土配合比：经深圳地铁公司根据项目研究结果讨论，决定采用水泥用量不超过 280kg/m³，突破当时规范最小水泥用量 320kg/m³的限制，为了进一步降低水泥用量以适应深圳地下水环境，在本试验中使用水泥用量 280kg/m³和 240kg/m³两种配合比以及模型试验的配合比如下：

混凝土①：水泥用量 280kg/m³，粉煤灰 90kg/m³（占胶凝材料总量 24.3%），水胶比 0.49；

混凝土②：水泥 240kg/m³，粉煤灰 180kg/m³（占胶凝材料总量 42.9%），水胶比 0.39；混凝土②中以 40kg/m³磨细矿渣等量取代粉煤灰；

混凝土③：水泥 180kg/m³，粉煤灰 180kg/m³（占胶凝材料总量 45%），比表面积不大于 400m²/kg 的磨细矿渣 40kg/m³（占胶凝材料总量 10%），矿物掺和料占胶凝材料总量的 55%，水胶比 0.42；

将试验墙分三段三次浇筑以上混凝土，浇筑安排如图 2 所示，浇筑方向由东向西，每小段之间为施工缝。

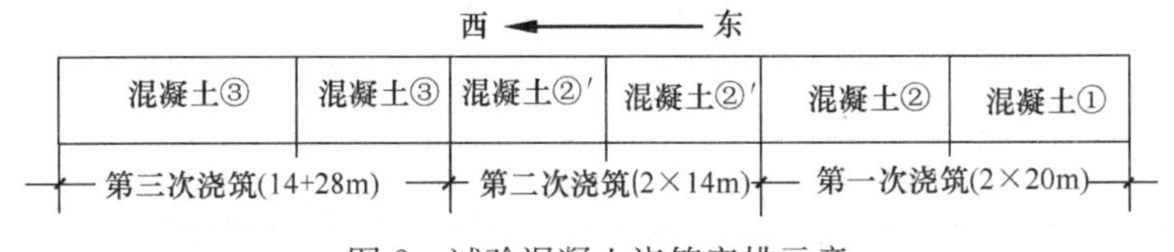

图 2　试验混凝土浇筑安排示意

2. 趾板和踵板用普通混凝土浇筑：从趾板和踵板以上，每段墙分两层浇筑，层间为冷缝。第一次浇筑为从趾板和踵板顶部往上 3.8m，第二次浇筑 3.8m 以上的 2.2m。第一浇筑层的混凝土按图 2 的安排，第二层的混凝土则全部采用普通混凝土，即混凝土①。

3. 全部采用 1.5cm 厚度的木模板，双排钢管加对拉螺杆固定模板，两侧采用钢管架及在地面打木楔用钢管做横支撑的方法防止模板的横向移动。

4. 第一段和第二段浇筑之前，第一段的全部和第二段东起第二小段在第一次浇筑层中埋设温度传感器和应力传感器。共埋设温度传感器 50 个、应力传感器 86 个，如图 3、4、

5、6 所示。各部分埋设数量见表 1 所示。

表 1　传感器埋设数量（个）

	第一次浇筑段		第二次浇筑段		第三次浇筑段		总计
温度	混凝土①20m	混凝土②20m	混凝土②14m	混凝土②14m	混凝土③14m	混凝土③28m	长度 110m
	12	12	14	无	12	无	50 个
应力	24	24	20	无	18	无	86 个

5. 混凝土拌和物坍落度控制为出机 160～180mm，运至现场为 140～160mm。现场取样的试件分别采用标准养护和温度跟踪养护的方法养护至指定龄期。

实验内容：

2000 年鉴定的地铁一期工程区间足尺模型所用高性能混凝土为 C30，本次实验要求强度等级为 C35，其他要求不变。按鉴定会专家意见，需增加构件抗疲劳试验。同时，根据地铁一期工程施工中的问题，还要进一步对施工缝的设置、浇筑高度、浇筑顺序、养护以及温控等进行研究。故此计划以下实验内容：

1. 高性能混凝土原材料选择和配合比调整。
2. 施工缝间距、浇筑高度对混凝土匀质性影响。
3. 模板、气候（温湿度、风速等）、养护等因素对混凝土温度发展的影响。
4. 混凝土初始温度、升温和降温速率以及最高温度对混凝土开裂性的影响。
5. 混凝土所用不同胶凝材料水化热和混凝土绝热温升检测。
6. 混凝土构件抗疲劳性试验。
7. 早期开裂情况及其分析。
8. 在标准养护条件和温度跟踪养护条件下混凝土性能指标检测。

四、现场试验及结果

1. 混凝土浇筑安排

表 2 所示为混凝土浇筑时间。

表 2　混凝土浇筑时间

浇筑次序	浇筑总长	下层混凝土浇筑时间	上层混凝土浇筑时间
第一次	长度（2×20）m	2002.10.25	2002.10.29
第二次	长度（2×14）m	2002.11.17	2002.11.23
第三次	（14+28）m	2002.12.09	2002.12.13

2. 第一次浇筑情况

1）试验目的和浇筑

浇筑两段各 20m 长，预期应至少有一段出现裂缝，以便分析影响因素。主要进行水泥用量、浇筑高度和浇筑长度（即施工缝间距）对墙体开裂的影响，对温度不加以控制，预埋温度传感器和应力传感器的目的是检测混凝土的温度及应力和混凝土开裂的关系。该段墙第一层浇筑高度分别为 3.8m 和 4m，浇筑长度为 20m，混凝土坍落度：140mm，入模温度：31℃，浇筑当天最高气温：27℃，最低气温 12℃，混凝土①浇筑时间：10 月 25 日下午

4：00左右到晚上 9：00 左右，混凝土②浇筑时间：晚上 10：00 左右到次日凌晨 2：00 左右，下料高度≤2m，下料口间距 2.5m；10 月 29 日下午浇筑第二层高度为 2.2m 的混凝土，直到 30 日凌晨；11 月 1 日同一天拆模。

2）拆模与裂缝情况

拆模前基本上没有进行湿养护，拆模从 11 月 1 日早上 7：30 开始，持续了两天。在拆模过程中 11 月 1 日上午 9：00 即发现两爿墙各有 1 条和 2 条裂缝。此后基本上每隔 1~2 小时继续出现 1～2 条，直到 11 月 4 日以后稳定。模板完全拆除后即开始向墙面浇水养护。共出现裂缝：混凝土①为 6 条，混凝土②为 7 条。位置大都在泄水管上方。裂缝长度分布大体在与高度相等的宽度范围内（圣维南区）呈自两端向中部长度增加的规律，在此范围以外则大体相等。见图 3 和图 4 所示。

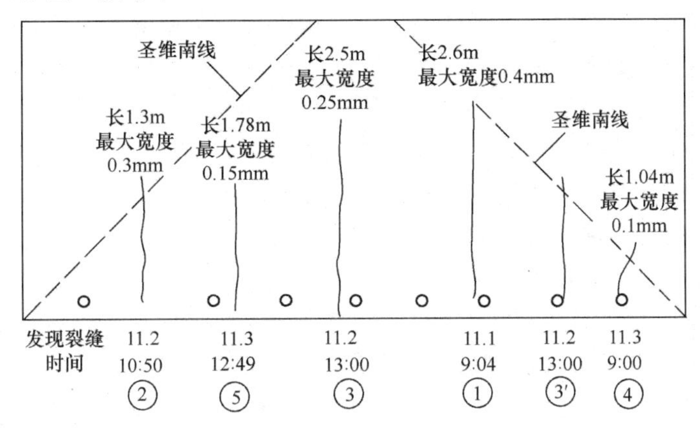

图 3　墙 1 裂缝出现时间、顺序和裂缝位置、长度以及最大宽度

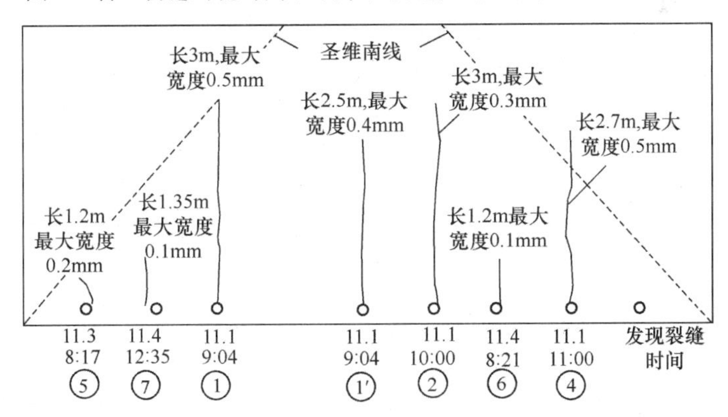

图 4　墙 2 裂缝出现时间、顺序和裂缝位置、长度以及最大宽度

①从墙的下端角部向中部引 45°斜线到顶端，称为圣维南线[1]。在此两条线之间的区域，从底部开始的裂缝向上发展，其尖端基本上落在线上，当二线不相交而分别交于顶端线上时，交点之间出现通长裂缝的可能性最大；如果二线交叉，即墙长小于两倍墙高时，则不会出现通长裂缝。图 7 和图 8 所示裂缝长度符合以上圣维南理论。

②裂缝基本上都出现在最下层的泄水孔上方。从应力传感器纪录的应力分布来看，墙的最低端拉应力最大，再加上泄水孔处的应力集中，裂缝就从底部泄水孔处产生并向上发展。

3）温度的影响分析

①从该混凝土温度发展来看，两爿墙混凝土中水泥用量相差 40kg/m³，但水胶比不同，水泥 280kg/m³ 的混凝土水胶比为 0.5，水泥 240kg/m³ 的混凝土水胶比为 0.4。现场浇筑的两种混凝土最高温升分别为 55.9℃和 55.7℃，并无显著差别，到达温峰的时间分别为：中部表面 24 小时，中心 26 小时；下部表面 22 小时，中心 27 小时，也差不多。在实验室中检测两种胶凝材料 96 小时的水化热分别为 201.19J/g 和 190.70J/g，差别也不显著。因此两爿墙开裂的情况也差别不大。

②我国有关标准规定，为了避免开裂，混凝土内外温差应不大于 25℃；从温差来看，以混凝土 1 为例，拆模前达温峰时，温差为 1.8℃～5.5℃；降温时温差为 0.2℃～1.8℃；拆模时混凝土内外温差为 6.6℃，混凝土表面和环境温度差为 0.3～3.3℃。温差始终都很低，可见裂缝不是因温差所引起的。

③两爿墙达到温峰后第二天开始降温，第一天降温速率最高为 8.1℃；第二天最高达到 9.8℃，这很可能是引起开裂的主要原因。

4）连续浇筑高度的影响

墙体底部受到基础的约束力由两端向中央逐渐增大，单位长度混凝土的高度越大，直接成为墙体在底边施加于基础的正压力也越大，因而影响该处产生的约束应力大小。从应力传感器得到的信息来看，在墙的高度方向，浇筑后五天内，由于温度的影响，顶部混凝土内部应力一直处于受拉状态，而在中部和底部均处于受压状态；可见浇筑高度越高，则处于拉应力的高度范围越大。所以浇筑高度和浇筑长度（施工缝间距）同样都不可忽视。

5）根据预设温度传感器分析混凝土内部温度历程表 3 所示。

表 3　第一次浇筑的混凝土①和混凝土②温度发展历程及特征分析

编号	水泥用量（kg/m³）	水胶比	浇筑时气温（℃）	传感器部位	混凝土最高温度（℃）		达温峰的龄期**（h）		温降速率（℃/日）					
									第一个 24h		第二个 24h		平均***	
					表面*	中心	表面*	中心	表面*	中心	表面*	中心	表面*	中心
①	280	0.5	26	下	48.2	53.7	13	18	4.4	6.2	5.3	5.7	4.7	6.2
				中	52.7	55.9	24	26	4.5	5.4	7.1	7.8	—	—
				上	45.8	47.6	13	13	7.8	8.1	6.2	6.6	—	—
②	240	0.4	26	下	46.9	52.2	18	21	5.9	6.6	5.2	4.7	—	—
				中	51.8	55.7	22	27	4.2	6.7	9.8	9.6	—	—
				上	46	48	20	20	9	9.2	7.0	7.2	—	—

* 混凝土表面向内 5cm；
** 自浇筑开始计；
*** 从达温峰到拆模。

由表 3 数据分析可见，两爿墙混凝土温峰到达得均很早，表明强度发展也会很快；但是混凝土表面和中心的温差均不大；最低 1.8℃（墙 1 顶端），最高 5.5℃（墙 1 底部），平均 3.7℃；而降温速率太快则是主要问题，最高达 9.8℃，这应当是引起开裂的主要因素。

6）应力传感器数据分析[2]

此次试验共埋设钢弦式应力传感器 86 个，埋置原则遵循地基上长墙开裂有序性理论[2]。一般认为，地基上长墙水平法向应力在长墙中部最大，当超过抗拉强度时，在长墙中部会出现第一道裂缝，长墙由一块变为两块，同时应力重新分布，形成新单独墙，如果单独墙中部应力仍然超过抗拉强度，则从中部继续开裂，如此循环，见图 5。

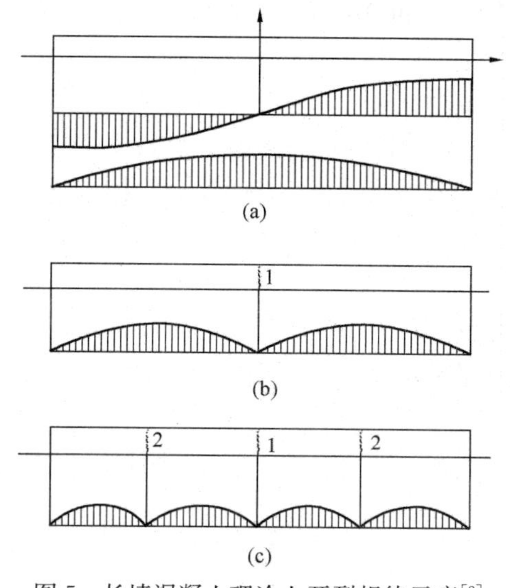

图 5　长墙混凝土理论上开裂规律示意[2]
(a) 主要应力分布；(b) 一次开裂；(c) 二次开裂

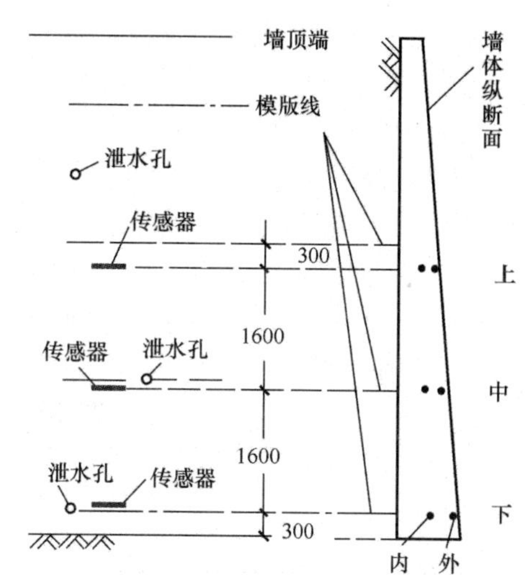

图 6　应力传感器位置举例示意

①因墙体混凝土浇筑在已硬化的趾踵板的混凝土上(见图 1)，当墙体混凝土收缩时底部约束最大，而上部为自由端，约束最小，故按图 6 的传感器位置，分析下部、中部外侧(距表面向内 5cm)和中心点，如图 7 所示的实例。

②应力传感器所给出的是变形值。假设弹性模量一致，可计算出应力，故在此用变形进行分析。由于初始标定值误差较大，取变形相对稳定时的值为零点，进行相对比较如图 7 所示。

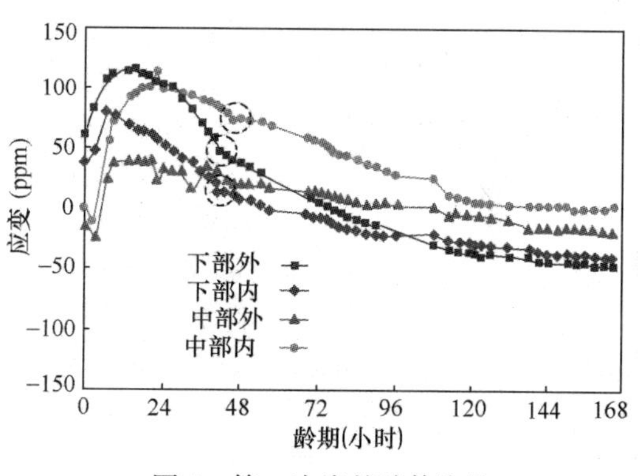

图 7　第一次浇筑墙体浇筑至 7 天内混凝土应变的发展

③图 7 所示变形趋势和温度传感器所显示的温度变化趋势是一致的，在 24 小时到 48 小时之间出现突变点(图中圆圈内)，说明此时有开裂点，对照表 3 及上述温度分析，此时(表 3 所示第二个 24 小时)降温速率最快，内外分别为 9.8℃ 和 9.6℃，可断定在拆模时就发现的裂缝应当是在拆模前就已出现。拆模后再相继出现的裂缝，一种可能是拆模前的开裂释放应力不足，拆模后接触空气，因表面失水干缩而造成的；另一种可能是拆模后接触空气，表面降温速率增大所造成。

3. 第二次浇筑情况

1）试验目的及浇筑概况

第二次浇筑只用混凝土②，改变浇筑长度(即施工缝间距)以了解浇筑长度对墙体开裂的影响，同时对混凝土温度加以控制，预埋温度传感器和应力传感器，检测混凝土的温度及应力与混凝土开裂的关系。该段墙第一层浇筑高度为 3.8m，浇筑长度为 14m。混凝土坍落度为 160mm。11 月 17 日下午 15：00 左右到 16：30 左右浇筑，当天最高气温为 21℃，最低气温 10℃，用冰水使入模温度为 15.5℃。下料高度≤2m。下料口间距 2.5m。11 月 23 日下午浇筑第二层高度为 2.2m 的混凝土；11 月 25 日同一天拆模。

2）根据第一次浇筑的分析，确定采取以下措施：

①控制混凝土入模温度低于浇筑当天的气温。

②在混凝土浇筑前就向钢模板表面浇水冷却模板，浇筑时停止浇水，浇筑完毕后立即继续在模板上浇水以抑制混凝土的升温，约 15 小时后（即浇筑的第二天早上），松动模板，继续通过在模板顶部安装的多孔水管，不断向模板内的墙身和顶面浇水进行养护和控温。拆模后至今无任何裂缝。

混凝土温度分析表 4。表 4 表明混凝土内部温度很低，中心最高温度＜40℃，达温峰龄期为 38 小时，降温速率＜4℃/天。到拆模时混凝土的降温速率平均为 2～4℃/天。

表 4　第二次浇筑的混凝土②温度历程和特征

水泥用量（kg/m^3）	水胶比	浇筑时的气温（℃）	传感器的部位	混凝土最高温度（℃）		达温峰时的龄期**（小时）		最大内外温差（℃）	温降（℃）/日			
									第一个 24h		第二个 24h	
				表面*	中心	表面*	中心		表面*	中心	表面*	中心
			下	36	39	31	34		3.4	2.9	1.8	3.5
240	0.42	21	中	37.8	39.7	38	38	4	2.8	3.1	3.2	4.0
			上	31.5	31.9	28	28		3.2	2.9	1.2	1.9

* 混凝土表面向内 5cm；

** 自浇筑开始计。

4. 第三次浇筑情况

1）试验目的及浇筑概况

第三次用混凝土③分两段浇筑，第一段浇筑长度为 14m，浇筑高度增加到 4.5m，同时对温度加以控制，预埋温度传感器和应力传感器，检测混凝土的温度及应力与混凝土开裂的关系。混凝土浇筑时间为 12 月 9 日下午 12∶24 左右到 16∶34 左右。下料高度≤2m。下料口间距 2.5m。混凝土坍落度 160mm。入模温度约 12℃，浇筑时气温为 13℃，12 月 13 日下午浇筑上部高 2.0m 的混凝土；12 月 16 日同一天拆模。第二段仍用混凝土③，于 12 月 17 日 15∶02 开始浇筑，浇筑长度为 20m，混凝土温度控制相同，无预埋传感器。入模温度平均 15.5℃，现场气温 21℃，坍落度平均 170mm。

2）混凝土养护和控温方法与第二次浇筑时的相同。

3）由于气温低，入模温度低，并早期控制了温升，混凝土初凝缓慢，但初凝后很快就达到终凝，并增长强度。

4）拆模后未发现任何裂缝。2011 年 11 月到现场观察，浇筑后经历 11 年而并未发现裂缝。

5）温度历程及分析

第三次浇筑的第一段混凝土温度历程分析如表 5 所示。因向模板表面浇水，传感器接触水后，量测的数据波动很大，但内外温度差、降温速率温峰到达时间等数据的规律仍可看出，即混凝土最高温度很低，混凝土内外温差以及混凝土与环境温度差均很小，降温速率很低，受环境温度影响有时降温速率为绝对值很小的负值。从图 8 可见几乎看不出明显的温峰。

表 5 第三次浇筑第-段混凝土温度特征

水泥用量（kg/m³）	水胶比	浇筑时的气温（℃）	传感器部位	混凝土最高温度（℃）		达温峰时龄期**（小时）		最大内外温差（℃）	温降（℃）/日			
									第一个 24h		第二个 24h	
				表面*	中心	表面*	中心		表面*	中心	表面*	中心
180	0.42	13	下	23.6	28.1***	70	65***	4	2.1	5.1***	−1.9	0.6
			中	24	23	70	76		1.1	0.26	−1.1	0.5
			上	24.3	22.6	73	76		4.3	3.0	−3.3	−4.0

* 混凝土表面向内 5cm；
** 自浇筑开始计；
*** 此处传感器损坏，故该数据不准。

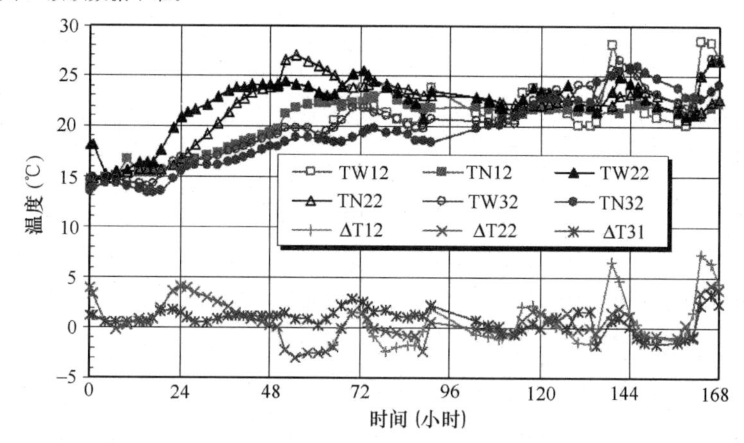

图 8 第三次浇筑混凝土内部温度发展情况

T—温度；W—外侧（保护层内 5cm）；N—内部（横断面中央），

Δ—温差；后面数字代表传感器所在的列和行

5. 从第一次浇筑和后两段浇筑时的混凝土应力影响的因素分析裂缝控制措施的机理

混凝土早期的开裂是因混凝土在约束条件下的收缩产生的应力超过实时混凝土的抗拉强度而造成的。则开裂与否的条件主要是应力，控制应力的影响因素，即可控制开裂；即使早期未出现可见裂缝，如果内部有积蓄的应力，在使用阶段还有可能因环境因素而引发为不可见裂缝的扩展，成为侵蚀性物质侵入的通道，而威胁混凝土结构的耐久性。

1）以第一次浇筑的混凝土①为例，根据所浇筑构件特点，按一定的假设，用下式计算任意时间混凝土的收缩（自收缩＋干缩）值：

$$\varepsilon_y(t) = \varepsilon_y^{\circ}(1 - e^{-bt}) M_1 \cdot M_2 \cdot M_3 \cdots\cdots \cdot M_n$$ [3]

式中符号：$\varepsilon_y(t)$——龄期 t 的收缩值；ε_y°——标准状态下极限收缩值；b——经验系数；t——指定龄期；M_1，……，M_n——与混凝土所处环境条件、水泥品种和细度、骨料品种、水灰比、构件尺寸、配筋率、振捣质量等有关的修正系数。

编号为 TN1 测点温度应力和以上计算的（自收缩＋干缩）应力的对比如图 9[3] 所示。

由图 9 可见，在混凝土硬化初期（7 天前），温度应力占主导。

2）以第一次浇筑的混凝土①和第二次浇筑的混凝土②作对比，示于图 10[3]。

由图 10 可见，入模温度高于气温和低于气温时，混凝土在升温阶段均处于压应力下，后者温度曲线的零应力点比前者的几乎晚一天，发展到拉应力阶段，应力稳定后的最大值比前者约小一倍。这就是两次浇筑的混凝土中水泥用量相差不大，而第二次浇筑的混凝土无裂缝的主要原因。

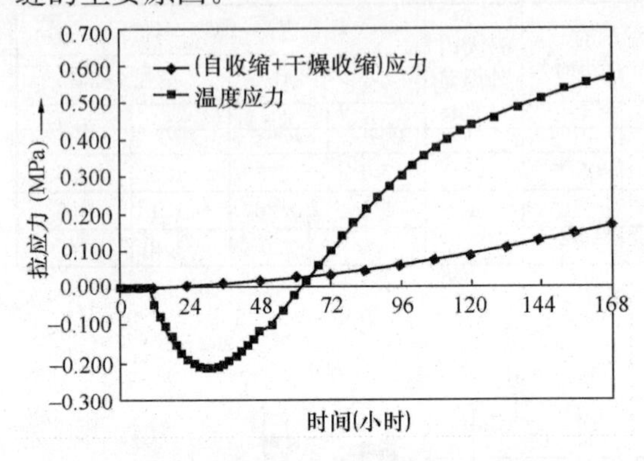

图 9　TN1 点温度应力和（自收缩＋干缩）应力

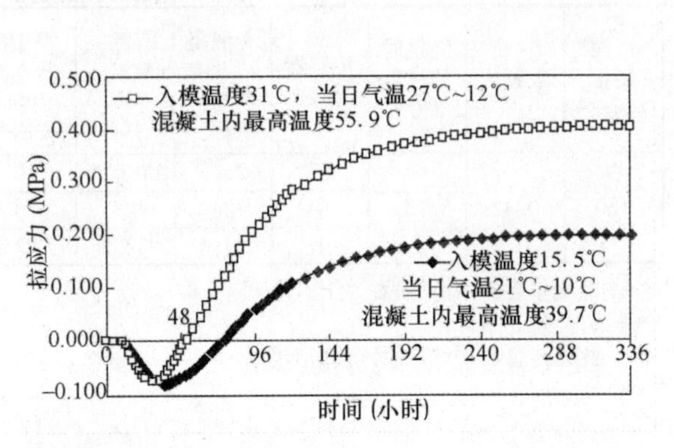

图 10　入模温度对混凝土内部温度应力的影响

3）表面浇水养护开始时间和水温的影响

模板内的混凝土浇筑后的内部温升可使表面温度也高于气温，拆模后立即浇水，会使表面温度降低过快而引起冷缩，在温度很高的内部混凝土约束下产生拉应力，可能造成开裂。本实验中第二次和第三次浇筑采取浇水预冷却的措施，则是在温升开始前就人为地降低环境温度，抑制混凝土内部的温升，则有效地防止了裂缝的出现。图 11 所示为混凝土表面开始浇水时间对混凝土中最大拉应力的影响。

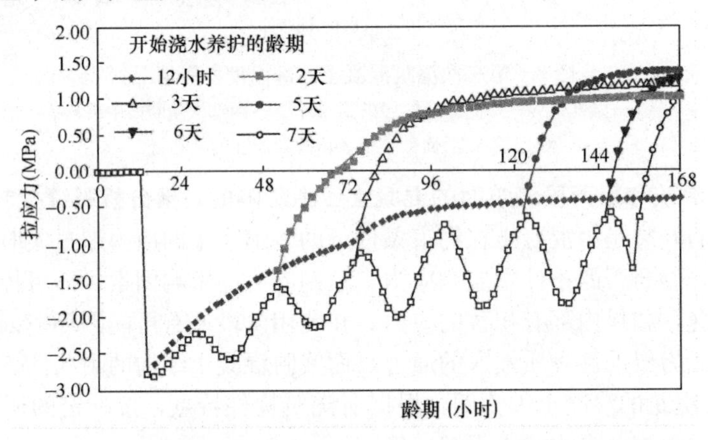

图 11　开始浇水的龄期对混凝土内部应力的影响[3]

从传感器记录的温度历程来看，各段墙混凝土温峰到达时间与入模温度有关，入模温度越高，达到温峰时间越早：第一次浇筑混凝土①最快为 13 小时；混凝土②最快为 18 小时；第二次浇筑最快为 28 小时，第三次浇筑则为 70 小时。由图 11 可见，龄期 12 小时尚未达到温峰，此时开始浇水，外部混凝土处在压应力状态，随温升的提高压应力下降，但因同时抑制了内部混凝土的温升，则尽管压应力下降，却一直处于压应力状态；当龄期两天开始浇水时，已超过温峰期，应力很快就超过零点而转变为拉应力，浇水开始得越晚，拉应力也随之增大，5 天以后混凝土已降温较多，浇水对最大应力的影响就不再增大。

浇水不仅要早，而且水温对应力也有影响。表 6 是龄期 5 天开始浇水至 7 天，养护水温对 7 天表面应力的影响。

表 6　养护水温对 7 天表面应力的影响

养护水温（℃）	8	12	16	20	平均上升（MPa/℃）
内部最大应力（MPa）	0.281	0.302	0.323	0.345	0.0055
外部最大应力（MPa）	1.908	1.403	0.898	0.392	0.1263

在表 6 的数据中可见，提高养护水温度时对内部应力影响很小，只有少许增加，而明显有利于降低外部混凝土的应力。

五、试验结果和建议

1. 本次试验取得不同胶凝材料组成（不同水泥用量）和配合比的混凝土立方体抗压和轴心抗压强度、跟踪养护试件强度、不同尺寸试件抗压强度发展对比、劈裂抗拉强度、受压弹性模量、疲劳性能等力学性能和抗碳化、抗氯离子扩散、抗渗、抗硫酸盐侵蚀、混凝土绝热温升和胶凝材料水化放热、不同条件下的干燥收缩等物理性能 400 多组，现场实测混凝土内部温度和应力数据 2 万多个，对适用于深圳地铁混凝土结构耐久性的混凝土配合比及其特性有了进一步较全面的了解。

2. 对厚度较大的墙体混凝土浇筑中影响混凝土结构抗裂性的因素及施工控制进行了试验和分析，表明平均厚度为 60cm 的混凝土墙体一层浇筑高度控制在 4.5m 以下是合适的；而当混凝土入模温度很低、降温速率很慢时，施工缝间距达 20m 是可行的。

3. 试验结果表明混凝土的入模温度和升温、降温速率以及合理的养护是控制混凝土结构裂缝的首要因素，而只控制混凝土内外温差是不够的；为今后的墙体施工提供了参考。

4. 试验表明混凝土②（水泥用量为 240kg/m^3，水胶比 0.39）各项性能指标均优于混凝土①（水泥用量为 280kg/m^3，水胶比 0.49），建议今后在深圳地区地下工程中采用。至于混凝土③，因水胶比稍大（0.41），除抗硫酸盐侵蚀性较好外，其他性能未显出明显优势，鉴于混凝土可持续发展的需要，建议今后对耐久性要求更高的合适部位进一步降低水胶比进行试验研究。

5. 施工时正值深圳的秋天，第二次和第三次浇筑时最高气温已低于 20℃，加上施工中的养护方案的实施，以致第三次浇筑的混凝土没有出现显著的温峰。此项经验是否适合于夏季施工，需要进行研究。

参考文献

[1] F. S. Rostasy, T. Tanabe & M. Laube. Assessment of External Restraint, see: Avoidance of Thermal Cracking in Concrete at Early Ages, Report of work of RILEM Technical, edited by R. Springenschmid, Munich1998.

[2] 王铁梦. 工程结构裂缝控制[M]. 北京：中国建筑工业出版社，1999.

[3] 许小荣，混凝土长墙中温度变化及其对内部变形发展的影响. 北京：清华大学工学硕士学位论文[D]. 2004.

自评

1. 本文介绍的工程试验对象是中等厚度的长墙，试验表明，混凝土是可以不裂的。裂缝不是只要水养护就能控制的，混凝土的原材料与配合比、构件尺寸、形状、施工方案、浇筑成型工艺等都有影响。施工和浇筑成型工艺中的关键是养护，而对养护的观念应从“及时浇水养护”转变为早期温度与湿度的控制。浇水也有技术问题。鉴于当前水泥熟料早期强度较高、细度过细，即使构件最小断面并不很大（例如 30cm）也会因水化放热较快使内部温升较大，形成内外温差而产生拉应力，因此早期温度控制更加重要。

2. 本工程试点所采取的控制裂缝措施中最有效的是控制浇筑温度低于当时的环境温度，夏季施工尤其要控制混凝土浇筑温度，如果也能低于气温，则裂缝也可控。第一次浇筑未提出特别的养护控制要求而开裂，是在夏季；第二次与第三次施工中有养护的控制，但已处于秋季，二者环境有别。因此还应当再进行夏季施工裂缝控制的试验。

3. 由工程实践来看，现代混凝土的养护应包括早期温度控制和保温两个内容，不仅要根据混凝土中的温差，更重要的是根据降温速率的控制来确定拆模时间。温度控制应注意控制拌和物入模温度，并在混凝土升温时采取降温措施以控制温升；降温阶段要注意保温以控制降温速率。

古今之成大事业大学问者必经过三种之境界。“昨夜西风凋碧树，独上高楼，望尽天涯路”。此第一境也。“衣带渐宽终不悔，为伊消得人憔悴”。此第二境也。“众里寻他千百度，蓦然回首，那人却在，灯火阑珊处”。此第三境也。

——王静安《人间词话》

论文之十二

混凝土施工中裂缝的控制*

廉慧珍
清华大学土木水利学院

1 混凝土结构产生裂缝的原因及控制的必要

混凝土是一种非均质的复杂多相混相材料，在其微观结构相组成之间主要的结合力是范德华力。因此其抗拉强度远低于抗压强度。当混凝土内部产生拉应力超过其抗拉强度时，就会产生裂缝。因此，混凝土发生开裂的条件就是：在约束下变形产生的拉应力超过实时的抗拉强度，也就是说必须同时考虑三个条件：变形的大小、约束的程度、实时抗拉强度。不受约束的自由变形不会产生应力；抗拉强度足以抵抗所产生的拉应力时则不会开裂。也就是说不能笼统地认为收缩必然开裂。所产生的应力大小和实时的弹性模量有关，和能够松弛应力的徐变有关；是否引起开裂还和混凝土的抗拉强度有关。

凡是组成良好并经适当捣固和养护的混凝土，只要内部孔隙和裂缝尚未形成相互连接而直达表面的通道，则基本上是水密性的；在使用中，结构的荷载以及大气环境的影响如冷热交替、干湿循环，可使这些内部微裂缝发展并传播，成为环境中侵蚀性介质浸入的通道[1]。早期裂缝控制的意义在于，已有裂缝的扩展比新生成裂缝容易。

可能引起开裂的变形主要是收缩，影响最大的早期收缩如下：

1）干缩。停止养护后，环境相对湿度低于100%，混凝土干缩即开始；在干燥的空气中，收缩会持续进行，甚至在28年后仍能观察到一些变化[2]。对于普通混凝土，28天收缩约40%，3个月收缩60%左右，180天收缩约70%，1年收缩平均75%，完全收缩的时间可长达20年。完全干缩值为10000微应变，F. M. Lee曾实测到4000微应变[2]。影响收缩的主要因素是骨料的品种和用量。当骨料品种一定时，单方混凝土中骨料用量越大，即浆骨比越小，则干缩越小。骨料的“骨架”作用即在于此。当水泥(或胶凝材料，以下同)用量不变时，水灰比(或水胶比，以下同)越大时，浆骨比越大，干缩也越大。因此混凝土的配合比中应当尽量减小用水量。

2）温度收缩。随着水泥实际强度的提高、比表面积的增大，水化热也相应较大，再加上因要求混凝土具有较高早期强度而使用较大的水泥用量，使现在用于厚度为30cm的混凝土构件也需要控制内部温度的变化。混凝土温度每下降15℃时，收缩约150微应变。例如抗压强度约30MPa的混凝土，其弹性模量为约30GPa(按我国结构设计规范)，则在约束下可产生弹性拉应力约4.5MPa，而30MPa的混凝土的直接抗拉强度约为2.7MPa。常有工程中的混凝土拆除模板时就发现已产生裂缝，显然是由温度变形所致。

---

* 原载中国工程院土木水利与建筑学部工程结构安全性与耐久性研究咨询项目组《混凝土结构耐久性设计与施工指南》的论文汇编。

3）自收缩：自收缩是在与外界无水分交换情况下因水泥水化消耗浆体内部自身的水分而产生的。自收缩从混凝土初凝就开始产生，在1天以内发展最快，3天以后减慢，此后就发展得很缓慢了。自收缩不同于化学收缩，但由化学收缩引起。化学收缩的原因是水泥和水发生水化反应，产物的固相体积增大，而与反应前水泥与水的体积之和相比则减小，故也称化学减缩。化学收缩在初凝前导致整个体系体积减缩，在初凝后导致体系产生孔隙，而对体系体积无影响；自收缩则导致毛细孔收缩而产生体系的收缩。胶凝材料（包括水泥和活性掺和料）的活性越大、水灰比越低，则自收缩越大。例如根据安明喆的试验[3]，w/c＝0.36时，3天自收缩约100微应变，而w/c＝0.275时，3天自收缩可超过270微应变。

按目前的标准方法（GBJ 82—85）检测水泥或混凝土收缩值，检测不到全部的自收缩值，而是停止养护(1天或3天)后的干燥收缩和一小部分自收缩值。而所测不到的那部分自收缩值恰恰是影响早期开裂的重要部分。水灰比越低，这部分所占比例越大[4]，而混凝土的总收缩几乎和水胶比无关。在目前大量使用较低水灰比和较大水泥用量的混凝土中，早期收缩最重要的就是温度收缩和自收缩；如果拆模较早而养护不当，则早期还可产生较大的干缩。

由于近年来混凝土所用的水泥强度高，尤其是早期强度高，混凝土水灰比较低，使混凝土温度变形和自收缩变形较大，即使早期未开裂，已产生的应力未消除，在后期使用阶段有时因外界条件如急剧的温度和湿度的变化，又会有新的应力生成，与已有应力叠加后如果超过混凝土实时的抗拉强度，就有可能在原有不可见微裂缝处扩展成可见的裂缝。因此控制混凝土早期内部的应力尽量减小，才是提高混凝土耐久性最重要的环节。为了减小早期内部应力，就要减小温度变形和自收缩变形，同时尽量避免高早强以降低早期弹性模量，增大早期徐变。

2 裂缝控制总则

影响开裂的因素很复杂，往往不是单一因素造成的。控制裂缝也不只是施工人员和混凝土生产者的事，而是涉及包括设计、混凝土及其原材料生产、施工甚至监理和业主(开发商、房主或政府主管)在内各方面的责任。因此需要各方共同努力解决，但是混凝土的施工，包括混凝土原材料的控制、混凝土的制备和现场施工的各个环节，则对于控制早期裂缝、减小后期开裂倾向、保证实现设计的混凝土结构耐久性是至关重要的。不能把施工看成什么人都能干的事，相反，需要知识面很宽的、能运用哲学思想(例如能根据具体情况，具体分析，具体处理工程中的问题)的管理人员和技术人员共同筹划、决策和把关。

混凝土施工中影响混凝土质量的并不只是养护的问题；养护也不只是浇水保湿的问题，而是包括模板种类、浇筑方式、浇筑顺序、振捣方法等的选择，以及混凝土内部温度的控制、拆模时间和方式等等各方面的内容。由于现代水泥成分中较高水化速率的组分因素增加，即使不是早强水泥的品种，水化放热速率也都加快，加之，为耐久性而设计的混凝土水胶比低，混凝土的自收缩变形和温度变形都会较大，上述施工的各环节就更加重要。

每项工程施工前，应针对不同工程的特点、环境、施工季节、条件，由监理(必要时可还有甲方代表参加)和技术人员按照设计要求，参考本《指南》和有关混凝土结构施工验收规范，共同制定具体保证措施和实施计划。

混凝土的制备应当密切配合混凝土的施工，提供混凝土合适的流变性能和浇筑温度，并且应当做好售后服务，跟踪混凝土施工中重要阶段的质量控制。国内已有混凝土生产企业做到了“混凝土生产和施工现场浇筑及质量控制的一体化”，这是很值得提倡的。但是要做到这一点，在经济核算和责任方面必须有相应政策性的调整。

3 混凝土的制备和运输

混凝土的生产者应当改变只按强度要求购买原材料的观念，而应更关心水泥的抗裂性能以及与抗裂性有关的指标以及含碱量，美国垦务局的 Burrows 根据现场监测和其他科学家的试验认为即使所用骨料没有碱活性，含碱量超过水泥质量的 0.6%时，也会因为促进水泥增大收缩而降低抗裂性，K_2O 比 Na_2O 的影响更大[5]。更应关心所供应的水泥质量稳定性(对主要性能指标，要监测其标准差)。水泥的实际强度不应超出标称强度太多，如果考虑储存，则一般要求富系数约 1.13 倍即可。如果立即使用，则富系数为 1.10 即可。高强度等级的水泥由于比表面积大，强度不易保持。例如某工程 C40 混凝土使用的 42.5 硅酸盐水泥，复验 28 天实际抗压强度可达 64MPa，C60 混凝土用 52.5 水泥，复验 28 天实际抗压强度则为 67MPa，二者相差无几。这种水泥超强度太多，抗裂性很难控制。此外，购进散装水泥的温度常常很高，甚至可达到 90℃，入仓后散热较慢，对混凝土的早期抗裂性不利，需要与水泥厂协商解决，或自行解决配制混凝土前水泥的散热。

任何工程必须根据本工程原材料、工艺条件、施工水平，选择合适的配合比进行试配，当原材料变化时还要进行调整，不能使用现成的“配方”。

应根据试配优选的配合比检测混凝土的抗裂性能和收缩值。由于混凝土自收缩在初凝时就开始，即使没有条件检测，也应在终凝后尽早测定初长。试件的养护条件应与现场实际养护条件相当。例如现场构件湿养护 7 天，则试验时试件也应湿养护或密封养护 7 天，则 7 天内所测收缩可包含一部分自收缩值，撤除养护后继续检测的收缩值主要是干缩。除按标准制作检测抗压强度的试件外，要检测 1 天的抗压强度。有条件时，最好能测定直接抗拉强度，以便控制混凝土内部应力始终低于其抗拉强度。

重要工程混凝土性能宜在现场制作模拟试件钻心取样进行检测。

热天施工的混凝土可掺入适量缓凝剂和引气剂，以延迟温峰的出现，并抑制初期强度的发展。但不可过于缓凝，否则可能引起后期的开裂。

掺用粉煤灰后混凝土拌和物黏聚性增加，达到相同流动度的流动速率减小，但与容器的黏附力下降，而且触变性明显，国内外都有经验表明，大掺量粉煤灰的混凝土拌和物坍落度大于 100mm 即可泵送[6]。坍落度不要太大，有利于控制裂缝。但这必须有级配和粒形良好的骨料为前提。为了方便浇筑和振捣，对中等强度的混凝土，目前宜控制在(140±20) mm。坍落度可用减水剂调整。

热天用的砂石料场应当有棚子以遮盖直射的阳光和雨水，冷天要保温，避免其中的含水结冰。在砂石料场取料时，应取用距底部以上 300mm 以上的砂石。含水率以饱和面干计，每天检测应不少于两次，如下过雨，则应增加检测次数。

搅拌机称量装置应定期校核，并经过试验控制下料的冲量。

运送混凝土拌和物的输送车在装料前，筒体应湿润，但不得积水。

4 混凝土配合比设计

4.1 混凝土配合比设计原则

按耐久性设计应首先满足低渗透性的要求。按工程设计的氯离子扩散系数要求或抗渗性指标确定氯离子扩散系数指标，水胶比一般不大于 0.45。

掺入掺和料时混凝土的水胶比应低于无掺和料的混凝土的水灰比，胶凝材料总量应稍大于设计相同强度等级传统混凝土时的水泥用量，以保证良好的施工性，提高混凝土的耐久性。对不同强度等级的混凝土，在目前我国骨料条件下，胶凝材料总量一般不少于 350kg/m³，C50 以上混凝土的不大于 500kg/m³。

砂率按混凝土施工性调整。为不严重影响混凝土弹性模量，对现市售的骨料，砂率也不宜大于 45%。

由于胶凝材料中各个组分密度相差较大，宜采用绝对体积法进行配合比的计算。至少第一盘试配料要采用绝对体积法。混凝土拌和物应有最小的砂石空隙率。

试配后应检验其强度是否满足设计要求。检验应按配制强度进行。混凝土配制强度：

$$f_{cu} = f_{cu0} + 1.645\sigma \tag{1}$$

式中 f_{cu}——混凝土配制强度，MPa；

f_{cu0}——混凝土设计强度，MPa；

σ——标准差，若无统计资料档案，设计强度为 C50 以下时，σ 取 5.0MPa，设计强度为 C50 以上时(含 C50)，σ 取 6.0MPa。

按计算出的配合比进行试拌，检验其施工性：调整其坍落度和坍落流动度，观察其体积稳定性，测定混凝土的表观密度，调整计算容重和各材料用量。

4.2 配合比计算步骤

1）按工程所要求的耐久性，确定目标氯离子扩散系数，选择水胶比。

2）按照施工条件确定施工性要求。一般，泵送时坍落度可为(140±20)mm，坍落流动度(400±50)mm。

3）强度等级为 C30 左右时，胶凝材料总量变动于 350～430kg/m³。

4）根据步骤 1）初选的水胶比和步骤 3)初选的胶凝材料总量计算用水量。

5）计算砂石用量。用砂浆填充石子孔隙乘以砂浆富余系数，列出下式：

$$V_C + V_M + V_W + V_S = p_0 \cdot k \cdot V_{0G} \tag{2}$$

按绝对体积法列出下式：

$$\frac{C}{\gamma_C} + \frac{k}{\gamma_M} + \frac{W}{\gamma_W} + \frac{S}{\gamma_S} = p_0 \cdot k \cdot \frac{G}{\gamma_{0G}} \tag{3}$$

式中：V_C、V_M、V_W、V_S——分别为每立方米混凝土中水泥、掺和料、水、砂的密实体积；

V_{0G}——每立方米混凝土中石子的松堆体积，L；

C、W、S、G——分别为每立方米混凝土中水泥、水、砂、石子用量，kg/m³；

γ_C、γ_M、γ_W、γ_S——分别为水泥、掺和料、水、砂的表观密度，kg/m³；

γ_{0G}——石子的松堆密度，kg/m³；

p_0——石子的空隙率，%；

k——砂浆富余系数，k=1.7～2.0（根据流动性不同要求调整）。

根据式（2）、（3），即可用计算出砂石体积，再根据砂石表观密度，计算砂石用量。

6）按胶凝材料总量掺高效减水剂试拌，进行坍落度和坍落流动度试验；测定拌和物表观密度，调整配合比，校验强度。

4.3　简易绝对体积法

吴中伟提出简易绝对体积法[7]，经北京建筑材料研究院实验验证[8]，表明对砂石来源稳定的搅拌站，使用该方法有简便易行的优点。其基本原则是要求出砂石有最小的混合空隙率，按绝对体积法原理计算，步骤如下：

1）按设计或甲方提出耐久性要求的氯离子扩散系数。

2）求砂石混合空隙率 α，选择最小值：先按石子级配情况设定砂率：如石子级配较好，可设砂率为 35%～40%，石子级配不好则砂率可加大，但不宜超过 45%。按砂率换算成砂石比，将不同砂石比的砂石混合，分三次装入一个 15～20L 的不变形的钢筒中，每次用直径为 15mm 的圆头捣棒各插捣 30 下(或在振动台上振动至试料不再下沉为止)，刮平表面后称量，计算捣实容重 ρ_0（kg/m^3）；测出砂石混合料的混合表观密度 ρ（kg/m^3），一般为 $2.65g/cm^3$左右。计算砂石混合料的空隙率 $\alpha=(\rho-\rho_0)/\rho$，最经济的混合空隙率约为 16%，一般为 20%～22%，24%左右则是不经济的。

3）计算胶凝材料浆量。胶凝材料浆量等于砂石混合空隙体积加富余量。胶凝材料浆富余量 ΔV_P取决于工作性要求和外加剂的性质和掺量（可先按坍落度为 180～200mm，估计为 8%～10%），试拌后，可按坍落度减小到(140±20)mm 调整用水量和外加剂掺量。则浆体积 V_P为：

$$V_P=\alpha+\Delta V_P(L/m^3) \tag{4}$$

4）计算各组分用量。设 1 份胶凝材料中掺入粉煤灰量为 f 份，表观密度为 γ_F，磨细矿渣掺量为 k 份，表观密度为 γ_K，水胶比为 w/b，水泥用量为 c 份，表观密度为 γ_C，水 w 份，$f+k+c=1$，则 1 份胶凝材料的体积为：

$$V_B=\frac{f}{\gamma_F}+\frac{k}{\gamma_K}+\frac{c}{\gamma_C}+\frac{w}{1} \tag{5}$$

则每 L 浆体中胶凝材料用量 b 为：

$$b=\frac{1}{\frac{f}{\gamma_F}+\frac{k}{\gamma_K}+\frac{c}{\gamma_C}+\frac{w}{1}}(kg/L) \tag{6}$$

$1m^3$中胶凝材料总量 $B=V_P\times b$，水泥 $c=B\times c$（kg），粉煤灰 $f=B\times f$（kg），

磨细矿渣 $K=B\times k$（kg），水 $W=c\times$（w/c）（kg），

骨料总量 $A=(1000-V_P)$(kg)，砂 $S=A\times$砂率(kg)，石 $G=A-S$(kg)。

因引入浆体积富余量，总体积略超过 $1m^3$，所计算的各材料用量总和需按实测的表观密度进行校正。

5）调整　按 15L 筒试配的砂石量+以上胶凝材料、水各量×1.5%，掺入外加剂试拌，测坍落度和流动度，如不符，则调整富余量或外加剂掺量，达到要求后，再装入筒中称量筒中混凝土和多余混凝土拌和料质量，求出混凝土表观密度，并校正各计算量。富余量±1.5%。

在以上基础上，经多次试拌，求得符合要求的合理、经济的配合比。

4.4 简易绝对体积法计算混凝土配合比举例

1）某地下工程，地下水中SO_4^{2-}最大含量为1000ppm，Cl^-最大含量为1000ppm。要求混凝土耐蚀系数≮0.85，抗渗等级≮S8，设计强度等级为C30，氯离子扩散系数应为$500\times10^{-14}cm^2/s$左右。由搅拌站集中供应混凝土。选择原材料为：GB 175—2000的42.5普通硅酸盐水泥，密度为$3.1g/cm^3$，复合掺入粉煤灰和磨细矿渣共计60%。粉煤灰为Ⅱ级，需水量比为104%，烧失量7.76%；磨细矿渣比表面积为$3700cm^2/kg$，碎石压碎指标9.8%，针片状颗粒6.1%，堆积密度$1520kg/m^3$，表观密度$2.66g/cm^3$；细骨料：Ⅱ区中砂(中粗砂)，细度模数2.9，堆积密度$1370kg/m^3$，表观密度≮$2.65g/cm^3$

2）初选目标坍落度为(180±20)mm；水胶比0.38～0.42。

3）计算：石子空隙率$=\left(1-\frac{1.52}{2.66}\right)\times100\%=42.9\%$，砂空隙率$=\left(1-\frac{1.37}{2.65}\right)\times100\%$
$=48.3\%$

砂石混合空隙体积$\alpha=0.429\times0.483=0.21m^3$

假设胶凝材料浆体富余量为10%，则浆体体积为$320L/m^3$。

选择水胶比为0.4；

掺入粉煤灰和磨细矿渣共60%，粉煤灰和磨细矿渣混合密度为$2.5g/cm^3$，则

$$\frac{\text{胶凝材料用量}}{\text{浆体体积}}=\frac{1}{\frac{0.4}{3.1}+\frac{0.6}{2.5}+0.4}=1.3kg/L$$

则胶凝材料总量$=320\times1.3=416kg/m^3$。

4）按前述步骤计算出配合比，掺入液体高效减水剂2.5%，试拌中因发现有泌水现象，调整胶凝材料总量为$420kg/m^3$，并确定水泥用量为$180kg/m^3$，水胶比0.40。按前述步骤调整各材料用量后，拌和物坍落度为220mm，坍落流动度为550mm，成型检测混凝土各项性能见表1和表2。

表1 混凝土力学性能

抗压强度（MPa）				28天轴心抗压强度（MPa）	28天抗折强度（MPa）	28天劈裂抗拉强度（MPa）	28天弹性模量（GPa）
3天	7天	28天	90天				
22.3	38.2	50.7	53.5	39.9	6.33	2.95	38.9

表2 混凝土耐久性指标

抗蚀系数	抗渗等级	ASTM C1202法测量结果及评价（库仑）	氯离子扩散系数（$10^{-14}cm^2/s$）
1.14	＞P12	1126/很低	161/中等

实测温升较快，约1天即达高峰，为了控制早期强度，调整了坍落度为140～160mm，增掺了缓凝剂，掺入引气剂4%。

5 施工

5.1 关于浇筑和振捣

1）混凝土运到工地后应立即检测坍落度，并尽快浇筑。如发现坍落度不足，不得擅自加水，应当在技术人员指导下用追加减水剂的方法解决。

2）浇筑温度：夏季浇筑混凝土应降低温度，至少应比当天最高气温低10℃。但混凝土浇筑温度太低时，受环境较高温度影响的表面硬化较快，内部温度升高时产生膨胀，会使先硬化的表面受拉而开裂[9]。因此在夏季，不仅要降低浇筑温度，而且要采取措施(例如避免上午浇筑，冷却模板，避免阳光直射于混凝土表面等)避免混凝土表面受气温影响而先于内部硬化。冬季要提高混凝土浇筑温度，则混凝土内部温度高于气温，内部成熟快，产生膨胀时，表面仍有一定塑性，可变形而不裂，而当混凝土降温时，在表面产生压应力，而有利于抗裂。冬季浇筑温度应不低于10℃。

3）图1[9]所示不同气温下、不同浇筑温度、不同厚度构件的混凝土在约束条件下最大应力水平和最大温差的关系。由图1可见，当混凝土浇筑温度 T_i 降低10℃时，相应地使积聚的拉应变减小约70微应变，占极限应变相当大的部分。

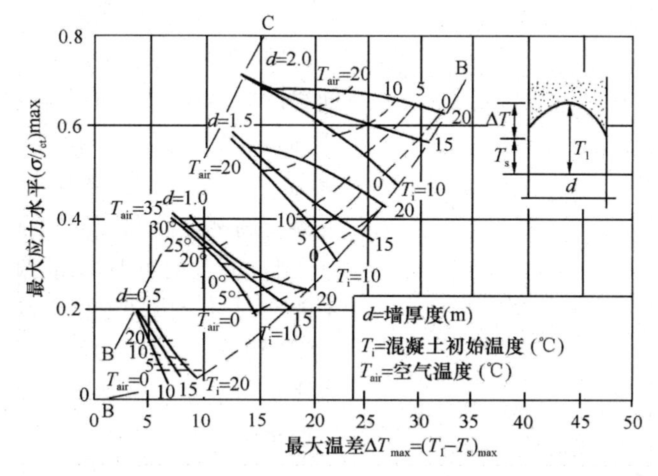

图1 不同气温下、不同浇筑温度、不同厚度构件的混凝土在约束条件下最大应力水平和最大温差的关系

4）在气温高的热天，不宜在上午浇筑，以免在混凝土升温恰值气温最高时而加剧。

5）长墙或板的施工缝的间距应视构件尺寸而定。高宽比大于2的墙，上部一般不会出现裂缝，但是应注意分层浇筑时，下层高度要大于上层高度，否则，上层拌和物会增大，对与基底接触面的正压力而增加约束应力。混凝土浇筑高度不宜超过2m，如必须超过，则必须用串筒等辅助下料；每层混凝土一次性布料不宜超过1m。

6）泵送混凝土下料位置相隔应当小于3m，每层下料位置应当交错，如图2所示，以保证均匀。

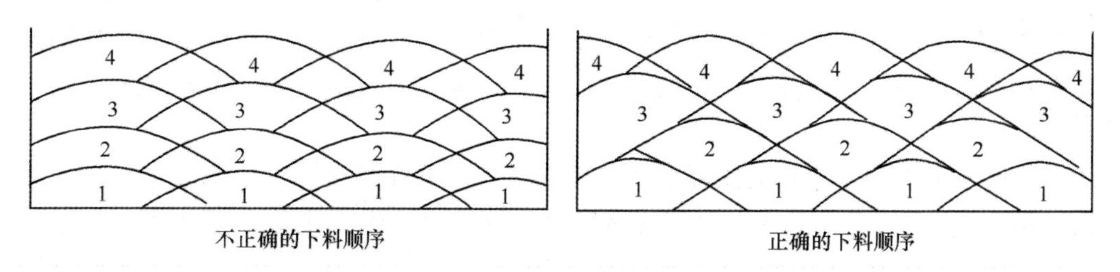

图2 下料顺序示意

7）不正确的浇筑顺序会造成可以避免的约束和不均匀的沉降。例如梁和柱或板和墙同

时浇筑，会因沉降不匀在交接处产生裂缝；相反，采取恰当的浇筑顺序会减少开裂，如大面积的板当使用膨胀剂时，采取“跳仓”方式浇筑可减少开裂。所以不同构件浇筑前应认真规划浇筑顺序。

8）应当正确进行混凝土拌和物的振捣，使用振捣棒时绝对禁止用振捣棒横拖赶动混凝土拌和物。否则必然造成离下料口远处砂浆过多而开裂。

5.2　关于养护

1）对于板，浇筑后立即覆盖，避免塑性开裂。当混凝土表面“收水”过快时，会结成一层硬壳，而内部则凝结变慢。在干燥有风的条件下，硬壳会开裂并脱落。实践证明，如能及时覆盖，就能避免发生这种情况。

2）尽早开始湿养护。墙、柱等在拆模前应及早松动模板浇水，或是用透水性模板或吸水性模板。

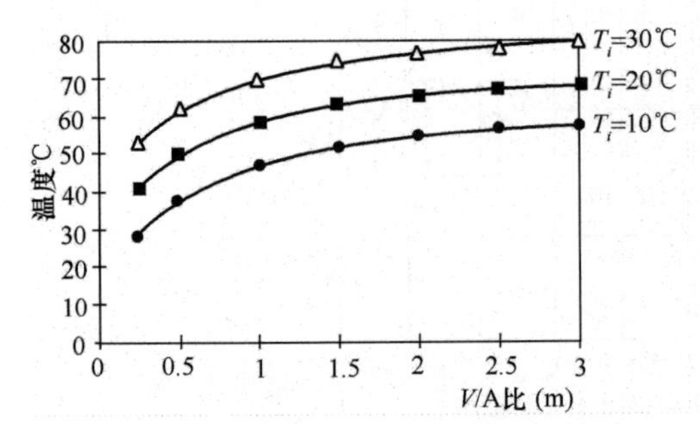

图3　作为浇筑温度、体积/面积比 V/A（m）函数的混凝土最高温度。硅酸盐水泥用量350kg/m^3，假设浇筑温度与气温相同

3）在最小断面大于30cm的构件中，早期温度应力引起的开裂往往占大部分（至少60%），因此温度控制很重要。首先应降低浇筑温度，浇筑温度和构件体积与面积的比值和混凝土内部最高温度的关系参见图3[9]。

4）拆模时间应视混凝土内部温度而定，不能在混凝土内部温度最高时拆模，尤其不能在混凝土内部温度最高时拆模后立即浇凉水，以避免对混凝土产生热震。拆模后注意保温，以避免降温速率太快。

5）避免间断浇水。

6）混凝土在相对湿度低于100%时开始失去毛细水，在相对湿度低于65%时，开始失去凝胶的吸附水；凝胶越多，体积越不稳定。硬化的混凝土水泥浆体需要有一定量的未水化颗粒稳定其体积[5]。因此浇水周期既要足够，又不宜随意延长。但目前主要矛盾是湿养护不足。

7）应在混凝土还处于塑性时开始冷却表面。夏季使用钢模板时，可在浇筑时同时向模板表面浇凉水，以推迟混凝土温峰时间，并降低温峰；混凝土内部达到温峰后开始降温时则应控制降温速率，避免在混凝土升温以后尤其是在温度最高时拆模，更不能立即浇凉水。冬季尽量使用导热系数小的模板，以减小混凝土中心和表面的温差，必要时，应采取灵活保温措施。对于厚度超过30cm的墙、柱、基础地板等中等体积和大体积混凝土结构，夏季施工时应尽量降低入模温度，在混凝土达到温峰前应在模板外（对墙、柱）或覆盖的塑料薄膜上面（对板）浇水降温，到达温峰以后的降温阶段应采取保温措施以降低降温速率，必要时可用热水养护。

图4[10]所示为混凝土典型的内部温度发展曲线。在刚浇筑很短时间的第Ⅰ阶段（约3～6小时）温度还没有上升，基本上保持浇筑温度；第Ⅱ阶段开始升温，但因混凝土尚处于塑性而内部为零应力，直到温度为 $T_{1,2}$ 时，混凝土内部开始产生压应力；第Ⅲ阶段混凝土持续升温，但由于徐变和自收缩的影响，在达到温峰前，压应力就开始下降；第Ⅳ阶段混凝土开始降温，当压应力下降为0时，混凝土仍然为温度很高的 $T_{2,3}$；第Ⅴ阶段内部应力由压应力

变成拉应力；在温度到达 T_c时，混凝土开裂。$T_{1,2}$称为第一次零应力温度，$T_{2,3}$为第二次零应力温度，T_c为开裂温度。开裂温度越低，混凝土抗裂性越好。这个图可以指导施工期间混凝土的温度控制，即尽量在第Ⅰ、Ⅱ阶段冷却混凝土，减小升温速率和温峰值，第Ⅳ、Ⅴ阶段要采取灵活保温措施控制降温速率。对于尽量在第Ⅰ、Ⅱ阶段冷却混凝土，我国已有单位实施，其根据可参考下面的一段引文[7]：

按传统的观点，通过在金属模板上洒水进行外部冷却，因通常必然增加混凝土体内的温差，会被认为是一种不适当的方法。然而，想起早先在本文 9.2.3 和 9.3.5 说过的关于表面层“有利的”压缩力，证明外部冷却可自然地减小断面的平均温度，并减小表面成熟度的发展，但是表面冷却的最重要影响是使早期表面混凝土温度适应（或低于）环境的温度。证明温度引起的混凝土表面预应力能由洒水过程的持续和适时来控制。

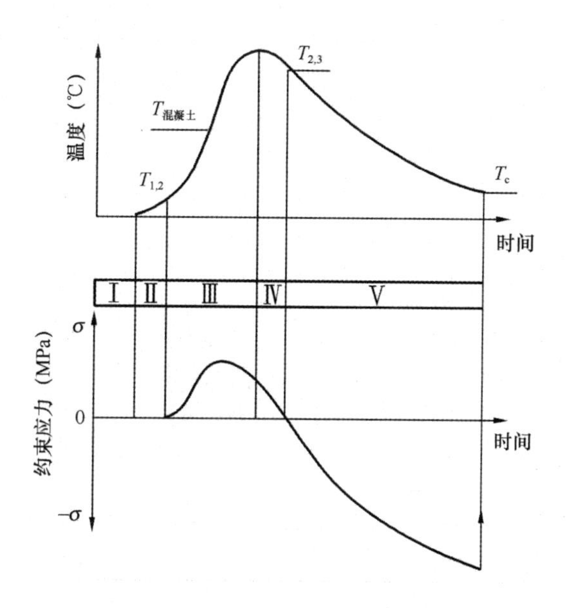

图 4　混凝土典型温度发展曲线

8）在混凝土降温阶段应控制降温速率不超过 2℃/d。

9）关于温差的控制，当混凝土内外温差和混凝土表面与大气的温差不大于 15℃时，混凝土就不会开裂。

10）为避免早期裂缝，应尽量控制抗压强度 12h≯6～8MPa，24h≯10～12MPa。

11）掺用大量矿物掺和料时应特别注意保湿养护，养护时间至少需 7d。

6　结论

1）裂缝控制是限制环境中侵蚀性介质进入混凝土结构的第一道防线，控制裂缝尤其是早期裂缝，对保证混凝土结构达到设计要求的耐久性，有重要意义。

2）裂缝控制需要建设主管、开发商、设计人员、材料供应商和施工承包商的共同努力，但是施工过程的各环节对混凝土成型质量和裂缝控制尤为重要。

3）除从观念上、计划上改变追求高早强外，还应尽量控制减小 2 天内强度增长速率，对竣工时间长的大型工程（如高层建筑的底层结构），尽量延迟强度验收期限。

4）配制混凝土时应选用开裂敏感性小的水泥（低 C_3A、C_3S，低碱、低比表面积）、抗裂性好的矿物掺和料（磨细矿渣如掺量小于 70%，其比表面积不宜超过 400m^2/kg，尽量避免使用硅灰，必须使用硅灰时，应与至少 30%的大掺量的粉煤灰复合），以及尽量选用热膨胀系数小的粗骨料（例如与花岗岩相比，石灰岩的热膨胀系数就较小）。选用配合比时尽量减少水泥用量和胶凝材料总量，提倡使用引气剂。

5）施工中应重视采取正确的施工浇筑顺序，严格禁止违反操作规程的浇筑和振捣方式，重要工程应有在线测定混凝土温度和应力的措施，根据实测结果调整养护措施。夏季要注意降低混凝土入模温度，并尽量提前在混凝土处于塑性的阶段开始采取降温措施，避免横跨断面的温差；在混凝土降温阶段，无论夏季、冬季，都要注意采取合理保温制度，避免混凝土内部降温太快；避免拆模时产生热冲击，要尽早开始湿养护，并避免间断浇水，不得在混凝土内部温度达高峰时开始浇凉水，浇水周期要足够。

6）提倡混凝土供应商和施工承包商联合实行混凝土的生产、浇筑、养护（包括温湿度控制）等一体化的施工。这不仅有利于裂缝的控制，也有利于施工质量的控制，应当是建筑工业集约化生产的方向，建设主管或开发商应当支持这一措施，并在经济政策上作相应的调整。

参考文献

[1] P. K. Mehta, Durability——Critical Issues for the Future , Concrete International, 1997.

[2] A. M. Neville. 李国泮，马贞勇译. 混凝土的性能[M]. 北京：中国建筑工业出版社，1983.

[3] 安明喆. 高性能混凝土自收缩的研究. 清华大学博士学位论文，1999.

[4] 宫沢伸吾，田澤栄一，左藤剛，など. 铁筋拘束 による超高强コンクリートの自己収缩応力. コンクリート工学年次论文報告集，1993(15)1[5].

[5] R. W. Burrows. The Visible and Invisible Cracking of Concrete, ACI Monograph No. 11, see: International Symposium on Sustainable Development of Cement and Concrete Industry. Ottawa, Canada, Oct. 1998.

[6] P. K. Mehta. Durability ——Concrete Technology for Sustainable Development－A overview of Essential Principles.

[7] 吴中伟，廉慧珍. 高性能混凝土[M]. 北京：中国铁道出版社，1999.

[8] 阎培渝，姚燕. 水泥及复合材料科学与技术——吴中伟院士从事科教工作六十年学术讨论会论文集[M]. 北京：中国建材工业出版社，1999.

[9] S. Bernander, Practical Measures to Avoiding Early Age Thermal Cracking in Concretes, see: Avoidance of Thermal Cracking in Concrete at Early Ages, Report of work of RILEM Technical, edited by R. Springenschmid, Munich1998.

[10] Phlipp Holzman. Optimization between early strength and low risk of ckracking, see: Conferrence Documentation of Our World in Concrete & Structures Vol. XIX, 23－24 August 2000, Singapore.

自评

1. 该文原作为《混凝土结构耐久性设计与施工指南》的论文汇编而发表。从搅拌机卸出的混凝土材料只是拌和物而并非最终产品，只有通过浇筑、振捣、收面、养护的工艺而制成构件用于结构，才能评价其质量，因此在上述《指南》中针对现代混凝土的特点选择几篇论文以供参考。由于混凝土不仅有原材料波动的影响，而且其微结构成长和性质具有对环境条件极强的依从性，针对不同工程需要、不同原材料、不同的使用环境和施工条件，在一定的原则要求下，可能会有若干有效的对策，因此对施工技术不宜制定规范做统一的规定。《指南》则可以提供各种具体技术供选择，有利于发挥施工人员的专业知识、经验和智慧。该文章中对于配合比设计方法只是一种建议。

2. 对于硅灰的使用，所述“尽量避免使用硅灰”指的是低水胶比的大体积混凝土，需要控制高活性组分产生自收缩，不能理解为一概不能用。任何材料都有最合适使用的场合，都有最适合于解决某种问题的作用。近年来在工程中发现，少量的硅灰可以改善低水胶比混凝土的施工性——缓解泌水，降低黏度。

3. 这里说的裂缝控制指的是早期尤其是初期可见裂缝的控制。混凝土结构劣化以致失效的极限状态是钢筋的锈蚀，钢筋锈蚀的条件是氧气和水分的提供，无宏观缺陷而致密的混凝土即使在有化学腐蚀性的环境中，其受腐蚀而劣化的过程是很缓慢的，因而可以很好地保护钢筋。但是如果混凝土出现裂缝，水和氧气得以进入，则劣化和腐蚀钢筋的速率就会很快。从能量的角度来说，已有裂缝的扩展比新生成裂缝容易，因此控制早期裂缝是为了避免或减少长期受环境作用而出现可见裂缝。混凝土裂缝的控制不止是施工的问题，也不是只靠配合比就能解决的，而是涉及设计、材料和施工，甚至与建设方的要求都有关系。

4. 这里所指的早期裂缝是指混凝土初凝后在施工期间所产生的温度变化、自收缩和干缩引起的裂缝。至于初凝前的塑性阶段产生的沉降裂缝和塑性收缩应当是最容易通过施工中工艺可控的，例如及时覆盖并在失去流动性后进行二次搓面。这种裂缝也应包括在早期（或初期）必须控制的裂缝中。控制早期裂缝的意义在于早期裂缝是后期在环境作用下的开裂源。

5. 可见裂缝不一定有害，不可见裂缝不一定无害，重要的是裂缝的部位、形态、尺寸等性质。裂缝即使不可见，仍然会有初始应力的积聚而成为荷载或环境作用下引发开裂的因素。裂缝是侵蚀性介质侵入的通道，因此是影响混凝土结构耐久性的主要因素。

一个人光溜溜地到这个世界上来，最后光溜溜地离开这个世界而去，彻底想起来，名利都是身外物，只有尽一个人的心力，使社会上的人多得他工作的裨益，是人生最愉快的事情。

——邹韬奋

论文之十三

对“高性能混凝土”十年来推广应用的反思*

廉慧珍
清华大学土木水利学院

摘　要　高性能混凝土不是混凝土的一个品种，而是达到工程结构耐久性的质量要求和目标，是满足不同工程要求的性能和具有匀质性的混凝土。高强不一定耐久，高流动性也不是任何工程都需要的，也不是只要有掺和料就能高性能；混凝土的质量不是实验室里配出来的，而是优选配合比的混凝土由生产、施工和管理人员在结构中实现的，开裂了的就不是高性能混凝土，除了特殊（如临时性结构）外，没有什么混凝土结构不需要耐久。针对不同工程特点和需要，对混凝土结构进行满足具体要求的性能和耐久性的设计，比笼统强调“高性能混凝土”的名词要更科学。

关键词　高性能混凝土，误区，反思，耐久性，匀质性

1　前言

自从1994年清华大学向国内介绍和发起应用高性能混凝土以来，高性能混凝土得到较多的应用，对混凝土耐久性的重视有所加强了，粉煤灰、矿渣等矿物掺和料的使用增多了，预拌混凝土更普遍了，混凝土拌和物的施工性能得到了改善。但是，近年来在国内外却发生较多使用“高性能混凝土”的结构开裂，特别是早期开裂的问题。而混凝土的裂缝正是在使用阶段环境侵蚀性介质侵入的通道。尽管在工程进行过程中，这些裂缝得到了修补，也仍然可成为混凝土提早劣化的隐患。于是不免产生一个问题：开裂了的混凝土还能是高性能的吗？反过来，我们又产生一个问题：当初提出“高性能混凝土”的概念，是强调混凝土的耐久性，那么，什么混凝土不需要耐久呢？究竟应如何理解高性能混凝土？

实际上，近年来混凝土的开裂成为国内外的热门课题，并不是“高性能混凝土”所独有的，只是“高性能混凝土”也没有逃脱当前较普遍存在的混凝土结构开裂的命运罢了。也就是说，某些“高性能混凝土”的开裂带有普遍意义上的原因。

对高性能混凝土的定义，或含义，至今没有统一的理解。就我国的情况来看，十年来由于对高性能混凝土的理解存在若干误区，造成“高性能混凝土”使用上的盲目和混乱。大多数对“高性能混凝土”的理解是“高强”、“高流动性”、“掺用矿物掺和料”，而用现行规范在实验室制作的小试件，以简化的条件和强制的过程进行的“耐久性”试验结果，又同该混凝土在实际结构中的表现不一致。存在这种现象并不仅是使用者的责任，许多科技工作者本身的认识也是在不断发展的，对已有认识的不断深化和修正是正常的、必需的。本文的目的是和大家共同探讨几个问题，以使我们的主观世界更加符合客观实际，不断促进科技进步。

---

* 原载《混凝土》2003年第7期。

2 误区之一：高性能混凝土必须高强度

高强混凝土技术的发展对建设事业起着重大作用，使得过去只能用钢结构建造的高层、大跨和恶劣环境中的结构物可以使用钢筋混凝土，不仅节省了能源的消耗（承受1000t死荷载、高1m的柱子用混凝土与用钢相比，所消耗的能源可减少约57%[1]），而且相对于大气对钢的锈蚀来说，混凝土更耐久。因此，100年来，尤其是近50年，混凝土的强度不断提高，是科学技术不断进步的体现，并促进了建设事业的发展。然而，世界上不存在完美无缺的事物，有一利必有一弊，有所得必有所失；任何技术都有其适用的范围。目前大多数人仍把高性能和高强联系在一起。甚至有人盲目追求混凝土的高强、“超高强”以至“特超高强”，并以此为“水平”的标准。西方国家报道的许多开裂的“高性能混凝土”，其实都是高强度的混凝土。提高混凝土的强度可以减小构件断面，但是结构物不仅需要强度，还有刚度的要求，例如对于一定高度的柱子，有一个最小断面的要求，低于这个最小断面就会发生失稳的破坏；也就是说，当满足达到这个最小断面的柱子所要求的强度时，再提高强度是没有意义和抗压强度的关系[2]的，反而会造成资源的浪费和高强带来的开裂敏感性等问题。

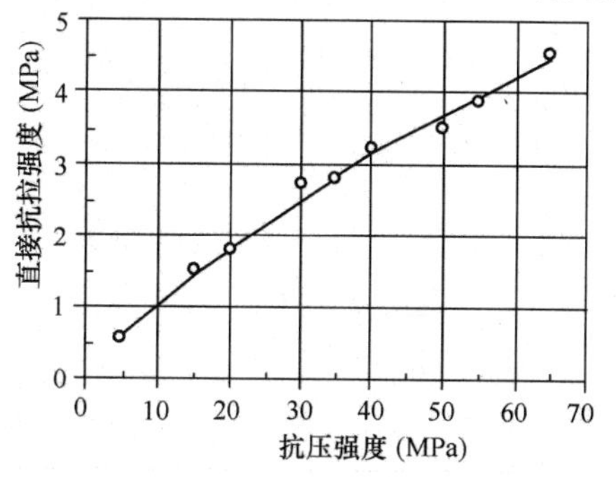

图1 混凝土直接抗拉强度

由于混凝土是一种非均质的“混相”体，其内部各相之间的作用力主要是范德华力，抗拉强度远低于抗压强度。如图1[2]所示，当混凝土抗压强度为10MPa时，抗拉强度和抗压强度的比值（以下简称拉压比）约为1/10；抗压强度为30MPa时，拉压比约为1/12；抗压强度为60MPa时，拉压比约为1/15；抗压强度为90MPa时，拉压比则约只有1/18。这样，就使得混凝土的强度越高，其脆性越大，断裂韧性越小(见图2)，抵抗突发荷载(如地震、爆炸)和疲劳(如高耸结构承受的风荷载，道路承受的动力荷载)的能力越差。

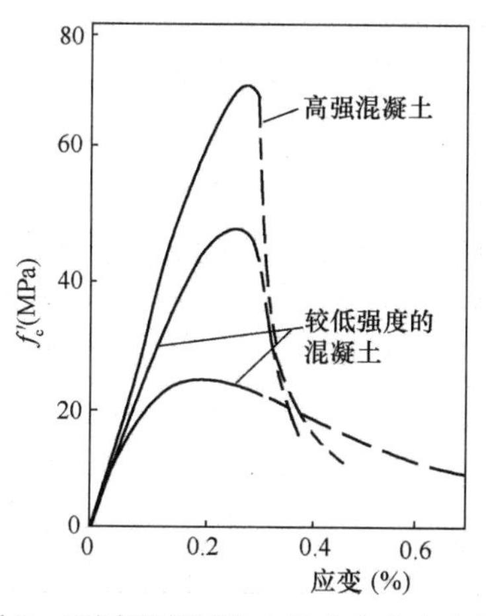

图2 不同强度混凝土的应力应变曲线

由图2可见，随着混凝土强度的提高，破坏以前出现的塑性变形越不明显，越会出现突发性破坏。图3中梁的延性比随混凝土强度的提高而下降速率提高，从26MPa提高到63MPa时，尽管箍筋率增加了约60%，梁的延性比却下降了约80%。

此外，高强混凝土一般使用强度等级高的水泥，同时，为了满足施工性的要求，水灰比很低的高强混凝土必须有较大的水泥用量，尤其是不掺用或掺量小的矿物掺和料时，会使混凝土中水化温升很高，控制不好时会产生较大的温度应力，再加上低水灰比造成的较大自收缩，可增加混凝土的开裂敏感性。

鉴于高强混凝土的优点和缺点，在土木工程学会高强与高性能混凝土委员会制定的《高强混凝土设计与施工指南》中关于高强混凝土的设计，最高考虑到C80，并建议使用钢管混凝土来提高构件延性比。问题是，由于人们以强度作为

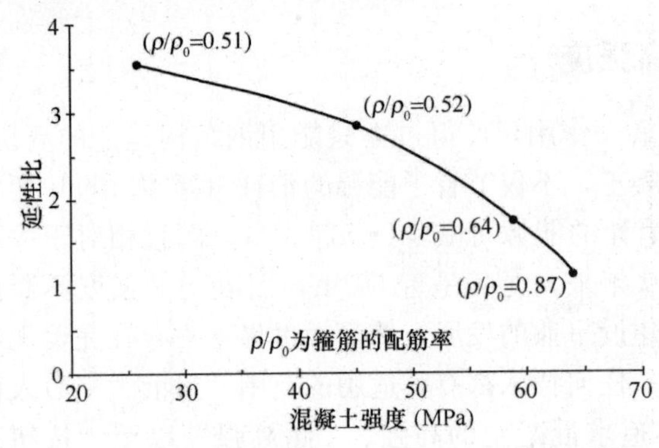

图 3 采用不同强度混凝土无负筋梁弯曲变形延性比

衡量混凝土水平和质量的定势，往往从设计到混凝土生产，再到施工，混凝土的实际强度层层被提高，并非工程实际要求那么高的强度。大多数工程所要求的是 C30、C40 这样中等强度等级的混凝土。正如吴中伟所论述图 3：“如果强调高性能混凝土必须在 C50 以上，则必然大大限制高性能混凝土的应用范围。大量使用的钢筋混凝土建筑物如低层和多层房屋以及高层房屋的上层构件，又如海工、水工，尤其是在不利环境中的结构物大体积混凝土等，对强度要求并不高，……但对耐久性要求却很高，……高性能混凝土不只是高强度的，而是包括各强度等级的……。”[4]

3 误区之二：高流动性就是高性能

提高流动性同样有利也有弊。混凝土拌和物的流动性从 10 年前普遍的 70～90mm 发展到现在大量预拌混凝土的 180～200mm，甚至已经有自密实的混凝土的浇筑，这也是混凝土技术的一种进步：可减轻振捣的劳动量，推动预拌混凝土的发展，泵送高度已可达 300m，并大大减少了“蜂窝”、“狗洞”等质量事故，可提高混凝土的匀质性。目前很多人以高流动性为“高性能混凝土”的特征。影响混凝土流动性的因素是用水量和高效减水剂，而影响相同流动性混凝土用水量的主要因素是骨料的质量。我国目前骨料质量非常差，如北京和深圳，自然状态砂石的空隙率接近 50%，则混凝土就要有较多的含浆量才能满足流动性的要求。目前北京预拌混凝土的用水量许多都在 175～180kg/m^3。较大的用灰量和用水量组成了较大的浆骨比是混凝土收缩增加的又一个因素。目前随着不适当要求大坍落度混凝土拌和物的要求，使许多工程中发生混凝土泌浆和离析的问题。如果能将混凝土用水量降低到 150kg/m^3 以下，则抗裂性就会得到较大改善。国内外的实践表明，掺入大量的粉煤灰后，因其存在大量玻璃微珠而能减少混凝土的需水量，同时可增加拌和物内部黏聚性而大大改善混凝土泌水、离析的性能，而却可减少拌和物与泵管之间的黏附力而增加可泵性。P. K. Mehta 在美国加州大学伯克利分校一项加固工程中，密配筋的基础和剪力墙的 C30 和 C35(约相当于我国混凝土的 C40)混凝土掺粉煤灰 50%和 60%，在施工中实测拌和物的坍落度为 128mm 时，与坍落度为 180mm 的普通混凝土的泵送性能相当，不离析或泌水[5]。实际浇筑时坍落度仅为 100mm。拌和物泵送性能良好，而且这样的混凝土浇筑的构件均有很好的抗裂性。此外试验和实际工程都表明，坍落度太大的拌和物还不利于混凝土温升的降低。

因此，不能把流动性一概作为混凝土拌和物“高性能”的指标，而应当根据不同工程特点，注重拌和物的施工性能。坍落度的选择要服从于混凝土的匀质性和体积稳定性要求。

4 误区之三：限制掺入粉煤灰

从排斥粉煤灰到如今“Ⅰ级”粉煤灰供不应求，是我国混凝土技术的又一进步。许多人认为掺粉煤灰的混凝土就是“高性能混凝土”。但是大部分的掺量不超过 20%，有的规范还规定掺量不大于15%。这种现状实际上仍然是20年前对掺和料水泥的指导思想的延续。基于对矿物掺和料的认识不足和对水泥混凝土科学技术的分解论，建筑材料的教科书至今仍以“调节水泥标号”作为矿物掺和料的一项作用：6%～15%的矿物掺和料等量取代硅酸盐水泥，则水泥标号降一等。也就是说，认为掺入矿渣或粉煤灰必然影响强度。实践证明，以简单取代的方法在使用硅酸盐水泥的混凝土中掺粉煤灰，在掺量不大于20%时，不仅对强度影响不大，对其他性能影响也不大，和普通硅酸盐水泥差别不大。必须正确认识粉煤灰的作用，否则不增加掺量，如何发挥粉煤灰的潜在优势？

尽管许多人接受粉煤灰对混凝土耐久性的作用、降低混凝土温升、减小收缩和抗裂的作用等观点，但潜意识里仍然是怀疑的。这也是不无根据的，但这种根据恰是从分解论的研究方法中得出的。现在讨论三个问题。

4.1 关于强度

粉煤灰对混凝土的贡献对水胶比特别敏感。如图4[6]所示的粉煤灰掺量为40%的碾压混凝土中，粉煤灰和水泥分别对混凝土强度的贡献与水胶比和龄期的关系。在7天，当水胶比约为0.7(体积比约为1.7)以上时，水泥的贡献大于粉煤灰的贡献(粉煤灰的贡献为0)；28天粉煤灰的贡献虽然增加了，但当水胶比约为0.7(体积比约2.2)时，粉煤灰的贡献还是为0；而后期，则粉煤灰的贡献超过了水泥的贡献。重要的是，可见到无论哪个龄期，粉煤灰的贡献随水胶比的降低而增长的速率比水泥的快，也就是说粉煤灰对水胶比更加敏感。如果在较高水胶比下以粉煤灰等量取代水泥，所得到的强度，尤其早期强度当然会下降很多。

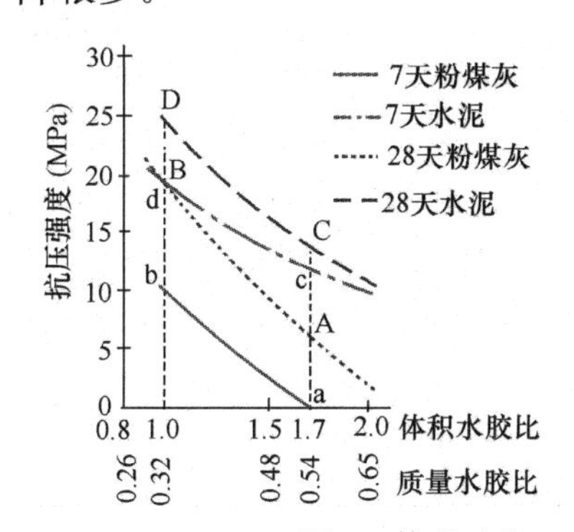

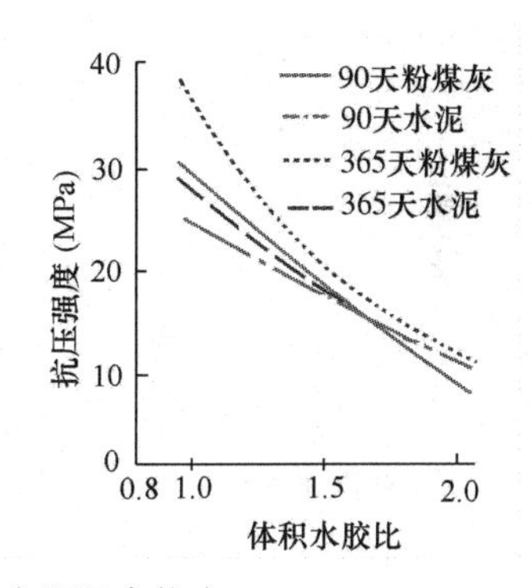

图4 体积比为1∶1的粉煤灰和水泥混合物中
水泥和粉煤灰分别对强度的贡献随水灰比和龄期的发展

英国Dunstan在Abrames1912年建立的强度-水灰比两维关系图上加了一根粉煤灰掺量的轴，成为强度-水胶比-粉煤灰掺量的三维关系图（图5)[6]，一定强度水平上的混凝土水胶比必须随粉煤灰掺量而改变(如图中的阴影区即水胶比-粉煤灰的等强度面)。因此，掺40%粉煤灰和10%矿渣粉的混凝土抗压强度28天的平均可达到52.7MPa，3个月超过60MPa[7]。

所以从耐久性出发，粉煤灰掺量直到60%对混凝土强度都是可以接受的。但使用方法要得当。同时，粉煤灰的掺量越大，对潮湿养护越敏感（见图6)[8]。另外，粉煤灰对温度也很敏感。跟踪实际构件中的温度进行养护实验的结果表明，实际构件中的较高温度对粉煤灰较有利而对硅酸盐水泥较不利，实验室标准条件下养护的试件强度，掺粉煤灰的低于构件中实际强度，而不掺粉煤灰的则相反。因此用现行标准制作的试件强度不反映现场实际构件混凝土的强度[9]。

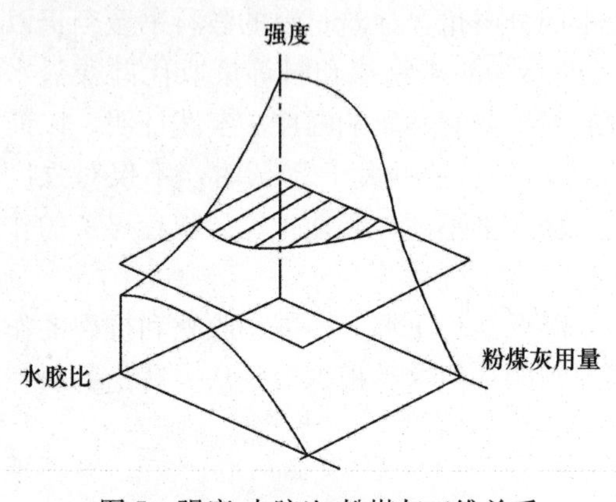

图 5　强度-水胶比-粉煤灰三维关系

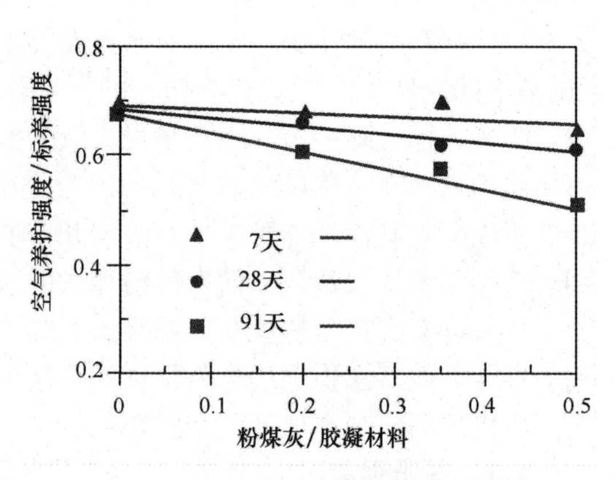

图 6　养护方式不同时粉煤灰掺量对混凝土强度的影响

4.2　关于收缩和开裂

怀疑掺粉煤灰后会加剧混凝土的收缩和开裂，所以不敢多掺。多数把掺量控制在 20%左右，掺用的目的仍主要是经济效益。哈尔滨工业大学工程硕士研究生刘旭晨作了一项有意义的试验，他分别掺 30%和 60%粉煤灰与纯水泥配制同强度等级的混凝土各制作三块 3m×1.5m×0.2m 的板，各用三种不同养护制度进行养护：①浇筑 1h 后覆盖薄膜；②浇筑 8h 后覆盖薄膜；③始终不覆盖。观察混凝土板面开裂情况。其结果是，1h 后覆盖的都没有裂；8h 覆盖掺 60%粉煤灰的没有裂，其余的都裂；始终未覆盖的在 16h 后全都开裂，而且掺 60%粉煤灰的最大裂缝长度和宽度最大，但是却还没有凝固；裂缝宽度大于 0.5mm，裂缝的数量，无掺和料的占约 37%，掺 30%粉煤灰的约占 55%，掺 60%粉煤灰的约占 64%。这些差别的原因应当是三种混凝土对水的敏感性的差别造成的。混凝土的密实度取决于水胶比，密实度低的混凝土 C 在塑性阶段，裸露的表面水分蒸发后，内部的水分能及时向表面泌出作为补充，因此不易出现塑性收缩裂缝；很密实的混凝土内部的水不易泌出，当表面水分蒸发后，无水分补充的来源，就会产生塑性裂缝。水胶比越低，混凝土的密实度越大。这就是当成型后不覆盖的混凝土 C 在还没有凝结时就出现裂缝的原因。但粉煤灰在 28 天以前基本上不参与水化反应，掺入不同量的粉煤灰后，混凝土中的水灰比（水和水泥的比值）实际上不同程度地增大了。为了进行分析，分别计算三种混凝土中的水胶比和水灰比，见表 1。

表 1　粉煤灰掺量不同时混凝土的水胶比和水灰比

编号（粉煤灰掺量%）	粉煤灰掺量	水胶比	水灰比
A（0）	0	0.5	0.5
B（30）	30	0.41	0.59
C（60）	60	0.36	0.85

由表 1 可见，掺粉煤灰后尽管水胶比减小了，而水灰比则加大了，混凝土 C 中因其水灰比很大（0.85），在 8h 覆盖之前，仍有大量的水供其消耗，因此成型后 8h 再覆盖也可不开裂。而混凝土 B 的水灰比相对较低，水泥水化较快的混凝土也容易失水，所以 8h 后再覆盖已为时过晚。现行规范中对混凝土收缩的测试是成型一天（注意：并无覆盖）后拆模测初长，这对于需要加强保湿养护的粉煤灰来说是不公平的。凡是在工程中掺用了粉煤灰而开裂

的混凝土都和养护有关。

4.3 关于抗冻性

用现行标准的方法检测掺粉煤灰混凝土的抗冻性或抗路面盐冻剥落性变差。M. D. A. Thomas“对20多个现场（主要是公路工程结构物）进行了调查，这些工程的粉煤灰混凝土一直是暴露在使用化冰盐的冻融环境之中。有些工程已经经历了20多个冬天，但调查混凝土都没有出现剥落现象，使用性能非常好，而有些从现场取得的一些粉煤灰混凝土样品进行实验室实验的结果却不好，这说明用现行对普通水泥混凝土进行的抗剥落性能试验方法不适合用于评价粉煤灰混凝土。……显然粉煤灰混凝土的快速试验和现场条件下的表现缺乏一致性。大量现场的粉煤灰混凝土在严酷条件下正继续提供着良好的服务，而同一混凝土样品在实验室里试验中却表现不好。”[10] 抗冻性也如此。粉煤灰对试件制备和养护条件更为敏感。制作试件条件没有考虑掺粉煤灰前后混凝土中水化动力学的变化，因而相同龄期掺和不掺粉煤灰的试件的微结构不同。掺粉煤灰的混凝土尽管水胶比低了，而水灰比却增大了，在28天以前的密实度相对较低，只有经过较长龄期，才能形成更加密实的微结构；实验证明，掺粉煤灰的混凝土的微结构适合于在较高温度下发展，20℃的标准养护温度对掺粉煤灰的试件比对不掺粉煤灰的试件不利；另外，掺粉煤灰的混凝土凝结慢，如果抹面结束后泌水还在继续，会对抗剥落性能带来很不利的影响[10]。因此目前需要建立针对掺用矿物掺和料的混凝土的、更接近现场条件的实验方法作为工程合格的评定。

并非掺矿物掺和料就一定是高性能，矿物掺和料使用不当也会出现问题。例如粉煤灰在28天以前基本上不参与反应，水和硅酸盐水泥的比值（水灰比）增大（尽管掺粉煤灰后降低水胶比），混凝土早期孔隙率随粉煤灰掺量增大而增大，碳化并非到后期才发生，而是一旦拆除模板，相对湿度下降到70%以下就开始，在以碳化引起钢筋锈蚀的一般环境中，碳化速率显然增大。此外只要掺入30%的矿物掺和料。胶凝材料中的SO_3就被稀释到使混凝土的需水量增大、与外加剂相容性差、早期强度低、收缩增大、后期弹性模量降低等等。

5 怎样理解高性能混凝土

自从20世纪80年代末美国和日本相继提出高性能混凝土的概念以来，对高性能混凝土仍无被各国普遍接受的确切定义，各个国家不同人群有不同的理解[1]。大多数承认单纯高强不一定耐久，而提出“高性能”则希望能既高强又耐久。可能是由于发现强调高强后产生的弊端，1998美国ACI又发表了一个定义为：“高性能混凝土是符合特殊性能组合和匀质性要求的混凝土，采用传统的原材料和一般的拌和、浇筑与养护方法，往往不能大量地生产出这种混凝土。所指特性为：易于浇筑，振捣不离析，早强，长期力学性能，抗渗性、密实性，水化温升，韧性，体积稳定性，恶劣环境下的较长寿命”[11]；1998年ACI和1990年美国NIST和ACI定义的区别是，后者把早强列入“特殊性能组合”可选性能之一，而不作为必要的规定而强调。

经过反思，结合ACI的定义，由本文上述分析，应当理解为，高性能混凝土不是一个混凝土的品种而是强调混凝土的“性能”（performance）或者质量、状态、水平等表现，或者说是一种质量目标。对不同的工程，高性能混凝土有不同的强调重点（即“特殊性能组合”）。吴中伟说：“高性能”显然是针对过去不高或“中”、“低”的性能提出的。在不同时期，“高”和“低”的分界线是大不相同的。随着科技的飞速进步与要求的不断提高，今日的“高”必然会被新的“高”所代替而成为来日的“低”。用“高”、“低”来命名科技产品，尤其是命名像混凝土这样的大宗材料，是不科学的与难以持久的。现在“约定俗成”，待使用一段时期后，希望找到一个更科学、更合理的新名词[5]。这段话就是这个意思。如果还用“高性能”的名词的话，则应当认识，高性能混凝土是整个工程全部环节协调、配合共同得到的耐久的可持续发展的混凝土，不是只要有配合比就能生产的，而是由包括原材料控制、拌和物生产制备与整个施工过程来实现的。由于要求混凝土结构的耐久性，混凝土首先就必须是体积稳定的、匀质的；开裂了的当然就不能称之为“高性能”。可持续发展的含义则是最大限度地节省能源、资源、保护环境，并且耐久性本身也包含可持续发展的意义——减少修补和拆除的建筑垃圾和重建的能源、资源消耗。可以断定，除了个别临时性结构，所有混凝土结构都应当符合这样的质量目标。因此，为了避免误导，针对各个不同工程特点和需要，对混凝土提出满足具体要求的性能和耐久性的设计，是否比笼统强调“高性能混凝土”的名词要更科学一些，更能促进我国混凝土技术广泛的进步？

参考文献

[1] Oskar Beijer，Energy consumption related to concrete structures ACI Journal Nov. 1975.

[2] A. M. Neville著. 李国泮，马贞勇译. 混凝土的性能[M]. 北京：中国建筑工业出版社，1983.

[3] 363 Committee，ACI，State of the Arts Report on High Strength Concrete，Chapter 6，1994 .

[4] 吴中伟，廉慧珍. 高性能混凝土[M]. 北京：中国铁道出版社. 1999.

[5] P. K. Mehta. Ggreening of the concrete industry for sustainable Development. Concrete International. July 2000.

[6] M. R. H. Dunstan. Fly-ash，as the Fourth Composition of Concrete，Proceedings of Second Canmet International Conference on Fly Ash ， Silica fume，Slag and Natural Pozzolans in Concrete，1987.

[7] 深圳地铁一期工程科技成果鉴定文件：高性能混凝土在深圳地铁一期工程中应用的实验研究，2000.

[8] M. N. Haque，M. K. Goplan，Strength Development of Inadequately Cured High Fly Ash Content an Structural Concrete，Cement and Concrete Research，Vol. 16，1986.

[9] R. K. Dhir，M. R. Jones，PFA Concrete：Influence of Simulated in Situ Curing on Elasto-Plastic load Response，Magazine of Concrete Research，1993，45，No. 163.

[10] M. D. A. Thomas. Laboratory and field studies of salt scaling in fly ash concrete. Frost Resistance of Concrete. Proceedings of the International RILEM workshop. E & SPON 1997.

[11] Nicholas J. Carino and James R. Clifton ， High-Performance Concrete：Research Needs to Enhance its Use，Concrete International，September 1991.

[12] Henry G. Russell，ACI Defines High-Performance Concrete，Concrete International，February 1999.

后记

1. 该文发表后，覃维祖老师指出文中对高性能混凝土定义中的引文不确切。又查对了一下原文特此更正。这是我的疏忽，特此向编辑部及读者道歉！现更正如下：

1）原文 p. 13 第一栏，第 5 节“怎样理解高性能混凝土”中第 3 行参考文献编号应为[4，12]；同页第二栏第 4 行编号［11］参考文献遗漏。（本文第 5 节“怎样理解高性能混凝土”中的第一段）

2）原文 P. 13 倒数第三行“……希望能高强又耐久。”后面关于美国对高性能混凝土定义的叙述不尽确切，现做以下修正（本文第 5 节“怎样理解高性能混凝土”中的第一段）：

……希望能高强又耐久。1990 年由美国国家标准与技术研究所（NIST）和混凝土协会（ACI）主办了第一届高性能混凝土的讨论会，定义高性能混凝土为："靠传统的组分和普通的拌和、浇筑、养护的方法不可能制备出的具有所要求的各种性质和匀质性的混凝土。作为实例，这些性质可以包括如：易于浇筑和振捣而不离析、增强的长期力学性质、高的早期强度、高韧度、体积稳定性和在严酷环境中的长寿命"[11]。1999 年 2 月的"Concrete International"发表布了由美国 ACI 的技术活动委员会（TAC，Technic Alactivities Committee）提出的作为 ACI 对高性能混凝土的定义："高性能混凝土是符合特殊性能组合和匀质性要求的混凝土，采用传统的原材料和一般的拌和、浇筑与养护方法，未必总能生产出这种混凝土。"在此定义下面的注释为："高能混凝土是某种特性适合于特殊工程应用和环境的一种混凝土。对于一定的工程应用可考虑选择的特性实例，如：易于浇筑，振捣不离析，早强，长期力学性能，抗渗性、密实性，水化温升，韧性，体积稳定性，严酷环境下的较长寿命"。并指出，由于高性能混凝土许多特性是相关的，改变其中之一，通常造成其他一个或多个特性的改变；因此，对预定的工程，配制混凝土中必须要考虑的一些特性，必须在合同中逐条明确规定[12]。1999 年 ACI 的定义是 1990 年讨论会定义的发展，强调混凝土的工程针对性以及匀质性。定义中没有提到如何达到高性能混凝土，因为有许多途径达到定义中的最终目的[12]。

2. 参考文献补遗：

Nicholas J. Carino and James R. Clifton. High-Performance Concrete：Research Needs to Enhance its Use. Concrete International，September 1991.

Henry G. Russell. ACI Defines High-Performance Concrete. *Concrete International*，February 1999.

3. 鉴于编辑将我文中的掺和料和拌和物中的"和"字都改成了"合"。在此做以下说明：

据说目前有关标准规定用"掺合料"，这在汉语用词上是有误的。"和"在汉语中有五个读音：hé（如温和、和平、共和），hè（如一唱一和）、hú（如打麻将和了）、huó（将粉状物加水搅拌成糊状、膏状等物，如和面、搅和、拌和、和稀泥、掺和等；huò（量词，指换水的次数，如衣服已经洗了三和水了）。掺和料和拌和物的"和"的读音显然应当是第四个，有搅拌的意思，与混凝土的和易性的"和"字是一样的意思。而"合"字则是由部分结合成整体的意思，如结合、汇合、配合、混合、化合。掺和同拌和的"和"都不应写作"合"。

自评

1."误区之三：限制掺入粉煤灰"是针对论文发表时（2003 年）工程中的问题而言的。那时多数是因不敢用而限制，当前的倾向却是认为掺矿物掺和料的就是高性能。实际上掺矿物掺和料也未必就是高性能，任何材料都有其使用的范围和使用方法，使用不当则会适得其反；不同材料在不同场合使用有不同的利和弊；

2. 文中多处提到"粉煤灰在 28 天以前基本上不参与水化反应"，这个结论是在实验室的标准条件下得出的，而在工程中的混凝土在早期内部温度都是高于常温的。粉煤灰恰恰在温度较高的环境中可快速反应，论文之七的实验表明，掺 60%粉煤灰的混凝土的 7 天强度在 40℃时比 20℃时的几乎可提高 10MPa，但是比起硅酸盐水泥和矿渣粉，粉煤灰早期的反应速率还是很低的。该分析在原则上是对的。

3. 第4节的最后一段中有“在以碳化引起潜在钢筋锈蚀的一般环境中”，这样对一般环境的定义不确切，容易使人误解为只要碳化就引起钢筋锈蚀。实际上碳化和钢筋锈蚀是不同步的。一般环境应当是指没有冻融循环氯盐和其他化学侵蚀性介质，只有碳化和因空气和水分的作用引起钢筋锈蚀的环境。

4. 刘旭晨做了一项有意义的试验，但在本文4.2小节中对其试验结果没有叙述清楚，鉴于该试验针对“有人反映在施工中粉煤灰掺量大了的混凝土容易开裂”的问题，在此特作较详细的叙述。

表评1 试验用混凝土配合比

编号	粉煤灰掺量（%）	水胶比	水灰比
A	0	0.5	0.5
B	30	0.41	0.59
C	60	0.36	0.85

注：水胶比：水与胶凝材料总量的质量比。
水灰比：水和水泥的质量比。

试验方案：用三种粉煤灰掺量模拟平面型构件，制作面积为1000cm×200cm混凝土板各三块，各以三种方式养护，一昼夜后观察。配合比见表评1，试验方案见图评1。

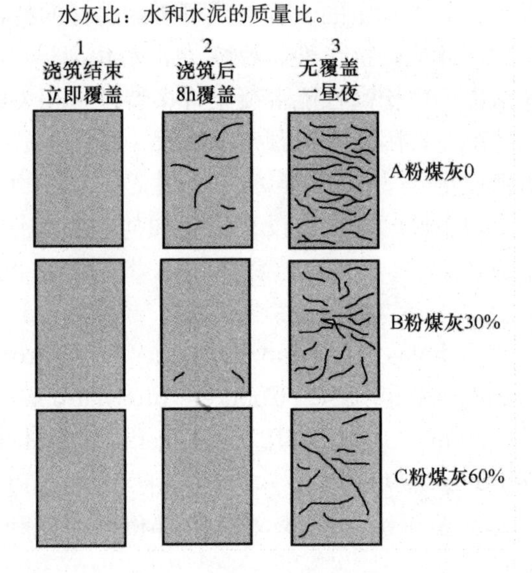

图评1 试验方案及结果示意图

试验结果及分析如下：

①浇筑后立即覆盖的都没有开裂。

②浇筑后暴露于空气中8h再覆盖的A-2有裂缝8条，粗而长；B-2有裂缝2条，细而短；C-2没有裂缝。原因是粉煤灰掺量60%的混凝土实际水灰比很大（见表评1），表面水蒸发后，内部有足够的水向表面迁移。掺粉煤灰少些的混凝土B-2实际水灰比相对较低，而且因掺粉煤灰会增大拌和物的表观黏度，泌水性差，覆盖前在塑性阶段已经产生了少量塑性收缩裂缝，8h后才覆盖已晚。

③一直没有覆盖的都有裂缝：A-3裂缝有24条，最长850mm；A-2裂缝有18条，最长600mm；A-3裂缝有11条，最长1150mm。在数量和尺寸上综合分析，三块开裂程度相当。原因是因在空气中暴露一昼夜后，粉煤灰掺量60%的混凝土C-3尚未终凝，表面持续失水收缩。但因已经初凝，拌和物稠度增大，失去向表面补充水分的能力。

该项试验表明，对如道路、场坪、筏式基础（大型基础底板）等结构部件浇筑后立即覆盖保湿养护的重要性。掺大掺量粉煤灰的混凝土在初凝前短时间内，可比不掺粉煤灰的混凝土有较高抗塑性收缩开裂的能力，但也要尽早覆盖，如果一直暴露于空气中，粉煤灰掺量越大，塑性开裂的可能性越大。

论文之十四

对“高性能混凝土”的再反思*

清华大学土木水利学院建材研究所　廉慧珍

摘　要　对于混凝土，性质(property)是在实验室中可用标准条件和方法检测出而量化的、反映混凝土本征特性的实验指标(并非真值)，如强度等力学性质；抗冻性、抗渗性等物理性质以及抗化学侵蚀等化学性质；性能(performance)是无法量化和检测的，而是只有用到具体工程中才能表现出的行为，如施工性、耐久性、匀质性等，对工程特点和服役环境具有很强的针对性。高性能混凝土的性能主要指的是后者，对前者则需根据工程需要予以选择。因此高性能混凝土不是混凝土的一个品种，不是只要有配合比就能生产出来的，而是由包括原材料控制、拌和物生产制备和整个施工过程来实现的，是由整个工程全部环节协调、配合而共同得到的、耐久的、可持续发展的混凝土。

关键词　性质；性能；高性能混凝土

前言

吴中伟先生是国内最早(1992 年 6 月)提到 HPC(High Performance Concrete)的人，当时他把 HPC 译成“高功能混凝土”，也有人译成“高性能混凝土”[1]。1993 年下半年，我和吴中伟先生讨论对 Performance 的翻译问题，认为“功能”和“性能”都不确切。Performance 一词在一般英汉词典中，被解释为“性能、表现、演出”等；在英汉科技词典中，又解释为“(运转)性能、(操作)效能、表演”等。实际上，该词含有操作和运行的表现和效果之意，是一种动态的现象。HPC 译成“高效能混凝土”比较贴切些，其含义是混凝土在施工、硬化及在所处环境条件下的使用过程等各阶段都满足技术要求的效果和表现，为此达成了一致意见。随后，陈肇元、覃维祖、陈恩义等几位老师和我一起，先后编辑了三册《高强与高效能混凝土译文集》，并发表了文章《高效能混凝土及其研究现状》(作者：陈恩义、廉慧珍)[1]、《高效能混凝土实现的可行性—国内外 HPC 材料研究现状及问题》(作者：廉慧珍)[2]；在 1994 年建工出版社出版的《混凝土工程新技术》一书中，收入了《高效能混凝土(HPC)》(作者：陈恩义、廉慧珍)[3]，并在全国性会议上作介绍。后来，有些工程技术人员提出，“高效能混凝土”叫起来觉得别扭，希望暂时叫‘高性能混凝土’。我们接受了这个意见，从《高强与高效能混凝土译文集》第四册开始，就改成了《高强与高性能混凝土译文集》。此后十几年来，人们在工程实践中，逐渐接受“高性能混凝土”这一术语，混凝土结构的耐久性要求也逐渐深入人心，反映出我国混凝土技术发生的变化。然而，任何事物有利必有弊，当注意到一种倾向时，还要注意到另一种被掩盖的倾向。虽然业内都承认“高性能混凝土”是“以耐久性为目标”代替“以强度为目标”的传统，但是在实践中，却出现一些误区，认为“高强”、“高流动性”、“掺用矿物掺和料”就是“高性能混凝

* 原载《混凝土世界》2010.6。

土”，造成在“高性能混凝土”使用上的盲目和混乱，在工程实践中不断出现很多与“高性能混凝土”的名称不相符的问题。例如有人发表的《高性能混凝土开裂问题的分析》、《高性能混凝土收缩开裂分析》等文章。由于不同国家文化的差异，有些词确实不好翻译，如实在无法找到与本国语言确切相符的词汇，可采取音译，例如日文就把“high performance”音译成ハィポルフォルマンス，而对混凝土，日本原来就从英文中音译成コンクリート。又如北京奥运会期间，曾对中文“加油!”的英译公开广泛讨论过，最后还是用了音译“jiayou!”而被外国观众接受并使用。科学技术是无止境的，不断认识，不断修正，才能“与时俱进”。2007年我在《混凝土》上发表《对“高性能混凝土”十年来推广应用的反思》以后，通过工程实践、思考、交流、讨论，感到正确的思维方法和观念确实比技术更重要。有人说，什么名词并不重要，关键是质量。这样说也不错，问题是如今“高性能”已被滥用了，被当成商业运作、申请项目时“忽悠”领导、某些企业标榜自己的“水平”和某些官员们显示政绩等的一种工具，“内涵”反而被忽视了，结果“高性能”成了包装的筐，凡是对自己有利的东西就往里装，某些人甚至还将其误认为是“创新”的代用词。这种状况的负面效应最直接的是造成国家大量财力、物力的浪费，因此有必要进一步再反思，也希望业界同仁也都反思一下，以纠正当前这种浮躁的风气，这对于混凝土科技的进步和工程质量的提高，将会产生积极意义。

1 对“performance”一词的理解

2008年北京奥运会期间，一次电视报道一位记者采访某个外国教练，问他：“你如何评价你的运动员?”他回答：“He has good performance.”2009年，我国残疾人艺术团到欧洲巡回演出，一次电视报道一位记者采访当地一位观众，问她：“你认为我们的演员怎么样?”她说：“Their performance is good。”目前我国对员工普遍采用的“绩效考核”的“绩效”就是从英文的“performance”译过来的。英汉辞海中的解释是：“进行或实行某事的行为或过程，某一行动的实现”，“要求，允诺，或请求的满足和履行”，“在戏剧中扮演角色的表演”，“执行的能力，取得预想结果的能力”……总之，与我国当前所述“性能”的涵义相去甚远。

在我国，传统教科书中基本上用的都是“性质”(properties)，包括力学性质、物理性质和化学，没有performance的概念。现在看来，性质应当是能够量化而在实验室的标准条件下用标准方法检测出的、反映混凝土本征特性的指标(但不是真值)，例如强度等力学性质，抗冻性、抗渗性、导热性、热胀冷缩等物理性质，抗碳化性、抗化学侵蚀等化学性质，等等。而“performance”则是无法量化的，是只有用到工程中才能表现出来的行为。例如耐久性，必须针对具体工程和所处环境。著名英国混凝土专家A. M. Neville撰文指出：“不存在一般含义上的耐久性……。混凝土在一种条件下耐久了，在另一些条件组合下就可能不耐久。……说某一混凝土是‘耐久的’或‘不耐久的’是错误的。”[4]又如施工性，在实验室中只能量化和检测出流动性，而施工性则是混凝土在输送、浇筑、振捣、抹面等过程中才能表现出的行为；再如匀质性，试件尺寸越小匀质性越好，检测出的强度值也越高，在实验室中成型的小试件是反映不出混凝土浇筑形成结构后的匀质性的。所以，如果理解正确，还叫高性能混凝土不是不可以，问题是如果当成一个混凝土的品种就大错而特错了。因为“高性能”不是仅靠搅拌站就能做出来的。高性能是一种质量，是要靠从工程业主到设计、施工，直到混凝土搅拌站都来共同承担的一种符合工程要求的质量。

从1993年我们开始介绍、宣传和推广“高性能混凝土”，到1996年吴中伟邀请我与他合著《高性能混凝土》，经过不断地研究、思考和实践，直到1999年该书正式出版。近年来不断地反思，感觉对“高性能混凝土”的认识应更趋向于科学的思考。10年前吴先生就提出：“中等强度的混凝土也应当能做到高性能”（现在看来，即使是低强度的混凝土，例如C20、C25，也应当而且能做到高性能），“‘高性能’显然是针对过去‘不高’或‘中’、‘低’的性能提出的。在不同时期，‘高’和‘低’的分界线是大不相同的。随着科技的飞速进步与要求的不断提高，今日的‘高’必然会被新的‘高’所代替而成为来日的‘低’。用‘高’、‘低’来命名科技产品，尤其是命名像混凝土这样的大宗材料，是不科学的与难以持久的。现在‘约定俗成’，待使用一段时期后，希望找到一个更科学、更合理的新名词”[5]。

2　对当前混凝土存在问题的原因分析

推广“高性能混凝土”以来，混凝土工程中出现较普遍的主要问题，一是硬化后开裂和拌和物泌浆造成结构表面出现砂线、麻面；二是对矿物掺和料，从不敢掺到近年来为降低成本而盲目地掺，造成与质检单位和施工单位的矛盾和纠纷。

在将一种新的概念引入伊始，认识可能有一定的误区是正常的。1994年陈恩义和我在《高效能混凝土（HPC）》一文中所介绍的法国、美国、加拿大、英国和日本一些有代表性的学者们的观点，尽管观点不尽相同，但有一个共识则是“HPC的高效能与高耐久性能具有相同的意义，而不是高强度”。也就是说，耐久性比强度更重要。但是除日本强调高流动性（不需要振捣）外，大多数仍看重“比普通混凝土高的强度”[3]。R. W. Burrows在他的《混凝土的可见与不可见裂缝》中，所列举的美国严重开裂的“高性能混凝土”示范桥[6]，实际上都是高强混凝土。因为美国在一开始时，多数人持“高性能即高强”的观点。直到今天人们仍在不停地追求混凝土强度的提高，于是误导了生产的水泥比表面积几近$400m^2/kg$，现在还有增大的趋势；熟料的C_3S达58%左右，有人曾追求60%～80%；这是混凝土开裂的根本原因之一。混凝土强度不适当的提高，尤其是早期强度的过分提高，正是“为什么过去很少见混凝土的开裂，现在却很普遍”的主要原因。Withy在1923年用比表面积为$231m^2/kg$、C_3S为30%的编号为7M的水泥浇筑混凝土，测试抗压强度，28天为21MPa，25年增长到52MPa，还有增长的趋势；1937年用比表面积为$380m^2/kg$、C_3S为57%的编号为I的水泥(和现在的常用水泥的相当，当时叫做快硬水泥)浇筑的混凝土，28天抗压强度是36MPa，10年以后强度开始倒缩，25年从10年时的54MPa倒缩到45MPa[6]。混凝土强度的提高使混凝土的应用范围大大增加，但是，因为要降低水灰比，又要保证流动性，必然要增大混凝土的浆骨比。就我国骨料质量的现状，现今C60混凝土的浆骨比一般大于38：62，超出了P. K. Mehta所给出的浆骨比35：65[7]的限值，该限值是平衡高强混凝土在强度、施工性和体识稳定性之间相互矛盾的需要，浆骨比是影响混凝土体积稳定性的主要因素。

尽管我们在《高性能混凝土》一书中强调“中等强度的混凝土也应当做到高性能”，还强调了“高流动性”。这实际上是一种“矫枉过正”的认识。当时(1994年前后)，北京市混凝土结构的“蜂窝、狗洞”现象很普遍，验收时抽查中“一抽一个准”。分析其原因，当时的建筑工程公司取消了固定工人的编制，只保留技术人员，对从农民中招收临时的一线工人进行技术管理。当时全国只有很少数的预拌混凝土，混凝土的流动性不大(多为坍落度 70mm～90mm)。由于振捣的劳动强度大，技能培训又不到位，欠振、漏振必然造成混凝土的缺陷。日本推行流态混凝土也是出于这个原因，而通过强调混凝土的流态杜绝了此类缺陷。现今我国的混凝土也因流动度大，已经做到了不出现蜂窝、狗洞等严重的缺陷，同时，工人的劳动强度降低，施工效率提高，建设速度加快。然而，凡事有利就有弊，我们当时的误区在于没有想到“过犹不及”的规律在拌和物流动性的问题上也存在着。现今的施工人员越来越不懂混凝土了，当今的混凝土水胶比比较低、掺用高效减水剂和磨细的矿物掺和料，拌和物黏聚性较大，对泵压的要求提高了，施工队为了降低成本，长期不更换、不标定混凝土泵，甚至还一直使用地泵，为了泵送省力，他们一味要求增大混凝土坍落度。混凝土运到工地，大多数施工队都在现场随意加水，甚至混凝土拌和物坍落度已超过 220mm，工地上还要加水。有一个包工队队长说“不加水太糨”，他们不知道现场加水不仅强度和密实性会下降，而且后加的水会造成混凝土匀质性的进一步降低。为了增大流动度，搅拌站不得不多加高效减水剂，而对于低水胶比的混凝土，拌和物的保水性对高效减水剂掺量非常敏感，再加上水泥太细，泌浆的现象就很容易发生。近年来，混凝土墙、柱表面的砂线、麻面出现得很多，成为新的问题。混凝土表面质量对混凝土结构耐久性有极重要的影响，不少工地对上述缺陷“一抹了之”，造成钢筋的混凝土保护层质量的隐患。此外，实验表明，坍落度太大时，浆骨比要较大，即用水量要大，外加剂用量也大，则释放出的自由水也多，这时很容易泌浆，而且混凝土温升也会有所提高，干缩也会大，开裂敏感性也就会增大。

3　反思的结果

1999 年 2 月的“Concrete International”发布了由美国 ACI 的技术活动委员会(TAC，Technical activities Committee)提出的作为 ACI 对高性能混凝土的定义为：“高性能混凝土是符合特殊性能组合和匀质性要求的混凝土，采用传统的原材料和一般的拌和、浇筑与养护方法，未必总能生产出这种混凝土”。在此定义下面的注释为：“高性能混凝土是某种特性适合于特殊工程应用和环境的一种混凝土。对于一定的工程应用，可考虑选择的特性如：施工性，早强，长期力学性能，抗渗性、密实性，水化温升，韧性，体积稳定性，严酷环境下的较长寿命等”。并指出，由于高性能混凝土许多特性是相关的，改变其中之一，通常造成其他一个或多个特性的改变，这也是混凝土的复杂之处；因此，该定义的说明还要求“对预定的工程，配制混凝土时必须考虑的特性，必须在合同中逐条明确规定”[8]。定义中没有提到如何达到高性能，因为有许多途径可达到定义中所述最终目的。该定义强调混凝土的工程针对性以及匀质性，这是对混凝土工程最重要的质量要求。

对“符合特殊性能组合”的要求不难理解，这就是对工程特点、所处环境和施工工艺等有很强的针对性，强度有不同的要求；也不一定都要求高流动性；掺和料的选择则和使用环境有关系。

匀质性是当前在我国最不被重视的问题。以强度为例，强度是保证结构承载安全的重要性质。在力学上，强度就是极限应力，即单位面积上能承受的最大荷载。对均匀的材料来说用受力面积去除破坏荷载值，就等于该材料的强度。而混凝土这种高度非匀质的材料，材料内部组成在空间分布不均匀，并存在许多不可见的原始缺陷，而且因为内部各处温度、湿度的差异（这种差异会因组成与配合比的不同而不同）造成处处强度不同。在实验室的条件下用标准方法检测出的抗压强度值，实际上并不是垂直于荷载方向无数平面上无数个点应力的平均值，混凝土最后破坏是由最薄弱区域的断裂引起的。对于结构设计来说，并不直接就使用该强度值，而是有个取值，即根据不同结构的功能引入不同的安全系数。构件尺寸越大，缺陷存在的几率越大，匀质性越重要。例如某重点工程浇筑混凝土拆模后，发现表面是“花的”：有一块、一块的墨绿色，如果没有注意到掺入矿渣粉后都会出现的这种暂时颜色现象，很难发现该构件的混凝土是否均匀。强度的真值无法测出，但如果没有混凝土内部温度的影响，可断定一定的混凝土在结构中的总体强度(测不出)会低于实验室检测出的同龄期标准强度值。工程中对构件混凝土匀质性的检测，可使用回弹法，但目前基本上还无验收匀质性的规定。

我们在一开始介绍时也曾把 HPC 当成了“一种很有发展前途的性能好而节能的新型混凝土”[3]，这是个误区，至今仍普遍存在。从上述 ACI 的定义和原文词意可知，高性能混凝土并不是一个混凝土的品种，而是强调混凝土的“性能”(performance)或者质量、状态、水平等表现，或者说是一种质量目标。对不同的工程，高性能混凝土有不同的强调重点(即“特殊性能组合”)。如果还用“高性能”的名词的话，则应当认识，高性能混凝土是整个工程全部环节协调、配合共同得到的耐久的可持续发展的混凝土，不是只要有配合比就能生产出来的，而是由包括原材料控制、拌和物生产制备与整个施工过程来实现的。由于要求混凝土结构的耐久性，混凝土首先就必须是体积稳定的、匀质的；开裂的当然就不能称之为“高性能”。可持续发展的含义则是最大限度地节省能源、资源、保护环境；混凝土结构耐久性本身也包含可持续发展的意义——减少修补和拆除的建筑垃圾和重建的能源、资源消耗。可以断定，除了个别临时性结构，所有混凝土结构都应当符合这样的质量目标。因此应当针对各个不同的具体工程的特点和需要，对混凝土进行满足具体要求的性能和耐久性的设计与施工，这才是根本。诚然，任何事物有利就有弊，有得必有失，但“高性能混凝土”既然不是一个混凝土的品种，当然就不存在利弊的问题。至于得失，是做任何事情都会遇到的，混凝土工程的复杂性需要具体问题具体分析并具体对待。

当注意到一种倾向时，必须同时注意另一种潜在的倾向，这样才能使我们对混凝土这种复杂体系的认识不断提高。

参考文献

[1] 陈恩义，廉慧珍．高效能混凝土及其研究现状[J]. 混凝土与水泥制品，1993.

[2] 廉慧珍．高效能混凝土实现的可行性——国内外 HPC 材料研究现状及问题[C]. 高强与高效能混凝土研究课题讨论会(黄山)，1994.

[3] 陈恩义，廉慧珍．高效能混凝土(HPC)[M]. 北京：中国建筑工业出版社，1994.

[4] A. M . Neville，Consideration of durability of concrete structures：Past，present，and future [J]. Materials and Structures，March 2001.

[5] 吴中伟，廉慧珍．高性能混凝土[M]. 北京：铁道出版社，1999.

[6] R. W. Burrows. The Visible and Invisible Cracking of Concrete. ACI Monograph No. 11.

[7] P. K. Mehta，P. C. Atcin，Basic principles of preparation for high performance concrete [J]. Cement，Concrete and Aggregate. 1990 winter.

[8] ACI's Definition of HPC [J]. Concrete International. Feb 1999.

对两次反思的自评

1. 反思了两次，是因为感到第一次的反思不彻底，在第一次反思中，指出当下对高性能混凝土的误区是认为高性能必须高强、高流动性、掺矿物掺和料。对其中的高强和高流动性的利弊基本上都做了分析，对矿物掺和料，则着重针对了当时对粉煤灰的限制做了阐述，却未能分析为什么掺矿物掺和料也不一定就高性能。现在普遍使用矿物掺和料主要是粉煤灰和矿渣粉，而实际上能用作矿物掺和料的材料很多，不同的矿物掺和料也都有各自的优缺点与在不同场合使用的利弊，都必须正确使用。这几年在使用矿物掺和料中出现的问题，实际上都源于使用不当。由于有是否正确使用和施工质量问题，对实验室试配后而由搅拌站生产出来的拌和物不能判定是否高性能。

所谓反思不彻底。根本问题是对“performance”这一词汇的理解得还不够，也没有明确承认自己也曾把“高性能混凝土”当成混凝土的一个品种（见论文之一及其反思以及吴中伟、廉慧珍著《高性能混凝土》第一章 第二节）。尤其是第一次反思，似乎认为误区都是别人的误解，而没有反思自己误导的责任。我们在一开始的宣传和推广中，也是强调高流动性和矿物掺和料的使用。在当时的情况下这样强调并没有错，只是不应和“高性能混凝土”挂钩，而且没有料到会有“矫枉过正”以致“物极必反”的另一种倾向。这类似于UEA膨胀剂在我国受到褒贬的命运：UEA膨胀剂20多年前推广应用伊始，全国年产量只有几百吨。因为是研制人员亲临现场指导使用，效果很好；随着的需要的扩大，而且表面上似乎生产技术难度不大，很快地，UEA的生产就“一哄而上”，产量飞速上升。结果，生产的人一知半解，使用的人更是不懂，使用中出现很多问题，纷纷反映“用了不如不用”，于是乎又“一哄而下”，这就是“物极必反”。当时一位膨胀剂的专家曾和深圳一位混凝土公司总工对话。

问：“张总啊，你认为UEA怎么样?”

答：“是个好东西，但是要看谁用?”

问：“谁用?”

答：“明白人!”

答得好！没有弄清楚就先用起来，必然会导致物极必反。现在的一位膨胀剂专家用了7年的时间，在对传统各种膨胀剂的机理潜心研究之后，提出了一种新概念的膨胀剂，可以不依赖于外部水而产生膨胀，可以使膨胀不会在后期倒缩。现在他还在继续更深入的研究。经过又一次否定之否定，我们或许会得到一种不开裂的水泥。历史的发展总是会因为人们对世界认识的不断更新而遵循着否定之否定的规律。

当然，任何事物在一个时期针对一定的问题有所强调，这也很正常。从第一次反思到再反思是一个再认识的过程。

2. 对客观事物或者现象的认识“知其然”后，必须还要“知其所以然”，才能继续前进；因当初“不知其所以然”而在实践中产生误区，出现问题甚至错误时，又必须修正，才能继续发展。学习最重要的是思考，而反思就是思考的一种方法，符合否定之否定的规律，是人类自我能动地认识自然界的规律。害怕批评、害怕反思是一般人害怕否定自己的一种心态。其实否定之否定不是为了否定，恰恰是进入一种新的肯定的过程。思维方法和观念的转变比技术更重要。

3.20年来对“高性能混凝土”的推广应用，确实也推动了我国混凝土技术的进步，但是因为这个理念是从国外引进的，中外文化的差异使我们对原文的理解和翻译不恰当，这是产生误区的根源。中国的术语“混凝土”原来是从日本移植过来的（日文写成汉字“混凝土”，音读こんこりど，与现在的音译片假名读音很相似）。顺便一说，我国有不少词汇来自日本，例如干部、支部、企业等。从中文来看，“混凝土”这个名词创造得不错，形象而贴切，只是不知为何后来改成了片音译的假名コンクリート。朋友赵筠告诉我，他的朋友张金喜教授曾在日本留学，有一次请他的日本导师来华交流。那位日本专家到中国后感到很惊讶：居然中国还在用“混凝土”的词汇！日本年轻人都已不知道这个词汇是什么意思了。日本是一个在科学上理智的国家，对外来语都用音译。想必他们考虑到毕竟由于东西方文化的差异，意译很容易发生不确切而误解。当初他们对 HPC 是用音译，现在连这个音译的名词都已不再提及。可能也因为悟到 HPC 不是混凝土的一个品种吧？近年来我国有人把国外的 UHPC 介绍到中国，译成“超高性能混凝土”，而赵筠看到日本就总体的力学性能而言，这一假定也适用于混凝土材料。有关的文章是称之为“超高强纤维增强混凝土”(Ultra High Strength Fiber Reinforced Concrete，简称 UFC)，这倒是对国外论文理解后的一种意译。尽管并未反映出其涵义，亦即其性能和以该性能为目标的技术路线、配制、生产的特点等，至少可以作为一个品种，能反映其本质上的特点。这种翻译比较恰当，举个例子：“硫铝酸盐水泥”用的就是水泥的一种命名法，也不能从术语上反应其内涵，只表明它是一个水泥的品种。

4. 在写《再反思》的时候，我查阅和思考过原文辞书的释义。明白了 property 和 performance 的区别。近来又查阅了中国的《现代汉语词典》对“性质”和“性能”的解释如下：

性质：一种事物区别于其他事物的根本属性；

性能：机械或其他工业制品对设计要求的满足程度。

由以上的解释可见，我在《再反思》中的理解是对的：性质，反应本征特性。对不同混凝土来说，是固有的，无所谓好的与不好的，可以在实验室的标准条件下检测出量化的指标而供选择；性能是满足设计要求的程度，所以有好和不好的区分，难以在实验室量化，只有用了才知道。因此，高性能混凝土不是配出来的，甚至成型为构件后，在经历时间和环境的考验之前，不能断定其好坏，当然就不是混凝土的一个品种。

5. 为了避免一开始就对“超高性能混凝土（UHPC)”产生误区，我邀请赵筠撰写一篇有关的综述，我们用了大量时间，几次反复研究讨论，最后认为，这完全不同于传统的高强混凝土和钢筋混凝土，也不同于传统的钢纤维增强混凝土，更不是原来的“高性能混凝土”的“高强化”。其主要特点是可得到应变硬化的性质，以提高构件的延性比。钢筋混凝土或钢骨混凝土* 是在混凝土浇筑成构件后制成的，而 UHPC 则是在混凝土配制时形成的，恰恰是一个混凝土材料的新品种。UHPC 将来有望替代或部分替代钢筋，成为钢-混凝土共同工作的又一种新的模式。我想，现在正在研究和推广应用这种新材料的人们，可能还未必有我们现在这种感悟。希望这篇将在《混凝土世界》连载四期近六万字的详细综述，能尽量对避免我们可能进入一个新的误区起到一种预防的作用。

---

* “钢骨混凝土”，又称“劲型混凝土”，即以型钢或钢筒（圆形或方形）做骨架，外包混凝土保护，一般多用于高层建筑的柱子。

6. 近来有些人纠结于要不要还是用“高性能混凝土”这个术语。其实我并不固执地一味要求“取消”这个词，这只是个人意见，现在我不再用，别人谁愿意用并没有什么限制。只要正确理解，不再进入误区，叫个什么名字，似乎也不相干。只是正像 A. Neville 在他的第四版《Properties of Concrete》中所说：“高性能的并不是一种创新材料”，“‘高性能’一词带有明显的广告色彩。原来称为‘高强混凝土’，但很多情况下，要求的是高耐久性，……”(刘树华、冷发光、李新宇、陈霞译；《混凝土的性能》)。因为不是在实验室里能配出来的，也不是搅拌站能生产出来的，一听见这个名词，就感到像是一种“忽悠”。在国外，并没有什么“取消”一说，大多数已不再用是一种认识的自然结果，没有必要讨论。

人类需要善于实践的人，这种人能由他们的工作取得最大利益；但是人类也需要梦想者，这种人醉心于一种事业的大公无私的发展，因而不能注意自身的物质利益。

——居里夫人

维持一个人的生命的事物，是他的事业。

—— 爱默生

第四部分　思维方法与观念的转变比技术更重要

在我退休后的二十年间参与工程建设、项目论证、事故诊断、技术咨询等实践中，不断地使我想起学生时代学过的哲学，越来越感到辩证法的规律在工程中随处都有体现，使我的收获远远大于我退休前知识的总和。在工程中，各种因素之间几乎没有单调增减的关系（即使有，也只在一个范围里），大都存在一个极值，即“物极必反”。看到工程中充满对立统一的规律，能创造条件使对立的因素得到统一；掌握事物的量变和质变的关系，做到适度；善于利用否定之否定的规律，使那些过去因条件不具备而中止了的技术又得到新的发展……，这就是方法问题。方法比技术更重要。这不是说技术不重要，技术是“技”和“术”，任何事物，无技术就无法实施；但方法更重要，方法是“道”，是指导技术和决策的方向和原则。

方法通常是指为完成某事的途径、步骤、手段。广义地说，是为解决问题的经常性的和系统性的、详细的、逻辑有序的计划和行为。方法有不同层次。思维方法是最高层次的方法，指的是对客观事物的反映和分析、概括、认识、判断的方法，决定人的一切外在行为表现和对一切外在行为的认知深度。观念是思维活动的结果，可以简单化地说是“看法”，是人们在自己的实践中对客观世界的一种结论性的认知。正确的技术决策取决于正确的观念，正确的观念取决于正确的思维方法，树立正确的思维方法才能不断更新知识，不断更新知识才会有真正的技术进步。可以简单地概括思维方法、观念和技术，即想法、看法和做法。

人们常常因已有知识和自己的经验而形成思维定势，主要表现在传统思维、从众思维、机械思维。我们每个人几乎都或多或少地会产生过这样的思维定势。思维定势就是不思变的思维方法，是科学思维方法的对立，阻碍我们与时俱进。

北宋时期王安石变法受到保守派的反对和中伤。有一次宋神宗对王安石说：“外边的人说，现在朝廷以为天变不足惧，人言不足恤，祖宗之法不足守，这是什么道理?”王安石说：“陛下自己管理国事，没有流连享乐，这就是惧天变。陛下征询、采纳人们的意见，岂是不恤人言？然而人言固有不足恤者，如果合于义理，人言又何足恤？以人言为不足恤，并不是错误。至于祖宗之法不足守，本来就应当如此。仁宗在位四十年，数次修敕（宋朝新定的法令）。如果法一定，子孙就当世世遵守，祖宗为什么还屡次变改?”这“三不足”说正好概括了王安石变法的思想和气魄（蔡美彪，朱瑞熙，李湖，卞孝萱，王会安著《中国通史》第五册，第158页，人民出版社，1978年出版，1993年再版）。转变思维定势才能真正地实行改革。畏惧天变、恤人言、守祖宗之法也恰好是当下人们思维定势：

其一，传统思维，或称习惯性思维，是用曾经成功的经验证明来思考问题，只相信自己所熟知的和自己的经验，“别人没用过的我不用”，“我一辈子都这样干的”……。例如楼盖得越高越好，企业做得越大越好……；又例如对强度的盲目追求，认为强度高就是高质量，就是优质的；快就是水平高的；等等。这种实例是很多的。总之，已经形成的习惯不能改变。例如在建筑材料的教学和工程的使用中，对材料含水率的检测和计算都是用干基，即将试样烘干前后的质量差与材料烘干后质量之比作为材料含水率。这是在物理概念上和逻辑上的错误。但是人们都说“规范就是这样规定的，这么多年一直都这样做，不能改。”这就是定势。世界是发展的，人对世界的认识是前进的，存在的不一定都合理。曾经被认为合理的，现在也不一定合理。“熟知并非真知”（黑格尔《精神现象学》）；“经验是一条我们曾经沿着它到达今天的道路，可是世界在昨天早晨变了，经验还会是通往明天的阶梯吗?”（王宏甲《中国新教育风暴》）。

举个例子来说转变传统思维对创新的意义：

“两点决定一条直线，过直线外任一点有一条，而且只有一条直线与该直线平行。”这是传统欧氏几何学中的5个公理之一 —— 平行公理，2000多年没有人怀疑过。当很多学者试图证明但都没有成功时，1815年，23岁的俄国数学家罗巴切夫斯基大胆地提出质疑：“这个公理是否可能并不成立?”他创造性地提出一个假设：“过直线外的任一点，至少有两条直线与该直线不相交”，以此代替平行公理，结合其他4个公理进行推理，推出一系列完全悖于欧氏几何的结论，他称之为“新几何”。罗巴切夫斯基对自己的理论坚持了30年，不仅所有俄国数学家都反对和嘲讽他，而且一位英国著名数学家在没有认真阅读原为俄文的新几何论文时，就武断地说：“我认为，任何时候也不会存在与欧几里得几何本质上不同的另外一种几何。”只有高斯理解“新几何”，把它称作“非欧几何”，但在一片反对和指责声中，他也不敢公开支持罗巴切夫斯基。直到罗巴切夫斯基死后12年（1865年），意大利数学家贝尔特拉米找到适用于罗氏非欧几何的曲面，使数学家们理解了非欧几何的存在。如今非欧几何得到很大的发展，除了贝尔特拉米的曲面几何外，还有射影几何、拓扑几何、分形几何以及爱因斯坦用于创立广义相对论的黎曼几何，等等，也大大促进了其他学科的发展。我们长达2000多年坚持欧氏平面几何，是因为我们涉及的几何问题常常是在较小的空间范围内，将其看作平面处理，误差小到难以察觉，于是我们就习以为常地认为欧氏几何是唯一正确的。但如果在在地球上画一个边长几百公里的“三角形”，其内角之和就不等于180°。如果不转变传统思维，我们不仅离不开地球，连地图也无法画出来。

其二，从众思维，即人云亦云，没有自主的分析和判断，亦步亦趋地跟进，或一知半解，想当然而误判。一哄而上，又一哄而下。例如对膨胀剂的褒贬、对聚丙烯纤维的迷信、对待工程质量责任问题，等等。有的则盲目攀比，赶时髦，跟潮流，认为“外国的就先进”，“大家都这样做就没错”。从众思维即“跟风”，国内外都普遍存在，而在我国尤甚。盲目必盲从，盲从必会付出代价，这不仅会放慢前进脚步，而且会造成很大的损失。

其三，机械思维，表现在对上级指示、领导意图、规范，知其然而不知其所以然，不管对错，不管本地实际和工程的具体情况，一律照章办事，照本宣科；认为标准和规范都是法律，报上和书本上写的都正确，唯上、唯本本是从，只要是有条文规定的、领导指示的，明知不对也照办不误。这种思维方法主要是由于我国历史上长期的等级制度不容独立思考和求异而形成的。人们形成这样的思维定势，主要是出于对自身利益的保护和对责任担当的回避。

上述三种思维定势没有绝对的界线和区分，其相互之间是有关连的，都是受僵化的、不冒风险的意识所支配而造成的。思维方法的定势具有后天获得性。人之初的天性具有强烈的好奇心、求知欲和想象力。人在成长过程中受到环境的影响，譬如包办、呵护过度，或者遭漠视，受训斥和压制，会使他们失去自由探索和发展的空间而逐渐泯灭了天性，形成依赖的惯性。这种惯性是可以改变的。首先是要认识到转变是必要的。有的学生在听我讲《科学研究方法》的课后对我说："听你讲了以后，我才知道，原来书上写的和杂志上登的并不一定都对呀!"这就是转变思维定时的开始。"什么?"，"为什么?"这是我们小时候最多的简单提问。长大后没有问题了，谁说了都信，就成了思维定势。现在应当先问自己，先思考，反复思考。已故教育家蒋南翔去世前留给年轻人的最后嘱托就是"脑要用，腿要动。"说的不仅是养生之道，也是一种思维方法。用脑是思考，动腿是调研、实践。

科学的思维方法是辩证思维、创新思维。转变思维定势为科学思维，要相信存在的不一定都合理，外国的不一定都先进，专家和权威不一定都正确。凡事不要盲目接受，都要问一问"为什么?"养成对任何事物都从两个方面去思考的习惯，"当注意到一种倾向时，必须还要注意另一种被掩盖了的倾向"，就会促进我们不断地去求索真理。不是为了突破而突破，而是产生问题才需要突破，突破才会有发展。但是怀疑要有根据，不能是怀疑主义。而是"对任何真理或知识，都只有经过自己头脑的独立思考、批判检验之后，才加以肯定或否定，决定取舍"，"在承认客观性和客观世界可认识性的前提下，强调认识的相对性。科学的现状是不可能有终极意义的"（《爱因斯坦文集》第一卷）。因此思维方法也是一种能力。建立科学思维方法的前提是承认客观世界是变化的、互相联系的，人对客观世界的认识是发展的。

合作的前提是相互需要，专长互补。合作的保障是相互尊重，相互信任。美国人亨利·福特曾说："To come together is a start; To stay together is an advancement; To work together is a success."香港中文大学吴奇院士很好地把它意译成："相聚是缘，同行是乐，共事是福。"

摘自清华大学化学系教授、张希院士2014年9月3日为清华2014级研究生新生所作的"开学第一课"

一、关于思维方法和观念

论文之一

水泥基复合材料科学研究中的辩证思维*

吴中伟　廉慧珍　（清华大学土木系）

我在1999年10月25日庆祝吴中伟教授从事科教工作60周年的科学讨论会上介绍了吴先生的学术思想。今年元月吴先生在病中建议将该文修改发表，并同意与我共同署名。吴先生于2000年2月4日不幸作古。哀悼之余，遵照吴先生遗愿完成此文，以资纪念。

——廉慧珍

以水泥作为基材复合而成的材料称为水泥基复合材料，例如各种混凝土、砂浆、钢丝网水泥、玻璃纤维增强水泥，等等。其中用量最大的是各种混凝土，尤其是以砂石为骨料的普通混凝土。与钢材、木材、塑料、砖瓦等常用建筑材料相比，混凝土尽管有单位质量大、导热系数大、抗拉强度低等缺点，其优点却是无可比拟的：原材料来源广泛，可满足宽范围强度等级的要求，可成型为任意形状，适应不同条件下的施工；从耐久性来说，比钢、木和塑料耐火、不生锈、不腐朽、不老化；从原材料的开采、加工，直到最后用于建筑物或构筑物，其全过程所消耗的成本是最少的。因此混凝土作为大宗的建筑材料而被广泛使用。但是由于生产量和用量庞大，混凝土长期以来是资源、能源消耗和环境污染的“大户”。生产和使用中劳动强度大，质量波动和随意性较大，常被认为是“低技术”。为了把混凝土从经验技术转变成由材料科学和力学基础的科学技术，从而有效地发挥其效用，吴中伟在40年代末提出了“混凝土科学技术”的概念。与其他各类学科一样，混凝土科学技术中充满了辩证法，混凝土科学技术研究也必须由符合辩证唯物主义的学术思想来指导。正如钱学森先生所主张的“对各种事物进行从定性到定量分析与综合集成来较准确地把握复杂事物的现象与本质、微观与宏观、部分与整体、固定与发展的辩证联系，有所前进，有所创新”。

1　水泥基材料科学研究的整体论与分解论的结合

科学研究的分解论，或称还原论，即将科研对象分解成小单位，进行量化研究。其优点是可排除复杂环境的干扰，进行本质的研究，缺点是容易脱离整体。整体论在早期则常依靠直觉和经验认识问题，缺少精确性。钱学森的“综合集成”思想主张从宏观上把握事物的整体，运用分解论对事物的系统、层次、结构、功能进行具体深入量化研究，以避免整体论的模糊性[2]。

长期以来，水泥、混凝土、钢筋混凝土构件和工程结构的研究被划分属于不同学科，在不同领域的不同层次（level）上或对不同尺度（scale）的对象进行。一般，水泥的研究属于

---

* 原载《混凝土》，2000年第4期。

微观和亚微观（或称细观）层次，主要用化学的和物理化学的方法进行研究，如研究熟料烧成和水泥的水化；混凝土的研究主要在“宏观”（粗观）层次上用物理的和力学的方法进行；钢筋混凝土构件的研究则完全在宏观层次上用力学方法进行。例如，过去一直认为混凝土中骨料是惰性的，水泥是活性的，为了研究水泥的水化，大量研究工作着重于水泥熟料单矿物的水化及其主要的水化物 —— C-S-H 相和钙矾石相。已有的研究成果对工程的技术进步起过很大的作用。但实际上，水泥，特别是混凝土，是一种复杂的、非均质的多相体，水泥浆体、混凝土，以至钢筋混凝土构件的行为都不能用其中各组分单个行为的简单叠加来表征。换句话说，不同层次上研究的结果都不能简单地外推。例如图 1 所示，在相同水灰比（水胶比）下，砂浆和混凝土中掺入硅灰后的表现是不同的。混凝土的行为不仅取决于水泥单独的性质和行为，这是由于现在众所周知的界面的存在所引起的。如果把不同层次各自研究的结论简单地放在一起，就会发生各性质及各因素之间的相互制约或矛盾。

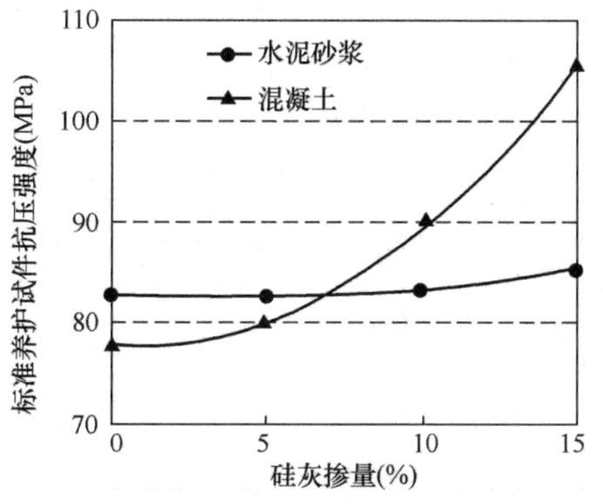

图 1　相同条件下硅灰对水泥砂浆和混凝土强度的不同影响

反过来，如果没有细观层次的研究，或者不了解细观层次的规律，则混凝土技术就不能脱离经验性的束缚。过去，在各层次研究分离的情况下，本来应是整体的水泥-混凝土-工程结构，由于分属不同领域，造成其工程技术人员“隔行”。在当前进入高性能混凝土时代之际，就会出现矛盾而影响工程质量和混凝土科技的进步。例如，随着建设的发展，工程结构设计人员要求提高混凝土强度，提高混凝土的强度则要求水泥提高标号（为了工程的进度和经济的缘故还要求提高水泥早期强度）；而不懂混凝土的结构设计人员并不知道混凝土的强度是怎样提高的，混凝土的生产者也不知道水泥标号（或早期强度）的提高是采取了什么措施，而这种措施会反过来对混凝土的其他性质和工程有什么影响。水泥的生产者并不了解混凝土技术的进展，不知道水泥的性质如何与混凝土技术相适应。结果就发生使用外加剂的混凝土流变性能的问题、大体积混凝土的温度应力问题、收缩问题、混凝土的长期性能问题，等等。水泥最终要体现到工程上去，向工程渗透，才能符合工程实际；反之，结构工程设计和施工技术人员必须深化水泥/混凝土的知识，才能知道如何对水泥提出全面而正确的要求。

用“中心质效应假说”[4]可把水泥基复合材料的不同层次联系起来。见图 2 所示。

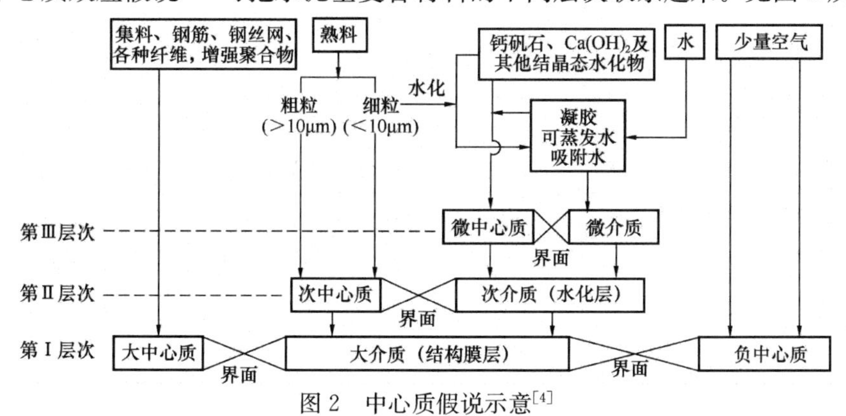

图 2　中心质假说示意[4]

在图2中，水泥基复合材料的每一层次包容了下一个层次，各级中心质是分散相，分散在介质（连续相）中，形成上一级的介质。各层次之间通过界面联系成整体。各层次的行为是相互影响的，故需对各层次的行为进行研究，但是下一层次的行为是在上一层次中产生的，与其独立的行为不尽相同。例如，混凝土的行为受水泥行为影响，但水泥的行为是在混凝土中产生的，因此应当把水泥放在混凝土中进行研究，把混凝土放在钢筋混凝土构件中研究，钢筋混凝土构件应当放在工程中研究。各领域、各层次的研究相互渗透、吸收、干预，各层次的研究方法互相借鉴、交叉，从系统论的观点进行研究，使各种因素得到对立的统一。

因此，水泥基材料的研究应当从宏观到微观，从整体到局部，注意组分（局部）之间的相互作用与联系。在进行局部研究与分解研究时，不忽视整体，而以提高整体性能为最终目的；必须联系实际，符合实际条件，试样的制备、实验条件与方法必须力求接近实际[5]。应用科学的目的是促进社会生产和经济的发展，因此目前水泥基材料的研究应在宏观、粗观、细观三个层次（level），或尺度（scale）进行，以宏观为主，再向微观深入的思想[2]。其中，nm尺度是微观层次；μm尺度为细观（submicro，或亚微观）层次，研究熟料的矿物及水泥浆体的组成、结构（如水化物、孔、界面过渡区）、物化性能；>mm的尺度的为粗观层次（macro，吴中伟认为原来称作宏观不妥），研究材料的组分、配比、物理性质与行为；而过去未被重视而不列入科研范围的混凝土及其配筋构件的使用环境、经济、能耗、生产工艺、应用场合等（可统称为环境，circumstances）的研究，则为宏观（grand view）层次。这就是说，不仅要强调水泥基材料本身各层次研究和各组分生产的整体性，而且还要把这个整体放到宏观环境中进行研究。以混凝土耐久性的研究为例：混凝土耐久性研究的高度复杂性在于原材料的复杂多样，混凝土本身组成结构的复杂与不断发展变化、环境条件的多样性与综合作用[2]，如果单从水泥化学来研究，就容易强调某单一因素而产生片面的结论。研究混凝土的耐久性是为了保证钢筋混凝土结构长期使用的安全性，必须与工程密切结合，从结构在使用状态下和使用环境中的行为来研究混凝土的耐久性。

基于宏观-细观的系统观点，高性能混凝土应当成为“绿色混凝土”[7]才符合可持续发展的战略。“绿色”的涵义为：①节约资源、能源；②不破坏环境，更应有利于环境；③既满足当代人的需求，又不危及后代人满足其需要的能力。“绿色混凝土”，无疑应是下一世纪混凝土科技的发展方向。

2 认识本质，因势利导

一切材料的用途都不是万能的，都有其合适的应用场合和范围。有利于某种场合的材料，对另外的场合也可能有害。一般来说总是要扬长避短，因时、因地、因事制宜。但是有利作用和有害作用在不同的条件下是可以转化的，把握事物的本质，因势利导，就能实现这种转化。在科技发展史上，有很多科学家以这种变化的观点、发展的观点治学，通过变害为利的创造发明为人类造福。膨胀剂的发明是因势利导、化害为利的一个范例[7]。许多混凝土工程破坏是因产生膨胀所致，因此膨胀被认为是混凝土的一大祸害。1892年W. Michaelis发现，波特兰水泥混凝土受硫酸盐侵蚀而膨胀破坏的主要原因是生成钙矾石，故称之为“水泥杆菌”，并在试验室中合成了该矿物[8]。此后，不少人对这种有害的“水泥杆菌”的组成、结构、特性及其生成并膨胀的条件进行研究，创造条件使其在混凝土的早龄期有控制地产生膨胀，补偿混凝土的收缩，或产生自应力，使混凝土密实，减免了裂缝的生成。这样，就产生了膨胀水泥、膨胀剂。利用大的膨胀能造成的破坏力研制成无声爆破剂，又是一项因势利导的创新。化害为利的转化是一种创新，是基于对“为害”的本质的认识，因此必须首先从宏观到微观，研究“利”和“害”产生的机理和转化的条件。

又如，在早年的教科书中一直认为在混凝土中水泥是活性的，骨料是惰性的，其间的界面是个薄弱的面，甚至无法改变。碱-骨料反应启发了人们，骨料和水泥浆体之间的关系不是不变的。用水泥熟料代替石子制作混凝土，抗压强度达到了126MPa，表明含有活性组分的骨料对混凝土的作用可以是有害的，也可以是有利的。

1972年法国G. Farran提出界面过渡层的概念，把对界面的认识从“面”发展到“层”。而“中心质假说”则认为各级中心质和介质之间存在一个比过渡层厚度大的“效应圈”。按“过渡层”的观点，认为骨料和浆体过渡层结构疏松，对混凝土性能不利，要提高混凝土的性能，应当尽量减小其厚度；按“效应圈”的观点，则认为界面的效应有不利的也有有利的，不利的效应会引起混凝土破坏，有利的效应可以改善或提高混凝土的匀质性和变形性能。用钢丝砂浆外圆锥形试件研究钢丝网水泥砂浆中的大中心质效应，经试验得出钢丝有效效应范围，指导了钢丝网水泥砂浆制品的研制和设计。

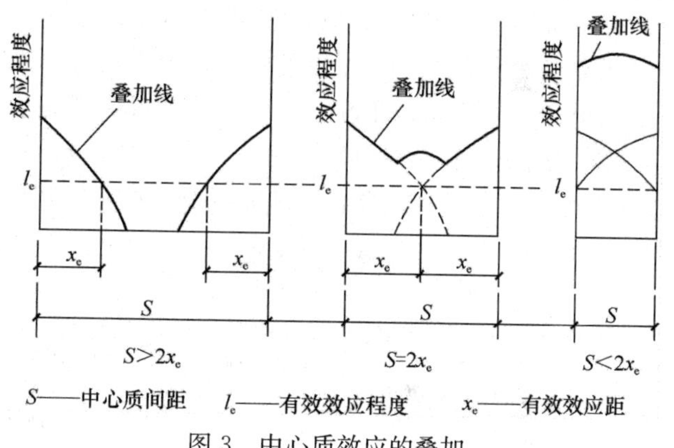

图3　中心质效应的叠加

中心质效应是可以叠加的，如图3所示。以往对水泥水化的研究认为，水泥水化程度越高，混凝土的性能越好，按“中心质效应假说”，未水化的熟料颗粒（次中心质）比水化物颗粒的强度高、体积稳定性好，其有利的效应叠加所形成的中心质网络有利于混凝土的匀质性，因此在高强混凝土中，虽然水泥水化程度较低，但低水灰比造成的低孔隙率和未水化熟料颗粒间距的减小，增加了中心质效应的程度，提高了混凝土的匀质性。活性细掺料如硅粉、矿渣、粉煤灰等的掺入，可增加中心质数量进一步减小中心质间距，提高效应程度，从而提高混凝土的匀质性。

对孔的研究是因势利导的又一例。传统认为孔在混凝土中是有害的。“中心质效应假说”中把孔也称作“中心质”，其涵义是把孔缝看作是混凝土中不仅必然也是必需的组分。冠以“负”字是为了区别于实体颗粒，而其作用则不仅是负的，也有正作用。例如引气剂对混凝土抗冻性的改善。有的孔只有正作用，例如孔径小到一定程度的孔。吴中伟提出“孔级配”的概念，并根据孔的不同作用对孔进行分类（见图4）。提出提高混凝土性能的数学模型[9]。

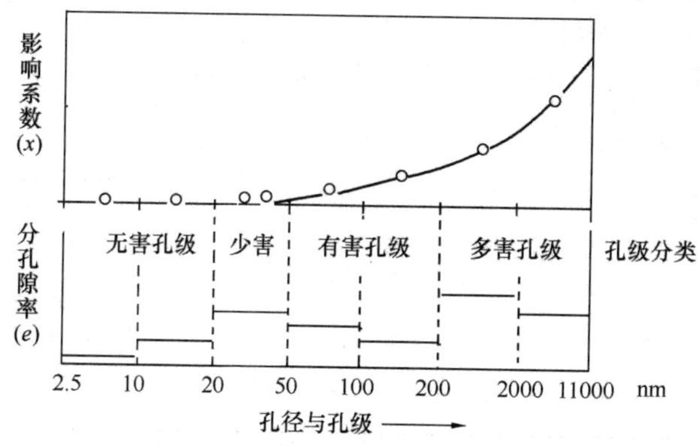

图4　孔分级、分孔隙率与影响系数的关系

优质混凝土应当有尽量小的单位质量和尽量高的性能，其与孔级配的关系为：

$$\begin{cases}\sum e_i \cdot X_i = \min & \text{性能最好} \\ \sum e_i = \max & \text{单位质量最小}\end{cases}$$

式中 e_i 为第 i 级孔的孔隙率，X_i 为第 i 级孔的影响系数。

按此数学模型进行混凝土材料的设计，可尽量多地使孔由有害转化为有利。

3 继承和创新的关系

科研创新是促进经济增长的主要途径，今后要讲究集约型经济增长方式，现在的标准是依靠科技进步的贡献率超过 50%，因此科研创新应成为不断努力的方向[2]。科技的发展总是从实践开始的。实践中提出问题，通过试验研究上升到理论，反过来又指导实践。但是每一次创新则必然来源于继承，因此必须十分重视信息。不仅要重视当前最新的信息，而且应善于掌握本领域已取得的成果，避免在同一水平上重复，这样才能早、快、新。

继承是为了创新。各学科都非常重视科技发展史的研究，就是要研究科技进步的规律，以便更好地继承和创新。科技发展和社会生产的发展相辅相成，各学科发展之间也相辅相成，因此往往是否定之否定，不断前进。例如“绿色混凝土”、环保型胶凝材料，就是从古罗马时代以石灰-火山灰为胶凝材料的混凝土——波特兰水泥——掺矿物细掺料的水泥——掺用细掺料的高性能混凝土——高掺量粉煤灰混凝土的否定之否定。其发展过程可如以下描述：

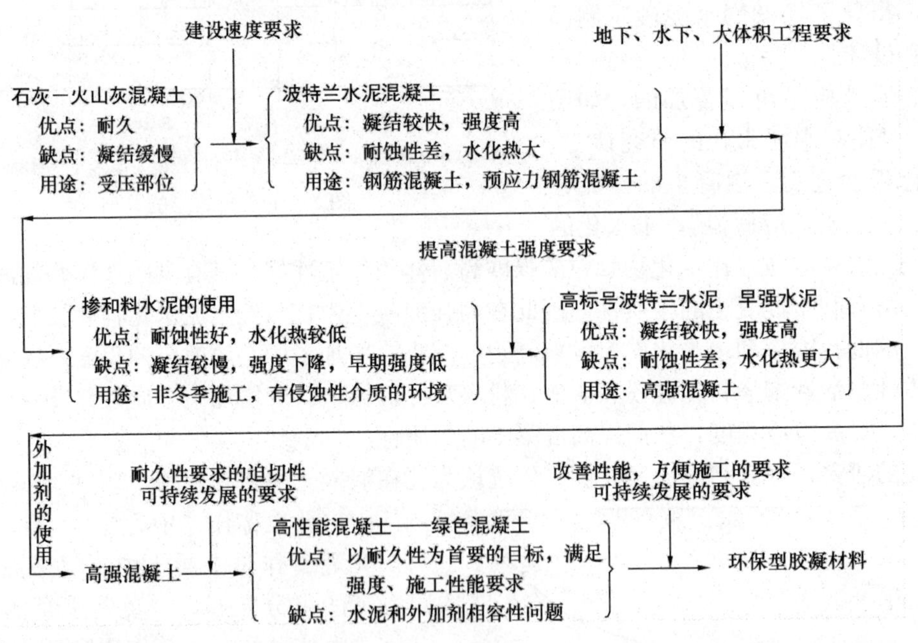

环保型胶凝材料可以用优质细掺料取代 50%以上的硅酸盐水泥[10]，就是借鉴了历经 2000 年之久至今还完好能用的古罗马石灰-火山灰混凝土的经验，结合当代混凝土技术特点和工业生产条件，根据可持续发展的战略的创新。

只有认真研究前人的工作，才能继承和发展。创造来自于勤奋，要勤读书，勤思考，勤动手实践。读书不仅要溯源，而且要善于“挑剔”，也就是说要批判地继承。研究工作要着眼于解决问题，因此读书不是“从文献缝里找题目”，必须密切联系实际的需要、实际的条件，抓住疑点和“闪光点”，把握创新的时机，不断在应用中积累和再创造[6]。

创新不是盲目发展，不是臆造，必须符合客观规律，实事求是。例如，纤维增强混凝土是水泥基复合材料的很好的发展，但由于弹性模量的差异，就不可能用玻璃纤维代钢制作水泥和混凝土构件。又如，补偿收缩混凝土的研究使我国在自应力管中的应用达到了世界领先水平，但由于混凝土技术的发展和工程条件的变化，必须继续对膨胀剂作用的条件进行试验研究；由于试验室对膨胀剂膨胀性质的检测与工程实际差别较大，不要盲目过度加长设缝间距，对不设后浇带的做法应当慎重采用。

创新不可避免会有一定的风险，但必须有充分科学依据。任何成功的技术都不能绝对地肯定，盲目地延伸；人类对客观事物规律的认识是无穷尽的，科学研究不会一劳永逸，在认识的一定阶段，由于条件的变化，出现一些曲折是正常的，因此又绝对地否定也是错误的。只有吸取前人成功的经验分析发生曲折的原因，不断结合新条件进行研究和实践，才能不断深化认识，不断创新，不断取得科技进步。

参考文献

[1] 吴中伟．怎样做好混凝土工程，科技出版社，1951年．

[2] 吴中伟．科研工作的选题、方法与创新，清华大学校庆学术报告会报告，1996.4.

[3] Goldman A，Bentur A. Bond Effects in High Strength Silica Fume Concrete，quoted by Bentur in paper tided The Role of the Interface in controlling the performance of High Quality Cement Composites. In Advances in Cement Manufacture and Use，eddied by Gartner E，published by Engineering Foundation，New York. 1989.

[4] 吴中伟．混凝土中心质假说，1959年北京水泥科学讨论会报告，1961年该报告论文集．

[5] 吴中伟．混凝土科学技术的反思，混凝土与水泥制品——吴中伟在祝贺其70寿辰时的学术报告（1987.7），1988.6.6.

[6] 吴中伟．绿色高性能混凝土与科技创新，建筑材料学报，(1)No.1998.9.

[7] 吴中伟，张鸿直．膨胀混凝土，北京：中国铁道出版社，1990.10.

[8] 鲍格 R H. 波特兰水泥化学，杨德骧，朱祖培，黄大能译．北京：中国工业出版社，1963.

[9] 吴中伟．混凝土科学技术近期发展方向的探讨，中国硅酸盐学报，(7) No.3，1979.8.

[10] 吴中伟．环保型高效胶结材，混凝土，1996.No.4.

自评和进一步的认识

1. 很多人说“混凝土没有科学，只有试验和经验。”曾有个混凝土搅拌站的总工委屈地告诉我，有一次在他们供料的工程中发生了纠纷，对方老板训斥她说：“你们有什么了不起的？不就是和泥的吗！”什么是科学？有人说能用数学式子计算的才是科学。按照《辞海》的条目，科学是关于自然、社会和思维的知识体系，适应人们生产斗争的需要而产生和发展，是实践经验的结晶。有人说“混凝土是实验科学”，不错，实验是科学技术研究的重要手段。实际上物理和化学都离不开实验，其理论也来自实验。物理学家沈致远就说过：“物理学是实验科学，一切要由实验来判定。”（沈致远：《科学是美丽的》上海教育出版社，2012年。引自书中“薛定谔猫的生与死”）他对科学的定义是：“科学是一种理论知识体系，它是人类对于客观世界的正确反映，是人类认识世界和改造世界的社会实践经验的概括和总结，同时，科学又是为社会实践服务的。”

2. 吴中伟在上世纪40年代末针对我国的情况提出了“混凝土科学技术”的概念，对混凝土学科的发展起了重要的推动作用。他在1961年提出的“中心质效应假说”早于西方国家关于混凝土的界面过渡层理论。界面过渡层理论指出混凝土中界面层是薄弱环节，中心质效应假说则认为效应有正有负，对传统认为不利的孔，也指出其有利的一面。按中心质效应，可以解释水胶比远低于水泥完全水化理论水胶比的高强混凝土强度来源，即“化学组成通过结构决定物质的性能”（唐有祺，《结晶化学》，人民教育出版社，1964年1月第8次印刷，p.126）。H. F. W. Taylor在1997年提出一个孔隙率-不同尺寸、不同结晶度和不同密度粒之间级配的等强度曲线（H. F. W. Taylor，DISCUSSION ON PAPER “MICRO－STRUCTURE AND STRENGTH OF HYDRATED CEMENT” BY R. FELDMN AND J. BEAUDOIN，C. & C. R.，Vol. 7，1977 pp 465～468），和“中心质假说”有共通之处，也符合结构决定性质的理论。从经验总结出规律，再用于指导实践，就加入了人类的知识体系，就是科学。做得出，未必能算得出；算得出也未必能做得出。正像有的人计算出他所做的混凝土结构使用寿命为108年，如此之准确显然是不可能的。用传统经典数学无法计算的事物恰恰是1＋1≠2的复杂混沌体系、模糊系统。正像从欧氏几何到非欧几何一样。从精确计算发展到模糊理论和不确定性科学，无疑是一种进步。这种进步是随着相邻学科的发展而发展的。如果没有现代的高速、大容量的计算机和高级的计算机程序语言，也不可能将天气预报提高到现在这样的准确度。

3. 2012年12月美国《物理学通报》刊登了麻省理工学院（MIT）混凝土持续性中心（Concrete Sustainability Hub）的论文：《水泥的纳米结构和纳米力学：多分散的胶状聚集体》（Nanostructure and Nanomechanics of Cement：Polydisperse Colloidal Packing），报道了MIT混凝土持续性中心的研究人员对水化的水泥产生强度的机理研究的进展。认为“组成水泥基本纳米尺度单元的硅酸钙（C-S-H）粒子具有一种无序的几何排列，而不是科学家们长期假设的有序晶态结构”，“C-S-H粒子本身也是无序的。形成任意尺寸的粒子并非均匀的圆形，而这种纳米尺度单元的尺寸多样性，导致一种致密的、无序团聚的粒子群，其所形成的是较强的水泥浆体”。他们发现，“在水化过程，通过静电力聚集的C-S-H粒子群产生一种低密度凝胶，当将其成型为一个致密聚集体时，就成为具有不同局部密度的刚性颗粒”。他们采用蒙特卡洛模拟法建立从3nm到35nm的粒子所组成具有不同聚集密度的粒子群模型，和纳米压痕实验结果建立其相关关系，发现是“尺寸多样的水泥水化物纳米粒子优化的聚集密度赋予混凝土强度。”这不禁使我想起吴中伟的“中心质假说”。重读先生1959年发表的论文，对照他1973年又发表的同类论文，发现后者的中心质模型对前者的做了修改。1959年的中心质模型如下：

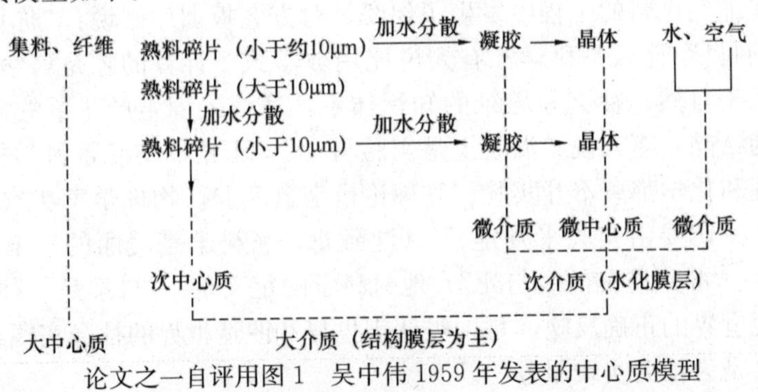

论文之一自评用图1　吴中伟1959年发表的中心质模型

由上图可见，吴中伟认为，组成次介质的微介质是含水和空气的凝胶，微中心质则是“由凝胶发育成的晶体，逐渐成为水泥石的主体”。这是20世纪50年代盛行的以前苏联科学院院士 П. А. Ребиндер（列宾杰尔）为代表的观点。正是上述MIT论文所述“……而不是科学家们长期假设的有序晶态结构”所针对的观点。吴中伟在1978年修正了这个模型如所选本文图2所示，把“AFt 相、$Ca(OH)_2$以及其他晶态水化物”归为微中心质。1960年T. C. Powers发表的论文《水泥浆体的物理性质》中明确指出，水泥水化物的强度和体积稳定性都不如未水化颗粒的，因此在水泥浆体中必须保持一定量的未水化水泥颗粒。实际上这就是“中心质假说”模型中的次中心质。1988年H. F. W. Taylor把由他自己命名的“C-S-H凝胶”修正成他自己最初命名的包含纳米C-S-H粒子聚集体和其他结晶态产物的“水泥凝胶”，这相应于“中心质模型”中包含微中心质和微介质的次介质。那么水泥水化后强度的来源就可以描述为：**成型密实的水泥浆体水化后，生成由纳米级粒子群聚集成水泥凝胶（次介质），其内部因多种尺寸的粒子级配而紧密接触，产生范德华力引力，把粗的、结晶态的、本身强度高的其他粒子（次中心质）胶结在一起组成具有强度的整体；这就是硬化的水泥浆体（大介质）；水泥浆体和各种集料、钢筋等大中心质通过界面（效应圈）而互相联系，组成整体，这就是混凝土。**

由以上分析可见吴中伟对水泥混凝土微结构的理解是大大超前的。遗憾的是因吾辈之悟性未到，而使其“埋在深山无人识”！

4. 实际上我们已经很长时期地接受了H. F. W. Taylor早在几十年以前的观点，即：水泥水化产物C-S-H粒子不是晶体，在常温下也不会发育成晶体，而是近程有序、远程无序（ACI《Smart Briefs》关于MIT研究时报道所说：“研究人员发现，C-S-H粒本身也有点无序……”应当也有此意）。因此对x-射线的衍射只出现很弱的弥散峰（见论文之一的自评用图2）。图中为了和C-S-H做对比，只显示$Ca(OH)_2$衍射三个特征强峰中的主强峰（$d=2.62Å$）和第三强峰（$d=1.8Å$），而未显示次强峰（$d=4.89Å$）；而且为了能看见C-S-H的弱峰，未显示出远远超过图中显示的$Ca(OH)_2$峰的全高度。Taylor说：“对C-S-H正确的描述是无定形的产物。把它设想为具有高于实际存在的结晶程度是错误的。……广义地说，C-S-H表示任何无定形和半结晶的C-S-H（我注：例如蒸养和蒸压的水泥水化产物）”。1979年H. F. W. Taylor的孔隙率-粒子级配-强度关系的等强度曲线模型（见本论文集第一部分“二”之图2）可对“中心质假说”做形象的解释。而“中心质假说”的研究正是辩证唯物主义和专业结合的结果。

5. “中心质假说”和Taylor的孔隙率-粒子级配-强度关系的等强度曲线模型及其实际应用表明，混凝土中的主要内聚力是范德华力，即分子间的引力。粉体颗粒越细，比表面积越大，颗粒间引力越大，斥力也随之增大，当势能最大时，应力和斥力相平衡，合力为0，此时粒子间的距离即范德华半径。粒子间距离小于范德华半径以后，斥力开始主导（见论文之一的自评用图3）。涉及水泥的元素范德华半径都不大于0.3nm。在混凝土中无论水泥水化产物生成的C-S-H还是矿物掺和料反应产物，其粒子之间的距离都不会小于范德华半径。从这个角度来看，掺矿物掺和料后，可作为次中心质，取代出“宝贵的”熟料，使更多的熟料水化，增加微介质；一部分矿物掺和料与$Ca(OH)_2$反应后，还可增加微介质。使全部颗粒的总级配得到改善。而不必追求所有矿物掺和料的化学活性。

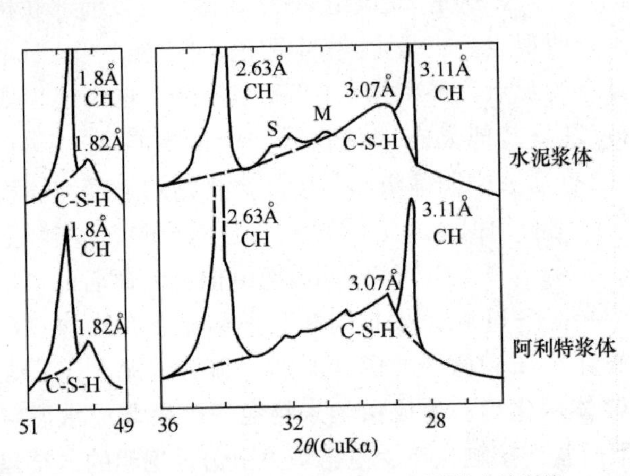

CH—Ca(OH)₂；M—单硫盐；S—阿利特＋贝利特

论文之一自评用图 2　水灰比为 0.65、水化 1 年的阿利特和水泥浆体局部 x-射线谱图

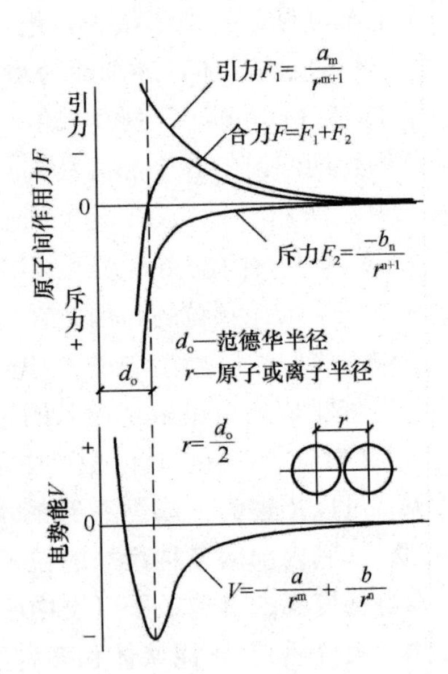

论文之一自评用图 3　范德华力图解解

6. 根据 Taylor 的理论和 MIT 最新研究，通过上述分析，将中心质假说补充如自评用图 4 所示。补充后的中心质假说可以从整体论解释混凝土强度的来源，作为从整体论解释矿物掺和料作用的混凝土微结构模型。

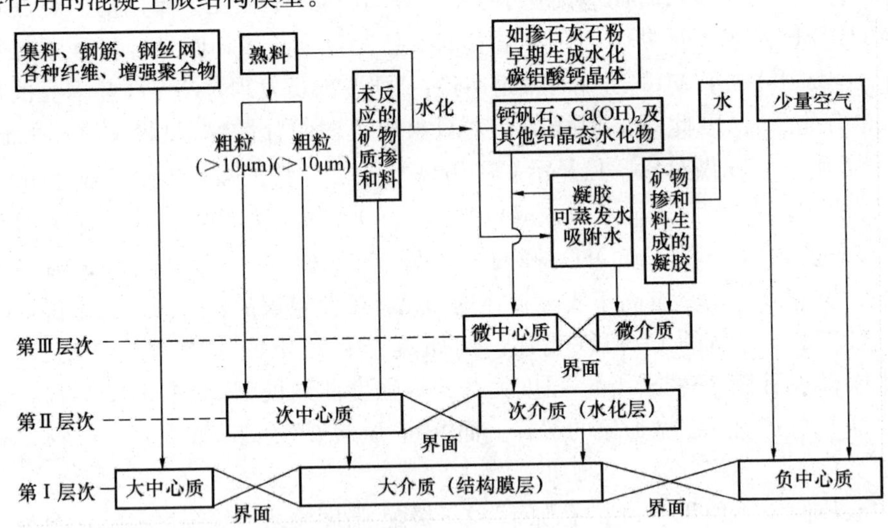

论文之一自评用图 4　对中心质假说的补充

7. 文中的第 3 节“继承、发展和创新”倒数第二段的提法是不错的，但是两个例子都不全面。不能用玻璃纤维代钢制作水泥和混凝土构件，主要不是因弹性模量的差异，而是因为玻璃纤维的耐碱性差。像硫铝酸盐水泥这样的非硅酸盐水泥才能用玻璃纤维。国外有的工程用于代替钢筋的 FRP（纤维增强树脂），用的就是玻璃纤维，但是玻璃纤维并不接触水泥。对混凝土的增强，不同纤维的作用是不同的，现在市售聚丙烯纤维因弹性模量太低，极限伸长率太大，不能用于抵抗因温度应力和干缩应力引起的开裂，有效的是用于缓解混凝土大面积水平构件的塑性开裂，和减少喷射混凝土施工时的回溅率。而对硬化后混凝土的裂缝

控制则要求纤维极限变形率小、单丝抗拉力大，外形比要较大，等等；关于膨胀剂的超长无缝施工，也需要对膨胀剂做更深入的研究，并由掌握了使用技术的人来用。这只是两个例子，主要是为了说明创新必须有科学根据，行动不能走在研究的前头。

在学习和生活中，工作的最重要的动力是工作中的乐趣，是工作获得结果时的乐趣以及对这个结果的社会价值的认识。

——爱因斯坦

较高级复杂的劳动，是这样一种劳动力的表现，这种劳动力比较普通的劳动力需要较高的教育费用，它的生产需要花费较多的劳动时间。因此，具有较高的价值。

——马克思

论文之二

混凝土结构耐久性的系统分析*

廉慧珍　（清华大学土木工程系）

1　系统分析的意义

分工是生产技术发展的产物，并且推动了生产技术的进一步发展。但是随着分工的精细和现代科学技术的快速发展，高层次上的相互联系与作用又显得非常必要。学科的交叉和渗透，甚至艺术和工程这种看似不相干的门类之间的互相渗透，也是现代生产与科技发展的必然。土木工程包括建筑、水利、道路、桥涵、港湾以及各种地上和地下的基础设施构筑物的特殊结构，它们有着共同的力学和材料学基础，只是工程性质、功能、荷载作用条件和所处环境有差别。随着生产的发展，由于这些差别而造成了专业科学技术的分工，而且越分越细。同样，结构的设计、材料和施工也因各自研究的深入而分离，但是这种由于分工精细而相互割裂的情况越来越不能适应工程的需要。例如，使用不同材料有不同的设计理论和方法，以及不同的施工方法；即使同一种材料，也是在不断地发展和变化。设计和施工方法必须跟随材料的变更或更新而变更或更新。“一个人当然不可能具备所有必要的知识，因此设计者可以简单地咨询材料专家，然而，除了少数特例外，大多数结构工程师对混凝土行为都缺乏必要的了解；材料专家又多是一些纯粹的科学家，对结构的问题缺乏了解，结果他们不知道什么问题要回答，而设计人也不知道要问些什么”[1]。这种知识分割的现状越来越不能适应当前工程技术的发展。“混凝土技术是进行结构设计的基本知识”[1]；施工主要是把各种材料组合成满足质量要求的结构；材料制备和使用要符合工程性质和功能的需要。从方法论来说，就是要把人的认识、研究和涉及的对象当作一个系统，分析其各个组成部分之间及其与外部环境的相互关系，研究对象的系统性质和运动规律。以往，结构工程师注意力集中于防止钢筋的腐蚀，所以许多规范中对腐蚀性介质的限制只有 Cl^-，而忽略了混凝土本身受环境中腐蚀性介质作用而引起的劣化。混凝土结构规范和材料规范的分离，使我国至今还没有一部按整体论编制的混凝土结构耐久设计和施工（包括制作）性规范。在实践中还造成例如图 1 所示的后果。

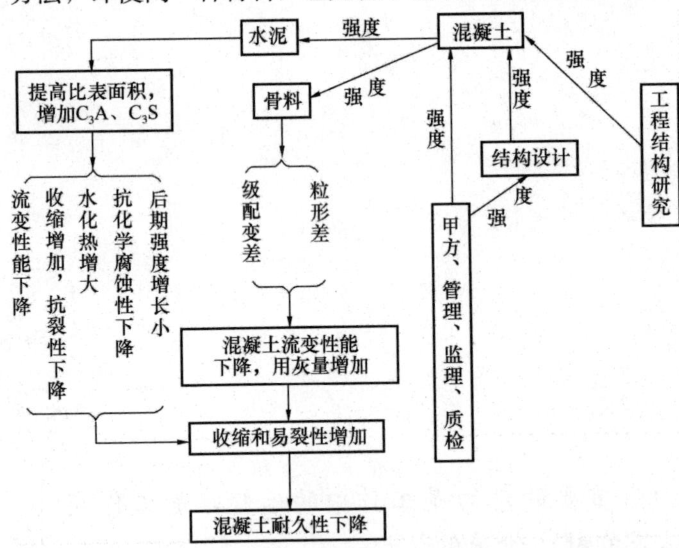

图 1　行业分离造成混凝土结构耐久性下降的一例图解

* 在中国工程院土木水利建筑学部《混凝体结构耐久性设计与施工论坛》的发言，北京 2002.11。

长期以来，混凝土材料的耐久性是以抗渗性和抗冻性作为主要衡量指标的，其出发点是几乎所有的环境因素对混凝土的腐蚀都首先由于腐蚀性的因素进入混凝土内部。因此混凝土的密实性和混凝土本身的抗力最重要。抗渗性反映混凝土的密实性，抗冻性反映混凝土的抗力。随着严酷环境中混凝土结构工程量的增大、混凝土组成材料的变化和对工程所处环境复杂性认识的提高，对混凝土材料抵抗不同环境作用能力的实验研究和检测也增加了。然而，在实验室用现行标准方法所检测的所有“耐久性”指标，只是混凝土材料本身的固有性质在一定条件下的反映，并不等于混凝土构件或结构的耐久性。单独就混凝土材料的耐久性的讨论没有意义（当然可用作横向的相对比较）。耐久性是混凝土必须用在的结构中并处于一定环境和条件下才能发生的行为，而这种行为的产生又是系统中各元素和系统与环境之间相互联系、相互作用的表现，因此混凝土耐久性的研究和设计必须用系统分析的方法。

对系统认识的观点即系统论。系统分析的方法就是从整体全局出发，对系统内外各种联系及其规律性加以辩证的分析，找出符合目的的最佳方案，以评定指导人们行动的方针、计划和措施。系统论是一种科学认识和社会实践的方法论，即从系统的观点出发，研究不同层次中的各要素之间和各层次系统间的运动规律而对研究对象进行优化处理的方法。系统论的基础是辩证唯物主义，即世界是物质的，物质是相互联系的而不是孤立的；是发展变化的而不是静止的；是全面的而不是片面的；物质世界是无数事物相互依存、相互制约、相互作用的过程形成的统一整体。

2　系统和系统的性质

系统是相互作用和相互联系的各个部分所组成的、有特定功能的有机整体。其主要性质包括：整体性、层次性和开放性、结构决定性质、无序到有序转化（变化的、发展的，达到动态平衡的趋势）。其中整体性是最重要的，将在下面再分析。

系统是有层次的，不同层次上都存在系统，所有系统都是较大系统的一部分，系统之间也有互相依存、互相作用的关系，因此还具有开放性。构成系统的元素必须是相互联系（相互依存、相互对立）、相互作用（相互影响）的，它们互相联系和作用的方式就是系统的结构，“物质的成分通过结构起作用”。所以没有相互联系和相互作用的不是系统中的元素。例如土木的结构工程大系统包含结构的设计、材料和施工三大元素。使用不同材料或不同用途的、功能的结构则是结构的分类，而不是这个系统中的元素。材料是土木工程的子系统，材料中的混凝土由胶凝材料（包括各种添加的组分）、骨料和水三个次级子系统组成。胶凝材料、骨料又由再小一些的系统组成……，一直可以分到最简单的组分，如图 2 以硬化的混凝土为例所示意。

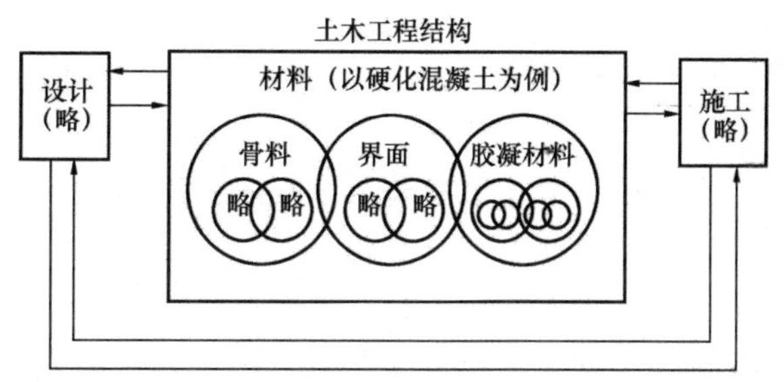

图 2　土木工程结构系统的层次性示意

耐久性不是这个系统的元素，而是这个系统的目的。但是作用于这个系统的大气，即环境作用条件，则存在着系统，并且与其所作用的系统形成了更大的系统。

系统论的观点可以指导我们对复杂的问题在不同层次上和层次间进行分析，逐步使无序向有序发展。如数据的离散就是无序。控制发生变化的条件，从不同低层次向较高层次逼近，就可以总结出规律，向有序发展。例如影响混凝土耐久性结构的因素非常复杂，没有单一因素引起的混凝土结构的劣化，但分析出各因素的作用因数 k_i 后，总会因其相互联系和作用而有：$D=f(k_1, k_2, k_3, k_4, k_5\cdots\cdots)$ 的关系。

3 整体论（Holistic methodology）

整体论是系统论的精髓。整体性是系统最重要的性质，即由部分构成整体时，出现组成部分所没有或对组成部分来说无意义的性质，同时失去组成部分单独存在时所具有的某些性质。这个性质就是：整体不等于组成整体的个体简单算数和。例如上述混凝土结构耐久性和各环境作用元素之间是个复杂函数关系，而不是加和关系。整体论就是从全局出发研究整体的规律性，确立这个函数。与整体对应的是分解论（Disaggrega-tire view），也有人称之为还原论（reductivity）。分解论认为组成部分或主要组成部分的性质可代表整体的性质，其方法是把复杂的体系中的主要因素抽象出来进行简化，进行科学假设和研究。如对水泥单矿物水化的研究。水泥化学家认为混凝土的性质就是由胶凝材料决定的，研究水泥，就可以了解混凝土，而水泥中又是硅酸盐凝胶起决定性作用。但实际上，水泥浆体的行为和混凝土是不同的。因为水泥作为一个元素在混凝土的体系中，与骨料和水的相互作用并不是简单叠加的。如图 3 所示，硅粉在水泥砂浆和在混凝土中的行为是不同的。实验室进行的分解实验往往与实际相差甚远，最典型的就是根据实验室中检测损坏的混凝土中既有活性骨料，所用水泥又含超过标准的碱，断定为碱-骨料反应所致，但拆除的混凝土损坏处并没有检测出碱-骨料反应的产物。Burrows 在美国克罗拉多州青山坝做了 104 块混凝土面板，进行了 53 年的观测，在其中使用了含高碱水泥和活性骨料的面板开裂的裂缝中没有检测到碱-骨料反应的产物[3]。分解研究可以发现事物的本质，以指导对整体现象规律的认识，因此在工程科学技术发展的一个阶段，分解研究是必要的，而且已有研究成果对工程的技术进步起过很大的作用。但实际上，水泥，特别是混凝土，是一种复杂的、非均质的多相体，水泥浆体、混凝土，以至钢筋混凝土构件的行为都不能用其中各组分单个行为的简单叠加来表征。换句话说，不同层次上研究的结果都不能简单地外推。分解论往往导致“只见树木不见林”。例如关于混凝土的耐久性，实验室检测的抗冻性、抗渗性、抗碳化性等指标只反映一定原材料和配比及工艺下所表现的混凝土材料本身固有的相对性质，而与实际的混凝土结构的条件和所处环境差别很大，所以结构耐久性不等于混凝土各项“耐久性”指标之和。

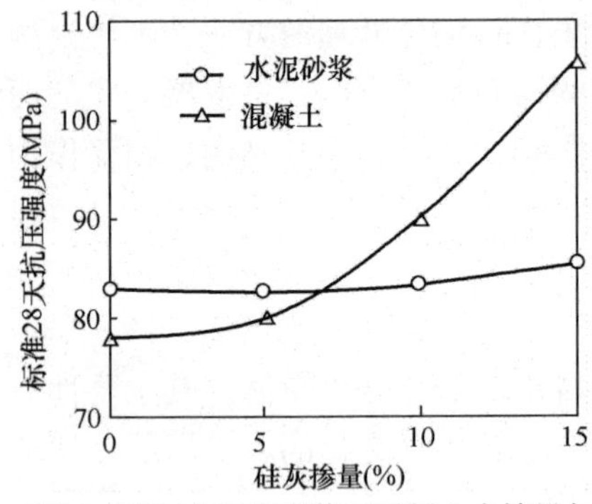

图 3 硅灰分别在砂浆和混凝土中掺量与 28 天强度的关系（水胶比相同为 0.33）[2]

分解研究的结果和整体行为有联系，但是不能真实地反映整体的行为。如上述实例，混凝土材料耐久性不等于混凝土结构工程耐久性，其原因一是条件的简化，忽略了环境的交互影响，而实际上混凝土耐久性并不等于各简化条件作用的叠加；二是过程的强制性，例如混凝土结构的碳化是在 CO_2 浓度不到 0.04％的大气中缓慢进行的，而实验室试验则是在 CO_2 浓度为 20％的条件下快速进行的，这两种动力学过程并不相同；又如使用在很苛刻的条件

（碱浓度、温度等）下检测出骨料有碱活性，而在实际工程却不一定发生碱-骨料反应；三是试件状态的差别，实验室十几厘米的小试件处于自由状态，而结构中的混凝土则处于约束状态；实验室试件处于“标准条件”，最多是自然条件，而结构中的混凝土则要经历复杂的温度历程。有人为了改善耐久性分解论研究的缺点，采取了“多因素影响”的试验。但却仍脱离不了过程的强制性。仍然存在自由的小试件和结构中混凝土状况的差别，而且，各因素交互的作用也并不等于同时的作用或叠加的作用。换句话说，方法仍是分解论，至多加一、两个因素，是不会有本质改变的。

又如，当前评价混凝土结构强度的方法是在浇筑前用同样的混凝土预留 150mm×150mm×150mm 或 100mm×100mm×100mm 的小试件，分别在（20±2）℃、相对湿度 90%以上的条件下和在现场“同条件”养护。实际上因为与实际结构混凝土所处条件和状态并不相同，不同混凝土强度发展是不同的，这种强度并不等于结构内部混凝土的强度。现仅就标准养护试件和结构内部混凝土强度发展来分析这种差别（见图 4）[3]。对于无掺和料的纯硅酸盐水泥混凝土，由于混凝土内部温升的影响，结构内部混凝土强度在早期稍高于标准养护试件的强度，随龄期的增长，则结构内部混凝土强度就越来越低于标准养护试件的强度；对于掺粉煤灰的混凝土，则结构内部混凝土的强度始终高于标准养护试件的强度；混凝土强度等级越高，以上差别越大。也就是说按标准养护试件评价结构混凝土的强度，当无掺和料时是不安全的；而粉煤灰掺量大于 20%时，则是保守的。这就是因为忽略了结构混凝土和标准养护温度的差别、对象尺寸的差别，忽略了不同组分混凝土对温度的敏感性差别。同样，实验室各项混凝土耐久性指标的测定也没有考虑由于这种差别而造成的混凝土内部微结构变化对结构耐久性的影响。

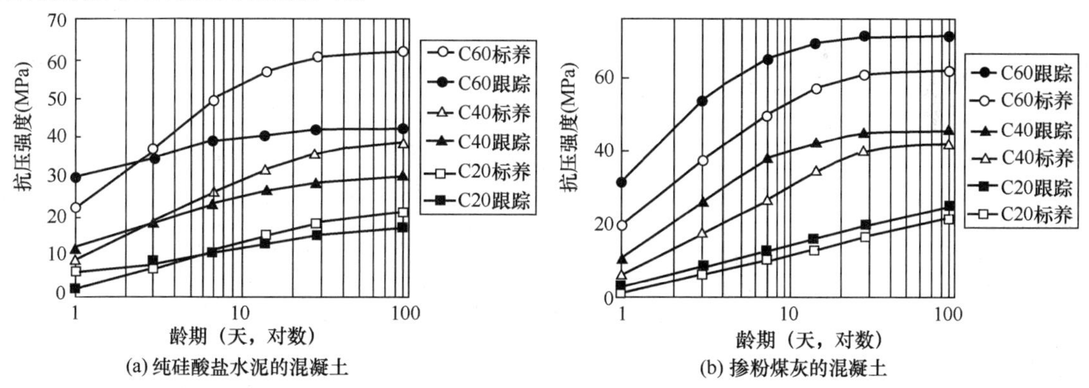

图 4　不同强度等级、纯硅酸盐水泥的混凝土和掺粉煤灰的混凝土内部强度的发展

注：①标养为标准养护，跟踪为用实测实际结构中混凝土的温度养护；

②各强度等级混凝土的水灰（胶）比，无掺和料的为：C60－0.437，C40－0.561，C20－0.766；掺粉煤灰的为：C60－0.357，C40－0.465，C20－0.593，粉煤灰掺量分别为 21%、28%和 37%。

4　对结构物耐久性和寿命评估的整体论

混凝土结构耐久性是指钢筋或预应力钢筋混凝土结构抵抗环境中各种因素作用而保持正常使用功效的能力。没有由单一因素造成的混凝土劣化，必然是几种因素的交互作用。但是，也并不是在任何地区、任何场合下相同影响混凝土耐久性的因素都起作用，在不同气候条件下和工程部位，其中必有一个起主导作用。同样的因素，结构部位（是在地下还是地

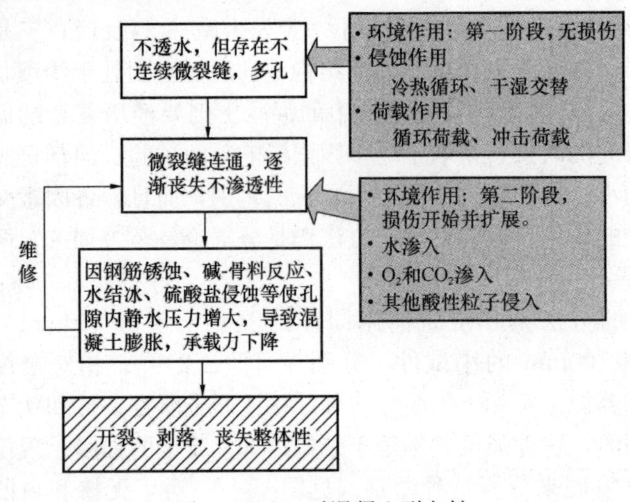

图 5　Mehta 对混凝土耐久性整体论的混凝土耐久性模型[5]

面，接触雨水的程度等)、地下水位高低、气候等条件不同时，对混凝土的作用是不同的，引起混凝土劣化的主导因素在不同条件下是不同的。不能把在不同地区试验的结果看成普适的规律，应当根据系统的整体性，分析系统中各元素之间对立统一的关系，找出主导元素。

按照 P. K. Mehta 的观点[4]，如图 5 所示。结构物寿命第一阶段：凡是组成良好并经适当捣固和养护的混凝土，只要内部孔隙和微裂缝尚未形成相互连接而直达表面的通道，则基本上是水密性的；使用期间结构的荷载以及大气环境的影响如冷热循环、干湿交替，可使这些内部微裂缝发展并传播；结构物作用第二阶段：混凝土丧失水密性，变成水饱和，有害离子侵入。由于有害离子作用而连续膨胀、收缩，开裂，重量损失，渗透性增加；修补、加固后仍可用；结构寿命第三阶段：上述过程反复进行，直到修补不如重建时，则废弃。这个过程的前提是开始服役时早期无可见缺陷。如果早期产生可见裂缝，则劣化会加速。

从整体论的观点分析，混凝土早期产生可见裂缝的原因及其对混凝土耐久性的影响如图 6 所示[5]。

由图 6 分析可见，建设速度和混凝土耐久性之间的矛盾是基本的矛盾。现在不少工程都将自己的“混凝土 3 天强度达 70%、7 天达 90%以上甚至 100%”引以为豪。而这恰恰是混凝土结构早期开裂和耐久性下降的主要原因。这样的混凝土，即使早期由于采取某种措施而

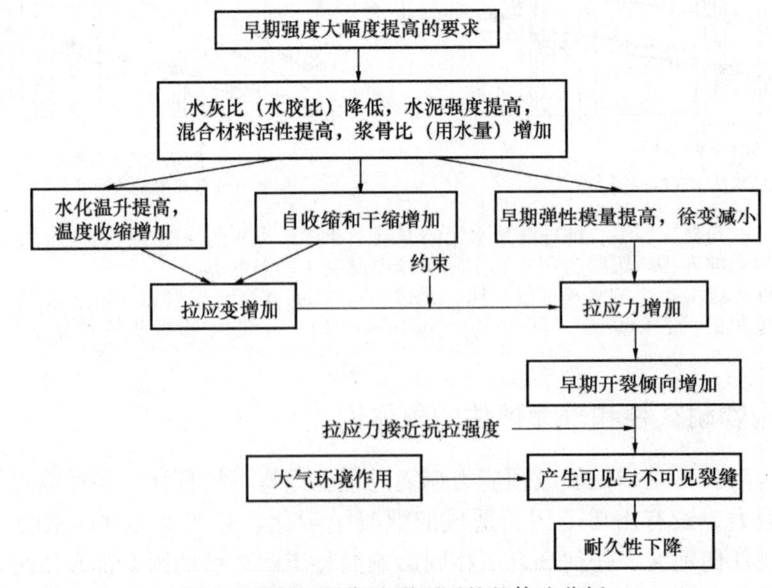

图 6　混凝土早期开裂原因的整体论分析

没有开裂，但却掩盖了不可见的微裂缝的存在。在恶劣气候下（特别是干湿交替）会加速混凝土的劣化。因此从整体论出发，混凝土结构密实，同时控制混凝土内部非荷载引起的应力尽可能地小，是提高混凝土结构耐久性的根本措施。然而，正如人的出生要优生优育，但不能不允许他们生病一样，还需要经常体检和保健，在他们生病时还要治病。直到他们的生命尽头。混凝土结构也有“生老病死”，进行耐久性设计只是优生优育的措施而已，混凝土结构的寿命应当包括使用期间进行定期检查、诊断、修补、加固。

对于引起混凝土结构劣化因素，整体论认为，在一个系统里，所有元素构成一个整体，如果存在的元素没有相互作用，就不具有系统的功能。例如对于碱-骨料反应来说，限制骨料碱活性同时又限制水泥含碱量是不必要的，因为二者同时存在才是发生碱-骨料反应的必要条件，但是还必须有充分条件，碱-骨料反应才会发生。其充分条件就是水。也就是说，从整体论出发，碱-骨料反应必须同时具备充分必要的条件：含碱量、活性骨料、水三者缺一不可。而水分的保持还必须是持续的，混凝土才可能因碱-骨料反应而劣化。在常年大部分时间相对湿度都很低的干燥环境下，即使混凝土中含碱量和骨料碱活性都大，碱-骨料反应也不会是引起混凝土结构劣化的主导因素。国外最早发现碱-骨料反应破坏的实例是水坝，因为水坝是最具有水这个充分条件的，在三峡水利工程的建设中曾有人认为应当考虑碱-骨料反应限制水泥的碱含量，经过大量论证，认为三峡所用石子不具有碱活性，必要条件不具备，最后不予考虑。这种决定应当是正确的，但是却忽略了碱对开裂的影响。其他如冻融、硫酸盐侵蚀以至钢筋的腐蚀，也都以水为充分条件。因此从整体论出发，混凝土劣化的根本原因是在大气作用下混凝土内部微裂缝扩展而丧失水密性。控制早期裂缝很重要，但是控制早期混凝土内部应力则更加重要。因为早期即使因施工的努力能控制住可见裂缝的发生，而如果混凝土结构内部积蓄的应力仍然存在，混凝土的干燥收缩期很长，使用阶段时期环境条件的骤然变化（干湿、冷热）会增加内部的应力，促使已有的微裂缝发展成可见的裂缝。所以早期不裂不等于后期也不裂，控制早期内部应力以减少内部微裂缝和应力的积蓄才是根本的。

如何从对混凝土耐久性的实验室试验研究走出来到实际工程混凝土结构耐久性的评价，这是新时期对科技工作者（不仅材料的研究，也包括设计和施工）的挑战。

5 耐久性设计的对立统一规律

矛盾无处不在。任何给工程带来好处的技术，同时必定还存在潜在的不利因素。混凝土耐久性设计中也存在各种既对立又有统一性的因素。在混凝土施工性、强度和耐久性等主要性质和水灰（胶）比、浆骨比和砂石比等配合比主要参数之间也存在着矛盾。在其他因素相同时，水灰比大的混凝土强度较低，密实度较差，但不易开裂；水灰比小的混凝土密实度高，但因自收缩大而易于开裂；当水胶比一定时，混凝土浆骨比大时，流动性好，但收缩大，易裂；其他条件相同时，砂石比大的混凝土施工性好，强度稍高，但弹性模量稍低而会稍增大收缩。增加水泥比表面积，可以提高混凝土早期强度，但是同时还会提高混凝土的温升和增加收缩。其他如建设速度（要求提高早期强度）和混凝土的抗裂性，施工性能和混凝土耐久性所要求的低浆骨比（低用水量）等也都有类似的矛盾。对立统一的含义就是矛盾双方互相依存，互相对立，在一定条件下互相转化，而绝不是调和。是要从整体出发，①根据工程特点、条件和要求，抓住主要矛盾。②透过现象抓住本质，重点解决本质的问题。③有所得必有所失，局部服从整体；力求代价最小。

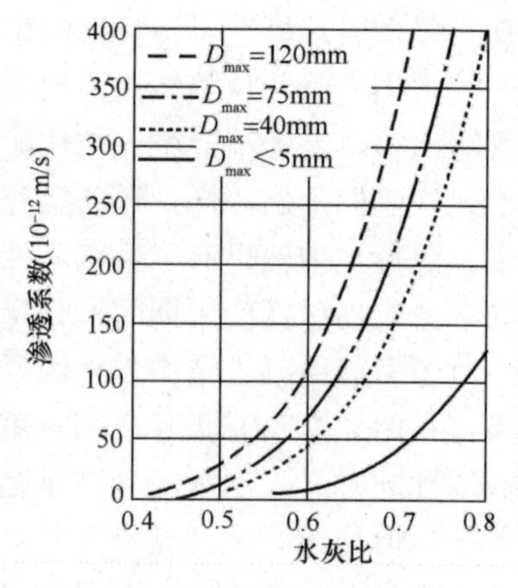

图 7　不同水灰比下骨料最大粒径对混凝土渗透系数的影响[6]

实现混凝土的结构耐久性不仅需要设计、材料和施工的配合，也需要甲方和监理一起来处理各种矛盾。为此，设计既要提出混凝土的原材料和配合比主要参数及施工控制的要求，还要考虑到尽量减小对施工控制的不利因素。例如，混凝土使用的石子最大粒径较大时，弹性模量较大，体积稳定性好，但是渗透性较大（见图 7 所示）；石子最大粒径小时虽然混凝土渗透性低，但抗裂性不好，石子粒径又不能太小，比如最小也不能小于 10mm，因此设计要求施工中混凝土使用的石子最大粒径都能通过钢筋间隙，则所设计构件钢筋最小净间距就不能小得连 10mm 的骨料都通不过去。而目前工程中这样小的钢筋净距并不少见。保证骨料最大粒径小于钢筋净间距的 3/4 的要求，首先就得在设计中来平衡。设计耐久的混凝土，重要的是提高混凝土的密实性，所以要限制混凝土最大水胶比，又因为要控制收缩变形，就要限制硅酸盐水泥的最大用量。这样受限制的参数对混凝土施工控制的影响，应当由材料工程师来解决，例如可以采取掺用矿物掺和料、使用减水剂、缓凝剂和引气剂等措施来调节，而设计并不能迁就施工，就需要施工方认真分析各个互相矛盾的因素，做好矛盾的转化，并采取措施如控制浇筑混凝土的各阶段的温度、控制 1 天以内的强度和弹性模量等，进一步保证达到设计的要求。甲方和监理应当接受对某些构件或部件延迟验收日期的观念。……不论哪一方是矛盾的主要方面，在处理类似上述旨在保证混凝土结构耐久性的的措施方面，都必然有得和失的对立统一。

正像医生不能照搬药典来给病人开药方一样，标准、规范、指南等都只提供基本要求和原则的指导，而不能提供保证，解决工程问题的复杂性不亚于医生看病，需要工程技术人员正确地运用科学方法论，不断实践，总结经验，丰富和更新知识，尤其是本系统的相邻、相关知识。混凝土材料的知识是结构设计和施工的基础，设计和施工不跟随材料科技的发展，必然会出问题；同样，如果材料的研制和供应商如果不懂得工程（结构和施工），施工工程师不懂得结构也不能保证耐久性设计的实现。

参考文献

[1] Adam Neville. Concrete Technology ——An Essential Element of Structural Design, Concrete International, July 1998.

[2] Goldman A, Bentur A. Bond Effects in High Strength Silica Fume Concrete, quoted by Bentur in paper titled The Role of the Interface in controlling the performance of High Quality Cement Composites. In Advances in Cement Manufacture and Use, eddied by Gartner E, published by Engineering Foundation, New York. 1989.

[3] R. W. Burrows. The Visible and Invisible Cracking of Concrete, ACI Monograph No. 11, Published by ACI, Farmington Hills, Michigan, First Edition, 1998.

[4] Dhir R K, Jones M R. PFA Concrete : Influence of Simulated in situ Curing on Elasto-Plastic load Response, Magazine of concrete Research, 1983(45)163.

[5] Metha P K. Durability——Critical Issues for the Future, Concrete International , July 1997.

[6] Ruettgers A E, Vidal N, Wing S P. An Investigation of the Permeability of mass Concrete with Particular Reference to Boulder Dam. Journal of ACI. 1935. 3～4.

自评

1. 当时我发言的时候，题目是混凝土结构耐久性的哲学问题，立刻遭到会场上的质疑："怎么现在还讲政治?"我说："哲学不是政治，是科学，而且是指导一切科学的科学。西方国家也在运用哲学，也在研究辩证法。"爱因斯坦说过："如果把哲学理解为在最普遍和最广泛的形式中对知识的追求，那么，哲学显然就可以被认为是全部科学之母。"

2. 关于耐久性的涵义，用 A. M. Neville 的话说："不存在一般含义上耐久的混凝土：如果我要为花园工具室修个地板，打算用上一年就推倒，某种质量的混凝土就'耐久'了，但这种混凝土用于一座大桥、一条隧道或一座大坝中也许就不耐久了。所以，说某一混凝土是'耐久的'或'不耐久的'是错误的"，"混凝土在一种条件下耐久，在另一些条件组合下就可能不耐久。仍然是没有本质上耐久的混凝土"。"有时混凝土不耐久，是因为用途发生了变化，例如本来所设计存放纸的仓库地板，却改成放化学品了"。（Adam Neville. Consideration of durability of concrete structures：Past，present，and future.《Materials and Structures》March，2001.）

3. 混凝土工程是个高度复杂的非均质多相体系，其各个环节（原材料选择、质量控制-试配-成型工艺）之间与每个环节中各个元素之间有着很复杂的关系。耐久性又是一项比强度更重要的性能，但却是必须在用到工程中在随时间的延续而变化的环境作用下所表现出的行为，无法确切定量地预测。如果没有哲学思想的指导，没有系统方法，如何保证混凝土工程的质量?

4. 在"4"中用必要条件和充分条件说明碱-骨料反应的条件，确切地说应当是：碱、活性骨料和水三者缺少任意一个都只是发生碱-骨料的必要条件而不充分；碱、活性骨料和水三者同在，才是充分必要条件。不能说水是充分条件，因为发生碱-骨料反应必须有水参加，但只有水也不能发生碱-骨料反应。

5. 在"5"中提出"设计既要提出混凝土原材料和配合比主要参数及施工控制的要求，还要考虑到尽量减小对施工控制的不利因素。"这句话的后半句要比前半句更重要。当然，混凝土结构要进行耐久性设计，必然会涉及材料和施工，但是在行业严重隔离的现状下，设计一般很难正确地提出原材料和配合比主要参数。设计能做的应当是提出技术性能指标的要求；尽量减少会对混凝土凝结硬化过程中产生约束的条件和应力集中，并避免给施工造成难以克服的困难，则更是设计者对保证结构耐久性本分的贡献。

6. 混凝土工程，尤其是以外加剂的使用为特征的现代混凝土及其工程中，充满对立统一的因素，需要技术人员用辩证思维去处理，分析和掌握主要矛盾和矛盾的主要方面，创造条件达到统一而满足工程要求。从自评表 1 中所列为混凝土配合比主要参数的变化对混凝土性质与性能影响的矛盾情况可见，任何因素的变化对混凝土都有利也有弊，而没有绝对有利或绝对不利。只要抓住主要矛盾，并分析矛盾的主要方面和矛盾转化的条件，创造转化的条件，就能保证达到主要的目标。该表所表示的只是在原材料和工艺不变的条件下，混凝土材料配合比的各参数（骨料粒径除外），调整其中任一项都会影响其他，必须结合原材料和工艺条件综合考虑调整的措施。所谓统一并不是"抹稀泥"，而是使矛盾双方的主次地位发生变化，取利弃弊，化害为利。

论文之二自评表1　配合比参数的变化对混凝土性质的影响（最后一列为非配合比参数）

配合比参数	水胶比		浆骨比		砂石比		骨料粒径	
增减*	↑	↓	↑	↓	↑	↓	↑	↓
拌和物流动性	**↑**	↓	**↑**	↓	**↑**	↓	↓	**↑**
抗压强度	↓	**↑**	**↑**	↓	**↑**	↓	↓	**↑**
弹性模量	↓	**↑**	↓	**↑**	↓	**↑**	**↑**	↓
抗渗性	↓	**↑**	**↑**	↓	**↑**	↓	↓	**↑**
干燥收缩	↑	↓	↑	**↓**	↑	**↓**	**↓**	↑
自收缩	↓	↑	↑	**↓**	↑	**↓**	**↓**	↑
开裂敏感性	**↓#**	↑#	↑	**↓**	↑	**↓**	**↓**	↑

* 原材料和工艺不变的条件下，固定其他参数，只改变一个参数；除第一行增减外，其他箭头灰色底纹中的为有利，无灰色底纹的为不利。

\# 因表面塑性收缩所致；水胶比较大时，表面失水会因内部水分向上迁移而得到补偿，水胶比很低时，拌和物黏聚性大，内部水分向上迁移困难，表面失水后，如果不能及时得到补充，就会产生塑性开裂；对于硬化混凝土：在原材料选定、浆骨比和砂石比一定的条件下，当水胶比大于0.3时，随水胶比的增大，干燥收缩增大，自收缩减小，收缩总量基本上不变，但是收缩是一种相背变形，在约束条件下，粒子间趋向于分离。水胶比大时，内部的约束小，所以开裂敏感性反而小；水胶比减小后，内部约束增大，就容易开裂。这就是强度等级低的老结构物开裂少的原因之一。

7. 除了原材料和工艺的影响外，混凝土各项性质之间也存在对立统一的规律，自评表2举例给出原材料条件和混凝土性质对工程质量影响的利弊，由此可见混凝土工程的复杂性。设计人员也需要用系统论观念决策。

8. 自评表2只是举例说明混凝土工程的复杂性，不能简单化地照此处理，必须对具体问题具体分析。除上述各因素外，加上矿物掺和料、高效减水剂和各种其他外加剂，更增加了复杂性。混凝土的复杂性堪与人体的复杂性相比，我们同样知之甚少。西方国家因建设量的饱和，科学家和工程师已经有时间去细致、深入地研究混凝土的科学问题，同时也因工程问题少了，出现了一些没事找事的“研究”。我国混凝土的用量已居世界第一，工程中有丰富的课题需要研究，有必要和有能力在对混凝土这样复杂体系的研究上做出更多自主创新的成果。关键是要有正确的思维方法和观念。

9. 从混凝土材料本身来说，之所以复杂，一是原材料不能提纯，品位和成分波动；二是混凝土的微结构的形成和发展具有时间和环境的依存性，而环境随时间的变化又是不确知的；现代混凝土更加复杂，是因为使用高效减水剂后，虽然有标准和规范，但是影响减水剂具体效果的因素就很复杂，如水泥品质（矿物组成、细度、助磨剂、石膏品质、混合材等）、添加顺序、水胶比、矿物掺和料和骨料品质（级配、粒径、粒形、含泥量）、拌和时的温度、环境温度和其他现场条件，等等。再加上一些非技术的原因，如因商业保密的缘故而无法选择和控制原材料，使得科学优化配料难以实现。监督、监测和管理者们、混凝土拌和物的使用者们对此并不了解。

论文之二自评表 2　水泥与混凝土品质差别对工程质量利和弊的影响

水泥	强度	高	利	相同水胶比的混凝土强度高	比表面积	大	利	水化快，早期强度高	含碱量	大	利	促进C_3A和C_3S水化，早期强度高；影响与外加剂相容性
			弊	水化热大，长期强度可能倒缩			弊	水化热大，易裂，后期自愈性差或无			弊	后期收缩大，开裂敏感性大，在合适条件下有潜在碱-骨料反应性，长期强度可能倒缩
		低	利	水化热低，抗裂性好，强度长期持续发展		小	利	水化热低，抗裂性好，强度长期持续发展		小	利	无碱-骨料反应的风险；开裂风险小
			弊	早期强度较低			弊	早期强度较低			弊	使用高效减水剂时易泌水
混凝土	强度	高	利	可减小构件断面，可加快施工进度，早期强度高	坍落度	增大	利	适合泵送，方便于成型工艺，节省劳动力	掺引气剂		利	可改善施工性；提高抗冻性和缓解一切对混凝土产生膨胀性的破坏；保持抗压强度不变时，可提高抗折强度
			弊	温升大，抗裂性差；脆性大，断裂韧性差；拌和物黏度大，后期增长率低			弊	离析、泌水倾向大；混凝土温升有所提高；拌和物性能可预测性差				
		低	利	开裂敏感性小；强度持续增长期长		减小	利	可降低开裂敏感性			弊	对施工有更高的要求，增加施工控制的工作量和成本；引气剂的质量和混凝土的匀质性影响工程质量
			弊	早期强度低；抗渗性差			弊	泵送阻力增大；振捣成型工艺要求较高				

10. 当前的预拌混凝土供应商不仅有面对原材料的无奈，还要把所生产的拌和物交给那些不懂混凝土的供应商去进行成型工艺的操作，自己无法对最终产品的质量负责，甚至不得不满足某些无理的要求，例如现在有的拌和物坍落度已经达到 240mm，就是这样出来的，以至于拌和物离析、泌水的问题时有发生，成为结构物耐久性的隐患。混凝土对这种复杂无法承受也不应该承受。

论文之三

思维方法和观念的转变比技术更重要之一
——打破专业藩篱是时代发展的需要和必然*

廉慧珍　覃维祖　（清华大学土木水利学院）

今后，一生中只从事一种职业的人越来越少，……今天专家是高级人才的代称，未来各领域的顶尖人才将更多地表现为那些善于打通专业藩篱、学科壁垒的“通才”。

通识教育之重要不仅在于拓宽知识面，更在于培养通识能力，这是比综合能力更高级的认识力。大成功首先得益于准确地认识和决策，再运用综合能力去实施才会取胜。

——王宏甲：《中国新教育风暴》[1]

一、专业的藩篱——当前土木建筑工程领域专业隔离的现状

先举一个实例。有一座商业大厦（现今，这样的建筑物并不是个别的），建筑师按要求将底层设计成大空间，故柱子少而间距大，不需要大开间的上部20几层商业楼则要增加柱子。从力学上来说，任何结构物的死荷载与活荷载都由墙、柱或墩、桩等支撑构件传递给基础（可以有不同形式），最终再由基础传递给地基。地基是上部结构的“根”。墙或柱等传递垂直荷载的构件是必须“生根”的。由于建筑设计是最先满足业主需要的（功能、视觉）环节，长期形成建筑设计说了算的惯例。适应因现代人类活动需要而出现的底层大开间、大跨度的高层建筑，结构设计发明了转换梁来承受上部无法“生根”的柱子。该转换梁实际上起上部柱子的“基础”作用。本实例中的结构工程师就是这样“遵”建筑师的“命”而采用了转换梁。因“负担很重”，该转换梁被设计成断面为高3.5m、宽2.5m的庞然大物（虽然在结构上这也是梁）。梁内配置一根工字梁，工字梁上下各三层ϕ30钢筋，钢筋最小净间距只有几毫米（见图1所示）。建筑师认为，如何实现我的设计，是结构设计的事；而结构工程师也不考虑这样的断面和配筋如何施工。施工单位将这种密集的钢筋骨架绑扎完毕后，不知如何才能让石子最小粒径也有5mm的混凝土通过只有5～6mm的钢筋间隙。最后由混凝土专家勉强用不需振捣的自密实混凝土来解决。说“勉强”，是因为即使不需振捣的混凝土从梁的侧面工字梁腹板的位置靠泵压灌入模板，密集的钢筋仍然无法被混凝土很好地包裹，只能进去一些砂浆。至于上下表面的保护层，则只能从外面抹上去，其密实性与均匀性就不能不令人生疑了。

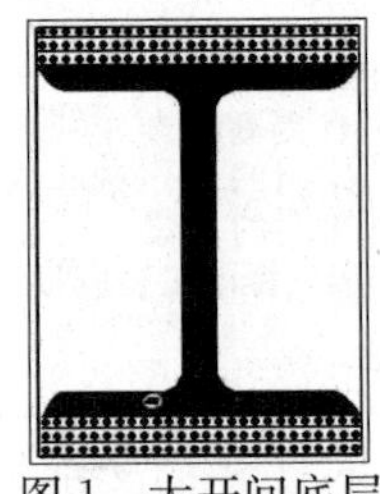

图1　大开间底层柱支撑的转换梁配筋示意

同一领域中专业划分过细之后，逐渐形成专业的藩篱，造成从建筑设计、结构设计、设备设计到施工和材料各专业形成了“指令”的阶梯，这个“指令”就是你“要配合我”，“满足我的要求”。各专业之间没有相互沟通，没有相互了解，没有相互交流，缺少相互支持与合作。长期以来，建筑设计着重点是满足功能和视觉的要求；结构设计要按照建筑设计着重

* 原文由编辑改成《建筑技术需要打破专业藩篱》为题发表于《建筑技术》2005年第一期。

从强度上保证安全性；施工则保证设计的实现；对材料的要求，由施工者说了算。各行其是，一级管一级。实际上处在一定环境中的结构工程，在正常情况下，几乎没有因承载力不足而破坏的，原因主要在材料和施工。以混凝土结构为例，过去在混凝土强度不高，技术较简单的情况下，尽管对不懂材料的设计和施工通常没有明显的影响，也还是出现过因材料和设计不匹配而发生问题。如1971年我国当时的机械工业部颁布了一项规定，禁止在湿热环境下使用高铝水泥。这一规定的出台源于盛产高铝水泥的南方某城市一项工程事故。当时设计人员只知道高铝水泥早期强度高，并不知道它只有温度在25℃以下才能稳定，超过25℃的潮湿环境中则会因水化物结晶转化而强度急剧下降。该城市常年气候湿润，四季温暖。在结构主体浇筑混凝土数日后，所有混凝土表面用手就能抠掉。类似的错误后来在北京也出现过，在用含膨胀组分的浇筑水泥灌注预制梁、柱节点时，使用了电热法加速硬化，结果所灌注的混凝土成了“豆腐渣”。更早的例子如在钢筋或预应力混凝土中使用氯化钙做早强剂等。这些都是因为设计和施工人员只认强度而不了解材料其他物理、化学性质所致。而对于“用膨胀剂防裂为什么有时反而裂得更厉害?”“掺用合成纤维就能防裂吗?”这样一些问题，则恐怕连材料工程师也不能正确地解答。在现今混凝土材料更加复杂、混凝土结构耐久性问题日益突出的情况下，“结构设计人员不了解混凝土，就不是一个真正完全的设计师”[2]。

由于专业的藩篱和传统观念的影响，从业主开始，对结构物主要就要求承载力，对材料就要求强度；而施工方在加速资金周转的利益驱动下，也迫使混凝土早期强度不断提高。混凝土主要原材料水泥和骨料的生产也就应市场的需求而变化。100多年来水泥生产技术的发展，主要体现在其活性不断提高、强度发展加快。而加快强度发展速率主要靠增加水泥熟料矿物中C_3A和C_3S的含量和粉磨细度。由图2可见，由于专业的隔离，水泥的使用者和水泥生产者都不知道这样的技术措施除提高混凝土强度发展速率外，会带来什么负面影响；而目前粗骨料质量越来越差的原因，也与用户对强度的要求有关。在现有绝大多数落后的骨料生产技术条件下，岩石强度越高，破碎后的粒形越差，针、片状颗粒也越多，以致公称连续级配的石子实际上几乎没有5～10mm颗粒存在，这是目前混凝土用水量和胶凝材料用量居高不下的主要原因。但是只要有人买，石子生产商绝不会顾及对混凝土品质的影响。

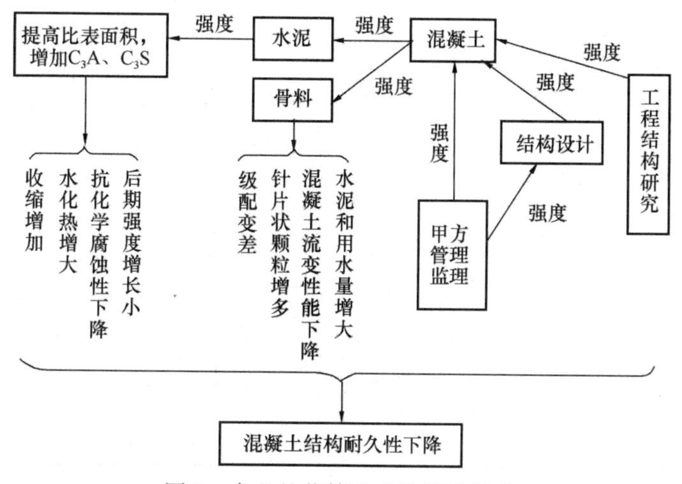

图2　专业的藩篱造成的后果示意

现状正如 A. Nevill 所述："一个人当然不可能具备所有必要的知识，因此设计者可以简单地咨询材料专家，然而，除少数特例外，大多数结构工程师对混凝土的行为都缺乏必要的了解；材料专家又多是一些纯粹的科学家，对结构的问题缺乏了解，结果他们不知道什么问题要回答，而设计人也不知道要问些什么"[2]。工程中出现问题时，不知怎样才能找到真正的原因和怎样从根本上来解决。

不仅如此，国内新兴的房地产商也参与打造了这种专业藩篱。为了某种防范需要，有个不成文的规定：水泥由业主（甲方）供应。而业主对混凝土和水泥的了解更少，更不知道为避免混凝土早期开裂影响耐久性，除了强度之外，水泥应具有哪些品质。

二、打通专业藩篱是时代的要求

分工是人类生活和生产的需要，显然任何个人都不能靠自己独立的活动满足自身的全部需要，必须借助于他人的劳动和产品。最早形成分工的因素是不同的性别、年龄、能力以及所处环境条件的差异，这就是在生产力低下原始自然经济社会的自然分工。随着生产发展和商品出现，生产和劳动的分工逐渐发展成社会分工。随着生产的发展，历史上发生过三次社会大分工：第一次是畜牧业和农业的分离；第二次是私有制出现后农业和手工业相分离；第三次是商人阶层出现后城乡分工。随着生产的发展，社会分工也不断发展，逐渐形成了各个行业。工业革命促使行业精细分工，同时带来工业时代的分科教育。教育的分科越来越细，来源于二次世界大战后航天技术的兴起，使美国向培养顶尖科学家的"精英教育"发展。当前的行业或专业的分科应当说是教育造成的。最早，在美国，工程只分成军事工程(military engineering) 和民用工程（civil engineering)；后来，民用工程如机械、电气、化学等逐渐分出去，最后只剩下土木工程，就沿用了原来民用工程（civil engineering）的名称，因此现今土木工程包括房屋、道路、铁路、运输管道、隧道、桥梁、航道、堤坝、洞库、港口、机场、电站、海上采油平台、给水与排水等所有需要"大兴土木"的工程。随着科学技术的发展，又逐渐出现许多新兴学科。前苏联教育的分科则比欧美国家更细。1952 年，我国高校院系调整主要是"学苏"的结果。在院系调整以前，土木系的学生毕业后可从事土木工程的任何工作，具备很宽的知识面；院系调整后，把土木、水利分开，土木工程则划分成结构、建造（construction，或称施工)、道路与桥梁、给水与排水、暖通、工程测量等专业。由于在工程中发现建筑材料和制品对于土木工程的重要性，在 1958 年增设了建筑材料专业，这也是"学苏"的结果。

分工是科技和生产发展所必需的，精细分工有助于学科的深入。但是客观规律往往是过犹不及。当科技和生产发展到更复杂的程度时，精细分工造成的知识面狭窄和知识的缺陷就凸显出来。越是复杂的科学技术，越需要学科的交叉；越是学科交叉，越是能产生出创新的技术和产品。因此打通专业藩篱是当今时代的要求。

划分过细的专业培养出的人才缺少什么？在我国 10 年浩劫使教育处于停顿状态的年代，美国的教育从 20 世纪 70 年代以后开始返回"注重社会化应用"，开始重视工程教育。近些年来，国内有些著名大学工民建专业毕业的学生，擅长于数学、外语和计算机，却在毕业时还不知道房子是怎么盖起来的。他们到设计院里去做设计，认为只要会用计算机软件就能从事"设计"了。建筑材料专业培养的学生，则力学基础不如"工民建"的，物理化学基础又不如"材料科学"的，更不知道房子是怎样盖起来了。……他们缺少的是工程的知识，缺少一切与工程有关的知识和相互之间的知识的渗透。更重要的是“准确地认识和决策”的“通识能力”[1]。他们在面对工程中的问题时，常常忘记了已经学过的最基本的道理，而被一些人云亦云的思维方式所左右。就好像已有的学识被那些思维方法蒙住了。例如对“大掺量矿物掺和料会降低混凝土 pH 值”的担心，就是忘记了在中学就学过的什么是 pH 值[3]。

为什么过去专业隔离造成的问题并没有像现在这样突出呢？这就是因为“知识经济时代更要注重的不是‘知识爆炸’，而是‘知识换代’”[1]。现代科学技术和生产的发展，越来越需要不仅使各工科专业知识的交叉和渗透，甚至也需要文科和理科这两个传统上认为是风马牛不相干的学科交叉和渗透，更何况密切相关的土木工程大系统内部中的各个专业！如果道路设计者懂得如何布设现今比过去复杂得多的地下管线（上下水、电、暖、通讯、煤气等），能与相应专业交叉合作，就不会发生道路常见“大开膛”的现象了。在建筑工程中，混凝土材料及其原料的生产发生了很大变化，设计人员和施工人员并没有关心，他们的知识和观念还是很久以前得到的，技术措施还是基于那时对材料的了解。因此他们就不理解为什么过去只有水工的大体积混凝土结构才需要控制温升，而现在厚度只有 30cm 的墙也需要控制温度[3]？他们也不理解，既然在钢筋混凝土中混凝土承受压应力，那么在楼板设计中取消了受压区的架立钢筋会造成什么后果？墙的构造筋放在主筋内侧会出现什么后果？反之，由于缺少结构的知识，混凝土生产者常常在工程中出现问题时无法回答用户提出的质疑；由于不懂水泥，当他们遇到“泡在水里的混凝土也会开裂”的现象时也无法解释。

现代科技需要专业之间的密切合作，进而知识的相互交叉。电视中曾播放过一个节目，叙述一位腿部残疾的运动员在世界残疾人运动会上创下 100m 短跑 11 秒 3 的成绩，他所用的假肢是由三个国家的科技人员共同完成的，因为该项技术涉及生物、力学、电子、机械和材料等专业。跨学科研究是科学本质的必然表现，也是当代科学的新特征[4]。不仅本学科各专业需要交叉、渗透、合作和融合，而且自然科学与社会科学的结合也是必然的，且意义更加重大。众所周知的画家达·芬奇同时也是个数学家，集自然科学知识和社会科学知识于一身。近代科学先驱者培根和迪卡尔是首先在科学观和方法论上取得了突破，才推动了科学的进展；经济学家运用自然科学的理论和方法获得了创新的成果，如数学家和经济学家合作创立的对策论、乔治斯库洛根的《熵定律与经济过程》，等等。可以说，时代需要人的知识结构重组。学科交叉是当代科学不可阻挡的强大潮流，正在摧毁几百年来形成的科学知识结构，也冲击着人们头脑中的科学专业化观念[4]。

三、我们应该做什么，能够做什么和怎么做？

● 混凝土技术是进行结构设计的基本知识[2]　材料是工程的基础。不同材料有不同的设计理论和方法，以及不同的施工方法；设计和施工方法必须跟随材料的变化或更新而变化或更新。土木工程中所有的专业或工程实体都与材料有关，任何材料和技术都会有利有弊，有得有失。工程中的主要问题是如何处理使用材料时所遇到的各种相互矛盾的因素，取利弃弊，进而创造条件化弊为利，最大限度地做到得多失少。其中作为结构材料的混凝土则是用量最大宗的、与结构安全性和耐久性最密切相关的材料。一方面，当前土木工程技术人员在学校里所学的混凝土知识只是皮毛[5]；另一方面，现在高校很多有关混凝土材料教材的内容，除了标准以外基本上多年没有更新，知识和观念都已陈旧。如 Mehta 说：“缺乏混凝土技术的整体论教育、缺乏研究和实践的整体观是造成近来几个重要工程项目延误，付出高昂代价的主导原因。从 Gerwick 和 Idorn 两人（他们都与混凝土业联系密切）和我来往的通信表明，缺乏经验的设计师制订的不符合实际的规程常常是一个出问题的原因"[6]。混凝土是一种能用简单技术制作的复杂体系，用于处在环境复杂条件下的结构，绝不仅仅是通过计算就能保证承载安全的，一旦丧失耐久性，就会危害结构的安全。而结构的耐久性主要就涉及混凝土材料。不能脱离耐久性考虑混凝土结构的安全性。因此结构设计人员应当具备这样的复杂体系的知识，并不断更新。

● 重视继续教育　曾任教育部长的何东昌说过："人一生只有 20％的知识是在学校取得的。由于人对自然的认识是不断发展的，知识需要不断更新"。王宏甲在他的书中说道："终身学习是以个人自主的姿态开拓心灵容纳世界的能力，是以不断更新的知识保卫我们生活的边疆"[1]。终身学习，在口头上无人反对，但是人们往往不自觉地对自己不理解或不熟悉的事物产生不同程度的抗拒，而且，在当前商业竞争激烈的情况下，往往也造成企业和个人的短期行为。事实证明，由那些短期行为所取得的效果常常不能补偿它所带来的长期的损失。

我国自实行了注册工程师（结构、监理等）资格考试制度，最近将举办注册建造师认证考试，这对促进技术人员的继续学习会起到积极作用。考前辅导是进行继续教育的好机会，希望注意不要把这宝贵的学习时间用来进行"应试教育"。有远见的企业家应当把人力培训放在重要的位置，爱护和鼓励那些暂时不够应试资格的技术人员学习的积极性，并且能为他们创造条件。目前重视对员工培训的企业太少，老板们总担心经过培训的员工留不住，担心他们水平提高了就会"跳槽"。有的时候，员工在工作中遇到问题时，需要查阅一些资料，这应该属于工作的一部分，但是往往遭到阻拦。今年 5 月的《环球时报》发表了一篇关于美国通用电气公司培训工作的报道，其中说道："美国通用电气公司认为，为员工提供良好的培训是企业留住人才最好的方式。因为只有员工感到自己在企业里不断进步、受到重视、能体现自己的价值，他们才愿意留下"。这是关于凝聚力和人力培训相互关系的辩证法。美国通用电气公司各层次的管理人员，包括最高层的，都是自己培训出来的，而且他们也为其他企业培训或输送人才，其结果是他们的实力影响遍及全世界。

"必须为那些负责设计和施工混凝土结构的人们提供充分的受教育机会。很显然，如果我们想要明天的材料科学家和工程师们能应对本文所描述的挑战（本文作者注：即所引用文献中关于自然资源的保护、工程材料的耐久性和环境污染……混凝土耐久性科学、质量保证和服务寿命预测试验方法，以及混凝土技术教育中存在的不足之处等），就必须在让他们掌握扎实的基础理论的同时，还有足够的现代混凝土加工技术的试验室和现场经验。"[4]

● 首先是要有正确的思维方法　正确的技术决策取决于正确的观念，正确的观念取决于正确的思维方法。采用正确的思维方法学习，才能正确地认识并运用客观规律。现在从各种渠道得到的信息量是很大的，公开发表的论文真有铺天盖地之势；专家也多得很。但是因为每个人都有他们自己知识和实践范围、环境、条件和历史，对某一件事物或问题，有的有共识，有的却会见仁见智。如果没有正确的思维方法，可能从不同来源的信息得到互相矛盾的结论，而无所适从；如果看书、听讲时自己提不出问题，就不能算作学习。带着问题学，学了发现问题，才能真正学好，否则就是人云亦云的盲从。最高层次的思维方法就是哲学。运用辩证法，具体问题具体分析，才是正确的思维方法。存在的不一定都合理，外国的不一定都先进，专家权威不一定都正确。

● 贵在主动　土木工程是一个涉及知识面宽，内涵广泛、门类众多、构成复杂的多分支综合体系，具有很强的综合性。建造一项大型的土木工程设施一般要经过规划、可行性论证、勘察、设计和施工等程序，需要运用工程测量、工程地质勘察、水文地质勘察、土力学、工程力学、工程结构、工程设计、建筑材料、建筑设备、工程机械、建筑管理、建筑经济等学科和施工技术、施工组织等领域的知识，以及电子计算机、工程试验和力学测试等技术。就绝大多数人来说，我们并不需要，也不可能通晓土木工程所有专业知识和技术，但是至少要建立在很多情况下会涉及到跨专业问题的理念。例如设计者、施工者和混凝土技术人员时常会遇到需要了解工程所在地区土质和地下水的情况。

具有丰富的跨专业知识的专门人才在历史上有，现实也存在，这并不是可望而不可及的。关键在是否有一种内在的需要，尤其是同一领域。古埃及第一个金字塔是在公元前 2630 年，由当时第三王朝大臣英霍泰普（Inhotap）设计并主持建造的。他同时是建筑师和工程师[7]。罗马时期索菲亚大教堂的设计者希腊人恩塞缪斯（Anthemius）是当时（公元 537 年）最伟大的建筑师和工程师，同时还是几何学家和物理学家[6]。成为巴黎象征的艾菲尔铁塔（1889 年），其建筑、结构设计和施工主持人，是承担过 20 年钢结构设计与施工的法国工程师艾菲尔（G. Eiffel）[6]。20 世纪 50 年代和 60 年代，就有不少人从工业与民用建筑专业毕业后从事建筑设计，或建筑学专业毕业后从事结构设计、施工的。现在建筑材料方向培养的研究生也有其他专业的生源，这在一定程度上可以扰动已有的专业藩篱。但是从实际工程来说，建筑工程本来就是一项大的系统工程，各专业都不能缺少。打通专业藩篱，各专业都应当主动，主动向相邻专业请教、学习，主动配合，知己知彼。有条件时，主动造就自己成为以我为主，通晓全土木工程知识的人才。

参考文献

[1] 王宏甲．中国新教育风暴，北京出版社，2004 年 8 月．

[2] A. Neville. Concrete Technology——An Essential Element of Structural Design. Concrete International, July, 1998.

[3] 廉慧珍．思维方法和观念的转变比技术更重要，之一：传统思维和从众思维对混凝土技术进步的影响，《商品混凝土》2004 年第 2 期．

[4] 金吾伦．当代科学的新特征，中国青年报，1999 年 5 月 16 日，第 3 版．

[5] P. K. Mehta. Concrete Technology for Sustainable Development——An Overview of Essential Principles. International Symposium on Sustainable Development of Cement and Concrete Industry, October, 1998, Ottawa..

[6] P. K. Mehta. Concrete Technology at the Crossroads——Problems and Opportunities. Concrete Technology, Past, Present and Future. ACI SP144-1(1994).

[7] 罗福午．土木工程史话：做整体结合建筑结构施工的建设者，《建筑技术》第 33 卷第 8 期，2002 年 8 月．

自评

1. 60 多年前“院系调整”时专业过细划分所形成的专业隔离，对土木工程的影响最为显著。发展到现在，不仅不同学科之间难以沟通，而且同一行业体力劳动和脑力劳动分家也有发展。西方国家由施工部门做设计的施工图，则很容易使施工与设计结合起来。而我国，刚从校门进入设计院的年轻人“连一根梁都不会设计”，而且没有施工经验，自然施工图就错误率很高。我刚开始听到这种情况，总有点耸人听闻的感觉，怀疑是否为个例？后来我问设计院的人，得到的回答是："是这样。因为会不会，关系不大，计算机里有现成的设计软件。"难怪有人对我说："现在都在把人培养成机器人"（罗福午语）。混凝土过去由施工单位自己试配，自己拌制，自己成型，制成构件，而现在搅拌站制备出混凝土的拌和物交给那些不懂混凝土的人去完成成型工艺，这已经成为当前能否保证混凝土工程质量的关键问题。

2. 水利、公路、铁路、建筑等同属于土木工程领域，其所用的混凝土，尽管各有其特殊要求和使用条件，但在材料的本质上和基本性质方面是相同或相通的。各专业有各自的长处和短处，相互隔离、各行其是的结果很不利于工程技术的发展。对此似应有进一步的论述。

3. 文章对解决行业隔离的问题提出的建议很好，但是实际上看来，实施的难度很大，其难度主要不在于技术问题，而是利益问题。1952 年院系调整时，那么大的动作，解决得那么快，就是因为那时没有什么利益的纠纷。现在则不然。实际上，从长远出发，可以通过整体的系统分析，平衡各方利益，先在有条件的部门做试点，逐步解决。一旦试点有所突破，利益达成新的平衡，进一步就会迅速展开。

4. 这篇文章现在的题目是我原始的题目，在《建筑技术》刊登时被编辑修改成"建筑技术需要打破专业藩篱"，这当然也无可非议，只是针对性窄了。需要打破专业藩篱的不止是建造行业，而且还需要和其他学科的交叉。

思考与实用的结合，就能产生明确的概念，就能找到一些简便方法，这些方法的发现激励着自尊心，而方法的准确性又能使智力得到满足，原来枯燥无味的工作，有了简便方法，就令人感到兴趣了。

——卢梭

论文之四

思维方法和观念的转变比技术更重要之二
——传统思维和从众思维对混凝土技术进步的影响*

廉慧珍 （清华大学土木水利学院）

1 前言

对同样一个问题、同样一件事，不同的人有不同的思维方法，就会有不同的行为和结果。可能许多人听说过一个关于推销鞋的故事[1]：一个推销员到一个岛上推销鞋。他发现这个岛屿上每个人都赤脚，没有穿鞋的，于是认为，这里的人都不穿鞋，谁买鞋呀？马上通知厂里“不要运鞋来了，这个岛上没有鞋的销路”；第二个推销员来到同一个岛上，一看，哇！每一个人都没穿鞋啊，如果每个人买一双鞋，会有多大的销路呀？马上打电报回厂：这个岛屿上鞋的销售市场太大了，赶快运鞋来。当然，第二个推销员要想让岛上的人都买他的鞋，肯定要做一番宣传、说服工作，还要请人家试穿一下。但是结果，就会因他的行改变了岛上人的生活质量而大开销路。所以，正确的技术决策取决于正确的观念，正确的观念取决于正确的思维方法。

如果只注重单纯的技术而没有正确的思维方法，仍然办不好事，反而会造成相反的后果。举两个混凝土养护的例子：南方某个工程地下的混凝土墙，遵照现行混凝土规范中“拆除模板后要及时浇水养护”的规定，施工人员在混凝土浇筑两天拆除了模板后，立即向墙面浇凉水，但是同时却听见混凝土开裂的声音，发生了严重的开裂。这是因为施工人员只考虑在混凝土具有一定的强度后就可以拆除模板，也注意到必须及时养护；却没有分析应当“及”什么“时”，没有想到，现在的混凝土由于早期强度高，浇筑后两天正值温升最高，“及时”浇凉水使表面温度骤然下降。产生很大的瞬时应力而开裂；另一个例子是，为了防止混凝土板的早期裂缝，目前普遍采取的养护技术就是浇筑后立即覆盖塑料膜，然而，如果是在炎热的日射环境下用于基础底板，这种覆盖就会造成“暖房效应”而促使混凝土内部温度升高。孤立地看这两个例子中的养护技术都不能是错的，但是恰恰错在没有“具体问题具体处理”的思维方法，因循了一种思维定势。

在思维方法上，思维定势有三种：传统思维（或习惯性思维）、从众思维和机械思维，而辩证思维、创新思维才是科学的思维。如果不转变思维定势，就不可能接受新技术；即使道听途说了某些技术，也不能科学地运用。

鉴于篇幅所限，在本文中只就传统思维（或习惯性思维）和从众思维的定势对混凝土技术进步的影响谈一些看法，望与大家讨论。

2 传统思维

环境和条件都在变化，生产技术都在不断发展，如果思维方法毫无变化或变化甚微，就

* 原文以《思维方法和观念的转变比技术更重要之一》刊载《商品混凝土》2004 年第 2 期。

不可能取得技术的进步。目前对混凝土技术进步影响最大的传统思维是“强度第一（甚至唯一)”。例如：

▲ 以强度高低作为水泥和混凝土“好”、“差”之分，认为“抗压强度高的混凝土，其他性能都好”；“早期强度高的水泥和混凝土是好的水泥和混凝土”

这种思维方法使人们对不断提高混凝土强度，尤其是早期强度，孜孜以求。甚至单纯追求什么“超高强”、“特超高强”。实际上，不同材料都有其本身的特点。混凝土是一种黏弹塑性材料。中等强度混凝土在荷载作用下，断裂之前有一段塑性变形过程。强度越高，这个过程越短，则在超载或震动、冲击荷载作用下的安全性就越差。高强混凝土几乎没有这种过程而会突然断裂，如图 1[2]所示。由图 2 可见，混凝土抗拉强度和抗压强度比值随抗压强度的提高而下降。当抗压强度为 20MPa 时，拉压比约为 1/11；抗压强度为 40MPa 时，拉压比则约 1/13；抗压强度为 60MPa 时，则拉压比减小到约为 1/15。

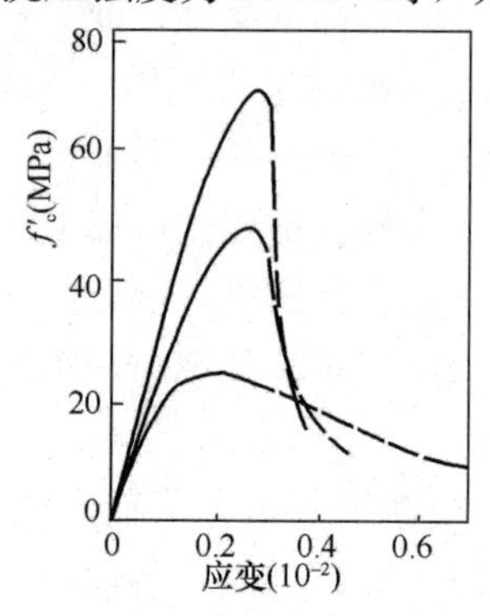

图 1　不同强度混凝土应力应变曲线

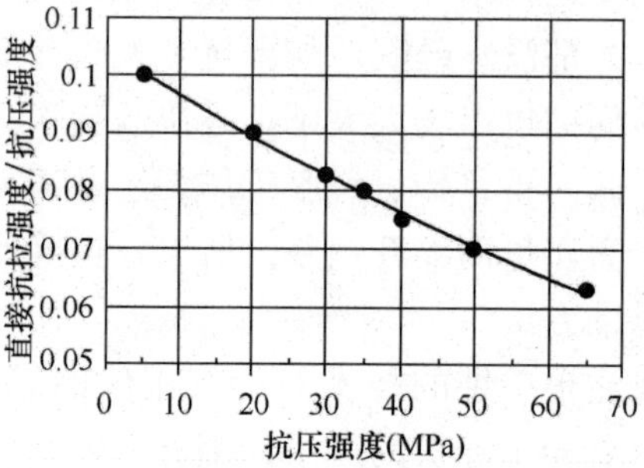

图 2　不同抗压强度混凝土的抗拉强度与抗压强度的比值[2]

图 3 表明在该实验中混凝土抗压强度从 26MPa 提高到 63MPa 时，梁的延性比下降的程度大于抗压强度提高的程度。图 4 表明混凝土抗裂性随抗压强度的提高而下降。

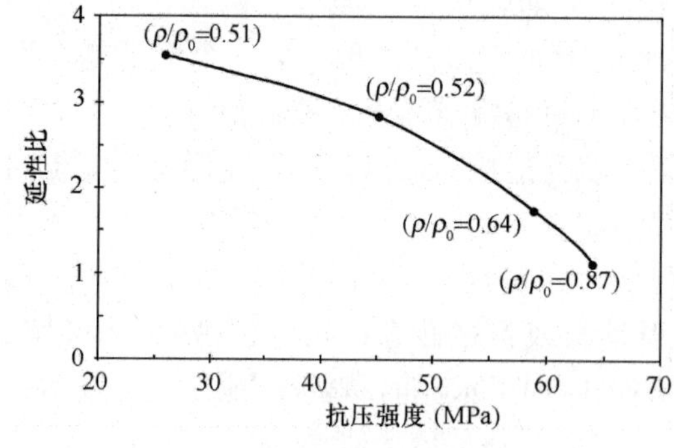

图 3　混凝土强度与无负筋梁延性比的关系
图中 ρ/ρ_0 为钢箍配筋率，增加 ρ/ρ_0 在一定范围内可提高结构延性比[3]

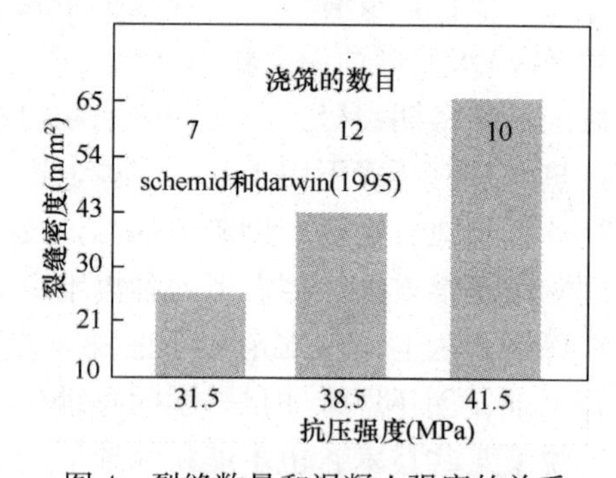

图 4　裂缝数量和混凝土强度的关系（图中抗压强度是按美国标准检测的大约是我国标准强度的 0.83 倍）[4]

高强不一定耐久，也不一定安全；抗压强度高，其他性能未必也高。为了某种目的而超过工程的需要任意提高混凝土抗压强度是有害的。

▲ 主张“高强混凝土必须用高强度水泥”，“建议取消 32.5 水泥”

其实这已经是一个“古老”的问题了。在 70 年代初，我国水泥标准最高强度的水泥是 500＃的，相当于今天的 32.5 水泥。当时配制高强混凝土时，许多人表示不解，因为人们概念中从来就是“水泥强度宜为混凝土强度的1.5～2倍”。直到今天，仍有人坚持这个观念，是因为他们的思维定势是把“强度”当成了绝对值，而不了解或忘记了无论水泥还是混凝土，强度都是在一定条件下——用一定水灰比和配料，成型为一定尺寸的试件，在一定温度和湿度下，养护到一定龄期——所检测的。同一种材料，条件不同时所测出强度值必然不同。

水泥的强度是按标准在固定水灰比下检测的。70年代检测的500＃使用的水灰比大约是0.36（依标准稠度用水量和掺和料而变化），现行水泥标准修订前检测水泥强度使用的水灰比是0.44，而当前使用的是0.5。混凝土的水灰比则是根据需要而改变的。如果混凝土的水灰比减小到低于检测水泥所用的水灰比，则混凝土强度就有可能超过水泥的强度。过去没有减水剂时，如果混凝土的水灰比和检测水泥强度所用水灰比一样，则水泥用量就会很大。因此从施工性和经济性考虑，水泥强度要比混凝土强度高出较多。现在，高效减水剂已经被普遍使用，混凝土的水灰比多低于0.5，当水灰比减小到0.32以下时，用水灰比0.5检测的32.5水泥配制出C60混凝土，不仅可能，而且在水泥标准修订前已用于工程（那时的425＃水泥）。混凝土的强度不再受水泥强度的制约。

水泥标准修订后对水泥强度要求高了，水泥厂多采取提高C_3A和C_3S以及比表面积的措施来提高水泥的强度，同时带来另一方面的结果则是水化热增大、流变性能变差（和外加剂的相容问题）、抗化学腐蚀性下降、收缩增大、开裂敏感性增加。而且早期强度高而后期强度增长率减小，甚至会倒缩（见图5所示）[4]。不同水泥厂生产的相同强度等级和品种的水泥，甚至同一水泥厂不同批量的水泥，其与外加剂的相容性和开裂敏感性可能会有很大的差异。由于现行水泥国标对水泥的强度和比表面积只规定下限而不规定上限（西方国家标准中是有规定的），有的水泥厂的产品实际强度比标称强度高出许多，被大多数用户认为是“好水泥”。活性越大的水泥储存性能越差，因此常有42.5水泥和52.5水泥的28天实测强度差不多的现象，出于价格的考虑，42.5水泥比52.5水泥销路好。但是强度的大幅度提高，增加了水泥的易裂性。

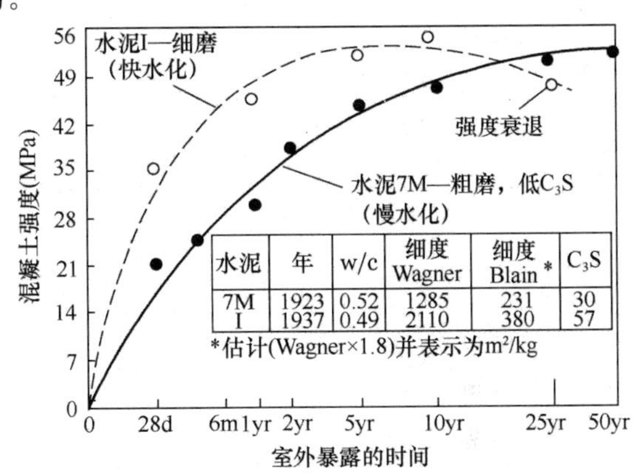

图5 采用快硬水泥的混凝土10年后强度倒缩：1937年按特快硬水泥生产的水泥I与现今水泥的平均水平很相似。在对依阿华州劣化的公路路面钻取芯样的一项研究中，Lemish和Elwell（1996年）也发现10～14年强度倒缩而得出结论：强度增长慢的混凝土与性能良好相关[4]

因此单纯以强度评价水泥和混凝土的“好”“坏”是传统思维造成的误区。应当把这种观念转变成：

① 水泥强度的高限比低限还重要；抗裂性比强度更重要；最重要的是质量均匀；

② 好的混凝土是具有满足工程特定要求的各项性能和具有最大匀质性，体积稳定而不开裂。

▲“纯的水泥混凝土是真货，掺用掺和料的是假冒伪劣的；硅酸盐水泥和普通硅酸盐水泥是正品，其他都是次等产品”

20多年前，我国水泥尽管标准列有硅酸盐水泥，实际上市售的大多数是普通水泥，硅酸盐水泥基本上用于出口。用掺和料主要是的是为了调节标号，增加产量。掺入掺和料由于是简单的等量取代，都以降低早期强度为代价。而且，由于矿渣比硅酸盐熟料易磨性差，造成矿渣水泥易泌水、抗渗性和抗冻性差；火山灰水泥由于需水量大而影响混凝土的流变性和抗冻性；粉煤灰水泥由于粉煤灰含碳量大、质量不稳定，基本上不受欢迎。因此，形成了人们排斥矿物掺和料的思维定势，在对提高混凝土强度的需要日增的情况下，硅酸盐水泥和掺和料低于15%的普通水泥成为市场的主流。推广高性能混凝土以来，由于对混凝土耐久性认识的提高，磨细矿渣、分选粉煤灰及其使用技术的发展，使矿物掺和料的使用日益增加，但是就社会上多数人来说，由于对矿物掺和料的本质、特性和使用技术的不了解，排斥矿物掺和料的思维定势仍在影响着混凝土技术的进步。事实是，在当前水泥强度不断提高的情况下，如果没用矿物掺和料，混凝土结构耐久性的问题将无法解决。在英国，粉煤灰已被列为混凝土除水泥、砂石和水以外的第四组分[5]，将矿物掺和料叫做“辅助性胶凝材料”，混凝土的水灰比改为“水胶比”。科学家预言，总有一天，即使粉煤灰价格比水泥价格高，也必须使用。现在这已成为事实：在距离美国西部大陆约 4000km 的太平洋上一个安静的小岛上，要使用从印度进口的手工精雕细刻的梁、柱，建一座美丽的庙宇。为了保证 1000 年的使用寿命，在软地基上 36m×17m×1m 的筏型基础混凝土采用了 C_3S 为 14%、C_3A 为 1% 的水泥，粉煤灰掺量近 60%。因附近岛屿都没有粉煤灰，必须从美国西海岸的火电厂运来，费用是每吨约 200 美元，是硅酸盐水泥价格的 3 倍。该混凝土已于 1999 年浇筑了 $760m^3$。当 2000 年 7 月发表介绍文章时，仍未发现任何裂缝[6]。我们排斥假冒伪劣产品既针对其假冒，更重要的是针对其伪劣。矿物掺和料对混凝土的贡献能补偿硅酸盐水泥的劣势，如降低混凝土温升、提高抗化学腐蚀的能力、后期强度的持续增长，以及粉煤灰的抗裂性。正确使用矿物掺和料不仅无任何假冒问题，而且对于性能得到很大改善的混凝土何谈伪劣！

对多数混凝土供应商来说，使用矿物掺和料主要的出发点是降低成本，前提是不降低强度。目前市场上的磨细矿渣比表面积都定位在 400～$450m^2/kg$，这主要是按经济原则优化的：矿渣比表面积越大，活性越高。因强度第一的思维定势，用户希望要高活性的，但是生产成本高，供应商利润低；在这个定值以下，售价低，则供应商利润也低。而比表面积为 400～$450m^2/kg$ 的产品从活性到价格，用户能接受，供应商客能得最大利润。尽管比表面积超过 $400m^2/kg$ 的矿渣自收缩随其掺量而增大，并且在目前多数人能接受的掺量下（例如 30%，最多 40%）并不降低混凝土的温升，而强度和降低成本对搅拌站的诱惑是很大的。细矿渣当掺量大于 75%以后，混凝土的自收缩和温升才会随掺量的增加而减小[7]，西方国家允许矿渣在水泥中的最大掺量是 90%[8]，新加坡用于海底隧道混凝土的水泥中矿渣掺量是 75%。在我国现行标准和规范中，这样的掺量是不允许的。我国有的水泥厂已采用熟料和矿渣分磨的技术生产矿渣水泥，尽管标准允许矿渣掺量为 20%～70%，为了保持在标准规定的 0.5 固定水灰比检测的强度为 42.5 以上，实际上矿渣掺量不会超过 40%。从标称强度来说也并不是“次等产品”，而从抗化学腐蚀来说，更优于硅酸盐水泥和普通水泥。

恰恰相反的是“现代水泥性质的变化对混凝土强度发展以及对硬化水泥石微结构带来的影响是侵蚀性离子在混凝土中的侵入性（penetration，包括水的渗透性、离子扩散性和毛细孔的吸附性——本文作者注），因此也是一定条件下混凝土耐久性的控制因素”[9]。水泥强度和耐久性没有直接关系，因为“现代水泥厂的生产改变了，对耐久性产生了不良影响。”[9]

显然用是否用掺和料来评价水泥质量的等级是由于不了解水泥标称强度和水泥对混凝土

强度的贡献是两回事。关键是当前水泥的生产和检测与混凝土性能根本没有关系。一旦水泥厂按混凝土的需要生产，并按混凝土的规律检测，二者才会发生关联。

顺便一提：有人主张取消水泥标准，而直接用混凝土来检测，这又走向了另一个极端。任何产品都必须有标准，否则无法控制产品质量的均匀。比方说药品，如果说其本身没有标准，全要有由人的服用效果来检测行吗？

▲“石子越多混凝土强度越高”，“石子只是填充材料，只要强度高就行”

这又是一个“强度唯一”的传统思维定势，以为骨料作为混凝土的骨架，就是起强度作用。这是一种误解。骨料的骨架作用主要是稳定混凝土的体积。纯的水泥浆体硬化后收缩过大，无法用于结构，必须有骨料对水泥浆体的收缩起约束作用，即使水泥不要钱而骨料很贵，骨料在混凝土中也必须占据大部分体积。至于骨料强度对混凝土强度的影响则随混凝土拌和物浆骨比的增大而减小，只有水泥用量特别小（例如只有 160kg/m^3）的干硬性混凝土（贫混凝土，无坍落度，需高频强力加压振捣或碾压才能成型）骨料的强度才重要。对于当前坍落度在 100mm 以上的泵送混凝土，骨料的强度并不影响混凝土的强度。

传统观念认为骨料粒径大的混凝土的强度高，事实恰好相反，由于界面的影响，混凝土强度随石子粒径的增大而下降，水灰比越低，影响越大（图 6）[2]，混凝土渗透性随石子粒径的增大而增大（图 7）[2]。传统观念还认为砂率大（石子少）时混凝土强度会降低，实际上还是由于界面的影响，骨料总用量相同而砂率增大时，强度有所提高、渗透性减小、弹性模量下降，收缩增大。骨料的表面组织（表面粗糙程度）影响混凝土强度的原因也与界面有关。

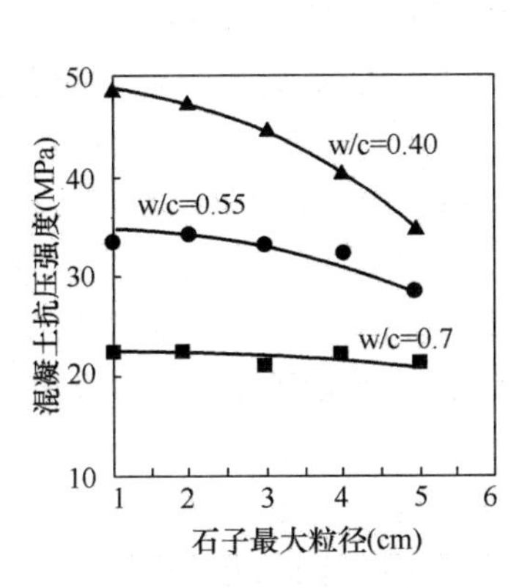

图 6　石子粒径与混凝土抗压强度关系[2]

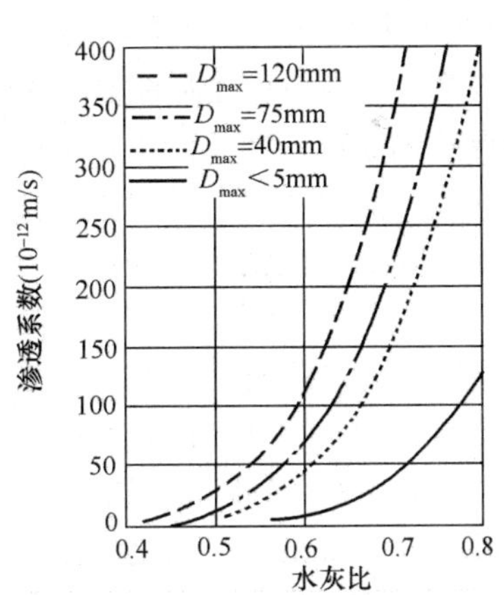

图 7　石子粒径对混凝土渗透性的影响[2]

从骨料性质来说，影响混凝土强度的不是骨料强度，而是骨料的形状和表面特征，对抗弯强度比对抗压强度影响更大些[2]。

鉴于混凝土的体积稳定性是混凝土质量的重要性质，则尽量减小混凝土的浆骨比就成为提高混凝土质量重要的重要措施。

用作混凝土骨料的岩石平均强度都超过 100MPa，即使是强烈风化的低强度花岗岩，其抗压强度也达 80～100MPa；非常坚硬的粗粒花岗岩强度可达 180～200 MPa[10]，因此，不必把强度作为对骨料质量的要求，至少不是唯一。骨料质量重要的是粒形、颗粒级配，吸水率和线膨胀系数等性质。其中石子的粒形甚至比级配还重要。砂子的细度模数并不说明级配，细度模数大而级配很差时，影响更大。

由于普通骨料加工简单，价格低廉，而且看不到骨料对混凝土结构耐久性不可缺少的重要作用，人们除了强度以外不关心骨料的质量和消耗，生产和运输、使用中浪费也严重，加重了天然资源的负担。忽视骨料质量的传统观念还表现在，许多年来已经不再有人根据构件尺寸、钢筋最小间距和保护层厚度来确定石子最大粒径了，设计人员也不考虑所设计钢筋密集程度能否允许现在最小的石子通过，以致常发生钢筋无法被混凝土足够地保护，水平钢筋底部由于混凝土沉降而留下水隙甚至形成水囊，为结构服役期加速钢筋的锈蚀留下严重的隐患。

骨料的质量差已经成为我国混凝土质量的重要障碍。当前骨料生产的现状基本上是经营分散，生产方式落后，几乎无人执行砂石标准。所供应标称连续级配的石子基本上都是10mm以上单粒级的，粒形也很不好。这就使我国混凝土的用水量居高不下，混凝土的裂缝问题存在普遍。造成这种状况的主导原因应当说是市场的需求。也就是说用户对骨料质量的只重视强度，大多不懂技术的供应商只要能卖出去，就不会求进取而提高骨料质量。

3 从众思维

从众思维是一个很普遍的现象。对众说纷纭的看法不加分析，盲目跟从。这种思维定势不仅在人类社会普遍存在，而且动物也有此现象，比如羊群效应。从某种意义上来说，也是正常的，但是往往是有害的，用到科学技术上，更是不利的。混凝土结构工程中的从众思维的形成，很多是出于想当然，或者把局部现象当成整体的规律。这里举两个常见的实例：

例1：担心掺粉煤灰、矿渣后混凝土的pH值下降

众所周知的是混凝土由于硅酸盐水泥水化释放出$Ca(OH)_2$而呈碱性，钢筋混凝土中的碱性环境可保护钢筋免予锈蚀。于是人们就认为掺粉煤灰、矿渣后不仅会稀释而且还要消耗掉$Ca(OH)_2$；人们又知道，pH值表明酸碱度，那么，想当然，$Ca(OH)_2$减少了，pH值必然下降，混凝土的护筋性就下降变差了。但是人们却没有想想什么是pH值？混凝土的pH值怎么测？pH值是[H^+]的负对数，用以表征溶液中H^+的浓度。既然表征溶液中H^+的浓度，则pH值与一定量溶液中的溶剂数量有关。在科研中为了了解掺矿物掺和料的水泥酸碱度的变化，测定pH值的方法首先是用该胶凝材料加水调和制取溶液。为了取得该胶凝材料完全水化后的$Ca(OH)_2$溶液，就要加大量的水，比如胶凝材料：水＝1：100或1：200；然后连续搅拌或震荡到水泥完全水化；用仪器检测溶液的pH值。只要能使水泥完全水化，加水量不同时，pH值当然不同。实验主要是对比矿物掺和料不同掺量的影响，没有绝对值。一般没有“某物质pH值”的说法，而是都对该物质饱和溶液而言，称作“某可溶物质饱和溶液的pH值”。

那么，混凝土是不溶性的固体，pH值并不是固体的性质。一般所说“混凝土的pH值”应当是指混凝土孔溶液的pH值。混凝土中的孔溶液时非常难以得到的，曾经只有极少的人用高压力挤出一些。因此可以说无法测定。但是可知硬化混凝土（尤其致密的混凝土）中毛细孔孔隙率是很低的，由于水泥水化的消耗和蒸发，孔溶液极少，只剩下孔壁吸附的水。不妨通过计算来分析一下：

为了计算方便，假定水泥用量为250kg/m³，按密度为3.15g/cm³计，则体积为80 L；w/c为0.48，则用水量为120 L。忽略骨料吸附的水，只计算水泥浆。在密封的情况下水化

程度为 50% 时，充水的毛细孔约为 67 L[11]，$Ca(OH)_2$在 20℃时的溶解度为 1.65g/1L，假定常温下每升水可溶解约 1.65 g，则 67 L 可溶解约 110.66g。

假设 1g 水泥完全水化，则吸收 0.227g 的水，水化物质量为 306.75kg，完全水化后水化物中 $Ca(OH)_2$约占全部水化物的 20%，则可粗略地认为水化率 50%的 250kg 水泥释放 $Ca(OH)_2$约为 31kg。

由此可推断，在实际混凝土中，即使 50%的胶凝材料被掺和料所取代，水化程度也很低(实际上水化可因实际水灰比的增大而加速)，所生成的 $Ca(OH)_2$也远比维持孔溶液 pH 值所需量多得多；而由于水的蒸发，孔溶液更加少，则只要有很少的 $Ca(OH)_2$，就能使孔溶液饱和，pH 值就不会下降。

注重混凝土的密实性、匀质性和延伸性，以及对由环境作用而引起的开裂的抗力，比担心 pH 值更有意义得多。

例 2：“混凝土结构早期开裂就是因为养护得不好”，“没有不裂的混凝土，出点裂缝没关系”

为什么过去没听说过混凝土结构裂得这样多，而近几年几乎成了普遍现象？公元 12～14 年罗马 Caligula 皇帝时期用石灰和火山灰以 1：2 配合，成功地用于建造了那不勒斯海港，现场观察，至今虽然被海浪磨光了表面，长满青苔，但混凝土却完好无损，数百米长的堤墙几乎无一裂缝。2000 多年后的今天，却有人形容，现在施工期间，地下结构“无墙不裂”。混凝土的养护固然很重要，但是从根本上来说是现在的水泥变了，混凝土也变了。

混凝土材料的特点是多孔而非均质，抗拉强度比抗压强度低得多。当由于各种原因引起的内部拉应力超过其抗拉强度时，就会开裂。如果能控制受约束的混凝土结构中拉应力始终低于其抗拉强度，则裂缝是可以控制的。过去一般只有水工的大坝这样大体积的混凝土才有因内部温升大造成过大温差应力而开裂的问题，但现今厚度只有 30cm 的混凝土墙早期因温度而开裂也相当普遍，“及时养护”都难以避免。还是因为现在的混凝土变了，因为混凝土的要求，水泥也有很大的变化。

对于混凝土的密实性，大家都很重视，但是如果“出点裂缝没关系”，密实性还有意义吗？从整体论的观点分析，凡是组成良好并经适当捣固和养护的混凝土，只要内部孔隙和裂缝尚未形成相互连接而直达表面的通道，则基本上是水密性的；在使用中，结构的荷载以及大气环境的影响如冷热交替、干湿循环，可使这些内部微裂缝发展并传播，成为环境中侵蚀性介质浸入的通道[12]。早期裂缝控制的意义在于，已有裂缝的扩展比新生成裂缝容易，而裂缝是一切导致混凝土劣化的因素侵入的通道。可以说，控制裂缝是得到耐久混凝土的关键。

混凝土结构耐久性和裂缝的关系、裂缝产生的根本原因可从图 8 来分析。

由以上分析可见，建设速度和混凝土耐久性之间的矛盾是主要的。现在不少工程都将自己的“混凝土 3 天强度达 70%、7 天达 90%以上甚至 100%”引以为豪，而且质检部门也这样要求。这恰恰是混凝土结构早期开裂和耐久性下降的主要原因。这样的混凝土，即使早期由于采取某种措施而有幸没有开裂，却掩盖了不可见的微裂缝的存在。在恶劣气候下（特别是干湿交替）会加速混凝土的劣化。事实上，近来许多搅拌站发现，7 天强度达到 100%的混凝土，到 28 天时强度基本上不再增长，甚至还下降。这是因为，事物都有其本身的成长规律，违反或扰乱了其规律，会影响质量，甚至会发生更严重的问题。拔苗助长固然错误，而三个月就长大的肉鸡不如过去一年才长成的家养鸡好吃，也是事实。比如人的生长，如果希望孩子出生后短时间就长成大人而给他喂食生长激素，则问题就严重了。混凝土内部微结构的形成和发展也如此。从整体论出发，混凝土结构密实，同时在早期控制混凝土内部非荷载引起的应力则同等重要。否则再密实的混凝土，一旦开裂也就成了不密实。

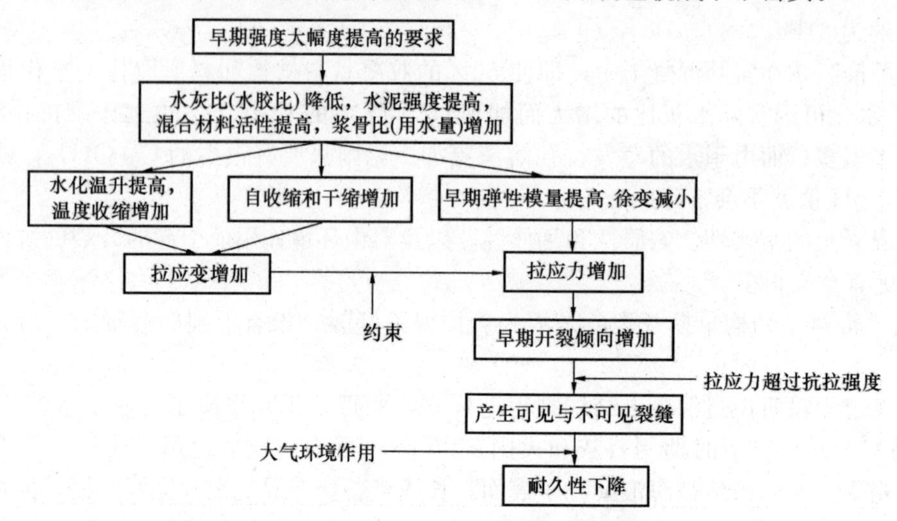

图 8　混凝土早期开裂原因的整体论分析

因此，混凝土结构裂缝的控制不仅是养护的问题，而涉及在设计中如何减小结构中的约束和应力集中、配筋如何配合裂缝的分散和便于混凝土浇筑，材料制作如何提高混凝土抗裂性能等问题。涉及施工的也不仅是养护问题。还包括模板技术、浇筑顺序、振捣方式、温度和湿度的控制、拆模时间，等等，而最重要的则是建设主管方的指导思想。主观意志要符合客观规律，才能使混凝土结构性能正常地发展。

参考文献

[1]　曾国平．创新思维与创造力的发挥．中央电视台 10 频道《百家论坛》2003 年 6 月 6 日(http://www.CCTV.com).

[2]　A. M. Neville 著．李国泮，马贞勇译．混凝土的性能(第一版)[M]. 北京：中国建筑工业出版社，1983.

[3]　ACI 363 委员会. State of the arts report on high strength concrete，1993.

[4]　R. W. Burrows. The visible and invisible cracking of concrete，ACI monograph No. 11，1998.

[5]　M. R. H. Dunstan. Pulverized fuel ash as the fourth component of concrete，ACI SP Proceedings of 2° international conference on utilization of fly ash，slag，silica fume and other pozzolana.

[6]　P. Kumar. Mehta and Wilbert S. Langley. Monolith Foundation: Built to Last a "1000 Years". Concrete International. July，2000.

[7]　E. Tazawa and S. Autogenous Shrinkage by Self Desiccation in Cementitious Material，9th international Conference on Chemistry of Cement，New Delhi，Sept. 1992.

[8]　BS 5382 Concrete Part 1 Guide to specifying concrete.

[9]　Adam. Consideration of durability of concrete structures: Past，present，and future. Materials and Structures. March 2001.

[10]　高等院校统编教材．混凝土学[M]. 北京：高等教育出版社，1983.

[11] T. C. Pwers. Physical Properties of Cement Paste，Fourth International Symposium on the Chemistry of Cement，Washington，D. C. 1960.

自评

1. 我也曾有过追求强度的误区。1974 年系里承接海军后勤部委托的 1000＃混凝土的研究任务。当时常用水泥最高强度是硬练标准 500＃（大约相当于现行标准的 32.5 等级）。当时我国尚无商品化的高效减水剂，只得用纺织业印染助剂 NNO 和 MF 代替萘磺酸盐甲醛缩合物的高效减水剂，当时每吨的价格分别是￥2600 元和￥3100 元，（那时我的工资是 62 元）。可见当年实现高强混凝土的困难。那时误认为配制高强混凝土需要高强度的水泥和高强的石子，所以热衷于寻找辉绿岩和致密花岗岩，总想“最好能有一种提高水泥强度的外加剂”……，因为没有商品高效减水剂，再加上技术水平低，自然也就没有做到 1000＃，只做到 700kg/cm^{2}左右。当硅灰的应用被介绍到国内时，我曾兴奋一时：觉得很神奇。直到我国有了自己的高效减水剂产品，并在全国推广泵送混凝土，才明白混凝土配制技术是原材料的协调。硅灰在提高混凝土强度的同时，也会出现别的问题。任何材料或技术，正确使用是有利的，使用不当就会有害；合适就好，不合适就不好。

2. 例 2 所指对养护问题的误区主要是想说明不仅是养护的问题，但叙述不够清楚，容易使本来就对养护重视不够的工地误解。养护仍然很重要，只是对养护也要转变观念：养护不只是浇水。如果“及时浇水”不适时，就会适得其反。对于现代混凝土，早期温度的控制更加重要，养护的内容应当包括温度和湿度的控制，温度控制和保湿同等重要。

3. 关于骨料强度的问题。过去使用低塑性混凝土时，浆骨比较小，骨料体积约占混凝土体积的 80％，骨料呈紧密堆积状态，混凝土硬化后荷载经骨料传递；如今的流态混凝土，尤其是泵送混凝土，拌和物中浆骨比较大，骨料呈悬浮状态，荷载主要由砂浆传递，骨料强度就不那么重要了。用轻骨料混凝土能配制 C50 以上的泵送混凝土，这也是其原因之一。但是对于高强混凝土，因浆体弹性模量和骨料弹性模量相近，界面较强，骨料强度的贡献会大些。

4. 骨料在混凝土中的作用主要是稳定体积。即使有朝一日骨料价格高于水泥价格，骨料还是必须作为混凝土中用量最大的组分。事实证明，由采石场供应单粒级石子，在实验室用最大松堆密度法进行级配至空隙率小于 40％，生产时按优化的级配分级上料，可以节省水泥约 20％，并可提高硬化混凝土的体积稳定性。但是前提是骨料的粒形要好。

5. 图 8 在本部分论文之二的图 6 中也使用过，后者在“早期开裂倾向增加”的文框下面使用的是“拉应力接近抗拉强度”，而本文图 8 则使用的是“拉应力超过抗拉强度”。国外有报道，实际上拉应力约达抗拉强度的 65％，混凝土就可能开裂。“拉应力超过抗拉强度”实际上是人们的推测，在覃维祖的博士生们用温度应力试验机的实验中也得到证明。因此用“接近”是对的。对其可能的原因分析如下：

①与抗压强度相比，混凝土的抗拉强度很低，直接抗拉强度一般难于检验，目前检测多为间接的方法（劈裂抗拉或抗折试验）；

②对实验方法和影响因素，抗拉强度比抗压强度敏感；

③混凝土强度检测值是在试件加载试验破坏后，所计算的受力断面上最大荷载的平均值。实际上，混凝土是高度复杂的非均质多相体，其力学性质在体内是不均匀的，受力后的

破坏是由最薄弱处断裂引起的，该处拉应力不可能超过所检测的抗拉强度值。

6. 很多人至今总是把矿物掺和料的使用当成假冒伪劣产品；而另一方面，又有很多人为了降低成本，盲目地什么都敢掺。这两种倾向都缘于对混凝土强度来源认识的误区。虽然按照吴中伟的“中心质假说”，矿物掺和料（或混合材）在混凝土中可以替代未水化熟料颗粒参与混凝土微结构的形成，但是必须是对混凝土无害。很多工业所使用的天然矿石都是与黏土质矿物共生的，其废弃物主要成分都会含有与水泥相近的化学成分，但是不能忽略其中也可能含有一些有害成分，未经试验研究和鉴定认证时，是不能随便用的。现在水泥厂32.5水泥的“乱掺”，需要教育和严格管理，不是不能管住的。因为“管不住”而取消该品种，可以说是因噎废食，是一种盲目崇拜标称抗压强度的传统思维。说严重些，也是管理者的不作为。简单的取消，仍然管不住那些“乱掺”的问题。

成功的人，都有浩然的气概，他们都是大胆的，勇敢的。他们的字典上，是没有“惧怕”两个字的，他们自信他们的能力是能够干一切事业的，他们自认他们是很有价值的人。

——戴尔·卡耐基

论文之五

思维方法和观念的转变比技术更重要之三
——再谈从众思维*

廉慧珍　（清华大学土木水利学院建筑材料研究所，北京，100084）

在这里引用《中国新教育风暴》中的一段话："当此经济全球化时代，有很多东西要学，但知识经济时代更要注重的不是'知识爆炸'，而是'知识换代'，这意味着有许多知识你已经不需要学和不需要像昨天那样学了。终身学习是以个人自主的姿态开拓心灵容纳世界的能力，是以不断更新的学识保卫我们生活的边疆"[1]。从众思维存在的普遍，就是因为人们脑子里那些已经"不需要学和不需要像昨天那样学"的知识保留得太多。尤其是土木建筑领域，陈旧而狭窄的知识正在束缚着我们的创造性思维和自主精神。因此，有必要再分析几个从众思维的实例，来促进以下我们思维方法的转变。

从众思维主要是人云亦云，凭想象盲目照办。其根源是对事物的分析"只认其一，不知其二"。例如，就像治病的药品一样，对什么人都适用而包治百病的药是没有的。而且最根本的是增加人体自身的抵抗力。任何材料也都有利必有弊。但是，为了要解决目前困扰人们的混凝土结构裂缝问题，有人总想找到一种灵丹妙药。结果是一哄而上，又一哄而下。曾经被绝对肯定的技术，又遭到简单的否定；绝对肯定是出于从众思维，简单否定也是出于从众思维。任何好东西使用不当，都有可能走向反面。矛盾的转化是有条件的。

例 1　膨胀剂的使用经历

按照辩证法的规律，任何事物都存在对立统一的规律：对立是绝对的，在一定条件下对立双方互相转化是相对的。膨胀剂的发明和应用的过程典型地符合辩证法的规律。最早发现的钙矾石是硫酸盐侵蚀混凝土的产物，因其产生膨胀而破坏混凝土结构，被称作"水泥杆菌"。当这种适当的膨胀发生在混凝土硬化初期的约束条件下时，就可以填充混凝土中的孔隙，增加混凝土密实度，进而可将多余的膨胀能（表现为内部的压应力，即自应力）储存下来，以补偿混凝土收缩或承载产生的拉应力（见图 1 的示意）。膨胀剂的发明正是运用辩证法化害为利的典型。对其他引起混凝土膨胀破坏因素作用的转化也如此。

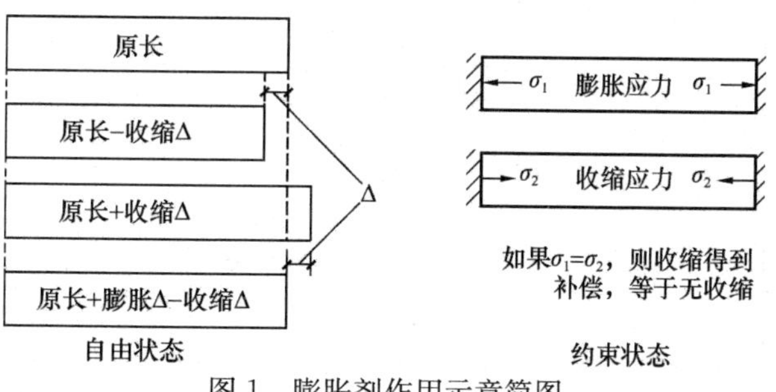

图 1　膨胀剂作用示意简图

我国科技工作者从上世纪 50 年代开始，就成功地研究和生产了各种类型的膨胀水泥，但是几十年后的今天，生产和工程客观条件都发生了很大的变化，许多问题是几十年前所不

* 原文《思维方法和观念的转变更重要之二》刊载《商品混凝土》2004 年第 3 期。

可能考虑到的。例如，在上世纪 90 年代以前，由于膨胀剂尚无大量生产，售价较高，补偿收缩混凝土主要只用于防渗层、抢修、锚固接缝等，构件尺寸都不大。在建筑结构构件（如梁、板、柱）中的应用没有本构关系的实验研究[2]。随着生产的扩大和推广应用的进行，形成了一种思维定势，认为只要掺膨胀剂就能减小收缩而不裂，而忽略了条件。为了防裂，无论墙、板、梁，还是大体积的基础地板，不管强度高低，膨胀剂被当成防裂的万金油。结果膨胀剂不再表现出“万能”。有人反映“掺了膨胀剂比不掺裂的还多”，甚至“掺膨胀剂的基础底板泡在水里也开裂了”的现象也是有的。一些设计单位、主管部门、施工单位因此又走向另外的极端，拒绝使用膨胀剂。由于缺少整体论的实验研究工作，盲目增加产量，而科学技术力量不足，反复是必然的，“物极必反”也是客观的规律。

现就其中需要考虑的主要的条件变化提出如下看法：

1. 水灰比

当时的实验是在 0.5 左右的常用水灰比下进行的。当水灰比从 0.5 减小到 0.45 时，膨胀水泥的膨胀率增加[2]。这是因为，较大水灰比的混凝土中较大的孔隙率可吸收较多的膨胀能，而较致密的混凝土才能产生较大的膨胀。当时没有水灰比小于 0.45 时的规律。现今的混凝土水灰比普遍小于 0.5，甚至 0.45。图 2 是清华大学阎培渝教授和他的研究生实验的结果。该图表明，掺膨胀剂的混凝土水灰比从 0.5 到 0.4，膨胀率随水灰比的减小而增大，与早先的结果相符，而水灰比进一步减小到 0.3 时，则膨胀率减小。即使从理论上来说，使水泥完全水化而毛细孔隙率最少的水灰比是 0.437[3]，水灰比低于 0.4 以后，混凝土中的水也是不足的。水灰比很低时，混凝土中的自由水很少，而水泥水化则随水灰比降低而加快，与反应时需要大量水的膨胀剂争夺自由水。同时，膨胀剂中重要组分 $CaSO_4$ 的溶出量随自由水的减少而减少。因此当水灰比低于 0.4 时，膨胀剂的效果会受到影响，膨胀剂会反应不完全而残留，在有水存在的环境中还可能继续反应而生成二次钙矾石。

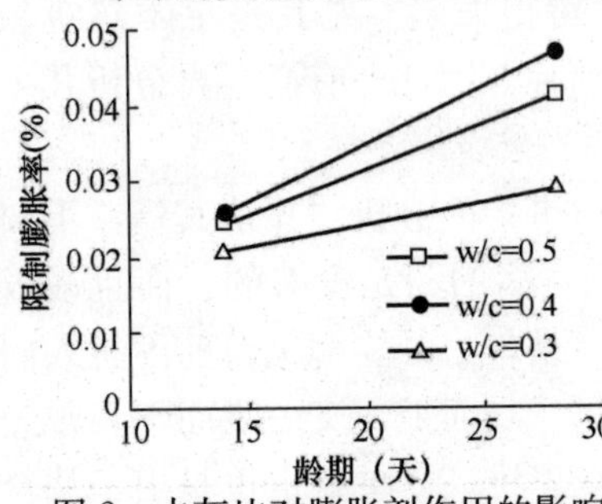

图 2　水灰比对膨胀剂作用的影响

2. 构件尺寸

由于水泥生产的变化和混凝土强度的提高，现代混凝土构件断面最小尺寸超过 30cm 时，如不采取有力措施，都像过去的大体积混凝土一样，内部会产生很高的温度。膨胀剂中的膨胀组分无论是钙矾石的生成还是 CaO 的水化都与温度有关。已生成的钙矾石从约 65℃开始脱水，同时在此温度下新的钙矾石不再生成；CaO 溶解度在冷水（10℃）和热水（80℃）中分别为 0.165g/L 水和 0.094g/L 水[4]。而厚度超过 80cm 的基础底板在常温下浇筑时，其内部温度都会超过 65℃。在混凝土降温以后，反应可继续进行。但如果内部温度很高，或构件尺寸太大，降温缓慢，在混凝土强度发展到较高时再反应，产生膨胀，可能就成了对结构不利的“延迟生成钙矾石”。因此控制大体积混凝土内部的温度对掺膨胀剂的混凝土来说很重要；掺膨胀剂的大体积混凝土最重要的是控制温升。控制温降则是对任何混凝土都至关重要的。“……除了（膨胀）能通过限制使混凝土内部建立一定压应力之外，其他如干缩、温度变化以及结构上的使水泥混凝土产生裂缝的各种因素，对膨胀水泥混凝土仍然同样起作用。在设计时必须注意到这点”[5]。此外，在对混凝土温度进行控制时还应考虑到，膨胀剂不仅不能降低混凝土的温升，而且在反应过程本身也发热。

3. 矿物掺和料的掺用

矿物掺和料对混凝土的膨胀有抑制作用，但是实验表明，不同掺和料的作用不同，对尚无实验根据的掺和料，需要在使用前做实验。

如图 3 所示，粉煤灰掺量对膨胀效果的影响是显著的。在一定掺量下（如 20%），养护方式的影响最大。这正说明使用膨胀剂的混凝土掺入粉煤灰后尤其需要加强水养护以补充水分。粉煤灰掺量较大时（如图 3 中的 40%），由于 28 天以前粉煤灰基本上不参与反应，尽管水胶比已相应降低以达到相同强度等级，而水和水泥的比值却增加很多，则密封养护和水养护的效果就差别不大了。因此掺膨胀剂的混凝土中粉煤灰掺量不宜太大，并加强水养护。

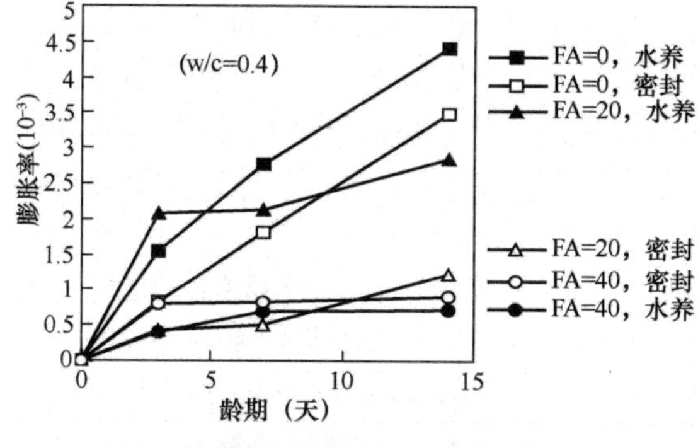

图 3　粉煤灰不同掺量和不同养护方式对掺膨胀剂效果的影响

图 4　天然沸石凝灰岩对掺膨胀剂混凝土膨胀效果的影响

混凝土龄期越长，粉煤灰抑制膨胀的作用越显著。但是其他品种掺和料的作用就未必与此相同。例如图 4 所示天然沸石岩的作用，显然是初期（7 天）对膨胀不仅不抑制，反而是促进的，并随掺量而增加；28 天才开始有所抑制；龄期越长，掺量越大，抑制效果越明显；SO_3越多时，抑制作用出现得晚些。对膨胀的这种早期促进、后期抑制的作用恰恰符合掺膨胀剂的目的，对混凝土结构是有利的。

实践表明，不仅水泥与外加剂有相容性问题，矿物掺和料也有，看来膨胀剂与不同掺和料的相容性，对现代混凝土的发展来说，也是需要研究的问题。

例 2　使用合成纤维的利和弊

大约在 10 年前，美国杜拉纤维公司开始在中国推广（说是推销也不无恰当）合成纤维时，实际上是“只说其一，未说其二”的。确实，纤维增强混凝土在美国、加拿大已有不少用于地面和场坪，表现出其在平板状结构中的抗裂作用。近年来国内针对混凝土大体积基础底板、墙、楼板和梁的开裂问题，也按照杜拉纤维的规格生产出合成纤维进行推广应用，由于每 m^3 混凝土只增加 40 元左右，目前使用量在不断增加。有人甚至还想用以代替膨胀剂！在已经使用的工程中，实际上是有的能不出现可见裂缝，而有的照样开裂。如果以为又找到了另一种万金油，再这样下去，有朝一日也会物极必反。

但是任何材料或技术都有利必有弊，各有各的用途，而且不同材料都有适合于其特性的使用方法。使用任何材料都需要了解其在使用中的利弊，采取有针对性的措施，取长补短。

1. 合成短纤维在混凝土中的主要用途

①由于变形模量低，在混凝土中掺入合成纤维，当混凝土受力破坏断裂后，纤维还能有较大的变形而使混凝土裂而不断，从而可提高结构的延性比，有利于在地震、爆炸等灾害情

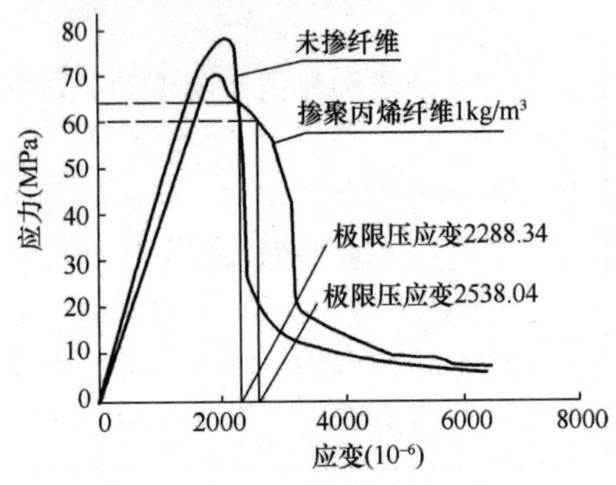

图5　掺和不掺聚丙烯纤维的高强混凝土应力-应变全曲线

况下提高安全性。由图5所示掺和不掺聚丙烯纤维的混凝土应力-应变全曲线可见，同样配比的高强混凝土掺入聚丙烯短纤维后，极限应变提高了约10.9%。有助于提高混凝土结构的延性比。

②用于平板结构，如场坪、楼板、基础底板、承台等结构的混凝土在终凝以前，尽管在一定条件下可产生很大的塑性收缩[6]，而塑性状态的混凝土弹性模量则极低，由塑性收缩所产生的应力就很小，但因为此时混凝土尚未产生抗拉强度，即使是很小的拉应力，也会造成开裂，产生塑性裂缝。在混凝土中掺入乱向的合成短纤维后，由纤维来抵抗这种塑性状态很小的应力，可减少或防止塑性裂缝的产生。此外，掺入合成纤维后，混凝土拌和物黏聚性提高，能阻止大颗粒的沉降，避免沉降收缩引起的裂缝。

③C20～C60混凝土弹性模量为20～35GPa，而目前市售聚丙烯纤维的弹性模量为3.5～4.5GPa，比混凝土弹性模量小一个数量级。因此合成纤维的极限延伸率比混凝土的大得多，例如混凝土极限拉伸变形为（150～200）$\times 10^{-6}$，而聚丙烯纤维极限变形率则为15%～20%，相差3个数量级。可见从理论上来说，用合成纤维提高硬化混凝土的抗裂性能是不可能的；也就是说，不能用于抵抗温度收缩和干燥收缩造成的开裂。

2. 使用合成纤维的利和弊

① 掺入合成纤维后，混凝土的抗渗性提高，但氯离子扩散系数增大，碳化深度增大，如表1所示（中关村建设集团中心实验室李玉琳的实验）。

表1　掺不同纤维的混凝土强度和碳化实验结果

名　称	掺量（kg/m^3）	抗压强度（MPa）	抗折强度（MPa）	渗水高度*（mm）	Cl^-扩散系数（$10^{-9}cm^2/s$）	碳化深度（mm）	
						28d	60d
不掺纤维	—	57.3	7.95	125	18.9	0.3	0.5
国产聚丙烯纤维	1	54.2	8.57	—	28.2	0.5	1.0
杜拉纤维	1	53.5	8.3	—		0.7	1.2
聚丙烯+粉煤灰				50			

*水压为1.5MPa。

这是因为介质在不同孔径中的传输机理不同。水在孔径大于100nm的混凝土毛细孔中传输以渗透为主，由于毛细作用而充水后，压力水的渗透符合达西方程；在孔径小于100nm时，水已无法在孔中形成黏性流（层流或紊流），介质是以一个一个分子或离子迁移的[7]，即由于吸附的扩散。当某种离子在混凝土内外存在浓度差时，能沿孔壁由高浓度处向低浓度处迁移，故以扩散为主，符合菲克定律。掺入纤维切断了可渗水的较大毛细孔通路，但却增加了纤维和水泥浆体界面小于100nm的毛细孔，氯离子扩散系数和碳化深度的增加即缘于此。在有侵蚀性介质存在的环境中使用合成纤维不利于保护钢筋。

② 拌和物黏聚性增加，坍落度减小，强度有所损失。如果混凝土配制技术跟不上，则混凝土拌和物就很不好用。

3. 合成纤维的指标和使用技术

目前无论是合成纤维的生产厂家还是用户，对目前市售合成纤维的质量只注重其分散性（这当然是很重要的指标）而对纤维在混凝土中的作用从众地深信不疑。目前极少有人研究符合混凝土要求的合成纤维指标。从纤维的作用来说，合成纤维的单丝抗拉力、纤度、长度和外形比（纤维的长度和纤度的比值，粗略地说即长细比）对混凝土的性质都有重要的影响。

①克黛拉力：从材料来说，合成树脂的抗拉强度可以很高，但是制成纤维以后，必须以单丝抗拉力评价。纤度是纤维粗细的量度指标，单位是“黛”，符号是D，即Denier。纤维长9000m、重1g，为1D。合成纤维的单丝抗拉力（以g / D计，简称克黛拉力）很低，而且在混凝土中的作用在当纤维是单向分布而且有足够长度时才有效。克黛拉力和纤维的纤度有关，亦即纤维越粗，克黛拉力越大。但是由于外形比的要求，纤维不能太粗，否则就会太长而影响拌和的性质。

②临界长度和外形比：如果纤维在基体中按平行于受力方向单向分布，纤维对基体抗拉应力的贡献与纤维的长度有关，如图6所示。图6中的临界长度$l_{拉}$的意思是当基体裂缝正好在纤维中部时纤维正好被充分利用的长度（图6b $l=l_{拉}$），如果裂缝不在纤维的中部，则短端就会被拔出。

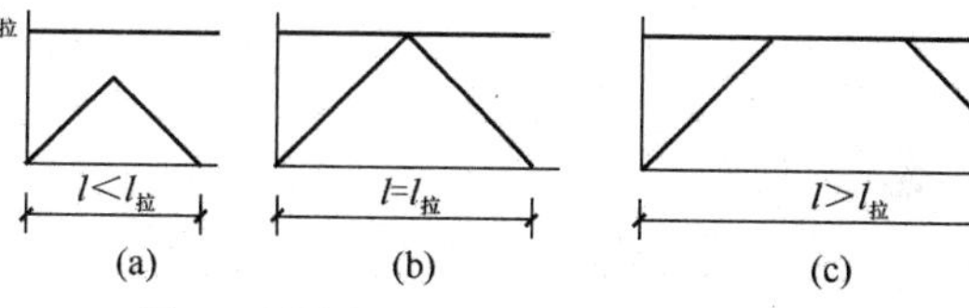

图6　纤维长度与纤维拉力效率的关系

如果纤维长度小于临界长度（图6a），当基体断裂时。纤维就会被拔出；由于短纤维在混凝土中是乱向分布的，一般总是要求纤维的长度比临界长度大几倍，以减小被拔出的机率。长度不足而在混凝土中又是乱向分布的短纤维，其效率只有0%～20%。要提高纤维效率而又不严重影响混凝土拌和物搅拌的纤维存在一个临界外形比。目前国产纤维都仿效杜拉纤维：长度18、19mm，纤度12～15D。试验表明，外形比主要影响工作性，对于提高混凝土拉压强度比与氯离子扩散系数来说，纤维的纤度比外形比更重要。增大纤度和长度能缓解掺纤维造成的氯离子扩散系数减小，并提高拉压强度比，如表2和表3（引自李玉琳的实验）所示。

表2　不同长径比的纤维对混凝土中氯离子扩散系数的影响

编号	长度(mm)×纤度(D)	长径比	纤维掺量(%)	纤维比表面积(mm^2/g)	Cl^-扩散系数(10^{-9} cm^2/s)
X55	空白	—	—	—	18.9
X56	19×15	349	0.1	1.79	28.2
X57	20×25	321	0.1	2.51	21.8
X58	24×25	385	0.1	3.01	20.8
X59	28×25	450	0.1	3.52	20.4

③ 混凝土的配合比：掺纤维的同时必须调整混凝土的配合比，以改善简单掺入时的损失。实验表明，掺用合适纤度和外形比的纤维同时掺用粉煤灰，对抗渗性和抗裂性的提高的效果都比单掺水泥时的明显。还要注意，因为纤维是被水泥浆体握裹而起作用的，所以适用于富浆的混凝土，混凝土中粗骨料粒径要小。

表3 掺不同纤度不同长细径比** 纤维的混凝土的主要性能对比

编号	纤维纤度(D)	纤维长度(mm)	长径比	纤维掺量(%)	坍落度/坍落流动度(mm)	28d抗压强度(MPa)	劈拉强度(MPa)	拉压比(%)
X55	—	—		—	235/ 600	74.0	3.40	4.70
X56	北京15	19	394	0.1	225/ 580	68.5	3.60	5.26
X57	北京25	20	321	0.1	220/ 580	69.8	3.95	5.66
X58	北京25	24	385	0.1	200/ 500	69.0	4.15	6.01
X59	北京25	28	450	0.1	200/ 500	69.5	4.30	6.19
X60	张家港13	19	425	0.1	180/ 450	62.0	3.0	4.84
X61	杜拉13	19	425	0.1	190/ 450	62.0	3.05	4.92
X62	北京15	19	394	0.3	220/ 450	65.5	3.85	5.88
X63	北京15	19	394	0.5	165/ 400	63.8	3.50	5.49
X64	北京25	20	321	0.3	190/ 450	62.5	4.10	6.56
X65	北京25	20	321	0.5	175/ 400	61.8	3.95	6.39
X66	北京25	28	450	0.3	190/ 400	62.5	4.25	6.80

** 按圆形截面、密度0.91g/cm^{3}计算聚丙烯纤维直径：13D为0.0045mm，15D为0.0048mm，25D为0.0062mm。

由以上二例的分析可见，任何材料都有其本身使用的规律，不可能是万能的，不能只听一面，人云亦云。正像医药一样，为什么俗话说“是药三分毒”？因为任何药都有使用的适应性和剂量、方法，用对了能治病，用错了能害人。西药的副作用哪怕只有万分之一的可能，也必须写在说明书中，为什么混凝土的各种外加剂不能在说明书中有其使用方法之外，还标明可能产生的副作用及注意事项呢？也可能是商业行为，也可能是商家自己也不清楚。那么，包括设计人员、施工人员和材料生产者在内的使用者就应当牢记：具体问题具体分析是最重要的方法。任何材料和技术的使用都会有得有失。如果能具体问题具体分析，针对具体问题采取整体的措施，即使扬长而不能完全补短，至少可以既能解决主要矛盾而又能尽量减少弊端。人不能只靠药物维持生命而最根本的是人的自身抗体，同样的道理，提高混凝土的抗裂性不能依靠外加剂，外加剂只能起一些辅助作用，最根本的还是要提高混凝土自身抗裂性能。

参考文献

[1] 王宏甲．中国新教育风暴[M]．北京：北京出版社，2004.

[2] 陈恩义．钙矾石类膨胀剂各组分作用机理及其应用的研究[M]．清华大学工学博士学位论文，1995.

[3] T. C. Powers. Physical Properties of Cement Paste，Proceedings of the Fourth International Symposium on the Chemistry of Cement，Washington，1960.

[4] 简明化学手册[M]．南宁：广西人民出版社，1978.

[5] 薛君玕，吴中伟．膨胀和自应力水泥及其应用[M]．北京：中国建筑工业出版社，1985.

[6] 岩崎训明．尹家辛，李景星译．混凝土的特性[M]．中国建筑工业出版社，1980.

[7] A．E．Шейкин，Ю. В. Чехвский，M. N. Бруссе. 胡春芝，袁效敏，高学善译．水泥混凝土的结构与性能[M]．北京：中国建筑工业出版社，1984.

自评

1. 任何产品的品质与供需双方都有关。供应方应当懂得用户的真正需求，而不是迎合落后；而用户的要求不科学或不合理时，也难以有高质量的产品。双方的水平都不高时，就会互相误导。高水平的供应方应当主动了解并使自己的产品适应用户并不懂得的实际需要；正如牛根生所说："培训顾客是销售的最高理念"。另一方面，用户自己应提出科学的要求，帮助供应方改善产品品质。

2. 文中关于对合成纤维性质的给出了正确的概念，例如表示纤维粗细程度的"纤度"，表示纤维抗力的"克黛拉力"，影响混凝土性质的"外形比"（即俗称的长细比）……，有利于合成纤维的正确使用。

3. 表 2 显示合成纤维会使氯离子扩散系数有所增大，可能因为纤维的掺入会减少混凝土中的大孔，却会增加纤维和浆体界面的小孔。提高纤维的长度、纤度和长细比可缓解此影响。尽管变化幅度不大（不超过 1 个数量级），却呈规律性变化。这个发现值得注意。

4. 不是所有合成纤维对抗干缩开裂都无效，现在国内外都已开发出其他具有良好抗裂性能合成纤维和无机纤维，应当做系统性的试验研究和科学的介绍。

5. 市售聚丙烯纤维合成纤维因单丝很细，极限变形太大，不能抵抗硬化混凝土的开裂，但是在拌和物中可以缓解沉降，而且可抵抗硬化前混凝土塑性收缩产生的很小拉应力，因此可缓解塑性收缩引起微细裂缝。对于道路工程的路面混凝土收面的工作量很大，常常在摊铺或浇筑后来不及覆盖和二次搓抹，掺用聚丙烯纤维不失为一种防止塑性开裂的措施。平面式构件的塑性裂缝初始宽度一般不会超过 0.1mm，深度多不大。没有经验的人往往会对其忽略，则在硬化后保湿不足，就可能成为后期的开裂源，工程表明，未及时处理的塑性裂缝发展最深可达 5～10cm。因为从能量的角度来看，有裂缝的发展比新生成裂缝容易。

6. 关于大体积混凝土中温度对膨胀剂生成钙矾石的影响，文中所述"已生成的钙矾石从约 65℃开始脱水，同时在此温度下新的钙矾石不再生成"，这是根据钙矾石的差热曲线得到的。差热试验是干热条件，而初期的大体积混凝土中是湿热条件，不能将前者的试验结论用于后者。

7. 有些误区是商业运作的片面性所致。现在正在研究并应用新型膨胀剂的专家低调、踏实、实事求是的作风，会为膨胀剂带来否定之定的新发展。不同矿物掺和料对膨胀剂的作用不同，与矿物掺和料和膨胀剂品种、成分、掺量都有关。例如用燃煤固硫渣由于含不同量的 $CaSO_4$、高活性 CaO 和 Al_2O_3，当掺量不同时，有的会抑制膨胀，有的却能促进膨胀或制成膨胀剂。必须深入进行机理，掌握其规律以正确使用并物尽其用。

郭保林评

在从众思维的"例 2 使用合成纤维的利和弊"中"1. 合成纤维在混凝土中的主要用途②"中弹性模量的提法不当。弹性模量的定义是指材料在弹性变形阶段其应力和应变比例的系数。不适合描述处于塑性状态混凝土的性质。对塑性收缩开裂的过程可解释为：因塑性状态混凝土的极限拉应变率很小，当塑性收缩变形达到实时混凝土极限拉应变时，混凝土便会开裂。

论文之六

思维方法和观念的转变比技术更重要之四
——对标准、规范的认识和使用*

廉慧珍　（清华大学土木水利学院）

1　前言

“无规矩无以成方圆”。产品与标准的关系、工程行为与规范的关系就如同方圆与规矩的关系。这就是标准、规范的严肃性所在。在计划经济年代，标准和规范具有绝对的权威。当时我国正处于建国初期，社会生产发展水平低下，技术落后，对标准和规范的更新没有很迫切的要求，因而一旦制定了标准和规范往往保持一、二十年不变。长此以往，执行监督和管理的人员只按本本、条条要求进行监督和检查，技术人员一切唯规范是从，过分依赖规范，视规范为法律，使他们的行为变成只对规范负责。即使在具体情况面前心存疑惑，也不敢越雷池一步，甚而至于，即使明明知道现行规范的规定已不再适用于当前的具体情况，也照办不误。一旦工程中出现问题，就可将责任推给规范。然而，改革开放使我们看到了外面正在飞速发展的世界，近年来我国工程建设也达到了空前的速度。工程特点、原材料的组分及其产生的性质都有了很大的变化。过去的结构多是简单的梁、板、柱，而现今的工程越来越复杂。例如，过去只有水工结构的大体积混凝土才有温控的问题，而现今房屋建筑中厚度只有30cm的墙体也会因未控制混凝土的温度而产生温度应力从而造成严重的开裂；又如，过去工程建设周期长，混凝土结构的强度低，现今施工速度加快，混凝土的设计强度提高，普通混凝土中水泥组分与细度同30年代的快硬水泥的组分与细度相当，而1937年采用当时的快硬水泥的混凝土10年后强度开始倒缩[1]；现在的混凝土组分比过去复杂得多……。使用思维方法和观念转变或更新之前制定的标准，机械地照章办事，常常引起工程中产生一些矛盾。再加上长期形成的土木工程领域各专业隔离的状况，造成各自按照自己的理解制订各行业或行业中某些分支的规范，相互之间难免有不协调之处。而在现今的绝大部分技术人员的观念中，凡是规范中白纸黑字的条文都是必须遵守的法律，遇到在同一工程中因具有不同技术要求采用不同规范而其内容互相矛盾时，使工程技术人员常不知所措。由于科学技术的飞速发展，我们常面对一些需要更新的技术；传统产品或技术已无法满足正在增加的复杂功能和在严酷环境中使用的工程的要求。新技术的推广常受到人们对标准和规范认识的传统思维的束缚而阻力重重。

我国有许多长期专门研究和编制标准、规范的专家，他们对规范有很深刻的见解。本人在这里仅就在工程中遇到的一些矛盾，从思维方法和观念的转变上就个人的理解略抒己见，与大家讨论。

---

* 原文为《思维方法和观念的转变比技术更重要之三》刊载于《商品混凝土》2005年第1期。由国家行政学院研究室收入《落实科学发展观的伟大实践》第三卷“改革、创新与发展”第二部分“增强自主创新能力”中（中国档案出版社出版），2006年5月。

2 标准和规范的性质

2.1 标准、规范是科学技术与实践相结合的产物

在我国《实施工程建设强制性标准监督规定》条文释义中说道："标准是以实践经验的总结和科学技术的发展为基础的，它不是某项科学技术研究成果，也不是单纯的实践经验总结，而必须是体现两者有机结合的综合成果。实践经验需要科学的归纳、分析、提炼，才能具有普遍的指导意义；科学技术研究成果必须通过实践经验才能确认其客观实际的可靠程度。……"

有人认为，科学就是计算、演绎、推理，而实践经验即使经过总结提高，也不是科学。这是一种观念的误区。科学是反映自然、社会等客观规律的知识体系，是实践的结晶。任何科学都是以实践为基础的。其一，科学技术总是从人类实践活动中发现现象和问题，通过不同的方法、不同途径进行研究的；其二，科学研究的最终目的是服务于实践的，通过研究，认识客观世界，按照世界客观规律改造世界。所以，不能把科学技术和实践割裂开来。计算、演绎、推理是科学，试验、观察、总结也是科学。用科学的方法发展成的工艺、操作技能就是技术。把人们在实践中的经验和现象、问题加以总结、分析，研究其规律，经过科学试验，上升到理论，然后又到实践中去检验，修正并补充，最后制订出对实践活动指导性的说明、建议和规定，这就是标准和技术规范。以上过程多次反复，不断地相互补充、促进、提高，就会使生产和科学技术不断进步。

因此，标准、规范要尽量反映科学技术的发展，但是永远是滞后的，并不代表也不可能代表最先进的技术。为此，建设部第 81 号令《实施工程建设强制性标准监督规定》的条文释义中还说，"标准与科学技术发展密切相连，标准应当与科学技术同步，适时将科学技术（作者注：似应为'科学技术的最新内容'）纳入到标准中去。科学技术是提高标准制定质量的关键环节。反过来，如果新技术、新工艺、新材料得不到推行，就难以获取实践的检验，也不能验证其正确性，纳入到标准中也会不可靠。为此给出适当条件允许其发展，是建立标准与科学技术桥梁的重要机制。"这才是实事求是的观点。正因为如此，不能用已有标准、规范去束缚新技术、新工艺和新材料的使用。当需要突破现有规范的规定时，可以也应当有试验研究的依据，试点应用和专家鉴定。工程技术人员都应当为其提供试点应用的条件，以促进科学技术的进步和建设质量的提高，并为标准、规范的修订提供依据。

2.2 标准、规范是妥协的产物和最低的要求

1）制订标准、规范的依据主要是已经发表或公布的、经过相当数量的实践检验过的信息和数据，其中也包括参考借鉴国外的经验，而现有上述资料往往因实验或实践时条件、环境、方法、材料等因素的差异而呈现离散性，必要时应当进行重演性试验验证。但是制定标准、规范基本上是案头工作，一般没有实验研究的经费支持，绝大多数只能基本上利用已有资料和数据进行分析、归纳、研讨，最多进行一些局部的试验，于是有的规定只能是由规范的制订者们协商，不可能准确无误而不可逾越。任何材料和技术都不能放之四海而皆准，对任何具体应用对象来说，都有利必有弊，再加上土木工程在客观上存在许多不确定的因素，因此对存在大量不确定因素的土木工程来说，标准的条文应当尽量避免过细、过死（例如对上下限的规定，上限不能太低，下限不能太高），如果规定过细、过死，则不免会形成一种

悖论。

2）标准的制订必须有若干人（应当尽可能地多）参与，广泛地征求意见。不同人有不同的经历、经验和条件，因而就有认识、观点和见解上的差异，难免会产生不同的意见。最后通过协调而采用一些大多数人都能够接受的意见，暂时舍弃那些争议最大而目前尚无法统一的意见。标准、规范既要有可操作性，又不能迁就落后的现状；既要纳入那些目前普遍尚无条件但能够创造条件实行的先进技术，又不能采用只有极少数人掌握的先进观念。因此要有原则地妥协，才能使所制定的标准、规范定稿。

3）标准、规范是最低的要求。技术人员应当根据具体工程的特点和条件（如所处环境，功能要求、重要性等）采取高于规范要求的措施，如果仍照搬标准和规范的要求，就会造成建设质量的问题。美国 ACI 混凝土结构设计规范的第一章、第一句话："本规范提供设计与施工……的最低要求"；美国公路部门 ASSHTO 桥梁设计规程的第一章第一节中写道："本规程无意取代设计人所具有的专门教育和工程判断的训练，仅在规程中规定为保证公共安全的最低要求。业主或设计人可能需要在设计中采用新的先进技术，或需对材料及施工质量提出更高的要求"。因此不能认为机械地照搬规范就万无一失了。

2.3 标准、规范是法律吗？

我国和建筑有关的法律、法规和标准、规范分为三类：(1) 法律、(2) 行政法规、(3) 技术标准。

1）与建筑规范有关的法律是由国家主席令颁布的，例如国家主席第 91 号令《中华人民共和国建筑法》、第 94 号令《中华人民共和国防震减灾法》、其他如《招标、投标法》、《合同法》、《标准化法》，等等；

2）与建筑有关的行政法规是国务院令或建设部令颁布的，国务院第 293 号令《建设工程勘察设计管理条例》，建设部第 68 号令《工程建设违法违纪行为处罚办法》、建设部第 81 号令《实施工程建设强制性标准监督规定》等，又如《建设工程质量管理条例》以及《强制性标准》或条文，等等。

3）技术标准按照《标准化法》以层次分类划分为：国家标准、行业标准、地方标准和企业标准。标准是标准和规范的泛指，包括产品标准、产品应用标准、质量标准、技术规范、技术规程等（凡国家标准都冠以 GB；行业标准按行业代号汉语拼音字母开头，例如开头字母，建筑行业用 J，市政工程用 C，水利行业用 SL，铁路行业用 T，交通用 JT，等等）。其中产品标准主要用于检验产品的合格性和匀质性。产品的国家标准一般针对全国通用性产品，例如水泥标准、砂石标准等。同一种产品的产品标准和应用（或使用）标准（或规程）不能相混。产品标准和产品说明也不是一回事。也就是说，产品标准只用于检验产品质量，而产品说明是对用户使用产品方法和注意事项的建议，产品标准和使用不一定有直接的定量关系。例如水泥标准中水泥的强度等级是统一按照水灰比 0.5 检测的。但是混凝土的水灰比是可以变动的，掺入不同掺和料时，要按强度要求在相同坍落度下进行比较，则低强度等级的水泥也可配制高强度等级的混凝土；或者说，不同强度等级的水泥可以配制出相同强度等级的混凝土。因此水泥用户对水泥关心重点应当不是它的公称强度，而是公称强度以外的性质，例如实际强度是否超出下限和上限，开裂敏感性如何，最重要的是质量均匀性。但是产品标准不仅用于检测产品质量的均匀性，而且也影响或决定产品本身的质量，也就会影响用户的使用效果。产品是为用户服务的，必须考虑这一点。地方标准是根据地方环境和资源特点制订的，或者地方作为对国家标准中未涉及的内容进行补充而制订的；企业标准用于非通用产品或试生产的新产品的检验。

与建筑规范有关的法律和行政法规不仅有法律效力，而且本身就是法律或具有法律性质；为了保障人民生命财产、人身健康，保护环境和公众利益，建设部还从国家标准中摘取了很少量条文作为强制性标准，也具有法律效力。凡是涉及技术的国家标准、行业标准、地方标准、企业标准则本身不具备法律效力，只有经过委托方和承担方（如建设方和实施方，即甲方和乙方）双方协商，共同选择，在合同中约定作为具体工程中使用时才具有法律效力。发达国家的这种合同的最后都附有双方契约的规范清单。制订标准、规范的目的是指导和制约工程中的行为，以保证工程质量。如果标准、规范变成了科学进步的障碍，就有悖于制订标准、规范的初衷了。从思维方法和观念上转变对标准、规范的认识，才能正确地使用标准、规范。现在，一旦工程中出现问题，当事人首先就是从成百上千、上万的各种标准和规范中去找依据，把标准、规范当成“救命稻草”，依赖标准、规范去对工程负责的这种推卸责任的现状，是长期计划经济的产物，对技术进步和工程的百年大计是很不利的。英国的标准（BS）在每一份的前言最后都声明：依从英国标准并不能免除其自身的法律责任。就是强调工程技术人员的责任感和创造性。

3 关于“与国际接轨”的问题

自从我国加入 WTO 之后，近年来，“与国际接轨”成了一些标准修订的理由和目标，从原则上来说，一般无可非议。但是问题在于对此问题也有个思维方法问题：“接轨”不等于机械照搬。例如在许多方面，国外也并没有完全统一在 ISO 下。对于土木工程来说，各国都有自己的标准：美国：ASTM，英国：BS，日本：JIS，德国：DIN，北欧：NT-BUILD，欧共体：EU，……。长期以来，美国人公开发表的文章一直是用英制单位，最新的 ASTM Designation C1157 中规定，国际单位制和英制都允许使用。近年来美国出版的图书中也有两种单位制并用的情况，例如美国已出版 14 版（每版都多次印刷）畅销 70 多年的名著《Design and Control of Concrete Mixture》，在 2000 年第 14 版中才在英制单位的数据之外，又另列出国际单位制单位的数据，以资对照[2]。又如美国的砂石粒径在本国仍用英制，对外时才换算成国际单位制数据。这是尊重本国的习惯和国情的表现。而这样做也没有违背“与国际接轨”。当然我们“接轨”后也是用自己的编号，但是值得注意的是，存在的不一定都是合理的。例如新修订后的现行水泥标准，因为要“与国际接轨”，造成了水泥实际强度* 比 20 年前提高了两级（约 20MPa）（见表 1 所示）。原来的意图是想用这种方法淘汰小水泥厂——水泥的强度越高，早期（3 天、7 天）强度和 28 天强度比值越大，在新标准中，因水灰比的增加，使同样熟料的水泥公称强度下降，而对上述比值的规定没有相应的下降，势必要增加早强的矿物组成。小水泥厂做不到就会被淘汰。但是因为标准中没有强度上限和细度（筛余量）下限的规定，那些难于提高熟料强度的水泥厂，为了生存，只能采用磨细的措施。现在市售水泥的 0.08mm 筛余量大都低于 5%，甚至还有低于 1%

* “水泥的实际强度”实际上是不存在的，因为在不同标准条件下的检测值是不同的，这里指的是用相同水灰比和相同条件检测的强度。例如 GB 175—63 用 0.36 左右的水胶比检测的 500# 水泥强度，大约相当于现行水泥标准用 0.5 的水胶比检测的 32.5 水泥强度。所以实际强度（也可叫做真值）是测不出来的。

的，有的已无筛余。大力提倡早期强度的结果，使不少混凝土强度增长的能力过早地耗尽，甚至到28天后发生倒缩。

表1　三次修订水泥标准的主要变化

	GB 175—63	GB 175—77	GB 175—92	现行标准
检验用水灰比	0.36左右	0.44、0.46	0.44、0.46	0.5
同熟料通用水泥标称强度	—	—	725	62.5
	—	—	625	52.5
	—	525	525	42.5
	500	425	425	32.5
	400	325	325	取消
其他变化		增加R型水泥	取消7天强度检测 加强早强意识	取消7天强度检测 加强早强意识
实际 C_3A	5%～7%	≥8%	≥8%	≤8%
细度	0.088mm孔筛余≤15%	0.08mm孔筛余≤15%	0.08mm筛余≤10%实际生产比表面积多为330kg/m²	0.08mm筛余≤10%实际生产都≤5%，比表面积超过350 kg/m²
最高强度等级（标号）	500#	525#	625#	62.5

“接轨”本无可厚非，但是不要把国外已发现问题而且正有改变趋势的旧事物“接”过来。西方国家也在批评现行标准与当前技术发展的不相容。过去检测水泥强度用0.5的水灰比是因为0.5是当时混凝土常用的水灰比，自从推广高强混凝土和高性能混凝土以来，用量较大的混凝土水灰比常低于0.5。而外加剂在高水灰比和低水灰比的拌和物中的表现是不同的，所以现在广泛使用外加剂的混凝土拌和物流变性能的问题突出了出来[3]~[5]。又如，矿物掺和料比硅酸盐水泥对水灰比敏感性大，水灰比越低，越敏感。我国水泥标准修订后，0.5的水灰比使掺用掺和料的水泥的公称强度下降更多，在当前用户的习惯认识上难以接受。

1992年美国公布了一个《Standard Performance Specification for Hydraulic Cement》(水硬性胶凝材料标准性能规范)，即对水硬性胶凝材料以性能的规范取代组分和掺和料掺量的规定（ASTM Designation C1157）。对于混凝土，在美国，混凝土基于性能的规范（performance-based specifications for concrete）的话题在当前混凝土业也日益受到重视[6]。

总之，不能关起门来“接轨”，而是要“知己知彼”，要很好地研究国外的前沿信息和国情。在标准制订过程中，要利用各种媒体发表制订标准的思维方法和观念，引起广泛的讨论，广泛发表意见，尤其要充分重视对不同意见的充分讨论。

4　当前标准、规范的问题

1）由于历史的原因和计划经济时期管理的分条、分块，长期形成的“行业”隔离，对于本属于一个领域的土木工程中同样的混凝土结构工程，建筑、水利、道路（铁路和公路也是分隔的）桥梁、港湾、市政（虽然也有道路和桥梁），其他如电力、冶金、化工等部门都有各自的标准、规范、规程或规定。由于工程特点和信息来源以及知识组成、认识观念的差别和编制时间的不同，彼此缺乏沟通，各自编制的方法和内容有不少差异，甚至有互相矛盾之处。例如，过去房屋建筑基本上没有因混凝土温升而发生的问题，现在，高层建筑、重载的结构物的基础底板、地下连续墙等比起水工的大坝，其体积小得多，甚至有些厚度只有30cm左右的墙也需要控制温度。但是目前有关大体积混凝土的规范除了水工以外，只有冶金建筑的相关规程中有一些规定。因水工混凝土有其特殊性：石子粒径大（最大150mm）、水泥用量很低，故其他“行业”就参照“冶金建筑”的规定：构件断面最小尺寸超过80cm的为大体积，并规定了水泥最大用量的限制以降低混凝土的温升。但是，因为大体积的基础筏板，往往还有防水、抗渗的要求，而有关防水混凝土的规范和上述大体积混凝土中对水泥用量的规定正好相反，使用者感到无所适从。又如不同行业的技术规范中对混凝土中含碱量就又有好几种不同的规定……。不同行业工程的结构各有其特点，但是工程结构的基本力学规律和结构所用材料的力学、物理、化学性质却没有本质上的差别。即使在同一行业中，混凝土材料的标准、规范也会因制订者各从自己所从事专业的角度出发而造成相互矛盾的规定。这种行业和专业隔离造成的标准、规范繁杂、混乱状态到了必须改变的时候了。山西省建筑设计院顾问总工程师胡德鹿在他编写的《胡德鹿说新规范》一书（未正式出版的参考书）中说得好：“目前我国经济体制正经历着由计划经济向市场经济转轨的巨大变化。通过投标竞争、承包、合同等形式打破行业封闭，统一的包括土木工程各行业的基本建设市场正在形成。而规范的不统一成为保护行业局部利益的不利因素，并可能阻碍技术进步，不利于我国设计单位在加入WTO以后参加建筑市场的竞争。……”

2）由于传统思维和从众思维，有时会把一些实践中的局部经验当成普适性规律，而使用标准的人如果以机械思维去照搬，就可能造成错误。例如对于混凝土的养护“拆除模板后要及时浇水”的规定，原则上不错。但是，有个工程在混凝土浇筑两天拆模后立即向混凝土浇水，结果当时就发生严重开裂。其原因就在于，按传统，过去混凝土内部温度达到峰值的时间是浇筑后的3～5天，而现今的混凝土可在两天左右温度最高（夏季可能超过80℃），突然受到凉水的冲击，因内外温度变化不均匀而产生很大的瞬时应力。只提供最低要求的规范可能会不够细致，但使用规范的人应当有正确的判断。另外，由于缺乏系统性的重演性试验与实践验证，有些标准或规范中所涉及的数据难免有不严谨之处。例如，在标准《粉煤灰混凝土应用技术规范》GBJ 146—1990中，通过对使用不同水泥配制混凝土中粉煤灰允许掺量的计算发现[7]，按照该标准，硅酸盐水泥和普通硅酸盐水泥中矿物掺和料允许掺量所占份额远小于在矿渣水泥中的，也小于火山灰水泥的。可见在制订该标准时的根据并不是很充分，是在一定程度上“拍脑袋”的结果。另一个例子是，《高强高性能混凝土用矿物外加剂》GB/T 18736—2002规定磨细矿渣含碳量不超过3%，其根据是什么？矿物掺和料的烧失量是在1000℃的马弗炉中灼烧后的失重量，而用于水泥混凝土的矿渣是1600℃的高炉炼铁排放的熔渣经水淬而成的粒化矿渣，一切能在1000℃下烧失的有机物和碳早已不存在，剩下的可能性只有碳酸钙了。但是在炼铁时加入的助熔剂CaO在高温下已熔融而以化合态存在于水淬矿渣中，由游离的CaO碳化形成碳酸钙的可能性不大。如果还有烧失量，那就只可能是外掺了石灰石或混入有机物等。当然3%的烧失量倒也无大碍，只是限制烧失量还不如限制含铁量。

3）由于思维方法和观念没有得到转变，尽管对有些新材料已经制定了相应标准，而检测的方法却仍然是用于检测传统水泥和混凝土的方法，因而常得到不符合实际工程情况的结果。现举例如下：

● 例1：使用膨胀剂的混凝土，在约束下才能储存膨胀能补偿收缩，无约束的自由膨胀会使混凝土体内疏松甚至开裂。因此按普通混凝土强度的检测方法，成型1天后拆模养护，此后处于自由状态，于是按龄期检测的试件强度比不掺膨胀剂的试件强度下降；改为带模养护后，试件强度就比不掺膨胀剂的试件强度提高了。实际工程结构中的混凝土都处于约束下，故带模养护反映的结果是合理的。

● 例2：过去，传统混凝土水灰比比较大，混凝土的收缩主要是干燥收缩，自收缩很小，可以忽略；现今混凝土水灰比较低，水泥活性又大大提高，则自收缩已不可忽略。自收缩在混凝土初凝就开始，1天可达全部自收缩的约三分之二[8]。按现行标准，试件成型后1天拆模测初长，则已经忽略了大部分自收缩。而水灰比越低，自收缩和干燥收缩的比例越大。对于大体积混凝土来说尤其不能忽略。

● 例3：在实验室检测混凝土性能指标的现行标准所规定的条件都是针对传统混凝土的，例如养护温度、龄期等。当使用较大掺量（例如大于20%）的粉煤灰时，粉煤灰在28天以前基本上不参加反应，即使为达到相同的28天强度而降低水胶比，水和水泥的比值（水灰比）也比不掺粉煤灰时的大；粉煤灰掺量越大，水胶比越低，而水灰比则越大。因此掺粉煤灰的混凝土早期内部孔隙率是有所增大的。但是60天以后，掺用粉煤灰的各种优势随着龄期的增长而凸显，这对于混凝土的耐久性很重要。都在28天检测混凝土的碳化、抗冻、抗渗等性能，对掺粉煤灰混凝土是不公平的。建筑工程中的大体积基础底板由于早期限制混凝土温升速率以控制开裂，常减小硅酸盐水泥用量，并采用缓凝剂，造成强度发展延迟，但并不影响竣工后结构的承载能力（反而有利于耐久性）。发达国家对这种需较长时间才达到额定上部荷载的结构混凝土的验收龄期规定为90天[9]。现行规范一律要求（强度）验收龄期不能超过60天，根据是什么呢？

以上只是新材料新技术和在传统思维下制订的规范之间不相容的少数实例。其他还有很多。正如P. K. Mehta所说[10]：“建筑规范……压制材料的再生利用。许多过时的规范限定工程必须使用一些特定范围的材料和配合比，而不是规定一些性能标准。例如许多联邦、州和都市的规范，常常限定混凝土拌和物里粉煤灰的最大掺量或水泥最小用量。用大掺量粉煤灰混凝土制备的高性能混凝土拌和物表明，指令性的规范是陈旧的，应该由以性能为基准的标准来更替”，“规范里像混合水泥中最大允许粉煤灰掺量，以及最小硅酸盐水泥用量这样苛刻的限制是陈旧的、是过去的遗物，应该废弃，而用性能导向的、足够灵活的标准来代替”。

5 结束语

土木工程技术问题是很复杂的，尤其是混凝土材料的性质，不仅影响因素太多，而且随时间变化，人们对这种材料的认识仍在不断发展。标准、规范不应是阻挡科学技术进步的障碍。采用新技术、新材料、新工艺需要突破现行标准、规范时，要有科学依据，经过试点和专家论证。当前实际情况往往“没有经过鉴定的不能使用”，同时“没有在工程中使用的不能鉴定”。其中的关键问题是工程技术人员能否为新技术提供试点工程。实际上已经有不少有识之士为许多新技术的工程试验提供过条件，正是他们在为标准的修订作贡献，在实践中不断为标准、规范的修订提供依据，促进科技不断进步。美国工程师协会制定的《工程师道德规范》[11]（Ethics，有人译成伦理守则）中就有这样的内容："工程师是创新和进步的发起人……，工程师经常根据科学和技术的发展更新知识提高能力……"。

由于标准、规范本身并不是法律，应当提倡工程开始前的"概念设计"和契约规范：——业主、设计、施工、材料、监理、质检等有关各方通过合同规定所采用的标准和规范以及现行标准规范未涉及的规定。而合同是具有法律效力的。

至于产品标准，只用于保证产品质量和匀质性，只能制约生产厂商，而不能要求用户服从。比方说，面粉出售时必须符合面粉的出厂产品质量标准，而不能规定用户对面粉的使用。产品标准制订者不能干涉用户对该产品应用标准的制订（尽管可以提建议）。但是产品是为了用户使用的，用以检验和控制产品质量的产品标准必须为用户着想，便于使用，并保证无害。所以生产者和标准制订者必须了解用户，了解产品和使用的关系对用户的影响，满足使用的要求。有人竟然宣称"我们水泥就是要高硅酸三钙、高强度、高能效，你们混凝土用得怎样那是你们的事，我们不管"。这种"大爷"作风显然违背市场经济规则。在现阶段产品的本质就是商品，商品的用户是上帝，产品必须为用户服务。工程中各环节往往具有多重身份：产品生产者本身同时也是用户，都要为用户着想。水泥厂是水泥原材料的用户，混凝土生产者是水泥的用户，建造者是混凝土的用户，建筑物或构筑物业主是该产品的用户，又以此服务于社会，最终的用户是广大人民群众。因此美国工程师协会指定的《工程师道德规范》规定："工程师是有责任保证科学技术与人类社会之间联系的公民……"。

存在的不一定都合理，国外的不一定都先进，专家、权威不一定都正确。有正确的思维方法和观念，不断更新知识，才能创造性，真正地对工程负责。

参考文献

[1] Richard. W. Burrows. The Visible and Invisible Cracking of concrete，AMERICAN CONCRETE institute monograph No. 11，ACI. 1998.

[2] Steven H. Ksmatka，William C. Panararese. Design and Control of Concrete Mixtures，14th edition，PCA，USA，2000.

[3] A. Tagnit，M. Baalbaki and P. C. Aïtcin. Calcium-Sulphate Optimization in Low Water/Cement Ratio Concretes for Rheological Purposes，Proceedings of 9th International Conference on Chemistry of Cement，New Delhi，1992.

[4] P. C. Aïtcin，C. Jolicoeur and J. G. MacGregor. Superplasticizers：How They Work and Why They Occasionally Don't，Concrete International，May 1994.

[5] P. C. Aïtcin. Cement of Yesterday and today ; Concrete of tomorrow，CEMENT AND CONCRETE RESEARCH，Sept. 2000.

[6] Peter Tayloy. Performance-Based Specifications for Concrete，CONCRETE NTERNATIONAL，August，2004.

[7] 廉慧珍，阎培渝．土木工程中的哲理和混凝土工程技术发展的关系，《建筑技术》2003. 1.

[8] 安明喆．高性能混凝土自收缩的研究，清华大学博士学位论文，1999.

[9] ACI committee 363. State of the Arts on High Strength Concrete，1994.

[10] P. K. Mehta. Greening of the Concrete Industry for Sustainable development. CONCRETE INTERNATIONAL. July，2002.

[11] Charter of Ethics of the Engineer，CNISF May 12，2001.

自评

1. 规范这个名词在英文中的对应词：一个是code，来源于拉丁文的codex，有法典、准则的意思。钢筋混凝土结构设计规范的英文名称就是：Code for design of reinforced concrete structure，因为涉及安全性。另一个是Specification，例如ACI 301 Specifications for Structural Concrete。我国技术规范都是Specification，连我国的英汉词典对Specifications的解释都只有"规格、详述、说明书"，和法律沾不上边。技术是发展的，技术规范必须及时修订，并且具体问题具体分析和对待，允许有条件和有根据地突破。故不是法律。

2. 第2.3条的"3)"关于技术标准的层次阐述不明确。当前有些人将此技术标准的层次理解为级别或等级，认为层次越高，越具有权威性，"下级服从上级"。这是误解。实际上层次越高，越是具有原则性，越是最低要求。各个地区乃至于各具体工程，都必须根据自己的具体情况制订出相应的地方标准或局部的标准，规定更高的要求。例如欧盟有欧洲标准，但是欧洲各国都有自己的标准；我的一个学生现在美国从事结构设计，他告诉我，美国的标准ASTM不是美国的国家标准，而是对各地方制订标准的指导，各个州甚至几乎每个具体工程都有适合于自己具体情况的自己的标准（应当说是根据具体工程的标准或规范的细则）。可以说是"老百姓说了算"。ASTM因其发布标准和规范的水平而具有权威性是众所公认的。

3. 第3条的"1)"中所述关于行业隔离造成规范的不统一，并引用胡德鹿的论点，分析得不全面。确实不少人反应，"现在对同样事物，不同标准、规范的规定不统一，使人无所适从。"而希望统一。这真是我们土木工程特有的问题。土木工程涉面很广，不仅行业种类有区别，而且行业内不同工程也各有特点，同样使用混凝土，会要求各异，如路面板和房屋楼板，要求就不相同；同样为大体积混凝土，央视大楼和面积200m×200m厚度最大处达9m的基础和水坝，对混凝土的要求就有很大差别；再者，我国幅员辽阔，东、西、南、北的气候条件、地质条件、资源条件的差别可称得上悬殊。对混凝土结构工程使用的混凝土技术性质如何统一！"统一"是有条件的，那就是在对混凝土本质和基本概念达到相对一致的认识条件下，统一对混凝土的基本要求，最低要求，以保证混凝土结构的安全和耐久。这种技术规范中的规定是一般性的（即普适性的）、原则性的、指导性的。行业标准、地方标准或企业标准的要求都应高于国家标准。建设者和工程技术人员可以在合同中进行选择或部分条文选择。问题不是"没有统一"的规范，而是没有统一的认识。因为制订标准的人员认识和技术水平的差异，是不断进步的，因此不能把技术规范孤立地看作法律而强制，而应当由建设者和技术人员交流、沟通，共同理解和选用适合于自己工程的规范与标准，由具有法律效力的合同来保证。

4. "和国际接轨"正如文中所述，本"无可厚非"，只是目的要明确：为什么接轨？接什么？怎么接？例如水泥，我国解放前没有自己的规范，解放前用的是英国的的路数，解放后学苏，和苏联不和后又改回"软练"（学英国），再后来要"与ISO接轨，"却"走的是欧洲标准"。可见一直在"和国际接轨"，而"国际"却也不是统一的。我国水泥产量已经达世界第一，我们有更多的话语权，为什么我们不能有自己原创的标准？原因只能说是对"水泥怎么使用"的问题没有想明白。美国1992年根据世界混凝土的变化发布了第一版ASTM

C1157《水泥* 标准性能规范》，至今已修订到 2011 年版。这是一个美国原创的标准，虽然未必完善，但是对我们应当有启发。

5. 由于规范是人定的，就必然和必须是妥协的产物，妥协结果的水平和参与人员的认识、经验与知识水平有关。例如，在对什么是水泥都没弄清楚而没有统一认识的情况下，硬要规定什么“最小水泥用量”；再如，并不知道现在的“包罗米公式”并非包罗米本来的公式，而是后人根据自己所处的条件而附会出来的，却规定必须使用；……。这些都是没有道理的。因此应当根据工程和科学技术的发展及时修订。每个技术人员都有义务为标准与规范的更新提供数据、经验和意见，使标准、规范不断提高水平。

郭保林评

“60d 龄期后的掺粉煤灰混凝土的优势随着龄期增长而凸显”，对于 w/b 不大于 0.45，在标养环境中的掺粉煤灰混凝土，这一规律是普遍存在的。而对于粉煤灰掺量超过 20%、w/b 介于 0.35～0.42、内部最高温度超过 55℃的结构中混凝土，采用跟踪养护，可在龄期 3d 左右达到甚至超过标养 28d 时的强度值。这种情况下，粉煤灰的技术优势不需要等到龄期 60d 后才逐渐发挥。”

生命的意义在于美好，在于向往目标的力量。应当使征途的每一瞬间都具有崇高的目的。苏联高尔基伟人的生平昭示我们，我们也能使自己的生命令人崇敬；当我们告别人生的时候，在时间的沙滩上留下自己的脚印。

——郎贯罗

* 原文是 Hydraulic Cement。在英文中，cement 并不只是水泥，而是胶结料或胶凝材料之意，包括水硬的（如硅酸盐水泥、掺混合材的硅酸盐水泥）和非水硬性的（如石灰、石膏等）。

论文之七

思维方法和观念的转变比技术更重要之五
——性能检验表现的差异缘于对变化了的材料使用了不变的方法*

廉慧珍（清华大学土木水利学院）

前言

对任何事物的评价都必须有一定的方法和指标。方法和指标是人定的，和人对客观规律的认识有关。随着科学技术的发展，方法会趋向于更科学；不同的方法自然会有不同的指标。用相同的方法所规定的指标也会因为是经过实践中统计得出的而发生变化。这样的事例有很多。如果我们的思维方法和观念不转变，用老方法评价新事物，必然会出现一些令人不解的现象，甚至会误事。科学研究的重要意义就在于不断提高对客观事物的认识，不断修正和更新已有的技术，不断前进。但是混凝土这一行业方法随材料的变化却步履艰难。分析其原因，一是人们认为从施工工艺来说，浇筑、振捣、养护是简单的过程，“在许多国家里，没有公认混凝土工需要培训、考试和认证。通常认为任何站在那里无事干的人都能直接就去浇筑或捣实混凝土”[1]；二是技术人员继续学习、接受新的信息不够，受传统思维方法和观念的束缚，普遍存在从众思维；三是土木工程安全性和经济性的影响很大，人们总习惯于保守地对待新技术，殊不知保守的并不一定是保险的，这也是观念的问题。实际上从材料科学的角度看，混凝土是用最简单工艺生产的最复杂的体系，更新观念才能不断认识这种复杂的体系，才能不致于出现混凝土结构提前劣化、“新的不如老的”的现象。

近十几年来，随着“高性能混凝土”的推广应用，混凝土技术不断发生变化。高效减水剂、膨胀剂、粉煤灰、磨细矿渣等有机、无机添加剂的使用日益广泛，混凝土的水灰比降低，商品混凝土发展迅速，……。同时也出现了一些困扰技术人员的问题，例如，“高性能混凝土的抗裂性”（这种提法是把高性能混凝土当成是混凝土的一个品种。其实，开裂了还能叫做高性能?），“商品混凝土更容易开裂”（开裂和“商品”有什么关系?），“掺粉煤灰的混凝土早期强度低，抗碳化性、抗冻性差”（是现场的混凝土吗?），等等；一方面许多混凝土结构提前劣化，另一方面在试验室检测性能很差的混凝土，在现场却有很好的表现；……。如此种种，实际上都是方法的问题——用不变的使用方法和试验方法评价变化了的材料。而方法是由观念决定的。

矿物掺和料已经成为现代混凝土一个必须的组分，这就更增加了混凝土体系的复杂性。听说最近某酸雨严重的城市有人在混凝土中掺粉煤灰和石灰石粉，进行抗酸雨侵蚀的试验，结果是掺粉煤灰的试件抗酸雨的性能变差，而掺石灰石的试件性能变好。众所周知，酸雨是大气中的 SO_2、CO_2 以及氮氧化物溶于大气中的水分（例如雨、雾）形成硫酸、碳酸和硝

---

* 原文《思维方法和观念的转变比技术更重要之四》刊载《商品混凝土》2005 年第 3 期。

酸，其对混凝土的侵害对象是$Ca(OH)_2$或碳化后生长的$CaCO_3$。例如：

$$H_2SO_4 + CaCO_3 \longrightarrow CaSO_4 + CO_2\uparrow + H_2O \text{ 或}$$

$$H_2SO_4 + Ca(OH)_2 \longrightarrow CaSO_4 + 2H_2O$$

因此，掺入石灰石后抗酸雨性能变好从化学上说是不可能的。至于掺入粉煤灰，则问题更复杂些，需要具体分析。现未得到该试验方法的资料，本不可妄加评论，但推断是采用了现行标准或规范中针对普通硅酸盐水泥而规定的实验室试验方法。

“混沌理论认为，在混沌系统中，初始条件的十分微小的变化经过不断放大，对其未来状态会造成极其巨大的差别。”[2]这就是拓扑学中著名的“蝴蝶效应”。什么是“蝴蝶效应”？

简单来说，“如问：两个眼睛的视敏度是一个眼睛的几倍？很容易想到的是两倍，可实际是6～10倍！这就是非线性：1+1不等于2”[2]。混凝土体系实际上也是个“混沌系统”，应当转变思维方法来认识这种非线性系统的特点。

在这里提出一些问题与大家探讨，至于解决的对策，尚需群策群力。相信只要思维方法和观念能与时俱进，可行的研究路线是会产生的，这需要基础科学和工程技术的结合。

1　关于现场混凝土强度的验收

① 本来试验室中对混凝土性能的“指标试验”只是用于混凝土材料优选时的相对比较，例如水泥混凝土的强度以28天计，是基于传统水泥混凝土强度发展规律统计的。在传统上，标准养护的普通混凝土强度发展以28天为100%，大体是3天约30%，7天约50%，三个月不到120%。28天以后增长就很缓慢了。取28天是因为7天以后就按周计算龄期。国外资料中28天（4周）以后的龄期计为56天（8周）、91天（13周），即缘于此。由于混凝土试件的离散性，实际检测时，“30天、2个月、3个月”和“28天、56天、91天”没有显著差别。过去认为，检测了28天强度，以后就保险了。但是，由于水泥的生产和性质的变化，以及混凝土早期强度的大幅度提高，现今的混凝土从7天到28天，强度增长很少，有的几乎不再增长，甚至有的还可能发生倒缩。即使28天没有发生倒缩，几年以后也可能发生[3]。那么现在规定在28天验收的强度还能保险吗？能保证混凝土结构在服役期间的设计承载力吗？

② 过去，人们认为混凝土的性能就是试验室中用标准试件检测的性能。对钻取芯样得到现场构件混凝土的强度值低于标准强度的结果，认为是钻芯技术造成的，而用一个统计的系数进行“修正”。在蒸汽养护中，人们早已经知道养护温度对普通硅酸盐水泥和有掺和料的水泥混凝土强度发展有不同的影响[4]。但是很少有人因此怀疑现场混凝土构件中的混凝土由于温度而产生强度差异。自从有人进行了“跟踪养护”[5]，按实测混凝土内部温度跟踪养护混凝土试件，才清楚构件中的硅酸盐水泥混凝土强度由于水泥水化热导致的温升而低于标准养护试件的强度值[6]，混凝土温度越高，初期强度越高，而后期强度相对越低（见图1所示）[7]。但是，至今我国对现场混凝土质量的验收仍

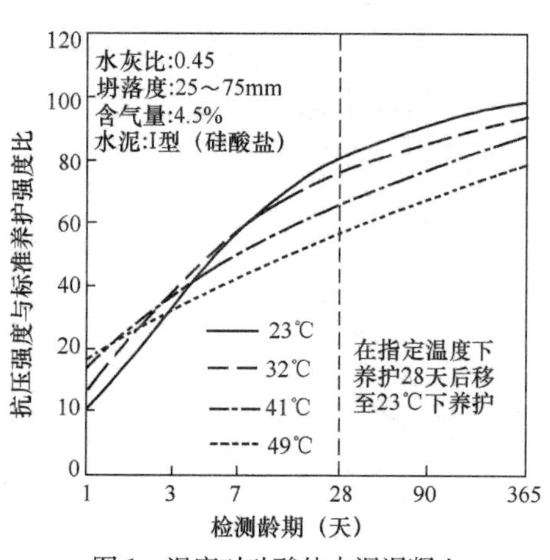

图1　温度对硅酸盐水泥混凝土强度发展的影响[7]

然以试验室标准成型和养护的试件为准。

③ 由于现场结构中混凝土内部温度高于标准养护的温度，这对纯硅酸盐水泥和有掺和料水泥的混凝土强度发展效果不同，用标准养护试件验收，对硅酸盐水泥混凝土是不安全的，对掺和料水泥混凝土则是保守的。在现行验收规范中，虽然为了更“准确地”代表现场混凝土的性能，增加了预留“同条件养护试件”的规定，但是也只是把试件放在现场结构旁边，用与结构相同的养护方式存放的小试件，仍然并不真是“同条件”。从北京航华大厦在底层 C60 混凝土柱子浇筑时实测的温度可见构件内部温度是不均匀的，强度也会有差别。如表 1 所示[8]。由表 1 可见，现场“同条件”养护的试件不仅不能代表柱子内部的混凝土，也不代表柱子表面的混凝土。实测表明，在重要构件中预埋温度传感器进行跟踪养护混凝土试件检测其强度，应当是趋近于实际的一种进步。目前可不妨从重大工程做起，待条件成熟后逐步推广。但是，对于处于严酷环境作用下的混凝土结构，从耐久性要求出发，钢筋的混凝土保护层质量更重要，所以还应当有保护层混凝土跟踪养护的数据，以验收保护层混凝土的强度。

表 1　不同养护条件下混凝土强度的发展[8]

龄期（天）	温度（℃）				试件抗压强度（MPa）			备注
	测温部位				养护条件			
	柱中心	保护层	柱表面	大气	标准	跟踪养护*	现场大气环境	
1	44	23	10	1.2	26.2	26.9	—	
2	49	31.2	12	0.7	45.6	48.5	33.2	
3	35	27	10.2	1.0	46.0	55.0	34.2	
5	18	14	6.5	−2.9	55.1	64.5	46.5	
7	11	−1	2.5	−2.9	56.0	65.0	49.4	7 天后将跟踪养护试件取出置于现场自然环境
14	0	−1.5	−4.5	−4	71.2	79.0	—	
28					71.3	79.8	67.5	
粉煤灰掺量 20%，C60，1995 年冬季施工，无防冻剂								

* 为柱中心跟踪温度的养护。

④ 混凝土内部由于硅酸盐水泥水化引起的温度升高，对掺入矿物掺和料的混凝土强度发展的影响，与对纯硅酸盐水泥配制的混凝土相反，强度不随温度和龄期下降，而是始终随龄期增长。如图 2 所示[9]，硅酸盐水泥掺粉煤灰与否，强度发展的差别比低热硅酸盐水泥的更显著，原因是低热硅酸盐水泥中的早强组分 C_3S 和 C_3A 含量低得多，本来早期强度就较低，掺粉煤灰后达到不掺的相同强度的龄期较长。硅酸盐水泥掺粉煤灰后，其强度增长率大得多。掺和料掺量越大，后期强度增长率越大（见图 3）[10]，在现今掺和料使用较普遍的情况下，仍用 28 天作为验收龄期显然有悖于按照混凝土强度发展规律确定标准龄期的初衷。不利用有掺和料混凝土

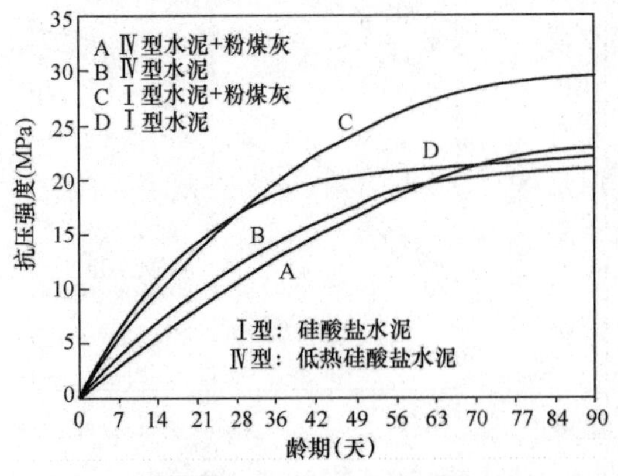

图 2　胶凝材料强度增长率（低钙粉煤灰）
(Samarin, Munn, and Ashby 1983)[9]

的后期强度，无疑是不经济的。更重要的是早期强度①低些有利于混凝土的耐久性。

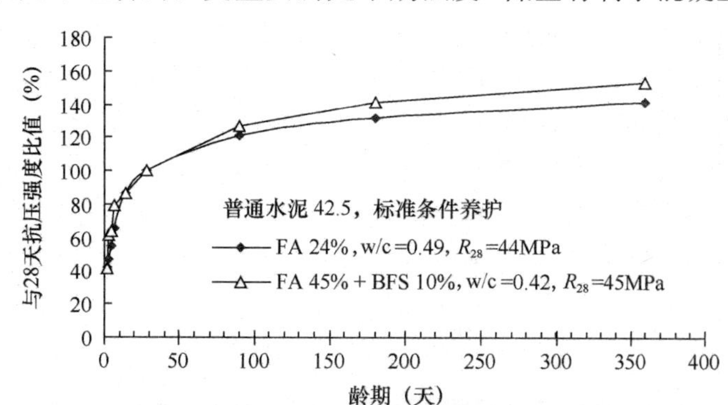

图 3　掺和料掺量不同的混凝土一年内标准养护强度发展

2　“混凝土耐久性”的实验室“指标实验”和现场混凝土结构耐久性表现的差异

为什么要把“混凝土耐久性”打上引号？在这里引用一段 Adam Neiville 的论述：“不存在一般含义上耐久的混凝土：如果我要为花园工具室修个地板，打算用上一年就推倒。某种质量的混凝土就‘耐久’了，但这种混凝土用于一座大桥、一条隧道或一座大坝中也许就不耐久了。所以，说某一混凝土是‘耐久的’，或‘不耐久的’是错误的。……实际上，有两个界定‘耐久性’的术语。……第一个，即预期的服务寿命，预期值可以规定为具体的年限。如大型隧道为 120 年；或工业建筑为 25～50 年；或一座家庭的木制房屋基础要 50～70 年。第二个术语涉及混凝土在预计暴露条件下的破坏过程。有的情况下，例如一座牛棚，牛群要呼出大量二氧化碳；另外一种，例如处于适宜气候中的一座桥梁，可能又受到了冻融循环的作用。还要重复说的是，混凝土在一种条件下耐久了，在另一些条件组合下就可能不耐久。……没有普适意义耐久的混凝土。”

这就是说，不能脱离在一定环境作用下的结构来评价“混凝土的耐久性”。传统上在实验室里进行的混凝土各种“耐久性”快速试验，只是一些在优选原材料和配合比时相对比较的“指标实验”[11]，而不能用以作为现场混凝土结构验收的标准。事实证明二者是难以相关的。其主要原因有三：

① 实验室的“指标实验”条件的简化　实验室研究的特点是为了弄清楚影响结果的因素及其规律，避免复杂因素的相互干扰，故常只改变一个因素、固定其他因素，即忽略环境中各因素的交互作用。“交互作用”不是“共同作用”或“耦合作用”，也不是“综合作用”，而是整体作用。环境（例如温度、湿度、流动的地下水等）是变化的，不同因素在不同时期对不同工程可能有不同的作用。各种作用不一定同时发生，交互作用的结果可能是加剧，也可能是削弱。例如夏季施工浇筑的混凝土，可能在秋、冬季由于冷热、干湿变化出现开裂，如果有水进入，可能带入侵蚀性介质加剧潜在的损伤危害；对于道路路面，夏季施工可以只设缩缝不设胀缝，而可避免常见的胀缝处造成的损伤；又如，当同时有镁离子和硫酸根离子存在时，潜在腐蚀性会加剧，而在海水中，因有氯盐的存在，会使因硫酸盐腐蚀生成的石膏和钙矾石溶解度增大而被流动的海水带走，则可缓解膨胀性的损害；在无流动水的土壤中，这

① 28 天以前称作早期，28 天以后为后期，7 天以前为初期（young age）。

种缓解就可能不存在。在实际工程中几乎没有由单一因素引起的混凝土结构劣化，但环境中的各因素如冷热和干湿的交替、冻融循环、化学腐蚀，等等，并不一定同时起作用，因此不会是简单的叠加。不同因素的作用有其各自发生的条件和机理。例如有人想模拟除冰盐的侵害而用盐水进行冻融实验。实验的结果当然是盐降低了混凝土孔中水的冰点而使冻融循环的腐蚀得到缓解，实际上这并不适用于除冰盐的情况，除冰盐的作用机理更为复杂。一方面，除冰盐直接接触混凝土表层，在冰雪融化时会吸收热量，使混凝土表层温度骤然下降到气温以下，例如理论上，以 40g NaCl 和 20g NH_4Cl 制成的除冰盐与 100g 冰雪混合后，温度可下降 30℃，使混凝土受温度冲击而开裂剥落；另一方面，反复使用除冰盐使渗入混凝土中的盐在水蒸发后浓度不断增加，达过饱和而析出结晶，产生结晶压力；对钢筋混凝土（如桥面板）更重要的是氯离子进入混凝土，会引起钢筋腐蚀，这种腐蚀要比盐水的冻融快速和严重得多。又如还有人想研究含盐的环境下冻融循环、干湿交替和碳化对混凝土的“共同作用”，所设计的实验为：把混凝土试件在含盐的水中浸泡一定时间达到饱和，然后冻融循环一定的次数，再暴露在空气中，自然升温并干燥和碳化。这个实验的意图是好的，但是冻融和干湿都必须有水进入混凝土的孔中，而碳化的条件是相对湿度 40%～70%；冻融、干湿的环境和碳化环境的条件是不同时存在的，除非春、秋大气相对湿度较低时碳化，冬季冻融。

② 实验室“指标实验”过程的强制性　在大气中，混凝土受环境作用的损伤非常缓慢，除非科学研究的需要，无论是预测还是验收，都不可能等待自然观测的结果。因此实验室中只能进行快速实验，办法就是设置严酷甚至极端的条件来强制加速实验的过程，这种强制过程的反应动力学和自然界工程所处环境中的差别很大。例如冻融循环实验，按照我国现行规范，无论是快冻法（冻 2 小时、融 2 小时一个循环）还是慢冻法（冻 4 小时、融 4 小时一个循环），相对于大气的温度变化来说，都是快速的。从室温降低到预定的负温（一般低于－15℃），冻结速率是 6～30℃/h，大气中的冻结速率很少超过 2℃/h[12]。另外，实验室快速实验的试件是先浸水饱和的，而在实际工程中不同环境和不同部位的混凝土有不同的饱水程度，并不是受冻结之前都含水饱和的。饱水试件的冻融循环实验是基于对混凝土损伤机理的认识而定的，这就是孔隙水结冰后体积膨胀 9%，因循环地冻融，使混凝土因疲劳而损伤。但是实际上，如果混凝土孔隙水饱和度低于临界值，则冰晶对混凝土孔壁不会产生压力。在密闭容器中（如混凝土中骨料的情况），临界饱和度是 91.7%[13]。混凝土抗冻性指数与混凝土中水饱和度的关系如图 4[13] 所示。可见当混凝土水饱和度＜85%时，混凝土是冻不坏的。水饱和度一旦达到临界值，就会很快被冻坏。可以认为 85%是临界饱和度。但即使在水中养护的试件，其残留孔隙也并非全部被水充满[13]。例如在水中养护 28 天的试件抗压实验破坏后，可见其内部因较干燥而颜色浅。强度越高这种现象越明显。因此，从水中取出的试件在第一次冻结时并不毁坏，随着冻融循环次数增加，内部水的饱和度因积累而提高，直到达临界值，很快就冻坏。现场混凝土除直接接触水而总是潮湿的构件如桥墩、水塔、水池等之外，很少有像实验室小试件那样的水饱和度。其他如实验室碳化试验条件是 CO_2 的浓度 20%、温度 20℃，比大气中 CO_2 的浓度（平均为

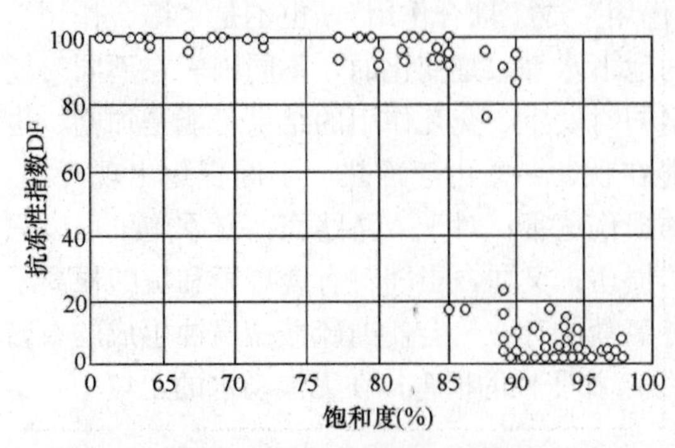

图 4　混凝土中水饱和度对抗冻性的影响[13]
抗冻性指数 DF＝冻融循环次数×0.6/300

0.03%）大三个数量级；硫酸盐侵蚀试验使用5%的硫酸钠溶液，在自然环境极少见；开发了目前使用广泛的氯离子扩散系数的RCM测定方法的唐路平报道，“从现场暴露了四年的混凝土断面检测计算得到的最大氯盐扩散值，还不到实验室里同一混凝土检测值的1/10”[14]，等等。

③ 实验室和工程结构中混凝土状态的差别　实验室试验所用试件都用标准成型的、无宏观缺陷的、自由的小试件，其内部微结构和现场混凝土在不同温度下形成的微结构不同；现场混凝土的微结构是在变化的温度下形成的，并且总是处于不同程度的约束条件下，内部混凝土则处于受三向约束的状态，而且常存在可见或不可见的裂缝；由于承受荷载，已有可见与不可见裂缝会发展，还会出现新裂缝。环境中的侵蚀性介质与这样不同状态的混凝土接触后，对混凝土的作用过程有差别，结果也不同。例如混凝土干燥收缩的测定，混凝土的干燥是由于失水，相同混凝土的试件尺寸不同时，失水率也不同，如图5所示[13]。

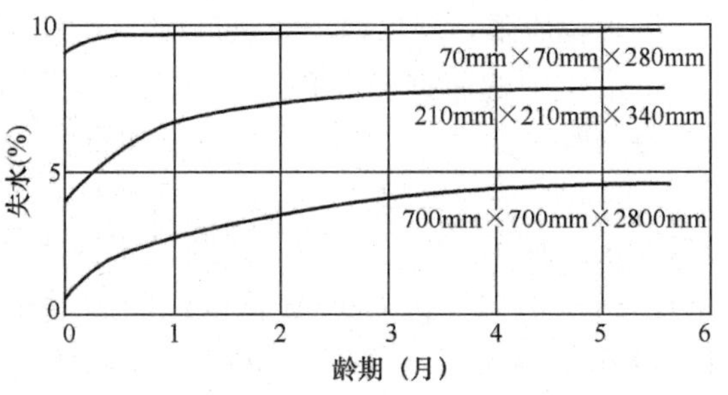

图5　试件尺寸对混凝土干燥失水率随龄期发展的影响

由图5可见，相同混凝土的试件（构件）尺寸越小，失水越快，失水率也越大。则推测其同龄期测出的自由干缩值也会越大。现行规范中检测混凝土收缩值所用试件尺寸为100mm×100mm×500mm，比环境温、湿度相同的现场混凝土构件失水率小得多，所测出的混凝土同龄期的收缩值当然也不同。试件（构件）形状对收缩值也有影响，构件体积和表面积的比值越小，同龄期收缩值越大。

有人为了建立实验室冻融循环试验结果和现场混凝土表现之间的相关关系，把同样的小试件放在现场环境下做对比检测，得出实验室冻融循环次数和现场混凝土寿命之间的关系。该试验仍然忽略了强制的快速和缓慢过程的差别、实验室制作的小试件和实际工程的构件的差别，因此仍然是不可靠的。对于混凝土这种复杂的混沌系统，任何企图用简单的公式建立实验室指标试验结果和复杂的现场条件下混凝土行为的相关关系都是徒劳的。对于4小时一个循环和8小时一个循环来说都是快速，ASTM C 671的方法才能算得上是慢冻法。ASTM C 671要求使用引气的混凝土试件，饱水度与预期或实测冬季开始时现场的混凝土饱水度一致，然后按所预期的现场冻融周期进行实验。所检测的是试件的长度变化。但该实验使用的也是小试件。小试件仍然不能真正反映工程实际的行为[15]。任何实验室快速试验方法用于评价建筑材料的基本要求，应能够可靠地反映该材料在现场使用的表现[16]。对重大工程，试件可实时取自现场预浇筑的实体混凝土，在发达国家已较多采用这项措施[11]。

3　对不同特性的材料用相同指标评价和实际情况的差异

从工程甲方和设计人员来看，现在混凝土的28天强度没有变化，但是他们一般并不了解28天强度没有变的现在的混凝土和过去的有什么区别。由于所用的水泥和骨料质量的变化以及外加剂的普遍使用，使得现代混凝土的配合比有了很大变化，加上矿物掺和料使用的日益普遍，混凝土的体系及微结构的形成过程都有不同程度的变化，因而尽管28天强度可以一样，如前所述，强度发展规律却起了变化，更重要的是其他性能表现也发生了变化。使用不变的方法和指标对现代混凝土进行试验并与传统混凝土进行对比，必然得到和现场表现不同的结论。M. D. A. THOMAS 对 20 多个公路工程现场结构物进行调查，所使用的粉煤灰混凝土一直是暴露在使用化冰盐的冻融环境之中。有些工程已经经历了 20 多个冬天。混凝土都没有出现剥落现象，即使有些从现场取得的一些粉煤灰混凝土样品进行实验室试验的结果不好，但现场使用性能非常好。THOMAS 的结论是，ASTM 的盐剥落实验方法不适合用于评价粉煤灰混凝土的抗剥落性能[15]。

因篇幅所限，现仅以粉煤灰混凝土为例讨论如下。

混凝土中的矿物掺和料，无论是具有潜在的水硬活性（如磨细矿渣），还是具有火山灰活性（如粉煤灰），都需要有石灰和石膏的参与，但是不同掺和料的活性差别很大，如图 6、图 7 所示为粉煤灰与硅灰分别和 CaO、石膏加水的混合物标准养护到龄期 1、7、28 天后对 x—射线衍射的结果。二图对比鲜明：活性很高的硅灰在 7 天时已大幅度地消耗了 CaO，28 天时除了因 CaO 碳化生成的碳酸钙之外，已无 CaO 存在，从 1 天到 28 天无定形 SiO_2 的散射干涉区变得高而宽，表明凝胶的产生。粉煤灰则除了由粉煤灰中的无定形 Al_2O_3 和 CaO、石膏生成很少量钙矾石和石膏消耗尽而转变成单硫盐外，石灰极少消耗。粉煤灰直到 28 天反应量极少，说明粉煤灰在 28 天以前几乎不参与 C-S-H 的生成，混凝土拌和物中的水基本上只供给硅酸盐水泥之用，相对于水泥来说，不能简单等量取代，即使取代后降低水胶比，实际的水灰比都会增大。表 2 所示为在粉煤灰不同掺量下，当不考虑粉煤灰的吸附水和早期少量钙矾石生成消耗的水时，不同水胶比下的实际水灰比计算值。

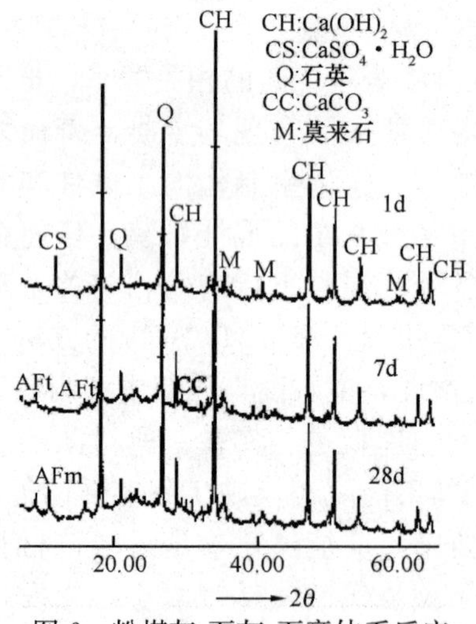

图 6　粉煤灰-石灰-石膏体系反应

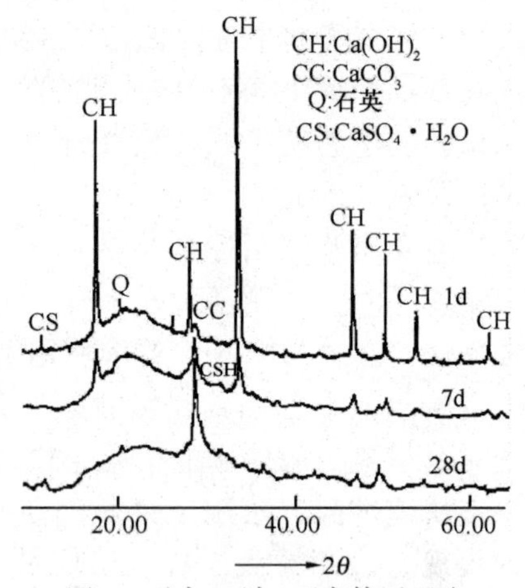

图 7　硅灰-石灰-石膏体系反应

表 2　粉煤灰不同掺量的混凝土在不同水胶比下的水灰比计算值

FA%	w/b	w/c	w/b	w/c	w/b	w/c	w/b	w/c
0	0.50	0.50	0.40	0.40	0.35	0.35	0.30	0.30
15		0.59		0.47		0.41		0.35
20		0.63		0.50		0.44		0.38
30		0.71		0.57		0.50		0.43
40		0.83		0.67		0.58		0.50
50		1.00		0.80		0.70		0.60

由表 2 可见，水胶比相同时，随粉煤灰掺量的增加，水灰比提高的幅度增大；即使扣除粉煤灰的吸附水和少量反应消耗水，水灰比仍然随粉煤灰掺量的增加而增大。在粉煤灰对混凝土减小孔隙率而增密的作用还没有发挥出来的 28 天以前，混凝土的孔隙率是增大的，一切和孔隙率有关的性质当然也就都不好。如果改变一下传统的方法，保持相同水灰比而不是相同水胶比来评价使用掺和料的混凝土，就能使粉煤灰有很好的表现。例如清华大学在深圳地铁工程中用大掺量掺和料制作的足尺模型，其配合比和主要性能指标如表 3 所示。

表 3　深圳地铁工程足尺模型试验用混凝土配合比和性能对比

水胶比	砂率（%）	浆体组成（kg/m^3）				抗压强度（MPa）			28 天弹性模量（GPa）	氯离子迁移电通量（库仑）	现行标准碳化深度（mm）
		用水量	水泥	粉煤灰	磨细矿渣	3 天	28 天	90 天			
0.57	0.45	208.8	278.0	90.0	—	12.4	39.1	47.3	31.11	2030	43.5
0.40	0.43	168.0	180.0	180.0	80	22.3	50.7	58.5	38.89	537	43.8

清华大学覃维祖教授的研究生成唯佳配制不同粉煤灰掺量的混凝土采用等胶凝材料浆体用量的方法降低水胶比，得到与不掺粉煤灰的混凝土一致的拌和物工作性和 3 天、28 天抗压强度，与等水胶比简单等量取代的方法比较如表 4 所示。

表 4　用不同配合比设计方法的混凝土掺不同量粉煤灰的比较

粉煤灰取代量（%）	抗压强度（MPa）						
	等水胶比（0.57）			等浆体体积（$0.31m^3 \pm 0.01\ m^3$）			
	计算水灰比	3 天	28 天	水胶比	计算水灰比	3 天	28 天
0	0.57	20	38.4	0.57	0.57	20	38.4
15	0.68	18.4	35.3	0.54	0.64	20.6	39.7
30	0.83	14.4	28.4	0.49	0.70	20.8	41.9
45	1.02	12.2	25.3	0.44	0.80	20.0	38. 6
60	1.41	7.8	19.3	0.36	0.90	20.6	40.1

用等胶凝材料浆体用量的方法，降低了水胶比后计算水灰比仍然增大，但却得到一致的 3 天和 28 天强度，其原因是计算水灰比并未扣除粉煤灰吸附的水和少量钙矾石生成消耗的水，也就是说，实际水灰比比计算值小。而且任何体系都是结构决定性质，混凝土的微结构比水化程度更重要，对混凝土强度和其他性质起决定性作用的不是水化程度，而是在不同条件下水化形成的微结构。

至于粉煤灰的“超量取代”，主观认为让超量部分“代砂”，而由于颗粒细度差别很大，并无法取代砂，实质上还是增加了胶凝材料总量；虽然降低了水胶比（与料单上的不同），但增加了胶凝材料用量（体积则更大）。不能不说是一种掩耳盗铃之举。

其他有关传统方法不能适应现代混凝土特性的实例很多，例如混凝土收缩值的测定，由于现代水泥的活性高、混凝土的水胶比低、早期强度高，传统方法测不出影响早期开裂主要的收缩：温度收缩和自收缩；又如传统观念认为硫酸盐腐蚀主要是化学反应的结果，实际工程调查发现盐结晶对混凝土的损伤更快、更严重，用传统的试验方法无法评价；再如对混凝土流变性能的测试，由于使用外加剂和矿物掺和料，混凝土拌和物具有明显的触变性，用传统的坍落度筒检测其工作性，与传统混凝土有很不同的表现。P. K. Mehta 在美国加州大学伯克利分校 Barker 会堂剪力墙和基础加固工程中所用混凝土，用坍落度筒检测的坍落度只有 100mm 的混凝土拌和物，其泵送效果竟与坍落度为 180mm 的传统混凝土泵送效果一样，这说明，即使方法不变，指标也得变，首先要改变的是观念。现代的混凝土拌和物按熟练混凝土工目测标准评价坍落度合适时，可能在浇筑时会流动性太大而严重离析。

结束语

1. 科学是人们对客观世界的认识，尽管混凝土是用简单的工艺制作的非线性复杂系统，也有其客观规律。传统的方法基于传统的认识水平，但是在一定的历史时期，就具体人来说，认识总是有限的，而且，客观事物总是在变、在发展，因此需要不断地加深认识和总结，认识和观念应当不断更新和发展。现存的不一定都合理。

2. 现在发现问题，正说明不变的方法和指标不能适应变化了的材料，首先是，是否认为应当改变。如何改变？

参考文献

[1] Neville A. M. . Consideration of durability of concrete structures: Past, present, and future. See: MATERIALS AND STRUCTURES, March 2001.

[2] 中国农垦信息网编录：蝴蝶效应，2003. 11. 3，www. chinafarm. con. cn.

[3] R. W. Burrows. The Visible and Invisible Cracking of Concrete, See; ACI Monograph No. 11.

[4] 沙凤丹．水泥强度快速测定，见：北京硅酸盐协会水泥制品专业委员会 1962 年学术年会論文报告汇编.

[5] Bamforth, P. B. . In-Situ Measurement of The Effect of Partial Portland Cement Replacment Using Either Fly Ash or Ground Granulated Blast-Furnace Slag on The Performance of Mass Concrete, See: Proc. instn. Civ. Engrs. , Part 2, 69, 777(1980).

[6] R. K. Dhir, M. R. Jones. PFA Concrete: Influence of Simulated in Situ Curing on Elasto-Plastic load Response, See: Magazine of Concrete Research, 1993, 45, No. 163.

[7] Steven H Kosmatka and William C. Panarese. Design and Control of concrete mixtures, 4th Edition, PCA Association , USA 2000.

[8] 李玉琳，钱选青，廉慧珍．北京航华科贸中心 C60 泵送混凝土的冬季施工[J]. 低温建筑技术，1996 年第 3 期.

[9] Use of Fly Ash in Concrete, Reported by ACI Committee 232, ACI 232. 2R-03.

[10] 深圳地铁车辆段挡土墙高性能混凝土工程应用验收报告，2004 年 6 月.

[11] Peter Taylor, Performance-Based Specifications for Concrete, See: CONCRETE INTERNATIONNAl, August 2004.

[12] S. mindess, J. F. Young 著. 方秋清等译. 混凝土[M]. 北京：中国建筑工程出版社，1989.

[13] Neville A. M. 著. 李国泮，马贞勇译. 混凝土的性能[M]. 北京：中国建筑工业出版社，1983.

[14] Sandberg. P and Tang, L. " A Field Study of The Penetration of Chlorides and other Ions Into a High Quality Concrete Marine Bridge Columns. " See: DURABILITY OF CONCRETE. In PROCEEDINGS OF THE THIRD INTERNATIONAL CONFERENCE, SP-145. 1994.

[15] Alexander M. Vaysburd, Chreistipherd D. Brown, Benoit Bissonnette, and Peter H. Emmons. "Realcrete" versus "Labcrete"— Searching for tests that give reliable results. See: CONCRETE

INTERNAIONAL, February 2004.

[16] M. D. A. THOMAS. Laboratory and field studies of salt scaling in fly ash concrete. Frost Resistance of Concrete. PROCEEDINGS OF THE INTERNATIONAL RILEM WORKSHOP. E & SPON 1997.

自评

1. 本文的题目说明变化了的材料不能用不变的方法试验和检测，也可引申到生产、加工、使用、验收等的环节。这是一个如何认识世界的思维方法问题。人对客观事物的认识应当随着客观事物的发展而变化，在这样的原则下才有可能产生创新。例如对构件混凝土性质验收的龄期一概定为28天或更长，都未必合适。应当考虑其服役开始的时间，尤其是耐久性指标，例如按现行标准，碳化实验的试件要养护到28天开始在20%的CO_2浓度下进行强制性碳化实验，而现场混凝土没有养护到28天才开始碳化的；海水中的桥墩也不是28天才开始接触海水的；冻融循环作用的开始和混凝土成型时的季节有关系，规定试件养护28天开始试验，也是没有实际意义的，等等。

2. 尽管实验室检测与现场性能不能相比，而作为混凝土材料的本征特性，可作为优化的根据，只是应当尽量模拟现场条件。

郭保林评

1. 应正视室内快速试验获得的数据与结构实体自然表现之间的差异。不能因室内快速实验方法不能真实反应有些混凝土的性质而全部否定，需要提出一个更加可信的方法。要么说明快速方法的适用性，尽量避免无用和实验值的误导。快速方法有其明确的用途，只是大家都不太重视适用范围，同时又随意外延测试值得作用。对混凝土实质性能把握能力的欠缺恰恰为这些存在缺陷的方法提供了很大的市场。

2. 图1所引用的图可能是经过重画而出现了错误。原图如下：

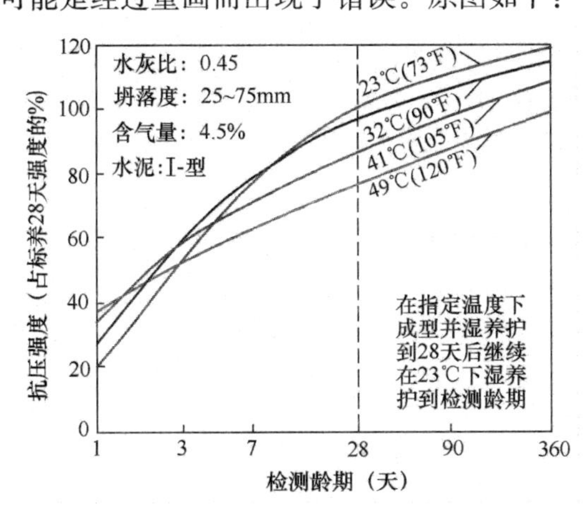

和廉老师沟通后，廉老师说："这是我读书不求甚解才出的错。"我觉得廉老师言重了。

论文之八

思维方法和观念的转变比技术更重要之六
——混凝土工程中的分解论和整体论*

廉慧珍（清华大学土木水利学院）

摘　要　整体论认为整体不等于组成部分个体的简单算数和，即由部分构成整体时，出现组成部分所没有的性质；分解论认为复杂系统各个组成部分的行为可以反映体系的性质，将体系简化成个要素，分别研究其行为和性质，可以代表整体的行为和性质。分解论对于研究事物的本质是必须的，但不是最终的，分解论研究的结果不能简单外推至整体，更不能用不正确的方法进行解释、归纳，甚至夸大；对混凝土工程这样的复杂体系必须按整体论进行分析和处理。而对基础性的研究需要采取分解论，现代系统论是将整体论和分解论相结合，如钱学森所说："我们所提倡的系统论，既不是整体论，也非还原论，而是整体论与还原论的辩证统一，是更高层次的东西，即我们的系统论既要包括整体论，也要包括还原论。"

关键词　混凝土工程；分解论；整体论；系统论；耐久性

1　树木和森林

人人皆知，森林由树木组成，但是并非种了大量的树木就能形成森林。多年以前，《中国青年报》曾刊登过一则惊人的报道，题为《绿色长城的垮塌》，内容摘要如下：

20几年投资上百亿资金营造了绵延千里的三北防护林，但是由于没有顾及当地的水文、气象和地貌的条件以及适地、适树、适草的自然规律，单一品种的杨树发生虫害时无法抵御，一倒一大片。许多杨树长成不死不活的"小老头树"，达不到防护效果。因此北京每年春天仍要迎接沙尘暴。

当然，此后我国已正在改造治理。

这样问题的出现，从根本上来说，是缘于一种分解论方法。

天然森林的特点是长期形成完整的、稳定的群落结构：茂密的林冠-灌木丛-草本层-落叶层-苔藓植被，相辅相成，形成一种大自然的发展的生物链，能保持水土和营养，保护生物的多样性和可再生资源，具有自我调节和抵御外来因素破坏的自我保护机制，在受损时有很强的自我修复及还原的功能。

人造林的失败就在于"南方沙家浜（杉木），西北杨家将（杨树），东南马家军（马尾松），东北莲花落（落叶松）"的单一品种：没有不同高度植物的互补，地表植被差，水土保持能力弱；单一树种无法给不同动物提供适宜的生存和繁衍的环境，生物多样性水平极低；缺少落叶改良土壤，营养循环过程被阻断，土壤营养日益匮乏；生态状况脆弱，缺少病虫害的天敌；随着树木的成长，单一元素在土壤中高度聚集易产生毒素。在狭隘的经济利益驱使

* 原文《思维方法和观念的转变比技术更重要之五》刊载《商品混凝土》2011.10。

下，巴西亚马逊雨林和我国云南、海南，“用经济林代替杂木林”，大量热带雨林被砍掉，种植橡胶、香蕉等经济作物。在种植橡胶树时，为了高产而严格清除林下植物，结果引起水土流失甚至山洪暴发的灾难。

人们处理任何事物都是基于对客观世界的认识。事物的发生和发展都有其客观规律，但是对客观规律的认识和研究的方法有关，而研究方法和思维方法有关。思维和研究的基本方法是整体论和分解论（又称还原论）。森林是由树木组成的，但是并不是种了树就一定成林，而是要根据“当地的水文、气象和地貌的条件以及适地、适树、适草的自然规律”规划建设。这就是整体论的方法，而分解论的方法则是“只见树木不见林”。然而，毕竟森林由树木组成，按自然规律认真研究每一种树木的特性和生长规律，保证成活率，是造林的先决条件。

20 世纪以来，相对论、量子论和系统论的出现，认为世界是非机械的、相互联系的、不可分解的有机整体。科学家开始看到整体论的重要性。这两种方法有什么区别、什么关系？如何取舍？这是涉及人们认识世界的重要问题。

2　混凝土工程中分解论和整体论

2.1　分解论（又称还原论，简化法，reductionism）

分解论是一种把复杂的系统（或者现象、过程）分解为其组成部分的过程。分解论认为复杂系统各个组成部分的行为可以反映系统的性质，将系统简化成各个要素，分别研究其主要行为和性质，然后将这些性质“组装”起来形成对整个系统的描述。例如，从研究水泥熟料单矿物水化来研究水泥的水化，用以指导混凝土材料的工艺；又如在研究影响混凝土性质的诸多因素中，每次只变动一个因素的研究方法，等等。分解论是研究本质，是科学发展所必须的，但不是最终的和整体的，其研究的结果不能直接用于整体。例如据 Goldman A 和 Bentur 报道，在 0.33 的相同水灰比下，硅灰掺量对砂浆标准养护试件的抗压强度影响很小，而对混凝土则影响显著，见图 1[1] 所示。按分解论的观点，水泥砂浆由水泥浆体和砂子组成，混凝土由砂浆和石子组成，砂子和石子都是惰性的，那么水泥砂浆的性质就能代表混凝土的性质，而按整体论来看，混凝土和砂浆的区别在于因石子的存在而多了一个组分相，即和石子的界面，如图 2 所示。尽管水泥砂浆也存在界面问题，而砂浆和石子组成混凝土后，浆体和石子的界面要比和砂子的界面在尺度上和作用上要大得多。当水泥中掺有不同混合材时，这种差别在宏观上就表现得更加明显。

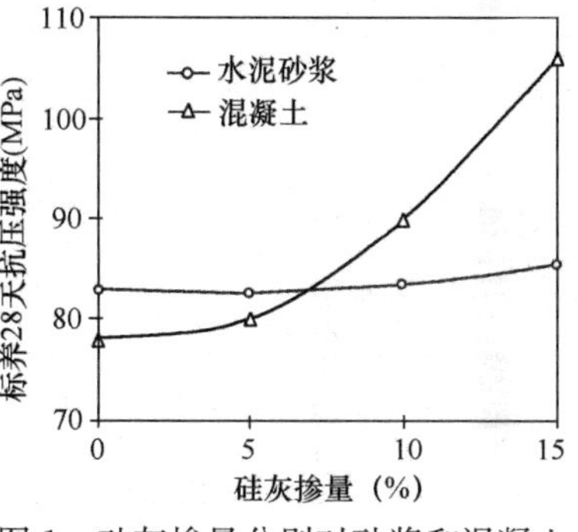

图 1　硅灰掺量分别对砂浆和混凝土 28 天抗压强度的影响[1]

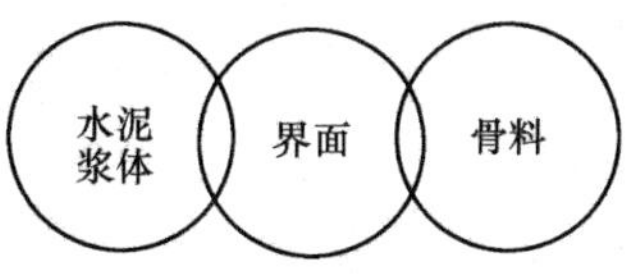

图 2　混凝土的组成

当混凝土已浇筑并硬化成构件后，就无法知道其内部究竟发生了什么。这时，就需要分解论了，而分解论方法是否可行，还是方法的问题。例如，当前水泥混凝土的检测都是按各自的标准条件进行的，将这样检测出的结果用于评价工程构件中混凝土的性质进行验收。这是一种以“静止的、孤立的”观念的分解。实际上，这样把混凝土结构构件分解成小试块后，其所处环境条件都不一样了，工程中的构件尺寸比实验室小试件的大得多，水泥水化产生的热量使混凝土内部的温度很高，和实验室标准条件下小试件实验的结果有很大的差别。

因为混凝土构件从表面到内部的温度是变化的，按照监测实际构件中的温度变换混凝土试件养护水的温度，则检测该养护条件下混凝土的强度，就可以反映实际构件中混凝土强度。如图 3 所示为水胶比为 0.4 的混凝土，分别掺入粉煤灰 0%、30%、40%、50% 和 60%，当温度≤20℃时，各组混凝土抗压强度均随龄期而增长；当温度达到 40℃时，掺粉煤灰的混凝土抗压强度均随龄期而增长，而不掺粉煤灰的混凝土的在 7 天以后增长减速；当温度达到 85℃时，掺粉煤灰的混凝土抗压强度仍均随龄期而增长，而不掺粉煤灰的混凝土则在 7 天以后，其抗压强度随龄期而下降。该试验应当也属于分解论方法，是一种用实验室小试件对结构构件中混凝土的模拟实验。当前较大尺寸构件混凝土内部温度远高于 20℃，对于不掺粉煤灰的混凝土，用实验室标准试验的强度验收是不安全的；对于粉煤灰掺量大于 30% 的混凝土，用标准试验强度验收则是保守的。

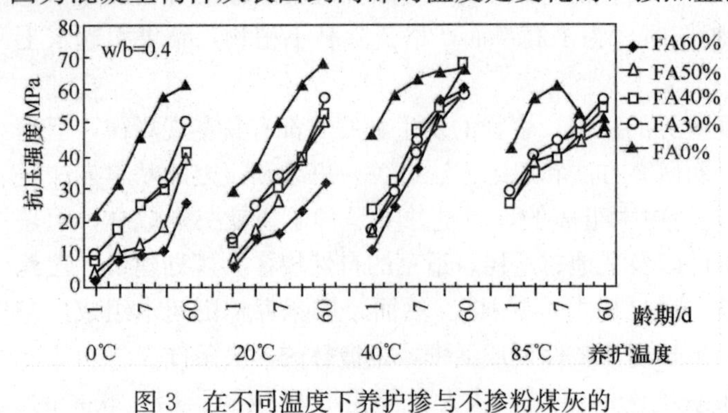

图 3　在不同温度下养护掺与不掺粉煤灰的混凝土抗压强度随龄期的发展

上述实验方法是在恒温条件下进行的，而实际构件并非恒温。实际工程中的混凝土温度不仅不同于标准条件的温度，而且在浇筑的初期，依混凝土的原材料、配合比、构件尺寸和环境温度而不同，随环境温度而波动。实测构件内部温度发展历程，跟踪变换养护水温度[2]，应该能更准确地了解构件混凝土的强度发展。虽然，一方面，养护温度的变换滞后于构件混凝土温度信号的传递；另一方面，试件内部温度的变换滞后于养护水温度的变换。由于只能如此“跟踪”而非同步，所测构件内部混凝土强度随温度的发展也是滞后的。然而，毕竟是一种整体论的思维，在定性上亦可以指导生产。Dhir R. K. 和 Jonesmann M. R. 配制相同强度等级的 C20、C40 和 C60 掺粉煤灰和不掺粉煤灰的混凝土，用跟踪养护的方法（temperature match condition）和标准养护方法检测其强度随龄期的发展，分别示于图 4 和图 5[3]。

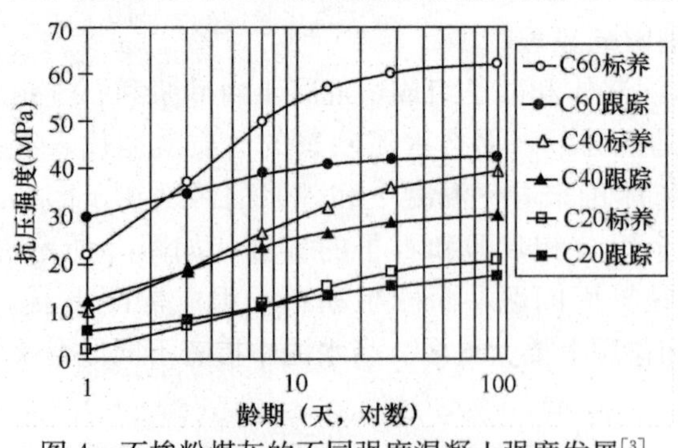

图 4　不掺粉煤灰的不同强度混凝土强度发展[3]

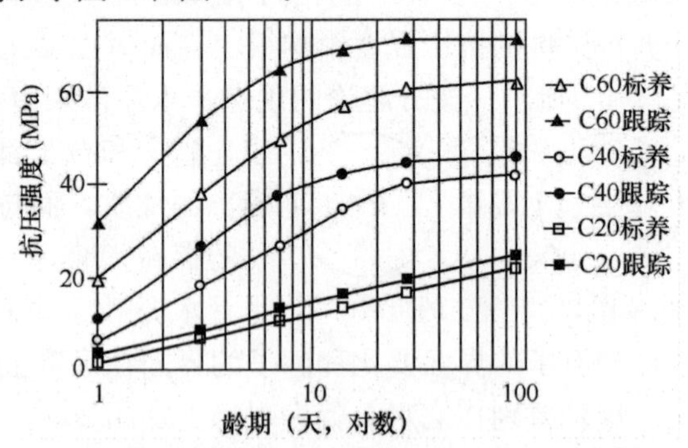

图 5　掺粉煤灰的不同强度混凝土强度发展[3]

由图 4 和图 5 可见，不掺粉煤灰时，C40 以上的混凝土在龄期为 3 天以前、C20 混凝土在 7 天以前，跟踪养护强度都比标准养护强度高，此后，则标准养护的强度都比跟踪养护的强度高，强度等级越高、龄期越长，高得越多；掺粉煤灰时，各强度等级混凝土跟踪养护的强度始终高于标准养护的强度，强度等级越高差别越大。在过去，混凝土不掺用矿物掺和料而强度等级普遍较低时，根据实验室标准条件养护的试件强度进行验收，由于结构设计时引入安全系数，影响还不是很大。现在混凝土强度普遍提高，更加之构件尺寸的增大、水泥水化发热量的提高等因素，分解论的验收方法，对混凝土不掺粉煤灰的结构会影响安全性，对掺粉煤灰的混凝土会影响经济性。

2.2 整体论（holism，又称机体论，organismic model）

整体论认为整体不等于组成部分个体的简单算数和，即由部分构成整体时，出现组成部分所没有或对组成部分来说无意义的性质，同时失去组成部分单独存在时所具有的某些性质。例如玻璃纤维布和树脂热压制成的玻璃钢，不再具有玻璃的性质（例如透明）和树脂原有的性质（例如可塑性）而成为一种新的材料。系统的整体性质或功能是由各组成部分相互联系和作用所造成。单个组成的性质无法解释整体的行为。例如，混凝土结构耐久性不等于混凝土各项“耐久性指标”之和。不同层次系统的规律不能简单地外推。其原因如下：

2.2.1 条件的简化——环境交互影响的忽略

环境的交互作用不一定都是加剧，也可能是减弱。例如硫酸盐和镁盐同时存在时，会加剧对混凝土的腐蚀，而在海水中的硫酸盐的腐蚀可因氯盐的存在而减弱；相同的交互作用在不同环境中也可以不同，例如硫酸盐与其和 C_3A 反应生成的水化硫铝酸钙，在海水中因氯盐的存在，其溶解度增大而被析出并被带走，混凝土可不发生膨胀；在不流动的含氯盐地下水中则不能被析出，就会引起膨胀。加剧的交互作用不是简单的叠加，不同介质作用一般不是同时发生的，不存在什么“耦合作用”*。例如钢筋必须在高碱性溶液中保持化学稳定。一般认为，混凝土碳化可减少混凝土孔溶液的碱度，对保护钢筋会不利。实际上，混凝土碳化和钢筋锈蚀的条件不同，不会同时发生，未必混凝土碳化就一定引起钢筋锈蚀。如图 6 所示，相对湿度 40%～70%是最适合碳化的条件，由图 7 可见，相对湿度低于 70%时，钢筋锈蚀速率极低；在相对湿度 95%～97%之间，几乎呈线性增大，相对湿度超过 97%之后，则又随相对湿度的增大而急剧下降，在相对湿度 100%时，钢筋锈蚀速率又极低；钢筋锈蚀率最大时，碳化则甚微；碳化深度最大时，钢筋锈蚀速率极低。这可解释北京拆除 50 年代建造的旧房碳化已达钢筋表面，钢筋却无显著锈蚀的现象。那些钢筋已锈蚀严重的构件，经查证都是当时为了赶工期而使用了氯盐作早强剂。（图 6、图 7 来自陈肇元讲座的引用）

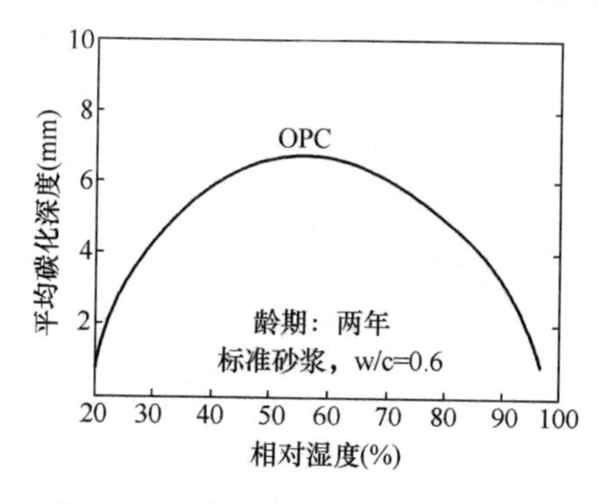

图 6 相对湿度和混凝土碳化的关系

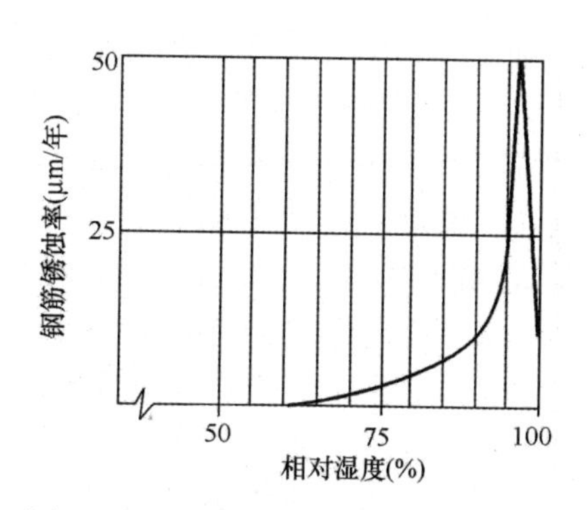

图 7 相对湿度和钢筋锈蚀速率的关系

* 耦合：根据全国科学技术名词审定委员会审定公布的定义，对电力学科为：两个本来分开的电路之间或一个电路的两个本来相互分开的部分之间的交链。可使能量从一个电路传送到另一个电路，或由电路的一个部分传送到另一部分。对通信科技学科为：两个电路或器件的接合或连锁。分“静电耦合”、“磁耦合”、“直接耦合”、“电阻性耦合”、“光耦合”等。从汉语词义来说，耦，即二人并耕的意思，衍伸为“成对”。

2.2.2 过程的强制性

混凝土在大气中的劣化是个很缓慢的过程，为了尽快得知混凝土抵抗环境作用的能力，作为优化时的对比，需要在实验室的标准条件下进行快速实验，这时的条件相对于自然条件，基本上都是强制性的，例如：

冻融循环：在实验室中的实验，冻结速率可大于 6℃/ h，最大可达数 10℃/h，而在大气中的冻结速率则很少超过 2℃/ h；在实验室中的实验，试件都为含水饱和，而在大气中，不同环境和部位有不同的饱水程度。饱水程度达到85%以前，永远冻不坏，只要水饱和度达到90%以上，便很快就会冻坏（见图 8）[4]。在影响混凝土受冻融破坏的环境因素中，严寒地区冬季因气温很低，冻后难融，而有的地区冬季气温并不比寒冷地区的低，而冻融循环次数有时可能比寒冷地区的冻融循环次数还多，则冻融循环次数会比最低温度更重要。在寒冷地区，避免混凝土的早期受冻更为重要。因为混凝土在早期（28天左右）内部含水饱和度高，或者因孔隙率较大而易为水饱和，这就是在将出台的《结构混凝土性能技术规范》中，做“寒冷地区混凝土应在受冻前混凝土的强度应≥10MPa，……开始受冻时的龄期应视当时气温而定，应至少提前 1 个月浇筑完毕”这样规定的理由。此外，按现行规范进行的抗冻融循环性试验，所用小试件是在实验室相对规范得多的条件下成型的，并标准养护至 28 天，在实际工程中，基本上没有养护 28 天的混凝土，而且，由于现场施工条件的影响，混凝土只有经过现场的浇注、振捣、收面、养护才完成最后构件，实验室的标准条件并没有考虑这种工艺的差别。

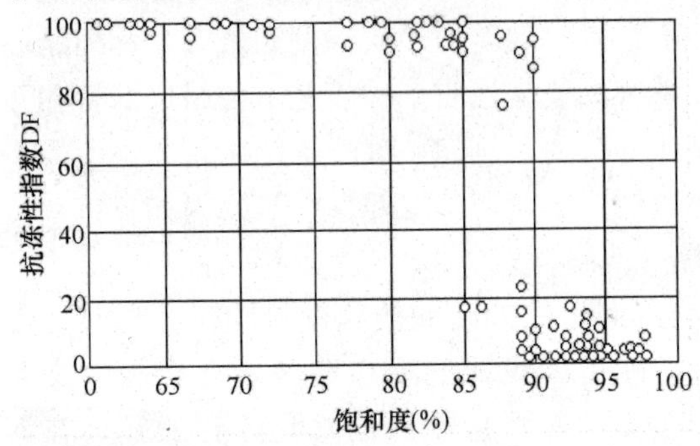

图 8 混凝土饱水程度与抗冻性指数 DF 的关系[4]

碳化：现行标准中规定混凝土碳化的试验是将标准养护至 28 天的试件在碳化箱中进行强制碳化，碳化箱条件是 CO_2 浓度 20%，温度 20℃，相对湿度（65±5）%，而大气中 CO_2 浓度平均约 0.04%，温度和湿度都是变化的，而且在实际工程中，没有养护 28 天的混凝土，不养护或养护龄期很短的混凝土自然碳化，和养护 28 天的混凝土强制碳化的规律完全不同。更何况对于当前普遍使用矿物掺和料的混凝土，不同品种矿物掺和料对混凝土在自然界温湿度变化条件下碳化行为的影响更加复杂，标准条件下的强制碳化无法反映该影响的规律。

碱-骨料反应：为了检测骨料的碱活性，常使用含碱量很大的水泥，例如人为地增加碱含量；为了检测掺混合材的水泥是否会发生碱-骨料反应，常使用碱活性很高的骨料，例如生产石英玻璃的下脚料。这种极端的条件在实际工程中往往是达不到的；即使混凝土中骨料具有很强的碱活性，同时有很大的含碱量，如果环境无水分的供应，也不会发生碱-骨料反应。碱、活性骨料和水三者同时存在是发生碱-骨料反应的必要而充分的条件。缺少其中任一项都只是必要而不充分。美国垦务局的 Richard W. Burrows 曾从 1946 年开始，在美国克罗拉多州的青山坝对使用 27 种水泥的 104 种混凝土胸墙面板进行了 53 年的观察、实测和研究。在浇筑面板时曾在坝顶每间隔 2.1 m 埋设黄铜插件，随后用尺寸稳定的钢尺和 30 倍显微镜进行量测（精确至 0.05mm）。观察中发现，使用高碱水泥的混凝土在仅 18 个月之后就开裂了，经 38 个月之后，出现了大量的地图状裂缝，但量测结果却并没有膨胀发生。由于 104 块混凝土板无一膨胀，可以断定这是一种干燥收缩现象[5]。在我国也发生过类似现象。

北京西直门的老立交桥在拆除后，检测开裂漏水处的混凝土，发现骨料有碱活性，水泥中碱含量也高，但是没有发现有碱-骨料反应的产物，其劣化的主要原因是干湿交替、温度变化、雨水渗透、除冰盐、冻融循环等对施工质量差的混凝土的交互作用。

氯离子在混凝土中的扩散系数：除了自然浸泡法以外，都是强制进行的——或者施加电场，或者真空饱盐。此外，由于条件的差异，有时还要进行一些假设，如氯离子扩散系数，无论何种方法检测，最后都是按 Fick 定律（在一定假设的条件下）计算出来的。Fick 定律主要的假设就是离子扩散的介质是均匀的，但即使成型规范的混凝土则也是高度的非均质体。自然浸泡法因为是在自然条件下的试验，被认为直观的就是可靠的，很容易被理性的人接受。但是用浸泡法试验的数据计算的氯离子扩散系数却仍然用 Fick 定律。混凝土不仅是非均质体，而且其微结构随环境（主要是温、湿度变化）和时间（龄期）而变化，具有不确知性，是从空间到时间上的不均匀。按照相同的方法得到的氯离子扩散系数用于优化混凝土材料是可行的，但是用于计算结构物的服役寿命则也只能是用于对比优化。有人以此计算出某结构物服役寿命为一百零几年，这样准确定量地预测混凝土结构耐久性反而是不科学的。

2.2.3 状态的差别

在实验室试验是用标准成型和标准养护的小试件，三向无外部约束、无宏观缺陷，考虑承载的作用时也是单向受压；而在结构中，内部混凝土处于三向约束和变化的温度下，存在可见与不可见的原始缺陷，而且承受荷载时，不同构件的不同部位应力状况不同。

试件尺寸不同也造成检测结果的差异，例如对混凝土干燥收缩性能的试验，因试件尺寸的不同，失水速率和失水程度都不同，则干缩变形率也不同（图 9）[4]；实验室用的标准试件尺寸一般使用 100mm×100mm×550mm，远小于结构工程中构件中的尺寸，所检测的混凝土干燥收缩值不能代表结构中混凝土的干缩变形。由于尺寸差别大，混凝土内部温度差别也大，会造成其他性质的差别。极端的实例是水工结构的大坝混凝土，由于其体积之大，水泥用量增减 1kg 都会对整体造价有很大影响，出于经济性的考虑，都尽量减小水泥用量；同时也为了尽量降低混凝土的温升而必须尽量降低水泥用量。为此，大坝混凝土使用的石子最大粒径会超过 40mm 甚至达 80mm、120mm、150mm。有时还要向已浇注的拌和物中抛填毛石。验收检测该混凝土强度和抗渗性时，则先剔除大于 40mm 的石子后才成型。实践证明这样的实验室检测结果与大坝实体混凝土的相去甚远。由于界面的影响，尤其是较高水胶比的混凝土，在理论上，骨料粒径越大，界面过渡层这个混凝土中的薄弱环节的厚度越大，因此，对于相同水胶比来说，混凝土的抗压强度和抗渗性随石子粒径的减小而提高；如果浆体量很少，而骨料粒径又很大，则骨料之间距离很小，界面过渡区几乎可以替代浆体，剔除大石子后混凝土的微结构和性质和坝体中的都会有很大差异。这就是分解论的结果。即使采取钻芯取样的检测，仍然有实验室混凝土和大坝混凝土之间缺乏相关性的问题，1984 年 J. M. Rapheal 就曾报道过一个无法解释取自大坝的芯样抗拉强度损失的疑团。经过漫长的调查之后，发现芯样在运往实验室期间，稍有干燥和表面裂缝所造成抗拉强度的下降高达 50%（资料来源为覃维祖讲座中所引用）。现在混凝土抗压强度和抗拉强度

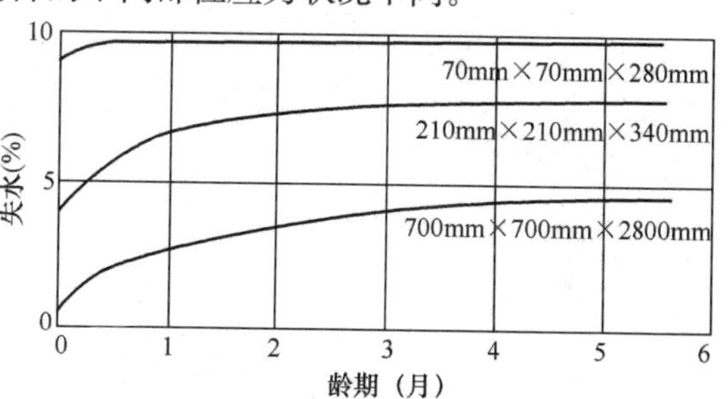

图 9　混凝土失水率和试件尺寸的关系

的关系和过去在较低混凝土强度等级时的不同，强度等级越高，抗拉强度与抗压强度的比值越小；实验条件对抗压和抗拉强度的影响也不一样，例如冻融循环后，抗拉强度损失比抗拉强度损失显著。

3 分解论和整体论的关系

现代科学技术活动规模扩大和工程技术复杂程度提高的速度空前，使自然科学、技术科学以及社会科学之间整体性联系的特点日益突出。该特点在土木工程领域也已显现——工程的功能复杂化对技术进步的要求、原材料的变化和资源的日益短缺对混凝土技术的挑战、观念和思维方法相对于时代发展的落后，等等，表现在土木工程中，尤其是混凝土这种复杂体，实验室分解研究的结果只反映混凝土材料本征特性的那些物理、力学性质指标都不能代表混凝土结构的服役行为，而且，在实验室，更无法量化和检测出混凝土那些只有在工程的活动中才能发生的行为、表现与效果（即英文中的“performance”），例如耐久性和施工性。这是由总是处在复杂变化环境中的混凝土内部微结构的复杂性所决定的。实验室分解研究的结果只反映材料相对的固有性质。在自然环境中几乎没有由单一因素造成的劣化，能给工程带来好处的措施，必然也会同时存在某些不利因素，必须因地制宜、因事制宜、因时制宜、因人（使用者的素质）制宜，因此分析和处理混凝土结构工程问题，必须用整体论观点和方法。

然而，不管有多么复杂，总是有规律的，需要不断地去探索、研究。当人们需要“知道究竟”的时候，就需要对复杂的事物进行剖析，去“刨根问底”，则研究机理是主要用分解论的方法。问题在于，不能把整体论和分解论对立起来而互相排斥，整体论和分解论相互结合，才能既能知其然而又能知其所以然；知其然才能确认其然。例如当萘系高效减水剂问世之后，正是科技人员研究了其成分和结构及其在水泥中的作用机理，人们才敢用，并逐渐了解了其使用的针对性而得以推广；对水泥水化机理的研究尽管至今还有很多问题不清楚，毕竟对混凝土的制备和应用起了很大的作用。如果行动先于研究，则必然会因盲目而走入歧途。对人体有害的“瘦肉精”曾经被大力推广就是一例。现在有人在推广某种产品时，为了狭隘的商业利益，只做“对混凝土性能改善”的宣传，而回避其成分和本征特性；对机理的“研究”只将一些推测的阐述或者皮毛的概念用来“点缀”蒙混。即使有证明在指标上是无害的甚至是改善的，对混凝土材料这种高度复杂的体系来说，也不能只看重短期的影响。整体论并不是只重表象不问实质，分解论也不是不研究元素之间的相互联系。

对于基础性研究，分解研究是必要的，是迄今自然科学研究的最基本方法。正如对人体的研究，不仅研究到细胞，还一直研究到基因，但是应当正确地简化，不能使用不可能存在的条件，并且不能做简单的外推，甚至夸大。例如在水灰（胶）比对混凝土收缩的影响时，从文献中查到有人说“水灰比对干缩的影响很大”，有人说“水灰比在 0.35 到 0.50 间变化对混凝土的干缩影响不大”，还有人认为水灰比可能间接地影响混凝土干缩，等等，莫衷一是。实际上这就是个方法问题。在试验时，所设条件不同会得出不同的结果。混凝土中骨料的线胀缩系数约是水泥浆胀缩系数的两倍，影响混凝土干缩的主要因素是骨料品种和用量。因此当骨料品种一定时，并在相同水灰比下，浆骨比是影响混凝土干缩的主要因素；对于水泥浆体来说，其干缩是由于失水而引起，因此当混凝土骨料品种和用量都一定时，水灰比是影响干缩的主要因素。试验时，就必须在水灰比、浆骨比（严格来说还有砂率）、骨料品种诸因素中，只改变一个因素而固定其他，才能知道各因素影响的程度；如果改变水灰比的做法是增减用水量而不变水泥用量，或增减水泥用量而不变用水量，则会同时引起浆骨比（体积比）的变化，试验结果不能说明各因素对混凝土干缩影响的程度。此外，与试验方法也有关——例如什么时间拆模测初长？从图 10[4] 可见，混凝土干缩值随水灰比增大而增大，而自收缩值则随水灰比增大而减小。自收缩在混凝土初凝后就开始，当试件硬化以后测初长时，已产生了一定的自收缩，此后再继续测长，就会得到干缩和自收缩造成的总收缩值，于是就出现“水灰比在 0.35 到 0.50 间变化对混凝土的干缩影响不大”的假象。当水灰比小于 0.3 时，自收缩更加敏感，继续降低水灰比时总收缩会增大。这个例子说明，分解试验的结论必须明确实验的条件，用于不同目的的解释。

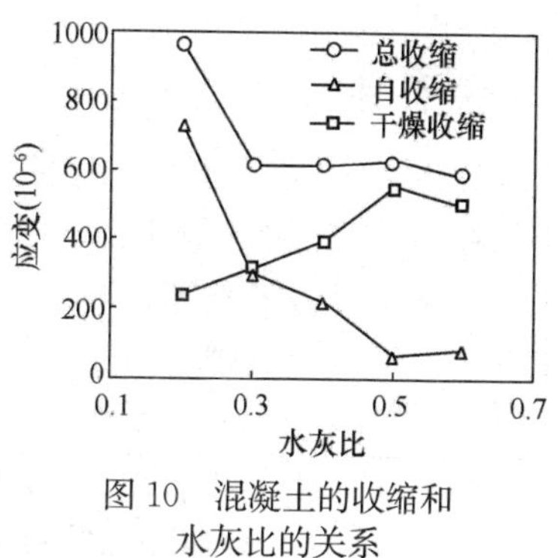

图 10　混凝土的收缩和水灰比的关系

自 1968 年美籍奥地利人、理论生物学家 L. Von. Bertalanffy 发表专著：《一般系统理论：基础、发展和应用》（《General System Theory：Foundations，Development，Applications》）后，开始形成了一个新兴的科学，把整体论和分解论结合了起来。

正如钱学森所说：“我们所提倡的系统论，既不是整体论，也非还原论，而是整体论与还原论的辩证统一，是更高层次的东西，即我们的系统论既要包括整体论，也要包括还原论。”系统论的基础是辩证法，把整体论和分解论割裂开是违背辩证法的，也就违背了系统论。不能认为只有在宏观层次上的研究才是整体论，实际上在物质的基本粒子研究中同样也需要整体论的思维，量子论就是最好的实例。不能认为分解论是科学的而整体论只是凭经验。什么是科学？科学的本质是人类对客观世界规律的认识，是一种知识体系。经验是经历和体验、验证的意思，无论是分解论还是整体论，经验都是重要的。牛顿发现万有引力，绝不是因为苹果掉到他头上这一次经历。同样，如果认为整体论是万能的而分解论是脱离实际的、无用的，就会丢掉某些可能是很重要的信息。分解研究时，要有整体论观念，整体研究时，要知其然又知其所以然。重要的是，研究的动力是人们了解世界的欲望和解决实际问题的需要，无论是分解论还是整体论研究，都不能从杂志缝里找题目，否则都会脱离实际。应提倡的是，重视基础性研究，按整体论观念用以指导工程的研究和实践。中铁十八局的杨雄利，在受 T. C. Powers 关于“不能使硅酸盐水泥完全水化”和 R. W. Burrows 关于混凝土养护的观点[5]的启发，对当时生产管片的养护工艺产生怀疑，经在生产实践中大量的对比试验，取消了原来工艺中无效或可能不利的环节，养护工艺优化后，取得了显著的经济效益和环境效益[6]。山东高速青岛公司的郭保林在青岛海湾大桥工程中使用了透水模板衬里，并对使用和不使用该模板衬里的混凝土进行模拟的对比试验，结果表明该模板衬里有既排水又蓄水、保水的功能，可基本消除毫米级的可见气泡，且可提高混凝土表面密实度，显著改善钢筋的混凝土保护层早期表面质量；使用不同来源的透水模板衬里的混凝土，与常规钢模板的相比，龄期 3 天时表面硬度提高率可达 40%以上，随龄期而下降，14 天下降到 30%以下，180 天一直到 540 天，基本稳定在 6%～8%。这是由于透水模板衬里的作用使混凝土表面水胶比降低，并得到充足的水养护，富集在表面的细水泥颗粒水化迅速而充分，“早长晚不长”而使表面增强稳定下来[7]。这种研究为严酷环境下重大混凝土结构工程的施工提供了可靠的依据。

结束语

1. 思维方法是哲学层次的方法，哲学是一门科学，是指导一切科学的科学，也是人们在实践中认识到的客观规律。用不同的思维方法会得到不同的结果。当前存在的传统思维、从众思维和机械思维等思维方法常会形成一种思维的定势，阻碍人们正确的认识和判断。运用辩证思维就可以较深刻地认识而遵循客观规律进行分析和决策。

2. 应当将分解论和整体论两种方法结合起来，而不是相互排斥，互相否定，当然也不是简单地从形式上的加合。这种结合的实质就是运用辩证思维分析和处理，就像是中西医结合一样，并不是用西医的手段诊断后再开出一些中成药方去治疗那样简单的拼装。对待混凝土这样复杂的体系，其复杂程度可以与人体相比，一个高明的中医一生中所开出的药方不会都是一样的，混凝土也不能一个强度等级都用一个配合比。现代“不确定性科学”理论应当也适用于混凝土。

3. 按整体论的观念，不能脱离工程而孤立地看待混凝土材料，首先要按工程需要优选和控制原材料，然后优化配制出具有合格实验室指标的混凝土拌和物，最后必须经过正确的浇筑、振捣、收面、养护等工艺，才能完成整个混凝土工程。三个环节互相影响、互相联系，每个环节都“好好做”，就能保证工程质量。离开工艺过程的混凝土拌和物是没有意义的。

参考文献

[1] Goldman A，Bentur A. Bond Effects in High Strength Silica Fume Concrete，quoted by Bentur in paper titled The Role of the Interface in controlling the performance of High Quality Cement Composites. In《Advances in Cement Manufacture and Use》，eddied by Gartner E，published by Engineering Foundation，New York. 1989.

[2] Swee Liang Mark（Astralia）and Kazuyuki Torii（Japan），Strength Development of High strength concretes with and without Silica Fume under the Influence of high Hydration Temperatures. Cement and Concrete Research，Vol. 25，No 8，1995.

[3] R. K. Dhir，M. R. Jones. PFA Concrete：Influence of Simulated in Situ Curing on Elasto-Plastic load Response，Magazine of Concrete Research，1993，45，No. 163.

[4] A. Neville. Properties of Concrete，Forth and Final Edition.

[5] R. W. Burrows. Visible and invisible cracking of concrete，ACI Farmington Hill，Michigan，1998.

[6] 杨雄利. 地铁管片养护制度优化的实验研究[J]. 混凝土世界，2010，10.

[7] 郭保林. 模板衬里改善表层混凝土质量的机理研究[J]. 建筑材料学报，2011，8.

自评

1. 分解论和整体论是当前科学学（即研究科学方法的学科）的两个学派，有各自的学术观点。绝对割裂来看，实际上各有利弊，不能相互排斥。

2. 有人说分解论和整体论无法结合，只能是微观研究用分解论，宏观研究用整体论。这样的说法是简单化了的理解。近代物理的研究证明，整体论观点促进了微观研究的进展。例如，量子力学把研究对象及其所处的环境看作一个整体，用量子态的概念表征微观体系状态，表明微观体系的性质总是在其与其他体系，特别是观察仪器的相互作用中表现出来，对决定状态的物理量不能给出确定的预言，只能给出物理量取值的概率。这就是整体论。这也恰好说明，人们通过观察、试验、实践对客观事物的认识与方法（包括手段）有关。任何理论都是从经验和实验中开始的。人们在实践或试验中当发现某种不协调（或者称矛盾时），引发思考和探究的兴趣，常常会在不自觉状态之中把分解论和整体论结合了起来，如量子论中薛定谔方程[①]、海森堡的不确定性（过去曾被译成测不准）原理[②]、玻尔的互补原理[③]、泡利的不相容原理[④]，等等，表明在研究微观世界的粒子物理中，也是整体论的观点（见本页附注）。

---

① 薛定谔方程（Schrodinger equation）：奥地利物理学家薛定谔（Erwin Schrodinger）导出波动力学建立的波函数（概率幅）而不是力学的量值，作为量子力学研究对象，描述微观粒子的状态随时间变化的规律。

② 不确定原理（Uncertainty principle 曾被译为测不准原理）：德国物理学家海森堡（Werner Heisenberg）认为，量测动作不可避免地会搅扰被量测粒子的运动状态，因此产生不确定性。即粒子的位置与动量不可能同时被确定，位置的不确定性与动量的不确定性遵守不等式 $\Delta x \Delta p_x \geqslant \frac{h}{2}$。式中 h 为普朗克常数。即粒子位置的不确定性×粒子速度的不确定性×粒子质量$\ngeqslant$普朗克常数。

③ 互补原理（Complementarity principle）：丹麦哥本哈根大学物理学家玻尔（Niels Henrik David Bohr）所提出："一些经典概念的应用不可避免地排除另一些经典概念的应用，而这'另一些经典概念'在另一条件下又是描述现象不可或缺的；必须而且只需将所有这些既互斥又互补的概念汇集在一起，才能而且定能形成对现象的详尽无遗的描述。"他用这样一种既对立又统一的原理解释光和粒子波粒二象性的互补。

④ 不相容原理（Exclusion principle）：是由奥地利科学家沃尔夫冈•泡利（Wolfgang E. Pauli）通过实验观察，总结原子构造时得到结论，认为在原子中不可能容纳量子状态完全相同的任意两个或两个以上的电子。

论文之九

面对混凝土工业可持续发展的挑战*

廉慧珍　阎培渝
清华大学土木水利学院

1　前言

同其他用于结构的建筑材料相比，混凝土具有不可替代的优势。由于大量使用或简单加工地方性原状天然材料，混凝土是耗能最低的材料（图 1）。能耗的计算包括从原材料开采、加工生产，直到构件完成制作[1]。100 年来，混凝土在人类生产建设发展中起着重大的作用。随着世界人口的增长、生产建设的进步和科学技术的发展，混凝土技术也在不断发展，用量也不断地增加。但是当混凝土用量大大增加以后，生产混凝土所需的水泥、砂、石等原材料大量消耗自然资源，破坏植被与河床的问题就日益显现。台湾目前已禁止开采生产水泥的石灰石，所用水泥都从境外购买；许多国家的优质骨料供不应求；在发展中国家，人类生活与生产所需的水资源正变得短缺[2]。当前全世界硅酸盐水泥工业每年向地球的大气层排放温室气体 CO_2 达 15 亿吨，约占总排放量的 7%[2]，我国在其中就占 40%多。1949 年以后，我国的水泥产量几乎呈指数增长（图 2），1997 年达到 5.1 亿吨，当时计划 2000 年产量为 5.5 亿吨，2010 年达到 8 亿吨。预计的水泥增长量为 2000 万吨/年左右。从 1997 年至 2002 年，仅过去 5 年，其间还经历了水泥标准修订所伴随的行业整顿，水泥产量却增加了 2 亿吨。照此速度，2010 年水泥产量将接近 10 亿吨。这将使用约 20 亿吨天然矿物原料，向大气排放约 10 亿吨 CO_2；这些水泥用于制备砂浆和混凝土，将消耗约 6 亿吨水和约 60 亿吨骨料，再加上生产、加工、运输将消耗的大量能源，对环境造成的巨大压力可想而知。

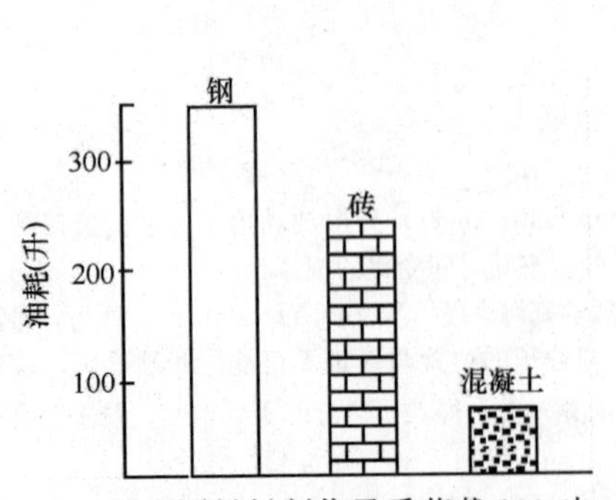

图 1　用不同材料制作承受荷载 100 吨、高 1 米的柱子消耗的能量[1]

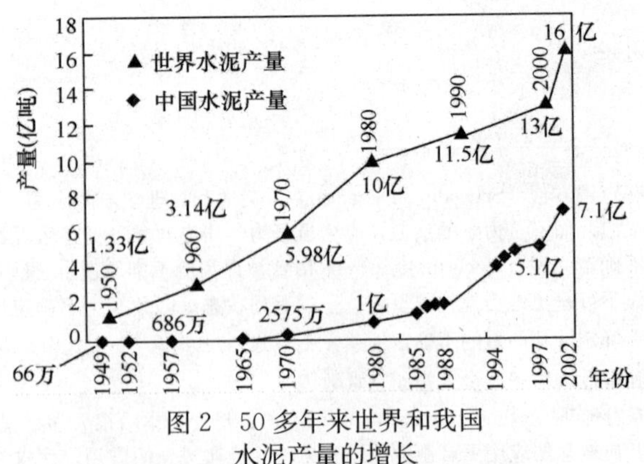

图 2　50 多年来世界和我国水泥产量的增长

上述论据都表明，当前的混凝土工业已面临是否可持续发展的挑战。钢结构建筑重新受

* 原载《建筑技术》2004.1。

到青睐，越来越多地用于重大工程建设，就是这种挑战的迹象之一。

我们可以采取哪些对策，来促进混凝土工业的可持续发展，以保持和发挥混凝土材料的优势，这就是今天我们要讨论的问题。

2 水泥工业可持续发展的出路

中国作为发展中的国家，大量基础设施的建设是必需的。混凝土仍然是最适合于大宗使用的结构材料，因此水泥产量的增长也是必然的，无法抗拒的。古代罗马2000多年前使用火山灰和石灰混合物制作的混凝土建筑物的耐久性告诉今人，减少硅酸盐熟料的用量（至少控制在现有用量）而大量掺用活性矿物混合材，既可满足建设对水泥的需求，又符合混凝土耐久性的需要，而且可以减少（至少不再增加）水泥工业向大气排放的CO_2，并大量利用工业废弃物，使水泥成为环境友好型材料。

实际上，除了P•Ⅰ型硅酸盐水泥外，我国现行水泥标准中对6个品种的硅酸盐类通用水泥都分别规定了其中混合材的允许掺量范围。例如P•Ⅱ的5%，普通硅酸盐水泥的6%～15%（活性混合材），矿渣硅酸盐水泥的20%～70%，火山灰硅酸盐水泥的20%～50%，粉煤灰硅酸盐水泥的20%～40%，复合硅酸盐水泥的20%～50%。但是除了普通硅酸盐水泥和一部分矿渣硅酸盐水泥外，其他含混合材的水泥因需求量小而很少生产。按现行水泥标准，水泥强度检验时一律采用0.5的水胶比；在高水胶比条件下，水泥强度随着混合材掺量的增加而急剧下降。而用户长期以来一直是以强度作为水泥主要的、甚至是唯一的质量指标。传统观念认为不能用低强度等级的水泥（含有较大量的混合材）配制高品质混凝土。随混凝土技术进步，低强度等级的水泥同样可配制出高强混凝土。例如清华大学、深圳港创建材公司和深圳地铁公司合作浇筑了一个足尺的深圳地铁区间段混凝土模型，所用胶凝材料总量中水泥只有180kg/m^3，其余有180kg/m^{3}粉煤灰和80kg/m^{3}的矿渣粉。现场留样检测28天混凝土抗压强度均超过50MPa。该混凝土中矿物掺和料约占胶凝材料的60%。如果按此比例生产水泥，其28天抗压强度是不会超过32.5MPa的。在统一的高水胶比条件下检验含掺和料的水泥，不能真实地反映矿物掺和料的作用；但在混凝土中如果简单地用矿物掺和料等量取代水泥，也是不正确的。以粉煤灰为例，如图3[3]所示，粉煤灰掺量对混凝土强度的影响不是线性的。在其他所有条件相同时20%以下的粉煤灰掺量影响不显著；此后，混凝土的强度随掺量的增加而明显下降；如欲保持强度不变，则需相应降低水胶比。在强度-粉煤灰掺量-水胶比之间存在三维相互影响关系。在等强度面上，可以找到混凝土的水胶比随粉煤灰掺量的变化关系。

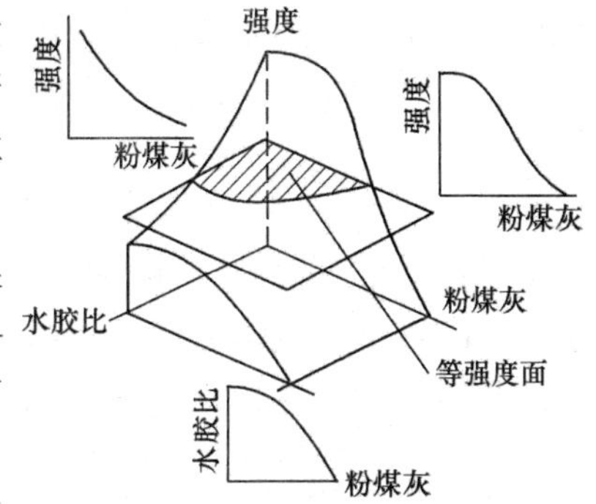

图3 强度-粉煤灰掺量-水胶比三维关系

目前水泥强度等级逐渐提高，使得低强度等级的混凝土的配制反而困难，往往导致混凝土拌和物泌水、离析严重。自从Malhotra及其同事们开发了大掺量粉煤灰（HVFA）混凝土，已有不少成功应用的实例，尤其适于低强度等级的混凝土。同传统混凝土拌和物相比，HVFA混凝土的拌和水用量要少1/3，结果是水胶比（w/b）较低，浆体总体积减少近16%。从而，直接与水胶比和混凝土中浆体比例有关的干燥收缩大为减小。同时，由于硅酸盐水泥用量大幅度减少，

HVFA 混凝土的早期水化放热量几乎减少 40%，因此大体积构件温度开裂的倾向大为减小[5]。Mehta 用 HVFA 浇筑了由配筋密集的剪力墙和一个 3.4m×1.8m 的大体积后张预应力钢筋混凝土基础组成的结构。剪力墙和基础的 HVFA 混凝土拌和物设计成坍落度 150mm、56d 强度 35MPa。此外，为了周转模板，墙体的混凝土设计成 7d 达到 20 MPa。保持混凝土内外温差在 25℃以内。大流动性的 HVFA 混凝土拌和物的水胶比为 0.32；为控制温升，剪力墙用Ⅰ型硅酸盐水泥 195 kg/m^3，基础用 160 kg/m^3；F 级粉煤灰 195 kg/m^3。由于 HVFA 混凝土优异的施工性，浇筑情况良好，墙和基础没有发现蜂窝和麻面。浇筑 9 个月后检查，墙体没有出现任何裂缝[6]。

矿渣水泥更是如此。英国和欧洲标准中允许矿渣最大掺量为 90%[7]，美国矿渣硅酸盐水泥（ⅠS 型）中矿渣最大掺量为 70%，矿渣水泥（S 型）中矿渣最低含量为 70%（需另加一定量的石灰）。我国水泥标准虽然允许在矿渣水泥中矿渣最大掺量为 70%，但极少生产如此高矿渣掺量的水泥。我国矿渣水泥销售情况不如普通水泥，用户宁愿购进普通水泥，在搅拌站掺加矿渣粉。这其中的原因之一就是，传统上矿渣水泥的生产是将矿渣和熟料以及石膏混合粉磨，由于矿渣比熟料难磨，成品中矿渣颗粒较粗，水泥磨得越细，矿渣和熟料间的细度差别越大，不仅不能发挥矿渣的活性，而且所配制的混凝土易于泌水，影响抗渗性和抗冻性。近来已有工厂将矿渣与熟料分别粉磨后再混合，大大改善了产品的性能。

由混凝土供应商在配制混凝土时掺入掺和料存在的问题是石膏含量的不足以及增加工序可能造成的管理问题。如果能将这道工序移到水泥厂，适应混凝土的规律来生产和检验粉煤灰硅酸盐水泥和矿渣水泥，或者分别控制熟料和混合材的性能及混合材的掺量，而由用户控制混凝土的强度，则可大大改善水泥工业的现状，逐步实现吴中伟生前曾提出过的减少或控制熟料产量但却增加水泥产量的目标，实现水泥工业的绿色化，并促进混凝土的绿色化。前苏联的 BHB 水泥[8]和清华大学的高性能胶凝材料[9][10]的技术路线应当可以借鉴。

3 砂石生产改造的紧迫性

20 多年前，我国已故混凝土专家蔡正咏曾说，中国混凝土质量不如西方的主要原因是砂、石的质量差。以石子为例，高质量的石子为大体等径粒形，级配良好，空隙率不超过 40%。我国石子生产的技术落后，20 多年前砂、石的空隙率一般都在 40%～43%，而对现今的砂、石质量说是糟糕并不过分；北京和深圳的砂石空隙率最好的也超过 45%，大多接近 50%，有的甚至超过。我国有混凝土用砂石标准，近年又经过修订，但是实际市售骨料的生产几乎都不执行。由于绝大多数采石场仍使用成本低廉的颚式破碎机，材质越硬（强度也高）的石料，破碎后针、片状颗粒越多，粒径小于 10mm 的颗粒几乎都是针片状颗粒，因此目前市售标称粒径 5mm 以上的石子，实际上缺少 5～10mm 粒级的颗粒。即使符合定义的针、片状颗粒数量在标准规定范围内，石子粒形也多数是多棱角的。砂子由于资源几近枯竭也无法讲究级配了。因此这几年来混凝土的用水量居高不下，一般都超过 175kg/m^3，C30 以下混凝土用水量更大，甚至达 200 kg/m^3。这是发达国家所没有的。美国混凝土用水量一般不超过 160 kg/m^3[4]，低强度（例如 25MPa）的混凝土也不会超过 180 kg/m^3（采用大掺量粉煤灰后，用水量可降到 130 kg/m^3 以下而仍具有良好的可泵性[4]）。用水量大时，水泥用量（胶凝材料）必然也大，以保证水灰比（水胶比）不变，在一定胶凝材料总量下，水泥用量必然也大，对混凝土的温升、变形也都会有影响；用水量少，意味着胶凝材料用量少，浆骨比低，混凝土结构开裂的倾向就会小。

另一个错误的观念是追求石子的强度。其实粒形比强度重要。构件受力后，材质强度高的针状和片状颗粒会因应力集中，承受过大的弯曲应力而提前断裂，从而降低混凝土的强度。还因粒形远离等径状，比表面积大，需水量就大，影响混凝土的施工性能。表1所示用相同配合比，分别以两种石子配制C60混凝土，测定其坍落度和28天抗压强度。其中1号石子结晶颗粒粗大，尽管粒形很好，大多为等径状，但岩体强度低；2号石子外观致密，岩石强度高，但针、片状颗粒明显地多。两组混凝土的水泥用量相同时，2号的比1号的需水量大，因而水灰比高，以致强度比1号的低，坍落度也比1号的低。如果2号的混凝土的坍落度提高到与1号的相当，则强度将会更低；如果使2号混凝土的水灰比与1号相同，同时增加用水量和水泥用量，2号混凝土的强度也会提高，但却会因浆骨比的增大而带来其他的问题：如混凝土温升提高，弹性模量降低，体积稳定性下降（收缩增加），费用增加，等等。

表1　石子对混凝土性质的影响

石子编号	石子特征	混凝土28天抗压强度（MPa）	坍落度（mm）
1	强度低，粒形好	71.3	195
2	强度高，粒形差	68.8	148

目前砂石质量降低的原因，首先是人们对砂石的认识有误，认为砂石级配不好不过是浪费点水泥而已；认为砂石价值很低而不珍惜。最近十年来我国建设规模快速增长，砂石需求量大，大量应运而生的小采石场，盲目占山采石，不仅产品质量低劣，而且生产过程中排出大量石粉、石屑，导致环境污染和资源浪费，再加上使用时的浪费也多，使天然砂石资源变得紧缺，造成了纯粹的市场控制——质量由卖方说了算，无法选择；想要求高质量，就会没有石子用而影响工程进度。

针对乱开山、乱采石造成安全事故频发和对环境的污染，我国有的地区已开始整顿砂石场，关停了一些规模小、设备落后、管理不善的作坊式采石场，加强了集中的大型采石场的技术投入。尽管其初衷只是着眼于安全，毕竟有利于环境保护，并为提高石子质量创造了一个良好的开端。砂石质量的提高，必须从严格执行砂石标准开始。为了生产出合格的以至优质的产品，现有生产方式必须改造。采取先进的破碎设备如反击式破碎机，还可采取不同类型破碎机分级破碎的方式，以改善石子粒形。

实际上即使在采石场进行了严格的级配，在装料、卸料和运输过程中，骨料在自重作用下也会破坏原来良好的级配，使用过程中再次装料、上料，使混凝土生产过程更无法保证骨料原有的级配。因此欧洲国家的石子都根据用户要求分级供应，用户在生产混凝土时按优化级配的比例上料。有条件的预拌混凝土厂可以自建、承包或定点采购，实现骨料的分粒级供应。目前我国有个别混凝土搅拌站已经或正在准备这样做了。实践证明，分级购进，级配上料，可以使掺用粉煤灰的混凝土用水量降低到100～130kg/m^3，仍有良好的可泵性。当然，这种骨料的价格应当高于原来无序竞争所盲目生产的劣质产品。优质优价的结果会避免人们因产品的“不值钱”而任意挥霍浪费而造成对环境不必要的压力。

4　减少混凝土的拌和与养护用水对生活用水的争夺

水是生命之源，水资源的严重不足已对人类造成威胁。混凝土是水资源消耗大户。我国目前水泥主要用于配制砂浆、灌浆料以及混凝土，每年约有3.5亿吨以上的水用于混凝土的拌和、养护和清洗。在保证混凝土各项性能的前提下尽量减少拌和水，不仅可以节省用水，而且能减小开裂的倾向。国内外都有实例表明，掺粉煤灰的混凝土，坍落度为100～130mm时仍有良好的可泵性[4]。假定每年7.1亿吨的水泥全部用于混凝土，按水泥用量为300kg/m^3计，则可生产23亿吨混凝土；如果每立方米混凝土少用100kg水，则每年混凝土行业可节省1亿吨水。目前绝大多数混凝土拌和与养护用水都是可饮用水，但实际并不需要如此洁净的水。在国外鼓励使用污水处理所得的中水作为混凝土拌和水[4]。目前出于生产成本考虑，已经有不少预拌混凝土厂循环使用设备清洗水，经沉淀后用于混凝土的拌和，成为减少使用生活用水的有效途径。地球上最丰富的水是海水，应当充分利用。当海水中所含可溶性盐不超过35000ppm时，可以用于拌和素混凝土[11]，尤其是掺大量矿物掺和料的混凝土；但严禁用于钢筋混凝土尤其预应力钢筋混凝土。当使用有潜在碱活性的骨料时，因海水中的钾、钠可能与其发生碱-骨料反应，所以使用有潜在活性骨料的混凝土不得使用海水拌和。与普通的混凝土相比，用海水拌和的混凝土早期强度会较高，但28天后的强度会较低。在配制时可用降低水灰比进行补偿[11]。用海水进行素混凝土的养护对混凝土不会造成损害。

混凝土的养护的目的是保持混凝土处于潮湿环境，洒水是最简便的方法，但是为了减少养护用水和工作量，已有人开发出若干有效的养护方法。例如蓄水性好的复合养护膜片，蓄水性模板，能蓄水的合成树脂毡片加塑料薄膜等。这些方法都能节省养护用水。但是还必须同时注意混凝土的降温问题，以免早期产生过大温度应力。

5 增强混凝土耐久性是混凝土业可持续发展的战略需要

混凝土结构的耐久性成为困扰当今世界各国的普遍问题。1991年在提交美国国会的报告《国家公路和桥梁现状》中指出，美国现存的全部混凝土工程价值约6万亿美元，而每年用于维修的费用高达300亿美元；英国1980年的建筑维修费用占建筑总费用的三分之二。许多发达国家每年用于建筑维修的费用超过新建的费用。美国联邦政府现在每年拨款50～60亿美元用于公路和桥梁的维修，但1998年ACI估计仅仅更换目前已损害的公路桥梁的混凝土桥面板就需800亿美元。我国目前也开始进入需要大规模修补或更换已建混凝土结构的时期。混凝土结构的提前劣化必然造成大量建筑垃圾和对混凝土原材料的大量需求与浪费。

必须从可持续发展的高度对混凝土结构的耐久性给予足够的重视。“如果我们能够生产出更耐久的产品，就必定能大量地节省材料。例如今天建造的混凝土结构物若不是现在的50年寿命，而是250年寿命，那么混凝土业的资源利用效率就能提高5倍。”[4]混凝土结构提前劣化的原因很复杂，发生早期开裂是重要原因之一。从根本上来说，混凝土属于多孔性的材料，精心制作并施工的混凝土可以有很好的水密性。如C30以上的混凝土，抗渗等级达到S8是很容易的事。但是在各种侵蚀性环境因素作用下，混凝土内部微裂缝会逐渐连通和开放，成为环境中的水、CO_2及其他有害介质侵入的通道，使钢筋混凝土开始劣化。如果混凝土结构在交付使用前甚至在浇筑的初期就出现裂缝，则会成为进一步开裂的源头，使劣化进程加速。早期裂缝产生的原因很复杂，单纯地追求高早期强度、以“混凝土强度7天可达到100%设计强度”为高水平施工的衡量标准是早期开裂的原因之一。任何事物，有所得必有所失。如果仅花半年时间盖起来的大楼的寿命只有20年，无论从经济上还是从保护资源和环境方面考虑，都是得不偿失的。

我们还需要重视大量建筑垃圾的综合利用。据报道，全球每年要产生超过 10 亿吨的建造和拆除废弃物[4]。大多数建筑废弃物作为混凝土的部分骨料的替代物是可行的，并在经济上也划算。二次世界大战末西方国家就已开始利用废弃的旧混凝土做混凝土的再生骨料，日本在旧混凝土再生骨料方面也做了许多工作，中国香港也制订了相关技术规程，开始在实际工程中使用。我国近年来由于混凝土耐久性问题的突出，以及旧建筑使用寿命终结，拆除的混凝土大为增加。但我国对废弃混凝土的综合利用率很低，大量建筑垃圾只能填埋，既占用了土地，又浪费了资源，对此应引起充分的注意。

最后还要说说节约能源的问题，减少熟料用量不仅减少 CO_2 排放量，而且也节省煅烧熟料的能源。但是用于混凝土的磨细矿渣等掺和料在磨细时能源消耗也很大，例如用传统球磨机粉磨矿渣，当比表面积达 $500m^2/kg$ 以上时，耗电超过 45 度/t；磨得越细，耗电越多。越细的矿渣虽然强度效应越高，但化学收缩和自收缩以及（掺入水泥后）水化热越大。所以从节省能源和混凝土耐久性考虑，矿渣不宜磨得太细。矿渣与熟料混磨的矿渣水泥中矿渣的比表面积只有 $250m^2/kg$ 左右，分磨时矿渣只要比表面积超过 $350m^2/kg$，即可提高其活性而没有副作用。目前矿渣比表面积居高不下的原因是单纯追求强度的传统观念所造成的市场需求所致。近年来出现的粉煤灰的磨细也是从强度尤其是早期强度出发的。尽管粉煤灰磨细后可能利大于弊，但却会增加粉磨的能耗。从混凝土对可持续发展的贡献出发，并且改变以强度为首要追求目标的观念，则应当按整体论规划混凝土及其原材料的市场。

参考文献

[1] Oscar Beijer. Energy consumption related to concrete structures, ACI Journal Nov. 1975.

[2] V. M. Malhotra. Introduction: Sustainable development and concrete technology, Concrete International. July, 2002.

[3] M. R. H. Dunstan. Fly-ash as the Fourth Constituent of Concrete Mix, Proceeding of Fourth International Conference on Fly Ash, Silica Fume, Slag and Natural Pozzolana in Concrete. Istanbul, Turkey, May 1992. SP-132-86.

[4] P. K. Mehta. Greening of the Concrete Industry for Sustainable development. Concrete International. July, 2002.

[5] V. M. Malhotra. High-performance high-volume Fly Ash concrete. Concrete International. July, 2002.

[6] P. K. Mehta. Heavily reinforced Shear walls and Mass foundations built with "Green" Concrete. Concrete International, August 2002.

[7] BS 5328: Concrete, Part 1: . Guide to specifying concrete, 1991.

[8] Podmasova S A, Babev SH T, Volkov YU S. New Low Water Demand Binders for High-Strength Concretes. Proceedings of International Conference on High-Strength Concrete, Norway. 1993.

[9] 廉慧珍，吴中伟. 混凝土的可持续发展和高性能胶凝材料[J]. 混凝土，1998，第六期.

[10] 廉慧珍，阮庆革，李玉琳. FK 系列高性能水泥的性能及其检测[J]. 混凝土，1999，第一期.

[11] Steven H Kosmatka and William C. Panarese. Design and control of Concrete Mixture, 4th printing, 13th edition, 1994.

自评

最后一段关于矿渣磨细问题的论述欠全面。水泥粉磨的耗电量目前约为 $45kW \cdot h/t$，如果矿渣磨细耗电量不超过此值，却可节省天然资源，是合理的。问题是现在有人用球磨机粉磨，磨到相同细度耗电量 80～100kW·h/t 时，则应当反对。国内现代化矿渣粉厂，耗电量约 38kW·h/t，这是应提倡的。不应一概否定矿渣的磨细。磨细后加大掺量（70%以上），即可降低混凝土温升和减小自收缩，最适合于大体积混凝土、地下、水下和海洋工程。

要记住，现在对人来说，如同需要空气一样，需要细腻的感情，而思想的细腻，智力的丰富，是它的源泉。

——苏霍姆林斯基

思想的伟大不在于能否容纳琐碎小事，而在于能否用自己的影响使小事变成大事。对小事漠不关心的人也不会对大事真正感兴趣。

——罗斯金

二、关于混凝土的原材料与配合比的思考

论文之十

水泥的品质和混凝土质量的关系*

廉慧珍** 梁文泉***

摘 要 在传统上，混凝土是按强度进行设计的，对混凝土的质量的最终标准主要是强度。因此混凝土生产者对水泥品质的要求也是强调强度；强度越高的水泥被认为质量也越高。如此的发展，造成近年来混凝土结构出现裂缝尤其是早期开裂的现象日益普遍。其原因很复杂。单从水泥来说，比表面积、矿物组成中C_3A、C_3S、碱含量的增加，热水泥的出厂，都增加了开裂的敏感性，降低了流变性能，是原材料中影响混凝土质量的主要原因。应当把抗裂性和与外加剂相容性作为水泥品质的重要要求，并限制出厂水泥的温度。

关键词 水泥品质，抗裂性，细度，矿物组成，含碱量

前言

水泥和混凝土的关系，可以比作食物和人的关系。食物被人消耗之后，应当变为组成人身体的各种必需的组分，不论近期还是长期都不应有什么有害的影响。并不是所有的人都清楚地知道自己应当对食物有什么要求。如果没有科学指导，那么双方都可能产生盲目性。水泥的强度，尤其是早期强度越来越高，虽然也是生产技术进步的一种表现，但也是一种盲目性追求市场的结果——即混凝土强度不断提高的要求。在传统上，由于人们对工程质量所注重的就是强度，自然对水泥的要求也主要是注重强度。尽管由于混凝土的耐久性问题开始显现，人们开始重视混凝土结构物的耐久性，但在实践中仍然把强度作为混凝土质量要求和验收的标准。尤其近两年来，混凝土施工中高效减水剂与水泥相容性不好的问题发生得比过去更多，地下连续墙和楼板甚至大梁开裂问题频频发生。其原因很复杂，涉及多方面，包括开发商、业主、建筑设计和结构设计、材料、管理。这些问题将另议。现单就材料本身来说，混凝土的质量不只是配合比的问题。配合比是与原材料性质相匹配的，质量差的原材料也很难做出高质量的配合比。因此有必要也从原材料找找原因。

暂不论骨料的品质，在原材料中，影响混凝土抗裂性的主要因素则是水泥。购进水泥时只检验强度（当然有时还可能复验一下凝结时间）是不能判断水泥对混凝土抗裂性影响的。如图1所示为两个不同厂家生产的相同品种水泥，B厂水泥的混凝土在约束条件下由于自收

---

* 原载《中国水泥》创刊号（2002.6.）、第二期（2002.7.）。

** 廉慧珍，女，教授，1933.11. 清华大学土木水利学院建材研究所，100084。

*** 梁文泉，男，副教授，1962.2. 武汉大学水力电力学院建材教研室，430072。

缩而产生较大的拉应力，使其对开裂敏感；A厂的水泥则应稍有膨胀而有较小的约束应力，抗裂性较好[1]。因此水泥的研究者和生产者应当除了关心按现行水泥标准规定的水泥性质外，更加关心水泥在混凝土中的行为，即对混凝土抗裂性能的影响。

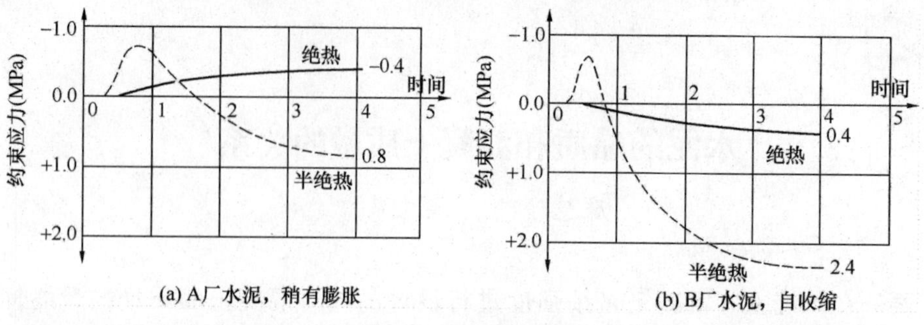

图1　用不同厂家的同品种水泥配制的混凝土在绝热和半绝热条件下的约束应力[1]

图1这种现象主要是随着水泥强度不断提高后才发生的。不同水泥厂家采用了不同的方法满足强度（尤其是早期强度）的要求，例如提高比表面积，增加C_3S、C_3A的含量等，我国有的水泥厂甚至还采用一些什么“增强剂”之类的措施（注意正像一些食品添加剂，短期无害，长期不一定安全）。由于建筑业的需求，现代水泥的组成和细度发生了很大变化[2]。美国从1920年到1999年70多年中水泥和混凝土主要参数的变化的趋势是水泥中C_3S含量从35%增加到50%～60%，比表面积从220m^2/kg增加到340～600m^2/kg，混凝土的水灰比从0.56～0.8减小到0.26～0.56[2]。水泥的7天抗压强度增长了几乎2.5倍[3]。近年来国外许多专家根据实际调查研究，对这种趋势提出了批评，指出当前混凝土结构不断增多的过早劣化现象主要原因是与此趋势有关。“20世纪混凝土业为满足越来越高的强度要求，不可避免地违背了材料科学的基本规律，即开裂与耐久性之间存在的密切关系。为了实现建设可持续发展的混凝土结构这个目标，有必要更新一些观念和建设实践。”[3]

我国水泥标准的修订的方针是“与国际接轨”，因此也是在按此趋势发展。回顾这段发展，分析其与混凝土结构耐久性的关系，会有助于我们更新观念，从关心强度转变到耐久性。从耐久性的角度评价水泥和混凝土的质量。

对我国水泥品质变化的简单回顾

对水泥标准的修订能反映出水泥品质的变化（不说“质量”而说“品质”是为了避免对当前产品水泥质量的褒贬）。修订水泥标准的人的出发点当然是通过修订标准提高水泥的质量，但是由于缺乏和水泥的服务对象——混凝土结构工程的联系，以至于忽视了水泥的品质对提高混凝土质量（不能只看到强度，更重要的是耐久性）的影响。20年来，我国水泥标准进行了三次修订。第一次修订的标准于1979年7月开始实施，第二次是1992年开始实施，第三次，即最近的一次是1999年开始实施。各次修订的基本出发点都是“与国际接轨”（尽管前两次还没有这个词，而实质相同），促进我国水泥生产工艺的改进和产品质量的提高。

第一次修订是将我国使用了20多年的“硬练”强度检验方法和标准改为“软练”强度和标准。这次变化较大，主要变化如表1所示。

表 1　水泥标准从硬练改为软练的主要变化****

变化因素	GB 175—63	GB 175—77
熟料的石灰 饱和系数（KH）	0.85 左右	0.90 左右
C_3A 含量	5%～7%	>8%
检验强度所用灰砂比	1∶3	1∶2.5
检验胶砂强度所用加水量*	$\frac{P^{**}}{4}+2.6$（普通水泥，约 0.36）	固定 0.44
用相同熟料的检验 28 天抗压强度的差别***	497 kg/cm² 613 kg/cm²	425 kg/cm² 525 kg/cm²
细度	4900 孔/cm² 筛余（0.088mm）	0.08mm 筛的筛余≤15%

* 只列出硅酸盐水泥和普通硅酸盐水泥的该量。
** 标准稠度用水量。
*** 以 GB 175—77 生产的 425＃和 525＃水泥为例，按统计计算。
**** 未列成型方法和试模形状和尺寸等重要的变化。

由表 1 可见，这次修订水泥标准的结果是增加了熟料中的 C_3S 和 C_3A 含量，水泥细度从比表面积平均 300m²/kg 增加到平均 330m²/kg，提高了水泥强度，尤其是早期强度，同时也提高了水化热。因检验强度的水灰比大幅度增加，减小了掺入矿物掺和料后强度的优势[3]。

第二次修订后的 GB 175—92、GB 1344—92 等强调了水泥的早期强度，28 天强度均提高了 2%，增加了 R 型水泥品种。该标准强化了 3 天早期强度意识，倡导多生产 R 型水泥[4]。普通水泥的细度进一步变细，从筛析法的<12%，改为<10%。

GB 175—1999、GB 1344—1990 等把强度检验的加水量改为 0.50，取消了 GB 175—92 中的 325＃水泥，水泥的强度进一步提高。业内人士都知道，水泥 3 天强度和 28 天强度的比值（反映水化速率）随水灰比的增大而减小，水泥标准修订后的水灰比从 0.44 增加到 0.5，但对 3 天强度的要求并没有降低，实际上提高了早期强度的要求，迫使水泥厂以提高 C_3S、C_3A 和比表面积来提高水泥的强度。某厂对 21 种来自不同厂家生产的熟料（包括大水泥和小水泥生产的）进行分析，C_3S 超过 60%的有 4 个样本（占总样本的 19%）、超过 58%的（含 60%以上的）有 10 个样本（占 47.6%）；有 17 个样本的 C_3A 含量超过 10%。大部分水泥细度超过了 350m²/kg。

综上所述，可见我国水泥各有关参数和性质变化的历程和趋势与国外相似。特点是增加 C_3S、C_3A、细度趋向于细，因而强度尤其早期强度不断提高。有一点我国标准和国际标准没有“接上轨”，那就是，国外，特别是欧洲国家的水泥标准对强度的规定不仅有低限，还有高限，而我国标准则只有低限没有高限。人们购买水泥认定“强度越高越好”，实际强度超过标称强度 20%以上的水泥被认为是“好水泥”。此外，上世纪 70 年代后期我国开始引进国外先进水泥生产的干法工艺，使水泥的含碱量提高，尤其使用北方的原材料的水泥含碱量普遍较高。GB 175—1999 对水泥中含碱量进行了限制，但只是出于对预防碱-骨料反应的考虑。这种变化的趋势虽然对混凝土提高早期强度有利，但却增加了混凝土的温度收缩、干燥收缩，再加上较低水灰比产生的自收缩，处于约束条件下的混凝土结构较大的收缩变形因高的早强而提高的早期弹性模量而产生较大的应力，而高早强又使能缓释收缩应变的徐变很小。于是开裂成为必然。

以下分别分析上述几个因素对混凝土抗裂性造成的影响。

水泥矿物组成的影响

众所周知，硅酸盐水泥主要的组成矿物有四种，它们的水化性质不同，在水泥中所占比例不同时影响对水泥整体的性质。表 2 所示为水泥中四种主要矿物的水化热，表 3 为四种主要矿物的收缩率。

表 2　传统水泥熟料四种主要矿物的水化热[6]

矿物 / 龄期	发热量（cal/g）			
	C_3S	C_2S	C_3A	C_4AF
3d	58±8	12±5	212±28	69±27
7d	53±11	10±7	372±39	118±37
28 天	90±7	25±4	329±39	118±22
3 个月	104±5	42±3	311±17	98±16
1 年	117±7	54±4	279±23	90±22
6.5 年	117±7	53±5	328±25	111±24

表 3　矿物熟料四种主要矿物的收缩率

矿物	收缩率
C_3A	0.00234±0.000100
C_3S	0.00079±0.000036
C_2S	0.00077±0.000036
C_4AF	0.00049±0.000114

由表 2 和表 3 可见，C_3A 的水化热是其他矿物水化热的数倍，尤其在早期。C_3S 的水化热虽然比 C_3A 的小很多，但在 3 天却是 C_2S 水化热的几乎 5 倍，因其含量在熟料中约占一半，故影响也很大；C_3A 的收缩率是 C_2S 收缩率的 3 倍，是 C_4AF 的几乎 5 倍。因此 C_3A 含量较大的早强水泥容易因早期的温度收缩、自收缩和干燥收缩而开裂。

水泥细度对混凝土工作性的影响

目前我国混凝土尤其是中等以上强度等级的混凝土普遍使用高效减水剂和其他外加剂。当高效减水剂产品一定时，水泥的成分（主要是含碱量、C_3A 及其相应的 SO_3 含量）和细度是影响水泥和高效减水剂相容性的主要因素。水泥细度的变化加剧了水泥与高效减水剂相容性问题。近两年时有发生高效减水剂的用户和厂家的纠纷。为此，天津雍阳外加剂厂丘汉用不同细度的天津 P·O 52.5 水泥和拉法基 P·O 52.5 水泥分别掺入不同量的 UNF-5AS，进行相容性实验。采用水灰比为 0.29 的净浆，分别在搅拌后 5 分钟和 60 分钟后量测其流动度，结果如表 4 所示。

由表 4 可见，随水泥比表面积的增加，与相同高效减水剂的相容性变差，饱和点提高，为减小流动度损失需要增加更多掺量的高效减水剂。不仅增加施工费用，而且可导致混凝土中水泥用量的增加，影响混凝土的耐久性。

表 4　高效减水剂与不同细度水泥的相容性实验结果

细度（cm²/g）	3014	3486	3982	4445	5054
饱和点（mm）	0.8	1.2	1.2	1.6	2
流动度无损失时的掺量（%）	1.6	2.2	1.8	>2.4	找不到

水泥细度对混凝土开裂的影响

在目前我国大多数水泥粉磨条件下，水泥磨得越细，其中的细颗粒越多。增加水泥的比表面积能提高水泥的水化速率，提高早期强度，但是粒径在 1μm 以下的颗粒水化很快，几乎对后期强度没有任何贡献。倒是对早期的水化热、混凝土的自收缩和干燥收缩有贡献——水化快的水泥颗粒水化热释放得早；因水化快消耗混凝土内部的水分较快，引起混凝土的自干燥收缩（图 2）[2]；细颗粒容易水化充分，产生更多的易于干燥收缩的凝胶和其他水化物。粗颗粒的减少，减少稳定体积的未水化颗粒，因而影响到混凝土的长期性能。

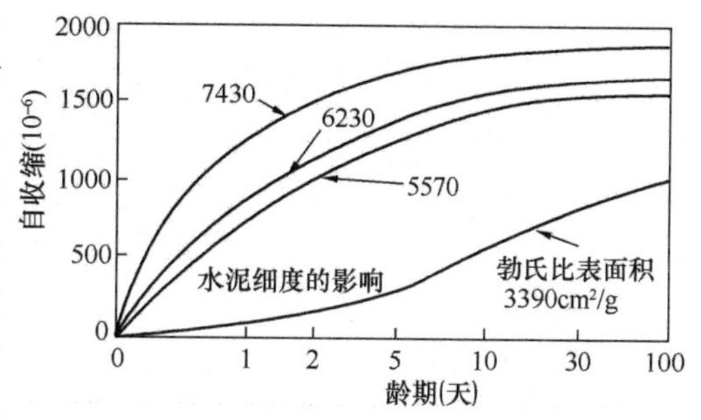

图 2　水泥细度对自收缩的影响

图 3 为 Burrows 引用的一个实例：在美国 1937 年按特快硬水泥生产的水泥 I 与现今水泥的平均水平的组成和细度相当，当时采用这种快硬水泥的混凝土 10 年后强度倒缩了（图 3 中的水泥 I）；而 1923 年使用粗水泥的混凝土，直到 50 年强度还在增长（图 3 中水泥 7M）[2]。水泥细度还会影响混凝土的抗冻性（见图 4）[2]。细水泥的易裂性可能与其低抗拉强度有关（图 5）[2]。

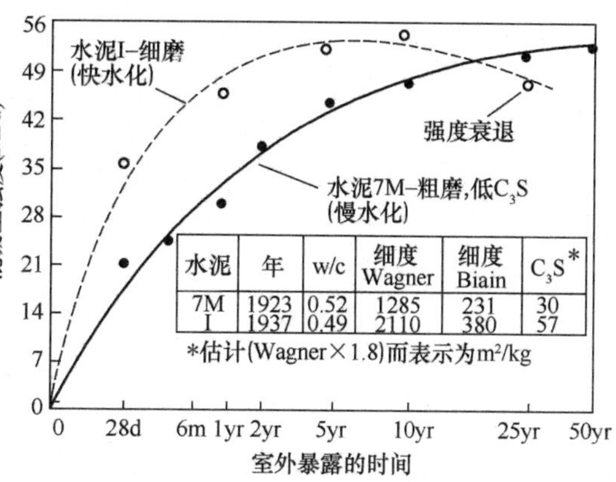

图 3　不同细度水泥的混凝土室外暴露的耐久性

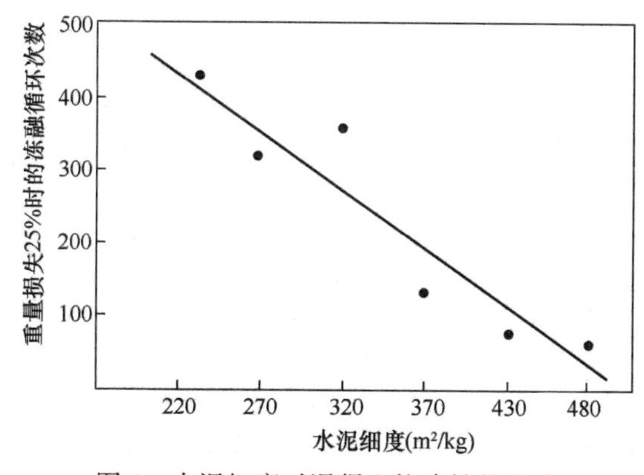

图 4　水泥细度对混凝土抗冻性的影响

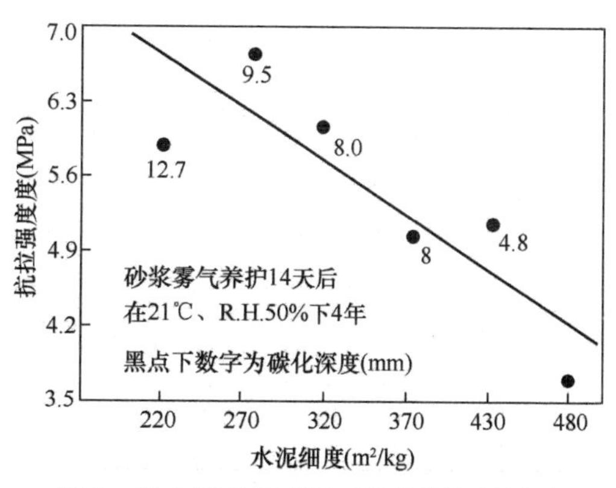

图 5　细水泥的易裂性对抗拉强度的影响

水泥中含碱量和混凝土开裂的关系

GB 175—1999 出于对预防碱-骨料反应的考虑对水泥中含碱量进行了限制。Burrows 在美国克罗拉多州的青山坝对 104 种混凝土的面板进行了 53 年的调查研究，发现开裂严重的劣化了的混凝土中，有的水泥含碱量高，但所用骨料并没有碱活性；还有的使用高碱水泥同时所用骨料也有活性，但是检测的结果却没有碱-骨料反应的产物，而混凝土却开裂而劣化了；低碱或虽高碱但低 C_3A 和低 C_3S 的水泥则完好。这表明碱能促进水泥的收缩开裂[2]。

图 6 为 Blaine 用环形收缩测定仪测定水泥中含碱量对水泥开裂情况的影响以及 1996 年调查的相应水泥混凝土状况的影响，图中的好和差表示抗裂性的好坏。在图中，注意当 Na_2O当量在 0.6 以下时混凝土状况的改善，还要注意水泥的细度和 C_3A、C_3S 影响。在图 7 中可见，用粗磨、低碱水泥时，引气混凝土可经受住 550 次冻融循环，但用细磨的、高碱水泥则经受不到 100 次循环[2]。

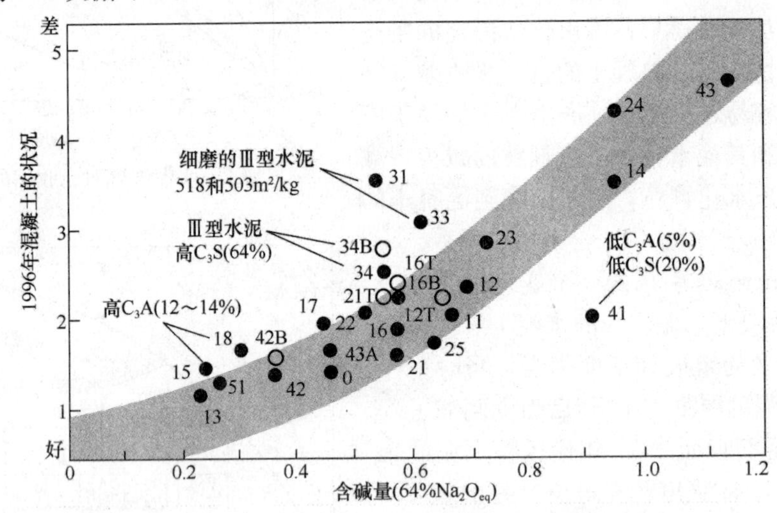

图 6　碱含量对混凝土状况的影响[2]（美国的Ⅲ型水泥是早强型硅酸盐水泥）

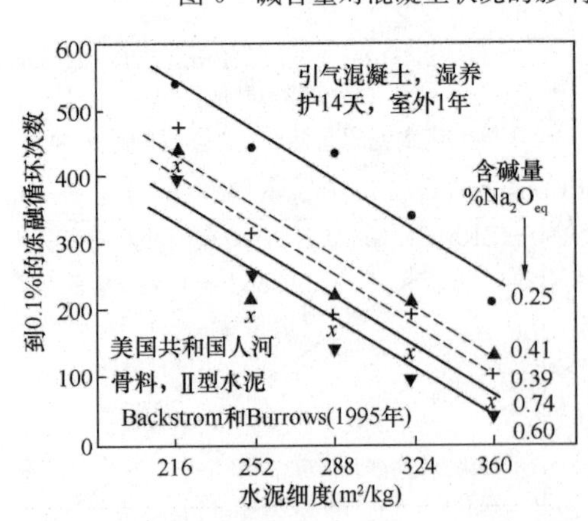

图 7　水泥细度和含碱量的交互作用

美国国家标准局对 199 种水泥进行了 18 年以上的调研，大量的发现是碱和细度、C_3A 和 C_4AF 的因素一起极大地影响水泥的抗裂性。即使水泥有相同水化率（强度）和相同的自由收缩，显然低碱水泥有内在的抵抗开裂的能力。当含碱量从低于 0.6% Na_2O 当量时，水泥的抗裂性明显增加，当进一步降低到趋向于 0 时，这种能力会进一步改善，尽管这一点是做不到的。由于碱-骨料反应必须在混凝土中有足够的含碱量、足够数量的活性骨料和足够的水分供应三个条件同时存在的情况下才会发生，并不要求任何情况下都限制水泥的含碱量，但是，促进混凝土收缩裂缝的生成和发展以至造成混凝土结构物的劣化，却是高含碱量对混凝土更大的威胁。不管是否使用活性骨料，必须将水泥中的含碱量减少到最少。

对水泥抗裂性评价和选择方法的推荐

用环形约束试验评价水泥或混凝土抗裂性的方法已有 60 多年的历史。世界许多国家的学者对钢环的材料、尺寸、信息收集和处理方法、评价指标，以至基于弹性力学的力学模型等都有研究，并分别用此方法研究过影响水泥和混凝土开裂敏感性的因素。Burrows 建议使用 Blaine 的方法评价水泥：开裂时间小于 1 小时的是很差的水泥，大于 15 小时的为优[2]。清华大学建材研究所覃维祖教授指导他的研究生对此方法进行了研究，结果表明用他们设计的材料和尺寸的环试验评价混凝土所用胶凝材料的抗裂性是有效的。使用热膨胀系数小的材料时，不仅可检测水泥的干缩开裂性能，还可检测由于温度收缩引起的开裂性能。

讨论和建议

1. 众所周知，凡是能提高混凝土早期强度的因素，都会影响混凝土后期强度的增长，目前在配制混凝土时都有较大的强度富余，以补偿这种后期强度的损失。这无疑造成很大的浪费。现在看来，问题远比此更严重，早期的高强度所带来的后患是混凝土结构物提早劣化。因此，除非工程有特殊需要，应尽量避免使用早强水泥。

2. 混凝土早期高强度的需求促使了水泥向高 C_3S 和高 C_3A、高比表面积发展，再加上混凝土的低水灰比、高水泥用量、超细矿物掺和料的使用，以及水泥出厂温度普遍过高，造成在约束状态下的混凝土因温度收缩、自收缩和干燥收缩和较高的早期弹性模量而产生较大的内部应力，早期的低徐变无法缓解这种应力，而产生早期裂缝；内部不可见的微裂缝在混凝土长期使用过程的干燥环境中继续发展，是混凝土提早劣化的主要原因。

3. 高含碱量的水泥会生成抗裂性能差的凝胶，加重混凝土后期的干燥收缩，所以不论骨料是否有活性，都应当限制水泥和混凝土中的含碱量。

4. 要有像管理食品添加剂那样管理生产水泥中的“增强剂”一类的措施，出厂时要注明主要成分，并有一年以上长期观测的数据证明其对混凝土长期性能的影响。

5. 应当改变以强度作为质量首要指标的观念，建议对水泥和混凝土品质增加抗裂性和与外加剂相容性的要求。

参考文献

[1] R. Springenschmid and R. Breitenbücher. Influence of Constituents，Mix Proportions and Temperature on Cracking Sensitivity of Concrete，PREVENTION OF THERMAL CRACKING IN CONCRETE AT EARLY AGE. Edited by R. Springenschmid. RILEM Report 15，pp. 6. Published in 1998 by E & FN Spon，11New Fetter Lane，London EC4P 4EE，UK.

[2] R. W. Burrows. The Visible and Invisible Cracking of Concrete，ACIMonograph No. 11，Published by ACI，Farmington Hills，Michigan，First Edition，1998.

[3] P. K . Mehta and R. W . Burrows. Building Durable Structure in 21st Century，CONCRETE INTERNATIONAL，March，2001.

[4] 王幼云. 实施水泥新标准对我国水泥生产工艺和产品质量的影响，中国建材院水泥研究所报告，1980，7.

[5] 王文义. 我国通用水泥新标准实施中的若干问题[M]. 北京：中国建材科技，1993，6.

[6] 岡田青、六車熙. コンクリート工学ハンドフック，改訂锌版，第一章セメント，朝仓書店，

1981，11.

自评

1. 100多年来水泥的生产追求的目标，一是降低能耗，二是高强，于是发展到现在的新型干法窑。但是，降低了能耗，却把碱留下固溶在熟料中；“耗能大户”的帽子却仍未能被摘除。强度提高了，而因 C_3S 和 C_3A 的增加却增大了早期水化热，提高了硬化水泥浆体的开裂敏感性。如果换一种思维方法，尝试一下，不要这么高的强度，是否能使能耗进一步有所降低呢？另一个不变的思维方法就是标准以前跟着苏联走，现在跟着ISO走。始终不能有自己的方式。有人问水泥界人士：“你们水泥除了能打混凝土（砂浆是一种特殊的混凝土），还能做什么?”既然如此，为什么不能按混凝土的需要生产，按混凝土的规律检测?

2. 文中所述水泥过细对混凝土的影响是与外加剂的相容性变差，增大混凝土早期收缩、硬化混凝土抗冻性下降。却未触及对结构耐久性更加重要的一个问题，即，硬化混凝土即使渗透性很低，也会不可避免地存在不可见的初始缺陷。在长期服役过程中，因和大气相对湿度相平衡，混凝土内部仍会有残留的水分，使水泥仍会继续水化，而使那些初始缺陷自愈；如果水泥过细，水化加速，水化程度增大，于是剩余的未水化水泥颗粒大为减少，会使原始缺陷失去自愈能力。在大气冷热循环、干湿交替的作用下，那些得不到自愈的原始缺陷会逐渐开展，成为提供侵蚀性介质侵入通道的可见裂缝。混凝土劣化就会开始。这就叫做“早长晚不长”，“早熟早衰”。

论文之十一

外加剂和混凝土工程质量的关系*

廉慧珍（清华大学土木水利学院）

1　外加剂的广泛应用是现代混凝土工程的重要标志

外加剂广泛应用使混凝土发生了巨大的变化，如表1所示。

表1　现代混凝土相对于传统混凝土的变化

项目	传统混凝土	现代混凝土
水胶比	0.5以上	0.15～0.6
强度	75#～400#	C15～C200以上
流动性	干硬性、低塑性、塑性	塑性、流态、自密实
骨料在混凝土中的状态	紧密堆积	悬浮
骨料对混凝土强度影响	大～很大	小～无
配合比选择	四组分、三要素，以强度为目标	多组分、四要素，以耐久性和施工性为目标

这样变化的原因首先缘于高效减水剂的使用：

1.1　改变了混凝土强度和水泥强度的依从关系，扩大了混凝土强度范围

水泥产品标准中规定水泥的质量均以水灰比0.5检验，为了施工性的需要，水泥强度必须是混凝土强度的1.5倍～2倍，小于1.5倍时将无法施工，大于2倍时则不经济。于是传统混凝土的水灰比必须大于0.5；使用了高效减水剂后，混凝土的强度不再受水泥强度的制约，因为水胶比范围可以从0.15（"超高性能混凝土"UHPC及其原创"活性粉末混凝土"RPC）直到0.6以至更高（用于垫层、回填等的低强度混凝土），自然混凝土的强度范围也就大幅度扩大了。

1.2　拌和物流动性范围扩大，丰富了施工技术

传统拌和物流动性从过去的无坍落度的干硬性贫混凝土、坍落度为10mm～40mm的低塑性到坍落度最大为50mm～90mm的塑性混凝土。浇筑和振捣的劳动强度大，难以避免"蜂窝"甚至"狗洞"之类的混凝土内部缺陷；现代混凝土拌和物的流动性范围大大提高，从道路等工程混凝土的塑性（坍落度为50mm～90mm）直到可以免振捣的自密实混凝土，坍落度最大不超过270mm，这样宽的范围极大地丰富了混凝土的施工技术，在高层、超高层、大跨度和异形建筑物和构筑物工程中都有了用武之地；同时大大增加了混凝土的品种，如高强泵送的轻骨料混凝土、自密实混凝土、超低水胶比的高密实度混

* 原载《混凝土世界》2012，1。

凝土等。

1.3 改变了混凝土强度和骨料强度的关系

传统混凝土中粗骨料相互之间距离小，骨料间水泥浆层很薄，属于紧密堆积，承受荷载时，荷载基本上通过骨料传递，随着拌和物坍落度的增大，粗骨料之间的料浆层增厚，粗骨料之间距离增大，骨料在混凝土中是悬浮状态的，荷载传递的过程既要通过骨料也要通过砂浆；混凝土受力后的破坏发生在最薄弱环节，对中低强度等级混凝土来说，最薄弱环节在界面和在砂浆中，坍落度越大，这个特点越突出。因此骨料强度对混凝土强度的影响随混凝土拌和物浆骨比的增大而减小，其减小的顺序是：干硬性混凝土（贫混凝土坍落度 0mm）→塑性混凝土（坍落度 10 mm～40mm）→塑性混凝土（坍落度 50 mm～90mm）→流态混凝土（坍落度 100mm ～120mm）→高流态混凝土（坍落度＞150mm）→自密实混凝土（坍落度＞240mm）。

1.4 扩大了水硬性胶凝材料（hydraulic cement）的涵义，改善了混凝土的性能

由于东西方文化的差异，常有一些词汇难以准确地互译。日文中有许多外来语都采取音译，就可避免一些误解。例如"cement"日文音译译成"セメント"，而我国译成水泥，多年来都被人们理解为在水泥厂生产出的具有水硬性的胶凝材料。实际上，"cement"在英文中泛指胶凝材料（包括气硬性的和水硬性的），水泥厂大宗生产的是 Portland cement，即硅酸盐水泥，所以才有 ASTM designation：C 1157 的"Standard Performance Specification for Hydraulic Cement"（水硬性胶凝材料标准性能规范）[1]，如按我国的翻译，岂不成了"水硬性水泥"? 美国这个标准的提出就是因为高效减水剂的使用，使得矿物掺和料作为水硬性胶凝材料组成的一部分有了依据。无论是西方国家还是我国，硅酸盐水泥标准中都包括掺混合材的硅酸盐水泥，在我国，人们的认识是掺混合材越多强度越低。现在因为有了高效减水剂，使矿物掺和料掺在混凝土中与普通水泥一起组成水硬性胶凝材料，矿物掺和料用量甚至超过硅酸盐水泥熟料用量，可以用低于水泥厂检验用的水胶比，得到任意强度的混凝土，而且可改善混凝土其他性能。因此现代混凝土以"水硬性胶凝材料"（现简称胶凝材料）代替"水泥"更加确切。

2 拌和物质量决定混凝土工程最终产品的质量

外加剂是通过拌和物性能的变化而影响混凝土质量的。鉴于现代混凝土拌和物性能的变化，必须转变传统观念。拌和物施工性是一种无法在实验室量化为单一指标而检测的性能，并非只是流动性，还有可泵性、填充性、抗堵塞性和凝结的性能，等等，第一位的是匀质性，混凝土拌和物质量指的就是满足特定工程施工性要求的匀质性。

虽然 ASTM C494/C494M 对其所定义的 7 种类型减水剂规定了减水率、凝结时间、抗折和抗压强度、干缩和抗冻融的详细要求，但减水剂的具体效果是依不同水泥、添加顺序、水胶比的变化、骨料品质的变化、拌和时的温度、环境温度和其他现场条件而不同的，既要保证产品达到所承诺的最低技术指标，并且均匀，还不能含有害于混凝土性能和操作人员健康的成分。对混凝土生产供应商来说，保证混凝土最终质量的是正确选择和使用好外加剂，以提供合格的拌和物。监督、监测和管理部门应摒弃机械思维和传统观念，不能用产品标准去限制、指责和惩罚用户，而应加强过程控制，以保证混凝土的最终质量。

影响混凝土工程质量的因素非常复杂，外加剂并非唯一因素，不能一概而论。但是当前混凝土工程中某些质量问题多半是由拌和物引起的，主要是掺用外加剂后才发生的，例如泌水、离析、凝结时间不正常等问题引起的墙、柱、墩等竖向构件拆模后表面出现砂线、麻面，甚至拌和物未充满钢筋的保护层而裸露钢筋（图 1）等问题；强度较高的混凝土有时还发生流动性的动力损失或者滞后泌水的现象；对于板、场坪等水平构件，发生表面因泌浆而硬化后“起砂”、塑性开裂或沉降裂缝的现象较为普遍；泌水还会造成一些看不见的隐患。拌和物沉降受到粗骨料或钢筋的阻挡，会在骨料或钢筋的下方因泌水而形成水囊，其中的水蒸发后即成间隙。骨料下方的间隙成为混凝土最薄弱的界面，而钢筋下方会引起钢筋的提前锈蚀[2]（图 2～图 4）。

图 1　混凝土未充满墙体钢筋的保护层而使钢筋裸露

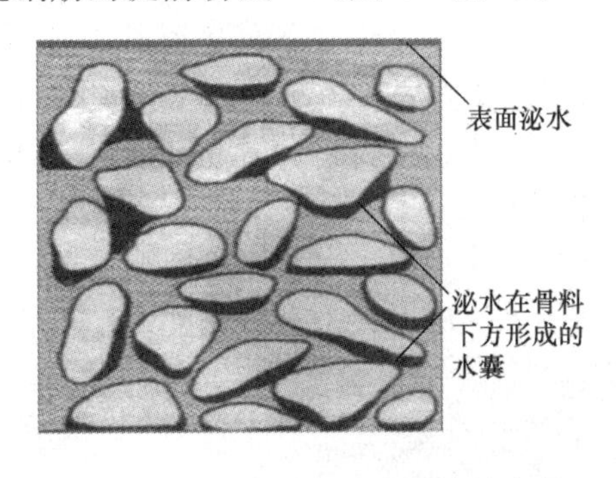

图 2　骨料下方混凝土水形成的水囊

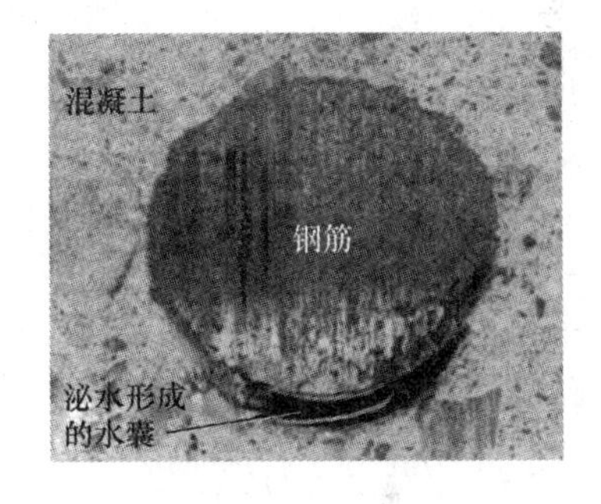

图 3　混凝土泌水在钢筋下方形成的水囊

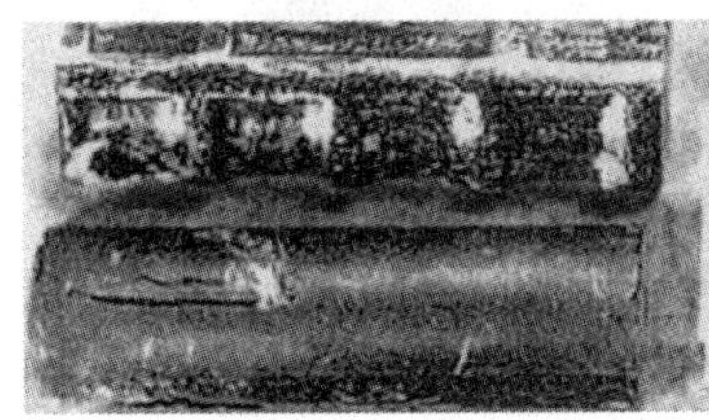

水平钢筋上方轻微锈蚀

水平钢筋下方严重锈蚀

图 4　东京大学现场暴露试验结果（引自覃维祖 ppt）

板面或路面混凝土的沉降受到钢筋的阻挡，还会引起表面的顺筋开裂（图 5），拌和物离析除影响混凝土结构匀质性外，还会造成堵泵等现象。

混凝土原始表面
钢筋阻挡混凝土沉降造成顺筋裂缝
沉降后混凝土表面
水平钢筋
钢筋下面混凝土囊形成的沉降裂缝

图 5　平板构件混凝土沉降

造成拌和物质量问题的主要原因是当前很多技术人员对现代混凝土的特性不甚了解，面对变化了的事物，必须转变传统观念。

3 现代混凝土拌和物特性及容易引发的问题

拌和物坍落度的动力损失、拌和物滞后泌水、高强混凝土表观黏度大还能否泵送？这些问题都涉及黏性、触变性、剪胀性等浓分散体系才有的特性，现代混凝土正是这样的体系。

3.1 低水胶比拌和物的黏性——水泥过细更增大拌和物的黏度

图 6 是常见四种类型流体的流变曲线[3]，其中 a 是牛顿体，为稀分散体系，剪应力 τ 与剪变速率 D 成正比，流变曲线是通过原点。实际使用的大多是浓分散体系，多属非牛顿体，其 τ 与 D 不成简单的正比关系，表观黏度（$\eta_a=\tau/D$）值也随 τ 而变化。b 是塑性体，即宾汉姆体，在这条曲线上，τ_L 为静切应力。当 $\tau>\tau_L$ 时，叫做塞流，只在容器边缘地区发生变形而产生滑动。中间未发生变化的部分仍按原来的结构形式向前运动。τ_y 为宾汉姆屈服（值）应力，τ_M 为层流剪切应力。当 $\tau>\tau_M$ 后，流动形式和牛顿体的完全一样看，体系中粒子间的结构被完全拆散，结构重组而平衡，塑性黏度恒定。从混凝土拌和物新性质来看，即属于这种塑性流体。在低水胶比的混凝土拌和物中，大小颗粒之间距离很近，相互碰撞、摩擦、吸附而产生流动的阻力。距离越近，颗粒之间抗剪切的应力越大，变形能力差；颗粒越细，颗粒之间的距离越近，这种作用也越大。因此，高强度混凝土拌和物剪切应力大，表观黏度大，但是和容器之间的附着黏滞性却并不大，仍可滑动，而且掺外加剂的低水胶比混凝土拌和物另一个特性就是触变性：看似很黏，却能在泵送和振捣下很好地流动。美国加州大学教授 P. K. Mehta 在工程中使用的大掺量粉煤灰混凝土，坍落度为 125mm，泵送浇筑时与坍落度为 180mm 的普通混凝土的效果相当[4]，用水量很少而不离析、不泌水。图 7 所示为该混凝土的流动特性，图 7 右显示所检测坍落度仅 100mm。[4]

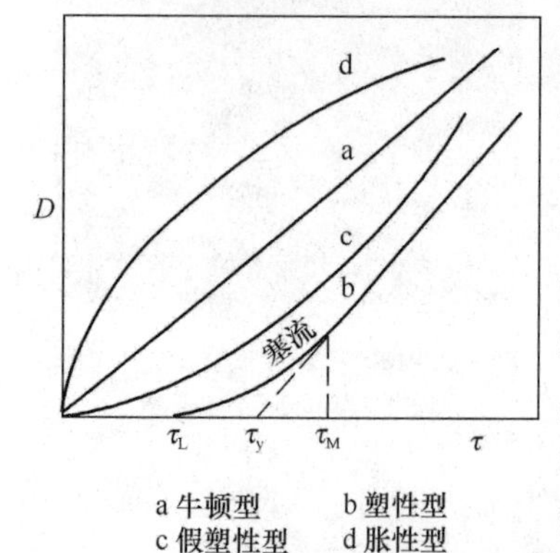

图 6 流体的四种类型

图 7 P. K. Mehta 使用大掺量粉煤灰混凝土泵送时的表现

高黏性液相可减缓颗粒间的相互作用，与颗粒一起平滑地流动，使颗粒的剪切变形局部化；同时，可大大约束由颗粒碰撞引起的颗粒变形，减少颗粒的接触应力。因此，尽管高黏性液相增大抗剪切能力，却可减少骨料相总体的抗剪切能力，从而阻止颗粒的离析与堵塞[5]。这就是高黏性拌和物具有的可泵性机理。但是低水胶比拌和物的黏性虽然不影响可泵性，却要求适当加大泵压。输送泵使用频繁后，泵压会因磨损而降低，目前许多混凝土泵的持有者只从利益出发，检修的频率过低、动力不够、不更新，当工程要求使用高强混凝土时，因黏度较大，泵送时只是需要适当增加一点泵压。如果发生堵泵，则其原因往往是离析造成的。

拌和物的黏性既和颗粒之间的距离有关，也和颗粒形状有关。在体积分数相同的情况下，颗粒尺寸越小，数量就越多，颗粒间距离越近，相互干扰的机遇越大；颗粒子越小，溶剂化后有效体积越大，溶剂化的需水量越多，自由水量越少，粒子移动阻力越大，因而黏度越大。颗粒子形状越不对称，阻力越大，浆体的黏度更大。高强混凝土的水胶比低，自由水量较少，现今的水泥颗粒过细，又加重了高强混凝土的黏度。为此，应当呼吁水泥厂配合，解决水泥熟料粉磨过细的问题。

3.2　拌和物的触变性——黏度大却不影响泵送和振捣的原因

拌和物在搅拌时成为流体，静置后体系中的粒子靠一定方式形成的网架结构而逐渐变稠、甚至变硬；施加外力时，在一定剪变速率下流动时被拆散，粒子在剪切应力作用下定向，剪切应力随时间而减小；这种表现叫做触变性。这种变化是可逆的，当剪变速率降低或停止时，被拆散的粒子靠布朗运动移动到一定的几何位置，又重新形成结构[5]。可用滞廻圈法检测评价触变性。方法是用旋转黏度计，均衡地变化转速，先从低到高，再从高到低，记录相应的剪切应力 τ，作 τ-D 图。上行线 ABC 和下行线 CA 组成月牙形的 $ABCA$ 环形曲线，称为“滞廻圈”（图 8）。

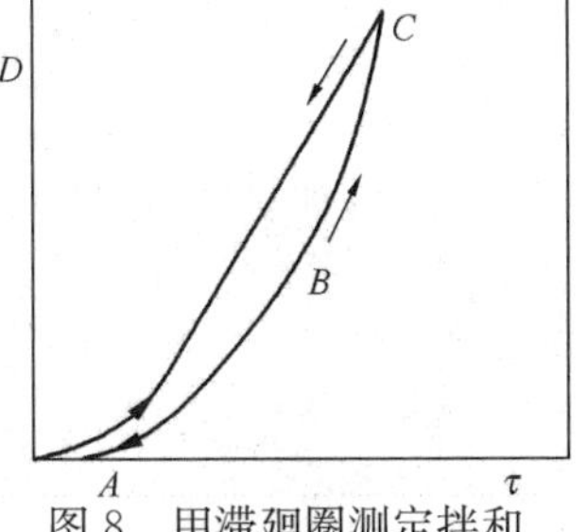

图 8　用滞廻圈测定拌和物的触变性

滞廻圈的面积可表示触变性的大小，面积越大，触变性越强。因为有触变性，表观黏度显得很大时却不会影响施工，反而在动力作用下能很好地流动并充满模型。现在的施工人员不认识这种具有触变性的拌和物，才会强行在现场加水。

3.3　拌和物的剪胀性——流动性动力损失的原因

在自由水很少的情况下，拌和物可能成为胀性流体，流变曲线为图 6 中的 d。静止时，粒子完全散开；搅动时，粒子重排而形成混乱的空间结构（图 9），会大大增大流动阻力，使表观黏度上升。表现在混凝土上，拌和物在过泵入模后完全丧失流动性，而无法振捣。这就是拌和物坍落度的动力损失。其机理是剪切稠化作用：表观黏度随切变速率增加而增大，称为剪胀性。具有剪胀性的胶体还有例如淀粉、芝麻酱、涂料等。

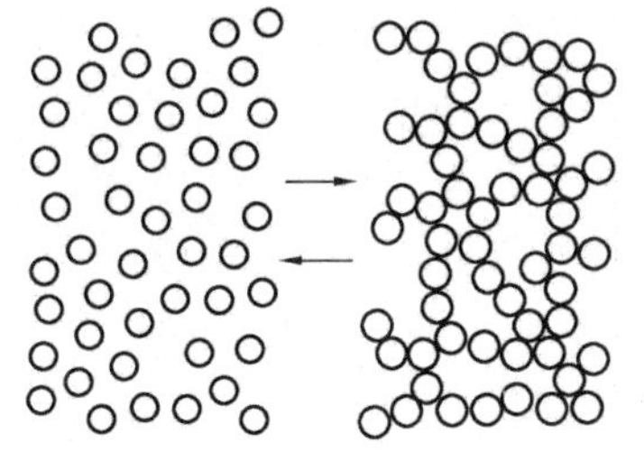

图 9　剪胀性机理示意

形成胀性流体的条件及解决的途径：

首先，体系分散浓度低时是牛顿流体，浓度高时是具有触变性塑性流体。出现胀性流体的浓度介于二者之间，最低浓度称为临界浓度，但是范围很窄；其次，是使用分散剂（例如水泥使用减水剂），使体系中颗粒分散很好的流体。

解决剪胀性问题的关键是调整拌和物的浓度，拌和物的浆骨比、砂石比、水胶比和掺和料品种及掺量都会影响拌和物的浓度。某桥梁工程的拉杆支柱 C60 混凝土出现了坍落度泵送后损失殆尽的现象。经检查分析，该混凝土原配合比中胶凝材料偏大（总量 548kg/m^3），胶凝材料中的矿物掺和料为 S95 矿渣粉（掺量 20%），砂子较细，这些都是需水量大的因素。高强混凝土的主要矛盾是低水胶比，应当以降低浓度为目标进行调整：按松堆孔隙率为 40%的目标级配石子，减少水泥和矿渣粉，代之以粉煤灰，以减小需水量；并减少用水量以减小水胶比以保证强度不变；增加聚羧酸高效减水剂用量（先稀释后用）以增加自由水量。

3.4　泌水和沉降——塑性收缩引发开裂的原因，“滞后泌水”的原因在于外加剂的选用

掺用高效减水剂的现代混凝土拌和物大多为流态，普通骨料的混凝土拌和物中固体颗粒密度大于拌和水的密度，颗粒所受重力大于拌和水的浮力，则颗粒会下沉，叫做沉降；同时拌和水或溶剂化了的微细颗粒向上或向与其接触的亲水性固体（如模板、粗骨料、

钢筋）迁移，叫做泌水或泌浆。混凝土硬化前的沉降和泌水引起的混凝土收缩叫做塑性收缩。

混凝土拌和物泌水有弊也有利：轻微的持续泌水有利于缓解塑性收缩引起的表面开裂；压力泌水有利于拌和物的泵送，也可以在拆模后得到平整的表面，但是泌水会给混凝土带来质量问题。

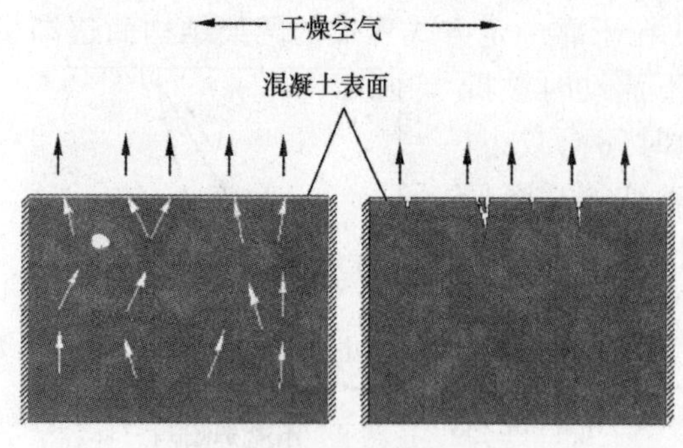

图 10　混凝土表面泌水引起塑性收缩的机理示意

塑性开裂形成的机理如图 10 所示。平板类的构件混凝土在干燥空气中，表面水分蒸发后，如果内部水分不断上升至表面继续蒸发，直到混凝土终凝，则不会发生开裂（图 10 左），这就是过去水胶比大的混凝土不出现表面裂缝的原因；对于黏度较大的拌和物，表面水分蒸发后，内部的水很难上升补充表面水，则表面干燥会引起收缩，内部仍处于湿润环境而不收缩，约束表面的收缩而产生收缩应力，在混凝土仍处于塑性状态时，不足以抵抗尽管是很小的拉应力，于是出现裂缝（图 10 右）。这种塑性裂缝属于发丝裂缝，往往不为肉眼所见，容易被忽略，但是这种裂缝是尖端朝下的楔形，长度一般不会超过 5mm，只要注意严密覆盖避免失水过快，在初凝前用二次抹面法消灭之，再继续进行保湿养护，即可无害；塑性裂缝往往数量很多，如果不予以处理，初凝后这些裂缝被固定下来，就会成为此后干缩裂缝的开裂源，甚至会开展而造成贯穿的通缝，后患无穷。黏性是现代混凝土拌和物的一个特性，塑性开裂与此特性有关，施工中务必予以充分重视。市售聚合物纤维因为单丝抗拉力很小而一般对抵抗干缩裂缝无效，但是对避免或减少塑性裂缝是有效的，当然也不能忽视湿养护。

沉降和泌水是相联系的，往往同时发生。不均匀的沉降会引发一些质量问题，如前所述。沉降和泌水引起的质量问题往往比较隐蔽，但当拌和物离析严重时，就会加剧沉降，在浇筑后造成构件混凝土分层，严重不均匀，成为安全隐患。

影响泌水和沉降的因素主要有：拌和物的浓度（水胶比、用水量、外加剂掺量）、颗粒的细度、凝结时间等。带浆的泌水会在水蒸发后在表面留下素浆或干粉层，使钢筋的保护层厚度减小。

近年来，使用聚羧酸系减水剂后的另一个常见现象是拌和物的“滞后泌水”：浇筑纵向尺寸较大的如墩、柱、墙等的混凝土时，泵前泵后一切正常的拌和物，泵送后 3～4 小时，在模板中的混凝土表面会出现一厚层泌出的水（浆）。这种问题的出现是由于所选用减水剂不当。聚羧酸系高效减水剂的优势在于可以通过分子设计和接枝共聚的工艺，达到使水泥颗粒束缚水按预期释放的效果。例如有时可以在拌和物出机时坍落度并不很大，运到工地开始浇筑时，可以达到预期的最大坍落度，这是人们所欢迎的。如果在热天施工，为减小坍落度损失，会需要束缚水缓释，而当同样的产品用于不需要束缚水缓释的季节或场合，可能就会发生类似这样的滞后泌水。

3.5　离析

离析主要指骨料和浆体分离，或粗骨料和细骨料分离，亦即“浆裹不住石”。

拌和物的形变能力（和流动性有关）是混凝土浇筑成型的首要条件，抗离析性是保证拌和物体积稳定、质量均匀的保证。拌和物中的自由水量是影响拌和物变形能力和抗离析能力的决定性因素：当自由水量不足时，拌和物形变能力差；自由水过量时，就出现离析的倾向。影响自由水量的因素还有总用水量（水胶比、浆骨比）、外加剂掺量、细粉料和细骨料的保水性以及粗骨料的粒形和级配等。在泵送过程离析的拌和物中，粗骨料形成拱架而阻塞流动，砂浆通过粗骨料间隙时被堵塞，只有浆体和水能通过砂子的间隙，于是发生堵泵现象。流动的拌和物堵泵的原因主要是拌和物离析。离析的拌和物即使没有堵泵而通过泵管浇筑到模板中，也会造成分层影响结构的匀质性。例如某工程已硬化的 C60 混凝土柱，发现如图 11 的表现，对几根柱钻芯检测其强度，上、中、下三个部位的强度有的分别为 51.5 MPa、19.2 MPa 和 56.6 MPa；有的分别为 17.2 MPa、27.2 MPa 和 67.6MPa。可见存在严重的匀质性问题。

图 11　拌和物离析加上施工不规范造成混凝土的分层

拌和物的变形能力和抗离析能力是相互矛盾的（图 12 和图 13），只有在一个合适的坍落度范围和水灰比，才能得到良好施工性的拌和物。施工现场擅自向拌和物中加水，以为能增大流动和变形能力，节省自己的力气，但是却同时会降低拌和物抗离析的能力，而且现场加水后在运输车的自落式搅拌机中，是不可能搅拌均匀的，更会加重拌和物的离析，还会降低混凝土的强度。

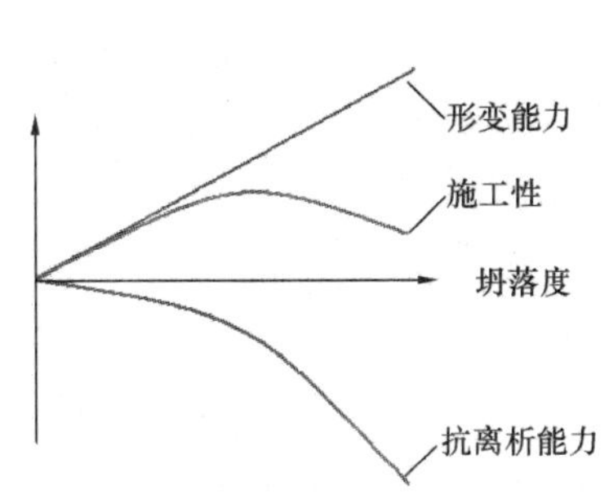

图 12　坍落度对拌和物形变能力与抗离析性的关系

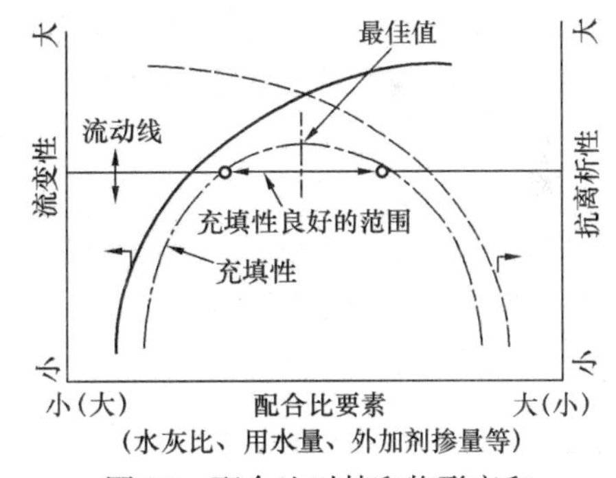

图 13　配合比对拌和物形变和抗离析性的关系

骨料质量对拌和物抗离析性的影响至关重要。在现行砂石标准中对石子粒形的规定偏宽，主要是迁就落后。尽管按针片状颗粒含量分成了等级，可方便用户选择，但是只要有人供应低价的石子，许多用户就会出于经济利益的要求而就低不就高。标准的导向作用不言而喻。

3.6　拌和物的坍落度损失——外加剂与混凝土原材料相容性问题的指标之一

由于水泥水化消耗和水分在空气中的蒸发，混凝土拌和物的坍落度都是要损失的，我们所指的是过快的非正常的损失，是外加剂和混凝土原材料相容性问题的表现。这个问题大家谈论很多，但是多问责于外加剂。这是认识上的一种误区。

外加剂的"适应性"提法是不妥的，徐永模早在 7 年前就已论述过这个问题[7]。"相容性"（compatibility）是双方配伍性质，是指两种或两种以上物质混合后，不产生整体性质劣化的能力；"适应性"（adaptability）则是单方的性质，是指随对方特性而改变自身的能力。减水剂和混凝土原材料之间的关系从减水剂来说，无论何种减水剂，其原材料纯度高，只要按在混凝土中的作用机理，经过科学的工艺过程，所生产的产品质量会是稳定和均匀的；而混凝土原材料复杂得多，如不加以控制，则影响就很大。凡是影响需水量和水泥水化速率的因素都会影响与外加剂的相容性。仅就水泥而言，凡影响与萘磺酸盐减水剂相容性的因素，同样影响聚羧酸系减水剂（只是其中SO_3的影响有些区别）。目前市售水泥细度过细、需水量大、熟料中C_3A含量大、含碱量过高或过低、SO_3只按凝结时间优化、出厂水泥温度过高，等等，很少有例外。凡此种种，都是影响外加剂在混凝土中正常作用的重要因素。在外加剂和水泥相容性的矛盾中，主导方面无疑是水泥。当前外加剂用于混凝土所出现的问题有增多的趋势，和混凝土所用如掺和料和骨料等其他原材料也有关系。外加剂只是一种辅料，而不是灵丹妙药，必须配合混凝土材料其他组成及配合比。混凝土是非常复杂的系统，甚至不同性能之间是相互矛盾的，首先要自己先做好，不能单向地要求外加剂适应自己的落后。

坍落度损失率和外加剂掺量及掺法也有关。追求简单无可非议，但是如果简单化地追求简单，则会适得其反。

对外加剂来说，复配是必要的，但不是万能的，根本问题是产品自身的配方和工艺。

4　现代混凝土拌和物的检测——对变化了的材料不能使用不变的方法

综上所述，现代混凝土拌和物所具有的特性不是坍落度所能表征的，坍落度只能检测无障碍的流动和无外力作用的静态流动性；掺外加剂所得到的较低水胶比、细掺和料的混凝土拌和物的特性在动态下才能表现出来，这就是“performance”。对于低流动性拌和物的填充性由变形能力控制，检测坍落度即可；高流动性拌和物的填充性则由抗离析的能力控制，坍落度不能反映施工性能；当配筋密集复杂，或泵管有弯头、模板形状变化时，检测抗离析性和抗堵塞性就很重要。很多人也发现坍落度法用于现代混凝土不合适，为此，多年来国内外都不断有人研究，也有很多检测技术[8]，但是因为都不如坍落度法那样简便而在我国实行起来都有难度，大部分都仍用坍落度法应付，个别用“倒坍落筒法”，也是静态的。图14是目前香港使用的方法，1984年，清华大学曾按德国DIN标准制作该装置并在研究中使用过。用高度为200mm坍落度筒装料并去筒后，用平板上的拉手提起40mm再自由落下，反复15次后，检测拌和物扩展度，表征在动态下的流变特性。拌和物离析、泌水、触变性、剪胀性（如果有）都可以有所表现。

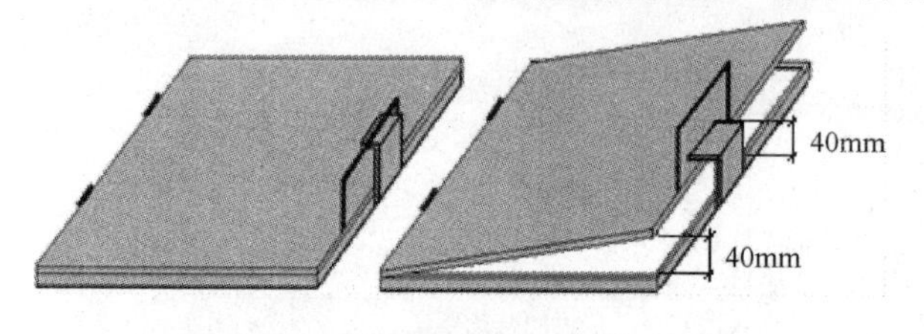

图14　流动度测定的振动式平板仪

此外，设计和施工人员、混凝土供应商都必须考虑：并不是任何混凝土都能顺利通过任何配筋状况的钢筋间隙的。如果钢筋最小间距、钢筋保护层厚度能和混凝土石子最大粒径相匹配，抗堵塞性（间隙通过性）可以不做检测，对于自密实混凝土则必须做。

5　结束语

1. 现代混凝土以外加剂的普遍使用为特征，外加剂通过改变拌和物影响混凝土的质量。变化了的材料不能用不变的观念和方法思考、使用和检测，需要清楚了解现代混凝土拌和物的特性，转变传统思维和观念。

2. 用产品“代”和“等级”评价减水剂是不合适的，因为任何产品都有利必有弊。没有好与不好的，只有合适与不合适的；匀质性是第一位的质量；售后服务也是一种质量，但不是什么问题都能解决的。世界上不存在什么包治百病的药。

3. 当前基础研究薄弱，机理不清楚，造成方法不科学，给工程带来诸多矛盾或隐患。例如热衷于使用不清楚成分和作用机理的“增效剂”、“高效剂”、“助磨剂”、“增强剂”，等等，它们未必不好，但是至少是不清楚，对混凝土强度无影响不等于对其他性质无影响；对混凝土质量眼前无影响，不等于未来无影响。许多人所做的“对长期性能的影响”，实际上最长期不过半年一载，或者只是在实验室条件下强制性检测的“实验室指标”而已。一个优秀的企业必须充分重视科学技术的研究，以“科学技术引领企业的发展”。

4. 混凝土企业不对水泥和砂石这样大宗的、主要而重要的原材料坚持自己选材的原则，却企图用掺量很小的外加剂解决自己的大问题，实在是本末倒置。混凝土企业应该自己首先做好，而不是单向地要求外加剂企业适应自己。

5. 为了尽量避免复杂性，应提倡混凝土企业使用在水泥厂优化后供应的复合水泥或混合材水泥。当然，这就对水泥企业提出更高的要求：按混凝土的需要生产水泥，这才是根本解决问题的途径。

6. 外加剂的生产者、销售者必须懂得混凝土，要有正确的观念指导，否则门槛易进，台阶难上，终究会随着社会的进步而被淘汰。

最后，引用美国垦务局工程师 Richard W. Burrows 在他的著作《The Visible and Invisible Cracking of Concrete》(混凝土的可见和不可见裂缝) 前言中的一句话来结束：

> 不要再没完没了地总盯在采用替代物、外加剂和钢筋及塑料增强等治疗办法上，而直接着眼于把石头粘在一起的胶结料——水泥吧![9]

参考文献

[1] ASTM designation C1157-00.

[2] T. U. Morammed，大即信明. Several Issues in Designing Durable RC Structures. Issues in Designing Durable Structures. Concrete International. July 2001.

[3] 冯绪胜. 胶体化学[M]. 北京：化工出版社，2005.

[4] P. K. Mehta. Heavily reinforced Shear walls and Mass foundations built with “Green” Concrete. See: Concrete International，August 2002.

[5] 侯万国. 应用胶体化学[M]. 北京：科学出版社，1999.

[6] 松岡康训. 超流動コンクリート，ンクリート工学 Vol. 31，No. 3，1993. 3.

[7] 徐永模. 关于超塑化剂与混凝土的“相容性”和“适应性”问题[J]. 中国建材，2004，第 10 期.

[8] 吴中伟，廉慧珍. 高性能混凝土[M]. 北京：中国铁道出版社，1999.

[9] Richard W. Burrows. The Visible and Invisible Cracking of Concrete，ACI Monograph No. 11，1998.

自评

1. 前几年很多人曾使用“现代混凝土”的名词，认为具有使用外加剂、矿物掺和料、大流动度、泵送，等等特征的混凝土就是“现代混凝土”。我因为没想明白究竟什么是现代混凝土，生怕又犯了从众思维的毛病，所以一直没接收这个提法，只用“现在的混凝土”、“当前的”或“当代的混凝土”等说法。这篇文章从外加剂对混凝土工程质量的影响，分析现代混凝土大不同于传统混凝土的特点，说明是外加剂改变了混凝土的一切。有人说掺矿物掺和料也应当是现代混凝土的特点。实际上，也是因为有了高效减水剂，才使混凝土掺入较大量矿物掺和料时，能把水灰比降低到0.5以下而发挥矿物掺和料的作用。1978年Malhotra在第一届国际混凝土外加剂会议上指出："高效减水剂的开发与应用是20世纪混凝土技术进展历程中一个重要的里程碑"。可以说现代混凝土是因外加剂的广泛使用而带来的所有特征的混凝土。因此，外加剂的使用才是根本的特征。

2. 文中所述触变性和剪胀性的特性，只是根据混凝土拌和物现存表象的理论分析。在应用胶体化学中，水泥浆体属于浓分散体系，现代混凝土因高效减水剂的使用而进一步提高了浓度。本文指出其除流动性以外的拌和物的表观黏度（apparent viscosity）与触变性(thixotropy)、剪胀性（dilatancy）等流变特性。其中剪胀性是指浓分散体系在浓度很窄的范围内，由于剪力的作用（如搅拌、泵送）而丧失流动性的性质。其表现是坍落度的动力损失，即经过泵管输送后丧失流动性，甚至无法振捣。水泥浆体本来就已经属于浓分散体系，掺入高效减水剂减水后，浓度更大。之所以在泵后发生剪胀性有两个可能的原因：一是拌和物运至工地又向拌和物中加水，改变了浓度，但又不可能搅拌均匀；另一是在搅拌机中搅拌不均匀，运输过程中搅拌车没有搅拌功能，亦即入泵前拌和物搅拌不足。有人发现一个现象，称之为"坍落度机内损失"：搅拌几十秒时坍落度达15～20cm，继续搅拌出机后，就只有几厘米了。还有人做搅拌时间的试验，发现搅拌时间从30秒到60秒，流动性提高，再继续搅拌到90秒，拌和物流动性反而变差。……这应当都是在浓度不合适的情况下发生的。调整了拌和物浓度后，即可解决。有人换了另一种减水剂，也可解决。是否减水剂的减水率越大，越容易使拌和物发生剪胀性？改换水泥也会解决，那么是水泥中的什么因素影响的？如果在拌和物泵前能预测，则可免去调试的盲目性。首先需要对机理进行研究，才能找到检测方法。目前有人在进行研究，还只是从拌和物本身着手，但是更重要的是开发一种能在实验室模拟泵送动力的装置，进行预测。

3. 萘磺酸盐类减水剂在我国已有近40年的使用经验，聚羧酸减水剂广泛使用至今不过4、5年，不断发现了一些和起初了解的不同的问题，需要生产者与使用者共同研究解决。

横眉冷对千夫指，俯首甘为孺子牛。

——鲁迅

论文之十二

砂石质量是影响混凝土质量的关键*

廉慧珍
清华大学土木水利学院

摘　要　我国混凝土中的水泥用量和拌和水比发达国家的至少多用了20%，不仅资源浪费严重，而且使硬化后的混凝土开裂敏感性大，威胁结构物的耐久性，而且因多用水泥而间接地严重影响减碳经济，其主要原因就是砂石质量太差，砂石质量差的原因是大多数砂石生产无人执行砂石标准，也无人监督管理砂石的质量，无人执行和监管砂石标准的原因是不认识砂石在混凝土中的重要作用，存在认为砂石取之不尽、生产砂石很简单、砂石没有技术问题等误区。砂石作为骨架作用主要是稳定混凝土的体积，即使将来砂石比水泥价格高，也不可不用。从可持续发展的目标出发，整顿砂石生产，加强标准的执行和监管已成为刻不容缓的问题。

关键词　砂石；粒形；级配；混凝土质量；减碳经济

产品标准是为了检验产品质量的标准，然而我国的砂石标准多年来已形同虚设，几乎无人执行。业内尽人皆知当前我国砂石质量普遍太差。石子松堆空隙率在45%以上（理想粒形和级配的石子松堆空隙率为38%），针片状颗粒超过10%（西方国家的限值为5%），砂子中含泥量最多的达20%，含石量超过25%……。砂石质量不仅是影响混凝土质量的主要因素，而且因此而成为影响我国减碳经济实施的重要因素。首先，使我国混凝土中的水泥用量、用水量和外加剂用量越来越大。一般来说我国混凝土水泥用量和用水量至少比西方国家的多用20%，若同国外的先进指标相比，可多用30%以上。假设混凝土年消耗量为20亿m^3，则少说每年多用水泥4亿吨，按掺入50%的矿物掺和料计，不算矿物掺和料加工的耗电，则至少多排放CO_2 2亿吨！更重要的是混凝土质量还因此而受到影响，当前工程中频繁出现的开裂问题，从材料本身来说，主要是不得不多用水、多用水泥的结果。当然，造成工程质量出现问题的原因并不是单一的因素，而且有时也不都是技术问题，但是目前我国基本建设中资源和能源的浪费和混凝土体积开裂敏感性增大的事实，主要的影响因素却是对砂石质量的忽视。砂石质量关系到建设工程的质量。

砂石质量太差的原因主要是无人认真执行砂石标准和对其质量的监督、检验，无人认真执行和监督、检验砂石标准的原因主要是对砂石的认识误区，对砂石质量认识的误区主要是不了解骨料在混凝土中的重要作用。

1　对砂石认识的误区

1.1　认为砂石资源取之不尽、用之不竭

北京的香山方圆160公顷，最高峰海拔557米，体积约4.5亿m^3，假定都是岩石，则

* 原载《混凝土世界》2010，8。

约有6.8亿吨。2009年我国水泥产量约16亿吨，约消耗砂石100亿吨/年。一年消耗掉的砂石就约相当于14.7座香山的体积！离北京最近的河北省三河县的石料场，和十几年前相比，已不见了好几座山头。近年来混凝土用砂石资源已经十分紧张，尤其是天然砂的无节制开采已威胁到我们环境的生态平衡，国家对河砂开采的控管是非常英明正确的。虽然可以使用人工砂，但是那也需要消耗天然岩石资源。天然资源不可再生，必须“省吃俭用”才能维持长久。

1.2 认为砂石生产技术简单，因而门槛低，有钱谁都能生产

砂石生产并非高科技，正因生产工艺简单，才能成为用量巨大的产品。但也必须生产出符合工程要求的合格砂石。首先，为用户服务，就必须了解用户的需要。砂石的生产者必须具备混凝土基本知识，了解和清楚砂石在混凝土中的作用和影响，以便按用户需要生产，并且能够培训顾客，指导顾客正确地选用；不同的岩石具有不同的物理性质和化学性质，砂石生产者还须具备一定的矿物岩石学知识，以便正确选择料源和加工工艺。例如由表1和表2可见岩石有不同吸水率和混凝土的线胀系数，会影响混凝土的施工和变形。例如，如果使用砂岩石子，就要比用石灰岩石子多用8kg/m³的水，水胶比必须保持不变，水泥也就相应增多。

表1 不同骨料的吸水率和收缩的关系

骨料	吸水率（%）	1年的收缩（%）
砂岩	5.0	0.12
板岩	1.2	0.07
花岗岩	0.5	0.05
石灰岩	0.2	0.04
石英岩	0.3	0.03

表2 不同岩石骨料的混凝土线胀系数[1]

骨料种类	热膨胀系数（10^{-6}/℃）
砂 砾	13.1
花岗岩	9.5
石英岩	12.8
玄武岩	9.5
砂 岩	11.7
石灰石	7.4

不同岩石有不同的热胀系数（表1）[1]，则在相同条件下会使混凝土有不同的热胀系数，也就是说混凝土在冷热变化、干湿交替环境下会有不同的变形性质，以致影响其耐久性。

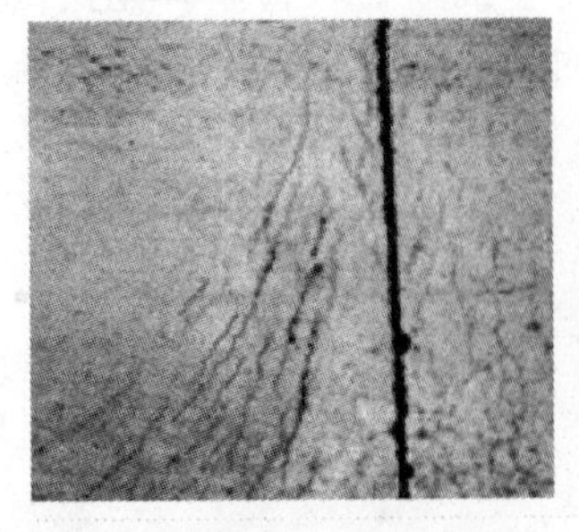

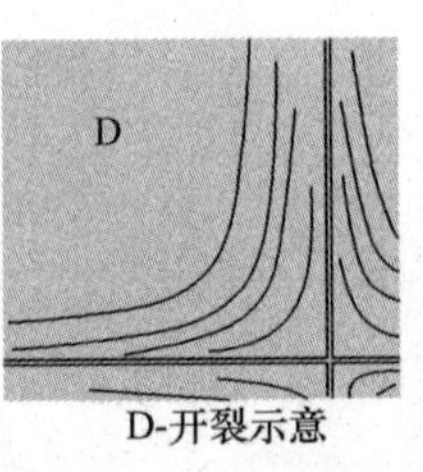

D-开裂示意

图1 混凝土路面平行于纵缝和横缝的D-开裂[2]

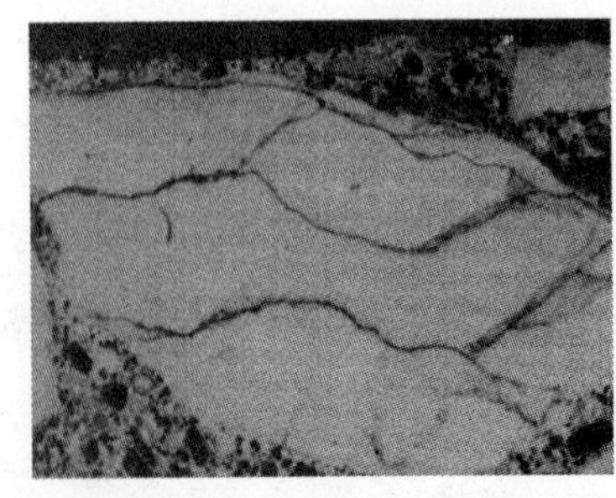

图2 石灰石骨料引起D-开裂[2]

不同的吸水率不仅影响混凝土的变形性质，而且还有可能影响混凝土的抗冻性。由于不重视岩石矿物学的基本知识，工程中出现问题往往很少考虑骨料是否会有什么影响，其实有时有可能和骨料有关，例如混凝土路面发生的D-开裂（图1左）[2]。叫做D-开裂可能是开裂的形状像英文字母D右半边（图1右）的缘故，是因混凝土中饱水的石灰岩骨料受冻融循环作用胀裂而引发的（图2）。

有一点岩石矿物学常识的人都会知道岩石大多由矿物组成，受力后可沿晶面断开呈现平整或光滑的表面，称作解理。中等解理（解理清楚）或完全解理的岩石（例如石灰岩），在使用颚式破碎机破碎时，会因解理而造成针、片状颗粒，增大混凝土拌和物的需水量。颚式破碎机价格最低廉，目前国内大多数作坊式的采石场仍然都在使用。这是造成石子粒形差的主要原因。

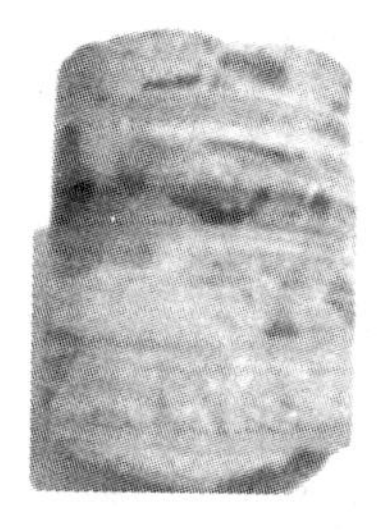

图3　芯样中可见粗骨料粒形很差，砂中含泥量很大，混凝土严重不均匀

图3为某工程抽检的芯样，由于骨料粒形和级配很差，砂子含泥量很大，必须使用较大量的水泥浆（水泥＋水）和减水剂才能保证拌和物有足够的流动性，结果造成拌和物的离析、泌浆，浇筑后严重不均匀。对这种结果，砂石供应商既不知道，也不关心。

因此，并不是任何人都能生产出合格的砂石，不懂混凝土，不懂岩石学和矿物学，只会无谓地消耗本该给子孙后代留一点的资源，而且还不断地为他们制造建筑垃圾。

2　砂石和混凝土质量的关系

我国在传统上把砂石称作混凝土的骨料，是因为岩石强度很高，对混凝土强度起很大作用，于是对砂石的选择主要看重强度，有解理的岩石，破碎的颗粒越小，针片状颗粒越多，致使公称粒径5～10mm的颗粒需水量很大，其结果，名义上的“连续级配”实际上其中5～10mm的颗粒几乎没有。由此也可以说，造成石子现状也是用户误导的结果。双方对砂石在混凝土中的作用的理解都有误。

2.1　骨料和混凝土强度的关系

过去，在主要使用塑性混凝土的情况下砂石强度确实会不同程度地影响混凝土的强度；上世纪50年代末我国曾经有构件厂使用过干硬性混凝土，水灰比低、浆体含量少，拌和物没有坍落度，用V-B稠度仪检测时，工作度为20秒以上，只能用于预制构件，在高频振动台上加压强力振捣，工人劳动强度大，噪声震耳，而且能耗很大，不久就不再生产了。这样的混凝土强度取决于石子强度和浆体—石子界面的粘结强度，石子对混凝土强度的贡献显而易见。当时，一般使用低塑性混凝土（坍落度10～30mm）和塑性混凝土（坍落度30～50mm、50～70mm、70～90mm）；上世纪90年代初开始使用流态混凝土（坍落度100mm以上）；石子对混凝土强度的作用随浆骨比的增大而减小。在泵送混凝土（坍落度150mm以上，当前普遍都超过200mm），石子在混凝土中呈悬浮状态，混凝土的强度基本上与骨料强度无关 。现在用颗粒强度很低的轻骨料（陶粒）已配制出高强（C50以上）泵送混凝土就是明证。在现代混凝土中，砂石在混凝土中的作用主要不是强度，但却是不可或缺的关键角色。

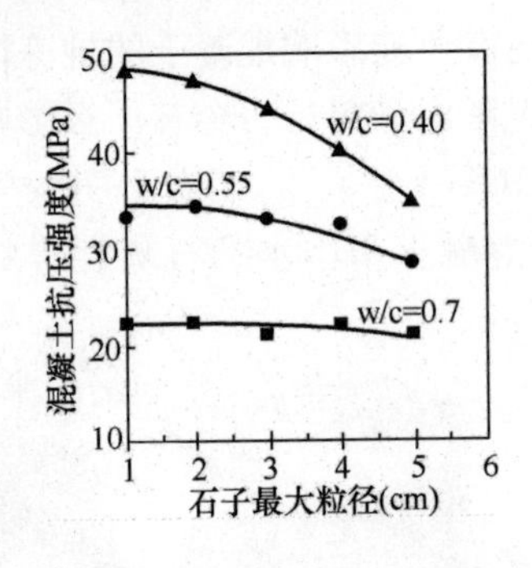

图 4 混凝土抗压强度和石子最大粒径的关系[3]

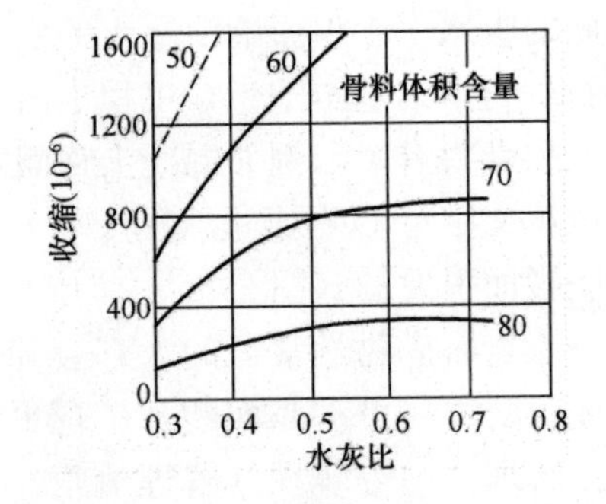

图 5 骨料含量对混凝土收缩的影响[4]

然而，当水胶比一定时，砂石用量和粒径影响混凝土中界面过渡区的厚度和数量，因此对混凝土的强度有影响。图 4[3] 表明，当水胶比很大时，石子粒径对强度的影响不显著，水胶比越低，影响越大。水胶比一定时，净浆强度高于砂浆强度，砂浆强度高于混凝土强度。

2.2 砂石在混凝土中的骨架作用主要是稳定体积

大多数混凝土用的普通岩石线胀系数为 $5\times10^{-6}\sim13\times10^{-6}$/℃，而硬化硅酸盐水泥净浆线胀系数为 $11\times10^{-6}\sim20\times10^{-6}$/℃，两者相差约 1 倍。如果没有骨料，水泥净浆硬化后会产生很大的收缩，稍有约束，就会严重开裂。

混凝土的收缩 S_c 与水泥净浆收缩 S_p 之比取决于骨料含量 a：$S_c = S_p(1-a)^n$（n 为经验系数，变动于 1.2～1.7，与骨料弹性模量有关）。由图 5[3] 可见，混凝土水灰比越大，骨料用量对混凝土收缩的影响越大。在图 6[4] 中可见，对混凝土塑性收缩影响的规律也是净浆＞砂浆＞混凝土，而混凝土水泥用量越大（骨料越少）影响越大。骨料对混凝土自收缩影响的规律也如此（见图 7）[4]。

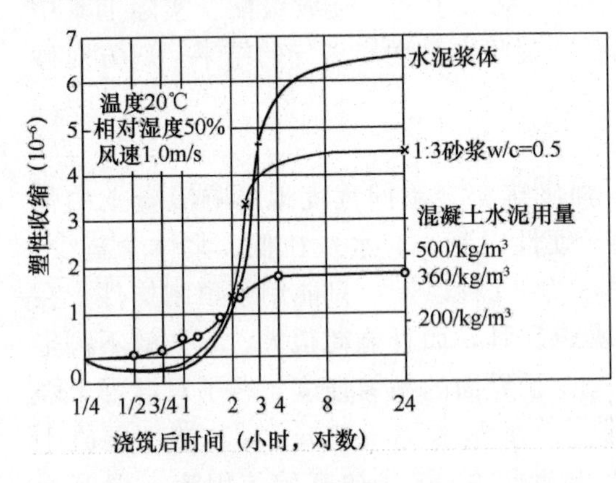

图 6 骨料对混凝土塑性收缩的影响[4]

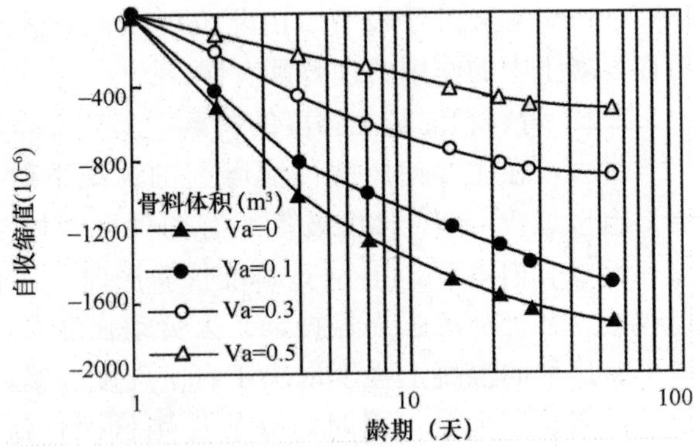

图 7 骨料用量对混凝土自收缩的影响[5]

2.3 骨料粒径和粒形对混凝土其他性能的影响

2.3.1 粒径的影响

图 8[3] 和图 9[4] 为骨料粒径与混凝土渗透性和抗冻性的关系。由图 8 可见，水灰比越大，骨料粒径对混凝土渗透性的影响越大，砂浆的这种影响最小；图 9 表明骨料粒径大会降低混凝土的抗冻性。

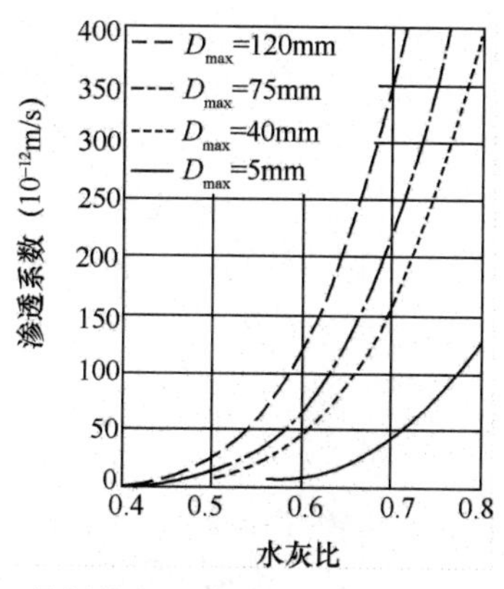

图 8　骨料粒径和混凝土渗透系数的关系[3]

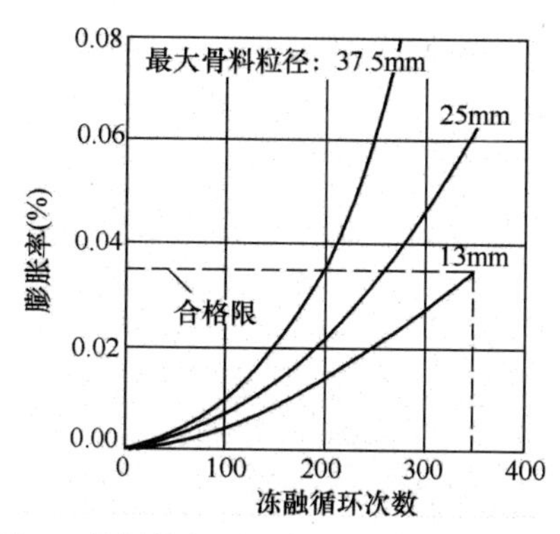

图 9　骨料粒径对混凝土抗冻性的关系[4]

2.3.2　粒形的影响

对于满足一定强度和密实性要求的混凝土来说，拌和物的施工性是保证混凝土最后质量的最重要性质。对混凝土施工性来说，骨料的粒形有时比级配的影响还要大。理想的骨料粒形呈等径状（即宏观球形）。表面粗糙程度相同时，等径状骨料的比表面积最小，需水量最小，可以同时满足施工性和强度等硬化混凝土性质。表 3 所示为某工程用不同骨料配制 C60 混凝土时的不同表现。用风化粗粒花岗岩的混凝土的坍落度为 195mm 时，28 天强度可达 71.3MPa，而乌石谷的石子属于致密石灰岩，强度高，但粒形不好，因混凝土拌和物需水量大，在满足混凝土强度要求的水灰比下，拌和物流动性较差。如要保证坍落度也达到 195mm，其措施一是增大用水量，即增大浆骨比，则既不经济，又会有更大的开裂敏感性；措施二是提高水胶比，则强度会达不到要求。在表 3 中，其所用水灰比降不下来，结果是强度达到 68.8MPa 时，坍落度才不到 150mm，显然不如使用强度较低但粒形好的的风化粗粒花岗岩。

骨料的粒形明显影响自密实混凝土的施工性，如表 4 所示，K53 和 K60 相比，K67 和 K61 相比，配合比相同，流动性一致，K53 和 K67（石子针片状颗粒为 5%）能很好地通过钢筋，而 K60 和 K61 只因石子中的针片状颗粒多了 2 个百分点（7%），遇到钢筋就被堵塞。

表 3　石子对混凝土性质的影响

所用石子	混凝土 28 天 抗压强度（MPa）	坍落度（mm）
深康风化粗粒花岗岩，强度低，但粒形好	71.3	195
乌石谷致密石灰岩，强度高，但针、片状颗粒含量大	68.8	148

表 4　骨料针片状颗粒对自密实混凝土抗堵塞性能的影响

编号	流动性		流动时间（秒）		配筋 L-流动充填性	充填性实验：通过钢筋流出一定距离时时间（秒）								
	Sl mm	*D* mm	流过 50cm	流过 65cm		150 mm	250 mm	300 mm	400 mm	500 mm	600 mm	800 mm	*T* mm	*L* mm
K53	260	695	6	17	√	2	4	6	7	15	22	44	260	>800
K60	260	690	5	9	×									
K78	245	660			√	2	5	7	11	15	24	55	245	>800
K67	265	670	4	8	√	2	6	8	12	18	30	53	260	>800
K61	265	665	6	16	×									

我国当前的混凝土“难做”和骨料粒形有很大关系。从图10中可见我国目前常用石子粒形质量的差距。图10中的（c）是针、片状颗粒形貌，可用其与日本和中国常用石子作比较。在日本石子（a）中找不到针、片状颗粒，而在中国的常用石子（b）中，则针、片状显而易见，即使在尺寸上还够不上（c）中的标准，也少见等径状。（d）是我国山东海瑞石料公司产品之一，和日本的产品已几乎无差别。可见这样的优质骨料我国不是做不出来。

(a) 日本的常用石子

(b) 中国的常用石子

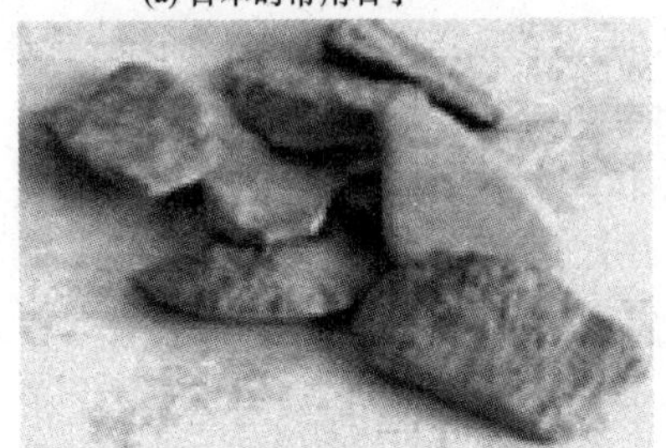

(c) 针片状颗粒

(d) 海瑞的石子

图10　石子粒形的对比

2.4　关于骨料的级配

颗粒尺寸由小到大连续搭配的比例，目的是不仅得到最小的骨料空隙率，而且也得到最小的骨料比表面积，但是在骨料的生产中只能从数量上做到级配，却无法做到整体产品级配的均匀性。因为砂石这种产品是散粒状的堆积物，在装、卸、运输等的动力作用下，原来不同大小的颗粒混合均匀的整体，会发生小颗粒向下移动而大颗粒留在表面的现象，堆积成锥形后，在表面的大颗粒会沿着锥形的斜面滚落。于是料堆的级配就失去均衡，混凝土生产中在这样的料堆所取骨料也就没有了级配。因此西方发达国家的砂石料都是分级供应，使用时按要求自行级配，按级配分级投料。规范中的连续级配不是对生产者而是对使用者规定的。为了保证满足混凝土的需要，我国砂石标准至少应当对石子明确规定分级供应。我国已经有一些混凝土使用了用单粒级两级配或三级配后，混凝土水泥用量减少了20%左右，但是如果粒形不好，也做不出小于40%的石子空隙率。

3　骨料质量和砂石标准

我国砂石质量问题产生原因比较复杂，但是除了体制问题、买方生存竞争问题等非技术性问题外，很重要的是观念问题：无论是卖方还是买方，无论是用户还是市场的调控者，对砂石的认识存在很大的误区，例如，一般产品标准是为了出厂的合格检验用的，带有强制性——不合格的产品不能出厂，但是我国砂石产品的国家标准居然是“推荐性”的，因而造成没有任何人执行砂石标准，也没有人对砂石质量进行检验。在技术人员申请修订砂石标准并改为强制性标准时，竟然遭到主管官员的不屑！

和其他技术标准与规范一样，制定砂石标准要需要先进而可操作，但又不迁就落后。所谓可操作是指只要努力就能做到，而不是对现状的无奈之举。例如过去曾有人在制订砂石应用标准时，既规定了级配标准，却又说明“如果级配不合格，通过实验证明不影响混凝土和易性时，也可使用”，又自己否定了自己——谁都知道，对不合格的砂石，只要增加水泥用量，就能“不影响施工”，这样严重迁就落后的条款本身就否定了标准的必要性！

颁布砂石标准后，应当像有水泥质检站负责检验出厂和混凝土质量监督站验收混凝土一样，设立各级砂石质量监督站。否则，标准形同虚设，目前现状就如此。那么为什么还要制订标准呢？为什么同类产品要有一致性标准呢？

当前砂石资源日益匮乏，尤其是天然级配良好的河砂，几近枯竭。用石子生产中的下脚料生产人工砂，符合节能降耗的要求。实践证明，只要正确使用，对混凝土不仅无不良影响，反而会改善拌和物的施工性能，尤其是天然河砂和人工砂混合使用，效果很好。目前已有人工砂标准，但用户大多不熟悉而不会用，应当尽快组织详解、培训，推广使用。

4 建议

（1）骨料质量对混凝土质量，以致对混凝土结构工程质量，尤其是混凝土结构的耐久性，举足轻重。没有骨料就没有混凝土，即使有朝一日砂石比水泥还贵，也必须使用；骨料质量影响混凝土结构工程的百年大计，不容忽视。实际上，不重视砂石质量，不仅因为无知，而且也涉及对社会和子孙后代的责任！天然资源不可再生，希望政府部门像调控水泥生产一样地调控砂石的开采和加工，免得当后人需要对我们留下的劣质结构物进行修补加固或重建时，却已是资源枯竭。

（2）混凝土的配合比中充满了对立统一的因素，尽管自从有了高效减水剂，缓解了若干矛盾，但仍然存在对立统一规律的问题。其中对骨料用量来说，在水胶比一定的条件下，增大浆骨比时，拌和物流动性增加，混凝土弹性模量减小，体积稳定性下降，开裂敏感性增大，强度和抗渗性提高；浆骨比减小时，拌和物流动性减小，强度和抗渗性会下降，但弹性模量会增大，体积稳定性增强，易裂性减小；对骨料粒径来说，粒径大小和骨料用量大小的规律一致。平衡这些矛盾靠的是技术人员的水平。

（3）提高砂石质量也需要形成买方市场，砂石用户首先要转变观念，不能为图便宜或能欠账等一己私利损害社会的利益。

（4）对砂石质量和生产及管理要求的建议

- 改变生产工艺，以保证粒形，例如除初破外不能再用颚式破碎机；
- 供应单粒级石子，提高限制粒形的标准，例如针、片状颗粒至少应小于10%；
- 要求水洗骨料，带水供应（尤其是砂），以便于混凝土生产的质量控制；
- 改变繁琐的传统评价骨料级配的方法；
- 严禁在砂石中掺入泥土；
- 加强合同意识和管理部门的合同的管理，完善法律文件。

参考文献

[1] 潘志华. 泡沫混凝土的变形和开裂问题及其可能的控制途径. 泡沫混凝土协会官方网 技术研讨，2010.1.

[2] Steven H. Kosmatka，Beatrix Kerkhoff，Willian C. Panarese. Design and Control of Concrete mixtures，the first print of fourteenth edition. Portland Cement Association，2003. USA.

[3] P. K. Mehta，Paulo J. M. Monteiro. 覃维祖，王栋民，丁建彤译. MicoStructure，properties and Materials[M]. 北京：中国电力出版社，McGraw-hill Education(Asia) 联合出版，2008.

[4] A. M. Neville. Properties of Concrete. the sixth print of four and final edition，Pearson Education Limited，2000. England.

[5] 田澤栄一. コンクリートの自己収缩. コンクリート工学年次論文報告集，Vol. 14，No. 1，1992.

自评

1. 当前砂石质量差的严重性已为众所周知。根本的原因何在？如文章所述，主要是认识的误区导致观念的误区。其误区不仅存在于卖方，也存在于买方；更重要的是无人管理，未能形成产业。在第 3 节中说到曾有应用标准中既规定了级配标准，却又说明“如果级配不合格，通过实验证明不影响混凝土和易性时，也可使用”。这说明在制订标准的人那里也存在误区。懂得混凝土的人都知道，因为有了高效减水剂，只要不规定“最小浆骨比原则”，无论如何都能满足混凝土的强度和和易性要求。其结果就是对耐久性的牺牲。

2. 图 6 引自 A. M. Neville 的著作，因为英国不在混凝土中掺用矿物掺和料，图中“混凝土水泥用量”所对应的是骨料用量，“水泥”并非单指硅酸盐水泥，也包括混合材水泥，实指胶凝材料总量。因此该图中混凝土的水泥用量越大，则浆骨比越大，骨料用量越小，收缩就越大。所以砂石在混凝土中的主要作用是稳定体积，假如有一天砂石比水泥还要贵，混凝土也必须用，而且要多用。

3. 看不到砂石对混凝土结构耐久性的重要作用，宁可多用水泥，也不愿多加钱去买合格的砂石，这是买方的误区所在，不仅浪费水泥，更重要的是增大混凝土开裂的敏感性，影响混凝土结构的耐久性。市场的自由在于有人买就有人卖；既然再差也有人买，反倒没有人要求提高买价买合格砂石，当然也就没有人愿意去提高质量。这种恶性循环造成的后果将造成另一个恶性循环——砂石质量越差，工程质量越下降，结构物越提前劣化，修补和再建的资源越短缺。

4. 从卖方来说，把砂石做好并非难事，做不好的主要原因是舍不得投入资金更换先进的设备、引进先进管理技术。其实，健康市场的正常规律是优质优价。用户还是希望得到优质产品的。谁能把砂石的加工当成产业来做出品牌，做到使混凝土的骨料用量最大，谁的骨料销量也就越大，谁就能赚更多的钱。否则就会掉入上述恶性循环。

5. 对砂石生产和标准的建议：①优质骨料应为等径状粒形（作为学术研究，可以截头八面体即 24 面体为模型；作为生产，应合理定义与检测“针片状颗粒”)；②单粒级供应，由用户用不同粒级级配后空隙率为 38%～41%（降级最大不超过 43%)；③国家应从资源和环境保护的角度管制矿山，禁止作坊式的乱采，尽快将砂石生产产业化、现代化。

6. 文中几个数字计算有错：水泥是假设每年混凝土用量为 20 亿立方米、水泥用量 400kg/m^{3}计算的。且不说当年（2010 年）混凝土用量远不止 20 亿立方米，只就 20 亿立方米这个假设来说，文中的计算是错误的。混凝土中水泥用量以 300kg/m^3 计，如果多用 20%，就是多用了 1.2 亿吨。实际上文章发表于 2010 年，以 2009 年计，水泥产量为 16.46

亿吨。水泥是一种不能压库的产品，必须都用于混凝土及其各种制品，按平均用量 300kg/m³ 计，则混凝土总量应当不低于 50 亿立方米，比西方国家多用了应当有 3 亿吨之多。对香山体积的计算应当是约 3 亿立方米，以密度为 2.7kg/m³ 计共重约 8 亿吨，可生产石子 6.8 亿吨（按松堆密度为 1.6t/m³，级配良好计），则消耗 100t 砂石相当于 14.7 座香山。

假如我们毫不迟疑地实施对硅酸盐水泥、原生的骨料和淡水所需要的保护措施，仅通过采取这个行动，混凝土业就能几乎成倍地提高其资源的效率；同时，假如我们还开始建造今后几百年也无需更换或大修的混凝土结构；那么，从现在起，50 年后将产生显著减少混凝土消耗量的效果。到了 2050 年，如果发展中国家对建筑物和基础设施的需求已大部分得到满足，我们就可期望……看到混凝土业可持续发展的光明。

——摘自美国 P. K. Mehta，《Greening of the Concrete Industry for Sustainable Development》[J] Concrete International，July，2002.

论文之十三

矿物掺和料由谁掺好?*

廉慧珍（清华大学土木水利学院）

1　胶凝材料发展否定之否定的历程

由于可持续发展战略和混凝土结构耐久性的要求，矿物掺和料在混凝土中的掺用已日益普遍。人们对矿物掺和料的认识和使用技术水平正在提高。尽管由于认识的误区，矿物掺和料在混凝土中的掺量受到工程建设管理方和现行规范的限制，而在满足性能要求的前提下，大城市商品混凝土实际上已普遍掺用较大掺量的矿物掺和料。

人类最早发现，在有些石头（石灰岩）上用火烧烤食物后该石头会变得松软，遇水体积膨胀并产生很大的热量，因而开始了石灰的生产；后来又发现含有一定量黏土的石灰石烧制的石灰具有一定的水硬性，但是天然石灰石中黏土含量波动幅度大，产品不稳定，继而用人工添加黏土调配原料生产出现代硅酸盐水泥的雏形。从古代至今，用于混凝土的胶凝材料经历了否定之否定的发展过程，如图 1 所示。

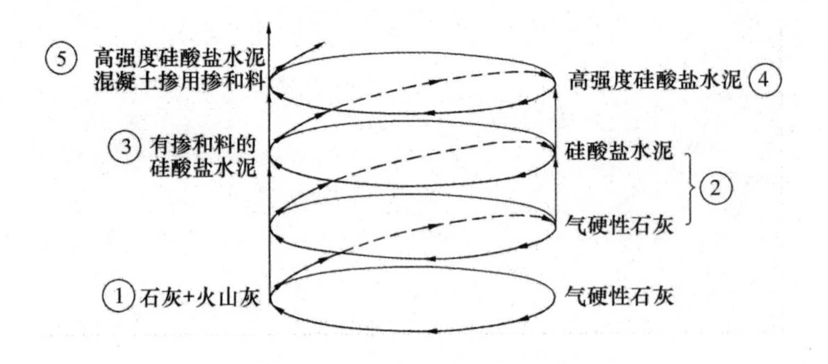

图 1　胶凝材料发展否定之否定的历程

—→第一次否定：从气硬性石灰的使用到与火山灰混合使用已得到水硬性

—→第二次否定：对第一次否定的否定：用黏土质石灰石烧制的水硬性石灰，发展到因天然黏土质石灰石成分不稳定而人工用黏土和石灰石配料煅烧的硅酸盐水泥

—→第三次否定：对第二次否定的否定：为调节硅酸盐水泥强度、降低水化热、提高抗化学腐蚀性的混合材料硅酸盐水泥

—→第四次否定：对第三次否定的否定：适应大跨度、高耸结构工程使用高强混凝土需要的高强度硅酸盐水泥

—→第五次否定：对第四次否定的否定：使用硅酸盐的混凝土耐久性问题凸现，发现“高强不一定耐久”[1]而在混凝土中掺用矿物掺和料

* 原载《中国水泥》2006 年第 7 期。本文所示各图除注明参考文献外，均为本文作者科研成果。

每次否定之否定都没有简单地回到上一次否定之前，而是向前发展了。第五次否定之否定之所以没有简单地回到混合材料水泥的使用，而是改为在混凝土中掺用矿物掺和料，原因有三：

① 过去含掺和料的水泥尽管有较低的水化热、较高的抗化学腐蚀性质，但用统一水灰比检测的强度，就成了以“牺牲强度”为代价。在以强度作为第一甚至唯一评价标准的传统观念支配下，用户无法接受。

② 矿渣和硅酸盐熟料硬度的差别使其共同粉磨后颗粒细度相差较大，磨得越细，相差越大，造成其中的矿渣比表面积太小（对于比表面积为 330m^2/kg 的水泥来说，矿渣比表面积大约只有 250m^2/kg 左右），不能发挥矿渣活性的潜能，不仅水泥强度低，而且因保水性差而不利于混凝土的抗渗和抗冻等性质。

③ 现在已有磨细矿渣的产品，尽管有的水泥厂也在将矿渣和硅酸盐熟料共同粉磨生产矿渣水泥，但是搅拌站自己掺用矿物掺和料则能降低成本获取更多些的利润。于是，“掺和料由谁来掺”转化成“钱由谁来赚”的矛盾。

有的水泥专家的意见认为，索性只生产高强度的纯硅酸盐水泥（当然可以提高价格来赚取利润），让混凝土有加大掺和料掺量的强度余地，掺和料的钱就由混凝土搅拌站去赚吧！有的混凝土专家也认为，在混凝土中掺用矿物掺和料比使用含掺和料的水泥，质量易于控制。然而，这只是短视的观点。社会发展需要多样化的产品供用户选择，以满足日益增长的各种不同需要。产品品种单一化则会更增加用户的麻烦和社会的不和谐，是不符合社会发展规律的。

对“掺和料由谁掺”的问题应当跳出各自眼前的利益来看待。一方面，从双方长远的发展着眼，不能只顾眼前而堵塞各自发展的前途；另一方面水泥和混凝土都是服务于社会的产品，掺和料究竟由谁掺好的问题，首先要从社会发展的利益出发；产品的经济效益是和社会效益密切相关的，没有社会效益的产品是没有生命力的。掺和料究竟由谁掺从根本上要服从结构工程的安全性和耐久性要求。

2 高强度硅酸盐水泥是混凝土掺用掺和料的需要吗？

2.1 从水泥厂来说，100 多年来水泥的发展主要围绕着两个目标：一是降低能耗和污染，提高生产效率，近年来我国水泥在这方面有了很大的发展；另一则是不断提高强度。对于硅酸盐水泥，提高强度的技术路线主要是增加 C_3S、C_3A 和比表面积。目前熟料中的 C_3S 最高的已达到 60%以上。有人想把 C_3S 提高到 70%～80%，这在生产上是无法实现的，而且是没有必要的。现代混凝土的强度已不依赖于水泥的强度，即使特殊用途的“超高强混凝土”如抗压强度达 200MPa 的 RPC（Reactive Powder Concrete，活性粉末混凝土），也能用现有强度的水泥配制出[2]。再者，高强混凝土的需求量是有限的，除了一些处于特殊严酷环境和特殊用途的结构工程外，目前绝大多数工程所用混凝土主要是 C30 和 C40；还有大量处于没有冻融和化学腐蚀性介质的一般环境中的结构只需要 C20 和 C25 混凝土，例如直径达几米的不冻江河上的桥梁墩柱、房屋建筑的基础垫层、跨度不大的楼板以及大坝，等等。即使在海水的严酷环境中，在某些情况下也有可能只需要很低的强度。例如日本 1998 年建成的明石大桥，全长 3910m，由于主跨长达 1990m，因稳定的需要，悬索塔处于海水中的塔基（桥墩）直径达 80m，浇筑于钢沉箱中的混凝土 91 天设计强度为 17.6MPa，配制强度为

24MPa[3]。高层、大跨结构构件采用高强混凝土可以减小构件断面，从而增大使用面积，减轻结构物自重，但是因稳定和刚度的要求，断面都有最小尺寸的要求，尺寸不可能无限地减小，强度也就不需要无限地提高，因此结构对混凝土强度的要求是有限的。而且混凝土抗拉强度与抗压强度比值随抗压强度的提高而几乎成直线地下降[4]，脆性增大。如图 2 所示，抗压强度 20MPa 时，拉压比约为 1/11，抗压强度 60MPa 时拉压比下降到 1/15。

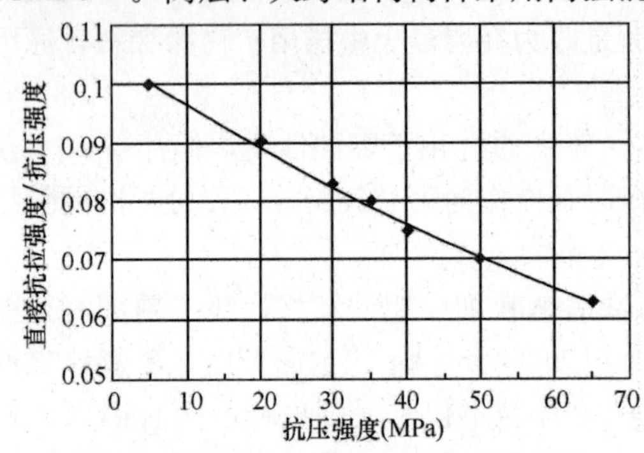

图 2　混凝土抗拉强度与抗压强度比随抗压强度提高而几乎成直线下降[4]

增大比表面积比起增加熟料的 C_3S 和 C_3A，相对容易做到，因此目前市售水泥的比表面积普遍很大，一般都达 350～380m²/kg。用筛析法检测，筛余量为 0 的并不鲜见，其比表面积最高的可超过 400m²/kg。这样的水泥给现今的混凝土带来了很大的麻烦，例如与外加剂和掺和料的相容性差，而且难以控制；水化热大、收缩大、混凝土结构开裂敏感性大、强度的后期增长率小，耐久性下降……。

2.2　水泥专家关于"只生产高强度的纯硅酸盐水泥，给混凝土加大掺和料掺量的强度余地"的意见是源于把"强度"当成了绝对值的思维定势，因为水泥强度是用固定的水灰比检测的，掺和料越多，强度值越低，要保持原有强度值就得提高熟料强度。现行水泥标准和 GB 175—77 中规定的 42.5、52.5 虽然公称强度相同，但实际强度已相差了一级。现代混凝土的强度不再依赖水泥强度就是因为高效减水剂的使用使混凝土的水灰比可以低于检测水泥强度所用水灰比，而水灰比是影响混凝土强度的最显著因素。上述明石大桥悬索塔基混凝土在试配时，曾使用 28 天强度为 32.4MPa 的低热水泥，掺用包括矿渣、粉煤灰和石灰石粉在内的复合掺和料 36.6%，91 天抗压强度为 42MPa，超过设计的 28 天配制强度 24MPa，最后改为对不同部位掺矿渣、粉煤灰总量分别 77%、79%、83%和 0.65～0.69 的水胶比，满足了设计要求。清华大学在深圳地铁足尺模型试验中使用 180kg/m³ 的 42.5 水泥，占胶凝材料总量 59%的掺和料，28 天最低抗压强度也超过了 50MPa；使用 40%粉煤灰和 10%矿渣、44%的熟料（GB 175—77 的 52.5）、6%石膏配制的水泥，混凝土 28 天抗压强度达71MPa……。以上种种都是水灰比（水胶比）的作用。

强度没有绝对值，也检测不出真值，而是在一定条件下（对水泥及材料来说，是用一定水灰比和配料，成型为一定尺寸的试件，在一定条件温度和湿度下，养护到一定龄期）所检测的。加大掺和料掺量时，只要不同程度地改变水胶比，就可达到不同强度的混凝土，现行水泥的强度已足够了。目前的 52.5 水泥和 42.5 的水泥 28 天实际强度差别并不大；龄期越长，差别越小，而且很不耐储存，强度越高，储存时强度损失越大。这表明，水泥强度的提高已经达到顶端，提高水泥强度的技术路线对于水泥的发展已不可行。因此，如果水泥品种单一化，则其自身的发展空间会越来越小。现代混凝土结构对水泥的需要首先不是高强，而是产品的匀质性，并有利于混凝土结构的安全性和耐久性。按混凝土结构的要求生产多品种产品才符合社会发展的需要。

3 在混凝土中掺用掺和料存在的问题

由于能源和资源的消耗和环境保护的需要，现代结构工程的耐久性问题越来越受到重视。适应可持续发展的战略要求和工程的要求，在现代混凝土中除水、水泥、砂、石四种传统材料之外多种有机和无机添加物质已成为必需组分，因混凝土密实性要求所用水胶比普遍减小。于是带来以下难以解决的问题：

3.1 复杂的组分增加了管理和控制的难度 因不同矿物掺和料有不同的特性和对混凝土的作用，复合掺用将成为发展趋势；不同功能的外加剂不一定都可以预先复配，有时需要几种外加剂分别向搅拌机上料。这无疑增加了原材料管理和上料控制的工作量。近十年间，在我国，把粉煤灰误用作水泥、膨胀剂误当成矿渣粉等的事故曾多有发生。其中有的是进料时就入错了料罐，造成混凝土几天都不凝固的事故，花费很大精力分析检查，生产和工程都受到影响。南方某城市有个搅拌站曾因这样的事故被迫停产1个多月。这是搅拌站应当可以解决的问题，但是管理的难度加大了。

3.2 搅拌的匀质性问题 预拌混凝土厂没有也不可能对复杂组分的胶凝材料进行预均化，水泥与掺和料是一起投入的，外加剂是溶于水一次投入的（为了减少坍落度损失，最好是分次掺入外加剂）。大多数搅拌站按照搅拌机的说明书设定搅拌时间为30秒。实际上这个搅拌时间原本可能是针对不用掺和料的传统混凝土设定的。十几年前土木工程学会的《高强与高性能混凝土设计与施工指南》就针对复杂组分和较低水胶比的混凝土需要均匀搅拌的特点，提出有别于传统混凝土的搅拌工艺，使用强制式搅拌机搅拌时间总计为150秒。如今只有很少数搅拌站的明白人把搅拌时间增加到60秒。他们说，“混凝土在搅拌车中还继续搅拌呢”。但是，搅拌车的搅拌机是自落式的，只起防止拌和物静止引起凝固的作用，对均匀搅拌无济于事。有人注意到掺矿渣粉的混凝土拆模后发现构件表面呈现不均匀的绿色（因矿渣粉引起，在空气中会逐渐消退），显然是搅拌不均匀所致。图3是以色列的R. Shalow和R. C. Reinitz堆搅拌时间和混凝土匀质性关系的实验结果，对没有任何掺入的混凝土，当搅拌时间少于75秒时[4]，随搅拌时间的缩短，混凝土抗压强度的离散性而增大，变异系数增大。

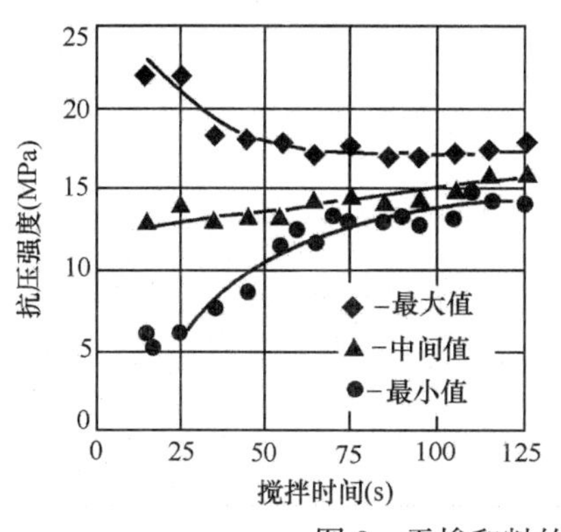

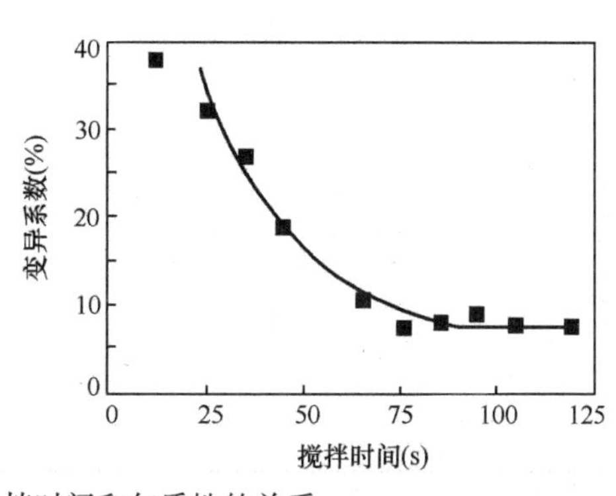

图3 无掺和料的混凝土搅拌时间和匀质性的关系

3.3 大掺量矿物掺和料使胶凝材料中SO_3不足 矿物掺和料的活性需要CaO和SO_3激发，故水泥标准规定允许矿渣水泥中SO_3最大掺量可达4%。而在混凝土中使用大掺量矿物掺和料会稀释水泥中的SO_3，掺量越大，SO_3越不足。因此混凝土早期强度低、凝结缓慢、收缩大。其影响如下：

3.3.1 强度 如图4所示为矿渣和粉煤灰总掺量为50%时，水泥砂浆各龄期抗压及抗折强度均随SO_3含量增加而提高；

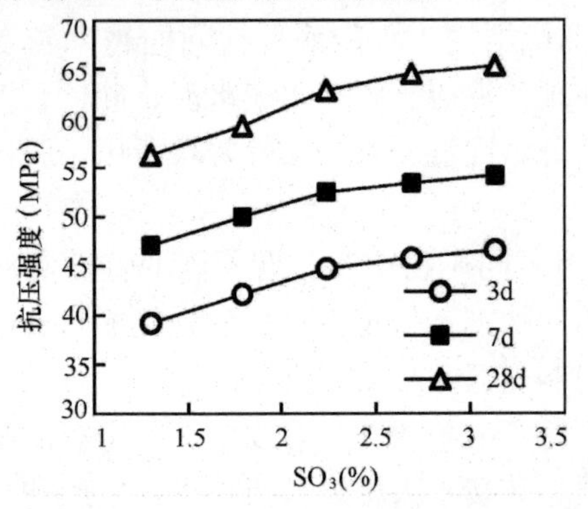

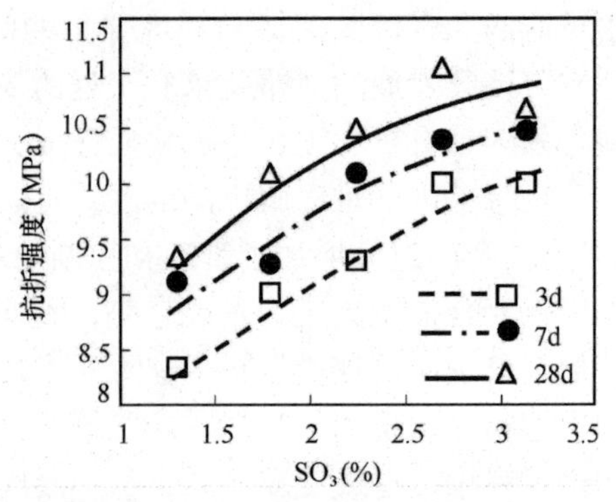

图4 SO_3含量对50%掺和料的胶凝材料砂浆强度的影响

3.3.2 变形 在水泥中掺入矿渣和粉煤灰总掺量50%，SO_3被稀释至1.3%，补充石膏后SO_3最大到3.15%。成型砂浆试件，在水中养护14天后，置于空气中。在不同龄期检测自由变形率，如图5所示。由图可见，水中养护到14天，各组试件均有膨胀，各组膨胀率之间差别不大，停止水养护后，各试件都随龄期产生收缩，SO_3越少的收缩越大。从在水中膨胀到在空气中收缩，其差值称为“落差”。“落差”越小，开裂敏感性越小。这个落差随SO_3的增大而减小：SO_3为3.15%的落差为0.037%；而SO_3为1.13%时，落差为0.050%；还可见到，SO_3为2.69%和3.13%的，无论干缩值还是干缩落差都已很接近，SO_3超过2.24%以后，并收缩发展稳定较早，干缩值都小于混凝土的收缩限，即0.02%。

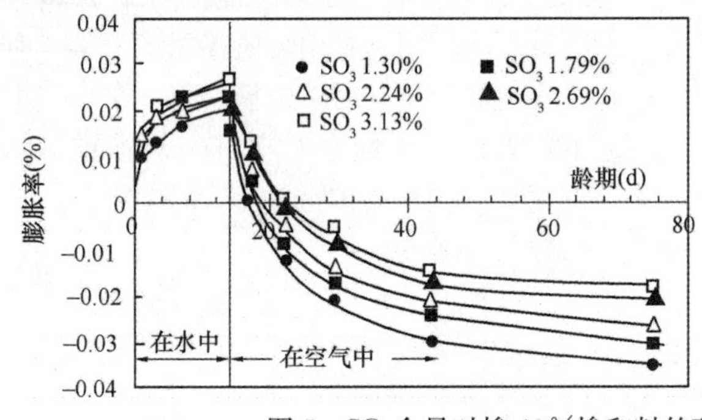

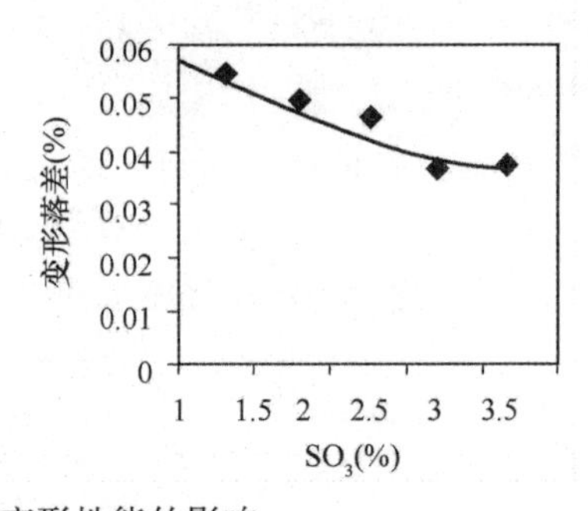

图5 SO_3含量对掺40%掺和料的砂浆变形性能的影响

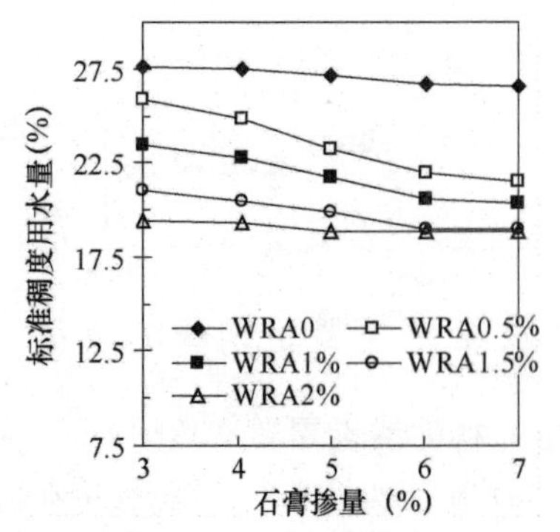

图6 石膏含量对掺外加剂混凝土的需水量影响

在传统上，生产水泥时对石膏的优化主要是为了调解凝结时间，基本上未考虑其他影响。在混凝土中掺入掺和料稀释SO_3的同时当然也稀释C_3A，但是石膏在有掺和料的浆体中的作用并不只涉及C_3A，对大掺量掺和料混凝土凝结时间的影响机理不同于和C_3A的关系。混凝土中掺和料只要掺量大于20%，则SO_3不足的影响就会有表现，掺和料掺量越大影响越大。

3.3.3 需水量 胶凝材料的需水量对混凝土的体积稳定性有重要意义，需水量小，混凝土中浆骨比小，混凝土的弹性模量高，收缩小。由图6可见，需水量在减水剂一般掺量下随石膏掺量的增加而减少，而对不掺减水剂的混凝土没有影响，减水剂掺量超过饱和点（即超过此掺量后，减水率不再增加的点，例如图5中掺量2%）后，就不再有影响。一般不会掺到饱和点。

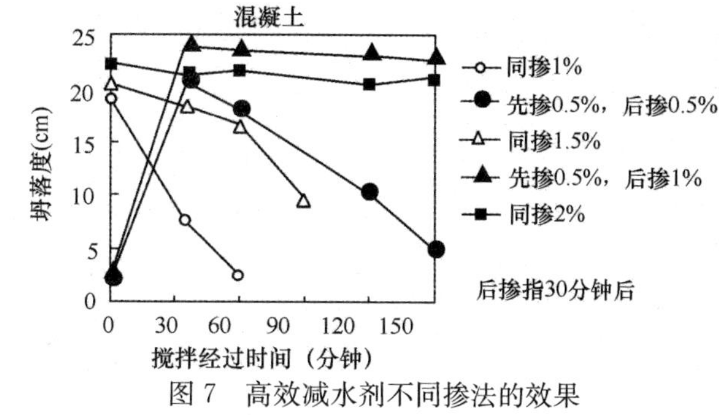

图7　高效减水剂不同掺法的效果

3.3.4　减水剂的掺法和效率

萘系高效减水剂的掺法有以下几种：

① 同掺法：将减水剂溶于拌和水中，一起投入干料中；

② 后掺法：先加一部分水拌和一段时间掺入减水剂；

③ 分次掺法：初始掺一半或三分之一，运送途中或到达现场再掺入其余部分；

④ 将粉状减水剂先掺入胶凝材料干拌，再掺入砂石干拌，最后加水。

图7所示为粉剂萘系高效减水剂掺法不同时拌和物流动性的变化。其中同掺法的流动性损失最大，掺量为1%时，流动度损失时间可以有所延长；掺量增大到1.5%，并且分次掺，可以在第二次掺后保持坍落度2.5小时不损失，掺量增大到2%时，则同掺法也可无损失。目前我国预拌混凝土高效减水剂采用的就是同掺法，其他各个方法都无法使用。

意大利V. A. Rossettihe和Curclo F, Cussino将水泥在生产时掺入超塑化剂共同粉磨，制成一种超塑化水泥（Special Superplasticized Cement，缩写为SPC），另外在使用水泥时掺入超塑化剂（称SpAD，相当于上述④）、将超塑化剂溶于拌和水，同掺入（相当于上述①的同掺法），用微型坍落度筒测定三种试样的坍落度经时变化如图8(a)所示；又检测三种试样的溶液中塑化剂的含量，如图8(b)所示。

如图8中所示，显然在水泥生产时掺入超塑化剂共同粉磨制成的超塑化水泥使用效率最高。

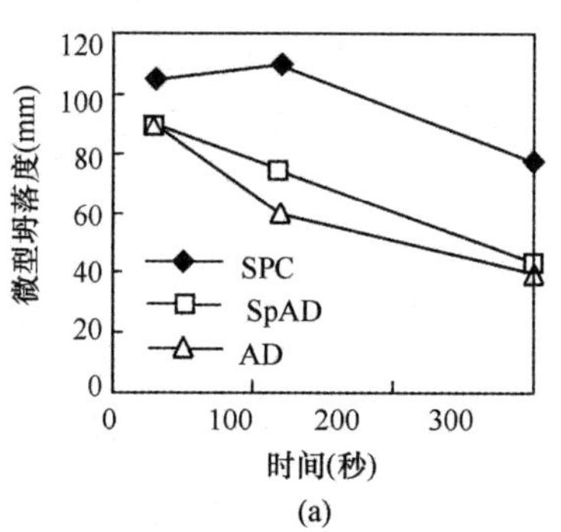

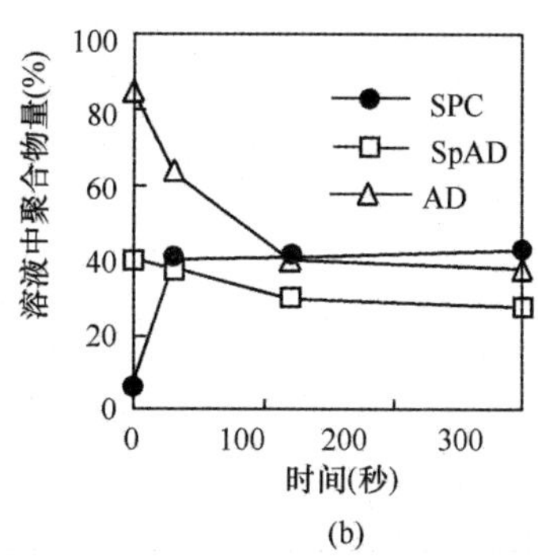

图8　水泥对高效减水剂的含量和浆体微型坍落度试验[5]

4　掺和料由水泥厂按照混凝土的需要来掺是下一次否定之否定规律的趋势

就组分复杂的现代混凝土的实际需要来说，把混凝土所需要用的胶凝材料还放回到工厂去生产，在共同粉磨中做到原材料的预均化，就能够不增加现行搅拌时间而大大提高混凝土的匀质性。搅拌站总体成本不会增加，反而对节约能源和资源、降低质量管理成本都会有明显的效益。从整体来讲，不仅可以实现水泥厂和搅拌站双赢，而且更重要的是有利于建设工程的质量和耐久性。在图1中按照已发生的否定之否定历程的规律预测下一次否定之否定的趋势示于图9，可见在水泥厂按混凝土需要生产掺用掺和料的水泥是符合客观规律的。

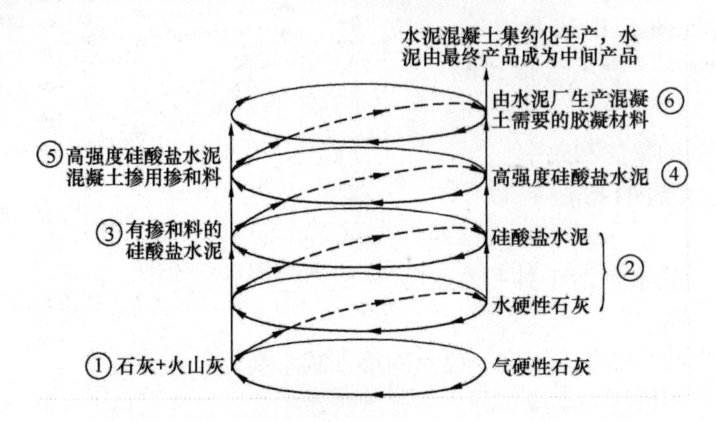

图9 水泥生产发展下一次否定之否定的趋势

创新才有出路，才能发展。不转变传统思维方法，就不会有创新。观念的转变才能有技术上的进步。世界在不断变化，为什么我们不能来一个“脑筋急转弯”，换一种方法来进行思维，转变一下传统的观念，适应世界的变化，采取一种适应当前社会可持续发展的生产模式呢？既然生产水泥时能把生料预热和分解放到窑外去进行，为什么不能把预拌混凝土中胶凝材料的均化生产放到搅拌机外去进行呢？建议搅拌站从提高资源效率、管理水平和混凝土质量的目的出发接受水泥厂的服务；水泥厂不要放弃混合材料水泥的生产，问题是如何生产，如何检测。将来的新型含掺和料水泥必定与硅酸盐水泥之外的传统的现行标准通用水泥不同，而是水泥和混凝土共同进步的产物。目前已经有一些水泥厂不仅建立了混凝土实验室，而且有的还收购了搅拌站，这就将使水泥由最终产品变为中间产品，更好地为生产建设服务，这是个很好的开端，但问题是不仅水泥厂的观念要转变，混凝土生产者也要从传统的、从众的和机械的思维定势中解放出来，才能对自身的需要有进一步的了解，否则，双方还会相互误导而产生新的问题，或者老的问题仍得不到解决，只是形式上的简单合并而已。其中水泥除按混凝土需要生产外，还必须按混凝土的规律进行检验，双方才能有共同语言成为一家人。

此外，任何事物都不要单一化，不要走千篇一律的极端，不能因此又全部变成一种生产模式。应当是可以有这样的生产模式，同时也还需保留其他方式生产的水泥，使我们的产品市场丰富多样，能满足日益增长的多样化的各种需求。

参考文献

[1] ACI 363委员会. State of the arts report on high strength concrete，1993.

[2] 曹峰. 粉煤灰活性粉末混凝土研究[D]. 北京：清华大学硕士论文. 1999，6.

[3] S. Kahima，M. Sakamoto，S. Okada，T. Iho and Nakagawa. Application of High Slag and Fly Ash，Low-Heat Cement to Antiwashout Underwater Concrete，Proceeding of Fourth International Conference on Fly Ash，Silica Fume，Slag and Natural Pozzolana in Concrete. Istanbul，Turkey，May 1992.

[4] A. M. Neville原著. 李国泮、马贞勇译. 混凝土的性能(第一版)[M]. 北京：中国建筑工业出版社，1983.

[5] V. A. Rossettihe, Curclo F, Cussino. Production Performance and Utilization of a Special Superplasticized Cement. Proceedings of 9th International Conference on the Cemistry of Cement, New Dehli, 1992.

自评

1. 矿物掺和料由谁掺好？这个问题在利益的驱动下被异化成“钱由谁赚?”的问题。大多数人对这个问题只从眼前的现象考虑。孤立地看，显然因为矿物掺和料现在的价格都低于熟料价格，谁掺谁得利。因此，搅拌站不肯放手，水泥厂偷着掺，结果是双方都面临产品质量问题。有的搅拌站，当知道了我说的这些弊病之后，就自己设法解决：在搅拌机前增加预混胶凝材料预混的装置，改造搅拌机叶片，在矿物掺和料中添加石膏，等等，但是仍然做不到水泥厂均化工艺的效果，此外，搅拌站仍然不了解所购水泥的组成，无法对水泥和矿物掺和料整体优化 SO_3 含量。如果购进熟料自己粉磨生产水泥，则会增加成本……；在水泥厂，因为不懂得混凝土的需要和规律，为了不降低水泥的强度，就过分地磨细，造成产品质量下降。直接影响工程使用效果和工程质量。

2. 矿物掺和料由谁掺好？必须从整体分析。当前一些水泥企业并购混凝土搅拌站，并不是从保证工程质量的全局出发，而是为了保障水泥产品的销路，结果形成了两层皮，没有解决根本问题，反而增加了新的矛盾。从根本上来说，这本来是一种适应当前生产力发展的生产关系创新举措，却被做成了夹生饭，再不改革，就会就会再走一次弯路，付出一次代价，错过一个成长的机会。马云说：“很多人输就输在对于新兴事物第一看不见，第二看不起，第三看不懂，第四来不及。”

3. 对于搅拌站搅拌不均匀的问题，图 5 所示为 A. M. Neville 第四版《Properties of Concrete》中仍然引用的，用以说明搅拌时间和拌和物匀质性的关系，不同机型要求的的最短搅拌时间不同，现在的双卧轴强制式搅拌机，至少也需要搅拌 60 秒。实验室试配采用的是 150 秒。

一个人对社会的价值，首先取决于他的感情思想和行动对于人类利益有多大作用。

——爱因斯坦

论文之十四

关于混凝土配合比选择方法的讨论之一
——当前混凝土配合比“设计”存在的问题*

廉慧珍（清华大学土木水利学院）
李玉琳（北京中宏盛建设工程质量检测有限责任公司）

摘　要　高效减水剂和矿物掺和料的大量使用，使用传统的混凝土配合比设计方法时，由于矿物掺和料与水泥的密度相差较大，计算混凝土配合比的假定容重法不再适用；现行以砂子绝干密度为基准的方法造成较大的质量管理难度；骨料质量成了影响混凝土质量的主要因素。改变矿物掺和料掺量时，应当保持浆骨比不变，以保持混凝土的体积稳定性。

关键词　混凝土配合比；假定容重法；砂石饱和面干状态的密度；骨料级配；等浆体体积

自从 1918 年美国的 D. Abrams 经 5 万多次试验，在“混凝土配合成分的设计”一文中提出混凝土的水灰比定则，认为可塑性混凝土的抗压强度完全受水灰比的控制，而与其他因素无关[1]。1932 年 I. Lyse 提出灰水比定则，认为混凝土的 28 天抗压强度和水灰比的倒数成正比，设混凝土 28 天抗压强度为 y，水灰比倒数为 x，则

$$y = -\mathrm{b} + \mathrm{a}x$$

其中系数 a、b 需根据原材料及工艺经试验确定。

我国 20 世纪 50 年代以来，长期使用 Bolomy 经大量实验数据统计拟合的公式：

$$R_{28} = R_c \mathrm{A}\left(\frac{c}{w} - \mathrm{B}\right)$$

该式试验时的条件是使用硅酸盐水泥、级配良好而清洁的河砂、粒形匀称的石子，系数 A、B 依石子品种而异；该式适用于坍落度为 30mm～90mm 的塑性混凝土，因施工性和经济性的要求，我国在使用该式时要求水泥强度 R_c 和混凝土强度 R_{28} 的关系为 $R_c = (1.5 \sim 2)R_{28}$。根据强度要求计算得出的水灰比与试配验证的结果相差可达 20%～30%。由于水泥标准的变化，JGJ 55—2002《普通混凝土配合比设计规程》修改了 Bolomy 公式中的系数 A、B。但是，我国自 1970 年代引进高效减水剂，直到 1980 年代末至今得以大量使用后，混凝土强度不再依赖于水泥强度，用 GB 175—77 水泥标准的 425＃水泥（相当于现行水泥标准的 32.5 等级）已能配制出 C60 的泵送混凝土。在本质上，混凝土主要还是由水泥、骨料和水组成的硬化体，但是其内涵已发生很大变化：总体强度水平高了，拌和物从低塑性发展到当前的泵送，流动性大大提高；原材料也有很大变化：水泥强度等级高、细度细，骨料粒形和级配差了，外加剂和矿物掺和料普遍使用，……。与此同时，多数人对混凝土却仍停留在

* 原载《混凝土》2009 年第 3 期。

以往的认识。为适应这种变化，在此讨论三个问题。

1. “假定容重法”对当前混凝土的配制已不再具有普适性

“假定容重法”本来是在绝对体积法的基础上产生的。混凝土配合比的原理是按照 $1m^3$ 混凝土拌和物由各原材料紧密堆积而成，即 $1m^3$ 混凝土体积等于各原材料绝对密实体积之和（即不计各原材料内部孔隙)。过去水泥、砂石的表观密度变化不大，所配制混凝土的表观密度变化也不大，因此为了简化试配，对水灰比为 0.5 左右的混凝土假定表观密度为 $2400kg/m^3$，对高强混凝土假定表观密度为 $2450kg/m^3$，试拌后实测差别不大。但是如今普遍使用较大掺量的矿物掺和料，例如粉煤灰表观密度为 $1.90 \sim 2.40g/cm^3$，磨细矿渣表观密度约为 $2.80g/cm^3$，与水泥表观密度的 $3.0g/cm^3$ 左右相比相差就大了，按上述假定的表观密度计算，则体积都会大于 $1\ m^3$，掺和料越多，大得越多。因此从根本上，还是应当使用绝对体积法。当然，正如任何方法都有一定的假设，绝对体积法的假设是忽略水泥水化所减少的那部分水的体积，但是，混凝土在新拌状态时，这部分水相对于混凝土的总体积来说是很少的。为了弥补这部分忽略水的体积，建议用绝对体积法计算时，不必计入搅拌时挟入的孔隙体积。

2. 用绝对体积法计算混凝土配合比时原材料密度的取值问题

材料的密度为单位体积的质量，量测出材料的体积和质量，即可计算。通常所说密度指的是绝对密实状态下单位体积的质量，对密实的无孔材料如钢铁，制成规则的几何形状，所量测的体积为绝对密实体积，或称作实体积；对含有孔隙的固体材料如砖瓦，如欲得出其密度，则可将其磨成细粉，与颗粒很细的材料如水泥、矿物掺和料一样，用比重瓶量测体积；磨得越细越接近真密度，因此只能磨到一定的粗细程度，求得近似密度值；一般凡是需要磨细来求得其密度的，统一磨到与水泥相同的细度。

对于砂石等散粒状材料是有微孔的材料，孔隙率都很小却不能忽略，其所含孔隙有能吸入水的开放孔，也有水进不去的封闭孔（包括 100nm 以下的开放但水进不去的孔)。组成绝对密实的混凝土拌和物中，砂石所占的体积，是能充水的气孔都充水到饱和程度而无表面吸附水状态的砂石颗粒体积的总和。这种状态就是饱和面干；这样求得的密度叫做表观密度，以区别于真密度。我国自 50 年代开始对混凝土进行配合比设计，其中的砂石计算就“以面干饱和状态中的材料为标准”[2]。而今，全世界只有我国是以绝干状态的骨料密度进行混凝土配合比的计算。这两种基准的配合比有什么区别呢?

饱和面干状态骨料所含的水既不影响混凝土拌和物的工作性，也不参与胶凝材料水化后微结构的组成。但是如果骨料不是面干饱和的状态，当含水率低于面干的饱和含水率时，就要从拌和水中吸收水；如果含水率大于面干的饱和含水率，则会增加拌和水量，对拌和物性能与硬化混凝土性能都会有影响。对于以绝干基试配的混凝土，目前我国在混凝土实际生产中，多采用炒干或烘干至恒重的方法求出砂子的实际含水率，依此扣除拌和水用量。结果是试拌的混凝土坍落度会小于预期值，又要调整配合比，造成质量控制的麻烦。当采取面干饱和基试配时，只要预先测得骨料面干的饱和含水率（其值与吸水率值相等)，再测出骨料实际的全部含水率，在生产中对拌和水“多退少补”即可，具有较好的质量可控性。

对骨料面干饱和状态的界定，实际上我国砂石标准中已有规定[3]。以往《建筑材料》教学中即按砂石标准明确测定石子表观密度的步骤：将一定量的石子在水中浸泡至少 24 小时，取出后放在拧干了水的毛巾上吸去其表面吸附水，肉眼观察表面无水的亮光，即为石子的饱和面干。称取 1kg 面干饱水的石子，用排水法测得其体积，即可求出石子的表观密度。砂石性能检测标准[3]测定砂子吸水率的方法中，对砂子面干饱和的界定方法为：将浸水饱和的砂子用吹风机吹干至表面开始变色，按规定方法装入一截头圆锥环（图2示意）中，向上提起截头圆锥环后，根据砂子的性状，即可判断其含水状态（图1示意）。其中面干饱和状态的含水率即吸水率。

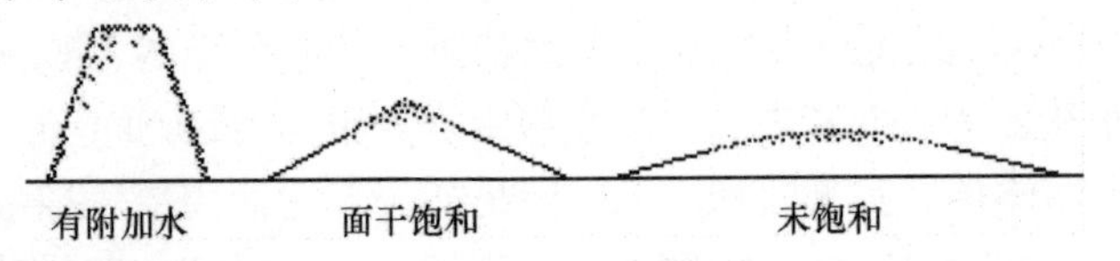

图1　砂子含水状态示意

图2　饱和面干测定用具示意

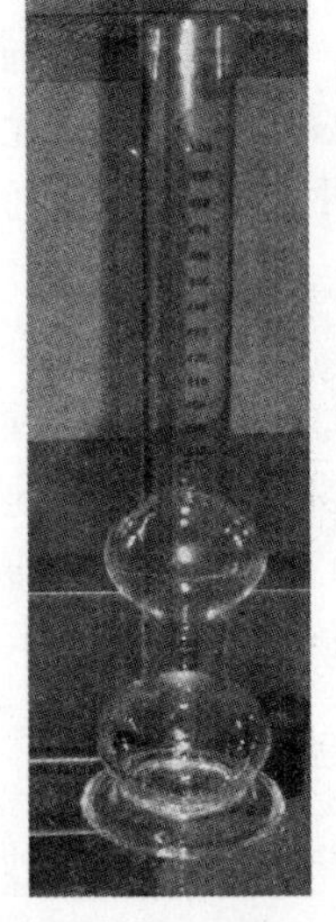

图3　在骨料下料口量测表面含水的比重瓶

称取一定量所得面干饱和状态的砂子，在图3所示比重瓶中测出其体积；将同等质量实际含水的砂子装入图3所示的比重瓶内，如果测得二者体积相同，表明二者都是面干饱和状态；如果所测得的体积大于或小于面干饱和状态砂子的体积，则体积的变化说明其含水量的变化，二者差值即需要“多退或少补”的水量。此项试验可在实验室也可在拌和楼的上料口进行，还可以用含水量传感器和电流表进行在线控制，而用此项试验作为校核。这样做的结果要求砂石进料必须是饱水状态，并且封闭储存，以防水分变化。对于严格质量管理和提高混凝土质量控制水平，这样做是非常必要的。

3. 骨料的级配和粒形成了影响混凝土质量的关键因素

骨料在混凝土中的骨架作用主要稳定体积。即使采石场生产的石子经过严格的级配，销售时经过装料、运输中的颠簸和卸料，再加上生产混凝土时的投料，就会大小颗粒分离而重新分布，失去级配。因此绝大多数国家配制混凝土所用的石子都采用两级配或三级配。例如德国，还在混凝土试配时将砂石一起连续地级配。我国目前市场供应的石子由于生产工艺落后，也由于大多数生产者的无知，无视砂石标准，号称连续级配，实际上小于10mm的颗粒极少，几乎没有。而且由于我国砂石标准中对针、片状尺寸颗粒限定要求过宽（实际上是牵就落后），使石子的粒形很差。如图4所示，我国粒形“合格”的石子中有的颗粒与针、片状颗粒的差别并不显著，而日本所用石子则各向径长差别较小，基本上是等径状的。

我国已故老专家蔡正咏在上世纪80年代初就说过：“我国混凝土质量不如西方国家的，原因就是石子质量太差。”但是那时我国石子随机取样的空隙率一般都在40%～42%，而理想粒形和级配的石子孔隙是36%～38%。现在，我国市售石子空隙率已达45%以上，甚至超过50%！这就使我国混凝土的水泥用量和用水量比西方国家混凝土水泥用量和用水量多用约20%。已经有一些搅拌站或工程采用了两级配的石子，混凝土的水泥用量减少了约20%。

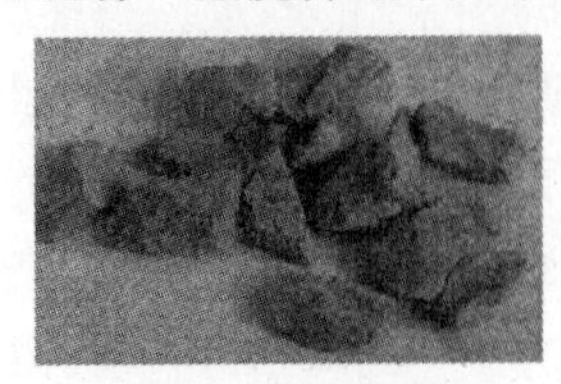

我国粒形“合格”的石子

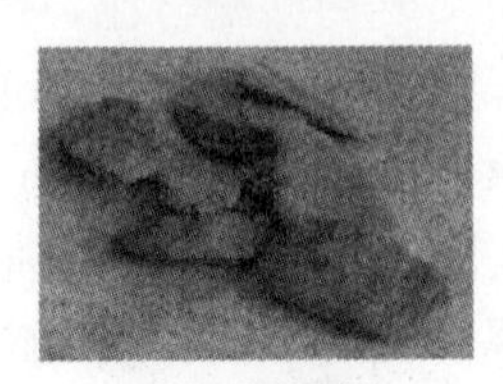

针、片状颗粒

日本常用石子

图4　我国目前常用石子与日本常用石子粒形比较

4. 掺用矿物掺和料的混凝土配合比计算的问题

多年来，人们对掺用掺和料的混凝土配合比的计算，基本上从等水胶比法（即简单等量取代）发展到超量取代法；也有人参照纯水泥混凝土的水灰比，计入掺和料后，在减少水泥的同时按原水灰比减少用水量，即等水灰比法。基本上没有人使用等浆体体积法。先将以上各方法分析如下：

（1）等水胶比法：掺矿物掺和料后的水胶比与未掺矿物掺和料时的水灰比值相同，即简单等量取代。因矿物掺和料密度小，使浆体体积变大，即浆骨比增大，例如，假定普通水泥密度为 3.0g/cm^3，粉煤灰密度为 2.2g/cm^3，当以粉煤灰简单取代 30%的水泥时，浆体体积就会增加 37 升。水泥加水硬化后的体积收缩是混凝土的特性之一，加入骨料制成混凝土后，由于骨料的温度变形系数比硬化水泥浆体的温度变形系数小一半多，则对混凝土起稳定体积的作用。浆骨比越小，硬化混凝土收缩值越小；浆骨比增大势必会对混凝土的体积稳定性有影响。此外，因粉煤灰反应速率和反应率低，混凝土早期浆体水灰比增大。例如假定有一原水灰比为 0.57 的混凝土，如果用粉煤灰简单取代 30%的水泥，水胶比仍为 0.57，忽略粉煤灰表面吸附水，则早期水灰比就会增大到 0.81，同时混凝土强度肯定下降；为了保持混凝土强度不变，将水胶比降至 0.5，则早期水灰比仍有 0.71。这样大的水灰比就会造成早期较大的孔隙率。如图 5 所示，水胶比为 0.35 的水泥浆体水化 1 天的孔隙率，无掺和料时约为 12%，掺粉煤灰 44%时约为 26%，掺粉煤灰 56%时约为 30%；降低水胶比时，上述孔隙率分别减小至约 3%、17%和 27%。早期孔隙率大是掺粉煤灰的混凝土早期碳化加速的主要原因。

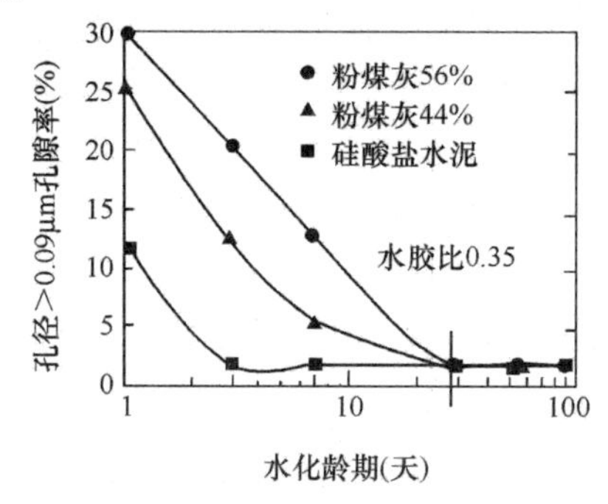

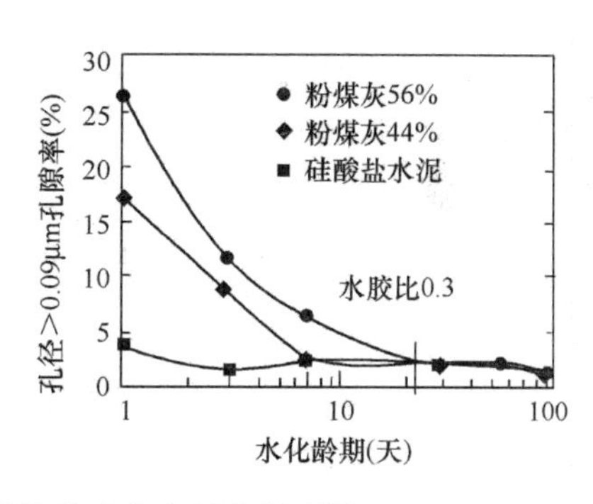

图 5　不同水胶比时粉煤灰掺量对水泥浆体孔隙率发展的影响[4]

而且，掺矿物掺和料混凝土的强度对水胶比更加敏感。英国的 Dunstan 研究表明，以 60%的体积取代水泥，水化 7 天时，水泥和粉煤灰对强度的贡献随水胶比的降低而增加，但是粉煤灰的贡献增加的幅度随龄期而增加显著[1]。如图 6 所示。

由图 6 可见，在龄期 7 天，质量水胶比 0.54 时，粉煤灰的贡献为 0；龄期为 28 天时，粉煤灰的贡献增大，但当体积水胶比约 2.2（相当于质量水胶比 0.7）时粉煤灰贡献仍为 0；随水胶比的下降，水泥和粉煤灰贡献差距明显减小；在龄期 90 天，当体积水胶比大于约 1.7（质量比约 0.54）时粉煤灰的贡献小于水泥的贡献，而在较低水胶比时，粉煤灰的贡献就超过了水泥的贡献；龄期 360 天后，则在任意水胶比下，粉煤灰的贡献都超过水泥的贡献。这说明粉煤灰作用比水泥作用对水胶比和龄期更加敏感，粉煤灰掺量越大，越需要减小水胶比。因此等水胶比法掺用的粉煤灰是无效的。

在图 7 中，中央的是 Dunstan 用以编制掺粉煤灰混凝土水胶比和粉煤灰掺量关系软件的

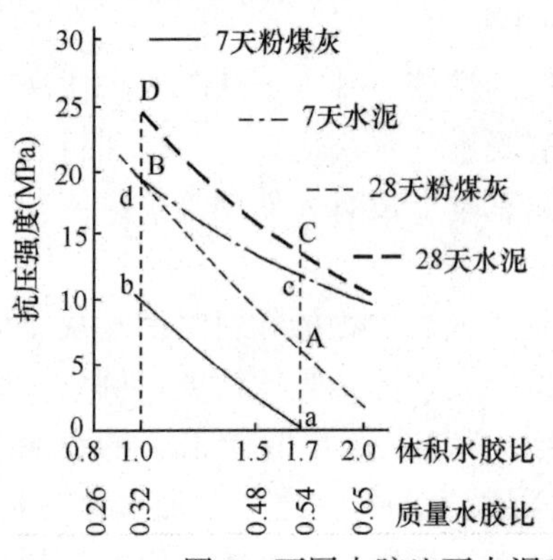

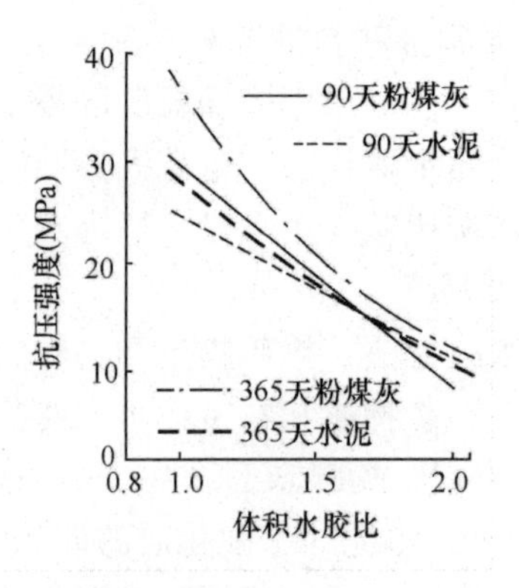

图 6 不同水胶比下水泥和粉煤灰比例为 40：60 时，水泥和粉煤灰对强度的贡献[5]

三维关系模型；①、②、③分别是一定粉煤灰掺量下强度-水胶比关系、一定水胶比下强度-粉煤灰掺量关系和一定强度下水胶比-粉煤灰掺量关系；②中虚线大体为粉煤灰掺量18%～20%，超过此值后，在相同水胶比下，强度随粉煤灰掺量增大而迅速下降。因此，掺粉煤灰时，不能采用不变的等水胶比，必须降低水胶比才能发挥粉煤灰的作用。

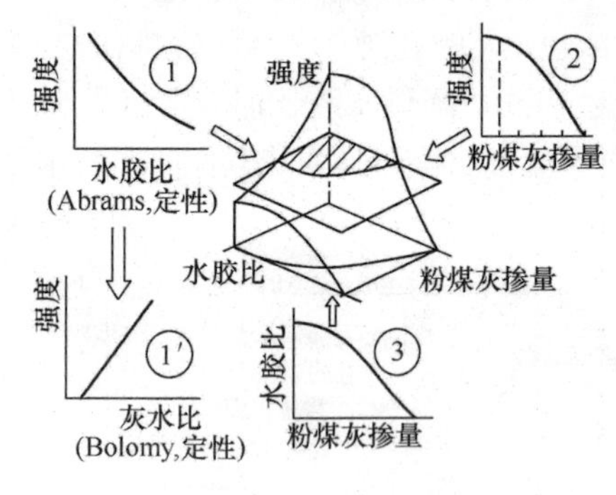

图 7 混凝土抗压强度-粉煤灰掺量-水胶比关系[5]

（2）超量取代法：由于对矿物掺和料的不了解，混凝土的设计与工程质量管理人员限制矿物掺和料的掺量，于是有关配合比的规范中提出粉煤灰的“超量取代法”，即，在能被接受的掺量范围取代水泥，另多掺一部分取代砂子。这只是一种计算而已，在数量上“代砂”，实际上因为细度量级的差别在功能上粉煤灰并不是砂，不可能“代砂”，仍然是胶凝材料，却因为“超量”而变相增加浆体含量、减小水胶比，但是，在形式上，并未公开实际的粉煤灰掺量和实际的水胶比，在客观上起了遮人耳目的作用。水胶比是混凝土配合比的三要素之一，在原材料相同的情况下，影响混凝土强度的主要因素是有效拌和水与包括水泥在内的全部粉细料的比值，即水胶比，即使掺入传统意义上的惰性材料如磨细石英砂等石粉。超量取代法不能用的原因，还在于对水胶比界定的混乱。例如有的搅拌站在胶凝材料中不计入超量取代的部分，声称掺粉煤灰前后的水灰比不变。已有实例表明，这种做法使得当工程中出现问题时，无法从所报的配合比上分析原因。有人认为掺粉煤灰后的混凝土抗裂性改善不明显，浆骨比增大是其原因之一。建议今后不再采用这种实际上增加浆骨比的计算方法。

（3）等水灰比法：基于某些人对水泥认识的局限性，把水泥厂生产的混合材水泥叫做水泥，而在搅拌站生产混凝土时掺的矿物掺和料不算在水泥中，简单地保持水灰比不变，减小用水量，降低水胶比，希望以此保证混凝土强度不变，但是这种做法的结果是水胶比降得过大，实际强度会远超过期望值。以粉煤灰为例，如果掺入粉煤灰后仍保持水灰比不变，则需降低水胶比。粉煤灰掺量越大，水胶比需降低越多。例如表 1 中，假定原始（表中 FA%为 0）水灰比为 0.50，当粉煤灰掺量为 20%时，使水灰比不变的水胶比应为 0.40，依此类推，粉煤灰掺量为 40%时，水胶比应为 0.30。这完全是忽略了粉煤灰的存在而计算出来的。实际上由于粉煤灰表面有吸附水，拌和物自由水并不像计算的那样大，则所需水胶比可以更大些。同时，这种方法的粉煤灰掺量是按等值量取代水泥掺入的，总胶凝材料质量不变，但因粉煤灰密度比水泥的小，粉煤灰掺量越大，总胶凝材料体积越大，水胶比降得太低时，会影响拌和物的施工性，就需要增加用水量（同时按水胶比增加胶凝材料用量），不仅会增加试配工作量，还会因浆骨比增大而影响混凝土的体积稳定性。

表 1　不同掺量粉煤灰的混凝土水胶比和水灰比的关系

FA%	w/b	w/c	w/b	w/c	w/b	w/c	w/b	w/c
0	0.50	0.50	0.40	0.40	0.35	0.35	0.30	0.30
15		0.59		0.47		0.41		0.35
20		0.63		0.50		0.44		0.38
30		0.71		0.57		0.50		0.43
40		0.83		0.67		0.58		0.50
50		1.00		0.80		0.70		0.60

（4）等浆体体积法：矿物掺和料密度小于水泥的密度，按质量掺入时，混凝土浆体体积会增大，按等浆体设计，可有利于保持混凝土的体积稳定性不变。覃维祖的硕士研究生成唯佳对 C30 混凝土的实验研究表明，按等浆体体积法与按等水胶比计算掺不同量粉煤灰的混凝土配合比相比，混凝土强度等级相同，而具有更高的抗渗性，如表 2 所示。

表 2　粉煤灰不同掺用方法对混凝土性能的影响

粉煤灰取代量（%）	等水胶比				等浆体体积			
	抗压强度（MPa）		$D_{Cl}\times10^{-9}$ cm²/s		抗压强度（MPa）		$D_{Cl}\times10^{-9}$ cm²/s	
	3d	28d	28d	84d	3d	28d	28d	84d
0	20.0	37.8	4.89	3.04	20.0	37.8	4.89	3.04
15	18.4	35.3	4.35	2.13	20.6	39.7	3.96	1.38
30	14.4	28.4	3.86	1.50	20.8	41.9	2.74	1.31
45	12.2	25.3	3.42	1.03	20.0	38.6	2.30	0.85
60	7.8	19.3	2.74	1.00	20.6	40.1	2.05	0.82

成唯佳按相同强度等级、相同流动性、相同浆体含量，进行粉煤灰掺量和水胶比关系的试验，如图 8 所示。图 9 为等水胶比时粉煤灰掺量与混凝土强度的关系。

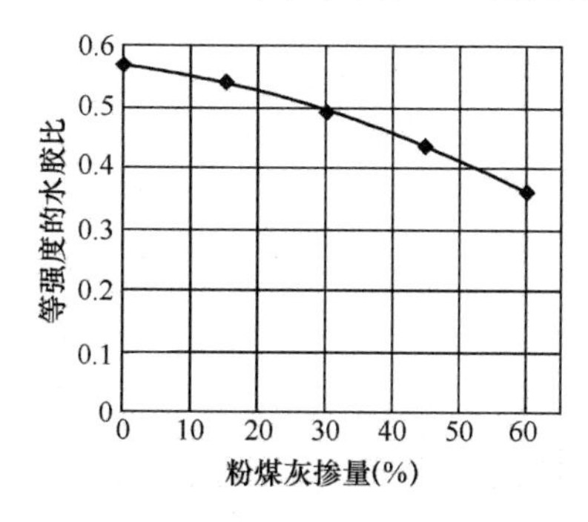

图 8　相同强度、流动度、浆骨比的混凝土粉煤灰掺量和水胶比的关系

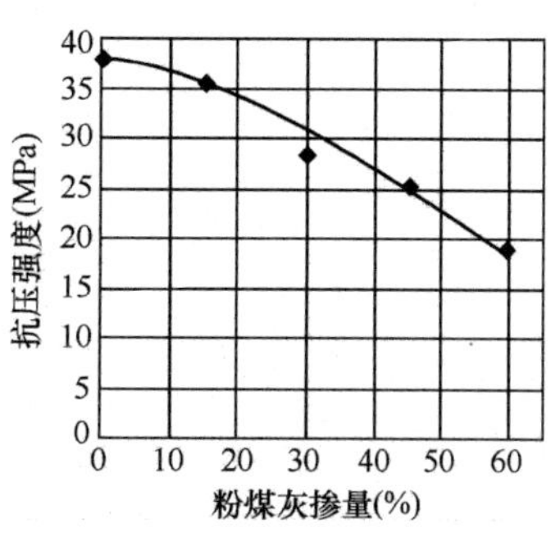

图 9　等水胶比混凝土的粉煤灰掺量与混凝土抗压强度的关系

图 8 与图 7 的③相比，图 9 和图 7 的②相比，规律是一致的。

结论

(1) 鉴于当前混凝土组分的变化，进行混凝土配合比的计算的假定容重法不再适用。建议改用绝对体积法；

(2) 以绝干基的砂子进行设计不利于混凝土质量的过程控制，饱和面干基才符合实际的客观规律；

(3) 以单粒级石子进行两级配或三级配，生产时分级投料，可得到满足施工要求的最小浆体总量，有利于工程的经济性和耐久性；

(4) 当水胶比或矿物掺和料掺量改变时，应使用等浆体体积法调整混凝土配合比，以保持混凝土的稳定性。

参考文献

[1] D. Abrams. Design of Concrete Mixtures，Bulletin 1，Structural Materials Research Laboratory，Lewis Institute，1918.

[2] 谭炳训．水泥与混凝土[M]. 北京：中国建筑工业出版社，1957.

[3] JGJ 52—92 普通混凝土用砂质量标准及检验方法[S].

[4] V. M. Malhotra. CANMET Investigations dealing with High-Volume fly ash Concrete，2006. 6.

[5] M. R. H. Dunstan. Fly-ash as the Fourth Constituent of Concrete Mix，Proceeding of Fourth International Conference on Fly Ash，Silica Fume，Slag and Natural Pozzolana in Concrete. Istanbul，Turkey，May 1992. SP-132-86.

[6] 成唯佳．基于现代混凝土设计理念的耐久性评价研究[D]. 北京：清华大学硕士论文，2005，6.

善人者，人亦善之。

——管仲

论文之十五

关于混凝土配合比选择方法的讨论之二

——关于当代混凝土配合比要素的选择和配合比计算方法的建议*

廉慧珍（清华大学土木水利学院）
李玉琳（北京中宏盛建设工程质量检测有限责任公司）

摘　要　当代混凝土配合比选择的内容实际上是水胶比、浆骨比、砂石比和矿物掺和料在胶凝材料中的比例等四要素的确定，以及在满足施工性要求的前提下依据紧密堆积的原理的计算方法。对于有耐久性要求的混凝土，这四要素的原则都能以混凝土结构耐久性设计给出的《混凝土技术要求》为根据来确定。调整配合比时，应采用等浆体体积法，以保证混凝土的体积稳定性不变。

关键词　混凝土配合比；水胶比；砂石比；浆骨比；掺和料；等浆体体积法

1　前言

多年来，多数混凝土工作者习惯于根据自己的经验，用最简单的方法计算混凝土配合比。混凝土的制作工艺看似简单，如果只满足强度和拌和物流动性要求，“怎么都能配出来”。但是对单方混凝土各组成材料质量和进行简单化的假定，实际上当原材料组成变化后，按假定的单方质量所计算的混凝土不是“胀方”就是“亏方”；只满足流动性要求的拌和物离析、泌水现象普遍。混凝土本来就是一种从细观到粗观都是高度非均质的多项复杂体系，由于原材料的变化和对混凝土性能要求的提高，现今的混凝土越来越复杂，现行简单化的方法已不再适用，需要进一步科学化。尽管混凝土配合比不能决定混凝土的最终质量，而得到高质量的混凝土配合比毕竟是当前人们所追求的。什么是“高质量”？按 1999 年 ACI 对高性能混凝土的定义，应是“符合特定工程性能组合和匀质性要求的混凝土”[1]。这个定义当然指的是最终质量，然而，高质量的配合比却是前提。“符合特定工程性能组合”是不言而喻的，匀质性则是很高的要求，不仅拌和物不离析、不泌水，而且硬化后也具有较高的体积稳定性，能较长期地抵抗环境作用引起的劣化，按照传统方法难以做到。任何新的观念和技术的提出，总是会遭到习惯思维的抵抗，只有改变思维方法和观念，才能使我们“与时俱进”。谁先做到谁领先。

现行当代混凝土的特点是普遍掺入矿物掺和料和高效减水剂。混凝土中水、水泥、砂、石四种原材料中增加了矿物掺和料，因此传统的配合比三要素——水灰比、浆骨比、砂石比，就成为水胶比、浆骨比、砂石比和矿物掺和料用量等四要素。配合比中需要求出的未知数由传统的 4 个变成 5 个。最后由各材料在满足施工要求的前提下紧密堆积的原理，用绝对体积法计算出各材料用量。不考虑外加剂占据混凝土的体积，则各组成材料的关系和性质及其作用和影响可用图 1 来描述。

---

* 原载《混凝土》2009 年第 5 期，更正于 2010 年第 9 期。

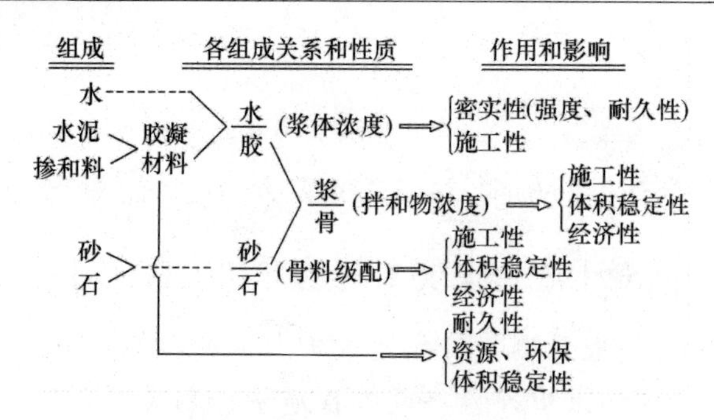

图1　混凝土各组成材料的关系和性质及其作用和影响

由图1可看出，混凝土配合比四要素都影响拌和物与硬化混凝土性能，当决定混凝土强度和密实性的水胶比确定之后，所有要素都影响拌和物施工性能。施工是保证混凝土质量的最后的和关键的环节，则考虑浆体浓度的因素、按拌和物的施工性能选择拌和物的砂石比与浆骨比，就是混凝土配合比选择的主要因素。其中浆骨比是保证硬化前后混凝土性能的核心因素。无论是改变水胶比，还是矿物掺和料用量，调整配合比时应使用等浆体体积法，以保持浆骨比不变。

我国混凝土年产量可占到全世界的一半，质量却相对落后。例如，全世界只有我国使用“假定容重法”计算混凝土配合比，也只有我国使用绝干基的砂石生产混凝土，造成我国混凝土质量控制的困难。由于混凝土原材料的波动和使用环境和条件的不确知性与不确定性，并不存在普遍适用的配合比，在本文中所介绍的方法只是一种推荐，相信会有更多方法可以使用，但是原理和原则应当是更科学与更便于质量控制的[2]。和施工技术一样，“混凝土配合比设计”不适于制订成规范。设计提出的《混凝土性能技术要求》所要求的也只是性能而不是组成（performance based specification）。希望能与大家为不断提高混凝土的技术与质量而共同努力，在实践中不断修正已有的知识、观念和方法，不仅在产量上，而且在技术和管理水平上尽快走向先进。相信目前“做不到”的事，只要大家齐心努力，外国人能做到的，我们也都能做得到。

为便于与读者沟通，给出所推荐配合比选择与计算的实例，供参考和讨论。

2　确定混凝土配合比的原则

2.1　按具体工程提供的《混凝土技术要求》选择原材料和配合比。

2.2　注重骨料级配和粒形，按最大松堆密度法优化级配骨料，但级配后空隙率应不大于42%。

2.3　按最小浆骨比（即最小用水量或胶凝材料总量）原则，尽量减小浆骨比，根据混凝土强度等级和最小胶凝材料总量的原则确定浆骨（体积）比，按选定的浆骨比得到1m^{3}混凝土拌和物浆体体积和骨料体积；计算骨料体积所使用的密度应当是饱和面干状态下所测定的。

2.4　按施工性要求选择砂石比，按《混凝土技术要求》中的混凝土目标性能确定矿物掺和料掺量和水胶比。

2.5　分别按绝对体积法用浆体体积计算胶凝材料总量和用水量；用骨料体积计算砂、石用量。调整水胶比时，保持浆体体积不变。

2.6 根据工程特点和技术要求选择合适的外加剂，用高效减水剂掺量调整拌和物的施工性。

2.7 由于水泥接触水时就开始水化，拌和物的实际密实体积会略小于各材料密度之和，则当未掺入引气剂时，可不考虑搅拌时挟入约1%的空气。

3 混凝土配合比四要素的选择

3.1 水胶比 对有耐久性要求的混凝土，按照结构设计和施工给出《混凝土技术要求》中的最低强度等级，按保证率95%确定配制强度；以最大水胶比作为初步选水胶比，再依次减小0.05～0.1百分点取3～5个水胶比试配，得出水胶比和强度的直线关系，找出上述配制强度所需要的水胶比，进行再次试配。或按无掺和料的普通混凝土强度-水灰比关系选择一个基准水灰比，掺入粉煤灰后再按等浆骨比调整水胶比。一般有耐久性要求的中等强度等级混凝土，掺用粉煤灰超过30%时（包括水泥中已含的混合材料），水胶比宜不超过0.44。

3.2 浆骨（体积）比 在水胶比一定的情况下的用水量或胶凝材料总量，或骨料总体积用量即反映浆骨比。对于泵送混凝土，可按表1选择，或按GB/T 50746—2008《混凝土结构耐久性设计规范》对最小和最大胶凝材料的限定范围，由试配拌和物工作性确定，取尽量小的浆骨比值。水胶比一定时，浆骨比小的，强度会稍低、弹性模量会稍高、体积稳定性好、开裂风险低，反之则相反。

表1 不同等级混凝土最大浆骨比

强度等级	浆体体积（m^3）/浆骨体积比	用水量（kg/m^3）
C30～C50（不含C50）	≤0.32/0.471	≤175
C50～C60（含C60）	≤0.35/0.538	≤160
C60以上（不含C60）	≤0.38/0.613	≤145

3.3 砂石比 通常在配合比中的砂石比，以一定浆骨比（或骨料总量）下的砂率表示。对级配良好的石子，砂率的选择以石子松堆空隙率与砂的松堆空隙率乘积为0.16～0.2为宜。一般，泵送混凝土砂率不宜小于36%，并不宜大于45%。为此应充分重视石子的级配，以不同粒径的两级配或三级配后松堆空隙率不大于42%为宜。石子松堆空隙率越小，砂石比可越小。在水胶比和浆骨比一定的条件下，砂石比的变动主要可影响施工性和变形性质，对硬化后的强度也会有所影响（在一定范围内，砂率小的，强度稍低，弹性模量稍大，开裂敏感性较低，拌和物黏聚性稍差，反之则相反）。

3.4 矿物掺和料掺量 矿物掺和料的掺量应视工程性质、环境和施工条件而选择。对于完全处于地下和水下的工程，尤其是大体积混凝土如基础底板、咬合桩或连续浇注的地下连续墙、海水中的桥梁桩基、海底隧道底板或有表面处理的侧墙以及常年处于干燥环境（相对湿度40%以下）的构件等，当没有立即冻融作用时，矿物掺和料可以用到最大掺量（矿物掺和料占胶凝材料总量的最大掺量粉煤灰为50%，磨细矿渣为75%）；一年中环境相对湿度变化较大（冷天处在相对湿度为50%左右、夏季相对湿度70%以上）无化学腐蚀和冻融循环一般环境中的构件，对断面小、保护层厚度小、强度等级低的构件（如厚度只有10～15cm的楼板），当水胶比较大时（如大于0.5），粉煤灰掺量不宜大于20%，矿渣掺量不宜大于30%（均包括水泥中已含的混合材料）。不同环境下矿物掺和料的掺量选择见GB/T 50746—2008附录B和条文说明附录B。如果采取延长湿养护时间或其他增强钢筋的混凝土保护层密实度措施，则可超过以上限制。

4 混凝土配合比选择的步骤

4.1 确认混凝土结构设计中《混凝土技术要求》提出的设计目标、条件及各项指标和参数：混凝土结构构件类型、保护层最小厚度、所处环境、设计使用年限、耐久性指标（根据所处环境选择）、最低强度等级、最大水胶比、胶凝材料最小和最大用量、施工季节、混凝土内部最高温度（如果有要求）、骨料最大粒径、拌和物坍落度、一小时坍落度最大损失（如果有）。

4.2 根据上述条件选择原材料

4.3 确认原材料条件：

➢水泥：品种、密度、标准稠度用水量、已含矿物掺和料品种及含量、水化热、氯离子含量、细度、凝结时间。

➢石子：品种、饱和面干状态的表观密度、松堆密度、石子最大粒径、级配的比例和级配后的空隙率。

➢砂子：筛除5mm以上颗粒后的细度模数、5mm以上颗粒含量、饱和面干状态的表观密度、自然堆积密度、空隙率、来源。

➢矿物掺和料：品种、密度、需水量比、烧失量、细度。

➢外加剂：品种、浓度（对液体）、其他相关指标（如减水剂的减水率、引气剂的引气量、碱含量、氯离子含量、钾钠含量等）。

4.4 混凝土配合比各参数的确定：

➢各材料符号：

质量：水泥C，矿渣SL，粉煤灰F，砂S，石G，水W，胶凝材料B，浆体P，骨料A，水胶比W/B。

胶凝材料的组成：水泥占胶凝材料的质量百分比α_c，矿渣粉占胶凝材料的质量百分比α_{SL}，粉煤灰占胶凝材料的质量百分比α_F。

密度：水泥ρ_C，矿渣ρ_{SL}，粉煤灰ρ_F，砂ρ_S，石ρ_G，水ρ_W，胶凝材料密度ρ_B。

体积：水泥V_C，矿渣V_{SL}，粉煤灰V_F，砂V_S，石V_G，水V_W，胶凝材料V_B，骨料体积V_A，将体体积V_P，浆骨（体积）比V_P/V_A。

➢按《混凝土技术要求》选取最低强度等级，并按保证率大于95%计算配制强度。

➢根据环境类别和作用等级、构件特点（例如构件尺寸）、施工季节和水泥品种，确定矿物掺和料掺量，根据矿物掺和料掺量，以《混凝土技术要求》的最大水胶比为限，调整水胶比（即水胶比随矿物掺和料掺量增大而减小）。

➢级配骨料，得到最小的骨料松堆空隙率。

➢根据骨料级配和粒形和《混凝土技术要求》中的混凝土强度等级要求的最小和最大浆骨比，以浆体与骨料绝对密实体积最小浆骨比的原则选定浆骨（体积）比，分别用浆体体积中的水胶比计算用水量和胶凝材料总量，用骨料体积中砂石比计算粗细骨料用量。

4.5 混凝土配合比各参数及材料用量的计算

➢按表 1 选定体积浆骨比 V_P/V_A (1)

➢混凝土拌和物总体积为 1m³，则由（1）可知 V_A 和 V_P，按级配骨料所用砂率和砂石表观密度计算砂石用量 S、G：

$$V_P = V_B + V_W \tag{2}$$

➢参考 GB/T 50746—2008 条文说明附录 B，根据环境条件和和构件尺寸确定胶凝材料组成：

$$B = C + F + SL \tag{3}$$

➢计算各材料占胶凝材料的质量百分数为：α_C、α_F、α_{SL} (4)

➢设各材料占胶凝材料的百分数为 β_C，β_F，β_{SL} (5)

➢由式（4）和式（5）以及各自相应的密度，计算胶凝材料的密度：

$$\rho_B = \beta_C\rho_C + \beta_F\rho_F + \beta_{SL}\rho_{SL} \tag{6}$$

➢为了消除难以测定的 β，将 $\beta_C = \dfrac{V_C}{V_B}$、$\beta_F = \dfrac{V_F}{V_B}$、$\beta_{SL} = \dfrac{V_{SL}}{V_B}$ 代入式（6）得：

$$\rho_B = \frac{1}{\dfrac{\alpha_C}{\rho_C} + \dfrac{\alpha_F}{\rho_F} + \dfrac{\alpha_{SL}}{\rho_{SL}}} \tag{7}$$

➢按《混凝土技术要求》选取试配用的最大水胶比 $(W/B)_1$，水的密度近似为 1，由式（7）已知胶凝材料密度为 ρ_B，计算体积水胶比：

$$\frac{V_W}{V_B} = \rho_B(W/B)_1 \tag{8}$$

➢由式（5）和式（7），计算胶凝材料用量 B 和用水量 W。

4.6 试配和配合比的确定

➢在所选用高效减水剂的推荐掺量的基础上，按混凝土的施工性调整为合适的掺量。

➢在《混凝土技术要求》最大水胶比的基础上，依次减小水胶比，选取 3～5 个值，计算各材料用量后进行试配，检测所指定性能指标值，从中选取符合目标值的水胶比，再次进行试配。

根据实测试配结果得出配合比的拌和物密度，对计算密度进行配合比的调整。

5 混凝土配合比选择实例

5.1 技术条件

某滨海城市地下水位为－2m，地下水中硫酸根离子和氯离子含量具有对混凝土结构中等腐蚀程度；商住楼地下两层，底层车库墙体厚度为 350mm，设计使用年限为 70 年，保护层厚度为 50mm，设计强度等级为 C40/P8；混凝土浇筑季节最高气温 33℃，最低气温 21℃。要求施工期间每次连续浇筑 100m³，宽度不大于 0.1mm 的纵向裂缝不多于 3 条。混凝土最大水胶比为 0.45，最小胶凝材料用量最小 320kg/m³，最大 450kg/m³；骨料最大粒径 25mm。混凝土坍落度 180～200mm，到达现场浇筑前坍落度应为 160～180mm。

5.2 技术要点

确认《混凝土技术要求》提供的工程所处环境为V-C级，对处于地下的350mm墙体热天施工来说，应按大体积混凝土考虑，以控制温度应力产生的裂缝为重点。

5.3 原材料选择

水泥：振兴P·O 42.5，已掺入粉煤灰20%，水化热262kJ/kg，密度$3.0g/cm^3$，氯离子含量≯0.6%，标准稠度用水量27%；

粉煤灰：0.045mm筛筛余量17%，密度$2.2g/cm^3$，烧失量4.5%，需水量比103%；

粗骨料：5～10mm和10～25mm以2∶8级配后，表观密度$2.69g/cm^3$，自然堆积密度$1620kg/m^3$，空隙率40%。

细骨料：筛除5mm以上颗粒的河砂，表观密度$2.6g/cm^3$，松堆密度$1432kg/m^3$，空隙率为45%。

外加剂：(略)。

5.4 配制强度的确定（略）

5.5 混凝土配合比参数选择

水胶比：按技术要求最大值选用W/B为0.44；

砂石比：按最紧密堆积原则，根据石子空隙率，选取砂率为40%，则砂石质量比为$S:G=40:60=0.67$；体积比为$0.67\times\rho_G/\rho_S=0.69$；

浆骨比：选择浆骨比为$V_p/V_A=32:68$，则$V_A=0.68m^3$；

粉煤灰掺量：按GB/T 50746—2008《混凝土耐久性设计规范》条文说明附录B，对V-C的环境作用有：

下限：$\frac{\alpha_F}{0.25}+\frac{\alpha_{SL}}{0.4}=1$，上限：$\frac{\alpha_F}{0.5}+\frac{\alpha_{SL}}{0.8}=1$

因拟掺入膨胀剂，为控制混凝土温升，不宜再掺入矿渣粉；则上述限定中，单掺粉煤灰的掺量限定范围为25%～50%，鉴于P·O 42.5水泥中已掺入粉煤灰20%，现选择粉煤灰掺量$\alpha_F=30\%$。

5.6 初步配合比计算

➢由材料条件知$V_A=\frac{G}{2690}+\frac{S}{2600}=0.68$，并$S:G=0.67$

则计算得$G=1080kg/m^3$，$S=724kg/m^3$

➢最大$(W/B)_1=0.44$ (1)'

➢B=C+F，则：

$$\rho_B=\alpha_C\rho_C+\alpha_F\rho_F=\frac{1}{\frac{0.7}{3.0}+\frac{0.3}{2.3}}=2.77 \quad (2)'$$

➢$1m^3$中$V_p/V_A=32:68$，则$V_p=0.32m^3$

$$V_B+V_W=V_p=0.32 \quad (3)'$$

$$\frac{V_W}{V_B}=\rho_B(W/B)_1=2.77\times0.44=1.22 \quad (4)'$$

由(3)'(4)'：$W=176kg/m^3$，$B=400kg/m^3$，由设定的粉煤灰掺量为30%，得知粉

煤灰用量为 120kg/m³。

5.7 初试配合比（表 2）

表 2 初试配合比

水	$(W/B)_1$	P·O 42.5 普通水泥	粉煤灰	砂	石	高效减水剂
176	0.44	280	120/30%	724	1080	略

5.8 试配、调整（略）

6 改变水胶比时计算混凝土配合比的等浆体体积法举例

按第 5 节的步骤另外分别计算出 $(W/B)_2=0.42$、$(W/B)_3=0.40$ 的配合比，与 $(W/B)_1=0.44$ 一起，共取得 3 组 (B/W)- 性能关系，以备优选生产配合比。

改变水胶比后浆体量发生变化，会影响到施工性，应按等浆体体积进行调整如表 3 所示。

调整说明：

- 水胶比 0.42 时，浆量减小了 8 升，可能影响施工性，如增加 9kg 粉煤灰和 4 kg 水，则浆体体积可增加到 323 升，可视为不变；
- 水胶比 0.40 时，浆量减小了 16 升，如增加 17kg 粉煤灰和 7kg 水，则浆体量增加到 323 升，可视为不变；
- 调整后质量水胶比不变，浆骨体积比不变，则砂石用量可不作调整，施工性不受影响；
- 如果调整水胶比后浆骨比减小，则拌和物体积会不足，从而影响施工性，可按新调整的水胶比增加浆量（即同时增加水和水泥用量）骨料用量而不改变浆骨比；
- 尽管浆骨比不变，而浆体浓度可能有变化，可视胶凝材料的需水性，调整减水剂用量。

表 3 改变水胶比后按等浆体体积进行调整配合比的计算

<table>
<tr><td colspan="4">原材料</td><td>胶凝材料
(B)</td><td>水
(W)</td><td>浆体数量</td><td>W/B</td><td>砂</td><td>石</td></tr>
<tr><td colspan="4">密度（g/cm³）</td><td>2.7</td><td>1</td><td>—</td><td>—</td><td>2.60</td><td>2.69</td></tr>
<tr><td rowspan="2">原配合比</td><td rowspan="2" colspan="2">$\left(\frac{W}{B}\right)_1$</td><td>用量（kg/m³）</td><td>400</td><td>176</td><td>576</td><td>0.44</td><td>724</td><td>1080</td></tr>
<tr><td>体积（m³/m³）</td><td>0.148</td><td>0.176</td><td>0.324</td><td></td><td>0.278</td><td>0.401</td></tr>
<tr><td rowspan="8">改变水胶比后的配合比</td><td rowspan="4">$\left(\frac{W}{B}\right)_2$</td><td rowspan="2">计算</td><td>用量（kg/m³）</td><td>400</td><td>168</td><td>568</td><td>0.42</td><td></td><td></td></tr>
<tr><td>体积（m³/m³）</td><td>0.148</td><td>0.168</td><td>0.316</td><td></td><td></td><td></td></tr>
<tr><td rowspan="2">调整</td><td>用量（kg/m³）</td><td>409</td><td>172</td><td>581</td><td>0.42</td><td></td><td></td></tr>
<tr><td>体积（m³/m³）</td><td>0.151</td><td>0.172</td><td>0.323</td><td></td><td></td><td></td></tr>
<tr><td rowspan="4">$\left(\frac{W}{B}\right)_3$</td><td rowspan="2">计算</td><td>用量（kg/m³）</td><td>400</td><td>160</td><td>560</td><td>0.40</td><td></td><td></td></tr>
<tr><td>体积（m³/m³）</td><td>0.148</td><td>0.160</td><td>0.308</td><td></td><td></td><td></td></tr>
<tr><td rowspan="2">调整</td><td>用量（kg/m³）</td><td>417</td><td>167</td><td>584</td><td>0.40</td><td></td><td></td></tr>
<tr><td>体积（m³/m³）</td><td>0.156</td><td>0.167</td><td>0.323</td><td></td><td></td><td></td></tr>
</table>

7 改变矿物掺和料掺量时计算配合比的等浆体体积法举例

如果已有一无掺和料的硅酸盐水泥混凝土的配合比，当掺入粉煤灰后，需用等浆体体积法，保持原配比的浆骨比不变，以保持混凝土的体积稳定性。

假定混凝土原配合比如表中所示，掺入粉煤灰30%，按粉煤灰特性，掺入粉煤灰的混凝土水胶比必须不大于0.5。计算配合比步骤见表4。

表4　等浆体体积法计算过程举例

原材料		水泥（C）	水（W）	FA（30%）	浆体数量	W/B	砂	石
密度（g/cm³）		3.10	1	2.4	—	—	2.61	2.67
原配合比	用量（kg/m³）	347	198	0	545	0.57	681	1151
	体积（m³/m³）	0.112	0.198	0	0.310		0.261	0.431
简单等量取代掺入粉煤灰	用量（kg/m³）	243	198	104	545			
	体积（m³/m³）	0.078	0.198	0.043	0.319			
掺粉煤灰后保持浆骨比计算	用量（kg/m³）	243	188	104	545	0.54	681	1151
	体积（m³/m³）	0.078	0.188	0.043	0.310		0.261	0.431
按耐久性要求水胶比为0.44	用量（kg/m³）	243	153	104	500	0.44		
	体积（m³/m³）	0.078	0.153	0.043	0.274			
按保持原浆骨比调整	需增加体积（m³/m³）	0.010	0.020	0.006	0.036			
	用量（kg/m³）	273	173	118	564	0.44	681	1151
	体积（m³/m³）	0.088	0.173	0.049	0.310		0.261	0.431

计算说明：

- 实测各原材料密度计算1 m³中原浆体体积：$V_p = V_c + V_w = 0.112 + 0.198 = 0.310\text{m}^3$；
- 掺入粉煤灰30%等量取代水泥后：

粉煤灰用量为347×0.3=104kg，实测粉煤灰密度为2.4g/cm³则浆体体积为104/2400=0.043m³；

水泥用量为347－104＝243kg，实测水泥密度为3.1g/cm³，则体积为243/3100＝0.078 m³；

- 浆体体积为：0.078+0.043+0.198=0.319 m³；
- 要保持浆体体积仍为0.310 m³不变，需减水0.01 m³，用水量从198 kg/m³减为188 kg/m³，则水胶比应为188/347=0.54；掺粉煤灰的混凝土水胶比应不大于0.5，并随粉煤灰掺量的增加而降低，现掺量为30%时，按耐久性要求设水胶比为0.44，用水量为347×0.44=153 kg。浆体体积为0.153+0.078+0.043=0.274 m³，则浆体体积不足。
- 为保持原浆骨比，需增加浆体0.036m³。按浆体中原比例调整，增加水（0.036/0.274）×0.153＝0.02m³，增加水泥（0.036/0.274）×0.078＝0.010 m³，增加粉煤灰（0.036/0.274）×0.043=0.006m³。

计算结果：

掺30%粉煤灰的混凝土配合比计算结果见表5。

表5　掺30%粉煤灰的混凝土配合比计算结果

材料		水泥	水	粉煤灰	浆体总量	水胶比	砂	石	拌和物表观密度
用量	质量（kg/m³）	273	173	117	560	0.44	681	1151	2392
	体积（m³/m³）	0.088	0.173	0.049	0.310	0.44	0.261	0.431	1

结果，胶凝材料总用量从347kg/m³增加到390kg/m³，但因用水量减少，故浆体体积不变，即浆骨比保持不变。

8　无论是经过优选还是经过调整得出的配合比，都必须再经过试拌。

参考文献

[1]　ACI TAC(Technical Activities Committee). Definition on High Performance Concrete,《CONCRETE INTERNATIONAL》, Feb. 1999.

[2]　廉慧珍，李玉琳．关于混凝土配合比选择方法的讨论之一：当前混凝土配合比“设计”存在的问题[J]. 混凝土，2009，第三期．

自评

1. 混凝土所使用的原材料会是是经常变化的，同一工程结构的不同部位可能处于不同环境，达到相同强度等级的混凝土拌和物可能会有不尽相同的配合比，论文之十四针对当前混凝土配合比设计中的问题，提出现代混凝土拌和物配合比设计的原则，目的是保证混凝土结构经久的安全性：①用体积法计算混凝土配合比；②选购单粒级粗骨料在实验室进行 2～3 级级配试验，以得到最大松堆密度的级配，生产时分级上料；③计算砂石体积使用饱和面干状态的密度；④在确定的水胶比下，采用最小的浆骨比；⑤掺入矿物掺和料时按等浆体体积调整配合比；⑥改变砂率时，保证浆骨比不变。

2. 论文之十五给出按上述原则进行配合比设计的一种方法和实例，仅供参考。各人都可以在合理的原则下，用自己的专业知识和经验，发挥自己的才能，试配出符合工程要求的拌和物。文中几次提到，按照结构设计要求的最低水胶比，上下变动，经试配确定拌和物满足强度要求的水胶比。如果设计文件未给出最低水胶比，只规定强度等级，则可酌情选择 3～5个水胶比进行试配。现代混凝土一般都掺用矿物掺和料，则常用强度等级混凝土的水胶比一般为 0.3～0.5 之间，选择范围并不大，强度等级 C40 上下的混凝土水胶比视矿物掺和料掺量，在 0.4 左右很容易选择。

3. 美国 ACI 曾花了很大力量，把“混凝土配合比设计”改成“混凝土拌和物配合比设计”，更具有科学性，因为我们的设计的确实是拌和物试配时各组分的比例。这个比例决定在经合理工艺成型并硬化后混凝土的性质。硬化后混凝土组分就发生变化，成为在细观以下各尺度上高度复杂的非均质多相体，不是我们能确切设计的出来的。

4. 混凝土拌和物配合比设计实际上就是根据工程要求选择原材料后各材料用量比例的确定，其中的核心是矿物掺合料掺量、水胶比、浆骨比、砂石比四要素的确定。其中水胶比如结构设计中没有最大水胶比的规定，对 C30 以上混凝土，可在常用的 0.3～0.5 取 3～5 个水胶比进行试配，做出强度—胶水比直线为参考，再继续试配确定。

宋少民评

第 3 节的 3. 4 中关于需控制掺和料掺量的一般环境条件的描述，语句欠通顺，建议修改如下：“对于一年中处于环境相对湿度变化较大（冷天处在相对湿度为 50%左右、夏季相对湿度 70%以上）无化学腐蚀和冻融循环的一般环境中的构件，对且断面小、保护层厚度小、强度等级低的构件（如厚度只有 10～15cm 的楼板），当水胶比较大时（如大于 0.5），粉煤灰掺量不宜大于 20%，矿渣掺量不宜大于 30%（均包括水泥中已含的混合材料）。”

论文之十六

关于混凝土配合比选择方法的讨论之三
——体积法计算混凝土配合比中原材料密度的问题*

廉慧珍（清华大学土木水利学院）
李玉琳（北京中关村建设集团中宏基公司）

摘　要　接受新疆农业大学葛毅雄教授的意见，《关于混凝土配合比选择方法的讨论》一文中体积法计算混凝土配合比，用加权法计算胶凝材料密度应当使用体积分数。用质量分数和体积分数计算值之差，随粉煤灰掺量的增大而增大，尽管只有 10^{-4} 之差，而从基本概念上是应当纠正的。饱和面干骨料所含的水，既不对拌和物和易性起作用，在混凝土硬化前也不参与水泥水化，而是属于骨料表观体积的一部分，因此骨料的密度必须使用饱和面干状态的测定值。河砂和人工砂的饱和面干含水率一般在0.8%～1.0%的范围。

关键词　混凝土配合比；体积法；加权法；胶凝材料密度；体积分数；饱和面干的骨料

用体积法计算混凝土配合比的重要参数是各原材料的密度。《混凝土》2009 年第 5 期刊登拙文《关于混凝土配合比选择方法的讨论》后，收到新疆农业大学葛毅雄教授来函，指出其中胶凝材料密度计算公式的错误。葛教授认真负责的精神令人敬佩，在此表示衷心感谢，并对原文中的错误进行更正。同时，在一年多的实践中，发现有的技术人员对使用“饱和面干骨料的表观密度”尚有不理解之处，并认为“难以操作”而有疑义。用体积法计算混凝土配合比的必要性是当前混凝土组成特点所决定的，混凝土是一种高度非均质的复杂体系，其复杂性首先就缘于原材料的波动性，因此在混凝土生产中除了需要控制原材料的稳定来源之外，希望在配合比计算过程中尽量消除和减少其他一些不确定的因素，其中主要的是骨料的密度，在此一并予以说明。

1　胶凝材料密度的计算

1.1　葛教授的意见摘要

……文中式（5）为

$$\rho_B = \alpha_C\rho_C + \alpha_F\rho_F + \alpha_{SL}\rho_{SL}$$

似与密度的基本概念相悖。

由于 α_C、α_F、α_{SL} 分别是水泥、粉煤灰、矿渣粉占胶凝材料的质量比，则胶凝材料的密度应按下式计算，才合理和精确：

$$\rho_B = \frac{1}{\dfrac{\alpha_C}{\rho_C} + \dfrac{\alpha_F}{\rho_F} + \dfrac{\alpha_{SL}}{\rho_{SL}}}$$

1.2　对葛教授所提供公式的推导验证

* 原载《混凝土》2010.9。

设 1m^3 混凝土中胶凝材料总量为 G_B，计算所用的各符号如表 1 所示。

表 1　计算胶凝材料密度的各符号

项目	1m^3 混凝土中胶凝材料			
	总量	水泥	粉煤灰	矿渣粉
质量	G_B	G_C	G_F	G_{SL}
质量比	1	$\alpha_C=\frac{G_C}{G_B}$	$\alpha_F=\frac{G_F}{G_B}$	$\alpha_{SL}=\frac{G_{SL}}{G_B}$
体积	V_B	V_C	V_F	V_{SL}
体积比	1	$\beta_C=\frac{V_C}{V_B}$	$\beta_F=\frac{V_F}{V_B}$	$\beta_{SL}=\frac{V_{SL}}{V_B}$
密度	$\rho_B=\frac{G_B}{V_B}$	$\rho_C=\frac{G_C}{V_C}$	$\rho_F=\frac{G_F}{V_F}$	$\rho_{SL}=\frac{G_{SL}}{V_{SL}}$

胶凝材料总量 $G_B=G_C+G_F+G_{SL}$

胶凝材料总体积 $V_B=V_C+V_F+V_{SL}$

胶凝材料密度

$$\begin{aligned}\rho_B&=\frac{G_B}{V_B}\\&=\frac{G_B}{V_C+V_F+V_{SL}}\\&=\frac{G_B}{\frac{G_C}{\rho_C}+\frac{G_F}{\rho_F}+\frac{G_{SL}}{\rho_{SL}}}\end{aligned} \tag{1}$$

将式（1）的分子和分母分别除以 G_B，则

$$\rho_B=\frac{1}{\frac{\alpha_C}{\rho_C}+\frac{\alpha_F}{\rho_F}+\frac{\alpha_{SL}}{\rho_{SL}}} \tag{2}$$

证毕。该公式成立。

1.3　用加权法计算密度的校验

在《关于混凝土配合比选择方法的讨论》中，按加权法，各组分对体系密度的贡献与该组分在体系中所占份额成正比；所以体系密度等于组成部分对体系密度贡献之和。

假设体系含有三组分，令其在体系中的份额（含量百分数）各为 a，b，c，则混合后的体系密度应表示为：

$$\rho_B=a\rho_C+b\rho_F+c\rho_{SL}$$

式中 a、b、c 为百分比，在此计算胶凝材料密度时，应当是体积比，故有：

$$\rho_B = \beta_C\rho_C + \beta_F\rho_F + \beta_{SL}\rho_{SL} \tag{3}$$

按式（3）推导，可得出同样结果，如下：

$$\begin{aligned}\rho_B &= \beta_C\rho_C + \beta_F\rho_F + \beta_{SL}\rho_{SL}\\ &= \frac{V_C}{V_B}\cdot\rho_C + \frac{V_F}{V_B}\cdot\rho_F + \frac{V_{SL}}{V_B}\cdot\rho_{SL}\\ &= \frac{1}{V_B}\left(\frac{G_C}{\rho_C}\cdot\rho_C + \frac{G_F}{\rho_F}\cdot\rho_F + \frac{G_{SL}}{\rho_{SL}}\cdot\rho_{SL}\right)\\ &= \frac{G_C + G_F + G_{SL}}{V_B} = \frac{G_B}{V_B}\end{aligned} \tag{4}$$

式（4）和式（1）完全相同，按式（1）到式（2）的推导，最后也回到与式（2）相同的式（5）：

$$\rho_B = \frac{1}{\dfrac{\alpha_C}{\rho_C} + \dfrac{\alpha_F}{\rho_F} + \dfrac{\alpha_{SL}}{\rho_{SL}}} \tag{5}$$

该式与葛教授所提供的公式相同，从基本原理上无误。

1.4 原文公式（5）的问题

原文中的公式

$$\rho_B = \alpha_C\rho_C + \alpha_F\rho_F + \alpha_{SL}\rho_{SL}$$

是符合加权法的，但是，错误出在系数 α_C、α_F、α_{SL} 使用了质量比。因为科学的配合比设计使用体积法，所以此处应为体积比，现设定体积比使用的符号水泥为 $\beta_C = \dfrac{G_C}{V_C}$、粉煤灰为 $\beta_F = \dfrac{G_F}{V_F}$、矿渣粉为 $\beta_{SL} = \dfrac{G_{SL}}{V_{SL}}$，则

$$\rho_B = \beta_C\rho_C + \beta_F\rho_F + \beta_{SL}\rho_{SL}$$

在实用上，α_C、α_F、α_{SL} 的值是现成的，而 β_C、β_F、β_{SL} 的计算就比较麻烦。从式（3）推导到式（5），就消除了体积比的系数 β，用质量比系数 α，则可方便于计算。

用体积比和用质量比计算结果的差异随掺和料的掺量而不同，现举例说明：

为计算方便，假设只掺粉煤灰，胶凝材料总量 G_B 为 400kg/m³，粉煤灰质量 α_F 分别为 25%和 50%，则水泥用量 G_C 分别为 300kg/m³ 和 200kg/m³，粉煤灰 G_F 分别为 100kg/m³ 和 200kg/m³，水泥密度 ρ_C 为 3.0g/cm³，粉煤灰密度 ρ_F 为 2.2g/cm³。胶凝材料体积 V_B 为：

$$V_B = V_C + V_F = \frac{G_C}{\rho_C} + \frac{G_F}{\rho_F}$$

$$\beta_F = \frac{V_F}{V_B}$$

从表 2 可见，粉煤灰掺量越大，α 和 β 差别越大，但是相对差别越小。

表 2 胶凝材料总量不变改变粉煤灰掺量时按质量比和按体积比计算胶凝材料密度的差值

粉煤灰质量份额 α_F（%）	粉煤灰质量 G_F（kg）	水泥质量份额 α_C（%）	水泥质量 G_C（kg）	按质量比 ρ_B	粉煤灰体积 V_F（L）	粉煤灰体积份额 β_F（%）	水泥体积 V_C（L）	水泥体积份额 β_C（%）	胶凝材料体积 V_B（L）	按体积比 β_B	质量比 ρ_B 和体积比 β_B 的差值
25	100	75	300	2.8	45.4	31.2	100.0	68.8	145.4	2.75	0.05
50	200	50	200	2.6	90.9	57.7	66.7	42.3	157.6	2.53	0.07

由表 2 可见：

①掺入粉煤灰后胶凝材料体积增大，掺粉煤灰 25%，胶凝材料体积增大 45.4L；掺 50%时增大 90.9L，因此使用矿物掺和料的混凝土必须使用体积法进行计算，否则会影响混凝土体积的变化，也会因浆体体积的变化而导致体积稳定性问题，进而影响混凝土结构的耐久性；按体积法计算时，即可调整浆体体积不变。调整的方法在拙文《关于混凝土配合比选择方法的讨论》中已有介绍。

②计算胶凝材料密度，按体积比计算比按质量比计算的小，粉煤灰掺量越大，差值越大；掺和料密度越小，差值越大。《关于混凝土配合比选择方法的讨论》原文所举一例中粉煤灰（密度为 2.3g/cm^3）掺量假设为 30%，用改正的公式计算密度值减小了 0.02 个百分点。尽管只有微小之差，而从基本概念上是应当纠正的。

想要避免这样计算的麻烦，最好的办法就是矿物掺和料由水泥厂掺，其密度直接可测。当然也可以在试配混凝土时，预先将胶凝材料配好后直接测其密度，只是每次改变组分时都需要重配、重测，会增大工作量，因此建议按本文公式（5）计算。

2 关于骨料密度

2.1 为什么体积法计算混凝土配合比所用骨料密度要用饱和面干状态进行检测

用作混凝土骨料的天然砂石属于多孔材料，骨料颗粒的自然体积包括其中的封闭孔和水能进入而在平衡湿度下不再出来的那些与外界连通的孔体积。使这些孔都充满水而表面没有附着水时就可测出骨料的自然体积。在混凝土的料浆中，含水饱和而面干的骨料中所含的水不会出来，既不参与水泥水化，也不参与混凝土的拌和；也就是说，既不改变水胶比，也不改变拌和物用水量。

以砂子为例，当前我国建筑工程用混凝土，在试配混凝土时，用假定密度法计算混凝土用水量，搅拌前先取砂样烘干或炒干，检测其含水率，然后在拌和水中扣除，再增加与所扣除水等量的砂：

调整的拌和水 = 计算拌和水量 − 烘干砂样测出的含水率 × 用砂量

调整的用砂量 = 计算用砂量 + 与所扣除水等量的砂

如果所检测的砂含水不饱和，则用这样调整的水拌和混凝土时，实际上砂子还要吸水，致使拌和水量不足，拌和物坍落度会减小，又需要再补充水，却无法知道需要补充多少水，只能一点、一点地试；如果所检测的砂子不仅含水饱和，而且还附有表面水，则根据砂子含水率调整拌和水后，就会多扣除砂子本来不参与拌和的那部分饱和面干所含的水，使调整更加麻烦。这样的控制就会使搅拌结束后的拌和物用水量变得不可知。实际上，在生产中调整用水量所需要知道的是骨料吸水率（等于饱和面干含水率），而不是骨料总的含水率，因此骨料的密度应当使用吸水饱和而面干状态下的表观密度。国外和我国的水利水电工程的混凝土都是这样配制的。

在日本，购买骨料都要求含水饱和以上，但水不能流淌，否则要被罚款。这样，在使骨料饱和面干时，即可用烘、吹等办法去除表面水，比对干燥骨料加湿处理操作方便。生产中所需要了解的是骨料饱和面干以外的含水率，求出骨料饱和面干体积（可称作标准体积）后，骨料实际的含水率便可从二者体积差得知，差值为正时，骨料含有表面附加水；差值为负时，则骨料不饱和，可在生产中多退少补；但是为了便于生产控制，当购买水洗过的骨料时，进料时含水都超过饱和，则与标准体积的差值均为正，生产投料过程用传感器在线监测骨料含水率所测出的只是骨料饱和面干以外的表面附加水值，直接传输到调水系统进行控制。如此可大大提高生产质量控制的水平。

2.2 如何界定骨料的饱和面干状态

有一定规模并生产规范砂石场供应的骨料，因需用水冲洗而往往都含有不同量的水。石子的饱和面干较容易界定，按标准方法，将石子浸没于水中 24 小时以上，取出后用拧干的湿毛巾蘸去表面水（使骨料表面不泛亮光为准）即可用排水法求出其自然体积；对于砂子，当购进的砂不含水或含水不饱和，可按砂石标准中检测砂子吸水率的方法检测饱和面干含水率——砂子的吸水率在数值上与饱和面干含水率相同；当前国内很多供应的砂子多为含水过量的湿砂，这实际上便于检测饱和面干含水率。

检测砂子饱和面干状态的模具和砂子典型的饱和面干状态见图 1 和图 2 所示。图 3 为砂子含水的三种基本状态。图 4 为含不同水率时砂子的状态及不同密度的砂含水率不同的状态。

图 1　检测砂子饱和面干状态的模具

图 2　砂子饱和面干状态

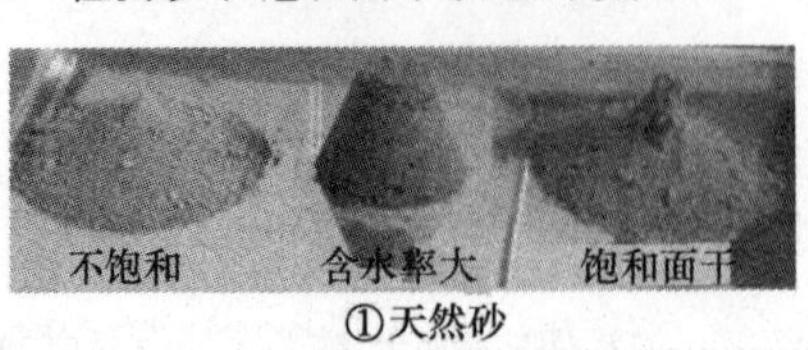

①天然砂

②人工砂

图 3　砂子三种基本含水状态

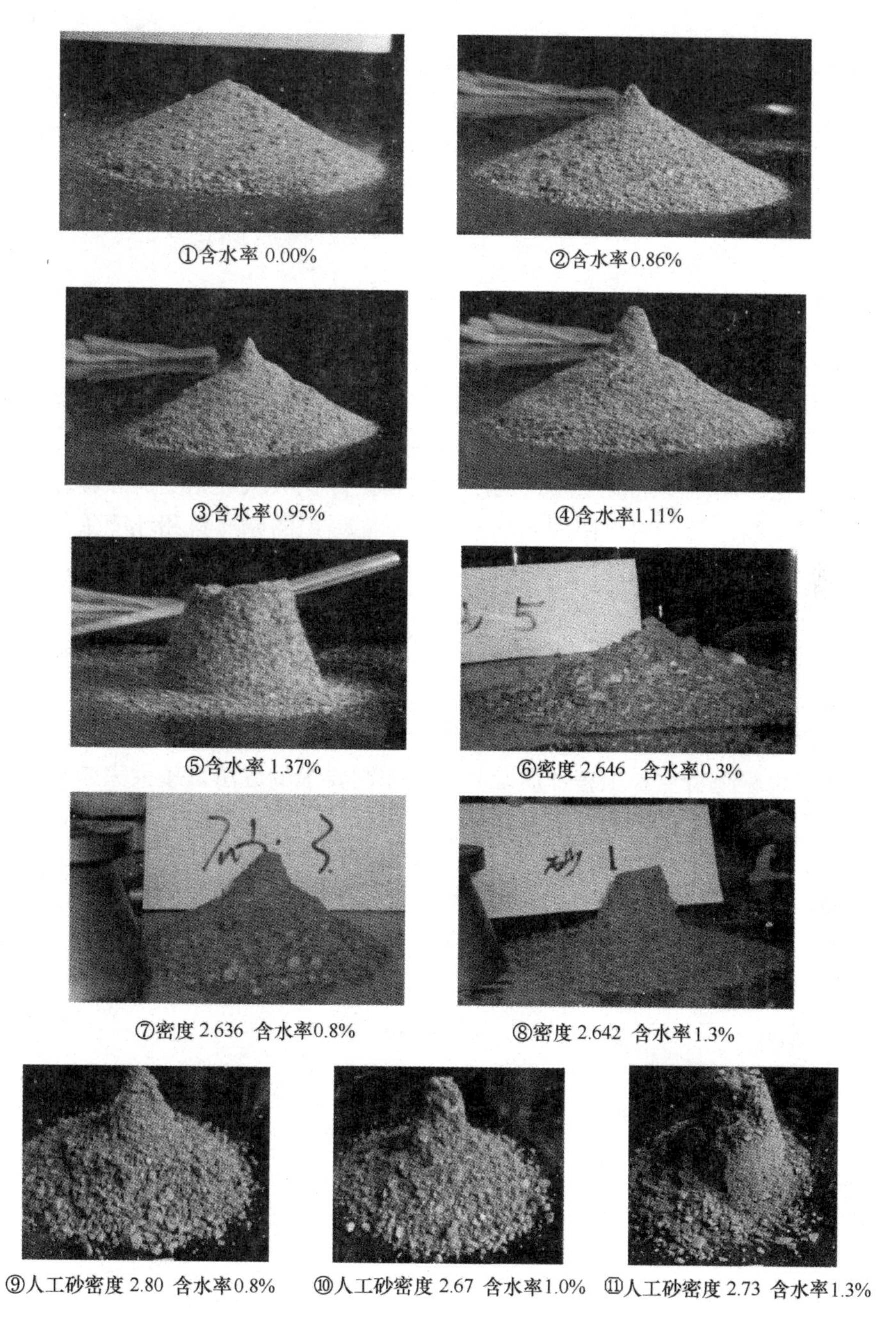

①含水率 0.00%　②含水率 0.86%

③含水率 0.95%　④含水率 1.11%

⑤含水率 1.37%　⑥密度 2.646　含水率 0.3%

⑦密度 2.636　含水率 0.8%　⑧密度 2.642　含水率 1.3%

⑨人工砂密度 2.80　含水率 0.8%　⑩人工砂密度 2.67　含水率 1.0%　⑪人工砂密度 2.73　含水率 1.3%

图 4　不同含水率砂子的状态及不同密度的砂含水率不同时的状态

因砂子总有些小于 0.075mm 的细颗粒，故只要含水，就可能在圆锥形堆的顶端出现一个附加的尖锥，这个尖锥的大小既反映含水量的大小，也反映含泥量的大小；由图 4 中的②、③、④可见，砂子饱和面干的含水率为 0.8%～1.0%，在此范围内波动影响很小，属于正常；当出现⑤的形状时，无疑远超出饱和面干。⑥、⑦、⑧是含粗颗粒较多的粗砂，密度在 2.60g/cm^3 以上，其规律也相同：⑥含水不饱和（0.3%），⑧含过量水（1.3%），⑦饱和面干（0.8%）；⑨、⑩、⑪人工砂的密度较大，为 2.67g/cm^3 以上，饱和面干状态的含水率也是0.8%～1.3%。

完全干燥的砂子在大气中自然堆放时，会吸收空气中的水分而达到与大气湿度相平衡，其含水率称做气干含水率，气干含水率基本上不超过0.5%。原始进料的砂，无论是破碎生产的人工砂，还是筛分的天然河砂，一般含水率为1%～3%，对检测饱和面干含水率是较方便的。当前北京的砂含水率多的也会达5%以上，甚至会有淌水的砂，从环保的角度，是应当禁止的。

3 结束语

当前在混凝土中使用外加剂和矿物掺和料已成为普遍，我国大规模基础设施建设正在大量消耗资源，从可持续发展的角度，大掺量矿物掺和料的使用是必然的，而且由于资源的日益短缺，开发新的资源也是需要的，那么，如继续沿用在混凝土组分较简单情况下所使用的传统配比计算方法已不利于混凝土生产的质量控制。体积法从理论上仍然合理，至少在实验室第一盘试配时应当采用。由于使用量巨大，受不能提纯的原材料和环境条件的影响，混凝土成为用最简单的工艺制作的最复杂的人工材料体系，努力加强基础性研究，尊重一线技术人员的专业经验，不断提高对混凝土的认识，这是混凝土技术人员负责任的行为，不宜制订“混凝土配合比设计”规范去束缚人们的创造力。然而，从满足结构工程耐久性和各项物理力学性能及施工质量的要求出发，则应当制订标准规定“混凝土配合比设计”的原则，使技术人员在正确原则要求下发挥各自的经验和专业技能，而不是具体限定“设计”的步骤。体积法是适应当前混凝土特点和工程性质及其质量控制要求的方法。望大家参与讨论。

说明：收到葛教授意见，我要求《混凝土》编辑部刊登我的答复和更正，编辑部要求我另写一篇文章，故撰此文，感谢《混凝土》编辑部予以刊登，给了我一个接受意见并更正的机会。

人类的全部历史都告诫有智慧的人，不要笃信时运，而应坚信思想。

——爱默生

论文之十七

评《普通混凝土配合比设计规程》(JGJ 55—2011)
——兼谈如何认识和编制及使用技术标准和规范*

廉慧珍（清华大学土木水利学院）

《普通混凝土配合比设计规程》(JGJ 55—2011) 不久前修订完毕并在2011年12月1日实施了。10多年来，我国混凝土技术和状况发生了巨大的变化，主要是因外加剂与矿物掺和料的普遍使用，使混凝土进入大不同于传统混凝土的现代混凝土时期，人们的思维方法和观念也随之发生了变化。例如，对水泥和混凝土强度的看法，过去大多数人都认为实验室检测出来的就是真值，混凝土强度不可能超过水泥的强度。1974年清华大学土木系应特殊工程需要承担过“1000＃混凝土”的研究项目，当时有位从事钢筋混凝土课程教学的教授问：“现在水泥标号最高的是500＃（折合为现在的水泥强度等级约为32.5)，怎么能做出1000＃（相当于现在的近C100)?”由于使用减水剂，1995年用当时的42.5＃水泥（约相当于现在的32.5水泥）已经配制出28天抗压强度达71MPa的泵送混凝土。实践逐渐使人们认识到，使用了高效减水剂后，现代混凝土的强度能超过混凝土中水泥的强度，其原因就是混凝土强度和水泥强度所使用的不是同一个标准检测方法。且不说检测所用试件型式和尺寸，成型所用水灰（胶）比的区别，更本质的是水泥强度检测使用的水灰比一律为不变的0.5，而现在的混凝土水胶比一般都低于0.5。也就是说，现代混凝土的强度已经不再完全依赖于水泥的强度，或者说，因为不使用同一标准方法检测，在普通混凝土范围内，就不能简单地将二者建立起什么定量的关系。这一点，十几年来，人们已经越来越明白，所以也就越来越没有人再去使用《规程》，如果说还有人用，也是为了应付施工（或质检部门）的要求，在配制混凝土后，再反推回去给出一个算式配比而已。然而，在全国范围内，现在仍有不少人（尤其是新手）“不知混凝土该怎样配”，其实这也是正常现象，就像现在工民建专业刚毕业的学生“连一根梁都不会设计”一样。在学校工程教育不足的今天，需要招工企业对刚毕业的大学生进行继续教育。工程技术问题尤其是对混凝土工程这样非常复杂的体系，是需要“传、帮、带”的。遗憾的是现在很多进行“传、帮、带”的人在当前社会大变革的形势下，思维方法和观念却仍然滞后。技术标准和规范的作用就是要在方法和观念上作引领，需要其具有先进性的意义就在于此。就此而言，仔细阅读现在这本修订的新《规程》（以下简称新《规程》)，感到其中确实存在若干问题。现就其中几个重点进行分析：

1　关于“混凝土配合比设计的原则”

什么是混凝土配合比设计的原则？对混凝土配合比拌和物的设计应当规定什么样的原则？矿物掺和料最大掺量限定的目的是什么？限定值和根据是什么？

---

* 原载《混凝土世界》2012.3。

要给出“混凝土配合比设计的原则”，这很重要。问题是什么是原则？给出什么样的原则？

原则的意思是行为的原理和准则，不涉及可以根据技术人员或管理人员的学识和经验能调整的技术和具体方法。例如 ASTM Designation C 1157《Standard Performance Specification for Hydraulic Cement》（水硬性胶凝材料的标准性能规范）的原则是“给出性能要求，而没有对胶凝材料成分或组成的限制”。我国当前尚不可能提出这样的原则，原因是当前我国仍以强度第一为原则，而强度是容易满足的，如果不规定胶凝材料组成和配比，就会更加混乱。从新《规程》各条文来看，其中的“基本规定”应当属于原则性的规定。然而原则既然是行为的准则，其条文必须符合科学逻辑，用词上就要严格，用“应”而不能用“宜”。

1.1 关于矿物掺和料

新《规程》总则第 3.0.5 条规定的第一句话是：“矿物掺和料在混凝土中的掺量应通过试验确定。”（根据什么进行试验？这涉及掺和料使用的目的。）这是一项原则，但在后面都使用的是“宜”，也就是说，这又并不是一个原则，但却列入了“基本要求”中（基本原则是必须遵守的）。

现在先分析一下《规程》中的表 2 和表 3“钢筋混凝土中规定的掺和料最大掺量”。为了分析的方便，按水泥现行标准规定的 P·Ⅰ硅酸盐水泥、P·Ⅱ硅酸盐水泥和 P·O 硅酸盐水泥中混合材最大含量分别为 0%、5%和 20%分别计入计算矿物掺和料总量，并分别列于第 4 列～第 7 列和第 10 列～第 11 列。将新《规程》中表 2、表 3 中的原规定的最大掺量值列于第 2 列、第 3 列和第 8 列、第 9 列（见表 1）。

表 1　对新《规程》中表 2、表 3 的分析

矿物掺和料种类	水胶比	最大掺量/%									
		采用硅酸盐水泥						采用普通硅酸盐水泥			
		允许掺量/%		计算掺粉煤灰后掺和料总量/%*				允许掺量/%		计算掺粉煤灰后掺和料总量/%	
		钢筋砼	预应力	钢筋砼		预应力		钢筋砼	预应力	钢筋砼	预应力
				P·Ⅰ	P·Ⅱ	P·Ⅰ	P·Ⅱ				
粉煤灰	≤0.4	45	35	45	50	35	40	35	30	55	50
	>0.4	40	25	40	45	25	30	30	20	50	40
粒化高炉矿渣粉	≤0.4	65	55	65	70	55	60	55	45	75	65
	>0.4	55	45	55	60	45	50	45	35	65	55
钢渣粉	—	30	20	30	35	20	25	20	10	40	30
磷渣粉	—	30	20	30	35	20	25	20	10	40	30
硅灰	—	10	10	10	15	10	15	10	10	30	30
复合掺和料	≤0.4	65	55	65	70	55	60	55	45	75	55
	>0.4	55	45	55	60	45	50	45	35	65	55
列号	1	2	3	4	5	6	7	8	9	10	11

*此处对硅酸盐水泥未规定型号。

根据表 1 分析提出问题如下：

①表中所给出的矿物掺和料最大掺量已属于“大掺量”范畴，但不知该表给出的限制有何根据？间隔 5 个百分点这样小的变动影响有多大？根据什么原理这样规定？第 4 列与第 5

列相比，第6列和第7列相比，第4列～第7列和第10列～第11列相比，水泥中混合材越多的，在混凝土中允许矿物掺和料总量倒越多。为什么要做这样的规定？理论上和实践上都无法解释。如果有实验依据，不知有多少数据支持？

②从表2和表3的编排来看，不知以水胶比0.4为界的原理是什么？现在业内人士基本上都已接受M. R. H. Dunstan关于“粉煤灰是混凝土的第四组分”[1]的观念，并且知道存在着尚不能普适化的水胶比-粉煤灰-强度三维的关系，也就是说当强度要求一定时，应当根据粉煤灰-强度的关系选择水胶比。而粉煤灰掺量是由混凝土服役环境耐久性要求、施工性能等来选择的。由于粉煤灰对水胶比更加敏感，掺用粉煤灰的混凝土水胶比必须随粉煤灰掺量而降低。按照新《规程》，在水胶比＞0.4的各个水灰比下，矿物掺和料最大掺量相对于水胶比≤0.4的混凝土，只减少5个百分点，而且固定不变，亦即可理解为水胶比大到0.6以上都只能限于该值。这样的规定对中等强度等级的结构混凝土是没有意义的，因为会影响矿物掺和料的贡献[1]；而对低强度等级的垫层之类用的C15～C20混凝土，更没有必要限定矿物掺和料的上限。因为现今的水泥强度都偏高（最低强度等级的水泥的最低强度也都超过用0.5水胶比检测的35MPa），当水胶比约达0.54时，粉煤灰对混凝土强度的贡献就为0了。于是要么因水胶比较低而强度多半会超标较多，要么增大水胶比而使拌和物严重离析、泌水。因此矿物掺和料掺量不能人为地去规定，试配最能解决问题。实际上，低强度等级混凝土中矿物掺和料掺量常需达60%以上（也可用32.5的复合水泥）才能配出来。

③第4列到第7列中对硅酸盐水泥未规定型号，如果不是因忽视而遗漏，则只能是规范编制者的观念问题，即认为P·Ⅰ和P·Ⅱ中已有混合材掺量的差别可以忽略。但实际上再允许掺入粉煤灰后所造成的掺和料总量就形成了悖论：如果混合材相差5个百分点可以忽略，那么表中所有相差5个百分点规定值都是毫无意义的。

④ 大掺量矿物掺和料的使用已成为现代混凝土的一个特征，符合可持续发展战略的要求，而且某些掺和料还可改善混凝土拌和物性能，改善混凝土耐久性。耐久性已成为决定矿物掺和料种类和掺量的重要因素。因此，尽管对房屋建筑来说“没有什么耐久性问题”（且不说这是对“耐久性”的误解），大掺量矿物掺和料的使用也已呈不可阻挡之势。但是任何材料都有利也有弊。在《混凝土结构耐久性设计规范》（GB/T 50476—2008）对环境的分类中，Ⅰ类环境所指“没有冻融循环和化学侵蚀性介质作用而只有碳化可能引起钢筋锈蚀的环境”，在全国不同的地区有不同的意义，例如在常年大部分时间相对湿度较低只有短暂湿热气候的北京，混凝土碳化问题的重要性一般主要不是潜在的钢筋锈蚀问题（由于气候的变化，这一点也变成不一定），而在于碳化收缩会与干燥收缩联合作用发生加重开裂的倾向[2]。一旦开裂，则在仅有不到一个月的湿热气候下，就有可能引起钢筋的锈蚀。对于在一年中的大部分季节处于湿热气候的深圳，则钢筋锈蚀就成为主要问题，但是一年中大约有2个月相对湿度在50%以下，碳化收缩与干燥收缩联合作用而发生的开裂倾向也是需要重视的。大掺量矿物掺和料的使用要求施工技术观念的更新（例如养护的概念和技术），为避免“Ⅰ类环境”中混凝土这种潜在的劣化因素，必须在设计和施工中采取相应措施，配合大掺量矿物掺和料的正确使用（如果要用的话）。在其他环境中，不同季节、不同构件的混凝土，对矿物掺和料有不同的要求，应根据耐久性进行选择和试验，当掺和料品种和用量不同时，还要调整配合比。配合比调整时，应保持浆体体积不能增大（这应是配合比设计的一项原则）。

⑤ 新《规程》条文说明第 3.0.5 为："规定矿物掺和料最大掺量主要是为了保证混凝土耐久性能。"此项说明似乎匪夷所思。因为，目前业内几乎尽人皆知，在混凝土中正确地使用矿物掺和料，目的就是为了改善混凝土体系中微结构以增加体系的密实性，提高混凝土抵抗环境中化学侵蚀性介质作用的能力。恰恰是那些严酷环境中的混凝土结构需要矿物掺和料去改善和提高其耐久性。如果因为条文中所规定的限值范围较宽，给某些执行者提供点参考，也未尝不可，只是这种说法令人怀疑："既然矿物掺和料的掺用搞不好会影响混凝土耐久性，那么掺用的目的是什么？"就是为了"废物利用"而降低成本？可以理解的是，在很久以前，使用矿物掺和料的混凝土的确出现过若干问题[3]，但那是对矿物掺和料的认识不足和使用技术不当造成的，至今也还有人用简单取代水泥的方法使用矿物掺和料，并用传统的水泥检测方法评价矿物掺和料性能。此外，不同矿物掺和料的种类和成分差异很大，导致性能差异也很大，人为规定一个固定的最大掺量缺乏科学依据。因此这样的限定值是不能作为"原则"来规定的。

⑥美国 ACI 211 委员会花了很大的力量把"混凝土配合比设计"改成"混凝土拌和物的配合比"[4]，其原因是混凝土搅拌站提供的不是混凝土的最终产品，而是拌和物。最终产品是经过浇筑、振捣、抹面、养护一系列工艺操作后形成的混凝土结构物。在当前材料与其工艺的行业隔离条件下，只有配合比是不可能保证混凝土最终产品质量的。只给出对矿物掺和料掺量限定的表格不能保证混凝土的质量，混凝土配合比不宜单独制订规范，而是应当在对混凝土性能要求中提出满足混凝土结构设计要求的配合的原则和工艺的配合。至于混凝土拌和物配合比的设计方法可以有很多，达到相同强度等级或再加上耐久性指标要求，方法可以千变万化，即所谓殊途同归。应当集中大家的智慧编制"指南"或"手册"，允许技术人员各显神通。有人说："'指南'没有权威性。"权威性不是指定的，更不是自封的，ACI、ASTM 都是民间学术团体，他们制订的标准由于质量高、适应性好而赢得美国工业界官方的信赖，不仅被美国各工业界纷纷采用，连美国国防部和联邦政府各部门机构也采用。我国土木工程学会 2004 年的标准《混凝土结构耐久性设计与施工指南》（CCES-01）及其修订版至今仍有不少重要工程在采用。一再强调"权威性"、"具有法律效力"，是对自己所制订规范的水平没有自信的表现。想当初原建设部要求把《混凝土结构耐久性设计与施工指南》改成"国标"时，编制《指南》的主持人陈肇元院士是经过反复推敲和协商后才同意的，但要求在题头加上代表推荐性的"T"。这是一种科学的负责任的态度。

1.2 关于骨料的含水率

对骨料含水率的限定怎么能做到？"长期以来一直……"能作为规范的根据吗？为什么在实验室试配时应当使用饱和面干的表观密度？

新《规程》第 3.0.2 条中规定混凝土配合比设计所采用的细骨料含水率应小于 0.5%，粗骨料含水率应小于 0.2%。不知为什么要做这样的规定？怎样才能达到这样的状态？新《规程》条文说明中对此的解释是："我国长期以来一直在建设工程中采用以干燥状态骨料为基准的混凝土配合比设计，具有可操作性，使用情况良好"。也就是说认为这样的含水率属于"干燥状态"。众所周知，骨料"干燥状态"基本上有三种——绝对干燥（绝干）、自然干燥（气干）、饱和面干（面干或日本所称表干）。可能上述规定指的是气干。实际上不同岩石的石子和砂子骨料在自然状态的大气环境（相对湿度是变化的）中的气干含水率变化很大。如果达不到此规定怎么办？实际上确实是“具有可操作性，使用情况良好”吗？

早年在学校教学实验中检测骨料密度时，都是在饱和面干状态下进行的。石子很容易做到饱和面干，只是因砂子做到饱和面干较难操作，后来使用的是气干状态。在混凝土配合比设计中，考虑到石子的饱和面干吸水率很小，而砂子含水不同时，称量会不同，不便于比较，约定砂子使用绝干状态的密度，在试配时测出砂子的含水率，在混凝土配合比的用砂量中扣除，加到用水量中。这样的结果，对所用砂子就是按绝干配料的（不考虑砂子的吸水率，即饱和面下的含水率），把砂子中实际的所有含水量全部算进拌和用水中。做过混凝土试配的人都有过这样的经验：在实验室试配时，使用绝干状态砂子配制混凝土，用绝干状态砂子试配计算的用水量拌和混凝土时，由于砂子吸水率的影响，会造成混凝土拌和物的工作性不能达到设计目标，于是需要尝试再增加用水量，一般为避免过加而需要反复数次测试坍落度，将最后坍落度合格了的拌和物用水量确定为最后的用水量。但是，因为上述用水量的调整是分数次添加的，用这样确定的最后用水量重新拌制一盘拌和物时，往往坍落度仍不够（在反复调整过程中会有一部分水的损失），就需要再试。在生产中的情形与此相似，只是由机械根据电流表自动调整代替人工控制调整，结果是拌和物出机后，拌和物中的用水量究竟有多少？很难有人说清楚。

有趣的是现在全世界除了我国建筑工程以外，都使用饱和面干状态下的骨料密度配制混凝土，要求骨料必须带水供应（但不得流淌），生产时所调整配合比中的用水量只须扣除骨料饱和面干之外表面吸附的水量。因为饱和面干的骨料所含的水在拌和物中是不会出来的（因为内外平衡），既不影响拌和物的流动性，也不参与水泥的水化，只在混凝土硬化后内部水量消耗后才会释放出来，反而会对界面有养护的作用。有人总说不会测“饱和面干”，但是却会做吸水率。其实饱和吸水率的数值与“饱和面干”含水率在数值上是相等的，检测方法也是一样的。绝干的骨料尤其是砂子，在拌和物中吸收拌和水的速率和在清水中吸水的速率不同，只要预先吸水饱和，可以不面干，在拌和物用水量中扣除骨料表面水后，即为“有效拌和水”。现在市场供应的砂子大多都带水，已为用饱和面干状态下骨料表观密度的使用提供了条件。提到“饱和面干”时，有人以为是要求把生产时上料的砂子先浇水再做成饱和面干，这完全是误解。正确的理解是：在试配时用饱和面干的骨料表观密度值计算体积或质量；生产时在拌和水中根据骨料含水状态多退少补。

“长期以来一直……”并不都是合理的，否则为什么还要创新、改革？“具有可操作性”的含义应当是努一把力就能做到，而不能成为迁就落后的同义词。

2　配合比主要参数 w/b 的确定

包罗米公式在80多年前就被发现其计算值和实际偏差可达20%之大，这是因为该回归直线方程的系数（直线的斜率和截距）被赋予了过于明确而狭隘的物理意义。而今面对混凝土原材料和混凝土技术所发生的巨大变化，必须转变思维方法和观念，否则就无法进步。

2.1　新《规程》规定用以下公式计算水胶比：

$$\frac{\mathrm{w}}{\mathrm{b}}=\frac{\alpha_{\mathrm{a}} f_{\mathrm{b}}}{f_{\mathrm{cu},0}+\alpha_{\mathrm{a}} \alpha_{\mathrm{b}} f_{\mathrm{b}}}$$

该公式并不能叫做“改进的”包罗米公式，经过移项后即可还原为原来的形式，并未改进：

$$f_{cu,0}=\alpha_a f_b\left(\frac{b}{w}-\alpha_b\right)$$

各符号意义也未变，只是改变了表达的字母，其中α_a和α_b即原公式的A、B，意义也未变，只是改变了数值。而这样数值的改变在我国也已经有过数次，其中在1982年试行的《普通混凝土配合比设计技术规定》（JGJ 55—81）中是正式的一次，2000版的《规程》又改了一次，其余的改变多为不同地方标准所为。

2.2 现按新《规程》给出规定的数据举一实例：

混凝土设计要求强度等级为C35，按新《规程》表4标准差应取5MPa，则$f_{cu,0}=$ 43.225MPa，取43MPa；Ⅱ级粉煤灰掺量为45%，按表6取影响系数为0.55；矿渣粉10%，无影响系数；42.5水泥28天强度为50MPa，掺和后，$f_b=50\times0.55=27.5$MPa，按表5，使用碎石，α_a和α_b分别为0.53和0.2。按新《规程》所给公式计算：

$$\frac{w}{b}=\frac{0.53\times27.5}{43+0.53\times27.5}=\frac{14.575}{45.915}=0.317$$

南方含盐土质某地下工程试验段，在实际工程试配C35的混凝土时，使用与上述相同的原材料，水灰比为0.42，实测28天强度为45.1MPa，很接近配制强度，但是水胶比比公式计算的大得多（偏差超过30%）。根据实际经验，如果采用上述计算的水胶比，混凝土强度应该可以达到60MPa。

为了避免偶然性，再举该工程另一组粉煤灰掺量少些的配合比的混凝土实例，同以上设计要求的C35混凝土，原材料相同，粉煤灰掺量为24.3%，按新《规程》表6，取影响系数0.75，则$f_b=50\times0.75=37.5$，用与上述相同公式计算：

$$\frac{w}{b}=\frac{0.53\times37.5}{43+0.53\times0.92\times37.5}=\frac{19.875}{46.975}=0.42$$

上述工程试验段另一组试配时使用与上述相同的原材料，实际水胶比为0.49，28天抗压强度为44.0MPa，与设计的配制强度相当，而与按新《规程》公式计算出的水胶比值偏差约为14%。可见，掺和料掺量越大，偏差越大。

2.3 通过上述对比所显示的差异，有以下几点看法：

① 就包罗米(J. Bolomy)公式本身来说，是当时严格控制原材料质量，并经过大量实验数据拟合而成的直线关系。当时包罗米本人和后来的很多使用者都清楚：该公式的计算值偏差可达20%之大。这是因为他赋予了所回归直线方程的系数(直线的斜率和截距)过于明确而狭隘的物理意义——石子品种、水泥强度，而实际上，当水胶比确定后，浆骨比(或者说石子用量)对混凝土强度也是有影响的；水胶比和浆骨比都确定之后，砂率对强度也有影响；另外，相同品种、相同强度的水泥在混凝土的工程中可能会有不同的表现；相同强度而品种不同的水泥的影响更有差别。当然，1930年的包罗米没有考虑这些问题的条件。1932年，莱斯(I. Lyse)肯定了混凝土的28天抗压强度和水灰比的倒数成正比的关系，提出灰水比定则：

$$y=-b+ax$$

式中：y——混凝土28天抗压强度；

x——水胶比的倒数；

b——纵坐标轴截距；

a——直线斜率。

一个纯粹回归的直线方程，比起给出系数的物理意义和具体数值的包罗米公式具有普适性。普适性在于直线关系，其直线的斜率和位置则可根据不同实验者所在地区原材料条件和工艺条件由实验确定。对于混凝土这样复杂的体系可以而且应当允许生产企业按自己的原材料条件建立各有关系列曲线，例如英国就有粉煤灰不同掺量时强度-水胶比曲线（别的国家没了解过）；日本各搅拌站都有一系列砂子饱和面干含水率-表观密度曲线、拌和物在搅拌机中转动时电流变化的系列曲线等。目前我国已有一些搅拌站根据自己所常用的和变化的原材料建立了这样的直线关系系列曲线，使用时方便、有效。为什么非要坚守包罗米公式不可呢？

② 1930 年包罗米公式被提出至今已 80 余年，混凝土技术的变化是：因高效减水剂和矿物掺和料的使用，混凝土和水泥制品的强度不再受水泥强度的制约，强度、水胶比、拌和物流动性都可以在很宽的范围变化；拌和物中骨料从紧密堆积的状态变成现在的悬浮状态；配合比过去是三要素、四组分；现在是四要素、五组分……；过去在一定原材料和工艺条件下水灰比的选择是按照强度-水灰比的关系，现在是强度-水胶比-矿物掺和料量三维的关系……。面对变化，必须转变思维方法和观念，否则就会在陈旧的圈子里转不出来。有人说，现在还没有别的办法确定水胶比，只好还用包罗米公式算个基本数值再经试配确定。当然，对混凝土这种具有不确定性的、高度复杂的多相非线性体系，配制绝对是必须的。有经验的技术人员不必死套包罗米公式，就能很快地给出接近工程要求的混凝土水胶比的基本值，可见规律还是存在的。从事科学技术研究的部门是否可以和这些经验结合起来进行研究，找到其中的规律，总结、上升到理论，提出新公式的创意？

③ 新《规程》只是把包罗米公式中的水泥强度简单化地改成胶凝材料的强度，并引入了“矿物掺和料的影响系数”。而所谓胶凝材料强度是和水泥一样用 0.5 的水胶比检测出来的；“影响系数”也是在统一的水胶比下检测的。这对于水胶比可以在很大幅度变化的混凝土来说有什么意义吗？新《规程》中 w/c 的计算产生如此大的偏差的主要原因就是这个影响系数的不合理。目前在业内，几乎人人尽知掺入矿物掺和料的胶凝材料体系和硅酸盐水泥相比，对水胶比更加敏感[2]；矿物掺和料在对混凝土的作用中，物理作用要比化学作用更重要。检测矿物掺和料和硅酸盐水泥组成的胶凝材料的强度，首先要减小水胶比以保持原始浆体体积不变(即混凝土拌和物的浆骨比不变)，二是要保持相同流动度。这才能体现工程中混凝土的规律。所谓“粉煤灰不是水泥的简单取代物，而是混凝土的第四组分”[1]就是这个意思。“矿物掺和料的影响系数”的概念恰恰就是一种简单取代。变化了的材料不能使用不变的方法。

3 关于耐久性

什么是混凝土的耐久性？新《规程》对耐久性理解的偏颇。

新《规程》条文说明第 3.0.1 表明该《规程》修订的目的：“……强调混凝土配合比设计应满足耐久性能要求，这是本次修订的重点之一……”但是除了在前言中说明“本次修订的主要技术内容的第 2 点是：‘增加并突出了混凝土的耐久性的规定’和强制性条文第 6.2.5 条‘对耐久性有设计要求的混凝土应进行相关耐久性试验验证’”，此外基本上未见有关耐久性的规定。至于第 7.1 条抗渗性混凝土、第 7.2 条抗冻性混凝土，那是针对“有特殊要求的混凝土"。重要的是强制性条文第 6.2.5 条只是"对耐久性有设计要求的混凝土"而言，言外之意，如果没有对耐久性要求，本《规程》是不予考虑的。这里的问题是，什么是耐久性？有没有不需要耐久性的混凝土？

Adam M Neiville 说过[5]："不存在一般含义上耐久的混凝土。如果我要为花园工具室修个地板，打算用上一年就推倒，某种质量的混凝土就'耐久'了。但这种混凝土用于一座大桥、一条隧道或一座大坝中也许就不耐久了。所以说某一混凝土是'耐久'的或'不耐久'的是错误的"，"混凝土在一种条件下耐久了，在另一些条件组合下就可能不耐久。仍然是没有本质上耐久的混凝土"，"有时混凝土不耐久，是因为用途发生了变化，例如本是设计存放纸的仓库的地板，却改为放化学品了"。由以上引用的 Adam M Neiville 的几句话可见，第一，即使是打算只用一年的混凝土，一年就是该混凝土的耐久性，例如 Neville 所说要为花园工具室修的地板。可能有人会说"哪里会有用不到一年就坏了的混凝土？"现举一例：有一年北京一个物流公司仓库地面的混凝土用了不到一年，就损坏严重，分块的交界处碎裂、剥落，常使运货的叉车行走受阻。这就是说该地面混凝土耐久性不足。第二，混凝土的耐久性是在一定环境下的结构构件中表现出的性能（performance)，实验室检测的"长期性能和耐久性能试验"只是供混凝土拌和物试配时优选原材料和配合比用的"实验室指标"[5],[6]。按标准实验方法检测的"耐久性"只能代表不同原材料和配合比的混凝土相对的本征特性。同理，实验室的"长期性能试验"也是在实验室条件下强制的过程，实验时间再长也不过半年、一年。这一强制性条文其实没有什么意义："对耐久性有设计要求的混凝土应进行相关耐久性试验验证"是没有谁会不做的。至于新《规程》所"强调混凝土配合比设计应满足耐久性能要求"只是"应进行相关耐久性试验验证"和"有特殊要求的混凝土"，并未见混凝土配合比设计怎样"满足耐久性能要求"。

4　怎样制订和使用标准、规范？

科学技术的支持，反映最新成熟的研究技术规范的制订应当有科研成果，或借鉴、消化国外先进经验，主要由行业协会主管、负责，组织无功利关联的专家参与编制和讨论，审定。因各专家有各自的科研和工程实践经验和学术观点，必须最后达到求大同存小异，因此技术规范也是在较高水平上妥协的产物。

通过对《规程》几点分析，不免引发对如何制订和使用标准、规范的思考。

4.1　解读法律、法规、规程、条例、标准、规范、指南等的含义和区别

我国各方面的建设仍处于起步阶段，管理方面多借鉴已经发展了几百年的发达国家的做法，在法律、制度、管理上有很多不健全之处，这是社会主义初级阶段很正常的现象。社会行为的规则、制度或长期形成的约定，在社会发展中的重大意义是不言而自明的，其高级形式的体现就是法律、法规、标准、规范等。我国在制订这些规则时大体上都参考或借鉴了发达国家经验，其命名或用词也是和国际接轨的。为了更确切地理解，以下根据国外一些词典、辞书解释的词义解读如下。本人知识有限，希望与大家讨论。

在此顺便一提：我国一般人在外文的翻译中常常只根据英汉或汉英词典，而编纂这些词典的专家对专业词汇未必精通，时有误导之弊，最好使用外文的原文辞书或英汉双解的词典，并经讨论而确定。

法律（Law） 法律是一种人类在社会层次上公平的规则，是世所公认的公正不偏的权衡标准，是理性的体现，又是一个合同式的契约，代表政治上的正义。由国家制订或认可，并由国家强制力保证实施，以规定当事人权利和义务为内容的具有普遍约束力的社会规范。在我国，都必须经全国人民代表大会表决通过，由国家主席令或国务院令的方式颁布。

法规（Code） 法规是法律、法令、条例、规则、章程等法定文件的总称。来自拉丁文的 codex（圣书、典籍的手抄本，药典）意思是写在木头片上的文件，是最早制订的法律。可以由国家权力机构制订、颁布、施行，也可由地方或部门制订、发布、施行。

条例、规程、规章（Regulations and Rule） 都属于法规（Code）。在“Investorwords”网站中解释 code 为官方或当局为控制和管理而规定的必须服从的一种法律、规则、原则、条件或其他命令。对于那些直接涉及国计民生、公众安全的设计规范，如建筑物和构筑物（桥梁、隧道、港湾、水运、铁道等）设计的规范，尽管也涉及技术，但也属于法规，英文名称也是 code。所以钢筋混凝土结构设计规范的英文题名就是 Code for Design of Concrete Structure。又例如建筑设计规范属于建筑法规的一部分——建筑法规体系分为法律、规范和标准三部分，其中法律主要涉及行政和组织管理（包括惩罚措施），规范侧重于综合技术要求，标准则偏重于单项技术要求。技术规范的制订应当有科学技术的支持，反映最新成熟的研究成果，或借鉴、消化国外先进经验，主要由行业协会主管、负责，组织无功利关联的专家参与编制和讨论，审定。因各专家有各自的科研和工程实践经验和学术观点，必须最后达到求大同存小异，因此技术规范也是在较高水平上妥协的产物。

标准（Standard） “The free online dictionary（http：//www. bsigroup. com）”对 standard 的解释是“因其出色而被公认或广泛使用的实施或生产的某种事物”；“一种要求、优秀或成就的水平”；“优质的水平”，“旗帜、标杆，”例如军旗就叫 Standard。在技术层面上为：“以其他人评判或量测为准的某种事物公认的或核准的范例”；“公认的质量或数量的相对量值”；“在规定条件下确定的、表征的、记录的单位量级目标”……可见标准是一种量化的质量要求，例如目前正在征求意见的“小学教师标准”。标准多用于对产品（有形和无形的）质量的合格检验。用以检验出厂的产品是否符合生产者承诺的质量指标及其匀质性，并且无毒无害。国家标准指的是对全国同行业或同类产品普适性的标准，因此所规定的指标为最低要求，产品出厂检验合格未必能满足用户的要求，用户可以提出更高的要求或特殊的要求。地方标准是针对地方具体条件的，往往高于国家标准；企业的产品标准应高于行业标准，以适应用户选择的需要。与上述类似，凡直接涉及国计民生、公众安全的标准属于法规，是强制性的，例如食品药物、武器等的标准。技术标准一般不具独立法律效力。

规范（Specification） Specification 的原意是规格、说明书的意思。在中文字面上是规格和范例的意思。虽然“标准”也有范例的意思，但其指标可以量化。无法精准定量而形成标准的被称为规范。“Businessdictionary（辞书）”对 specification 的定义是：“对某种作业、行为进行定性的信息规定”，“对能得到满意的特定要求，或者顾客在货物、材料、方法、工序、服务、系统或工程上的要求而卖方必须交付的基本特性准确的表述，通常以双方（和/或独立证明人）都能量测其程度的方式书写，然而并不等同于控制极限（允许波动的范围）”，规范通常主要分为两类：①性能规范：符合已知的顾客要求，例如保持房间温度在一个规定的范围；②技术规范：表述个体单位性能的水平，又分为：（a）个体单位的规范，叙述由标称值和允差（偏离标称值的允许值）组成的个体性能边界参数；（b）合格的质量，叙述使大多数单位满意的限定值，但有一定比例的单位允许超过此限定值；（c）分布规范，为每个单位定义一个用于生产者监测生产过程的统计学的分布（就均差和标准差而言）。可见，规范是一种“约定”。我国技术规范使用的英文名称就是specification，规范条文主要告诉使用者什么是不可以做的，即对不合理或对公众利益有害的行为或操作的限定。基本上偏于原则上的规定，不讲道理、不做具体工艺和方法的规定，尤其禁止对具体产品的规定。因此，诸如具有丰富内容、影响因素众多的施工之类的技术就不宜制订规范，例如对施工只能规定湿养护的时间，而不规定养护的具体方法（对具体方法可以编制《指南》），目前我国只有施工验收规范而没有施工技术规范即缘于此。《混凝土拌和物的配合比设计》也属于此类。

技术规范的制订应当有科学技术的支持，反映最新的并经反复研究—实践检验—反馈—再研究而成熟的研究成果，或借鉴、消化国外先进经验。例如ASTM的技术规范制订的基础是ACI各专业委员会分别每3年发布一次的技术报告（Stats of the arts report）。因此技术规范永远滞后于但不能落后于科学研究，技术规范应具有相对的先进性—先进而可操作，又不迁就落后。技术规范主要由行业协会主管、负责，组织无功利关联的专家参与讨论、编制和审定，不能有行政的干预。因各专家有各自的科研和工程实践经验和学术观点，必须最后达到求大同存小异，因此技术规范也是在较高水平上妥协的产物。技术规范只规定为保证公众安全的最低要求，使用者可以根据具体工程的特点和条件按各自的专业知识和工程经验发挥创新精神，提出更高的要求。新材料、新技术、新工艺的采用，必须是有明确机理的，符合科学规律的，经实验室小试、向生产过渡的中间试验和工程试用，制订相关技术标准和操作规程，最后经专家鉴定，在相关主管部门备案。

指南或者使用手册（Guide or Handbook）“The freeonline dictionary”解释guide是“对方式、方法、途径的引领、指导，或建议，提供信息或指令的一种小册子”；“作为调节一种动作或操作的指示装置”；“为控制队列的矫直而安排在行进队伍两侧的纵队士兵，以指示方向或中心点”……可见，指南或手册就是指路的，给出在规定原则下满足一定要求的若干材料、技术、方法或别人成功的经验，供使用者根据自己的条件和需要进行选择。

4.2　标准、规范怎么用?

存在的不一定都合理，外国的不一定都先进，专家和权威不一定都正确。法律或属于法律范畴的其他文件也好，技术标准或规范也罢，是以人对客观规律的认识为基础的，也都有水平问题。如何使用和监管、裁定，更是有水平问题，这尤为重要。

不管是什么样的标准和规范，首先都具有严肃性，无规矩无以成方圆。各人按自己的意志所成的方圆只能个人自己用，凡是有人群活动的事物如果没有规矩就会乱套。打个比方：“游戏规则”并不属于任何法律，但重要的这是一种约定，只要参加某个游戏就必须遵守该游戏的规则，这是毋庸置疑的。

由以上解读和介绍可见，除建筑、结构设计规范（Code）以外，涉及材料、技术、工艺、方法等的技术标准和规范的本身都不具法律属性。在西方国家，工程合同中都必须有甲乙双方共同选定并契约的“规范清单”，合同具有法律效力，则合同中的“规范清单”所列标准和规范就自然与合同一起具有法律效力。为了执行的方便，在合同中还应随所选择规范而约定一个处罚条例。技术标准和规范中规定的指标是最低要求，在我国现行国家水泥标准和混凝土结构耐久性设计标准中，已列入此说明。国外发达国家都有此说明，例如

ASSHTO（American Association of State Highway and Transportation Officials，美国国家公路与运输管理者协会）桥梁设计规程的第一章第一节中写道："本规程无意取代设计人所具有的专门教育和工程判断的训练，仅在规程中规定为保证公共安全的最低要求。业主或设计人可能需要在设计中采用新的先进技术，或需对材料及施工质量提出更高的要求"。英国标准前言的最后一句话就有："依从英国标准不能免除其法律责任"。最低要求的意思是，保证品质要求最低的工程的安全性和最低等级耐久性的普适性要求，设计人员必须根据工程的重要程度提高要求。因此标准中的指标上限不能太低，下限不能太高，留有充分余地，用户可以根据需要，用性能（强度不是唯一的，也不一定是第一的性能）去确定和控制。通过合同，就可以避免对不同规范中规定的差异感到无所适从的问题，因为列入合同中的规范才是唯一在法律上有效的。合同是市场经济重要的法律手段，而且是契约双方互相沟通、交流、理解的过程，应当尽快改变我国目前土木工程尤其是建筑市场"合同"的单方强制性或者走形式的现状。一方面管理部门要从完善所提供的合同文本入手，引导合同双方正确制订和签约；另一方面合同双方要正确认识合同的意义而严肃对待，学会通过合同保护自身利益，"丑话说在前面"。无论如何，对法律和具法规属性的规定必须服从或遵守，对技术类的标准、规范、指南等，一经契约定入合同，双方都应严格遵守，遇到矛盾或意外情况时，双方应沟通洽商，并记录成文签字后作为合同的补充附件。当需要突破现有规范的规定时，应当有试验研究的依据、试点应用和专家鉴定。工程技术人员都应为其提供试点应用的条件，以促进科学技术的进步和建设质量的提高，并为标准、规范的修订提供依据。

结束语

存在的不一定都合理，外国的不一定都先进，专家和权威不一定都正确。法律或属于法律范畴的其他文件也好，技术标准或规范也罢，是以人对客观规律的认识为基础的，也都有水平问题。如何使用和监管、裁定，更是有水平问题，这尤为重要。当前我国正处于发展中的阶段，尚未实现范式*的转换，难免"公说公有理，婆说婆有理"。有不同意见是很正常的，但是一定要致力于符合时代要求的范式转换，首先需要思维的突破和技术的更新；如果把作为谋生手段而制订技术标准的范式，转换成为公益事业而制订技术标准的范式，则所谓思维的突破也就是思维方法和世界观的转变，技术的突破就是学习、了解、吸纳科学技术研究新的成果，既先进又可操作，但又不迁就落后。

下面引用原建设部第 81 号令《实施工程建设强制性标准监督规定》的条文释义中的一段文字作为结束：

"标准与科学技术发展密切相连，标准应当与科学技术同步，适时将科学技术（本文作者注：似应为'科学技术的最新内容'）纳入到标准中去。科学技术是提高标准制订质量的关键环节。反过来，如果新技术、新工艺、新材料得不到推行，就难以获取实践的检验，也不能验证其正确性，纳入到标准中也会不可靠。为此给出适当条件允许其发展，是建立标准与科学技术桥梁的重要机制。"

---

* 按照美国著名科学哲学家托马斯·库恩（Thomas. Kuhn）自 1962 年所提出和阐述，范式主要指的是从事某一科学的研究者群体所共同遵从的世界观和行为方式。

参考文献

[1] M. R. H. Dunstan. Fly Ash，as“The Fourth Composition”of Concrete，Proceedings of Second CANMET International Conference on Fly Ash，Silica fume，Slag and Natural Pozzolans in Concrete，1987.

[2] Adam M Neville. Properties of Concrete，Fourth and Final Edition. Pearson Limited，England. 2000.

[3] Richard w Burrrows. The Visible and Invisible Cracking of Concrete. ACI Monograph No. 11.

[4] Ken W. Day(澳大利亚). Manufacture of High Performance Concrete，来华讲座，1996. 7. 13.

[5] Adam Neville. Consideration of durability of concrete structures：Past，present，and future. Materials and Structures March 2001. pp114-118.

[6] 廉慧珍．思维方法和观念的转变比技术更重要—之五：混凝土工程中的分解论和整体论[J]．商品混凝土，2011，10.

附件更正

编辑同志：贵刊 2012 年第 3 期刊登我的文章《评普通混凝土配合比设计规程（JGJ 55—2011）》的表 1 中，本应按下式计算：

掺和料掺量＋(100－掺和料掺量)×现行水泥标准规定的混合材上限百分数

文中却直接以标准规定的百分数直接相加了。这是因我的粗心造成的错误，现予以更正如下：

表 1　对新《规程》中表 2、表 3 的分析

矿物掺和料种类	水胶比	最大掺量/%									
		采用硅酸盐水泥						采用普通硅酸盐水泥			
		允许掺量/%		计算掺粉煤灰后掺和料总量/%*				允许掺量/%		计算掺粉煤灰后掺和料总量/%**	
		钢筋砼	预应力	钢筋砼		预应力		钢筋砼	预应力	钢筋砼	预应力
				P·Ⅰ	P·Ⅱ	P·Ⅰ	P·Ⅱ				
粉煤灰	≤0.4	45	35	45	48	35	38	35	30	48	44
	>0.4	40	25	40	43	25	29	30	20	44	36
粒化高炉矿渣粉	≤0.4	65	55	65	67	55	57	55	45	64	56
	>0.4	55	45	55	57	45	48	45	35	56	48
钢渣粉	—	30	20	30	34	20	24	20	10	36	28
磷渣粉	—	30	20	30	34	20	24	20	10	36	28
硅灰	—	10	10	10	15	10	15	10	10	28	28
复合掺和料	≤0.4	65	55	65	67	55	57	55	45	64	56
	>0.4	55	45	55	57	45	48	45	35	56	48
列号	1	2	3	4	5	6	7	8	9	10	11

* 此处对硅酸盐水泥未规定型号，P·Ⅱ已有的混合材以 5%计；计算掺和料总量时，小数点以后的数四舍五入。

** 普通硅酸盐水泥中按现行标准规定的混合材最大量为 20%计，不考虑超标。计算掺和料总量时，小数点以后的数四舍五入。

由更正后的表来看，不能得出全都是“水泥中混合材越多的，在混凝土中允许矿物掺和料总量倒越多”的结论，例如第 4、5 列和第 10 列相比；但是第 6、7 列和第 11 列相比，第 4 列和第 5 列相比，第 6 列和第 7 列相比，则基本符合上述结论；其中有的高有的低，混乱无章，表明该表在制作时，“拍脑袋”的可能性很大。虽然也可能和我一样，在计算过程中有些失误，但是作为国家标准，则有失严肃性。按该标准在宣贯的通知时所说的“国标等同于法律”（通知的人在邮件中所述），则会使不明白的人接受误导而照办，明白人不知所措。

感谢清华大学阎培渝教授认真地阅读了该文指出上述错误。这种作风应当提倡。

我的原则是有话就说，有错必纠，希望和大家共同追求真理。

廉慧珍　2012.10.30

自评

1. 自从有了高效减水剂使混凝土的强度和水泥的强度不再有固定的依从关系，Bolomey公式怎么用？其实最早从前苏联引进的 Bolomey 公式已经不是原始的。原始的 Bolomey 公式如下：

$$S = \mathrm{K}\left(\frac{C}{V+A}\right) - \mathrm{K}'$$

式中符号应不是英文，S 为混凝土抗压强度，K 和 K′为回归系数，C、V、A 分别为水泥用量、用水量和含气量。

由该式可见，原始的 Bolomey 公式只是一个回归方程，和 Lyse 的二元一次方程是一样的：

$$y = \mathrm{a}x - \mathrm{b}$$

式中 y 为 28 天抗压强度，x 为灰水比，a 和 b 为回归系数，反应最早 Abrams 方程的本质。前苏联在用硬练法检测水泥强度时，把这个公式改造成和水泥强度与骨料质地有关的公式：

$$R_{Ц} = AR_c\left(\frac{C}{B} - Б\right)$$

式中符号为俄文字母，$R_{Ц}$为混凝土抗压强度，R_c为水泥强度，C、B 分别为水泥和水的用量，A、$Б$ 为和石子（碎石或卵石）有关的系数。

2. 我国引入该公式后一直用到现在。在我国水泥标准由硬练法改成软练法时，把 A、$Б$ 系数改成英文的 A、B，并修改过一次数值，2000 年又修改过一次，JGJ 55—2011 再一次修改。其实包括前苏联公式和我国几次的修改的系数，不过就是把水泥强度剥离出来的回归系数。如果制订标准的人弄清楚 Bolomey 公式的来龙去脉，就不会发生因使用了高效减水剂而“Bolomey 不再适用”的问题，或者直接使用 Lyse 的直线方程：$y = -\mathrm{b} + \mathrm{a}x$（式中 y 是混凝土抗压强度，x 为水胶比的倒数，a、b 为回归系数）。则问题就简单了。因为是回归的直线，当然并非所有的数据点都落在线上。影响回归系数数值的因素不仅是骨料质地（碎石还是卵石），还有骨料的粒径、用量和砂率，当然还有水泥（含其他胶凝材料）和生产工艺。因此应当由搅拌站按照自己的条件自行回归。统一规定是不符合实际的。

3. 在混凝土拌和物配合比设计中，核心的问题就是水胶比的确定。混凝土强度和水胶比倒数成正比的关系是已知的规律，目前大量使用的混凝土为 C30～C50，所用水胶比基本上为 0.3～0.5，可以选择设定 3～5 个水胶比试配得到相应的强度，在所得强度－水胶比关系图上选择所需强度相应的水胶比，用所选水胶比继续试配，即可确定水胶比。

4. 因工程条件、环境和原材料的复杂性，很多搅拌站试验人员一般都有调整配合比的经验。已经出现了配合比设计的各种有效方法，标准只宜规定混凝土拌和物配合比设计的原则，而不宜规定方法和步骤去限制一线人员的发挥他们的学识和才智，或者总结各个方法，制订出“指南”，供大家根据自己的情况选用。本文所举实例也只是供参考的一种方法，重要的是所述原则。

三、关于回弹法和混凝土强度的关系问题

论文之十八

质疑“回弹法检测混凝土抗压强度”*

廉慧珍（清华大学土木水利学院）

摘　要　材料的硬度和强度不是同一个概念。同一种匀质材料的硬度和强度之间有一定的相关性，而不同材料的硬度和强度之间不能建立相关的关系；同样水胶比的砂浆和混凝土是不同的材料，砂浆的硬度最多只可能与砂浆强度有一定的联系，而相同水胶比的砂浆强度和混凝土强度的关系却因浆骨比和砂率的不同而异；混凝土碳化层和该混凝土更是不同的材料，混凝土碳化层的硬度和内部混凝土的强度没有关系，再基于碳化层的硬度引进“折减系数”来推算混凝土的强度，在概念上是错误的。

关键词　回弹法；硬度和强度关系；碳化层；折减系数

1　什么是硬度?

严格来说，应当称表面硬度。

回弹仪是用肖氏硬度（Shore’s hardness）原理检测材料表面硬度的仪器。在有关混凝土的网站论坛中，发现有些人在概念上把混凝土的硬度和强度混淆了，以为硬度大的材料强度也高，回弹值就代表强度。尽管对业内人士澄清这个问题不免是画蛇添足，简单复习一下相关知识还是有益的。

表面硬度是指材料抵抗外来机械作用力（如刻划、压入、研磨等）侵入的能力，硬度很难测定和准确地表示，常用方法有三类：静压法，如布氏硬度、洛氏硬度、维氏硬度等；划痕法，如莫氏硬度；回弹法，如肖氏硬度。①对金属材料，多用静压法，以钢球或金刚石钻头在固定荷载下经一定时间压入受检材料表面的深度或压痕大小作为硬度值。例如布氏（Brinell）硬度 HB、洛氏（Rockwell）硬度 HR、维氏（Vecart）硬度 HV，其区别只是所用压头和标准荷载值的不同；②在地质学上多用莫氏硬度（Mohs’ scale of hardness），因1822年莫斯（Friedrich. Mohs）创立而得名。该法用10种标准矿物测定矿物的相对硬度，由小到大分为10级：滑石1，石膏2，方解石3，萤石4，鳞灰石5，正长石6，石英7，黄玉8，刚玉9，金刚石10。使用时作刻划比较得出相对硬度。例如某矿物能将方解石刻出划痕，而不能刻萤石，则其莫氏硬度为3～4，其他类推。莫氏硬度比较粗略，如虽滑石的硬度为1，金刚石为10，刚玉为9，但经显微硬度计测得的绝对硬度则金刚石的为滑石的4192倍，刚玉的为滑石的442倍；③肖氏硬度是一种回弹硬度，主要用于金属材料，方法是使一种特制的小锤或球从一定高度自由下落，冲击被测材料试样表面后，其回弹高度反映试样在

---

* 原载《混凝土》2007年第9期。

冲击过程中产生的应变能（储存继而释放），用以确定材料的表面硬度。这种仪器比较小巧，适用于现场使用，精度不高，但是方便。检测混凝土强度的回弹法用的就是肖氏硬度的原理，检测的直接读数应当是混凝土的表面硬度。

强度是混凝土在外部荷载作用下抵抗破坏的能力。不同材料的硬度和强度并没有固定的关系。例如金属这种各向同性的弹性材料，硬度和强度相关性较好；木材的硬度很低，但标准含水量的木材顺纹抗压强度则可从 20MPa 变化到约 100MPa。不同树种的强度差别大而硬度差别却较小。不同材料的硬度和强度的关系是不同的；一种材料的硬度和另一种材料的强度更是没有关系。混凝土强度是整体的表现，在整体观念上进行检测，而其表面硬度的检测则是在某些点上进行，其中的骨料和水泥浆体毕竟是两种不同硬度的材料，水泥浆体和混凝土由于粗骨料界面的影响，也是强度有区别的两种材料；水泥浆体的硬度和混凝土的强度是不能建立起关系的。我国使用回弹法已有近 40 年的历史。过去用于传统混凝土时，尽管回弹值离散性很大，而出现的问题尚未如今天这样突出。现在材料变了，还使用不变的方法，必然会造成一些突出的矛盾。例如凡是掺了粉煤灰的混凝土用回弹法测定的强度都不合格，某些质检站就增大碳化深度修正系数使其合格。这不禁使人想起“说你是时，你就是，不是也是”的童谣。在此先来质疑一下，希望引起讨论。是否应当否定这种检测方法是次要的，重要的是希望概念清楚。

2　混凝土是什么?

有个开发商在与混凝土搅拌站工作人员发生争执时训斥道：“你们有什么了不起的？不就是个和泥的吗!”这代表了人们对混凝土的认识，当前工程中出现的质量问题（尽管还不能叫做“事故”）都和这种认识有关。因此有必要在此重申一下对混凝土的认识。

混凝土是用最简单的工艺制作的最复杂体系。简单是必须的，否则不能成为最广泛使用的大宗建筑材料；但是复杂又是必然的，原因是：①原材料来源广泛而多样，成分波动而不可能提纯，所形成的微结构在不同层次上的多相、非均质，依配合比而离散；②微结构的形成具有环境（温度、湿度）和时间的依赖性；③水泥水化形成的复杂凝胶，在目前技术水平下难以测定。因此这样复杂的体系具有微结构的不确知性和性能的不确定性，使混凝土表现出“混沌体系”（非线性体系）的特征，可以说具有“蝴蝶效应”——事物发展的结果对初始条件具有极为敏感的依赖性，初始条件极小的偏差将会引起结果的巨大差异。

3　疑问

3.1　按以上所述的概念，现行技术规程的题目定为“回弹法用于检测混凝土的强度”[1]，即使能用，也只能是对混凝土强度的“推断”，说是“检测”是否欠妥？退而言之，对于当代的混凝土是否连“推断”也值得怀疑？

“回弹法用于检测混凝土的强度”的根据是认为混凝土的抗压强度和混凝土的硬度具有相关性。但是对于混凝土这样复杂的多相非均质材料来说，回弹值和抗压强度之间没有唯一的关系；不只是不同强度等级的混凝土没有相同的硬度-抗压强度关系，而且相同强度等级的混凝土也没有相同的组成和微结构；即使给定的混凝土，也会因骨料和基体之间的硬度不同以及骨料在矿物学上的变化而有不同的回弹值。合理的方法是对每一种混凝土都标定其强度-硬度关系，“……当用回弹值估计现场混凝土的强度时，必须和标定时的实验步骤与环境条件相似"[2]。把定到规范中的回弹值-抗压强度关系表格或公式作为通用标准是欠妥当的。规程规定在检测时要避开粗骨料而压在砂浆上，充其量这样得到的回弹值也仅是砂浆的，最多只能反映砂浆硬度和砂浆强度的关系。因界面的存在，在相同水胶比下浆骨比或砂率不同会影响混凝土的强度，因此，尽管砂浆是混凝土的一部分，砂浆硬度和混凝土强度却并没有固定的关系。从根本上来说，对于传统混凝土，回弹值对抗压强度只能起大体"推断"的作用，定义成"检测"实际上误导了对现场混凝土质量的评价，造成了有些人混淆了硬度和强度的概念。

3.2 混凝土碳化层和混凝土更加显然地是不同的材料，按前述"一种材料的硬度和另一种材料的强度没有关系"的原理，碳化层和混凝土总是两种材料吧，即使按不同碳化层厚度给出修正系数，仍然是把本来没有关系的两件事物硬拉在一起去对比。进一步说，材料表面硬度和材料的厚度有关系吗？材质相同的玻璃板和玻璃砖的表面硬度难道不同吗？同样材质的钢板和钢锭表面硬度应当也是一样的。按照碳化层厚度修正所测硬度推算出的混凝土强度是否荒唐？

混凝土中的$Ca(OH)_2$和潮湿空气中的CO_2反应生$CaCO_3$，称作碳酸盐化，简称碳化。碳化都从表面开始，逐渐向内部深入。碳化后的混凝土表面硬度会增大，也就是说碳化层是不同于水泥浆体、砂浆和混凝土的另一种材料。碳化层的硬度显然更不能用以推断混凝土的强度，于是规程中给出了按碳化层厚度取折减系数，以"修正"所测硬度推算出的混凝土强度。对于传统混凝土，强度高的在验收时(通常在28天)碳化深度不大，低强度等级的，因水泥强度过高，所配制的混凝土实际强度往往也超标。现今，掺入矿物掺和料，混凝土碳化后，酚酞试剂不显色的部分除了生成碳酸钙之外，还有未反应的矿物掺和料颗粒，则从整体来看，这时的混凝土及其碳化层和无掺和料时的混凝土及其碳化层又有了区别，尤其是在当前搅拌站的生产条件下，更增加了匀质性的问题[3]。对这样一种复杂体系，用简单的回弹法检测其强度有什么可靠性？

3.3 掺粉煤灰的混凝土碳化为什么会加速？

讨论这个问题的目的是说明碳化对混凝土的影响主要并不是强度，因为只要在掺用粉煤灰后把混凝土水胶比降低到一定程度，28天抗压强度无疑是会满足设计要求的，而且由于现场浇筑混凝土温度的影响，掺粉煤灰的混凝土实际强度总是会比标准养护的相同掺粉煤灰的混凝土试件强度高，并与碳化无关。

传统上认为，在混凝土中掺入粉煤灰后碳化加速是因为粉煤灰稀释了水泥中的$Ca(OH)_2$，那么，为什么掺用同样比例矿渣的混凝土碳化加速的程度会低得多呢？当然可能有人会认为是矿渣中含较多CaO之故。但是从矿相分析来看，矿渣中CaO主要为化合态，不会增加混凝土中$Ca(OH)_2$的含量，掺入矿渣似乎也会稀释$Ca(OH)_2$的浓度。传统认为碳化速率和环境中CO_2浓度有关，混凝土中$Ca(OH)_2$浓度减小时，相当于大气中CO_2浓度相对增加。这是一种概念的转移：按照Fick定律，一种物质在另一种物质中的扩散系数与其浓度有关，也就是说，CO_2初始浓度影响其扩散速率，并不等于影响碳化的速率和深度。不管$Ca(OH)_2$的浓度多少，在合适的湿度下，总是会和CO_2碳化反应的。按照现行有关规范，混凝土碳化性能的试验方法是：将试件养护到28天，在CO_2浓度为20%、温度20℃、相对湿度(60±5)%的碳化箱中碳化28天。这种方法对实际工程毫无意义，因为在实际工程中不会养护到28天。也就是说，现场混凝土的碳化都不会从28天才开始，而是停止湿养护后，混凝土表面层相对湿度下降到70%以下时，碳化就会开始。对于纯硅酸盐水泥的混凝土，碳化深度随水灰比的增加而增加，“水灰比0.4的混凝土碳化深度是水灰比为0.6的一半，水灰比为0.5的混凝土在一般条件下暴露10年，碳化深度为5～10mm[3]”；“水灰比为0.6的混凝土15年后碳化深度为15mm，而水灰比为0.45的混凝土，碳化深度为15mm时需要100年[4]”。也就是说，影响混凝土碳化性质的主要因素是混凝土的水灰比，水灰比是决定混凝土密实度的主要因素。而当掺用粉煤灰时，即使配制混凝土时能降低水胶比，使该混凝土28天强度保持与不掺粉煤灰时的一致，而其初期(例如3天、7天)强度还是低于不掺粉煤灰时的同龄期强度。从图1[5]可看出无论是掺粉煤灰还是磨细石英砂，浆体孔隙率均随掺和料的掺量的增加而增加。其中对混凝土强度有影响的是100nm以上的孔，规律亦然。由于用汞压力测孔法试验，与混凝土相比的试样尺寸太小，试验结果中可能会忽略了一些孔，尤其是大一些的孔。对气体或离子来说，在100nm以下的孔中也能在浓度差的驱使下进行扩散。

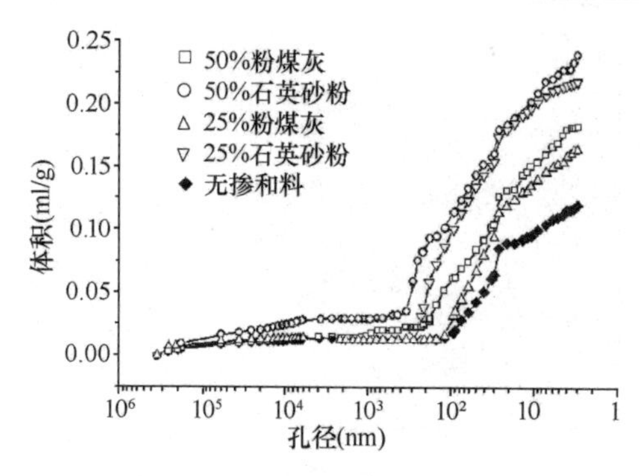

图1　掺矿物掺和料的水泥浆体养护7天孔隙分布[5]

在图2中，水化龄期应当是指在有水存在的情况下所经过的龄期，故可认为等同于湿养护的龄期。由图2可见，在一定的水胶比下，湿养护龄期越短，粉煤灰掺量越大的试件孔隙率越大；不同粉煤灰掺量的试件之间孔隙率的差别随湿养护龄期的增长而缩小；不同粉煤灰掺量的试件之间孔隙率无差别的湿养护龄期与水胶比有关，如图2中水胶比为0.35时，该龄期约在28天，水胶比为0.3时，则该龄期约为22天。对于纯硅酸盐水泥来说，在这样低的水胶比下，湿养护2天足矣，而对于掺粉煤灰的混凝土，尽管掺粉煤灰的前提是必须降低水胶比，实际工程中混凝土湿养护龄期一般不会超过7天，大掺量粉煤灰混凝土实际的碳化深度也会因孔隙率增加而增加。碳化本身不会造成混凝土劣化，但是$Ca(OH)_2$碳化后分子体积大约可收缩20%，如果先产生干燥收缩，随后再加上碳化收缩，可能在约束条件下产生开裂；更重要的是，钢筋在碱性环境下的稳定性会因碱度降低而受到破坏，引起锈蚀。对于混凝土的强度，则碳化前后并不会有太大差别，反而会因碳化而提高。对于保护层厚度很小、强度等级很低的混凝土，当无有效技术措施时，应当考虑的倒是大掺量粉煤灰混凝土早期孔隙率大而发生的碳化对可能引起钢筋锈蚀的影响，碳化后的混凝土不仅碱度下降，而且因碳化收缩，尤其是先产生干缩与继而碳化产生收缩的叠加，会使混凝土孔隙增多、增大造成表面开裂。因此，大可不必为按碳化层厚度的折减系数大小而担心混凝土的强度。

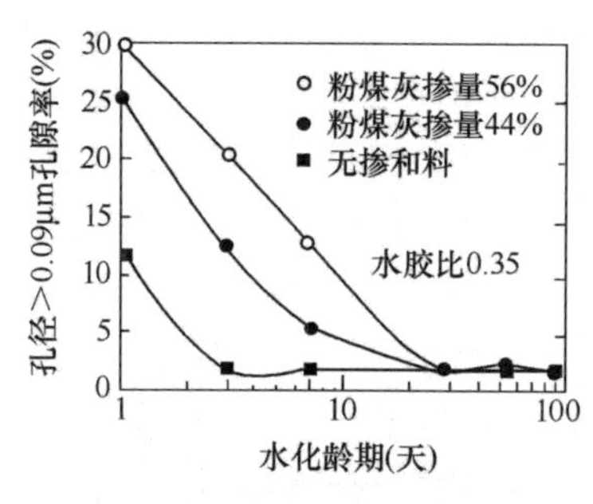

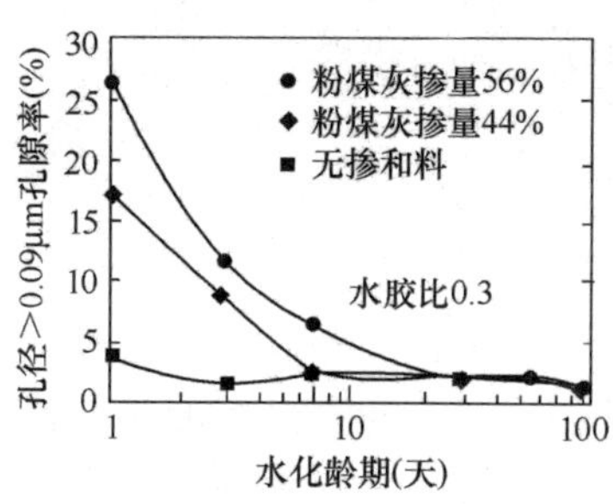

图2　不同掺量粉煤灰的水泥浆体碳化深度和水化龄期的关系[6]

3.4　工程上对碳化深度的检测和混凝土强度有关系吗?

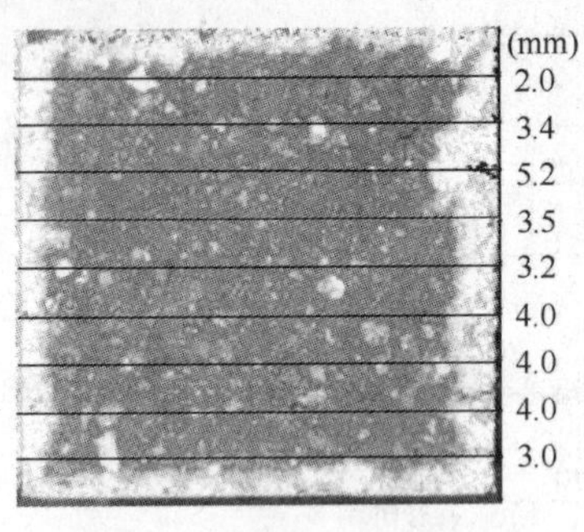

图 3　混凝土碳化前沿形状

由于混凝土材料的高度非匀质性，碳化前沿很难定量，如图 3 所示为一个 40mm×40mm×160mm 的砂浆试件在相对湿度为 50％的大气常温环境中碳化后横断面的酚酞显色，可见碳化区形状极无规则，充分显示了这种材料的非匀质性。显然，在取平均值时，选取测点位置和数量都会极大地影响计算结果。因此，取有限数量的测点时，不同时间、不同人的量测结果有很大的差异。测点数量越多，差别越小，而在实际工程中一般都是在构件上钻眼，滴入酚酞试剂，然后用卡尺量测不显色部分的深度，取 6 个点的平均值，作为碳化深度。这样的结果的代表性显然值得怀疑。而且，酚酞试剂在碱性下呈紫红色，在酸性和中性下无色，其变色范围为 pH＝8～10。$Ca(OH)_2$碳化后，pH 值可下降到 8.5。掺入粉煤灰后，$Ca(OH)_2$减少，酚酞无色之处并不都是 $CaCO_3$，还包含未水化的水泥和粉煤灰，还可能会有受大气中其他酸性介质(如酸雨中的 SO_2、工业排放和汽车尾气中的 NO_x等)作用形成的其他盐；还可能有未碳化的$Ca(OH)_2$核心；当然还有砂子和石子。因此，这个“碳化层”的硬度及厚度和混凝土的强度并没有关系，对于混凝土的强度来说是没有意义的。

4　不用“回弹法检测混凝土强度”，对工程中的混凝土强度如何验收？

在硅酸盐水泥混凝土问世之前，已经有古老的混凝土建筑和构筑物在世界上屹立了 2000 多年，例如至今仍供游人游览的古罗马万神殿，经历 2000 多年海浪冲刷至今仍完好无损、长数百米无一裂缝的那不勒斯海港，等等，尽管建造时没有硅酸盐水泥，使用的是以石灰和火山灰为胶凝材料的混凝土，却因“精心选择原材料，精心施工”[7]而有着如此优异的质量。实践证明，一般工程在实验室经过反复试配而优化的混凝土，到达现场验收合格，只要在现场不随意更动，而按合理的顺序浇筑，正确地振捣，并根据环境温度控制好入模温度和升温、降温速率，不要过早拆模，保证充分的湿养护，则混凝土的质量就不会有问题。因此过程的质量控制比“死后验尸”要重要得多。对于重要工程最好采用跟踪养护的技术进行监控和验收。因为现场混凝土构件的尺寸远大于实验室小试件的尺寸，现场混凝土构件依尺度大小和散热面积的不同，其内部的实际温度一般都不同程度地高于实验室内标准养护温度，则二者强度的发展也不同。跟踪养护即在混凝土内部一定部位（视需要控制性能的关键部位而定）埋设温度传感器，跟踪该所测温度调节试件养护池的水温。这样的试件强度可跟踪构件内混凝土实际强度。对于重大工程，可在现场预浇筑一个模拟实际构件尺寸的实体，预埋温度和应力传感器，并供结构运行期间钻芯监测其所需性能。如图 4 所示实例。高 330m 的北京国贸三期塔楼 A 工程在正式浇筑大体积混凝土底板以前，在工地现场预先浇筑了一个 4.5m×4.5m×4.5m 的足尺模型，以检验混凝土品质，观察结构内部温升、强度发展和应力分布情况，用于指导实际施工，取得很好效果[7]。

图 4　加拿大某工程现场预浇筑供测温和长期监测强度发展的 2.5m×4.0m×5.0m 实体[6]

参考文献

[1] JGJ/T 23—2001 回弹法检测混凝土抗压强度技术规程[S].

[2] N. Jackson. 卢璋，廉慧珍译．土木工程材料[M]. 北京：中国建筑工业出版社，1988.

[3] A. Neville. 李国泮，马国贞译．混凝土的性能[M]. 中国建筑工业出版社，1981.

[4] A. Neville，Properties of Concrete，4th & final edition，Pearson Education Limited，1st published 1995，6threprinted 2000.

[5] 阎培渝，张庆欢．含有活性或惰性掺和料的复合胶凝材料硬化浆体的微观结构特征[J]. 硅酸盐学报，34(12)：1491-1496，2006.

[6] V. M. Maholtra，CANMET Investigations Dealing with High-Volume Fly Ash Concrete，2006. 6.

[7] 栾尧，阎培渝等．大体积混凝土足尺模型内部的水化温升观察[J]. 建筑技术，38(1)：61-63 (2007).

本文在《混凝土》中刊登后的第二年，主持编制《回弹法检测混凝土抗压强度技术规程》的陕西省建筑科学研究院文恒武教授写了一篇对本文回应的文章。随后我又发表了一篇《讨论》的文章进一步阐述我的观点。现将文教授原文附录于此，供大家对照讨论。

附件：文恒武、魏超琪：对“质疑‘回弹法检测混凝土抗压强度’”一文中几个问题的看法*

本刊 2007 年第 9 期刊登了清华大学土木水利学院廉慧珍教授“质疑‘回弹法检测混凝土抗压强度’”一文，对“回弹法检测混凝土抗压强度”的方法提出了许多质疑，现就廉教授文中的一些观点和‘质疑’谈谈个人的看法，供同仁们讨论，以促进混凝土无损检测技术的健康发展。

一、回弹法发展史的回顾

1945 年瑞士史密特发明了回弹仪并获得专利，所以，世界上有些国家也把回弹仪称作“史密特锤”，它是借助于获得一定能量的弹击拉簧所连接的弹击锤冲击弹击杆，弹击锤连同弹击杆一同冲击混凝土表面后，弹击锤向后反弹，带动指针在回弹仪机壳的刻度尺上显示出回弹值。借助于回弹值，人们通过一定的经验公式计算，就可以获得被弹击的混凝土的抗压强度。上世纪 60 年代原天津建筑仪器厂开始生产标称能量为 2.207J 用于检测混凝土抗压强度的回弹仪，随后又相继开发出了用于检测黏土砖、砌体砂浆抗压强度和高强混凝土抗压强度的专用回弹仪。2002 年日本龟昌精机株式会社在天津投资成立贵昌精密机械（天津）有限公司主要生产回弹仪，山东乐陵回弹仪厂年产销各种规格型号的回弹仪近 20000 台，产品销往日本、新加坡、伊朗、香港、台湾等国家和地区，成为世界上最大的回弹仪生产厂家，浙江舟山博远科技开发有限公司等单位开发生产的数字式回弹仪，其技术水平已经处于国际领先。为了加强知识产权保护，我国与回弹仪有关的专利有 10 多个。

回弹法检测混凝土抗压强度的研究，上世纪六十年代由原国家建工部下达研究任务，陕西省建筑科学研究院会同中国建筑科学研究院、浙江省建筑科学研究院等有关单位进行了长期研究，其研究成果获 1978 年全国科学大会奖。在此基础上成立了以陕西省建筑科学研究院为主编单位的编制组，编制组会同全国建工、铁路、交通、水利等部门十多个单位参加，五十多个单位参与，分别对混凝土模板、水泥、外加剂、掺和料、配合比、养护条件、碳化等问题开展了专题研究。共取得实验数据近万个，通过建立不同的数学模型对实验数据进行处理分析，计算不同数学模型的误差，最终优选相对误差、相对标准差最小，相惯性最好的幂函数为基本数学模型，其相对误差≤±15% ，相对标准差≤18%，并用混凝土试块和芯样强度进行了验证。

1985 年编制成中华人民共和国行业标准《回弹法评定混凝土抗压强度技术规程》JGJ 23—1985。为了

* 原载《混凝土》2008 年第 7 期。

加强回弹规程的推广和应用，1986年建设部颁发了《回弹法规程管理实施细则》，并建立了回弹法全国管理组及各省市自治区管理小组，负责对回弹法检测人员培训考核发证，对回弹仪进行检定。为了规范回弹仪的生产，保证回弹仪的质量，国家城乡建设环境保护部于1988年4月，颁发了回弹仪产品的国家标准《回弹仪》GB 9138—88。

经过几年的使用，1992年建设部委托主编单位陕西省建筑科学研究院对《回弹法评定混凝土抗压强度技术规程》JGJ 23—1985进行了补充、修订和完善，修订的主要内容是进一步研究碳化深度对检测结果的影响。试验研究结果表明：在不同的强度区间，碳化深度对强度的影响并不完全是按照幂函数规律变化的，因而对不同的强度区间进行了修正。为了与已颁布实施的《混凝土强度检验评定标准》GBJ 107—87衔接，修订后的规程把“评定”改为“检测”，更名为《回弹法检测混凝土抗压强度技术规程》JGJ/T 23—1992。

众所周知，回弹仪作为混凝土强度检测的计量仪器，必须按时进行计量检定，才能保证检测的精度和准确性。为此，1993年国家技术监督局颁发了由陕西省建筑科学研究院编制的回弹仪计量检定规程《混凝土回弹仪》JGJ 817—93。规程对回弹仪的技术要求、检定方法、检定器械、检定周期、检定报告的格式进行了严格的规定。这就从根本上扭转了人们片面地认为，只要回弹仪的钢砧率定值达到80+2时，回弹仪就合格，就是标准状态的错误认识。回弹仪的钢砧率定值只是回弹仪的一个基本性能，是回弹仪处于标准状态的必要条件，而不是充分条件，只有回弹仪的指针摩擦力、拉簧刚度、冲击长度、起跳位置、脱钩位置等技术条件全部满足检定规程时，回弹仪才会处于标准状态，才能保证检测的准确性。

2000年主编单位对《回弹法检测混凝土抗压强度技术规程》JGJ/T 23—1992又进行了补充、修订。针对泵送混凝土的回弹法检测强度误差较大问题，修订中增加了泵送混凝土检测强度的修正值即附录B，2001年由建设部颁布实施。

我国幅员广大，地域辽阔，各地的气候、砂石、水泥、外加剂、掺和料等混凝土原材料及施工技术都有很大的差别，为了提高检测精度，《回弹法检测混凝土抗压强度技术规程》JGJ/T 23—2001第6.1.2规定：有条件的地区和部门，应制订本地区的测强曲线或专用测强曲线，经上级主管部门组织审定和批准后实施。各检测单位应按照专用测强曲线、地区测强曲线、全国统一测强曲线顺序选用测强曲线。

随着建筑技术的快速发展，泵送混凝土在我国得到了广泛的使用，泵送混凝土的特点是：流动性好、塌落度大，浆体含量高、砂率大，粗集料少、粒径小，一般均添加泵送剂和矿物掺和料，早期强度高。因此，全国许多地方相继编制了泵送混凝土地方测强曲线和专用测强曲线，从而提高了检测精度。

几十年来，经过广大检测工作者的不懈努力，《回弹法》这种适合我国国情的检测混凝土抗压强度的方法，已成为我国混凝土工程现场原位检测应用最广泛、最方便的检测方法；回弹仪已成为我国工程建设中的质量控制、质量监督、质量检测过程中必不可少的检测仪器，对提高我国工程质量无损检测水平、保证工程质量发挥了重要作用，我国已成为世界上回弹仪最大的生产国和使用国。

二、对“质疑”中几个问题的商榷

廉教授在文中写道：“发现有些人在概念上把混凝土的硬度和强度混淆了，以为硬度大的材料强度也高，回弹值就代表强度”。其实，在从事回弹法检测的人员中，大家的确认为混凝土的硬度和强度不是一个概念，混凝土的回弹值并不代表混凝土的强度，但混凝土的回弹值和强度是有一定的关系，回弹值越大其混凝土抗压强度越高。

廉教授认为：“混凝土的强度是整体的表现，在整体观念上进行检测，而其表面硬度的检测则是在某点上进行，其中的骨料和水泥浆体毕竟是两种不同硬度的材料，水泥浆体的硬度和混凝土的强度是不能建立起关系的”。笔者认为廉教授的看法是片面的。1824年阿斯普丁（J. Aspdin）发明了波特兰水泥后，水泥与混凝土的生产技术得到了迅速的发展。一百多年来，人们从微观、亚微观和宏观方面对混凝土内部结构和性能之间的关系进行了全面、系统的研究。尽管混凝土是一种非均质、各相共存的复杂体系，但是，混凝土的宏观强度理论是把混凝土当作宏观均质且各向同性的材料来研究的。回弹法是研究混凝土的表面回弹值与混凝土强度之间的关系的，它研究的对象就是混凝土，而不是混凝土中的砂浆。况且《回弹规范》

中规定在每个测区选择有效的16个测点，计算时去掉3个大的和3个小值剩余10个计算平均值，已经考虑了石子和气孔对回弹值的影响。当检测条件与测强曲线的适用条件有较大差异时，还可采用同条件试件或钻取混凝土芯样进行修正。那么，混凝土的回弹值和混凝土的强度之间到底有没有关系呢？几十年来，全国各地的许多专家和学者针对不同地区、不同的原材料、不同的配合比、不同的气候养护条件、不同的生产工艺条件下混凝土回弹值与强度之间的关系进行了系统的研究，利用数理统计的基本方法得出的关系式有：幂函数关系式、指数关系式、抛物线式、直线式等，分别用混凝土试块或混凝土构件的芯样进行了实验验证和误差分析；制订的地区和专用测强曲线有：泵送混凝土测强曲线、高强混凝土测强曲线、特细砂混凝土测强曲线、山砂混凝土测强曲线、管桩混凝土测强曲线、离心混凝土管测强曲线、水工混凝土测强曲线、港口工程、公路路面工程测强曲线，等等。难道能说混凝土的回弹值与强度之间没有关系吗？

廉教授在文中写道："规范的题目定为'回弹法用于检测混凝土的强度'，即使能用，也只能是对混凝土强度的'推断'，说是'检测'是否欠妥？"回弹法能不能称为检测混凝土抗压强度呢？请看我国目前的相关标准是怎用定义的，《建筑结构检测技术标准》GB/T 50344、《混凝土结构工程施工质量验收规范》GB 20204、《混凝土强度检验评定标准》GB/T 50107(替代GBJ 107—87，征求意见稿)等许多相关的标准规范都把"回弹法"一类的用于现场检测混凝土抗压强度的方法叫做"检测"。笔者最近查阅了1979年版上海辞书出版社出版的《辞海》，没有查到"检测"一词，与之相近的词有"检验"，其意为：用工具、仪器或其他方法对被检物体进行检查、测量看其是否合乎规格的过程。商务印书馆2005年版的《现代汉语词典》把"检测"定义为：(1)检验测定，(2)检查验看。其实，在业内"检测"这已是一个大家常用的词，人人都知其意。另外，不知道廉教授看过《回弹法检测混凝土抗压强度技术规程》JGJ/T 23没有，该规范是根据各测区的回弹值换算出测区的混凝土强值，再根据测区混凝土强度的换算值来推定混凝土构件的抗压强度值的，在附录F检测报告的格式中也表达得十分清楚，回弹法检测的结果是现龄期混凝土抗压强度的推定值。不知廉教授把规范中的"推定"叫做"推断"，这样一字之改有何意义？

廉教授在文中对混凝土的碳化及碳化对混凝土强度的影响讲得很多，混凝土表面的碳化对回弹法检测的影响是不容置疑的，尽管碳化会提高混凝土的抗压强度，但碳化对混凝土表面硬度的影响更大。众所周知，在回弹法检测中，碳化深度对检测结果有一定的影响，有时也会引起争议和纠纷，这其中的主要原因是用于测量碳化深度的方法有缺陷的缘故。目前用于测量混凝土碳化深度的方法是"酚酞法"，这是一个间接的测试混凝土碳化深度的方法，"酚酞法"测量的是混凝土的碱度，并不是碳化深度，而我们却把它当作混凝土的碳化深度（酚酞遇见碱变红），通常情况下，是没有问题的。但在实际的工程项目中，由于酸性脱模剂的使用、气候环境的影响、养护不当及外加剂和掺和料的大量加入等原因都可能会使混凝土表面"碱度"降低而出现"假性碳化"和"异常碳化"的现象，这正是回弹法要研究和解决的技术难点，现在有些单位正在尝试检测过程用砂轮机打磨掉碳化层的方法，以减少因碳化对检测结果的影响。

廉教授在文中写道："例如凡是掺了粉煤灰的混凝土用回弹法测定的强度都不合格，某些质检站就增大碳化深度修正系数使其合格。这不禁使人想起'说你是，你就是，不是也是'的童谣"。请问廉教授用"凡是"这么绝对的语言有何根据？根据我们每年检测的几百个工程项目大多是掺粉煤灰的混凝土，其结果并没有出现像廉教授所说的"凡是掺了粉煤灰的混凝土用回弹法测定的强度多不合格"的情况。况且《混凝土强度检验评定标准》GB/T 50107(征求意见稿替代GBJ 107—87)的基本规定：混凝土的强度等级应按立方体抗压强度标准值划分。回弹法检测的是现龄期的混凝土的抗压强度，其结果是作为处理混凝土质量问题的一个依据，而不是评定混凝土强度的。《回弹法检测混凝土抗压强度技术规程》JGJ/T 23总则中明确规定：当对结构的混凝土强度有检测要求时，可按本规程进行检测，检测结果可作为处理混凝土质量问题的一个依据，《混凝土结构工程施工质量验收规范》GB 20204和《混凝土强度检验评定标准》GB/T 50107(征求意见稿)也是规定其检测结果可作为处理混凝土质量问题的一个依据。

廉教授所讲的更改碳化修正系数，出具假报告的情况，确实存在，任何时候也都有可能发生，这和回弹法检测是没有关系的，其实你在压混凝土试块的时候也可以更改数据，更能使"说你是，你就是，不是也是"，况且不留任何后患，因为试块已经压碎了，无法复原，谁也无法查实。篡改实验数据是个人的职业道德和检测单位的管理水平问题，不是一个技术问题和学术问题，廉教授把它归结为回弹法检测的缘由是没有道理的。

廉教授提出的"'不用回弹法检测混凝土强度'对工程中的混凝土如何验收?"的问题。我国现行的标准、规范规定得非常具体、清楚，只要严格按照规范和标准去做，混凝土的质量是能得到保证的。那么在什么情况下需要对混凝土进行检测呢?《建筑结构检测技术标准》GB/T 50344 规定：在下列四种情况下，应该进行建筑结构工程质量的检测，1. 涉及结构安全的试块、试件以及有关材料检验数量不足；2. 对施工质量的抽样检测结果达不到设计要求；3. 对施工质量有怀疑或争议，需要通过检测进一步分析结构的可靠性；4. 发生工程事故，需要通过检测分析事故的原因及对结构可靠性的影响。有些地方的建设行政主管部门，针对当地的施工和管理水平相继出台了一些地方规定，例如，有的规定对结构的主要构件抽取一定比例进行原位检测。这些规定无疑都是符合我国国情的，对提高工程质量是非常必要的。如果廉教授把我国目前这种质量监管体系称为"死后验尸"的话，那么我认为这种"死后验尸"是非常必要的，当你搞不清楚其死因时，就必须"死后验尸"，这有什么不可以呢?

廉教授提出了"根据环境温度控制好入模温度和升温、降温速率"，"对重要工程采用跟踪养护的技术，在混凝土内部预埋温度和应力传感器"的方法。控制混凝土入模温度，在大体积混凝土和极端的冬季、炎热的夏季是很有必要的，但对于一般混凝土工程有必要花费很大的代价控制入模温度吗?在混凝土中预埋温度传感器和降温水管，这是控制混凝土升温、降温速率，减少温差，防止温度应力引起的混凝土裂缝而常用的办法。那么，对于一般混凝土结构要控制它的升温、降温速率那是劳民伤财，得不偿失，况且有些工程根本无法做到，也没有必要这么去做；在大体积混凝土中预埋温度传感器，用以监测混凝土内部温度的变化规律，从而确定冷却水管的进水量和进、出水口水的温差及养护制度，进而达到控制、减少混凝土裂缝的目的。十几年来我院已先后进行了 100 个工程大体积混凝土内部温度的监测，并研制出可以同时监测 100 多个点的大体积混凝土无线温度监测系统。在混凝土内部预埋应力传感器，不知其目的是什么?它能保证混凝土的抗压强度吗?应力传感器埋入混凝土中，在混凝土的浇筑阶段，由于重力的作用，它显示的是混凝土在该点之上的压应力，随着混凝土的不断凝结硬化，其压应力逐步消失，能够测到的仅仅是由于温度场而产生的温度应力。廉教授提出的这些办法仅是大体积混凝土测温，控制温度裂缝而已，与混凝土的强度检测毫无关系，更谈不到与回弹法有什么关系。

三、结语

作为一本用于混凝土抗压强度原位检测的推荐性方法标准，《回弹法检测混凝土抗压强度技术规程》JGJ/T 23 以其使用方便等特点，在我国工程质量控制、工程质量检测中已应用了 20 多年，它和其他标准规范一样，都有它的适应范围和使用条件，都需要继续不断的提高和完善。作为回弹法规程的编制组成员欢迎同仁们就回弹法的一些技术问题进行讨论，欢迎就回弹法检测中的经验教训进行交流，以利于提高我国混凝土强度检测的技术水平。

由于本人工作经验和水平有限，对廉教授的一些观点和见解谈了点自己的拙见，不一定正确，请同仁们多加批评指正。

参考资料（略）

论文之十九

对文恒武、魏超琪二位同志《对“质疑‘回弹法检测混凝土抗压强度’”一文中几个问题的看法》的讨论*

廉慧珍（清华大学土木水利学院）

时代在变化，过去我们熟悉的东西有些要闲置起来，而有些我们不熟悉的东西要尽快熟悉起来。

——毛泽东《七届二中全会上的报告》

经验是一条我们曾经沿着它到达今天的道路，可是世界在昨天早晨变了，经验还会是通往明天的阶梯吗？

——王宏甲《中国新教育风暴》

对一个理论的反驳（即对问题的任何认真的尝试性解决的反驳）始终是我们接近真理的前进一步。

——卡尔·波普尔《猜想与反驳——科学知识的增长》

摘　要　在过去条件下所造成合理的存在，在新的条件下就可能并不合理，需要根据新的认识进行修正。对混凝土这种高度不均匀的材料，按宏观上均匀整体评价其宏观行为是可以的，但是在厘米数量级的尺度上就不能认为它是均匀的。现行回弹法剔除 6/16 回弹数据所得数据只能属于砂浆，而砂浆表面硬度和混凝土强度之间没有固定的关系，因为混凝土是变化的。当代混凝土的“碳化层”更增加了回弹法检测混凝土强度的不可用性。回弹法用于评价混凝土结构匀质性才是各得其所。

非常高兴地看到文恒武、魏超琪二位同志的文章。文、魏二位无疑是使用回弹法评价硬化混凝土质量的专家，文章写得很好，又有理又有利，文笔也很好。

我一直希望在我国期刊杂志上不要只有一家之言，而是能形成一种讨论、争鸣的风气。人对客观世界的认识无止境，总要在前进中不断修正，不断更新，丰富认识。以往不能形成这种风气，原因之一是怕得罪人；原因之二是大多数人都有一种怕被否定的情结。可能还有别的原因，比如认为与自己无关不想管闲事。但是糟糕的是，很多人以为凡是期刊上刊登出来的文章，或出版的书中所说的都是对的。如果没有讨论和质疑，很容易误导人们的从众思维。现在二位发表自己的意见，确实令人高兴。“猜想与反驳”是重要的科学的研究方法。没有质疑，就不会有进步，不提问题就没有创新。交流、讨论的结果应当是相互修正，取得共同的“前进一步”，不存在谁是谁非的问题。存在的不一定都合理，外国的不一定都先进，专家和权威不一定都正确。在此就对二位的看法，也再谈一下看法，与大家交流。

1　为什么存在不一定合理？

1.1　存在总是在一定条件下形成的，而客观世界总是不断发展的，人类对客观世界的认识也是不断发展的；在过去条件下所造成合理的存在，在新的条件下就可能并不合理，需

* 原载《混凝土》2008.7。

要根据新的认识进行修正。发达国家的技术标准一般至少要 3 年修订一次，就缘于此。又如，过去只有像水坝那样的大体积混凝土结构，而如今体积比水坝体积小得多的“中等大体积混凝土”也出现早期的温度收缩和内部自收缩产生的应力问题[1]；再如，对于当前大多数有耐久性要求的混凝土，传统的抗渗性仪实验方法已不能评价其在含有腐蚀性介质环境下的抗渗性，于是出现电量法、氯离子在混凝土中的扩散系数法等评价方法；还有，由于高效减水剂的使用，包罗米公式除了表明抗压强度和水灰比倒数成正比的规律之外，已不能再用于当代混凝土的配合比设计，而且，已不必再要求“水泥强度等于混凝土强度的 1.5～2 倍”；如此等等，变化是绝对的，不变是相对的；符合发展规律的存在是合理的，反之则不合理。发展、变化，甚至由于发展和变化对已有结论的否定都是很正常的，但对已有结论做出的努力和成就仍然是必要的，是具有历史性意义的。而且一般不会是完全的否定，而是修正或限制条件或适用范围。例如上述现行抗渗仪实验法对低强度等级混凝土仍适用；包罗米公式可用于建立一定原材料的特定混凝土强度和灰水比关系（$y = ax - b$）；回弹法用于现场混凝土匀质性的评价简便而有效……。

1.2 混凝土是一种用简单工艺制作的复杂体系，其复杂的程度几乎可与人体相比。尤其是当代混凝土在混凝土中广泛使用外加剂和各种掺和料，水胶比较大幅度地降低。正像任何事物一样，有利必有弊，有得必有失。当代混凝土技术使混凝土强度总体上得到提高，使钢筋混凝土在大跨度、高耸的结构中得到应用；由于和易性大大改善，降低了施工的劳动强度，由于振捣不良而出现的缺陷大大减少，使施工速度得以大大提高……。但另一方面，由于组分增多，更增加了混凝土原材料成分复杂性和波动以及生产中质量控制的难度；其次，混凝土是至今各种材料中唯一不仅具有环境依赖性，而且同时具有时间依赖性的材料。其所处环境的不确定性和时间的进程造成其微结构动态的变化，也就会造成其工程行为的不确定性。除了由施工单位自己生产的混凝土（例如铁路建设）之外，就大量的预拌混凝土来说，由于商业利益的驱使，搅拌时间普遍只有 30 秒，很少数的增加到 45 秒；个别的因特殊需要，也最多只增加到 60 秒。对不掺矿物掺和料的混凝土来说，60 秒则是 $1yd^3$（$3/4m^3$）混凝土拌和物最少搅拌时间，而每增加 $1yd^3$，应增加 15 秒[2]。如果按一般搅拌站每次搅拌 $3m^3$ 计算，净搅拌时间应最少为 105 秒。现在这样复杂成分的混凝土用 30 秒这样短的时间是搅拌不均匀的。上世纪 90 年代初，中国土木工程学会曾制订《高强设计与施工指南》规定了一个复杂的搅拌流程，总的搅拌时间是 150 秒。当时这样规定是有实验和实践根据的。Shalow R 等人曾对只用硅酸盐水泥的混凝土搅拌时间和抗压强度关系的匀质性进行研究，结果如图 1、图 2 所示[3]。

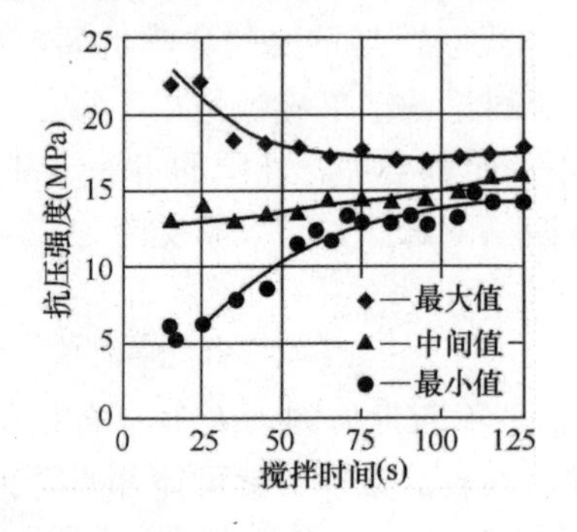

图 1　混凝土抗压强度与搅拌时间的关系

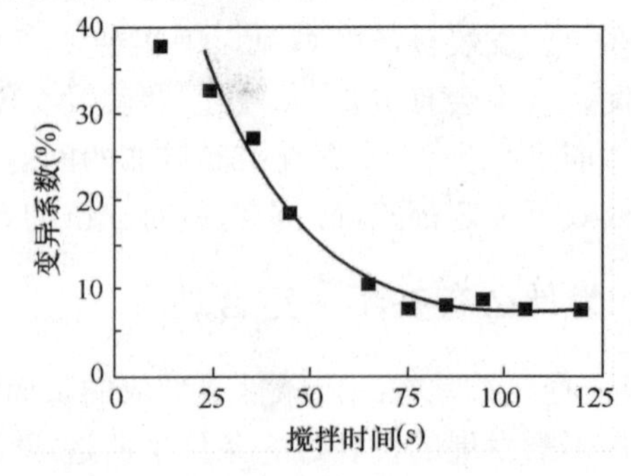

图 2　混凝土抗压强度变异系数与搅拌时间的关系

图 1、图 2 表明，搅拌时间少于 75 秒时，混凝土强度变异系数就很大。严格来说，搅拌时间应当在干料拌和 30 秒后，再加水搅拌 90～120 秒。因此现在的混凝土均质性更差。例如某工程混凝土构件拆模后发现表面有不均匀墨绿色的色斑。墨绿色是掺用矿渣粉造成的，而色斑的不均匀正是搅拌不足造成整体不均匀的表现。

1.3 "把混凝土当作宏观均质且各向同性的材料"，过去在混凝土组分不是这么多、水灰比比较大、总体强度等级比较低、生产和施工控制也没有现在这样复杂的情况下，还算可以，而现在情况正在起着很大的变化，混凝土的非均质性表现得更加突出。如今这种非均质性不仅是在于不同组分的、不同尺度的颗粒在空间上分布的不均匀，而且对不同体积的构件来说，内部和外部温度、湿度存在不同程度的差异。例如，强度高的试件在从养护室中取出擦干表面后试压破坏，总是发现中央部分是最干的，可见沿断面方向的湿度分布不一定是从中心到表面的梯度，不同强度等级的混凝土、不同组成的混凝土会有不同方式的分布；对不同组成的混凝土，因为早期内外温度差，使得在垂直于受力方向的各个断面上实际强度（即单位面积上的极限荷载）是不均匀的（即各点不同）。因此，由现行标准立方体试件得出的强度是各个承压面上各点强度的平均值，确实是"……整体的表现，在整体观念上进行检测"的，也就是说不是真值。用回弹法的主观意图虽然"研究的对象就是混凝土，而不是混凝土中的砂浆"，但是回弹锤和混凝土表面的接触面积很小，所接触的不是石子就是砂浆，这是不能否认的。即使"规定在每个测区选择有效的 16 个测点"（数理统计所需样本至少是 30 个，这里的 16 点的根据是什么?），也得不出"整体的表现"。"把混凝土当作宏观均质且各向同性的材料来研究"的"混凝土的宏观强度理论"是从混凝土结构或构件这样的宏观尺度来说的，用 150mm×150mm×150mm 立方体试件的抗压强度可做大体上的描述。而回弹法"在一个测区内"所测的 16 个点"去掉 3 个大值和 3 个小值剩余 10 个计算平均值，已经考虑了石子和气孔对回弹值的影响"，去掉了石子后剩下的不就是砂浆了吗（砂浆中也会有气孔）？在泵送混凝土中，水胶比一定时，石子越大或越多，弹性模量越大，收缩越小，而抗压强度则越低，渗透性越大。水胶比越低，这种影响越明显；一般来说在相同水胶比下，试件相同的水泥净浆强度比砂浆强度高，砂浆强度比混凝土强度高，所以严格地说三者不是一种材料。"水泥浆体的硬度和混凝土的强度是不能建立起关系的"是就普适性规律而言，并不是片面，而是笼统；相反，如果说"砂浆表面硬度和混凝土强度之间能建立关系"则必须有前提，即针对原材料和配合比以及成型、养护环境相同的混凝土。没有前提反而倒是片面的，是会产生误导的。

1.4 "推定"和"推断"并没有什么差别，只是个人语言习惯而已，但"检验"和"检测"似有所不同。"检验"是检查并验证的意思，而"检测"则有检查并量测的意思，也就是说，检验强调的是定性，检测则有定量的涵义。这个问题相对地并不重要，只是我国目前建筑工程的质检人员素质参差，许多人对回弹法检测强度的理解是后者。当然，当回弹强度不够时，都会要求钻芯，现在这种情况也很多。许多技术都和使用者的水平有关，对工程上使用的来说，"傻瓜"仪器装置似乎还没有。就连测坍落度这样简单的技术，也有人当面弄虚作假。现在质检的质量问题很复杂，有技术水平和理解水平的问题，也有很多非技术问题，当然不能都归咎于技术本身。我的原文中也并没有把那些非技术所固有的问题"归结为回弹法检测的缘由"。但是为了避免操作的误区，规范的语言应当尽量避免被钻空子。从回弹法的性质来说，改成"用回弹值估计现场混凝土的强度"[4]好像更确切些。

2 关于混凝土碳化问题的影响

回弹法在我国已经使用了30多年，过去人们对回弹法只是认为检测结果离散性大而已，不像现在这样纠纷之多。除了上述现在混凝土匀质性更突出的原因外，现在混凝土因掺入较大量矿物掺和料，碳化问题比过去突出，用酚酞试剂判断碳化深度，正如文、魏二位文中和我在原文中所述，确实有很多问题，而且由于过去对保护层质量的忽视，直到现在，不少人仍然只在关心结构表面视觉效果时才重视，有的工程在剔除表面“碳化层”后再检测回弹值，反而比未剔除时的还要高。更重要的是，“碳化层”确实是与混凝土本体不同的两种材料，“碳化层”的硬度和混凝土强度之间不可能有相关关系。因此用对“碳化层”的回弹值根据“碳化层”厚度修正的强度值的方法是不能用的。这是科学概念问题：不同材料的硬度与强度之间是没有关系的，材料的表面硬度和材料的厚度也是没有关系的。“凡是掺了粉煤灰的混凝土用回弹法测定的强度多不合格”的说法确实有失全面，不仅偏激，而且逻辑也有问题——前面用了“凡是”，后面却用了“多”。在这里接受文、魏二位的意见，并致歉。但是碳化问题毕竟使目前回弹法使用中的问题比过去的突出，本质上的原因就是对“碳化层”的认识问题。碳化检测方法还不只是二位所说的“假性碳化”和“异常碳化”现象，即使没有这些假的和异常的碳化，由于碳化前沿确定的困难，目前现场量测方法实在不能为准。混凝土比砂浆碳化前沿更加曲折，如图3、图4所示。当碳化深度较大时，打磨除去“碳化层”也不现实。

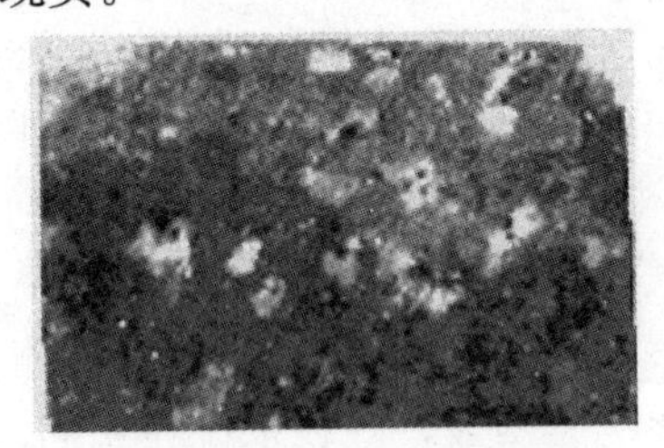

图3 混凝土墙体芯样碳化前沿

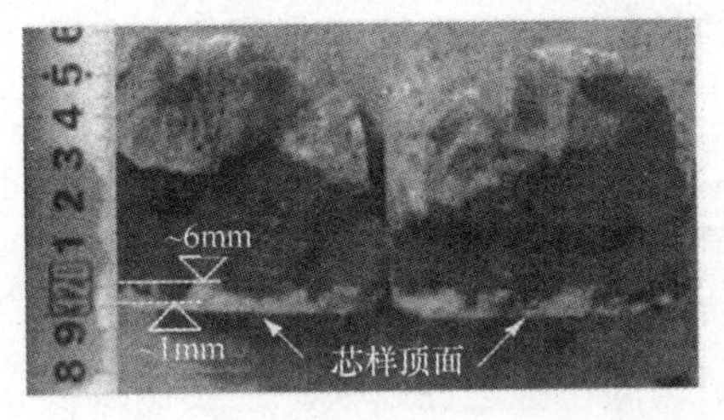

图4 现场浇筑模型混凝土芯样碳化前沿

3 科学技术发展和技术标准、规范的关系

3.1 科学主要解决“什么”和“为什么”的问题，技术则主要解决“怎么”的问题；科学研究注重的是事物的本质，技术研究注重的是从大量宏观试验中找出规律。科学研究是技术研究的基础，二者同样重要。宏观上的表象有其科学上的必然，但是不一定具有普适性。不管发明者是谁，在混凝土中使用回弹法总是从金属材料移植过来的，尽管工业上的金属材料也并不是理想的绝对均匀体，毕竟混凝土和金属材料的力学性质和均质性相差得太大。连金属的表面硬度都很难测准，何况混凝土？表面硬度的检测在金属工业中主要也是用来评价材料均质性、加工性，并不用于检测其强度。

3.2 按照我国《标准化法》的规定，和建筑有关的法律、法规和标准、规范分为三类：(1)法律、(2)行政法规、(3)技术标准。其中与建筑规范有关的法律是由国家主席令颁布的，例如国家主席第91号令《中华人民共和国建筑法》、第94号令《中华人民共和国防震减灾法》、其他如《招标投标法》、《合同法》、《标准化法》，等等；与建筑有关的行政法规是国务院令或建设部令颁布的，如国务院第293号令《建设工程勘察设计管理条例》、建设部第68号令《工程建设违法违纪行为处罚办法》、建设部第81号令《实施工程建设强制性标准监督规定》等，又如《建设工程质量管理条例》以及《强制性标准》或条文，等等。这两类本身就是法律，但都不涉及技术。涉及技术的产品标准、产品应用标准、质量标准、技术规范、技术规程等，无论国家标准、行业标准、地方标准和企业标准，其本身都不是法律，单独不具有法律效率。是国家的还是行业的、地方的，区别只在于适用范围，都不能用以进行强制性管理。技术标准（包括规范、规程、指南等）只有在委托方（甲方）和受委托方（乙方）经协商而列入双方合同的"规范清单"中才与合同一起成为有法律效力的文件。合同是双方互相信任、支持和制约的法律行为，是双方沟通和交流的过程。目前我国大部分人视一切技术标准和规范为法律是受约双方都缺少合同意识，也不了解技术规范的性质所致。于是经常发生不同技术标准中某些规定相互矛盾的问题。随着社会的发展，这种局面正在逐步改善。

技术标准本身不是法律的原因与其自身性质有关：

①任何技术都与使用条件有关，都有其适用的范围，而条件（时、事、地、使用的人）的变化是复杂的；没有放之四海而皆可用的技术，必须具体问题具体分析；

②技术标准总是滞后于科学技术研究的，但是"标准与科学技术发展密切相连，标准应当与科学技术同步，适时将科学技术（作者注：似应为'科学技术的最新内容'）纳入到标准中去。科学技术是提高标准制订质量的关键环节。反过来，如果新技术、新工艺、新材料得不到推行，就难以获取实践的检验，也不能验证其正确性，纳入到标准中也会不可靠。为此给出适当条件允许其发展，是建立标准与科学技术桥梁的重要机制。"（建设部第81号令《实施工程建设强制性标准监督规定》）；

③标准是妥协的产物和最低要求。标准是人定的，和人的认识水平有关；参与制订标准的个人会因其经历和认识的差异，难免见仁见智，共识必然是妥协才能得到；妥协的原则是先进而可操作，可操作而不迁就落后，因此必然是最低要求。

鉴于我国大众的传统认识和方便管理的目的，观念逐步转变，关系逐步理顺是需要并可能的。但是作为标准规范的管理和执行者，却应当有"天变不足畏，祖训不足法，人言不足恤"（温家宝总理谈改革、创新时引用王安石语）的精神。

3.3 顺便一说："阿斯普丁（J. Aspdin）发明了波特兰水泥"的说法是不确切的，尽管波特兰水泥的名称是阿斯普丁起的，他却不能称之为发明者，只是个申请了专利的聪明人而已。硅酸盐水泥是人类经历了上千年的生活和生产实践，不断总结、修正逐渐发展成的。硅酸盐水泥和混凝土发展的历史也证明"否定之否定"这一发展规律。了解这一规律，有助于正确对待今后科技的发展。

人类最早使用的胶凝材料是石灰，是从火的使用中得到启发开始的。有人发现经过火烧过食物的石头地面变得疏松、洁白、体积增大，加水后具有胶凝性，于是有了石灰的生产。后来，又有人在生产中发现当石灰石中含有较多黏土杂质时，烧出的石灰磨细后具有一定的水硬性，便出现用含黏土的石灰石生产水硬性石灰，但是天然石灰石中黏土含量极不稳定，逐渐悟到可以用黏土和石灰石配料并提高煅烧温度，能得到强度与水硬性更好的胶凝材料。在1813年法国维卡制得被称为现代波特兰水泥雏形的人工水硬性石灰石11年后，阿斯普丁取得了波特兰水泥专利。此后的100多年来，水泥生产工艺得到不断的改进。矿物掺和料的使用，也经过漫长的否定之否定过程。2000多年以前古罗马使用石灰和火山灰为胶凝材料的混凝土建造的结构物是对水硬性石灰的一次否定之否定，是现今概念的矿物掺和料使用的最早源头；后因其凝结缓慢而被硅酸盐水泥所替代，回到了水硬性石灰的本质——以黏土和石灰石为原料，又是一次否定之否定；随着生产的发展，硅酸盐水泥在具有化学腐蚀环境中表现出其弱点，混合材料开始用于水泥，硅酸盐水泥水化生成的$Ca(OH)_2$取代古罗马胶凝材料中的石灰，再一次实现了否定之否定。我国在20世纪90年代以前，水泥厂生产的水泥基本上都掺有混合材料，而混合材料水泥因公称强度低而不受用户欢迎；于是，由于高强混凝土的需要，硅酸盐水泥又成为主角，这一次否定之否定是以提高硅酸盐水泥高强和早强组分来实现的；20世纪90年代开始，出于混凝土结构耐久性的要求，又在混凝土中开始使用矿物掺和料，代替了在水泥生产时使用混合材料，这一次“转回去”的“上升”是由于高效减水剂的使用“否定”了混合材料水泥的公称强度。然而当前水泥混凝土的现状又出现新的问题，必然还会是否定之否定的发展，在此不予赘述。

由此可见，技术产生于人类对客观世界的认识，而认识总是不断前进的，继承和发展的关系是否定之否定的。因此“质疑”和“反质疑”对于发展来说是很正常而必须的。一般来说，事物的发展不会是对过去的完全否定，而是否定之否定，即螺旋式地上升。发展总是在前面基础上的前进。这一规律是普遍的，我国当前出现的问题在外国也曾出现过或目前也正在出现，国外专家也在质疑现行标准、规范中的问题，并逐渐修正。这是发展的必然。

4 关于过程控制中“跟踪养护”

陕西省建科院在大体积控温技术上的丰富的成果和经验不容置疑，希望你们促进和支持崔工著书，巩固你们的成果。这里所说的跟踪养护（temperature match condition）技术和大体积混凝土控温技术不是一回事，不是为了控温，而是用于评价现场构件混凝土性能。该技术的提出是因为，现今因水泥强度高、水化热大、混凝土水胶比较低，即使很薄的构件（例如厚度10cm的楼板）混凝土中的温度都不会是恒定的20℃，不论是高还是低，实验室检测的指标和现场构件中混凝土的性能差别都会较大。例如，当混凝土内部温度较高时，对不用或少用矿物掺和料的混凝土抗压强度，构件中的要低于实验室标准试件的，用实验室指标验收就会不安全；使用矿物掺和料的，则相反。跟踪养护是用埋设于构件内部的温度传感器所测出的温度调节混凝土试件养护水温度，则该试件抗压强度检测值可较好地反映现场混凝土内部实际强度。对一个构件只要埋设于中心点和保护层两点即可。甚至中心并不重要，重要的是保护层。对重要的工程，保护层比较厚时，只在钢筋表面埋一个即可，代价并不会很大。对一般结构，用与结构相同混凝土的随构件养护小试件（即所谓同条件养护试件，实际上条件也不相同）大体评价保护层混凝土也是可以的。保护层质量比中心强度重要。应力传感器是用于评价结构开裂性用的，一般并不需要，而且用目前的应力传感器技术上不成熟。

我在行文中常有对用语推敲不够而欠全面之处，以致会引起一些误解，在此再次感谢文、魏二位提醒，使我得以改正，望今后继续交流讨论，共同进步。

参考文献

[1] R. Springenschmid. Avoidance of Thermal Cracking in Concrete at Early Ages. Munich, Germany. 1998.

[2] ACI 304.1R-89. Guide for measuring, mixing, transporting, and placing concrete, ACI manual of Concrete Practice, 1994.

[3] A. Neville. Properties of Concrete, 4th & final edition, Pearson Education Limited, 1st published 1995, 6th reprinted 2000.

[4] N. Jackson. 土木工程材料，卢璋，廉慧珍译. 中国建筑工业出版社，1988 年.

自评

1. 我确实非常高兴文恒武教授能提出不同意见，因为不同意见的辩论是一种重要的科学研究方法，只有在辩论中才能不断发现问题，深入求索。

2. “凡是掺了粉煤灰的混凝土用回弹法测定的强度都不合格”确实说得绝对了，但是掺粉煤灰的混凝土标准养护试件和芯样强度合格，而用回弹法检测，该混凝土强度却不合格的实例很多。如何解释这种现象？

3. 为了尽量符合工程实际情况，我校博士生贾耀东曾做过一个试验：混凝土中掺粉煤灰 35%、水胶比 0.43、普通水泥 273kg/m^3，拆模后自然碳化 180 天，分析其表面碳化层 3mm 范围的成分，见自评用图 1 所示。结果表明，$CaCO_3$ 总量只有 6.5%，未反应的 $Ca(OH)_2$ 约为 2%。可见，掺入掺和料的混凝土表面“碳化层”遇酚酞试剂后不显色并非真正的完全碳化。大量尚未参与反应的粉煤灰和尚未水化的水泥，与少量碳化生成的 $CaCO_3$ 同时存在，遇酚酞试剂也不会显色，当然也就不会具有碳酸钙的硬度。那么，将其看作完全的“碳化层”而计算的强度再按“碳化层”厚度折减，就会使本来就不高的计算强度值更低。达不到设计强度也就成为可能。

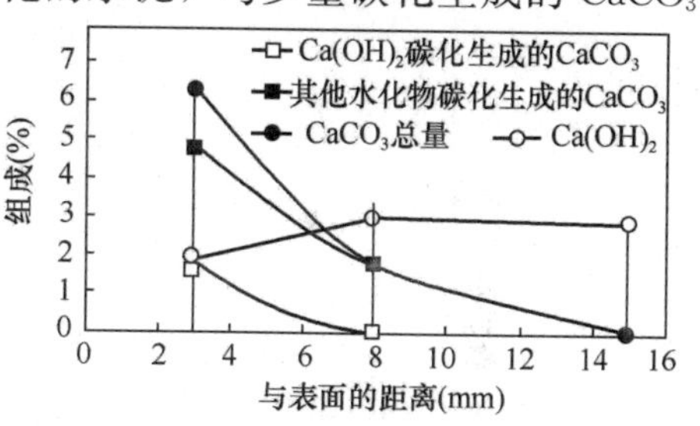

论文之十九自评用图 1　贾耀东分析的“碳化”表面成分结果

4. 在传统的感性认识上，认为碳酸钙硬度很大，这是对以碳酸钙为主要成分的石灰岩（俗称石灰石）的感受得来的概念，石灰岩的莫氏硬度为 7 左右，而化学纯 $CaCO_3$ 的莫氏硬度为 3（10 级的莫氏硬度中石膏硬度是 2）。那么，由 $Ca(OH)_2$ 碳化生成的 $CaCO_3$ 的硬度也不会靠近 7，而会靠近 3。也就是说，是否还应当怀疑一下即使真正完全碳化后的硬度需要折减么？此外，回弹法检测的是表面硬度，和被测物厚度无关。对于最接近材料力学假设的金属材料，也不用表面硬度去检测强度。表面硬度是和材料加工性质（切割、刻划、雕琢、刺入、压陷、磨耗，等等）相关的指标；强度则是抵抗在荷载作用下的断裂和永久变形的性质。这是两项不同的性质，由于检测方法不同，其间的相关性只是定性的。

5. 英国教科书中称：用回弹法“即使熟练的操作者，也不能很确定地预测强度。但是该实验简单而方便，所以常用于作为对现场混凝土初步的研究，例如用已知优质混凝土和可疑混凝土进行比较，以评价其匀质性。”

6. 从论文之十九自评用图 2 的混凝土回弹值和强度关系可见实验数据的离散性很大。所有数据点约有 130 多个，每个点还是 5 个读数的平均值。几乎没有一个点落在所回归的直线上（文献没有给出相关系数）。想来我国回归的方程不会比这个的相关性更好吧？

7. 2012 年 9 月，中国混凝土与水泥制品协会在西安举办了一个沙龙，讨论回弹法检测

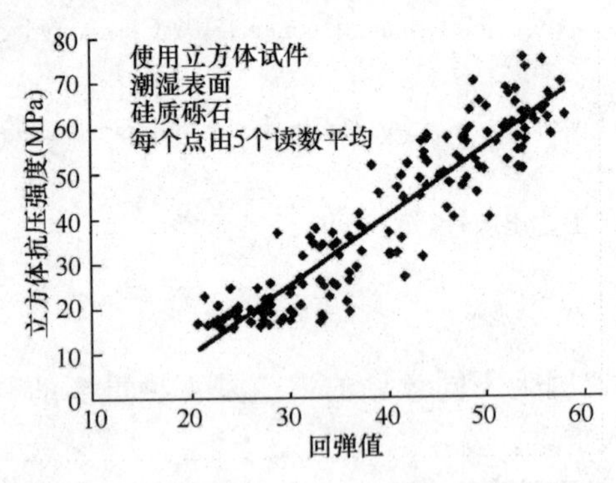

论文之十九自评用图2 美国伊利诺伊州立大学土木系学生数年试验积累的数据回归结果

混凝土强度的若干问题。与会专家们一致认为，过去，我国传统混凝土强度等级低，组分相对简单，操作者能按规范使用回弹法，没有出现过什么大问题。因为有了高效减水剂，现代混凝土比传统混凝土复杂得多。但是在当前还没有更有效的方法对结构混凝土进行验收的情况下，回弹法可以做为施工过程中的一种质量控制方法：在一定龄期内发现结构回弹值偏低时，可有足够的时间在结构验收前对该结构进行有效的养护予以补救；还可以作为一种对结构匀质性检测的手段；如果用于估计结构中混凝土强度，应当对回弹法的使用范围进一步缩小，在可使用范围内严格执行操作规程，特别要注意三个要素：使用条件、被弹结构条件、操作的规范性。如能在回弹前将回弹目标区域打磨光滑，所得到的数据就会比较真实。

书籍是思想的航船，在时代的波涛中破浪前进。它满载贵重的货物，运送给一代又一代。

——培根

四、主要的不是技术问题

论文之二十

中国混凝土向何处去?*

——写在中国混凝土与水泥制品协会预拌混凝土分会成立之际

廉慧珍（清华大学土木水利学院土木工程系）

摘　要　原材料性能指标的微小变化、质量的波动和环境的差异、操作的差异，以及时间的推移，都会造成混凝土微结构的变化而明显影响拌和物与结构物的宏观行为。至今还没有任何手段能够确切地了解混凝土内部在不同条件下究竟发生了什么、将会发生什么。混凝土具有微结构和性能不确定与不确知性，是一种混沌体系（非线性体系），初始参数微小的变化可能引起最终结果的巨大偏差。因此，不能用线性思维进行混凝土工程。做好混凝土首先需要转变观念：改变以强度为唯一目标的传统观念，高度重视耐久性；改变企业只为赚取最大利润而损害质量的做法，以获得社会的信任；按照工程需要和应用环境，选择和正确使用原材料；端正对混凝土质量的认识；提高管理水平，加强混凝土生产和使用的过程控制与服务。不仅需要技术创新，也需要在生产关系上创新。

中国混凝土与水泥制品协会成立了一个预拌混凝土分会，应当是一件好事。协会的水平代表行业的水平，也可代表国家的水平——管理水平和科学技术水平。国内外混凝土业界都知道，美国有个 ACI（American Concrete Institute 美国混凝土协会），还有个 ASTM（American Society for Testing Materials 美国材料试验协会），都是民间组织。ASTM 根据 ACI 的技术报告制订的材料标准也叫 ASTM（American Standard of Testing Materials 美国材料试验标准）。ACI 和 ASTM 不仅在美国具有权威性，而且在全世界都有很大的影响。如今我国混凝土和水泥制品行业也有了一级协会，希望能起到引导行业进入世界先进行列的作用。

协会是什么？协会是代表行业利益、为行业服务的民间团体；也是协助政府管理行业，规范行业行为的组织者。协会的功能主要是保护行业的利益，协调与其他行业的关系，提高行业的水平。

随着我国的改革开放，起步很晚的中国混凝土取得了飞速的发展。我国混凝土前辈黄大能先生 30 年前就有个愿望：在中国推行预拌混凝土。黄先生这个愿望现今已得以实现，我国混凝土工程已经蓬勃发展起来。然而，任何举措都有得必有失：当出现一种倾向时，必然会隐含着另一种倾向。我们在做我们的前人没有做过的事，尽管外国人做过，未必符合我们的国情，前进中出现一些问题是正常的。发现问题，才能去解决问题而继续前进。只能见阳

---

* 原载《混凝土世界》2009.10。

光，见不得阴影，就是讳疾忌医。科学地分析行业发展中出现的问题，找出原因并加以解决才是正确的态度，而当前的预拌混凝土确实存在一些问题。

一、现在的混凝土工程质量是进步了还是退步了？

随着我国机、电和电子工业的发展，适应预拌混凝土发展的计量、搅拌、运输等工艺得到大幅度的进步，高效减水剂的普遍使用和发展改变了混凝土的一切：高强、高流态、高程泵送、在狭窄空间的浇筑，等等，都已成为可能和现实；拌和物匀质性提高，施工方便，因振捣不善而造成的缺陷得以避免，极大地加快了建设速度。另一方面，从总体上看，从业人员的素质低，质量管理和控制水平差，以致混凝土结构的质量事故和裂缝比过去出现得多了，因质量而造成供需双方的纠纷多了。即使眼前没有发生问题，也已存在不少隐患：混凝土配合比的报告大部分失真；混凝土拌和料运到工地后，加水现象普遍；10 年前混凝土的骨料都用水洗，现在一部分砂子含泥量高达 7%，等等。

看来，不是进步不进步的问题，而是有得必有失！

二、水泥和混凝土的强度越来越高是进步了还是退步了？

1. 上世纪 70 年代水泥最高标号是 500＃，相当于现在强度等级的 32.5；现在 52.5 的水泥从强度来说相当于 20 年前的 625＃水泥。但是，过去规定水泥的储存期为三个月；而今不再提储存期，除了必要性外，重要的在于，储存期不到一个月后，52.5 水泥的强度就会和 42.5 水泥的强度相差无几，而且用户反映“不好用了”。

2. 有了高效减水剂，实现混凝土的高强已经不是难事，混凝土得以用到高层和大跨结构。人们对高强混凝土觉得还不够过瘾，时尚的是崇拜和追求“超高强”、“特超高强”。混凝土 7 天甚至 3 天，强度达到 28 天设计强度值的 100%，一般都能做到，但是后期强度不增长了，开裂敏感性增大了，对缺陷的自愈能力下降了，混凝土结构的耐久性又将如何保证？由图 1 和图 2 中可见强度不同的混凝土强度发展历程的差别。

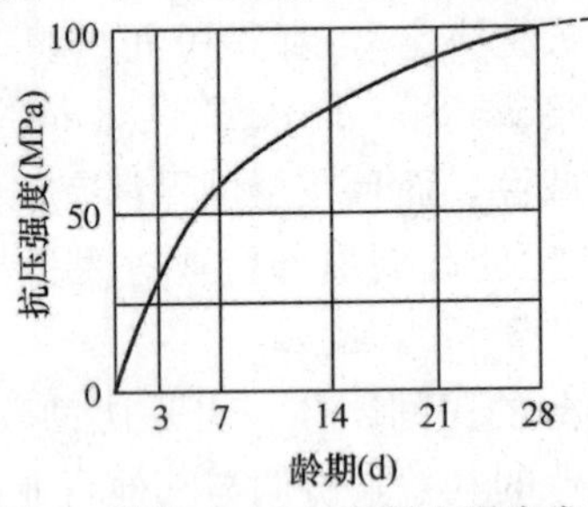

图 1　传统 30MPa 混凝土强度发展

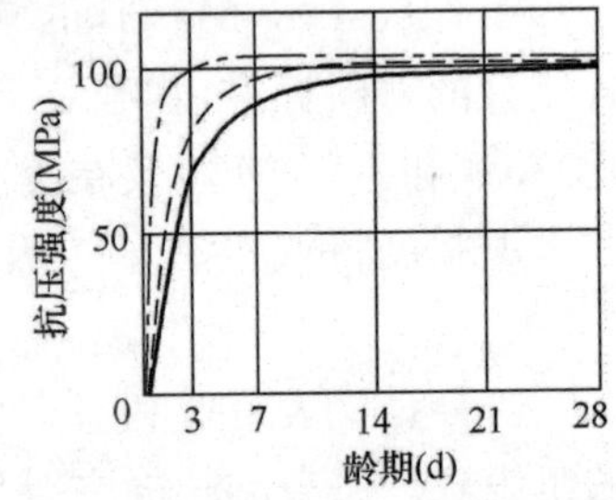

图 2　初期强度越高，发展余地越小

图 1 是传统的混凝土强度正常发展规律：以 28 天设计强度为 100%计，3 天约 30%，7 天约 60%。图 2 中，C50 以上的混凝土强度 3 天就可达设计强度的约 70%，14 天可达 90%；如果混凝土 7 天强度就达到设计强度的 100%，则到了 28 天就增长的很少，甚至几乎不增长；如果 3 天就达 100%，则 7 天还能长一点，28 天就基本不长了，甚至会倒缩。

美国的 Withy 在威斯康辛大学从 1910 年开始了 50 年水泥净浆、砂浆和混凝土的实验计划，浇筑了室内和室外混凝土。分别于 1910 年、1923 年和 1937 年成型了 5000 多个试件。Washa 和 Wendt 于 1975 年发表了这些试件 50 年观测的结果。结果是：1923 年用

Blaine 细度为 231m²/kg 的水泥配制的混凝土 28 天强度为 21MPa，25 年强度达到 52MPa；1937 年用当时的快硬水泥配制的混凝土 28 天强度为 35MPa，5 年达到 53MPa，10 年后强度开始倒缩，25 年强度就倒缩至 45MPa，比水化慢的混凝土 25 年强度还低。当时的快硬水泥与当今美国的 I 型水泥和 II 型水泥矿物组成与细度相当（C_3S 为 57%，勃氏比表面积为 380m²/kg，我国常用水泥也如此），见图 3[1]。

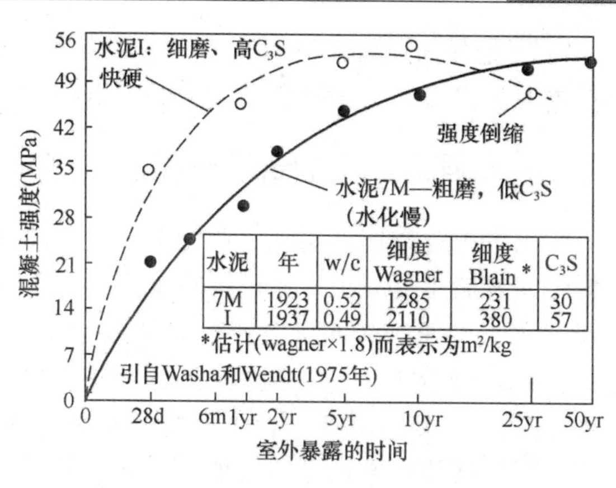

图 3　不同强度水泥配制的混凝土强度发展

3. 凡是提高早期强度的措施对混凝土后期性能都会有损害，这已是国内外专家的共识[1],[2]。前 RILEM 主席、德国混凝土专家 Rupert Springenschmid 证明控制混凝土 12 小时抗压强度不超过 6MPa，就可以避免混凝土的早期开裂[3]；我国混凝土专家黄士元经过试验得出结论：可用混凝土 24 小时抗压强度不超过 12MPa 来控制混凝土的早期开裂（见图 4）。

4. 高强是为了减小构件断面，例如高度一定的混凝土柱子，强度越高，柱子可做得越细，但是，柱子的高度与其最小断面尺寸的比例还受“压杆稳定”问题的限制（见图 5 的示意）。对一定高度的柱子来说，有个最小断面的要求，也就是说，对强度的要求是有限的。有人说，“目前房屋建筑应用的超高强混凝土达到 180MPa”，这有必要吗？

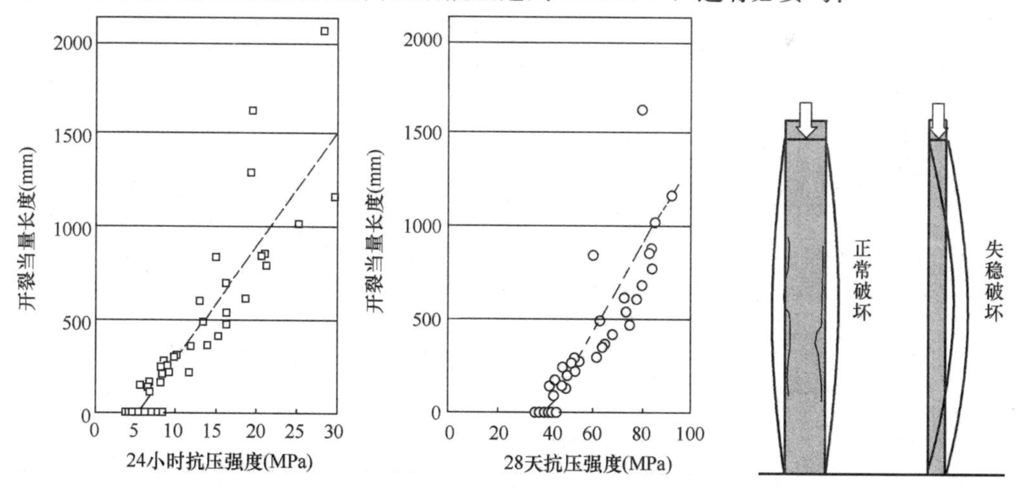

图 4　混凝土初期和早期抗压强度与开裂敏感性的关系　　图 5　压杆稳定问题示意

看来不是进步不进步的问题，而是有利必有弊，各有各的用途，需要具体问题具体对待。

5. 水泥现在“不好用”，应当说主要是用户误导的结果。由于提高混凝土的强度的要求，100 多年来水泥的发展目标除了降低能耗之外，主要就是追求强度。从煅烧工艺上提高熟料强度已取得成效，但是毕竟有限，当前水泥强度的提高还离不开粉磨得过细和五花八门的助磨剂、增强剂。于是在为混凝土提高强度的同时，也带来了损害耐久性的隐患。解铃还须系铃人！关键的是需要从用户这里转变传统的观念。

6. 无论如何，既然有利有弊、有得有失，我们总是应当看见并尽量避免或减少弊和失，取得更大的进步，尽量避免进一步退两步。

三、预拌混凝土的苦乐历程

摆脱手工业操作，实现工业化生产，是社会生产发展的必由之路和必然规律，因此预拌混凝土的发展是混凝土工业进步的重要标志。预拌混凝土推广的初期曾经“暴利”过一时，1996 年时的北京，按北京市公布的预算定额预拌混凝土出厂价格和 1995 年下半年《价格参考信息》公布增加的水泥价差及砂石价差计算，并取上限，包括原材料价格和运费、原材料 8%的损耗、厂站管理费（含人工费）及动力机械费、两税一费及计划利润，另考虑高强度混凝土所提高的技术含量而增加的管理费和附加利润，等等，所计算的 C60 混凝土价格为 570.64 元/m^3，C50 价格为 425 元/m^3，而实际上当时市场供应的价格从表 1 可见一斑。

表 1　1995～1996 年北京市部分高强预拌混凝土价格（元/m^3）

混凝土强度等级		C50	C60	备　注
使用工程	航华科贸中心		860	
	华普祥业大厦		1100	
	建威大厦		850	
	财税大楼		660	
	首都机场二期		790	
	恒基中心	570		
	大成广场	600		
	通讯中心	550		
	建威大厦	520		
根据北京市预算定额计算		437	587	未计附加利润
根据市场实际计算		387	495	未计附加利润
		425	570	计入附加利润

注：附加利润指的是因高强混凝土技术含量所增加利润。

在表 1 中，十几年前实际计算的价格所含利润至少有 20%，而市场销售后的平均（除去最高价 1100 元）利润超过 300 元/m^3，最高价的利润可近 500 元/m^3！1996 年上海宝钢申包混凝土公司 C30 混凝土成本约 300 元/m^3，售价为 660～680 元/m^3，利润约 50%。这在当时是很诱人的，使人们趋之若鹜，纷纷挤向这个“能发财”的独木桥，“有条件的上，没条件的也上”，潮起潮涌，泥沙俱下。现在大城市的搅拌站数量比那时翻了几倍，良莠不齐，造成“僧多粥少”，时至今日，由于恶性竞争造成部分搅拌站“难以度日”，混凝土价格一降再降，每立方米有几元的利润也好，赔本也撑着，结果造成预拌混凝土行业由乐变苦。现在使混凝土搅拌站苦恼和无奈的还有甲方的严重拖欠款，与监理、质检站的矛盾，规范的限制。

更重要的是原材料无法控制：不知道水泥里掺了什么，掺了多少；砂石已经基本上没有质量符合标准的产品出厂了，使强度越低的混凝土越难打，用水量（水泥用量）居高不下；以 C30 混凝土为例，用水量普遍大于 175kg/m^3，近年来甚至达到 200kg/m^3，C30 以下泵送混凝土普遍严重离析、泌水。混凝土耐久性问题确实潜在着隐患。

四、造成上述状况的原因很复杂，大多不是技术问题，首先是对混凝土认识的误区和利益的驱使

1. 有了预拌混凝土后，施工人员越来越不懂得混凝土了，施工中正确的浇筑和振捣也由此而被忽视。当前普遍使用高效减水剂和矿物掺和料的混凝土拌和物具有触变性，看似很黏的拌和物在振捣和泵送的动力作用下，仍会流动得很顺畅。现在浇筑混凝土的一线工人对此并不认识，误以为会难浇筑。以至于预拌混凝土运到工地后，几乎没有不往里加水的；如今浇筑混凝土很多是由“专业队”承包，对工程质量既不必关心，也没有责任，“打一枪换一个地方”，不让加水就撂挑子走人。施工人员不得不行此“加水”的无奈之举。然而混凝土工认为不加水的拌和物难打也有其道理：重要工程的实践证明，凡是使用刚购买的新泵，坍落度 130～150mm 的混凝土都能很顺利地泵送，而且不泌水、不离析，而今所使用的泵，为了省钱，往往不及时更换泵机，泵压自然会不足。混凝土生产和施工的各自独立，造成新的行业隔离，这种生产关系对生产力的影响已经很明显了。设想一下，混凝土搅拌站是否可以这样来做一种有偿的售后服务？——自己成立一个施工的“混凝土专业队”，招收有一定文化（例如高中毕业）的混凝土工，认真进行培训和管理，取代当前那些并不懂混凝土的“专业队”，在售出混凝土后，同时承包混凝土的浇筑、振捣和养护。

2. 人们认为搅拌站技术简单，投资少，门槛低。有的老板认为“混凝土有什么了不起？不就是和泥嘛！”门槛的形成，其实是由人的感觉而设置的。越需要投资大的，越令人感觉门槛高；越是不熟悉而新发展的，人们越会认为其门槛高。实际上任何专业都可能有很低的门槛，关键在于对品质和质量有怎样的要求。有的专业越不懂越觉得神秘莫测，进门后才知道不过如此。对混凝土，越不懂，胆越大；越懂得，胆越小。很多事是想做就能做，但是能做好就难了，关键是标准，是你心中所设的门槛。从技术来说，预拌混凝土的门槛并不低，而是因人们对混凝土认识的误区而设置的门槛过低。其实任何行业都有不同的门槛，关键是如何要求。计算机技术应当是高科技了，但是在一些大城市以“攒电脑”、“攒手机”为生的成千上万“打工仔”、“打工妹”，大多数并非计算机专业出身，最多是高中毕业。从工艺上，混凝土看似简单，简单得“……通常认为任何站在那里没事干的人都能直接就去浇筑或捣实混凝土”[4]；但这是客观的要求：必须使用地方可得的原材料和尽量简单的加工工艺，以适合于最大宗建筑材料使用的需要。否则，如果将原材料提纯、精加工，那就会使大多数工程承受不起，而只能用于“精品”的工程。也正因这种简单，就造成混凝土从宏观到微观各层次多相、非均质的复杂性。复杂得至今无法建立实验室指标试验结果和复杂的现场条件下混凝土行为的相关关系；原材料性能指标的微小变化、质量的波动和环境的差异、操作的差异，以及时间的推移，都会造成混凝土微结构的变化而明显影响拌和物与结构物的宏观行为。直到今天，还没有任何手段能够确切地了解混凝土内部在不同条件下究竟发生了什么、将会发生什么。因此混凝土具有微结构和性能不确定与不确知性，是一种混沌体系（非线性体系），具有“蝴蝶效应”——初始参数微小的变化可能引起最终结果的巨大偏差，即“差之毫厘，失之千里”。混凝土也有此现象，例如完全相同配合比的混凝土，使用针片状颗粒为 5%的石子时，拌和物可以顺利泵送而自密实；当石子针片状颗粒为 7%时，只增加 2 个百分点，就发生堵泵。因此，不能用线性思维进行混凝土工程。

五、矿物掺和料使用的误区首先也来自利益的驱使

混凝土生产人员在老板不断降低成本的压力下，对矿物掺和料，从过去的不敢用，到现在的“胆大妄为”——由于高效减水剂的普遍使用，混凝土的强度满足设计要求已不是难事，只要把水胶比降下来，不分场合，什么都敢掺、多少都敢掺。但是问题不在于强度。请看一例：

清华大学李飞用3种水胶比和普通硅酸盐水泥，分别掺入不同量粉煤灰，按胶凝材料与砂1∶3制作砂浆试件；成型1天后拆模，置于相对湿度（60±5）%、温度（20±3）℃的自然空气中，分别于3天、7天、28天、90天、1年，自端头切除约1cm后，用酚酞试剂检测其碳化（这里权且使用“碳化”的名词）深度。3种水胶比的试件在3天时基本上都无明显碳化表现（因为混凝土湿度下降需要时间），7天开始已显现。图6所示为自然碳化90天和365天的结果。试件断面为（40×40）mm^2，可见，当水胶比0.5时，掺20%的粉煤灰，1年的碳化深度平均就超过10mm；对C30来说，目前用得最多的水胶比为0.45左右，则1年的碳化深度估计可达10mm左右。对于目前厚度只有100～150mm的房屋楼板来说，其钢筋的混凝土保护层厚度一般只有15mm，如果粉煤灰掺量不是20%，而是30%、40%，甚至更多，则会是什么样的后果呢？当然这是比较极端的结果，因为同样水胶比和粉煤灰掺量，用于混凝土时，会比图6所示的结果好些，如果养护时间长一些，就会有更大的改善。

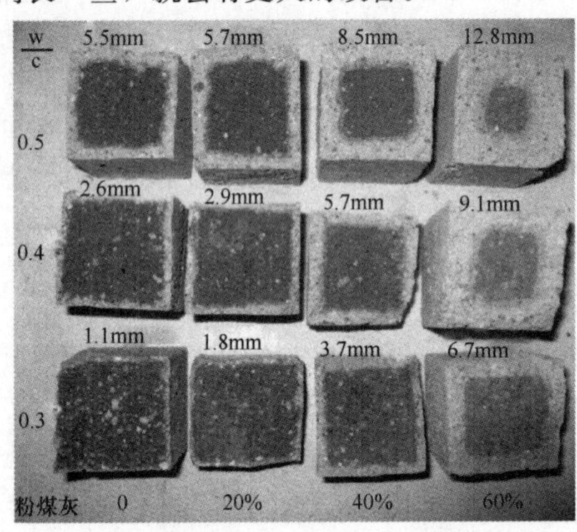

图6　42.5普通水泥1∶3砂浆粉煤灰掺量不同的试件自然碳化90天和1年的结果

以上均指处于《混凝土耐久性设计规范》划分的Ⅰ类环境的混凝土结构，因为碳化和钢筋锈蚀的条件是不同的，如果常年大部分时间环境相对湿度不超过70%，则对钢筋锈蚀而劣化的威胁就不大，但是碳化引起的收缩和干燥收缩的叠加会增大混凝土的收缩（见图7）[2]，而导致增加结构混凝土开裂的敏感性，影响结构物的适用性与增大混凝土劣化的速率。对于有冻融循环作用（II类）和海水、氯盐（Ⅲ类、Ⅳ类）及其他化学腐蚀性作用（V类）的环境，掺用矿物掺和料的混凝土早期（28天以前为早期，7天以前为初期）孔隙率随

矿物掺和料掺量的增加而增大，如果过早接触环境作用，则会提前遭到劣化，影响结构耐久性。因此，应根据环境特点、养护期长短、初次接触环境作用时混凝土的龄期、混凝土表面处理情况等条件确定矿物掺和料的种类和掺量，或者根据矿物掺和料种类及掺量确定养护制度及允许初次接触环境作用的时间。对于完全处于氧气不足的水下或地下的结构，则矿物掺和料可以掺到最大量。此外，不同的矿物掺和料有不同的特性和不同的作用机理，并不是怎么掺都行的。例如近来在我国刚流行起在混凝土中掺石灰石粉，使用者主要的目的还是为了降低成本，而对水化碳硫硅酸钙的腐蚀并不了解。当构件处于含硫酸盐的腐蚀环境（或者混凝土中含有反应剩余的石膏），有水的存在、温度低于15℃的条件，会生成没有胶凝性的$CaCO_3 \cdot CaSiO_3 \cdot CaSO_4 \cdot 15H_2O$（thaumasite 硅灰石膏），使混凝土软化。这个问题的第一次报道是美国公路研究委员会，直到上世纪90年代以后，在欧美及南非也发现不少此类破坏，人们才开始注意。2002年在英国首次举行“碳硫硅钙石型硫酸盐腐蚀”国际会议，表明对该问题的系统研究刚刚开始[5]。二十几年前我国原南京工学院张春梅在加拿大做访问学者时，曾做过用磨细石灰石粉提高混凝土早期强度的实验研究，证明掺量2%是有效的，可早期生成针状水化碳铝酸钙（$3CaO \cdot Al_2O_3 \cdot 3CaCO_3 \cdot 32H_2O$），提高早期强度。但因多量$CaCO_3$可能引起硅灰石膏的生成，加拿大规范规定石灰石粉掺量不超过5%。我国有些搅拌站尚无大量试验研究数据支持的情况下，只凭抗压强度，不分场合，将石灰石粉掺到20%以上，是很不慎重的。至于其他尚无规范依据的各种工业废渣都有其各自的利弊，更不能不问情由，只要抗压强度满足要求就敢用。有很多“好东西”也需要“会用”。

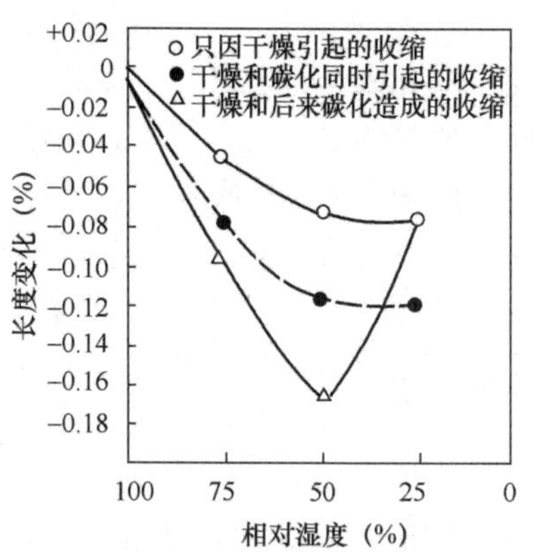

图7　碳化收缩和干燥收缩的共同作用

六、混凝土配合比设计只要原理清楚，原则明确，不需要定规范

我国现行《混凝土配合比设计》的国家标准仍然使用Bolomy公式，按强度计算水灰比。在美国有人曾建议以强度为基础确定混凝土配合比，“因为耐久性试验作为常规方法成本太高，而且也太费时间”，“这是很大的倒退，因为强度变化可能与耐久性变化很不一致。此外，利用比较细的水泥、某些外加剂以及特殊的养护方法所得到的7天或28天强度无论如何都不能代表混凝土的质量。”[2]而且，由于高效减水剂和矿物掺和料的大量使用，用了近100年的Bolomy公式已不再适用。据说目前我国《普通混凝土配合比设计规程》（JGJ 55—2002）正在进行修订。其实“混凝土配合比设计”首先是要明确配合比的原理和配制原则，至于具体设计方法、步骤不必制订成规范。由于原材料变化的影响很大，混凝土的配制很大程度上要依靠经验进行试配。在实验室工作多年、肯钻研、有经验的技术人员可以在合理的原则下发挥自己的专业才能，创造自己特有的有效方法。用规范规定一种方法则会限制技术人员的能动性和创造性。浏览国外发达国家的混凝土配合比设计方法，可发现其各有千秋。我们可以也应当从基本原理出发，在混凝土技术规范中规定混凝土配合比设计的原则。

当前我国配合比设计存在的问题主要是：

1. 在理论上，最大密实度的混凝土拌和物由水填充胶凝材料空隙形成水泥浆，再由水泥浆填充砂子空隙形成水泥砂浆，最后由砂浆填充石子空隙形成混凝土。这就是计算混凝土配合比的绝对密实体积法，简称绝对体积法。过去，混凝土强度等级范围较窄，在不使用高效减水剂和矿物掺和料时，这样计算出的1m^{3}混凝土的质量和实际试配出的拌和物质量大体吻合，约为2400kg，随强度的提高密度增加不大。为简便起见，使用“假定容重法”被普遍使用。本世纪以来，因矿物掺和料的普遍使用，混凝土强度范围增宽，不同矿物掺和料的密度和水泥密度的差别、低强度等级和高强度等级用水量的不同，还用“假定容重法”时，就使混凝土实际体积不准确，浆体体积也变化不定，会影响硬化混凝土的体积稳定性。目前全世界只有我国使用“假定容重法”，显然应当改变。有人说：“反正混凝土试配时都要根据实测调整密度，假定多少都没有关系。”需要注意的是总密度调整后，不能保证浆骨比不变。

2. 在计算混凝土配合比时，所使用的骨料，全世界只有我国除水工结构外，考虑的是绝干状态：试配时，将砂子烘/炒至绝干，测出其含水率，然后从所确定的拌和水中扣除砂子中的水。当含水未饱和的砂子投入拌和料中后，砂子又要吸水，于是拌和物流动性不足，又要调整水量，往往要调整数次。试配时如此，生产中质量控制的难度更大。调来调去，以致出场拌和物的用水量成为不可知。饱和面干骨料中所含的水不参与水化和混凝土微结构的形成，也不参与混凝土的拌和，是属于骨料本身的一部分，只会在混凝土硬化、自由水减少后才能出来对界面起养护作用。

3. 从全国范围来看，混凝土配制的水平差别很大，主要是概念不清楚造成的。例如我国曾有过规范规定粉煤灰的掺入使用“超量取代法”，原本是为了降低实际水胶比，但是却被现在许多人用来掩人耳目：超量部分用以代砂，则配合比中粉煤灰掺量就不会超过当前管理部门的规定——管理部门往往并不清楚为什么要掺粉煤灰、掺多少有什么影响，例如一律规定掺量不大于25%，他们中有人说，“粉煤灰不是水泥，所以不能多代水泥，代砂可以”。实际上从颗粒尺寸来说，粉煤灰确实不是砂。在水泥厂，掺混合材的水泥也是水泥，把混合材挪个地方到混凝土，怎么就不认了呢？粉煤灰不能代砂，因为所起作用不同。例如当水胶比不变时，灰砂比的变化和水泥的水化无关；而粉煤灰掺量变化则会影响水泥的水化。即使是砂，磨细到水泥的细度后也就不再是砂了。由于“代砂”的概念不清，有些人计算混凝土的水灰比（实际上应当是水胶比）时，就不把超量那部分粉煤灰计算进去，还有人把粉煤灰掺量按粉煤灰用量和水泥用量之比计算，这就造成配合比设计和质量控制的混乱。实际上不管算在哪里，粉煤灰总是会参与水泥浆体及其与砂石界面微结构形成的，而这个微结构和水有关，计算混凝土的水胶比必须计入全部水泥和矿物掺和料。此外超量取代法也忽视了浆骨比变化对混凝土质量的影响。

七、有人说：设计有安全系数，强度差一点出不了事

尽管强度已不再是现在配制混凝土的难事，也还是有人为了降低成本少用外加剂，多用掺和料而出现强度不够的现象，更多的是施工单位向已交货的拌和物中添加水，造成硬化后的混凝土结构芯样强度的问题也不是个别的。实际上我国结构设计的安全保证率或安全储备并不大。例如根据陈肇元查阅的文献，活荷载安全系数，我国为1.4，英国为1.6，美国为1.7；恒荷载安全系数，我国为1.2，美、英为1.4 。混凝土材料强度安全系数，我国为

1.35，英国为1.5；钢材强度安全系数，我国为1.08，英、美为1.15。从表2和表3可见我国结构设计对活荷载所取的标准值也是最低的（荷载=标准值×安全系数）。

表2　各国办公楼楼层设计活荷载标准值（kgf/m^2）（摘自陈肇元课件资料）

中国	美国	英国	法国、德国、俄罗斯	加拿大	日本	澳大利亚	意大利
150*	240	250	200	240	290	300	340

*2003年起我国新设计的办公楼楼层活荷载已改为200kgf/m^2

表3　不同建筑物的楼层活荷载标准值（kgf/m^2）（摘自陈肇元课件资料）

	体育馆	办公楼	办公楼门厅	公寓住宅房间	公寓住宅走廊	学校教室	学校走廊	剧院门庭	餐厅	防火通道
中国	230	150*	200	150*	150	200	250	300	350	同走廊
美国	480	240	480	200	480	200	390	480	480	480

* 2003年起我国新设计的住宅、办公楼楼层活荷载改为200kgf/m^2

从表2和表3可见，办公室设计的楼面活荷载（荷载=标准值×安全系数），我国只是英国的52.5%，美国的51.5%；计算构件承载能力时规定的材料强度安全系数，我国比英美的低10%～15%，按英美规范设计的办公室楼板构件，承受活荷载的能力比按我国规范设计的大一倍以上，承受恒载（自重）的能力高出约15%。

还有人说，“这么多年都这样干了，也没出什么事”，“只要强度够了就行，耐久性谁也看不见”。

黄士元教授曾说过：（只要结构设计计算正确）由于设计的承载力不够而毁坏的结构物几乎没有，但是由于缺乏耐久性设计而出现问题的结构物却多见。对我国既有混凝土结构使用寿命调查结果，大多未超过50年。尤其因使用除冰盐而遭到损坏的桥梁，使用寿命一般只有一二十年（见图8，资料来源于李克非讲课资料）。耐久性不足时，因环境作用而劣化，会威胁到结构的承载力。

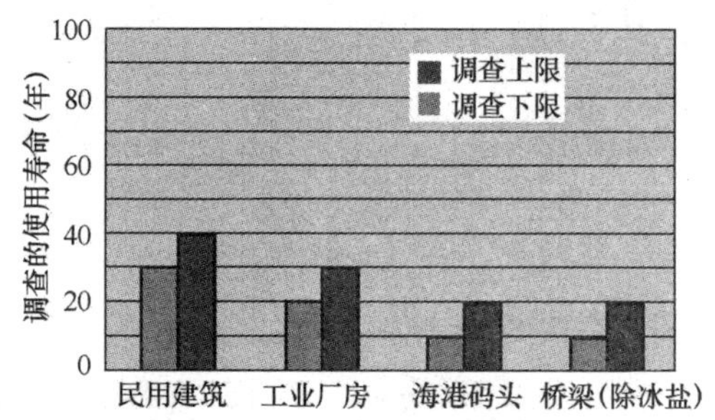

图8　既有结构物使用寿命调查

图9中所示的是不同结构在不同环境作用下劣化、损伤的实例，包括码头、桥梁及其栏杆和桥墩、盐渍土地区电线杆、道路、铁路、房屋等。其中（b）的上图所示是桥梁的仰视图，可见保护层的顺筋开裂，下图是开裂后剥落掉下保护层的混凝土，该桥只运行了8年；（d）的栏杆“烂根”是由除冰盐腐蚀造成的；（g）的电线杆在地面以上约1 m高度处胀裂，是因为地下水中的盐通过混凝土的毛细作用吸附上去，在空气中水分蒸发，盐的浓度不断增加而析晶造成的，这种腐蚀对墙、柱也有同样作用。（e）是宁波华光城住宅小区，1997年评比中获得建安质量第三名的奖励，2004年住户装修时发现楼板混凝土剥落，露出锈蚀的钢筋。这里只是举出一些典型的实例（资料除（g）由廉慧珍拍摄、（j）由覃维祖提供外，均来源于陈肇元）。

今年武汉长江大桥进行检查时发现，这座建于50年前的大桥完好无损，原因是当年施工质量的保证，当时所用的合格石子都经过冲洗。在十几年以前的重要工程骨料也都是洗过再用的，而现在除非重大工程，想用洗过的砂石简直是奢望。为武汉长江大桥而成立的武汉大桥局至今仍继承着他们的优良传统，这是保证我国混凝土工程质量的希望。

从太空看地球，见图 10，可见到中国区域几乎被沙漠颜色所掩盖，看不到绿色。图 11 所示，近 100 年人类向大气排放的二氧化碳直线上升。这当然并不都是建设工程造成的，但是结构物耐久性差是能源和资源最大的浪费，废弃的结构物还造成对环境的压力。

(a) 某地区北仓港码头运行11年

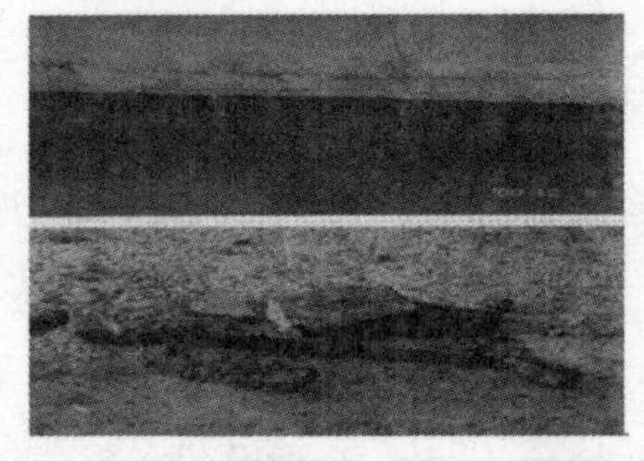

(b) 山东白浪河大桥运行8年保护层剥落

(c) 海水中的桥墩

(d) 某北疆城市的一座立交桥栏杆

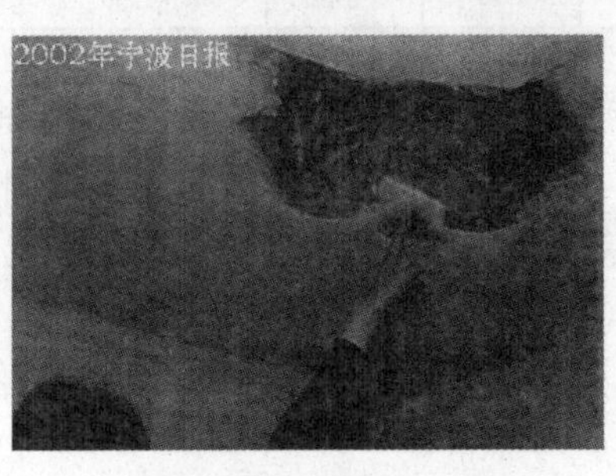

(e) 住宅楼楼板混凝土剥落露出锈蚀的钢筋

(f) 运行40年的北京美术馆地下室顶板钢筋锈蚀

(g) 西部地区盐渍土地面以上盐结晶破坏

(h) 某立交桥桥墩有落水管一侧钢筋腐蚀

(i) 快速铁路路肩的损坏

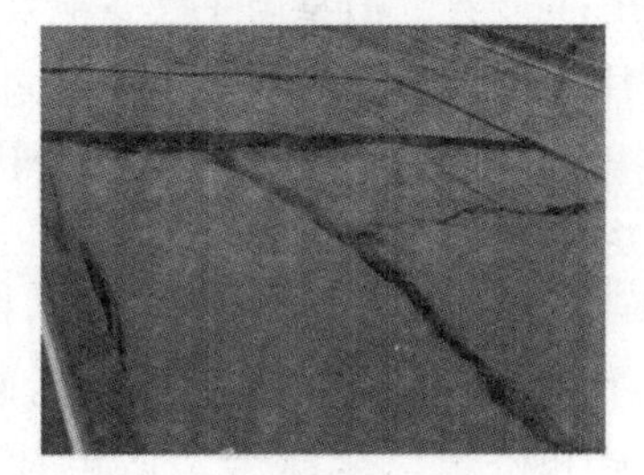

(j) 320国道运行10年

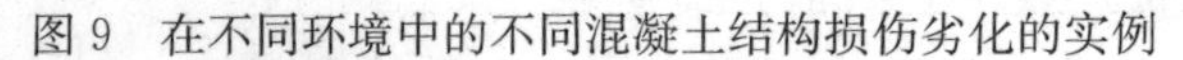

图 9　在不同环境中的不同混凝土结构损伤劣化的实例

图 10　从太空看地球——卫星拍摄分析的图片（引自互联网）

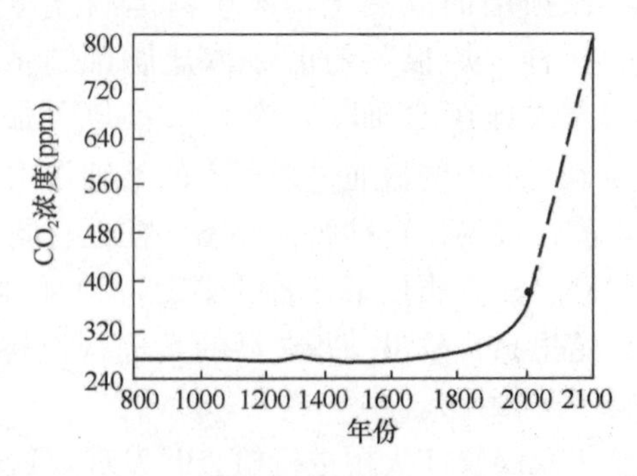

图 11　大气温室效应气体 CO_2 排放的增长趋势（引自陈肇元课件）

2002 年 Mehta 给出的数据是：全世界资源消耗，水泥 16 亿吨/年，砂石 100 亿吨/年，水 10 亿吨/年；每生产 1 吨水泥约排放 1 吨产生温室效应的 CO_2；混凝土结构耐久性下降提前劣化，由此造成每年 10 亿吨建筑垃圾和修补加固的资源消耗[6]。

我国去年水泥产量已达约 14 亿吨，如果全部用于混凝土，则我国混凝土目前产量可达约 40 亿 m^3/年（其中商品混凝土约 7 亿 m^3/年），骨料用量以 1600kg/m^3 计，则约消耗砂石 87 亿吨/年，约等于 11 座北京的香山。砂石资源的短缺是目前混凝土无法形成买方市场而造成砂石质量低下的原因之一，就连生产水泥用的石膏也很紧缺，有的地区石膏价格已接近水泥价格。“地大物博”的中国，资源也不是“取之不尽”、“用之不竭”的。

混凝土结构的质量一旦发现问题，建设的各方都有责任。混凝土的质量是涉及我国乃至人类可持续发展的大事。改变预拌混凝土现状是人类社会可持续发展的需要。

八、做好混凝土首先需要观念的改变

1. 改变以强度为唯一目标的传统观念并非忽视强度，而是因为过去太忽视耐久性。强度作为结构安全性的保证，仍很重要。目前强度不足的情况也说明耐久性同样不足。

2. 尽量降低成本是经营之道，节能、降耗、减排，则是社会可持续发展，如果只是为了赚取最大利润，损害了质量，则将失去社会的信任。

3. 按照工程需要和应用环境，选择和正确使用原材料，尤其是矿物掺和料，没有放之四海而皆准的材料和技术。例如欧洲标准中规定水泥应从已确认适用性的水泥中选择，应考虑施工条件、混凝土的最终用途、养护条件、结构的尺寸（考虑放热过程）、结构所暴露的环境条件、骨料与来源于组成材料的碱的潜在反应活性。又如对骨料的选择，欧洲标准中规定骨料的类型、级配和类别，如针片度、抗冻融性能、耐磨性、细度，选择时应考虑施工条件、混凝土的最终用途、结构所暴露的环境条件、混凝土表面终饰的要求（如暴露骨料或要用工具加工）、钢筋保护层和最小截面宽度等，可见考虑之细，值得借鉴。

4. 端正对混凝土质量的认识。质量是任何产品的生命，要想别人尊重自己，首先要自重。市场经济有两条铁律：一是优质优价—— 一分价钱一分货，买东西是这样，卖东西也如此。相互拼命压价的目的是想挤掉别人，让钱由自己赚，这种相互倾轧、恶性竞争所造成的后果不是谁死谁活的问题，对工程质量对社会造成的影响却是更深远的。只有由稳定的质量和诚信形成的品牌才有长久的生命力。其二，用户是上帝—— 按用户需要生产，市场需要什么，就生产什么，但是上帝未必知道什么是符合科学规律的真正需要，用户也需要培训。和谐社会需要买卖双方互相沟通、理解。

5. 混凝土结构的匀质性比强度更重要。我国目前的验收只重视强度，验收规范中的“统计方法”所规定的样本数也只有 10 组，还允许“非统计方法”。这种验收不仅不全面，而且忽视了匀质性。用回弹仪检测和评价结构的匀质性比“检测强度”具有科学性。当然这也属于黄士元教授所形容的“死后验尸”，重要的是加强过程控制。

6. 加强混凝土生产和使用的过程控制与服务。美国强调混凝土的试配，重要工程必须试浇筑，认为施工单位只要按试配优化配合比合格生产的拌和物正确地施工，质量不会有问题，注意力是对施工的控制。日本认为只要现场验证了混凝土拌和物的单方用水量在试配确定值允许偏差范围内，保证施工质量就能符合工程要求，因此对与现场留样的强度相比（28天龄期后才知晓），更重视对拌和物用水量的测定（保证强度的首要因素是水灰比）。日本现场检测混凝土用水量的技术约有七八种，官方要求强制执行的只有一种，日本对混凝土的质量控制是全方位、全过程的：从原材料选择、进场验收到存储、配合比选择与试配，从计量、搅拌、运输、浇注到拌和物验收，在各方面进行过程的控制。改变目前配合比计算的方法和生产质量控制的模式加强混凝土生产和使用的过程控制与服务。

7. 重视继续教育。因为客观世界在不断变化，人对客观世界的认识也应当与时俱进，任何人在一生中都需要持续地接受再教育。重视对自身的再教育和对员工的培训，是一个成功的管理者必备的素质。不要怕员工水平提高了可能会“跳槽”，当你这里能发挥他们水平的岗位已满员的时候，允许他们“跳槽”是应该的，很正常。全世界有1万多个CEO是美国通用电气公司培训而成长后正常“跳槽”的。这些人实际上都是美国通用电气公司的活广告。一个有发展前途的企业，一要有技术研究的力量，二要有培训的制度。

九、矿物掺和料由谁掺好？

目前预拌混凝土在掺和料和外加剂使用上存在如下问题：

1. 掺入组分太多，管理水平跟不上，上错料的问题时有发生。例如，去年北京就有人在一个工程中把膨胀剂当成了矿渣用。几乎每年都有把粉煤灰当成水泥的事情发生。

2. 搅拌机搅拌时间太短，混凝土的多组分无预均化工序，存在拌和物匀质性的问题。例如，掺矿渣的混凝土拆模后，表面呈现分布不均匀的墨绿色斑，虽然在空气中会消失，但是也说明搅拌的不均匀；还曾有人把掺加膨胀剂的混凝土试件泡在水里养护，其中竟有一个试件胀裂，也是搅拌不均匀造成的。

3. 使用大掺量矿物掺和料的混凝土 SO_3 不足，致使混凝土凝结慢，早期强度低，收缩大，与减水剂相容性差。现在的水泥中 SO_3 一般为2.2%～2.4%，假定矿物掺和料掺量为30%，则 SO_3 被稀释成不到1.7%。而 SO_3 是矿物掺和料的激发剂之一，水泥标准规定矿渣硅酸盐水泥中允许 SO_3 为4%！

4. 难以改变外加剂掺入方式，影响外加剂的效果和拌和物的质量。外加剂掺法有同掺法、后掺法、分次掺法等，其中同掺法是效果最差的。当前多数搅拌站是把外加剂溶于拌和水一起加入干拌料中，是效果最差的一种外加剂掺入方法。

上述问题是搅拌站很难解决的，这使混凝土的质量受到制约，如果矿物掺和料和外加剂由水泥厂来掺，就可解决这些问题。有三种路线：①水泥厂办搅拌站，按混凝土需要生产胶凝材料，并建立和培训专门负责混凝土浇筑成型、养护的专业队伍，从原材料到混凝土的生产和应用，实现集约化、一体化，不仅能提高混凝土结构的质量，而且可减少若干中间环节，从而降低成本、节约资源；由于所生产的胶凝材料是中间产品，可以不受水泥标准的限制；②搅拌站建立自己的粉磨站，按混凝土所需的胶凝材料的组成——熟料、矿物掺和料、外加剂、优化的石膏，预先在磨机中粉磨，既可达到预均化的目的，又可以不必因“不知道从水泥厂买来的水泥里都加了什么”而担心对混凝土配制的影响；胶凝材料的组成也可不受水泥标准的约束；③已有的独立粉磨站把生产水泥改成为搅拌站做胶凝材料“来料加工”，这种委托加工应不受水泥标准的制约，只需制订一个用于质量控制和质量检查的产品标准。

国外发达国家水泥厂办搅拌站很普遍。新加坡所用混凝土都是由水泥厂在销售水泥的同时，根据用户需要，提供经过试配优选的混凝土配合比。这符合水泥、混凝土发展否定之否定的规律（见图12），将是未来水泥混凝土发展的必然方向。目前我国已经有越来越多的水泥企业集团在并购混凝土搅拌站，但是仍然是“两层皮”，并未从根本上改变原来生产关系所存在的缺陷，进一步的努力可能会有脱胎换骨之痛。但似不可避免。

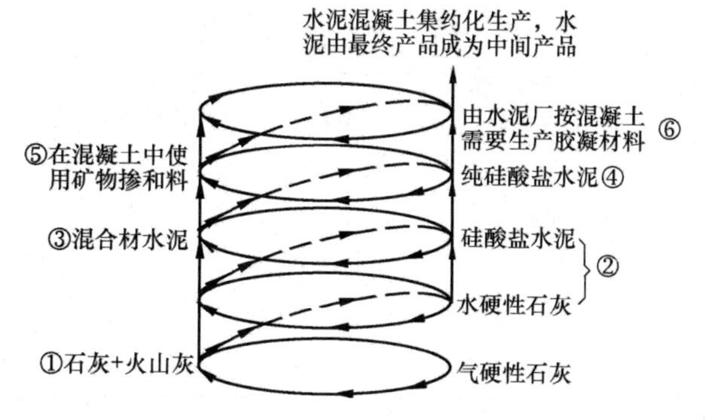

图12 按水泥混凝土技术发展否定之否定规律预测未来的水泥混凝土生产方式

十、做好预拌混凝土企业需要有什么？

1. 人是第一重要的，要留住人才，就要使他们感到自己在企业里不断进步、受到重视、能体现自己的价值，得到发展。

2. 有一批既懂设计和施工，又懂水泥和砂石的，有正确观念和先进技术的管理人员。

3. 有稳定的合格原材料来源。

4. 有切实的能够保证混凝土质量的严格的制度和措施。

5. 要想发展，就要做别人不做或做不到的事；做人之做易，做人之不做难，做人之难做更难。别人之不做和难做，正是自己发展的机会。

6. 做事先做人，做最好的自己。正像马云所说：欲望要有底线，否则是要付出代价的。

十一、对协会工作的建议

1. 定期出版发送内参简报，使政府相关部门能够了解行业的真实情况，必要时通过适当渠道和政府主管部门沟通，使他们及时了解行业动态和亟待解决的问题，争取政府主管部门的支持。

2. 定期提供技术发展报告，传播技术发展的信息，作为标准的制订、修订和技术决策的重要参考。

3. 促进行业团结一致形成买方市场，以控制原材料质量。

4. 规范行业行为，协助政府建立市场准入制度。

5. 制订行业道德规范，并监督执行。

6. 严格行业章程、制度的执行。

7. 除技术交流外，加强组织协会会员单位之间管理工作和思想观念、意见的交流。

8. 加强行业的继续教育。

十二、结束语

最后引用网上对当前H1N1流感话题的一个帖子作为结束语，以寄希望于引起行业同仁对混凝土发展中存在问题的足够重视：

“我们的世界、我们的未来，会有更多的‘非典型的’病例出现！我们生活的条件提高了，但生存的条件降低了：地球环境的恶化、自然界生态平衡的破坏、激素饲养的家禽家畜、食品添加剂养大的孩子，以后还可能看到满街跑的转基因人类！现在我们还活着，多做、多学点自我保护的、更能为后代有保障的事吧”！

参考文献

[1] Richard W. Burrows. Visible and Invisible Crack of Concrete，ACI Monograph No. 11，1997.

[2] Sidney Mindess and Francis Young. Concrete，2nd edition. 吴科如、张雄、姚武等译．混凝土[M]. 北京：化学工业出版社，材料科学与工程出版中心，2005，1.

[3] Rupert Springenschmid. Avoidance of Thermal Cracking in Concrete at Early Ages，Proceedings of Symposium held by RILEM，Munich，1998.

[4] Adam Neville. Consideration of durability of concrete structures：Past，present，and future. Materials and Structures March 2001.

[5] 胡明玉，唐明述．碳硫硅钙石型硫酸盐腐蚀研究进展[J]. 混凝土，2004. No. 6.

[6] P. K. Mehta. Greening of the Concrete Industry for Sustainable development. Concrete International. July，2002.

逆境是达到真理的一条道路。

——拜伦

当许多人在一条道路上徘徊不前时，他们不得不让开一条大路，让珍惜时间的人赶到他们的前面去。

——苏格拉底

论文之廿一

谁对工程质量负责?*

廉慧珍(清华大学土木水利学院)

摘　要　去年，北京市住建委大力检查建筑工程质量，受到广大群众的称赞。最近北京旧宫在建的不合格保障性住房被拆除，是迄今我国建筑工程中执法力度最到位的举措。对质量问题不讳疾忌医，及时发现问题，勇于暴露问题并采取果断措施解决问题，才能对工程质量的保证有所作为而使“前车之辙”成为“后车之鉴”。建筑工程质量问题的解决仍任重道远。建设单位、勘察单位、设计单位、施工单位(包括材料供应商)、工程监理和质检单位都对建设工程质量负有责任，必须明确责任关系，各负其责，负责到位。真正主导工程质量问题的主要不是技术问题。当前工程质量问题中并无高难科技问题，并非不能解决，体制的改革、执法的到位、人员素质的提高才是根本。

1　什么是工程的质量?我国当前工程有没有质量问题?

工程质量因涉及千家万户，涉及生命和财产，涉及公众形象……，其重要性不言而喻。当前我国正处于大规模基础设施建设时期，需要和正在消耗大量的资源和能源，则工程质量还涉及人类社会可持续发展的问题。近20年来资源消耗之迅速实在始料未及。且不说现在很难买到符合现行标准的砂石，就连现在那些质量很糟糕的砂石也越来越难买到了。照这样下去，难以想象我们的子孙后代还能看得见几座山……，更严重的后果是，消耗了本该留给子孙的资源和能源，却因为原材料无选择直到无法选择，以致滥用，造成工程质量的低下，而给未来留下大量的建筑垃圾。

什么是工程质量?工程质量指的是建成后的建筑物或结构物是否满足指定工程安全性、耐久性和适用性的要求。不同的工程有不同的功能，并可能处于不同的环境条件，因而具有不同的安全性和耐久性等级要求。适用性则是满足人的活动、行为和感官的要求。这些都指的是对工程最终成品的要求。混凝土工程的质量当然和混凝土的质量有关，满足指定工程的各项性能和匀质性要求就是混凝土的质量，但是混凝土并非最终产品，按指定工程要求选择原材料和配合比的混凝土拌和物，如果没有使混凝土均匀地密实成型的施工作为保证，也不可能保证工程质量。

正像厨师想做出高质量的蛋糕必须有高质量的面粉和鸡蛋，工程质量和材料质量当然关系密切，但是厨师的水平却是关键的条件;好材料是否能做出好产品，关键是工艺过程。混凝土不是最终产品，必须浇筑在模板里，密实地成型为构件，才是最终的产品。这样的产品，强度应符合设计要求，并且整体均匀，体积稳定，才是质量合格的。近年来工程质量问题确实增多。问题不仅是建设后十一、二年出现问题，而且有的问题是在拆模后还没交工就出现的:混凝土表面由于拌和物离析、泌水而出现砂线、麻面(图1)，不均匀的颜色(图

* 原载《混凝土世界》2011.01。

2)，裂缝（图 3），分层（图 4），没有振捣密实（图 6），柱子上下强度差别很大，甚至强度不够，等等；有的表面混凝土没有充盈钢筋保护层（图 5），却常常一抹了之，尽管装修后不再看得见，毕竟留下隐患。其实，近 20 年来，国内外对混凝土结构耐久性日益重视，主要是因为处于严酷环境的工程增多，如含有化学腐蚀性物质的地下铁道、海底或山体内、海底隧道等一些隐蔽工程，还有跨海桥梁以及近海采油平台等工程，不仅建成后会受到严酷环境的作用，而且在建设中的施工条件恶劣，保证施工质量的难度极大。水工结构、交通工程、铁道工程、西部工程等应当是需要重视并考虑耐久性设计的重点。然而，本来环境条件相对平和得多的城市住房建设工程，确实不需要什么高科技，只要按照规范做好，保证质量并非难事，如今质量事故却频发。

图 1　某桥墩表面混凝土的砂线（左）和麻面（右）

图 2　某工程柱子表面颜色不均匀

图 3　某工程拆模后墙面的水平裂缝

实际上，保证工程质量只需两个条件：一是责任到位，执法有力度；二是从业人员有良知和责任心。执法的力度是执法的质量，从业人员的良知和责任心反映国民素质。北京在建中的旧宫保障住房有 6 栋因混凝土强度严重不符合工程设计要求而被拆除，这是迄今我国建筑工程中执法力度最到位的举措，看来住房建设的质量问题解决有望。人们常说，要允许犯错误，越干事，越容易暴露问题，乃至错误；不干事的，永远不会出现错误。对质量问题不讳疾忌医，及时发现问题，勇于暴露，找到原因，公正处理，才能避免今后发生更大的错误，才能对工程质量的保证有所作为而使“前车之辙”成为“后车之鉴”。国民素质的提高绝非一日之功，因此确保工程质量仍然任重道远。

图 4　仰拍离析的柱子混凝土由离析造成的分层

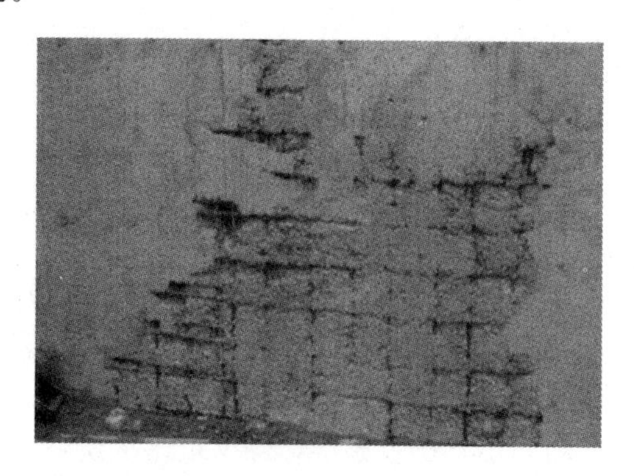

图 5　墙体拆模后的混凝土部分墙面钢筋无保护层

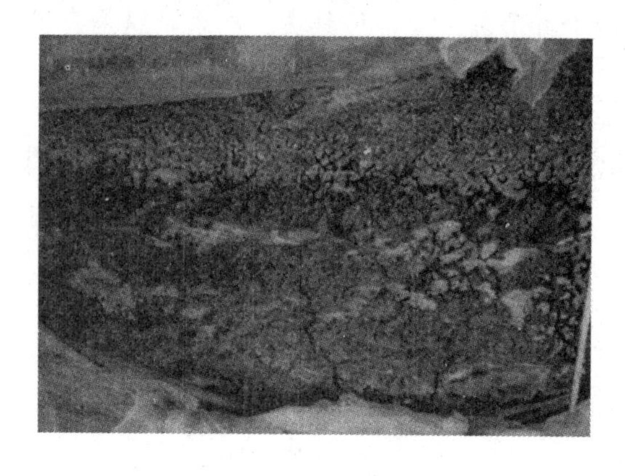

图 6　某工程钢筋混凝土柱子下部混凝土，右图中深色部分为混入的泥块

工程质量由甲方工程技术代表、工程设计、施工（包括材料）等环节完成并保证，而工程质量的检测由质检部门进行，工程质量的过程监督、管理的责任主体是工程监理人员；各个环节都有质量问题，除了自查之外，质检和监理的质量由有政府机构委托授权的质量监督机构管理。那么谁来监督质量检验的质量？质量检验的质量又是什么？是否还存在质量检验的质量问题？最终的检验当然很重要，但是更为重要的过程控制为什么不能实行？可见整个工程包括质量监督和检验，各个环节都存在质量问题。

2　工程质量问题谁负责？对谁负责？怎样负责？

什么是负责？负责就是承担应该承担的任务及其行为的后果。因此“负责”必然要涉及人和由人组成的部门、组织。然而，曾经遇到过这样一件事：某城市发生一起质量事故，有关部门组织专家进行了分析讨论，最后的结论竟然是对这起事故的发生“谁都没有责任”，而是“由规范负责”。这样的处理结果看似可笑，实际上却是普遍存在的现象。在计划经济年代，养成技术人员过分依赖规范的习惯和一切唯规范是从的行为准则，使他们的设计行为变成只对规范负责，而不是首先对工程负责。一旦工程中出现问题，责任就推给规范。在多数人的心目中，凡是“规范”或“标准”，就都是法律。照“规范”或“标准”办，自己就没有责任了，或者把规范或标准当作“救命稻草”，一旦出现什么问题，只要在任何规范或标准中找到与自己在工程中所作所为相符合的规定，就可解脱，而不负任何责任了。实际上，只有以“中华人民共和国主席令”或“中华人民共和国国务院令”颁布的法规和某些强制性的技术条例才是法律，凡是技术标准和规范，其本身都不是法律。因为，没有绝对正确的和不变的技术。如果只依赖于“由规范负责”，就会出现无人负责的结局，造成工程的损失和资源的浪费，而当事人就都会越来越有恃无恐。不能不说我国当前工程质量出现的问题与“由规范负责”有关。从事工程质量认证、检测、检查、监理等工作的人几乎无一例外地都这样说：“我们就是按（标准、规范、上级的指示和文件规定的）条文执行，条文如果有错，也执行。”这就是说，他们只是对规范“负责”，对领导“负责”，而不是对工程负责。

那么谁该对工程负责呢？在国外，工程项目的合同中都有一个“规范清单”，技术规范只有被列入合同的“规范清单”中，就与合同一起具有法律效力。按合同约定的双方责任检查判断，该是谁就是谁，谁该负多少责任，谁就要担当。我国具有法律属性的《建设工程质量管理条例》第三条明确规定：建设单位、勘察单位、设计单位、施工单位、工程监理单位依法对建设工程质量负责[1]。也就是说，工程各个环节都必须各负其责。本来，这是有国家明文规定的，不言而喻，然而现状并非尽如人意。

以房屋建筑为例，建设投资方（开发商）应当最关心建筑物的质量，也就是说对住房建设来说，开发商是责任的主体。我国确实也有较强实力的房地产开发企业十分关心自己的声誉，因而也就很重视工程质量，不仅聘请监理，而且有自己的技术人员队伍，在一个项目上能派出 100 多位自己的技术人员进行监管。但是更多的开发商为了应对巨大的开工量，应设的技术人员和管理人员常常只以人头充数，或是多处兼职，只能顾头不顾尾，处于应付的状态，很少尽到开发管理的首要职责，把全部责任推给监理。监理人员的培训和考试主要是规章制度条文的背诵而缺少理解，一方面，有责任心的和有技术水平的监理难以坚持原则；另一方面则更多的是根本不懂结构工程的非专业人员也能拿到监理资质，却不能尽到监理的责任。

第二，对于边设计、边施工、边完善手续的三包工程和工程的层层转包，建设部曾三令五申地禁止，然而屡禁不止，势必影响工程质量。尤其是房屋建筑，施工项目很大程度是通过关系、垫资和一定的经济利益才能得到。本来“低价中标”使得工程款已很紧张，常常是先保证自己所得之后就向下转包，到了三包、四包就很难维持，除了偷工减料，似乎再没有更好的办法。那些卖资质现象，更使得建设项目的质量责任形同虚设。正常的市场是以质论价，而我国现在多数商人则是以价定质，为了中标，不惜压价压到连自己都没有把握保证质量的程度，也硬着头皮“先拿到手再说”，结果是“拿多少钱干多少活”，怎么可能保证应有的质量！市场经济的规律是优质优价，用低等的价位得到上等质量的产品是几乎不可能的。

第三，设计是建筑工程实施的依据。而设计的依据就是设计规范。目前绝大多数设计人员也只对规范负责。他们希望规范“定得越具体、越细致越好”。西方国家基本上由建筑事务所进行设计后，再由负责施工的建筑工程公司做施工图。我国则绝大部分施工单位不具备这个能力。一位从事审图的老工程师说：“现在的施工图错误率很大，甚至达到 40%！”原因是在学校没有得到施工图的训练，刚出校门的工民建专业学生不知道房子是怎样盖起来的，最多只会设计一根梁，甚至只会使用计算机软件设计一个断面，当然就更不懂得混凝土，常出现照设计图纸施工的难度很大、混凝土难以浇筑等问题，有的难以保证钢筋的混凝土保护层质量，造成工程质量的隐患[2]，如图7所示为保护层厚度太小，水平钢筋太粗、太密，设计未对石子粒径提要求而造成的保护层混凝土质量问题；图8所示情况则更为多见。然而，基本上无人过问设计对混凝土结构质量有什么影响，只是要求不管设计成什么样子，施工也必须做出来，混凝土都必须能浇筑。这种体制对工程质量是不利的。

图7　保护层厚度太小，水平筋太粗，设计未对石子粒径提要求的保护层混凝土质量问题

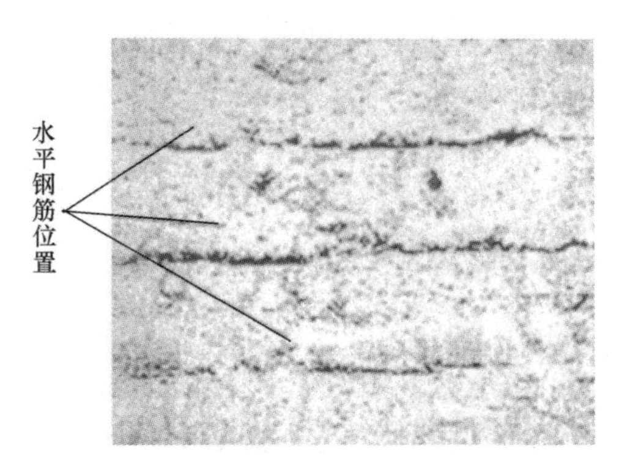

图8　保护层厚度和混凝土石子粒径不匹配，拌和物在水平钢筋下面沉降造成混凝土的开裂

第四，监理人员的混凝土知识最多也只是些皮毛，他们只会照“章”办事，对如图8出现的普遍现象大多默认其一抹了之的措施；质检部门大多数是“对来样负责”，而不管“来样”是怎么“来”的。明知试样强度不能代表结构混凝土的质量，但也出报告；监督人员往往用死条文管卡，例如水泥用量少1kg都算不合格。又如现在还在普遍使用明知不合理的回弹法检测混凝土强度，并且规定检测点不能离得太远，理由竟然是“泵送混凝土坍落度太大，上下强度不一样”，却不知这种“上下不一样”恰恰可以用回弹法检测结构混凝土的匀质性问题，等等。有的人打着“遵守规范”的幌子在工地上指指点点，实际上他们自己对规范也未必清楚。例如有的质检人员竟能说出“这里C30混凝土的强度只有C29”这样的大大的外行话！表明这些“掌管质量”的执法者们须接受继续教育的必要性。应当让他们清楚地认识他们的工作和工程质量的关系，清楚他们的责任。他们需要懂一点技术，虽不要求专，但应有一些基本概念。《建筑工程检测试验技术管理规范》（JGJ 190—2010）对检测试验的取样、制样、试验、检测数据和报告的真实性和准确性、见证管理、档案的建立等等都有明确的要求，实际上如果能一丝不苟地执行，就应该是尽责了。

混凝土质量对工程质量有至关重要的影响。鉴于当前工程的特点，对混凝土性能要求已不再仅是强度了，然而工程设计、施工中的各个环节的要求，甚至包括工程验收却仍然都是强度第一，甚至唯一，造成有些技术水平较低的混凝土搅拌站也只考虑强度，为了降低成本而盲目增大矿物掺和料用量，少数搅拌站甚至敢于掺用未经开发认证的标准以外的工业废弃物，有的试验量不足而连强度都不能保证。试验机长期不标定，实验操作不规范的情况也时有发生。至于混凝土质量当然对工程质量有至关重要的影响，鉴于其复杂性和篇幅所限，将另文探讨。

3　技术标准和规范能负责吗？怎样制订和使用标准？

目前，包括房屋、道路、铁路、运输管道、隧道、桥梁、航道、堤坝、洞库、港口、机场、电站、海上采油平台，等等所有需要“大兴土木”的工程，本来在学科基础上都属于一类，对工程质量的要求有相同的原则，即安全性、耐久性、适用性和经济性；但由于长期行业隔离的原因，各主管部门都有各自部门的设计与施工规范，目前我国涉及土木工程领域的各种各样规范、标准（包括专用设备、设施）约有上千种。往往会因标准和规范制订人的不同认识、经验以及观念的差异，造成几种标准或规范之间存在相互矛盾的规定。常发生不同规范对同一事物的规定相互矛盾的情况。因为把规范都当成法律，当不同规范相互矛盾时就不知所措了：以哪个为准？没有人敢做决定。其实出现这种问题的原因不在于规范太多而又互相矛盾，不同规范是针对不同用途的，如果像发达国家那样，行业协会、研究机构都可以制订规范和标准，供购销双方按工程特点和条件协商选择最适合的来共同遵守，问题也就好解决了。

事实上，技术标准和规范也有质量问题：是否符合工程实际？是否有科学根据？

技术规范的水平和质量与制订者的思维方法、认识水平和工程经验有关。例如，试验和工程实践证明，RCM 法检测氯离子扩散系数的离散性很大，最大可相差一个数量级，而《混凝土耐久性检验评定标准》（JGJ/T 193－2009）中的表 3.0.2-1（见表 1）用 RCM 法检测混凝土抗氯离子渗透性能（此处氯离子渗透性能的提法是概念性的错误——本文作者注）的等级划分，将氯离子扩散系数分成 5 等，其中Ⅰ等的标准值为 $4.5\times10^{-12}m^2/s$，Ⅴ等的标准值为 $1.5\times10^{-12}m^2/s$，5 个等级之间只相差不到半个数量级，每两个等级之间只相差亿万分之一！实验误差都比其大得多的多。绝不可能做到，严重脱离实际。请问制订者做过实验吗？这样的“拍脑袋”显然太离谱了。如果说“也做过试验”，则这种试验与操作者的水平及经验关系很大，例如用染色法检测氯离子迁移深度时，正像碳化前沿一样难以界定，量测时的微小差异可以引起计算结果的很大变化。在国外，为了制订标准，都需要三个以上地区（对于欧洲标准则是三个国家）用同样原材料在同一时间做平行的合格性验证试验。即使标准制订者自己做过试验，也需要反复对比、验证，更何况对这种精细量测的试验。上述实例的做法显然欠妥，如果再作为“法律”去宣贯和强制执行就更加不妥。

表 1　混凝土抗氯离子渗透性能的等级划分（JGJ/T 193—2009 表 3.0.2-1）

等级	RCM-Ⅰ	RCM-Ⅱ	RCM-Ⅲ	RCM-Ⅳ	RCM-Ⅴ
氯离子扩散系数 D_{RCM}（$\times10^{-12}m^2/s$）	$D_{RCM}\geqslant4.5$	$3.5\leqslant D_{RCM}<4.5$	$2.5\geqslant D_{RCM}<3.5$	$1.5\geqslant D_{RCM}<2.5$	$D_{RCM}<1$

规范的水平和质量还和怎样制订有重要关系。由于延续了计划经济年代形成的观念，我国从上到下绝大多数工程技术人员都把技术规范看成法律，也就自然形成了人们心目中技术规范的“等级”，认为国家标准等级最高，行业标准就位其次了（实际上行业标准也等同于行业的“国标”）。在这样的体制下，因为主管部门有规定，参加标准的制/修订对于企业和个人都可以作为晋级的资本，尤其是制订实验方法标准时还涉及实验仪器的推行，本来已与行政权力“脱钩”的研究单位，某些人仍一心追求成为权力机构的执法者，几乎包揽一切建设工程相关标准、规范的国标、行标“制订权”。制订标准时，向企业征集赞助费，参编标准是我国有关部委规定的企业业绩之一，则众多企业“乐此不疲”。而出钱的企业都能成为“编委”或“参编单位”，这样的规范必然要代表出钱人的利益，这未免有失公正。

美国 ACI（American Concrete Institute，美国混凝土协会）由各技术委员会专家工作组每 2～3 年分别提供一次相关领域的“现状发展报告”（State of the arts report on……），交由 PCA（Portland Cement Association，波特兰水泥协会）编写技术规范或标准的条文，最后由 ASTM（American Society for Testing and Materials，美国试验与材料协会）的相关技术分委员会专家组制订标准或规范。标准制订一直采用自愿达成一致意见的制度。标准制订由技术委员会负责，由标准工作组起草，广泛征求、吸收各方面的意见和建议。感兴趣的每个会员和任何热心的团体都能充分发表意见。委员会对提出的意见都给予研究和处理，经过技术分委员会和技术委员会投票表决，在采纳大多数会员共同意见后，由大多数会员投票赞成、批准，作为正式标准出版。

以上三个协会都是历史上百年的非营利性民间学术团体，分别都有数万名会员和专家。其中 ASTM 现有 33669 个（个人和团体）会员，22396 个主要委员会会员分别在 ASTM 的技术委员会下设的 2004 个技术分委员会中担任技术专家进行工作。参加了 ASTM 标准的制订工作的有 105817 个单位。

虽然 ASTM 标准是非官方学术团体制订的，但由于其质量高，适应性好，而赢得了美国工业界的官方信赖，不仅被美国各工业界纷纷采用，而且被美国国防部和联邦政府各部门机构采用。权威性不是被加封的，而是由其技术水平和服务质量所赢得的。我国也不乏一些认真负责的专家、学者，制订出的标准、规范有较高水平，但是也有一些没有足够工程实践经验和技术研究经历的人“根据××部的要求”（是官方要求的还是他们申请后被要求的?）去制订标准、规范，还自称“具有法律效力”，到处宣传他们的“权威性”。明白人都不会被那些存在问题的规范牵着鼻子走，但又囿于“国家标准”的威慑，因而不得不使用“阴阳配合比”。“阴阳配合比”的出现确实是违规的，但是如果现行标准规范能指导他们生产出合格产品，他们何必冒险?

各国制订规范都声明是最低要求，给技术人员和建设者留下充分发挥才能的空间，为此，现在美国正在讨论和推行“基于性能的规范”（performance based specification）[3,4]，英国标准（2004 年修订并入欧洲标准）中有“设计的混凝土（designed concrete）”和“规定组成的混凝土（prescribed concrete）”两种规定：前者是不规定组成，只规定拌和物和硬化后的性能；后者则规定组成，一般只用于 C16/C20[5]。使用规定组成混凝土，责任者是提出规定的一方。我国标准、规范的管理和制订不妨研究、学习和借鉴一下他们的做法。

4　当前工程质量问题主要不在于技术

如前所述，一般工业民用建筑工程质量的保证，比起那些严酷环境中的工程以及如奥运场馆工程、央视大楼等重大的公共建筑，从技术上来说，并没有什么难度。史书上记载庞大混凝土结构的古罗马万神庙能够屹立至今 2000 余年，除了材料和结构设计特点以外，更得益于对当时由石灰和火山灰组成的混凝土材料的“精心挑选，精心准备，精心搅拌”，基础和墙“用铁锤夯结实”[6],[7]，这是什么高技术吗？为什么现代人做不到呢？实在是“非不能也，乃不为也”。就以房屋建设中楼板裂缝问题来说，常出现在楼板中央的乱向裂缝甚至上下贯穿而漏水，这种裂缝基本上是平板结构混凝土的塑性裂缝发展而造成的。在塑性阶段，因混凝土沉降和表面水分的蒸发而产生的塑性裂缝非常细小，一般宽度不会大于 0.1mm，深度不会大于 3mm，因不易发现，很容易被忽略。从能量的角度，已有裂缝的扩展比新生成裂缝容易，所以塑性裂缝如不尽早处理，就会成为后期干缩条件下的开裂源。但这是最容易避免的：已经有不少施工的实践证明，只要在成型后立即严密覆盖塑料布，待初凝前后对表面进行二次搓抹，然后再覆盖至终凝后继续湿养护3～5天即可解决；如果舍不得劳动但舍得花钱，则在混凝土中掺入合成的聚丙烯短纤维，对避免平板构件塑性开裂是有效的。路面工程因难以实行上述二次搓面法，掺入合成短纤维则可以取得一定的效果。因此可以说，楼板这种裂缝是最容易避免的，却仍然未能避免，可见主要不是技术问题。

除上述诸多问题外，从根本上来看，大量的工程质量问题确实不在于技术。首先是人的问题。《建设工程质量管理条例》中所指建设单位、勘察单位、设计单位、施工单位（包括材料供应）、工程监理单位等的人，研究和制订规范的人，进行试验和质量检测的人……，所有有关人员的观念、知识和技术水平，尤其是责任心，决定了他们的行为和后果。原本应为一体的行业之间互相隔离，相互误导，由不懂混凝土的人向混凝土供应原材料，混凝土拌和物又交给不懂混凝土的人去完成生产结构构件的最后工序，“先天不足，后天失调”。这种生产关系已影响了当今工程建设的生产力的发展。

由于用量的巨大，混凝土结构制作工艺必须简单，所用原材料不能提纯，必须来源广泛，就必然成分波动较大，而正因为工艺的简单——多数人把混凝土看成是简单的“和泥”，根本就谈不上对具有对环境和时间依赖性的非均质多相体的混凝土有多深的认识。混凝土的制作没有什么高难技术，但是由于来自变化的原材料的混凝土，在工程中总是处在变化的时间和环境中，属于“混沌体系”，必须按“不确定性科学体系”[8]来对待，然而，从教育的源头就没有得到应有的重视，以至于大多数工程人员的混凝土知识只是“从包罗所有建筑材料的一门必修课中学到水泥混凝土的一些皮毛”，“改革的过程必须从大学开始”[9]。

对奥运场馆等重大工程项目，质量会受到高度重视，例如“鸟巢”钢结构的焊接是从造船厂调来的八级电焊工夜以继日地连续焊接，三天不下岗，完成任务后在结构上留下自己姓名，以示责任和自豪。有这样的责任心和自豪感、敢于担当的男子汉，质量能差吗？令人不禁想起建筑工人也曾有过的自豪：

从那海滨走到边疆，我们一生走遍四方，辽阔的祖国，万里山河，都是我们的家乡。
住着帐篷和草房，冒着山野的风霜，一旦盖起了高楼大厦，我们就再换一个地方。
面前是无尽的原野，身后是崭新的厂房，我们的生活就是这样，战斗着走向前方！

40多年前，刚就业的学徒工，需经过师傅传带至少一年，考核通过后方能转正定级为二级工。例如那时的墙砌体工程，必须达到四级工以后才能“上墙”。二级工也只能做辅助性的劳动。他们为了自己的前途，都自觉地自己出钱上夜校，努力地跟师傅学。施工单位也很重视创造条件对技术工人进行培养。当年这样成长起来的工人，经过不断的努力深造，很多已成为不同岗位的骨干力量，出色地发挥着作用。而现在，由于对混凝土的复杂性不了解，一线劳动力素质很差，他们没有师傅，没有人告诉他们什么是工程质量，没有人告诉他们应当对工程负有什么责任，也没有操作的规范，可以说是随心所欲。他们大多数不知道自己的前途是什么，于是向拌和物中加水，随意振捣，养护走形式，制作试块造假……，几乎没有什么规范性的操作。这一切都是怎么了？能怪他们吗？我国很多实验室研究的成果难以转化为生产力，主要原因就是工艺水平太差而难以实现。现在，缺少技能培养，现有职业教育师资水平差，技术培训走形式的多。技术如不被广大一线工人掌握，工程质量就是一句空话。

5　结论和建议

工程质量指的是建成后的建筑物或结构物是否满足指定工程安全性、耐久性和适用性的要求。在看到我国基本建设取得重大进展的成果的同时，还应看到工程质量确实也存在不容忽视的问题，以保证社会的可持续发展。工程质量问题产生的根本原因不是技术问题，首先是人的素质、观念和责任心，乃至执法的力度。

为改善和提高我国混凝土结构工程质量，提出以下建议：

1. 切实落实《建设工程质量管理条例》的规定，明确任何工程的质量由建设单位、勘察单位、设计单位、施工单位（包括材料供应）、工程监理单位（包括试验检测）共同负责。一旦出现问题，一律问责，一查到底。

2. 加强合同意识，以合同作为工程质量的法律担保。杜绝“霸王条款”，合同中必须有双方对质量要求和保证质量的承诺，以及违约的责任，必须附带双方认可的规范清单以及规范以外的技术措施要求。发生质量问题时禁止“私了”。

3. 整顿资质的挂靠和技术人员的多处兼职，所有总工、技术负责人、监理人员的信息必须由网络管理，坚决杜绝总工同时多处兼任和买卖资质证明。

4. 整顿当前无序的培训机构，提倡企业内部健全对各级技术人员有计划的系统培训制度（小企业可以联合组织），并与绩效考核相结合进行定期考核。考核者需经资格认证。

5. 行业协会是政府管理行业的助手，是行业与政府之间的桥梁，应重视、支持加强协会的工作和作用，不断提高协会的水平。企业培训师资可由协会组织学习与考核。

6. 尽快清理现有技术标准和规范，解决不同规范之间相互矛盾和其中某些规范无法操作的问题，提倡以性能要求制订标准和规范。建议由行业协会组织制订者认真学习国外先进经验，讨论、明确和规定制订标准和规范的原则。

7. 试点应用和推广检测混凝土拌和物用水量的技术，研究制订加强工程的过程控制的制度和措施；延伸混凝土搅拌站的售后服务，由搅拌站自己派出技术人员，自己招考工人，自己培训操作技工，并相对稳定。实现从原材料选择、试配、生产、输送、浇筑前检验、浇筑、振捣、养护（包括温度和湿度的控制，以控制早期开裂）直到完成混凝土结构构件的“混凝土工程”全过程。“自己生的孩子自己养”对保证最终质量是有利的。

8. 研究解决层层转包的问题，实事求是地调整定额。恶性低价竞标现象的出现是源于低价中标的政策，建议组织专人研究招投标的系统工程，科学地规范招标、投标制度。

以下引用一段宁夏回族自治区政府主席王正伟日前在区政府常务会上关于工程建设领域突出问题治理整顿的讲话，作为最后一段建议。

王正伟说[10]：“开展工程建设领域突出问题的治理整顿，必须要讲‘游戏规则’，针对工程建设的关键环节，以及恶性低价竞标、豆腐渣工程、超概算等突出问题，要探索设定更有效的制度和措施，防止公共资源交易领域腐败现象的滋生和蔓延。……制度要严格，不能大而化之……；评标要真科学，不能个人说了算……；操作要真阳光，不能走过场……；查处要真严厉，不能心慈手软……；诚信要真建立，不能让‘潜规则’大行其道。取信于民很重要的一条是政府诚信，政府带头把规矩立起来，并且带头遵守。没有细化、管用、易于操作的制度规则，‘潜规则’就会大行其道；有了规则不去遵守，‘潜规则’同样会大行其道。

我们必须用正规则勇敢面对‘潜规则’，遏制‘潜规则’，这是法治政府、向人民负责的政府必须勇于面对和着力做到的，也是对党的事业负责，对领导干部的爱护。”

参考文献

[1] 《建设工程质量管理条例》(2000 年 1 月 30 日中华人民共和国国务院令第 279 号).

[2] 廉慧珍. 建筑技术需要打破专业藩篱[J]. 建筑技术，2005 年第 1 期.

[3] ASTM Designation：C 1157-00 Standard Performance Specification for Hydraulic cement.

[4] Peter Taylor. Performance-based Specifications for Concrete，[J] Concrete International，AUGUST 2004.

[5] BRITISH Standard ，Concrete-Part 1：Specification，Performance，Production and conformity. BS EN 206-1：2000，Corrigenda Nos. 1 and 2 and Amendments Mos. 1 and 2.（The European Standard EN 206-2000，with the incorporation of amendment Al：2004，has the status of a British Standard)

[6] Grant M，The world of Rome，1960.

[7] 陈志华. 外国建筑史(十九世纪末叶以前)[M]. 北京：中国建筑工业出版社，1983.

[8] 于学馥，宋存义. 不确定科学决策方法[M]. 北京：冶金工业出版社，2003. 7.

[9] P. K. Mehta，Concrete technology for Sustainable Development—An Overview of Esential Principle，International Symposium on Sustainable Development of cement and Concrete Industry，Ottawa，Canada. Oct. 1998.

[10] 王正伟. 用正规则遏制“潜规则”. 在宁夏回族自治区区政府常务会上的讲话，《 人民日报 》07 版 . 2010，12，13. 记者 徐运平 周志忠整理.

有人说，知识就是力量。对我来说，知识就是幸福。有了知识，你就可以区别真理和谬误，可以分清高尚与渺小。当你了解到各个时期人们的思想行为时，你就会对发展到今天的人类产生同情和亲近的感情。

——海伦·凯勒

论文之廿二

对建材利废观念和行为的思考*

廉慧珍（清华大学土木水利学院）

建筑材料是一切材料中用量最大的材料，建筑材料产业也是天然资源消耗量最多的产业。除钢铁和塑料外，建筑材料所需要消耗的资源是地球上储量最大的含硅、铝、钙化合物的岩石。采矿、冶金等工业，在提取各自产品后所废弃的，恰恰是在另一方面为建筑材料提供原材料。因此从整体来说，“没有废物，只有放错了地方的资源”。既作为产品，就必然有质量问题。用为其他行业提供资源的观念改造自己的生产工艺，是现代工业发展的必须和必然。例如新建电厂的分级收尘正是为粉煤灰的利用提供方便；高炉矿渣的水淬也是资源化的措施；又如为了保证脱硫废渣利用的质量，在沸腾炉的排渣口增加冷却装置，等等。但是目前并不是所有工业企业都认识并能认真做到的，需要双方进一步沟通。建筑材料“利废”的目的是减轻建设对天然资源的压力，缓解工业废弃物对环境的压力，变废为宝，改善产品性能，或开发新产品，同时降低产品成本。但是，当前许多企业的“利废”首先或者主要是为了降低成本，并能得到有关减免税或给予贷款等的优惠政策支持，有的水泥和混凝土的生产厂，为了降低成本，或者在天然资源不足的情况下，什么“废弃物”都敢用，有把建筑材料当成“垃圾箱”的趋势。较多的则为了满足用户对工业“废弃物”活性的追求，过度采用细磨、煅烧等技术再加工，又增加能源的消耗；另一方面，也还有很多人不敢用。例如水工结构物规范中规定混凝土中粉煤灰掺量不得大于25%，路面混凝土干脆不允许使用矿物掺和料……。这两种倾向都不利于建材利废的科学性。

一、当前科学利废的障碍主要不是技术而是观念和利益

1. 障碍之一：由于对矿渣、粉煤灰一类工业副产品认识有误而禁止或限制使用

许多人把在混凝土中掺用矿物掺和料当成“瓜菜代”，认为掺用矿物掺和料是假冒伪劣的次品，是为了降低成本不得已而为之的措施。实际上，在严酷环境中，矿物掺和料的正确使用是提高混凝土结构耐久性必需的有效措施，即使粉煤灰和矿渣比水泥价格高，为了混凝土结构的耐久性，也要使用。P. Kumar. Mehta 和 Wilbert S. Langley 报道[1]，在距离美国西部大陆约4000km的太平洋上有一个安静的小岛Kauai，在岛上要用手工雕凿成非常美观的大理石建造一座庙宇（见图1），以供旅游观光。地下富含海水成分，并因无钻探条件而地质条件不明，为了避免

图1　太平洋Kauai小岛上一座观光用的庙宇[1]

* 原载《建筑装饰材料世界》中国混凝土专刊，2009年第三期。

建筑物的不均匀沉降和受化学腐蚀，采用不配筋的整块筏式基础；设计人员从古罗马用石灰-火山灰作胶凝材料的混凝土建造的万神庙得到启发，为耐久性的需要，混凝土的胶凝材料中使用含 C_2S 62%、C_3S 14%、C_3A 1%的硅酸盐水泥，掺 60%的粉煤灰。硅酸盐水泥价格约为＄75，而粉煤灰价格则约为＄200。

不敢或不愿用矿物掺和料的人所担心的主要是强度，这种担心是对强度的认识问题。原本矿物掺和料是用于水泥的混合材料。在水泥中使用混合材料是为了调节水泥标号（如普通硅酸盐水泥）和用于抵抗环境中化学腐蚀。在 1990 年之前，我国水泥标准中检测水泥强度的水灰比是 0.44 和 0.46，工程中大量使用的混凝土水灰比基本上大于 0.5，而且极少使用泵送工艺。那时大多数人们对配制混凝土的认识仍是“水泥强度应为混凝土强度的 1.5～2 倍”。混凝土强度受着水泥强度的制约。尽管因高效减水剂的使用使高强混凝土得以实现，而“高强混凝土必须使用高强水泥”的观念仍然主宰着许多人的行为。尤其是上世纪末我国修订水泥标准后，因检测水泥强度的水灰比改为 0.5，相同的水泥的标称强度下降了一个等级；标称强度相同的混合材料水泥早期强度则降低得更多。这是因为掺入混合材料的水泥比硅酸盐水泥对水灰比的敏感度大。如图 2 所示，掺 40%粉煤灰的水泥中，当水胶比为 0.54 时，粉煤灰对强度的贡献为 0，水胶比降低至 0.32，则粉煤灰的贡献比硅酸盐水泥的贡献增长得多。从图 2 中龄期 7 天和 28 天的两个四边形 abdc 和 ABDC，就可看出，bd<ac，BD<AC；龄期长比龄期短时，粉煤灰的贡献比硅酸盐水泥的贡献增长得更多。按现行水泥标准混合材料最大掺量所生产的混合材料水泥或复合水泥，用 0.5 的水胶比检测得到的强度值（即标称强度），使不明就里的多数用户（设计人员、施工人员、建设甲方，甚至多数混凝土生产商）不能接受。

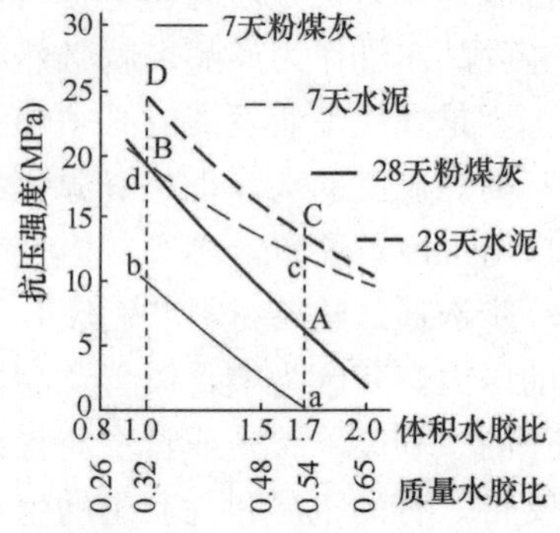

图 2 掺量 40%的粉煤灰与硅酸盐水泥强度分别随水胶比变化而发展的程度[2]

无论水泥还是混凝土，其强度都是在各自标准条件下检测的，都不是真值，真值是无法检测出的。水泥只是一种原材料，其本身的强度没有意义，也无法得知，必须通过加水成型为一种制品来评价，因此水泥标准中规定用一定灰砂比和一定用水量在一定工艺下成型为一定形状与尺寸的胶砂试件，并使用专用仪器设备按一定方法来检测。混凝土强度的检测则与此不同。除使用设备、操作方法、试件尺寸不同和养护条件稍有差异外，最重要的是用水量不固定，而是按设计要求变化。即使检测水泥的用水量和混凝土用水量相同，所得强度值也会不同，两者之间没有可比性。强度低的水泥也可能配制出强度高的混凝土。例如 1995 年在清华大学与住总集团技术开发中心合作研究中，使用 44%的硅酸盐水泥熟料和 40%的粉煤灰、10%的矿渣、优化石膏掺量为 6%，制作出的水泥，如果用现行水泥标准检测，其强度等级可能连 27.5 都达不到，但是在当时住总集团实验室配制 C60 高强混凝土，28 天抗压强度值达 71MPa（坍落度为 200mm）。原因就是高效减水剂使混凝土的水胶比降低到了 0.29。清华大学 1999 年在深圳地铁足尺模型试验中所用混凝土矿物掺和料总量达 60%，水胶比为 0.40，现场留样 28 天抗压强度最低的也有 50MPa；3 天强度最高的达 39MPa。7 年后钻芯取样检测，抗压强度最高达到 98MPa。因此，不要被水泥强度牵制，也不要因混合材料水泥的强度低而不愿用，也不要不敢在混凝土中用矿物掺和料。

人们的另一个顾虑是碳化问题。传统认为掺矿物掺和料加速或加深碳化是因为“混凝土

中 $Ca(OH)_2$减少”。这是一种误区，实际上 $Ca(OH)_2$的碳化与其含量无关。例如，磨细矿渣活性较高，应当比粉煤灰消耗更多 $Ca(OH)_2$，但矿渣却比粉煤灰有较好的抗碳化性能。影响混凝土碳化的因素，除 CO_2浓度外，主要是混凝土的密实度。矿物掺和料掺量越大、水胶比越大，混凝土早期孔隙率越大；湿养护期越短，孔隙率越大。现行评价混凝土碳化的标准是用养护到 28 天的试件，在 CO_2浓度为 20%的碳化箱中强制碳化的。实际工程中的混凝土构件基本上没有养护到 28 天的。一旦停止养护，钢筋保护层混凝土中湿度下降，就会开始发生碳化，并随着湿度的继续下降而加速，在环境相对湿度 50%左右时，碳化速率最快。降低水胶比、延长湿养护时间（或延迟拆模），可减少混凝土的孔隙率，提高混凝土的密实度。如图 3 所示，分别为粉煤灰 56%、44%和纯硅酸盐水泥浆体的孔隙率随（水养护）水化龄期的发展。当水胶比为 0.35 时，龄期 1 天孔隙率分别约为 30%、26%和 12%，水化龄期 28 天三条浆体孔隙率发展曲线交汇于约 2.5%；当水胶比降低到 0.3 时，龄期 1 天孔隙率分别减少到约 27%、17.5%和 3%，水化龄期 21 天三条曲线交汇于约 2.5%。水胶比 0.3 与水胶比 0.35 相比，三条曲线随养护龄期的增长而收敛得早，分别为 21 天和 28 天；水胶比降低到 0.3 时，粉煤灰掺量 44%，养护到 7 天抗碳化能力就可与纯硅酸盐水泥的相当。

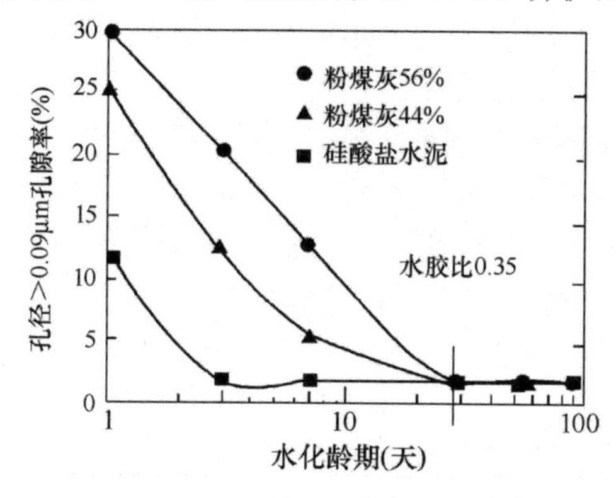

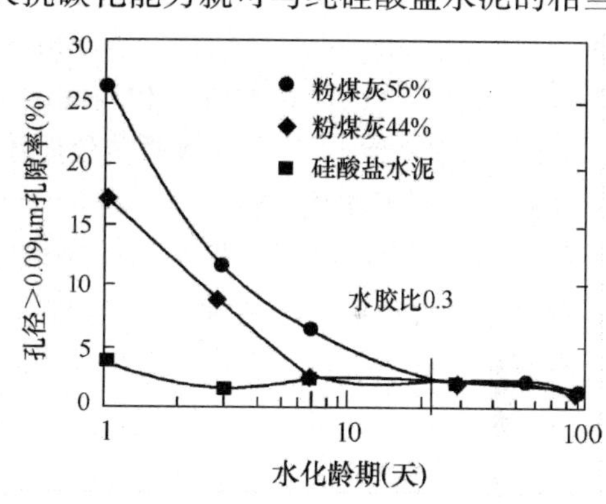

图 3　低水胶比硅酸盐水泥、大掺量粉煤灰浆体的孔隙率[3]

清华大学在深圳地铁足尺模型试验中，7 年后的钻取芯样表明，掺粉煤灰 24.5%、水胶比 0.57 的混凝土碳化深度为 13.7mm，而粉煤灰 41%、矿渣 19%、水胶比 0.40 的混凝土则碳化深度为 8.7mm。

以上表明，对用于较大体积的构件，降低水胶比、复掺粉煤灰和矿渣、适当延长养护期，则使用大掺量矿物掺和料时，不必担心碳化问题。

2. 障碍之二：把利益作为首要或唯一的目标，首先从降低成本出发而滥用。

就大多数企业来说，利废的目的主要是获取更多的利益——得到政府对利废的优惠政策和降低成本。当然这也无可非议，但是得到利益应当是结果，而不是目的，目的应当是为了我们的社会能够可持续发展。如果把利益作为首要或唯一的目标，就会走偏路。对不了解的东西不敢用，这是人之常情，但是也不能只懂得一点皮毛就胆大妄为地乱用。

当前有些企业（包括水泥厂和混凝土搅拌站）之所以什么都敢用，实际上就是因为一知半解之故。他们认为，水泥、混凝土的质量就是强度，有水泥（对水泥厂则是熟料）作基础，只要强度能达到，用什么都行。混凝土搅拌站的办法是降低水灰比，水泥厂的办法就是磨细（使用他们未必知道是什么成分并对混凝土有什么影响的助磨剂）。在他们的概念里，凡是工业废渣都能当火山灰质材料使用。殊不知不仅火山灰质材料和矿渣一类具有潜在水硬性的材料是不同的，即便目前按传统对火山灰质材料的定义都划分属于火山灰材料，来源不同时，其性质和在水泥中的作用也会有很大的差别。例如粉煤灰，其化学活性是很低的，最可贵之处在于其所含的玻璃微珠，甚至其中的少量莫来石和石英也都有用，而天然火山灰中凝灰岩和沸石化的凝灰岩就有很大差别，凝灰岩和浮石、火山玻璃碎屑也都有很大差别。至于不同的工业废渣，由于原矿石中除主要提取部分以外的成分复杂，使得如铜渣、磷渣、铬渣、锰渣、锑矿渣、硅锰渣、钢渣、铁合金渣……尽管主要成分都是 SiO_2 和 Al_2O_3，而其他以微量存在的某些成分或元素对混凝土是否有某种不利影响呢？人们对粉煤灰和矿渣的研究、认识和使用的历史和范围虽然也还不足，毕竟因经过较长期的使用和考验取得了一定的经验，对其他各种废渣的研究则甚少，最多也只是进行了强度试验。行动比研究超前的做法，是会有风险的。又例如当前兴起掺用磨细石灰石粉之风，也是主要从强度出发的行为。水泥生产者并不向用户说明掺了石灰石粉的水泥有什么适用性。例如抗冻性如何？抵抗低温下的硫酸盐腐蚀能力如何？在含地下水中含有侵蚀性 CO_2 的环境中有什么危害？主要用在哪里？怎样用？等等。

有些厂家为了得到国家对利废的优惠，但又不违犯质量监督部门的规定，竟然制造假报告，致使供应商和用户相互不信任，造成市场混乱。混凝土生产商不愿意购买复合水泥而自己掺用矿物掺和料的原因，除了成本和利润因素外，许多厂家表示“不知道水泥厂掺了什么东西”，新修订的水泥标准将 42.5 普通硅酸盐水泥中混合材料掺量的规定放宽到了 20%，但是仍有实际掺到 25%甚至更多的情况。这样给工程使用造成很大的麻烦和潜在危害。生产混凝土不管使用环境和构件特点，违背科学地滥用工业废料（品种和掺量），也使工程存在隐患。

混凝土这种混沌体系，本身就因其原材料的自然波动、施工控制以及对环境和时间等复杂因素而具有某些不确定性和不确知性，如果因供应商缺乏诚信，为自身利益而造假，不仅违背科学利废的原则，而且给工程又增添了不确知和不确定的因素，违背职业道德的原则。

观念决定技术路线，技术总是需要科学研究先行。科学利废就不能认为单有强度的保证就万事大吉，对工程来说耐久性和安全性同等重要，而耐久性不良时也会危及安全性。

二、建材利废也需要节能、减排

既然是利废，其本身是一种节省能源、资源和治理环境的行为，为了环境保护和更好地利废，可能需要消耗少量的能源去进行处理，其中如高炉炼铁矿渣的水淬、煤粉炉发电粉煤灰的分级电收尘、沸腾炉燃煤固硫的冷却排渣机，等等，都是按照用于建材的需要而在线处理废渣的工艺措施。但是如果在利用时又过分不恰当地增加处理的能耗或排放，就与利废的目的相悖。利废也需要节能、减排！

就当前的主流来说，人们的观念中仍因盲目追求强度和对活性的概念的误解，而导致对矿物掺和料活性的迷信。为了提高矿物掺和料的“活性”，当前所采取的主要技术路线就是一磨一烧。以矿渣和粉煤灰为例，对矿渣这种具有潜在水硬性的材料，磨得越细，其潜在活性越能得到发挥。过去矿渣水泥因矿渣和熟料共同混磨，硬度较大的矿渣难磨，比表面积为 $300m^2/kg$ 的水泥中矿渣的比表面积只有约 $250m^2/kg$，活性难以发挥，并使混凝土拌和物泌水，硬化后抗渗性和抗冻性差。如今采取熟料与矿渣分磨技术后，就解决了此类问题，而且也可提高水泥的强度。目前国内市售磨细矿渣比表面积多为 $400\sim450m^2/kg$。这项指标是某些大型钢铁厂引进国外超细磨设备和粉磨技术，按最大销售量和利润，经过投入产出比“优化”的产品规格。用该种粉末设备与技术达到这样的产品指标，耗电量约为 40 瓩小时/吨，与水泥厂生产普通水泥球磨机耗电量相当，但相对于水泥来说，省去了熟料的煅烧，对节能、减排也还是起作用的；问题是，有些人误认为只要一磨就可以卖钱，在地方上用球磨机粉磨水淬矿渣。球磨机的效率是很低的，将矿渣磨至比表面积为 400～450m²/kg，耗电量可达 100 瓩小时/吨。尽管由于电价相对于出售后的可得利润为生产商可接受，而这样大的能耗却有悖于节能、降耗、减排的目的！

另一方面，水胶比是决定水泥基材料强度的主要因素。自从高效减水剂大量使用以来，已能用调整水胶比来达到强度的目的。矿物掺和料对混凝土的作用，强度并不是一切。实验和工程都已表明，胶凝材料活性越高，水化产生的热量越大，自收缩也越大（见图 4）；从图 5 可见，当矿渣比表面积超过 400 m²/kg 时，掺量为 30％的胶凝材料水化热随龄期的进展有超过无掺和料水泥水化热的趋势，水胶比低时尤甚（见图 5）[5]。解决的办法是在分磨时，矿渣磨得细，而熟料磨得不要太细，所生产的水泥就可以得到较好的颗粒级配，长龄期性能可得以很好地发展；否则，最好矿渣比表面积控制在不超过 400 m²/kg。不少人反映掺磨细矿渣的混凝土后期干缩裂缝增多，已有人实验研究表明磨细矿渣掺量越大，长期干燥收缩越大（见图 6、图 7）[6]。

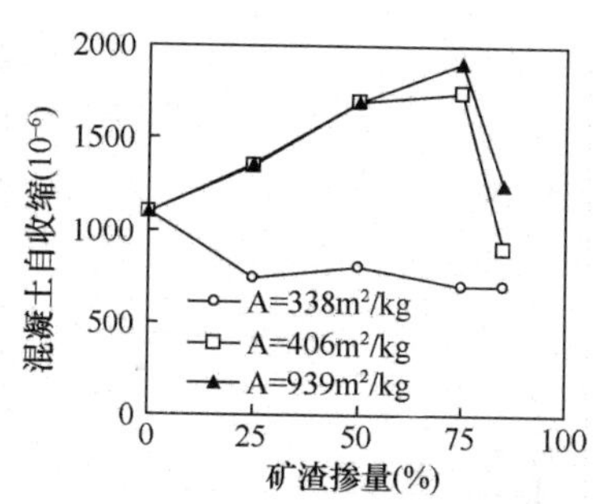

图 4　不同掺量的不同细度矿渣混凝土自收缩[4]

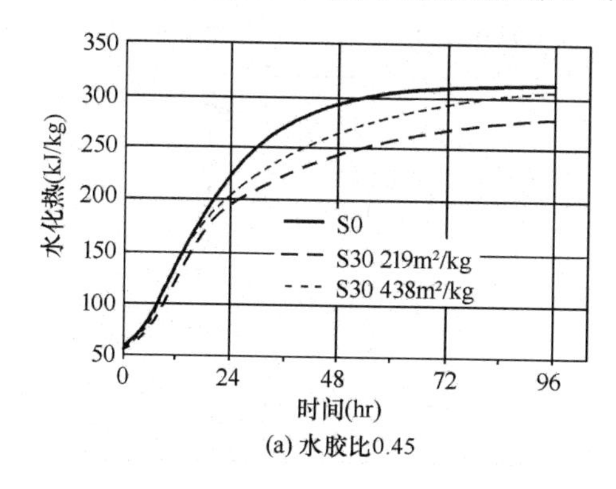

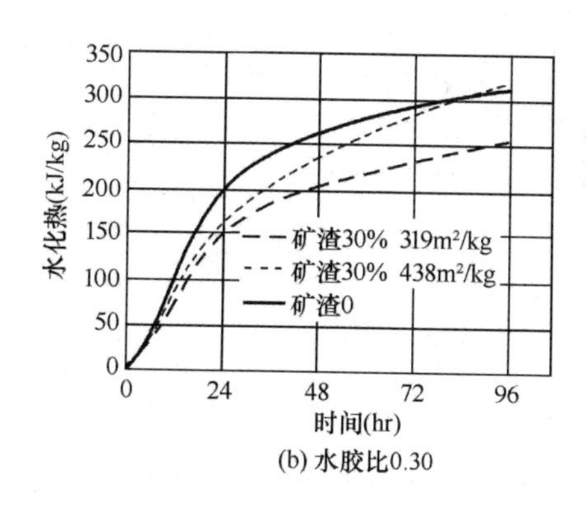

图 5　掺量 30％不同比表面积矿渣对胶凝材料水化热的贡献[5]

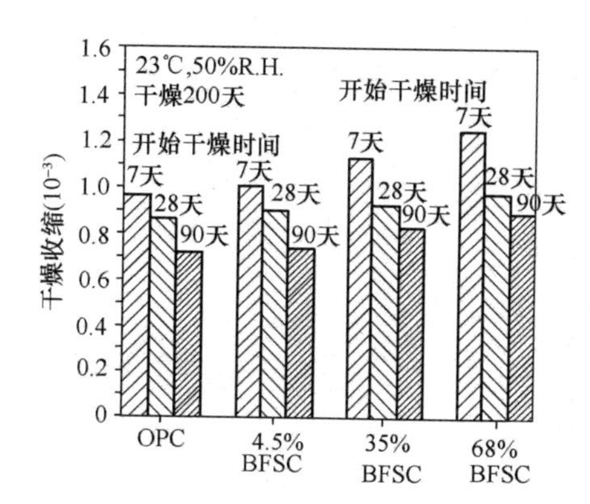

图 6　磨细矿渣掺量对砂浆长期干缩的影响

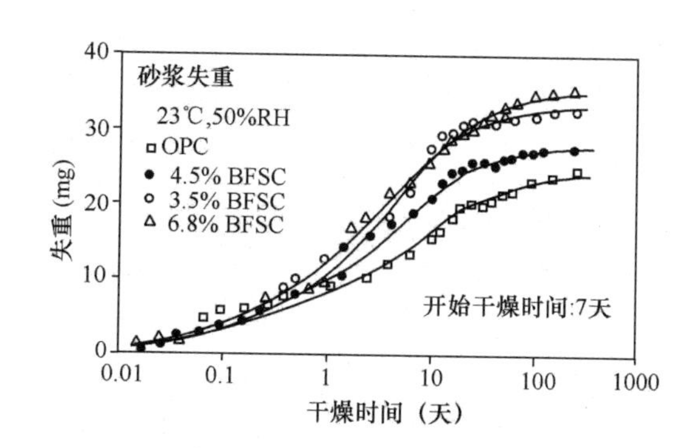

图 7　磨细矿渣掺量和砂浆干燥失重的影响

不少研究证明，粉煤灰本身的化学活性是很低的。建材院董刚检测在粉煤灰掺量为50%的浆体中不同比表面积的粉煤灰反应程度的实验结果表明[6]，水化60天时，比表面积为400m^3/kg的粉煤灰反应程度为11.42%；比表面积为700m^3/kg的粉煤灰反应程度为17.53%，比表面积增加75%，反应程度只提高了6.11个百分点（57%）。而反应程度也还只有不到20%。从图8可见，粉煤灰从比表面积326m^3/kg磨细至934m^3/kg，掺30%的水泥浆体水化热增大得很少。可见，比表面积增大近两倍，其早期反应活性还是很低的。由此可见尽管因比矿渣易磨的多，从能耗的角度来说却是得不偿失的。掺粉煤灰能降低混凝土温升的原因就是粉煤灰的化学反应活性低。在浆体中掺入粉煤灰，尽管减小水胶比（定义为水与胶凝材料总量之比），而因粉煤灰早期几乎不反应，拌和水主要都供应硅酸盐水泥的水化，水与硅酸盐水泥的比值（定义为水灰比）增大，可促进硅酸盐水泥的水化速率和水化度的提高[6]，粉煤灰中的玻璃微珠可取代所减少的未水化水泥颗粒作为浆体的微集料，以稳定浆体体积和保证强度。后期随龄期的发展，粉煤灰中玻璃碎屑与玻璃微珠表面活性部分参与水化反应，凝胶本体与玻璃微珠表面结合紧密，粉煤灰中各种微粒都可参与混凝土中的颗粒再级配，而不断增强。物质的化学成分通过结构起作用，体系的性质是由结构决定的。粉煤灰作用主要是通过改变混凝土的微结构来实现的。不同矿物材料有不同特点及作用，单纯迷信化学活性，就不能正确使用，反而使节能减排的原意适得其反。对粉煤灰来说玻璃微珠含量和烧失量（反应含碳量）比其细度、活性指数等指标更重要。

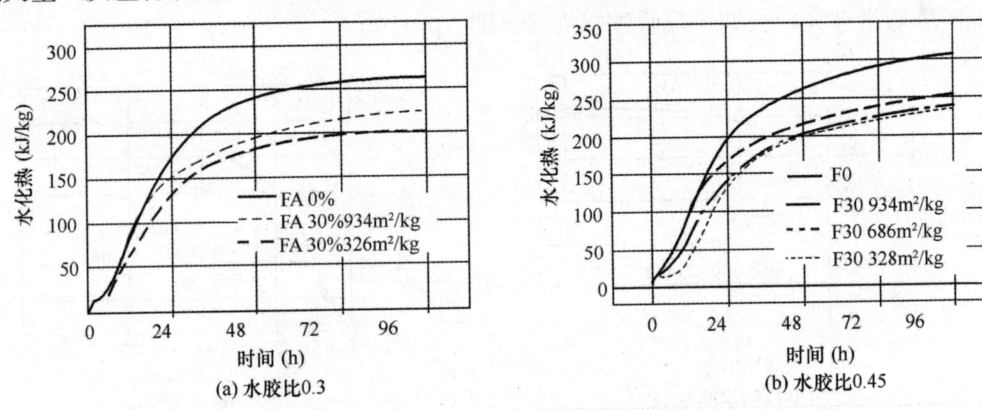

图8　掺30%不同比表面积粉煤灰对胶凝材料水化热的贡献

从“利废不能再增耗或增排”的原则出发，用单纯煅烧的方法提高如高岭土、煤矸石等的化学活性，应当都是不可取的——既增加能耗，又增加CO_2的排放。如果只有“烧”的办法，就应当有能源的再利用。例如目前国内的粉煤灰去碳装置，可将碳回收成民用煤粉；煤矸石可用以代替一部分黏土制生料生产水泥，其所含碳粉也可利用，等等。

三、如何看待利废的效益

1. 建筑材料利废的效益并不能只算经济账，首先应当是环境效益——缓解对环境的冲击，社会效益——节省天然资源和能源的消耗，参与循环经济的施行。只有不断转变传统观念，深化对产生利废效果机理的认识，才能高效地利废，因此在利废过程中对胶凝材料水化硬化机理的基础研究会产生学术效益。总之，建材利废是一项功在当代利在千秋的事业，是对人类可持续发展的贡献。单算经济账可能眼前盈利，但是节能毕竟是关系到人类社会可持续发展的大事。如果只从赚钱出发，采取增加能耗和排放的手段去“利废”，一旦赚到很多钱以后，却发现人类因能源和资源的匮乏和环境的恶化而无法继续生存时，就会悔之甚晚矣！相反，如果能换来环境的改善、天然资源和能源的节约和结构物的耐久，则应当“不因利小而不为”。我们赖以生存的根本不是钱，而是资源、能源和环境。正确利废产生环境效益和社会效益是第一位的，决不能违背节能、减排的方针。

2. 建材利废不能把自己当成“垃圾筐”，不管后果地什么东西都往里装。比方说，在天灾人祸造成经济困难年代的“瓜菜代”，用以“代”的也是能吃的、无害于人健康和生命的“瓜菜”，决不会容许食用观音土。同理，我们不能为利废而利废，也要以我为主，立足于有利于产品的质量。质量首先是产品的匀质性和无害性，而不能只用强度去衡量。任何利废的行为首先必须是研究，不仅进行强度试验，而且对任何无害性要进行试验和检测。国外水泥标准中的矿物掺和料只规定使用硅灰、矿渣和粉煤灰，我国新修订的水泥标准中也改为这样的规定。我们提倡建材利废，但首先提倡的是进行研究，至少要对其中的成分无害性进行检测证实。对不同的工农业废弃物都要研究其中除硅、铝、钙以外各种成分对例如水泥混凝土等建材产品性能的影响，有可靠的、足够量的数据，并经过工业试验和专家鉴定。

结论

1. 建材利废是“功在当代利在千秋”的事业，不能只算经济账，必须算能源账、环境账，勿以利小而不为。

2. 利废未必降低质量，利废的目的还应当是提高产品的质量，关键是正确的观念和技术。利废也需要节能、减排。

3. 任何行为都有得必有失，注意到一种倾向的同时要注意另一种潜在的倾向。对“废弃物”为了利用而进行处处理时，不能只算经济账，必须考虑是否增加了能耗？是否有新的环境污染？必须计算产品的总能耗，例如在水泥中掺磨细矿渣，所取代的水泥量并非实际的节能、减排量，必须扣除矿渣处理过程全部的能耗成本；例如使用煅烧路线的粉煤灰脱碳、煤矸石活化等，必须计算其能耗和废气的排放。应当鼓励和提倡能量的再利用。

4. 任何技术都有利必又有弊，没有好的和不好的，只有合适的与不合适的；不同矿物掺和料，有不同技术要求并处于不同环境中工程的适用性，不可盲目处之。利废也必须保证产品质量，但强度不是一切，甚至不是第一，不能只因能满足强度指标要求就乱用。既要讲科学地大量使用，又不能把自己当成“垃圾筐”。

参考文献

[1] P. K. Mehta. Monolith Foundation: Built to Last a “1000 Years”, Concrete International. July, 2000.

[2] M. R. H. Dunstan. Fly Ash, As “The Fourth Composition” of Concrete, Proceedings of Second CANMET International Conference on Fly Ash , Silica fume, Slag and Natural Pozzolans in Concrete, 1987.

[3] E. Tazawa and S. Autogenous Shrinkage by Self Desiccation in Cementitious Material, 9th international Conference on Chemistry of Cement, New Delhi, Sept. 1992.

[4] Feldman, R. F., Carette, G. G., and Malhotra, V. M.. "Mechanism of Development of Physical and Mechanical Properties of High-Volume Fly Ash in Cement Paste". Journal of Cement & Concrete Composite, Vol. 12. No. 4, 1991.

[5] 张士海. 胶凝体系对混凝土早期开裂的影响与评价方法研究[D], 清华大学硕士论文, 2002.5.

[6] 董刚, 陈志成, 陈益民. 比表面积和掺量对粉煤灰反应程度的影响[J]. 混凝土, 2008.7.

[7] 张庆欢. 粉煤灰在复合胶凝材料水化过程中的作用机理. 清华大学硕士论文, 2006.6.

自评

1. 第一节中"障碍之一"第三段的一句话"水泥只是一种原材料，其本身的强度没有意义，也无法得知，……"说法不确切，应为：水泥本身的强度并不是真值，而是在现行标准规定的固定水灰比及其他条件下检测的相对值，其意义仅供出厂产品质量的检测。水泥强度和混凝土强度没有固定的关系。

2. 同节中"障碍之二"提到关于石灰石粉的利用，这是因为我国很多地区使用石灰岩碎石，用生产石灰石碎石的下脚料生产机制砂，而机制砂的下脚料是石灰石粉。试验研究和生产实践证明，石灰石粉作为混凝土的掺和料或水泥的混合材，其中的$CaCO_3$在水泥水化的早期可与水泥中的C_3A反应，生成水化碳铝酸钙，提高混凝土的早期强度，并且可调节混凝土拌和物的施工性能，如缓解泌水，降低高强混凝土的黏性等。欧洲水泥标准已经把"石灰石硅酸盐水泥"列入CEM-Ⅱ中，最大掺量为25%。中国混凝土与水泥制品协会也制定了石灰石粉的产品与应用标准。有人质疑石灰粉的"活性"，因为"石灰石粉后期对混凝土强度没有贡献"。这是长期以来对矿物掺和料作用的片面认识造成的担忧。三十多年前，基本上混凝土不掺用矿物掺和料，作为水泥的混合材，在教科书中，把混合材分成活性的和惰性的（或称填充性的）。其中的"活性"指的是有自身潜在水硬性或火山灰性，也就是具有化学反应的性质。随着科技的进步，尤其是高效减水剂的广泛使用，对"化学成分通过结构决定强度"有了更深刻的认识，矿物掺和料的"活性"就不那么重要了。不同的矿物掺和料有不同的作用，可以按需要择优而用。石灰石粉主要成分是$CaCO_3$，而不是主要含铝、硅质的成分，当然不具有火山灰性，其自身也没有潜在的水硬性，所以也就不具有对混凝土后期强度的贡献。其活性主要表现在早期能取代$CaSO_4$而与C_3A反应生成水化碳铝酸钙，提高早期强度，其更重要的作用是可调节拌和物的施工性。至于生成碳硫硅酸钙型的硫酸盐腐蚀是有条件的，必须有水、低于15℃的温度和外部或内部硫酸盐三者同时存在，才是此类腐蚀发生的充分而必要的条件。缺少其中任一条件都是安全的。

3. 对节能、降耗、减排的决策本身也需要系统分析。例如，当前我国水泥产能过剩的根本问题是熟料的产能过剩。解决水泥产量过剩的根本措施是减少熟料产量，代之以矿物掺和料生产水泥，满足市场对水泥的需求。已知磨细矿渣和粉煤灰都是很好的矿物掺和料，石灰石、沸石岩等硬度很低（莫氏硬度3～4），作为水泥的混合材不会增加水泥粉磨的能耗。作为一种矿物掺和料资源，不仅可对缓解水泥产能过剩有利，而且还能改善水泥的性质。

4. 结论第4条指的是任何技术都有其适合的用途，用得合适就有利，用错了场合或方法，就会有弊。这也是工程中辩证的规律。

论文之廿三

解决混凝土结构工程质量的根本途径

——变革混凝土工程的生产关系*

廉慧珍（清华大学土木水利学院）

1. 混凝土工程生产关系现状及对工程质量的影响

混凝土材料不是最终产品，其最终产品是通过规范化的工艺制成的结构构件。混凝土工程是包括选择和控制原材料，经过试配、生产、运送，直到浇筑、振捣、收面、养护（控制温度和湿度）至指定龄期的全过程。完成最终产品的是混凝土工程。其中浇筑、振捣、收面、养护是混凝土工程中影响最终产品质量的关键工艺。脱离这关键工艺的混凝土拌和物不能保证最终产品的质量。

当前混凝土工程已被分离到不同行业，由不懂混凝土的人带有强制性地向混凝土供应原材料，生产出的混凝土拌和物又交给更加不懂混凝土的人去完成最终产品，信息不对称而缺损，混凝土工作者缺腿，其职责无法到位，难以确定混凝土工程的责任者。出现质量问题时，必然纠纷不止。这种水泥-混凝土-施工生产关系已经在阻碍混凝土工程技术和建设的健康发展，造成能源、资源的浪费和对环境的冲击，改革是必须的。

生产关系的实质是生产资料的占有和利益分配的关系。改革传统水泥-混凝土-施工的关系必然涉及各方各自的利益，因此生产关系的改革是一种革命性的变革，必然会遭到传统势力的反抗。生产力的发展是必然的规律，生产关系的具体形式必须适应生产力发展的要求。从人类历史发展来看，生产关系的变革符合随着生产力发展的否定之否定规律：从原始社会的无分工逐渐到自然分工——主要是家庭中按性别、年龄分别外出狩猎和家务的分工；此后又逐渐产生三次社会大分工：第一次农业和畜牧业分离；第二次手工业和农业分离；第三次出现商业，产生专门从事商业活动的商人。在这样三次社会大分工的过程中逐渐形成物质生产劳动和精神生产劳动以及体力劳动和脑力劳动的分离和对立。18 世纪（1765 年）开始的第一次工业革命和 19 世纪（1870 年）开始的第二次工业革命所带来的分工至今越来越精细，然而，20 世纪以来生产力的飞速发展，又需要集约化、规模化；科学技术的高度发展需要联合与传承，需要不同学科交叉。若干跨学科的新技术、新理论的出现，说明打破专业藩篱和学科壁垒也是现代社会发展的又一次需要和必然，否则必然会阻碍当代科学技术新的突破。当前混凝土工程的生产关系也面临着变革的问题。

100 多年以来，水泥和混凝土一直分属于两个学科系统，在生产中存在严重的行业隔离现象，水泥界的观念是：“我们就要高 C_3S、高强度、低能耗，至于对混凝土有什么影响，那是混凝土的事”，不了解除了强度之外，混凝土应当有什么样的质量要求；混凝土

* 原载《混凝土世界》2011.06。

界一味追求强度的提高，认为水泥强度高才能提高混凝土强度，并不了解水泥厂采取什么样的技术路线提高水泥的强度，也不了解这个技术路线对混凝土会有什么影响。在过去水泥强度和混凝土总体强度等级都较低、混凝土结构服役环境相对简单的情况下，水泥和混凝土相互之间的矛盾并不明显，而如今，随着社会与科学技术的发展，水泥业和作为用户的混凝土业之间的矛盾不断凸显。主要是因为高效减水剂的大规模使用后，混凝土强度不再依赖于水泥强度，但水泥强度却不断提高。原本因使用新型干法窑而降低能耗，却为了提高标称强度而又增加粉磨的能耗，以致当前用量仍然很大的C25以下混凝土无法配制。勉强配出的，则强度大大超过设计要求，或者严重离析、泌水，影响结构质量；同时因为水泥过细和因此而造成颗粒级配的不合理，使混凝土开裂敏感性增大、和外加剂相容性变差、抗拉强度下降、后期强度倒缩、耐久性变差……。这是水泥界不了解其所服务的对象所造成的弊端。

过去教科书中所说"配制混凝土时，水泥强度一般应为混凝土强度的1.5～2.5倍"，是针对当时用0.44/0.46的水灰比检测的水泥强度、而混凝土又不使用减水剂的条件而言的。现今，32.5水泥完全能配制出用于某种环境的C60混凝土，而用于C20混凝土时仍然强度太高，复合水泥用于C30、C40常用混凝土很好，但许多人不愿意用，认为复合水泥"强度低"、"混合材品种和掺量不透明"，这是混凝土界不了解水泥的生产而造成的结果；另一方面，他们自己反而为了降低成本而打着"提高耐久性"的旗号，在生产混凝土时盲目增大矿物掺和料用量，甚至只看强度而启用未经全面研究和工程考验的其他矿物粉料，最快显露的直接结果就是混凝土拌和物离析、泌水严重，混凝土碳化速率加快、碳化深度增大，造成施工方和混凝土供应上的矛盾、混凝土供应商和质检之间的矛盾。

在过去现场生产混凝土时，由于懂得混凝土，知道如何保证工程质量，施工方自己对混凝土质量负责；然而，从提高生产效率和工业文明的角度来看，预拌混凝土无疑是混凝土生产的一个进步，问题是有利必有弊：如今，基本上混凝土是脱离工程的，施工人员越来越不懂混凝土了，混凝土供应者也不管施工，施工方并不对混凝土负责。于是常出现施工单位的"无理"要求，擅自在拌和物中加水，改变配合比，等等现象，验收时纠纷不止。而混凝土供应商也只负责运到工地交货为止，并不管施工单位如何处理和对待混凝土的拌和物，难以确认对最终产品的责任。有人说："现在混凝土怎么打，实际上是民工说了算，随意加水的问题拗不过民工。"这是个真实反映的普遍现象。谁能解决？谁来解决？

2. 当前预拌混凝土生产存在的问题

行业隔离的结果造成双方相互误导，而从生产关系上来看，现在水泥和混凝土两个行业是各自独立的经济实体，双方都有各自占有自己的生产资料和分配方式与制度，亦即都有各自的利益立场。矿物掺和料由谁掺好？实质上是钱由谁赚的问题。现状必然是各不相让。然而，从发展生产力的角度，公平地说，目前由搅拌站掺用掺和料有如下弊端：

① 因为原材料种类多，常有人把粉煤灰错当成水泥，甚至有人把膨胀剂当成了矿渣使用，这类事故时有发生，一旦发生，损失就很大。解决此类问题就要增加管理的成本。

② 水泥厂将熟料与适量石膏共同粉磨而生产水泥（包括掺混合材的水泥和复合水泥）时，其中的石膏是进行整体优化的，搅拌站用水泥掺用矿物掺和料时，则不可能对石膏再进

行优化，因此矿物掺和料的掺入就会稀释石膏，而且也稀释 $Ca(OH)_2$，而石膏和 $Ca(OH)_2$ 恰恰对矿物掺和料有激发作用。于是，掺用掺和料的混凝土早期强度低、凝结缓慢、与外加剂相容性差、干缩大、长期弹性模量较低。

③为了迎合搅拌站解决这个问题的需要，有人办起矿物掺和料的生产厂（名为"复合胶凝材料"，不含外加剂），把不同矿物掺和料按一定比例（比例的根据是什么?）混合，再加入一定量的石灰和石膏（加入量的根据是什么?），当前一种"矿粉"这种有意无意混淆视听的产品，其中有的就是在生产矿渣粉时掺入了粉煤灰或石粉（什么岩石的粉?）和一定量石膏混合而成。石膏的优化必须和水泥结合，即对全部胶凝材料整体优化，目前市场的复合胶凝材料添加石膏或复合都是脱离水泥的、盲目的。且不论这种产品如何适应不同环境中不同工程的需要，即使这种产品能用，在搅拌站掺到水泥中后，也不能解决一个致命的难题——搅拌不均匀。水泥生产的质量匀质性是最重要的，因此从原材料、生料、熟料，直到水泥成品，每一道工序都经过均化，而混凝土搅拌站是无法做到这样均化的。传统混凝土拌和物达到变异系数稳定所需搅拌时间约为 75 秒[1]，现今掺入矿物掺和料和外加剂的拌和物，全程搅拌达均匀的时间（包括加水前的干拌）为 150 秒[2]。而目前我国预拌混凝土，通常搅拌时间都是 30 秒，个别最长的也不过 1 分钟，延长搅拌时间势必影响产量，所以至今基本无人能做到。

④由于混凝土搅拌站难以控制进厂水泥的来源，而难以从根本上解决水泥与外加剂相容性问题。西方国家控制预拌混凝土坍落度损失一般采取外加剂分次掺入的方式：在搅拌机中上料时先掺入一半，然后通过输送车上搅拌机中的传感器，感应拌和物黏度变化，而自动添加其余部分。我国混凝土厂商没有人舍得投资安装这种装置，一律将外加剂溶入拌和水中一起投入。这是外加剂使用效果最差的方式，不仅坍落度损失无法克服，而且也不经济——大量外加剂被骨料表面所吸附，使外加剂效率下降，不得不增加掺量。外加剂掺入最好的方式是在粉磨水泥时兼作助磨剂而掺入，外加剂由于被先吸附在水泥颗粒的表面，在混凝土加水搅拌后能缓慢释放而长时间保持在混凝土拌和物中的浓度，可提高流化的效果，减少坍落度损失，还可避免外加剂被骨料吸附而降低效率的问题。从外加剂效率来说，也是矿物掺和料在水泥厂掺好。

3. 当前水泥对混凝土不适应的问题及其对工程质量的影响

从水泥厂现状来看，其大宗的用户是混凝土，然而水泥生产并未考虑现今用户的特点和需要，例如，混合材比硅酸盐水泥对水灰比有更大的敏感性。随水灰比的减小，掺用矿物掺和料后水泥强度增高的幅度大于硅酸盐水泥的强度增高幅度[3]。按照现在人们的从众思维方法，对水泥质量的评价主要认为强度高的水泥才是好水泥。现行水泥标准把修订前的水灰比 0.44/0.46 增大到 0.5，则混合材水泥的表观强度降低得更多，使用户误认为"凡是有掺和料的水泥都不是正品，性能都不好。"实际上现在的混凝土强度已经与水泥强度不再是线性的关系了，用强度高低作为质量的标准已是落后的、不科学的观念！现行 32.5 水泥照样能配制出 C60 混凝土。传统观念认定的水泥强度，实际上是作为产品出厂合格检验所用的指标，只是水泥的标称强度（即以产品标准检测而称谓的强度），并非真值（真值是测不出来的）。水泥厂这种对强度认识的误区导致了不断追求用助磨剂使水泥比表面积已达 $400m^2/kg$ 以上，还要求助磨剂厂供应能增强的助磨剂，甚至有的还借助于工业盐来增强，对工程危害极大。

除强度以外，当前反应最多的有如下的问题：

①水泥与外加剂的相容性问题

使用现行标准加水量无法检测当前水灰比普遍低于0.5的混凝土中外加剂与水泥的相容性[4]，而且水泥厂不知混凝土厂使用的是什么外加剂，也无法在出厂前进行检验。这一方面表明现行水泥标准已不能适应混凝土的现状了，另一方面表明水泥和混凝土行业隔离、生产、检验分家的现状也已不能适应当前的混凝土质量控制的要求了，国外也存在如此的问题。

②水泥因粉磨得太细而需水量增大

上世纪80年代的水泥标准稠度用水量一般都在24%～26%，现在则普遍在30%左右。混凝土也提出希望水泥需水量要小，水泥厂采取的措施是在助磨剂中掺入少量减水剂。掺入减水剂后水泥所表现的需水量并非一定工艺下水泥的本征特性，造成水泥需水量大的颗粒细度和级配的不合理问题并未改变，混凝土在使用中反而因这种隐藏的信息而造成试配的难度。由于人们对大型水泥企业的信任，尽管我国水泥产能过剩，产量足够，却在市场上显得水泥供不应求，以致水泥厂不顾混凝土的需要而仍然我行我素。

③ 行业隔离所造成的信息不对称

信息不对称不仅影响交易，而且是对工程质量影响的最大问题。由于受了混凝土“强度第一”观念的误导，当前的水泥磨得太细，而且有越来越细的趋势，原本水泥厂为了降低能耗而使用助磨剂，现在使用助磨剂却主要为了磨得更细，以便掺入更多的混合材来降低成本。混合材掺多少、掺什么，都已和标准无关了，目标只是强度。其结果是水泥和减水剂相容性差，混凝土抗拉强度与抗压强度比值下降、抗冻性差、开裂敏感性增大，混凝土强度早期很高，28天以后就不再增长了；初期所形成的不可见内部缺陷因水泥在早期水化消耗几尽而后期没有足够的水化物去修复，更会增大在温、湿度交替变化下的开裂敏感性，加速混凝土劣化。即使在无化学腐蚀和冻融循环的一般环境下，也会如此。

混凝土从业者要求水泥不要太细但要早期强度高，这又是他们不懂得水泥的无理要求，不知道现在提高强度靠的主要手段就是磨细。在这个问题上，水泥和混凝土的从业者都应当知道一个常识，即“早长晚不长”，“早熟早衰”。速生的杨树、催肥的家禽家畜等如此，人也如此，混凝土也如此。

一些大型水泥厂因生料制备的需要，一般有自备石灰石矿山，从原岩的开采到逐级破碎到最后的细颗粒；另一方面，有的搅拌站还有自备矿山生产自用的石子和人工砂，产生数量很大的石灰石粉。双方都在浪费能源和资源。这也表明行业隔离的现状已经阻碍生产力的发展，阻碍科技进步，影响混凝土工程的质量和资源、能源的合理利用。

4. 混凝土生产关系的变革符合社会发展否定之否定的规律

人类最早发现，在有些石头（石灰岩）上用火烧烤食物后该石头会变得松软，遇水体积膨胀并产生很大的热量，因而开始了石灰的生产；后来又发现含有一定量黏土的石灰石烧制的石灰具有一定的水硬性，但是天然石灰石中黏土含量波动幅度大，产品不稳定，继而用人工添加黏土调配原料生产出现代硅酸盐水泥的雏形。从古代至今，用于混凝土的胶凝材料经历了否定之否定的发展过程，如图1所示。

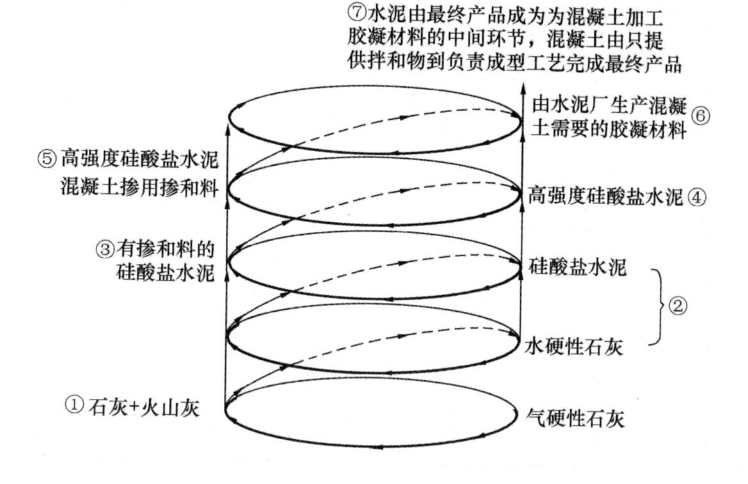

图 1 水泥生产发展否定之否定规律和趋势

图 1 表明：

⟶第一次否定：从气硬性石灰的使用到与火山灰混合使用以得到水硬性。

⟶第二次否定：对第一次否定的否定：用黏土质石灰石烧制的水硬性石灰，发展到因天然黏土质石灰石成分不稳定而用黏土和石灰石人工配料煅烧的硅酸盐水泥。

⟶第三次否定：对第二次否定的否定：为调节硅酸盐水泥强度、降低水化热、提高抗化学腐蚀性的混合材料硅酸盐水泥。

⟶第四次否定：对第三次否定的否定：适应大跨度、高耸结构工程使用高强混凝土需要的高强度硅酸盐水泥。

⟶第五次否定：对第四次否定的否定：使用硅酸盐水泥的混凝土耐久性问题凸显，发现“高强不一定耐久”[5]而在混凝土中掺用矿物掺和料。

⟶下一次否定：对第五次否定的否定：因在混凝土中掺矿物掺和料具有不可克服的缺点，而使用由水泥厂按混凝土需要生产胶凝材料，进而发展成水泥-混凝土-混凝土施工成型工艺一体化的生产关系的变革。

每次否定之否定都没有简单地回到上一次否定之前，而是向前发展了。第五次否定之否定也没有简单地回到混合材料水泥，而是改为在混凝土中掺用矿物掺和料，原因有三：

（1）过去含混合材的水泥尽管有较低的水化热、较高的抗化学腐蚀性质，但用统一水灰比检测的强度，就成了以“牺牲强度”为代价。在以强度作为第一甚至唯一评价标准和方法的传统观念支配下，用户无法接受。

（2）矿渣和硅酸盐熟料硬度的差别使其共同粉磨后颗粒细度相差较大，磨得越细，相差越大，造成其中的矿渣比表面积太小（对于比表表面积为 $330m^2/kg$ 的水泥来说，矿渣比表表面积大约只有 $250m^2/kg$ 左右），不能发挥矿渣活性的潜能，不仅水泥强度低，而且因保水性差而不利于混凝土的抗渗和抗冻等性质。

（3）现在已有磨细矿渣的产品，尽管有的水泥厂现在也在生产水泥时掺入矿渣粉生产矿渣水泥，而混凝土自己掺用矿物掺和料则能降低成本获取更多的利润。于是，“掺和料由谁来掺”转化成“钱由谁来赚”的矛盾。

有的水泥专家的意见认为，索性只生产高强度的纯硅酸盐水泥（当然可以提高价格来补偿利润)，让混凝土有加大掺和料掺量的余地，掺和料的钱就由混凝土去赚吧！有的混凝土专家也认为，在混凝土中掺用矿物掺和料比使用含混合材的水泥质量易于控制。然而，这只是短视的观点。社会发展需要多样化的产品供用户选择，以满足日益增长的各种不同需要。产品品种单一化则会更增加用户的麻烦和社会的不和谐，是不符合社会发展规律的。

对“掺和料由谁掺”的问题应当跳出各自眼前的利益来看待。一方面，从双方长远的发展着眼，不能且顾眼前而堵塞各自发展的前途；另一方面水泥和混凝土都是服务于社会的产品，掺和料究竟由谁掺的问题，首先要从社会发展的利益出发，产品的经济效益是和社会效益密切相关的，没有社会效益的产品是没有生命力的。掺和料究竟由谁掺从根本上要服从结构工程的安全性和耐久性要求。

就组分复杂的现代混凝土的实际需要来说，把混凝土所需要用的胶凝材料放回到工厂去生产，在共同粉磨中做到原材料的预均化，就能够不增加现行搅拌时间而大大提高混凝土的匀质性。搅拌站总体成本不会增加，反而在节约能源和资源、减低质量管理成本上都会有明显的效益。从整体来讲，不仅可以实现水泥厂和搅拌站双赢，而且更重要的是有利于建设工程的质量和耐久性。从图 1 中按照已发生的否定之否定历程的规律预测下一次否定之否定的趋势，可见在水泥厂按混凝土的需要生产胶凝材料是符合客观规律的。

5. 国内外水泥和混凝土生产关系现状

水泥厂办搅拌站在西方发达国家已有多年的历史，例如法国、德国、瑞士。在国外很少在搅拌站掺用矿物掺和料，水泥厂的水泥品种基本上都可以满足使用要求。只有加拿大因 V. M. Malhotra 多年提倡和推广大掺量粉煤灰[6]~[9]，将大掺量粉煤灰和磨细矿渣列入 2004 年颁布的混凝土标准中[10]。新加坡的双龙水泥公司生产磨细矿渣掺量为 70%～75%的水泥，销售时，该厂的中心实验室为用户进行试配，提供符合用户工程需要的混凝土配合比。日本的水泥是把熟料和不同矿物掺和料粉磨后按细颗粒级级配后混合优化石膏掺量而制成，混凝土不单独掺用掺和料；欧洲水泥标准允许矿渣水泥中矿渣掺量达 95%，也是由水泥厂生产而不是在混凝土中掺的。目前我国由于供需双方认识和诚信的问题，一般做不到。

上世纪 80 年代，苏联水泥科学研究院和混凝土与钢筋混凝土科学研究院合作进行低需水性水泥胶凝材料（BHB）的研究[11]，1991 年投入小批量生产。1993 年俄罗斯正式注册为 BHB（该符号为俄文，音译成英文为 VNV）水泥，已有数家水泥厂生产。BHB 的后缀数字代表该水泥中熟料的用量。其余除石膏外的组分为矿渣、粉煤灰、石英砂粉中的两种或三种。相对于普通水泥标准稠度用水量 25%～30%，BHB-100～BHB-40 的标准稠度用水量为 16%～20%。BHB 水泥熟料用量可减少达 50%～70%，但所配制的混凝土强度可达 80～100MPa，用于配制混凝土，具有低水化热和较好的耐久性、工作性等优点。该水泥中熟料取代量最多可达 70%，强度却比基准水泥的高。例如其中 BHB-50 水泥用量只有 350kg/m^3 时，虽然熟料只占其中的一半（175kg/m^3），混凝土强度却可达 55.9MPa。这种水泥无疑是可持续发展的。

V. Alunno Rossetti 等试验研究并在意大利一家水泥厂投产了一种特种超塑化水泥 SPC（Special Superplasticized Cement），该水泥是在意大利 525 型硅酸盐水泥生产时掺入超塑化剂而制成的，提高了超塑化剂（高效减水剂）的使用效率[12]；瑞典用中热水泥和硅灰掺入超塑化剂粉磨生产出一种强力改性水泥 EMC（Energetically Modified Cement），专用于高强和超高强混凝土，改性后的水泥比基准水泥的强度提高60%以上。可以用0.19的水灰比配制出170MPa超高强混凝土[13]。

清华大学廉慧珍从1994年开始与北京住总集团合作，进行了低需水量、低水化热、低收缩的环保型胶凝材料的研究，最后优选试样中熟料含量均为44%，所用矿物掺和料为粉煤灰和矿渣。试验研究结果表明，组分相同时，与在现场混合使用相比，有明显的优势。检测其性能如下：

标准稠度用水量：12%～21%，用于不同强度等级；

凝结时间：初凝3：00～8：00，终凝4：30～10：00，用于不同强度等级；

28天抗压强度：50～80MPa，用于C30～C80不同强度等级；

收缩：水中养护7天后继续在空气中养护28天的限制收缩≯0.02%，落差≯0.04%；

水化热：用溶解热法测定，3天≯250J/g。

以上表明该方式生产的水泥具有低水化热、优异的工作性、耐化学侵蚀性和长期强度的较高增长率的优点，尤其适用于较高强度大体积的基础工程。当时用以配制C60泵送混凝土（坍落度为180～220mm）时，28天抗压强度为71MPa。

1997年在北京窦店砖瓦厂试生产了120吨产品，同时送往国家建材质量检测中心、冶金部废渣利用质量检测中心、北京市建工院检测中心等单位检测，结果如表1所示：

表1 试生产产品送检结果

检测单位		国家建材院水泥检测中心		冶金部质检中心	北京市建工研究院第一检测中心			
编号					ZY-1	ZY-2	ZY-3	ZY-4
细度（80μm筛筛余%）		1.8		1.2	1.5	1.2	1.0	3.2
标准稠度用水量（%）		18		17.4	22	22	22	22
安定性（试饼法）		合格		合格	合格	未做	合格	合格
凝结时间（初凝/终凝 h：m）		7：35/9：25		3：28/5：40	3：02/5：12	2：02/6：13	—	—
胶砂强度折/压（MPa）	3天	4.9/28.9	5.3/30.5	6.7/41.4	5.0/35.0	6.2/47.2	5.7/42.4	5.0/33.9
	7天	6.2/40.2	7.0/47.0	8.6/58.8	6.5/50.7	7.5/58.2	6.6/53.3	5.7/44.9
	28天	8.26/64.5	9.27/71.3	11.1/84.4	9.4/67.8	9.9/70.0	8.9/64.0	8.3/64.1
储存1个月强度（MPa）	3天	—	—	—	4.5/30.2	—	—	—
	7天	—	—	—	5.7/41.8	—	—	—
	28天	—	—	—	8.1/64	—	—	—
*限制变形率（%）	水中14天	0.023		—	0.014	0.029	—	—
	转空气28天	−0.018		——	−0.014	0.022	——	——
水化热（直接法）（kJ/kg）	3天	111		—	—	—	—	—
	7天	176		—	—	—	—	—
**干缩（%）	7天	−0.05		−0.07	—	—	—	—
	14天	−0.09		−0.04	—	—	—	—
	28天	−0.099		−0.10	—	—	—	—
	60天	−0.095		—	—	—	—	—

* 模拟当时规定的施工中养护条件。

** 成型1天拆模后即开始试验。

1999年，清华大学土木系进一步与北京城建（集团）混凝土公司合作进行了生产和应用的试验研究[14]，试生产出200吨用于C50以下高性能混凝土的高性能新型胶凝材料，产品中磨细矿渣和粉煤灰总掺量为50%。硅酸盐水泥熟料用量为42%～44%。其主要性能如表2所示。

表2　试生产FK新型胶凝材料的主要性质

80μm筛筛余（%）	凝结时间（小时：分）		标准稠度用水量（%）	抗折强度（MPa）		抗压强度（MPa）		水化热（J/g）	
	初凝	终凝		3天	28天	3天	28天	3天	7天
1.1	4：59	7：00	17.6	5.2	10.4	33.7	67.6	144	180

该产品用于北京美林花园公寓2号楼地下一层内墙柱、地下二层Ⅰ段车库，共浇筑260m³ C40混凝土，坍落度200～220mm，扩展度≥500mm，1小时后坍落度损失率<10%，工作性能优异，水化热低，体积稳定性好，易于振捣，28天强度达到设计标号的130%～132%，共生产50m³ C30混凝土用于浇筑顶板，28天强度达到设计标号的124%～131%；拆模后观察，混凝土结构表面光洁、密实，无裂缝和蜂窝麻面，颜色纯正，外观质量良好。但因该项技术涉及生产关系的变革，如没有统筹规划，则以推行，故至今未得以产业化。水泥和混凝土的原材料及矿物掺和料地方性都很强，尚需有一定规模的统计性的系统实验研究和系统设计。

我国近年来也有一些大的水泥集团收购搅拌站、建立混凝土实验室的情况，但是被收购的搅拌站一般技术水平不会太理想。而且水泥厂和搅拌站仍然是“两层皮”，彼此独立经营核算，只是水泥可免去中间环节而稳定供应，可降低混凝土的成本。这毕竟是一种进步，但是尚未考虑搅拌站掺用掺和料所造成的质量不均匀、SO_3被稀释而不足以及外加剂效率等问题，仍然也还有投料管理的问题。而且目前我国混凝土技术水平虽有进步，总体上从业人员水平仍赶不上发展的要求。

需要说明的是，当前石子生产大多数仍然技术落后、对环境的粉尘污染严重，所生产的石子颗粒形状差、级配很差，成为影响混凝土质量的关键因素之一。实际上对于现代预拌混凝土，石子强度已经不再影响混凝土的强度，但是在当前传统思维支配下，用户仍然强调石子的强度，而在当前大多数采石场的技术落后的状况下，强度越高的石子，颗粒形状越差，破碎的颗粒越小，针状和片状颗粒越多。因此目前大多供应的名义上连续级配石子基本上没有5～10mm级的颗粒。这是目前我国混凝土用水量居高不下的主要原因。提高砂石质量在技术上并不是困难的问题，根本问题还是生产关系不顺所造成。如果水泥厂的石灰石矿山开采不要首先着眼于水泥生料，而是改变工艺，先生产石子，生产水泥时能利用生产石子的废弃物（不够时可取石子生产的中间产品作为辅助），则可提高能源和资源的效率，降低总体成本。

6. 建立从原材料制备-试配-生产-成型工艺一体化的混凝土工程体系设想

和其他产品不同，水泥最大宗的服务对象——混凝土是在人工材料中一种最复杂的、非均质、多相、多变的混沌体系，因此为了适应这个特性，水泥不能是最终产品，而应当是混凝土的中间产品；混凝土服务的对象是混凝土结构，从混凝土到混凝土结构的过程又是很复杂的，会受到环境和成型工艺操作的极大影响，因此混凝土也不是最终产品，而是结构物的中间产品。从水泥-混凝土-混凝土结构物，必须是一体化的连续生产，才能保证结构物这个最终产品的质量——安全的和耐久的。打破狭窄行业之间的藩篱是时代发展的需要和必然。这种生产关系的变革首先不是技术问题，而是观念的转变：

①合并水泥原料和混凝土骨料的石灰石矿山，采用先进破碎和筛分技术，生产分级的、粒形优良的单粒级石子，供混凝土在制备时按两级配或三级配分级上料（实践证明两级配石子后，混凝土用水量可减少约20%），用5mm以下的颗粒制作人工砂；用砂石下脚料制作水泥生料，不足的部分另外取分级的人工砂或5～10mm的颗粒来补充；

②通过优化的试验研究将多种矿物掺和料、水泥和外加剂纳入水泥的粉磨系统，生产符合不同混凝土性能和质量要求的低水化热、低需水量、低收缩的胶凝材料，与砂石、外加剂等混凝土其他原料一起作为中间产品，减少中间环节，提高资源和能源的利用效率，降低社会成本，加以建立混凝土成型工艺作为售后服务的机制，大幅度提高混凝土质量，实现对结构工程的优质服务。其中外加剂的生产也纳入本系统，以便控制外加剂和胶凝材料的相容性。其流程如图2所示：

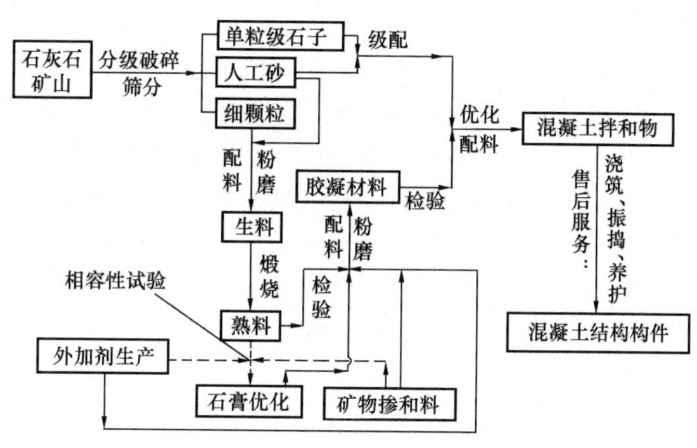

图2　作为混凝土工程的原材料制备-拌和物制备-成型工艺一体化生产模式

以上流程是把混凝土当作混凝土工程来做：总体控制原材料质量，水泥和砂石按混凝土需要生产，混凝土按工程需要试配、制备后，按合同与工程特点进行浇筑、振捣、养护后交货，即以对混凝土结构的“售后服务”取代目前施工单位对该工序发放的的分包，最终完成混凝土的全部工程。这时混凝土供应商对混凝土质量负责。

7. 混凝土工程生产关系改革预期节能、降耗、减排效果和社会效益

（1）节省现有水泥厂生料制作过程中的初级破碎能源，并减少采石场石粉的排放，总体减少石灰石的消耗；

（2）外加剂本身有很好的助磨作用，不同矿物掺和料也有相互助磨作用，可降低粉磨能耗并省去现有水泥所用助磨剂；

（3）可以通过在粉磨水泥中的优化配料和混凝土浇筑前的技术措施，提高混凝土表面密实度，解决大掺量矿物掺和料造成的早期碳化问题，将熟料消耗量减少到40%以下，达到减碳的目的；

（4）提高石子质量后，可减小混凝土水泥用量（用水量）约20%，增加节能、降耗、减

排的效果，并且可提高混凝土的耐久性；

（5）混凝土的所有原材料都直接按混凝土需要生产，可解决现有的水泥和混凝土、水泥和外加剂以及混凝土和外加剂之间的矛盾，形成和谐生产的关系，有利于社会稳定；

（6）当前施工人员越来越不懂混凝土，“混凝土专业包工队”人员素质差，任意向拌和物中加水、振捣不规范、养护不到位，严重影响混凝土结构质量，混凝土的售后服务机制可从根本上杜绝此问题的发生，保证工程质量，避免将来结构物劣化造成的资源、能源的浪费和建筑垃圾对环境的冲击。

8. 尚需研究解决的问题

改革混凝土工程生产关系的难度不在于技术，首先要转变观念，换一种思维方法，采取一种适应当前社会可持续发展的生产模式。既然生产水泥时能把生料预热和分解放到窑外去进行，为什么不能把预拌混凝土中胶凝材料的均化生产放到搅拌机外去进行呢？这个问题谁都能回答，最需要下决心第一个吃螃蟹要解决的关键问题就是如何解决各自的经济利益。不妨请一位经济学专家参与进行一下研究。

有不少人认为在技术上很简单。其实不然，尚有以下问题需要研究：

（1）如何做到不同工程需要的胶凝材料组成与性能优化和产品定型化？

（2）改变传统水泥从单一凝结时间优化石膏的方法，如何按多项性能优化石膏？

（3）如何建立符合混凝土规律的水泥性能检测试验方法和产品标准？

（4）如何使用成品胶凝材料进行混凝土配合比设计？包括程序设计和试验验证；

（5）如何实现混凝土生产与施工质量过程控制技术与自动化体系？

（6）如何进行水泥-混凝土-施工工艺流程、物料平衡、体系仿真，包括生产的过程测算和优化，以适应现代化生产的需要？

市场经济的特点之一就是产品多样化以适应不同的需要，不同的生产方式可以适应不同对象的需要。任何时候都不能“一刀切”地采取单一模式。上述设想可以现根据现有条件逐步实现。例如水泥集团收购、合并混凝土搅拌站只是个很初级的阶段，可以先从自有矿山的生产方式着手改造，为集团搅拌站提供优质骨料，培训、鼓励集团搅拌站先使用集团的复合水泥，然后再逐步进一步发展。

有生产关系的创新，才能保证技术的创新。

参考文献

［1］ A. Neville. Properties of Concrete，4th and Final Edition. An imprint of Pearson Education Limited. First published 1995，6th printed 2000.

［2］ 高性能混凝土施工指南.

［3］ M. R. H. Dunstan. Fly Ash，As “The Fourth Composition” of Concrete，Proceedings of Second CANMET International Conference on Fly Ash ，Silica fume，Slag and Natural Pozzolans in Concrete，1987. ［4］ Atcin.

［5］ ACI 363 委员会. State of the arts report on high strength concrete，1994.

［6］ Wilbert S. Langley，Georges G. Carette，and V. M. Malhotra. Structural Concrete Incorporating High Volumes of ASTM Class F Fly Ash. ACI Materials Journal/September-October 1989.

［7］ Andre Bisaillon，Michel Rivest，and V. M. Malhotra. Performance of High-Volume Fly Ash Concrete in Large Experimental Monoliths. ACI Materials Journal/March-April 1994.

[8] A. Bilodeau, V. Sivasundaram, K. E. Painter, and V. M. Malhotra. Durability of Concrete Incorporating High Volumes of Fly Ash from Sources in the U. S. ACI Materials Journal/January-February 1994

[9] V. M. Malhotro. CANNET Investigations Dealing With High-volume Fly Ash Concrete. CANNET Report.

[10] CSA Standard, A23. 1-04/A23. 2-04, Concrete materials and methods of Concrete Construction/Method of Test and Standard Practices for Concrete 2004.

[11] S. A. Podmasova, SH. T. Babev, YU. S. Volkov. New Low Water Demand Binders for High-Strength Concretes, Proceedings of International Conference on High-Strength Concrete, Norway, 1993.

[12] V. Alunno Rossetti, F. Curclo and Cussino. Production Performance and Utilization of a Special Superplasticized Cement, Proceeding of 9th International Conference on the Chemistry of Cement, New Delhi, 1992.

[13] J. E. Jonasson, V. Ronin. Energetically Modified Cement (EMC), Proceedings of International Conference on High-Strength Concrete, Norway, 1993.

[14] 北京城建集团有限责任公司混凝土分公司，清华大学. 环保型胶凝材料及其配制高性能混凝土的研究鉴定文件，1999.

自评

1. 从原材料质量控制，到试配、拌和物生产，直至浇筑、振捣、收面、养护，按混凝土工程的理念，整合原材料（水泥、砂石、外加剂等）、混凝土试配与生产、混凝土拌和物成型三个环节，使水泥、砂石、外加剂、混凝土搅拌站、施工中混凝土的成型工艺等五个分立的行业成为一体化的产业。符合可持续性发展的战略和当今我国企业转型升级的思路。不是各环节的简单大拼盘，而是从根本上变革生产关系，全过程进行投入产出的统一管理与核算。生产关系的变革本质上是打破原有生产资料所有和利润分配的制度，建立新的统一制度。因为是根本性的变革，就像人的出生一样，必然会经历一个艰难痛苦的过程。这个过程早晚会到来，否则生产力得不到发展，转型升级就会成为口号和形式。

2. 本文第 6 条提出“建立从原材料生产及质量控制-试配-生产-成型工艺一体化的混凝土工程体系设想”只是一种原则，具体的实施还需要按系统工程根据具体条件具体研究。例如因水泥企业有矿山而发展骨料的产业，这必然涉及资源利用效率问题，不能因使用高品位石灰石而与水泥争原料，宜利用矿山中的低品位石灰石生产骨料，等等。其他也一样，有规划的问题，管理的问题，也有大量技术问题。主要的困难在于人才的匮乏。

3. 广义地说，解决水泥和混凝土的生产关系主要的不是技术问题。说是利益的问题，并非否定对利益的追求，究竟市场经济的本质是利益。问题在于如何认识利益，怎样获取利益。归根到底，其实是思维方法和观念的问题。我们每个人都是双重身份，既是买者，又是卖者。如果卖者都只为追求自己的利益最大化而牺牲质量，那么，卖出低质产品的人也会买入低质的另一种产品。这还有什么利益可言？据说有一个从事建设的人曾担心自己参与建设的建筑物质量，说：“也许将来我的孙子会住上我们现在盖的房子啊！”如果人人都有这样一份担心，就会明白利人和利己是统一的，利人才能利己。就会按照客观规律公正地（不是公平）平衡（不是平均）各方的利益，使全社会不断进步，不断增加财富。

本来这本集子所选论文到2013年为止，现在看来，本文还是主要在技术上的论述，而这个问题解决之难，主要不是技术问题，而是观念问题。鉴于这个问题的重要性，在付稿之前再选入一篇2014年《混凝土世界》第一期刊登的文章，作为补充，希望能引起讨论。

思想上的努力，正如可以长出大树的种子一般，在眼睛里是看不见的。但，人类社会生活的明显的变化正发生于其中。

——列·托尔斯泰

科学家不是依赖于个人的思想，而是综合了几千人的智慧，所有的人想一个问题，并且每人做它的部分工作，添加到正建立起来的伟大知识大厦之中。

——卢瑟福

论文之廿四

从水泥和混凝土的关系对企业转型升级的思考*

廉慧珍（清华大学土木水利学院）

“转型升级”是当前企业家们的热门话题。我曾参加过某企业宣布开始“转型升级”的盛会。热闹过后，我没弄明白“转型升级”是什么意思：他们要转什么型、升么级呢？也就是说现在是什么型，要转到什么型？日前在网上看到原载《经济观察报》的一篇署名张斌的文章，文曰：“‘转型升级’在经济危机之后就成了一句口头禅，各种会议或者发言加上这个帽子，立即显得尊贵起来。听得越多，越是一头雾水。”接着他说，他从长三角走到珠三角，到处咨询，但是“遗憾的是，我仍然没有搞清楚。各种答案五花八门，纷繁复杂。唯一比较相似的一点是：各级政府都说不清楚怎么去做，而大多数正在做的企业是做了也说不清楚。”[1]

看来不是只有我闹不清楚。到网上百度一下去查查，对“企业转型升级的基本思路”有以下这样的答案（括号内是我的联想）：

（1）加强调整　促进产业结构升级（原来升级就是“调整”？什么是产业结构级别?）；

（2）自主创新　提高企业竞争力（毋庸置疑，调整不调整也得以此为目的）；

（3）提升服务　进一步优化企业发展环境（不升级就不能提升服务了吗?）；

（4）破解难题　努力强化企业要素保障（这一点看来是政府的事，要给政策）；

（5）齐抓共管　做好企业减负工作（这也是政府的事。但是谁和谁怎么齐抓？怎么实现共管?）；

（6）联合互补　着力提升企业应对能力（怎样联合？应对什么？是否联合起来力量大?）。

现在从混凝土本行来讨论讨论吧。

1. 从混凝土说起

（1）混凝土材料并不是最终产品，混凝土工程的产品是混凝土结构构件，是必须按照处于一定环境的特定工程的需要，经原材料选择、配合比确定、生产制备、输送，并通过浇筑、振捣、收面、养护等各工序的成型工艺而完成混凝土结构构件制作的全过程。通过这个全过程才能完成最终产品——混凝土构件。这样的全过程叫做混凝土工程。

（2）当前混凝土工程处于被割裂的状态：原材料品质和质量不可知，不可控，致使混凝土拌和物的试配存在盲目性，使用不合适的原材料无法生产出优质的混凝土拌和物；而这样的拌和物运送至工地，又交给那些不懂混凝土的农民工去完成混凝土的成型工艺。后果是随意在拌和物中加水，浇筑和振捣不规范，防止塑性开裂的收面工艺不及时，养护不到位，以至于出现问题时纠纷不止或者被掩盖而造成隐患。众所周知，任何材料必须经相应的工艺严

* 原载《混凝土世界》2014年第一期。

图 1　掺矿渣粉的混凝土因搅拌不均匀出现灰绿色的色斑

格加工生产才能保证产品的质量。我国一般机电（含电子）制造业水平长期难以赶超世界发达国家的水平，其重要原因就是工艺的问题。脱离成型工艺的任何材料都不可能实现预期的性质和质量。混凝土脱离成型工艺只是没有意义的拌和物。这样被割裂的生产关系已严重阻碍混凝土工程技术的进步，必须变革。

（3）现代混凝土由于普遍使用高效减水剂，可以将水胶比降至很低，因而得以掺用大掺量的矿物掺和料。然而由此而带来一些质量问题如下：

①管理问题　掺入组分多，上错料的现象时有发生，这一点通过增加管理成本，搅拌站应当能够解决；

②拌和物匀质性问题　这是任何产品最重要的质量标准，而现行绝大多数搅拌站使用的搅拌机一般因搅拌时间短，拌和物大多搅拌不均匀，当掺用矿渣粉时，混凝土硬化后会出现灰绿色的色斑，该色斑在空气中会逐渐消失，但却反映出拌和物不均匀的问题（见图 1）。水灰比越低，匀质性越差；掺入减水剂和矿物掺和料，更会增加不均匀性。图 2[2] 显示掺硅灰的低水胶比净浆搅拌时间不同时颗粒分散的情况。

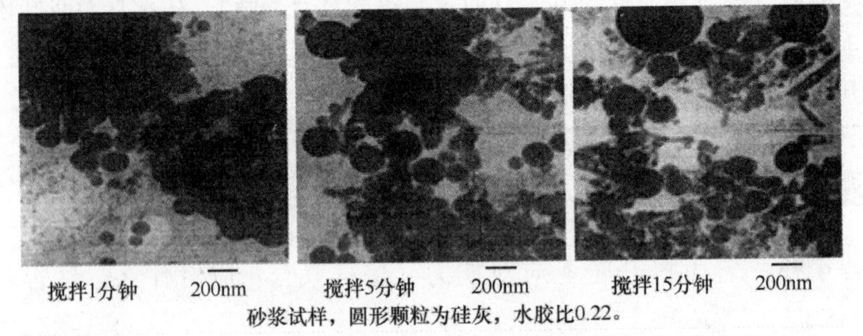

图 2　掺硅灰的低水胶比净浆颗粒分散程度随搅拌时间的变化

③SO_3不足问题　水泥厂生产的水泥不论掺多少混合材，都是整体优化石膏掺量的，搅拌站的混凝土拌和物掺矿物掺和料后，水泥中的SO_3就会被稀释，于是往往出现与外加剂相容性问题，混凝土强度低，掺减水剂的拌和物需水量大，收缩大等问题（见图 3～图 6）。

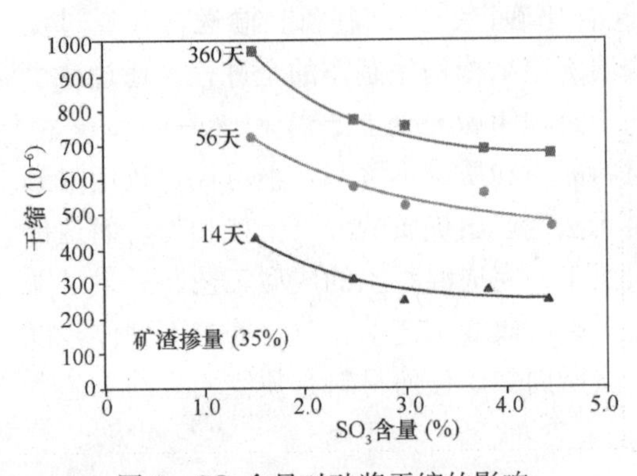

图 3　SO_3含量对砂浆干缩的影响

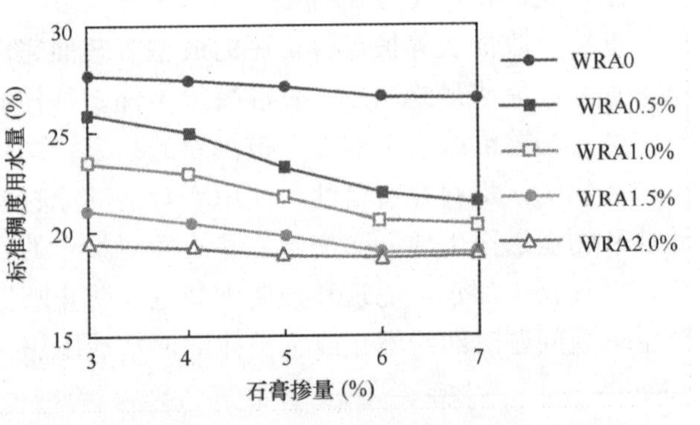

图 4　石膏掺量对不同掺量减水剂拌和物需水性影响

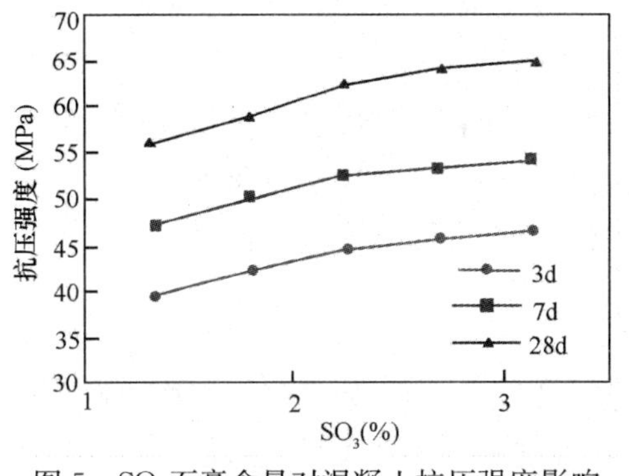

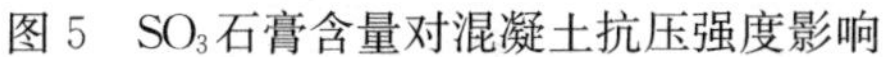
图 5　SO_3 石膏含量对混凝土抗压强度影响

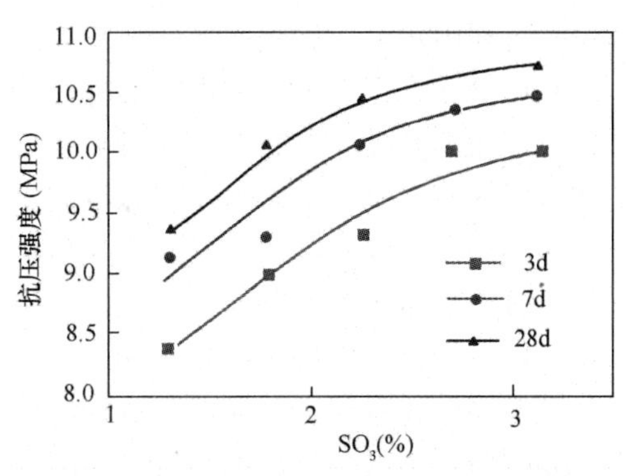

图 6　SO_3 石膏含量对混凝土抗折强度影响

有人为了解决此问题，就在矿物掺和料中先掺入石膏。但是石膏的影响必须是针对全部胶凝材料的，在并不知道水泥中已经掺入的是什么石膏，掺了多少，也不知道现在要掺入矿物掺和料中的石膏品种和品位的情况下，掺量的根据是什么?

④高效减水剂的掺法问题

在搅拌站把高效减水剂先溶于拌和水，然后再一起倾入干料中。这是一种外加剂使用效率最低的一种掺法（与水同掺法）。图 7 所示是萘系减水剂不同掺量和不同掺法对混凝土拌和物坍落度经时损失的影响。可见同掺法的坍落度损失最快，除非掺量达到饱和点。

意大利 V. A. Rossettihe 和 Curclo F，Cussino 将水泥生产时掺入超塑化剂共同粉磨，制成一种超塑化水泥（Special Superplasticized Cement，缩写为 SPC），另外在使用水泥时掺入超塑化剂（称 SpAD），将超塑化剂溶于拌和水，同掺入干料中（AD 试样），检测三种试样的溶液中塑化剂的含量，如图 8[3] 所示。

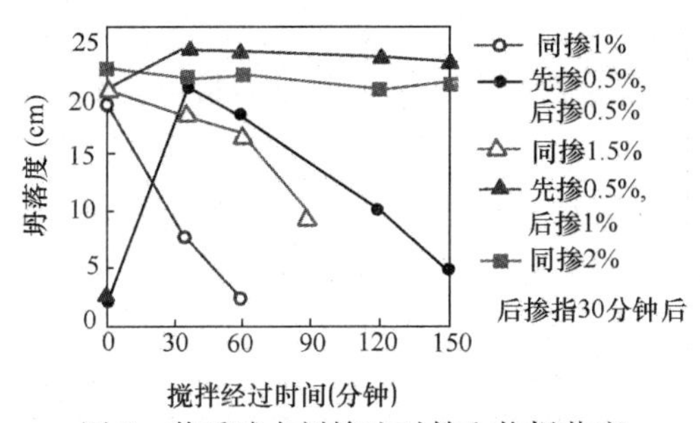

图 7　萘系减水剂掺法对拌和物坍落度经时性的影响

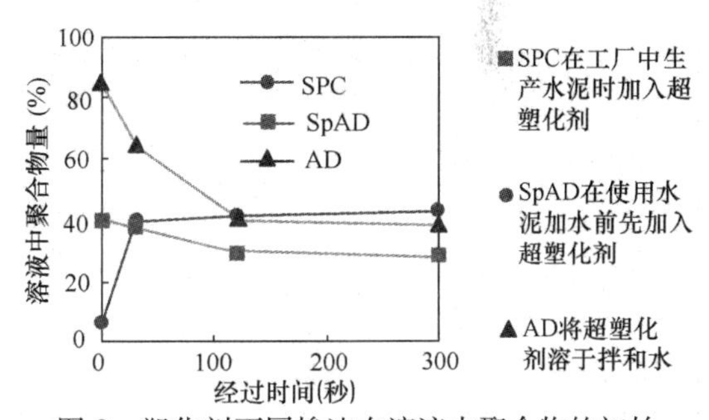

图 8　塑化剂不同掺法在溶液中聚合物的初始浓度及其随时间的变化[3]

在图 8 中所见为：将塑化剂溶于拌和水中时（AD），溶液中聚合物的初始浓度很高，约 2 分钟即下降一半；将塑化剂先与水泥混合（SpAD），初始浓度不高，但随时间变化不大；与水泥共同粉磨的（SPC），初始浓度很低，很快即上升，此后未见下降，塑化剂在溶液中的浓度随时间的变化，反而仍稍有增高。表明外加剂的效率，上述 AD 恰为当前搅拌站普遍使用的掺法，损失最大，效率最差。

由以上可见，矿物掺和料由搅拌站掺用，有难以克服的弊病；而由水泥厂掺用，则可得到以下改善：

（1）可整体优化掺和料的品种及颗粒度；

（2）可提高管理水平，实现工业化管理；

（3）可利用水泥厂均化的条件，提高均化程度；

（4）可整体优化 SO_3；改善胶凝材料（新水泥）的物理力学性质；

（5）外加剂可代替助磨剂，并可直接解决外加剂与水泥相容性问题，提高其利用效率；

（6）搅拌站可以集中精力控制砂石质量，着力进行现代化生产和信息化管理的水平，并研究产业链向下的延伸，实现混凝土工程全过程的优化，保证最终产品的质量；

（7）可从整体上提升节能、降耗、减排的力度；

（8）可降低整体成本，包括交易成本。

可见矿物掺和料在生产水泥时掺用为好。目前之所以做不到，就是“矿物掺和料由谁掺好?”的问题异化成了“钱由谁赚?”的问题。也就是说这是个生产关系的问题——生产资料和资本占有与利润分配的制度问题。正如肖耿所说[4]：“改革就是改变游戏规则。……改变游戏规则就一定会有人受益，有人受损。……判断必定会受自己利益与立场的影响，屁股指挥脑袋，容易有偏见”。

2. 水泥和混凝土粗放型经营的现状和转型升级的必要性

（1）社会经济的转型升级应当是指社会经济体制的改革和生产力的提升。在资本原始积累阶段，生产力低下，经济的发展以粗放型为主，即靠资金和劳动力的高投入，不惜高能耗，却换来相对低的效益和对环境的污染。另一方面积累了一定资金的中产阶层许多人想脱离那些风险大的赚钱方式而继续赚大钱，看好我国大规模基础设施建设中的土木工程混凝土市场，纷纷将资金投向工艺简单、投资门槛较低的混凝土搅拌站，据不完全统计，全国混凝土搅拌站大约已有 7000 个。其中大量搅拌站管理水平和技术力量都较差，不乏靠买“配方”生产的，几近乎作坊的水平。在建设规模大的部分城市，搅拌站过于集中，以致粥少僧多，面对市场竞争的压力，竞相压价，为了生存，不得不低价购进毫无质量可言的劣质骨料。现在落后的检测理念和手段又只认强度，谁都知道只要把水胶比降下来，强度是最容易做到的事。这样，那些蒙混过关的劣质产品就又有了藏身之地，结果是出了问题则纠纷不止；勉强过关却留下隐患，而且造成资源和能源的浪费、环境的污染。这样下去，将会耗尽大量优质资源却给后代留下一堆需要处理的垃圾，而到那时，用于修补或再建的资源和能源从哪里来？这种初级阶段无序状态能够存在的原因是预拌混凝土搅拌站的归属不明。也就是说实际上“没有人管”。在计划经济年代，混凝土和施工是一体的，自然就归原建设部管，那时的关系是明确的；转变到市场经济后，行政管理技术的状况至今仍未理顺，还归现在的住建部管。但是由于预拌混凝土与施工行业分离，随着科学技术的发展和人们对混凝土材料认识的提高，以致于住建部主持制订的某些标准对混凝土企业而言是如鲠在喉，非但没有推进行业技术的进步，还多处阻碍先进技术的推广。

（2）水泥行业的人从来都以自己的“科学”而看不起“没有科学”的混凝土：“我们水泥就要低能耗、高 C_3S、高强度，至于混凝土怎么用，有什么问题，那是你们混凝土的事”。但是谁都知道，水泥如果离开砂石就无法“成气候”。水泥的用户就是混凝土，那么就要了解用户的需要，甚至要比用户更加了解自己的产品对用户使用会有什么影响，为用户服务。混凝土学科属于一种复杂的不确定性科学，其复杂性的根源就是水泥——水泥水化以及水泥浆体与骨料的界面结构对时间和环境非线性的依赖性。而水泥化学近 30 年来进展如何？水泥如何赋予混凝土以强度？新型干法窑到头了吗？还需要如何发展？耗能大户的帽子怎样才能摘掉？为什么出厂按标准检验合格的水泥也有人很不满意？水泥企业有多少做到了“精细化管理”？美国在1992年制定的ASTM C 1157《水泥标准性能规范》，现在我查到的1998年以后修订版就有98、00、03、08a、09、10、11等7个，该规范只规定性能要求而不涉及混合材品种（成分）和掺量，中国有几个人对此清楚？2012年12月，美国麻省理工学院混凝土持续性中心在世界著名的《物理学通报》发表了关于水泥水化研究的论文，为什么没有听到国内学术界的反响？……现在我国水泥界潜心研究水泥和混凝土基础科学的人已经很少了。关注降低成本、增加产量和销量，这本是正常的要求，问题是采取什么手段。由于没有正确认识水泥和混凝土的关系，不按科学规律发展，盲目的结果致使产能过剩，技术发展举步维艰，无论如何地向上下游延伸去寻找出路，也都带有一定的盲目性。

（3）当生产关系的发展滞后于生产力的发展时，就对生产力起阻碍作用。使生产关系适应生产力的发展，才能解放生产力，先进的生产关系推动生产力的发展。我国改革开放以来第一次的转型升级是从计划经济转成在计划经济指导下的市场经济，进而是国营企业改造成国有股份制，对我国的生产力发展起了很大的推动作用。30多年来，我国经济和科学技术上都有了很大发展，但是国家希望我们有更多的自主创新的成果。水泥和混凝土现在仍处于技术落后而产能过剩的状态，与当前对基本建设质量的要求不相适应。转型升级是必要的。现在的转型应该是从劳动力密集和资金密集的粗放型转为以技术发展生产力的集约型，升级为长线持续发展的经济实体。无论是联合改造、合并重组还是向产业链的上下游延伸，都不应当是个形式上的问题，主要的都是生产关系的改革。

（4）华新集团混凝土板块副总裁刘效峰概括混凝土企业的发展为以下几个阶段：

①起步阶段：小规模、垄断经营、高利润；

②发展阶段：无序竞争、应收款上升、大企业进入；

③规模化阶段：并购小企业、大规模企业出现；

④一体化阶段：产业链进一步上下延伸、规模扩张；

⑤成熟阶段：大企业间的合资、合并，低利润、低应收款。

这个概括很符合实际，实质上也是生产关系问题。只是成熟阶段还应加上“高质量，高诚信，树品牌。”

3. 当前水泥企业并购混凝土搅拌站后出现的问题及对策

近年来，具有管理、经营和资金实力的水泥集团企业纷纷并购混凝土搅拌站，生产混凝土用的骨料和外加剂，成立混凝土公司或板块统一管理，实现产业链的延伸。这是好事，是进行当前企业转型升级的第一步。但是由于事先的准备不足，目的和目标有误，有些并购者的老总很快就发现自己步入了“陷阱”，遇到不少棘手的问题，感到“原来混凝土并不那么简单！”首先，所有并购搅拌站的集团的动机主要是为解决销售渠道问题，对整体产业的发展缺乏深刻的认识，不了解混凝土从技术到管理经营上的复杂性和风险，也不了解混凝土从业者的难处，更不了解混凝土用户的状况。于是只做了一个大拼盘，忽视了生产的改造和生产关系的改造，水泥和混凝土仍然还是两层皮。真正做到生产关系变革的指导思想是要模糊水泥-混凝土-施工的界线，把混凝土搅拌站做成水泥企业的一个车间，把混凝土工程的质量作为企业最终产品的命脉，在整体上进行管理与投入和产出、资本和利润的核算。只有真正做到生产关系变革后，才能解放生产力，顺利进行技术改造，实现技术的创新。

水泥和混凝土一体化以后，真正实现混凝土工程的整体化才能使混凝土工程跨越一大步，符合生产力发展的规律，甚至把混凝土做成精品。因此还有很多工作要做，例如骨料的优化，按混凝土的需要优化水泥的生产，按混凝土的规律检测水泥的质量，混凝土配制技术与拌和物质量评定方法的改进，混凝土拌和物成型工艺的优化，等等。

其实，当前这种水泥企业并购混凝土搅拌站的出发点只是解决对市场占有的问题。肖耿说[4]："什么是市场？市场就是产权的交换，有人买、有人卖就形成市场。企业也是市场的一种表现形式。企业是由职工、所有者、出资人、供应商与消费者等共同形成的合约网络。各方与公司法人建立自愿的合约关系，形成现代公司这种市场化的经济组织。"目前我国的市场经济还很不完善，混凝土及其相关产品行业的市场尤其很不正常，远未形成健康的网络。其核心问题是生产关系，而且不只是水泥和混凝土之间的问题，涉及混凝土工程整个产业链。

改革是一个制度的建立与创新过程。生产关系的变革是最难的、最根本的改革，改革之难在于必然触及利益，而"触及利益比触及灵魂还难"；改革必然要打破人们熟悉的、习惯了的传统，进入需要研究和建立的新的程序。一位朋友曾说："一切革新都会招致孤立，创造性始终包含有做人们所不熟悉的事情的意义。它需要有一种能抛弃人所共知的事情的意志。"也就是说，首先要转变观念。

4. 对当前进行企业转型升级的建议

（1）企业的转型升级是和国家经济的转型升级是有关系的。转型升级的过程，最重要的是政府转变职能的转变，构筑恰当的产权基础设施与宏观环境。过去三十年来，我国不断地在进行改革，如企业从行政单位变成真正意义上的现代公司，就是我国一个重要的制度改革。而所有的改革，政府都在为市场提供涉及产权界定、产权交易、产权纠纷的调解与裁定等三个领域的产权基础设施[3]。企业的升级与此有关。

（2）由行业协会通过制订技术标准和规范、信息资讯、教育等手段对行业进行引领导向，规范行业的行为，保护行业的利益，解决有关的纠纷，这是发达国家的惯例。政体分开以后，那些清楚政府政策、掌握先进技术信息、熟知行业动态和发展方向的、公正的行业协会，应当是企业的主心骨。协会有水平问题，需要不断完善，加强行业协会的协调自律功能，形成政府、协会、企业三位一体的良性互动。涉及混凝土工程的各个行业协会应当像美国的混凝土协会（ACI）与波特兰水泥协会（PCA）、美国材料与实验协会（ASTM）一样，打破专业藩篱，各司其职，精诚合作，同做贡献。我国当前有关混凝土的协会是历史原因造成的分散甚至混乱，在转型升级的新形势下需要资源整合、力量集中，否则难以担当应有的重任。

（3）混凝土工程范畴的转型升级，首先就是生产关系的变革，实现从原材料到成型工艺完成 混凝土结构构件/部件的一体化。路线可有多种，如：

①由优势企业兼并劣势企业，或具有产业优势的混凝土企业实行兼并联合，组建集团：上游解决稳定的硅酸盐水泥熟料（也可以是水泥成品，但条件是组分透明，诚信合作）和优质骨料的来源，自建粉磨站进行胶凝材料预加工。这种方式等于把水泥厂的粉磨车间搬来，对于不懂水泥的混凝土从业者有一定的困难。下游解决售后服务的问题，对混凝土最终的产品负责，而不是只对拌和物负责。中间环节发展现代化管理技术。

②由有条件的水泥企业集团并购混凝土搅拌站，向产业链上下延伸，整合生产关系，发挥集团优势，全面进行技术改造，提高科技含量。

③实力不足的混凝土搅拌站可委托独立的粉磨站“来料加工”，提高胶凝材料的匀质性。当然这一步不属于转型升级之列。

（4）全行业共同关注和致力于人才的培养和继续教育。协会掌握高级人力资源，聚集了有实力的专家和名师，企业配合协会建立人才培养和管理体系与制度，有利于提升培训的水平。

5. 生产关系变革的思想准备

（1）不要简单化地看待生产关系的变革，新的生产关系建立像人的出生一样，无论顺产还是难产，都会经历一个艰难痛苦的过程，这个过程迟早会发生，最大的困难在于人才。

（2）任何事物有舍才有得。当前很多尤其是年轻人都有想做“老大”的心态，这也是对既得利益不放手的原因。例如有个搅拌站总工自己做得比较难，被并购以后，发现“没有得到多少好处”，反而“挣了钱不能都归自己，挣不挣都得上交管理费”，于是跳槽了。这就是“两层皮”的结果。当然这是此人心态和认识问题，不认识并不是所有的人都适合于当“第一把手”，但和“两层皮”和生产关系不无关系。

（3）生产关系的变革不是即得利益的再分配，而是重新建立一个适合于生产力发展的新的制度。在设计改革路线时，不仅要考虑现有利益集团能够接受的极限，更重要的是要考虑对长远的、可持续的发展的影响。也就是说决策人能看到多大范围和多远，是成功改革的关键。这是一项系统工程，需要水泥工业的、混凝土材料的、施工的、管理的、经济学的和工业工程等各方面的专家共同研究策划，真正做成完整的高质量混凝土工程，也不是谁都能做得到的。

（4）现在造成这样两层皮的现状，成了夹生饭，是简单化的结果，再不改革，就会再走一次弯路，付出一次代价，错过一段成长的机会。

参考文献

[1] 张斌．所谓“转型升级”．《经济观察网》2010-08-20.

[2] 日本清水建设(株)橋 大介，山崎 庸行．超高强コンクリート，《コンクリート工学》特集・のコンクリート製造技術/4.6 No.3 1993.3.

[3] V. Alunno Rossetti，F. Curclo and Cussino. Production Performance and Utilization of a Special Superplasticized Cement，Proceedings of 9th International Conference on the Chemistry of Cement，New Delhi，1992.

[4] 香港经纶国际经济研究院资深研究员及学术研究副总裁、香港大学荣誉教授 肖耿．改革就是改变游戏规则．FT 中文网・中国政经，2013，11.

自评

1. 无论如何，从根本上说，资本原始积累到一定程度时，经济的转型升级确实必要——从主要靠资金和劳动力的投入“做强、做大”的粗放型经济或产业，转变成靠技术水平和人员全面素质提高的集约型经济或产业；由资源消耗型向资源节约型、环境友好型转变，而不是盲目追求“做强、做大”。“强”的本质不是规模，盲目做大未必能做强。强靠的是品牌，品牌的实质是质量和信誉，是口碑。因此首先是做好，才能做强，规模应在市场需求下应运而生。这是转型的主要内涵。转型的目的是进一步提高生产力，生产力发展最主要的障碍是落后的生产关系，生产关系的变革才是根本的创新和转型，转型的目的是升级，目标是升级到长线持续发展的经济实体。好一强一大是相辅相成的。

2. 生产关系的变革因涉及利益而属于根本性的变革，不可能自发地产生，但是可以在科学的指导下平稳过渡。水泥-混凝土-施工之间生产关系的改革是企业转型升级的重要步骤，是一个涉及经济学、管理学和技术领域的系统工程，尽管主要不是技术问题，也必须有技术的支撑，在能得以完善。这不是一般技术人员所能做的事，必须由高瞻远瞩的高层管理者从大局出发统筹解决。

3. 只有必然触动利益的生产关系改革，才能真正做到转型升级；改革不是利益重新分配，而是建立更合理的新秩序，使市场健康发展。领导者能看到多远，企业就能走多远。在实行改革时，必须跳出自己的天地，站在高处，看到本行业的改革对行业、社会、国家乃至人类发展的意义。水泥-混凝土-施工之间生产关系的改革只是我国当前改革很小的一部分，但是道理是一样的。

中国人有一句老话："不入虎穴，焉得虎子。"这句话对于人们的实践是真理，对于认识论也是真理。离开实践的认识是不可能的。

——毛泽东

结束语 我还有什么问题希望研究？

问题一 水泥混凝土强度发展规律现状和问题

1. 在传统上，水泥和混凝土都以 28 天作为抗压强度和其他性质的标准龄期。其原因有二：其一西方人习惯使用 7 和 7 的倍数（7 天、28 天、56 天、91 天），实际上 28 天和 30 天并不会有多大差别；其二，过去混凝土强度等级较低，龄期越长，发展越慢，呈对数曲线方式增长，到 28 天以后，增长就很慢了（见图 1）。为了工程进度的要求，就按世俗习惯选择了 28 天。

2. 以 28 天强度为目标强度（100%）计，如图 2 所示，过去用 GB 175—63 水泥的 300# 混凝土，3 天强度约为 30%左右，7 天约为 65%；现在的 C30 混凝土 3 天强度约为 50%，7 天约为 75%；C60 混凝土 3 天约达 70%，14 天可达 90%以上；高强混凝土当 7 天就达到目标强度时，28 天几乎不再增长；如果 3 天就达到目标强度，28 天更是无增长，甚至有的还低于 100%。

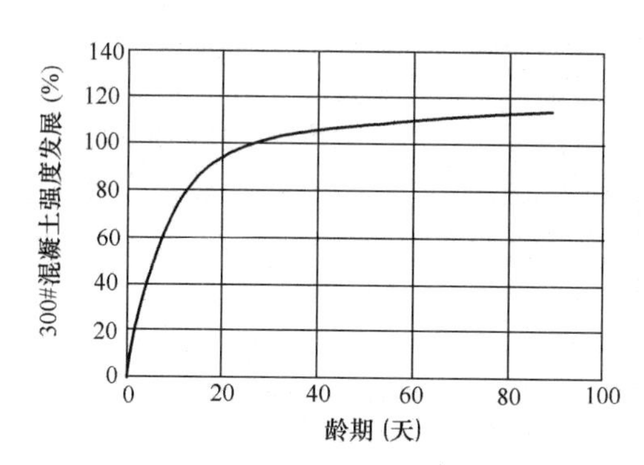

图 1 GB 175—63 水泥的 300# 混凝土（相当于现在混凝土 28MPa）强度发展

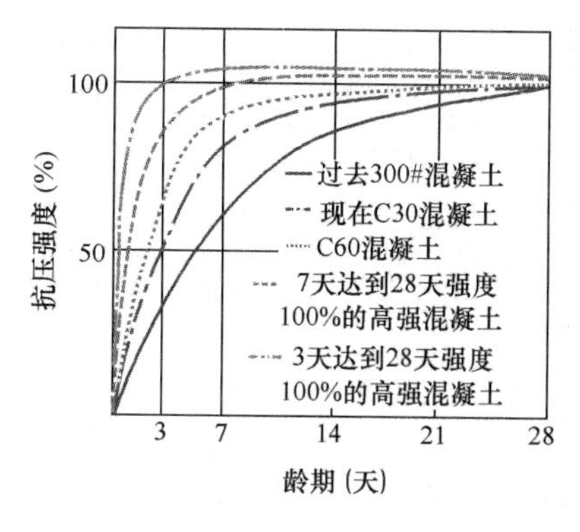

图 2 混凝土不同初期强度和 28 天强度的关系

3. 1910 年 Withy 在威斯康辛大学开始了 50 年水泥净浆、砂浆和混凝土的实验计划，浇筑了室内和室外混凝土。分别于 1910、1923 和 1937 年 3 个不同时间成型了 5000 多个试件。50 年的结果由 Washa 和 Wendt 于 1975 年发表（当时 I 型水泥的还没到 50 年，故只有 25 年的数据），如图 3 所示。其中 7M 水泥的熟料 C_3S 为 30%、水泥细度相当于 Blaine 比表面积为 231m^2/kg，1923 年用这种水泥配

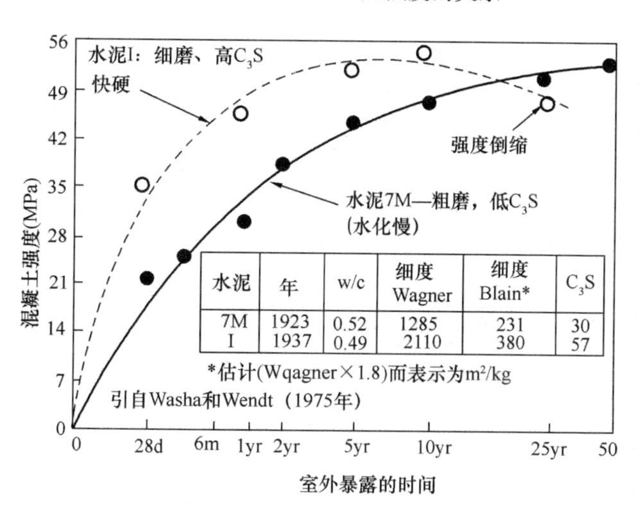

水泥	年	w/c	细度 Wagner	细度 Blain*	C_3S
7M	1923	0.52	1285	231	30
I	1937	0.49	2110	380	57

图 3 不同强度的水泥配制的混凝土强度发展

制的混凝土，28 天强度约为 21MPa，此后一直增长，10 年达约 46MPa，50 年后达到约 53MPa；I 型水泥是 1937 年的快硬水泥，熟料 C_3S 含量为 57%，水泥 Blaine 比表面积为 380m^2/kg，这与目前的通用水泥的指标一致，用以配制的混凝土，28 天强度约为 36MPa，10 年强度达 54MPa，此后强度开始倒缩，25 年的强度下降到约 46MPa，低于 7M 水泥的混凝土同期强度。二者水灰比相当（0.49 和 0.52）。

4. 从以上情况看来，水泥混凝土的强度发展和生物生长的规律一样：也会“早长晚不长”；强度越高，早期强度越高；早期强度越高，后期增长率越低。混凝土成型后始终处于生长——劣化——自愈+生长——劣化——自愈+生长的平衡循环之中，早期以生长为主，晚期以劣化为主。一旦以劣化为主，就会表现出强度倒缩的现象。那么这个转折点在哪里？这将决定如何预测结构物的服役寿命。

5. 我国大规模基础建设开始较晚，经历 50 年以上工程使用考验的混凝土很少，即使有，强度等级也多在 30MPa 以下。极少有人做过 1 年以上的观测，只有个别人对海洋环境中的混凝土做过 10 年的观测。对不同环境中不同混凝土的分析需要很长时间、大量统计性的数据。

6. 近年来水泥粉磨过细又增加了促使“早长晚不长”的因素，将使混凝土初始缺陷得不到自愈，成为服役期间环境的作用的开裂源而提前劣化。

7. 由以上分析，我很想知道以下的问题：

①20 世纪 20～30 年代所用早强水泥中熟料的 C_3S 含量与细度和现在通用水泥的一样，28 天强度 36MPa 的混凝土强度会在 10 年以后开始倒缩，那么现在大量使用的 28 天配制强度都达到 40MPa 以上的混凝土，是否会随早期强度的提高而提前倒缩？会提前到多少年？

②业内人士都知道掺入矿物掺和料的混凝土强度后期增进率大，那么是否会延迟混凝土强度倒缩的时间？规律如何？和强度等级有什么关系？强度等级相同时，和矿物掺和料掺量以及水胶比有什么关系？

③现在有人发现高强混凝土的强度 7 天或 3 天达到目标强度 100%，而 28 天只在 100%线上下波动而不再增长，这是在标准养护条件下的表现，28 天以后怎样呢？在自然条件下又会如何？现在水泥太细，会“早长晚不长”，那么是否会加速这种混凝土强度的倒缩？

④对预应力钢筋混凝土来说，因张拉预应力的需要，混凝土强度现在都被要求为 C50 以上，如果混凝土强度后期倒缩，那将非常危险。

⑤混凝土硬化初期的温度影响混凝土强度的发展，在浇筑后立即进行蒸汽养护，其后期强度会发生倒缩，还会影响混凝土的耐久性。原因是较高的温度加速水泥的水化反应，水化产物大多是结晶度差而多孔的水化产物。（参考 A. M. Neville，《Properties of Concrete》4th edition，Chapter 8）。压蒸条件下则水泥水化产物以稳定的硬硅钙石晶体为主（H. W. F Taylor，来华技术讲座，1980 年，北京），因而不会发生强度的倒缩（参见同上 Neville 的著作）。但是在我国已有发现，压蒸后强度达到 100MPa 的混凝土，其抗冻性耐久性指数只相当于低强度等级的水泥。那么，是否也有必要对蒸汽养护和压蒸养护的混凝土做长期系统实验研究？

⑥混凝土结构设计 100 多年来依据的一直是混凝土 28 天强度，而现在 28 天强度不可靠

了，结构设计该怎么做?

8. 有必要对现代混凝土强度发展规律做系统试验研究，其目的是：

①确认不同强度等级的现代混凝土强度发展与劣化的规律；

②研究影响混凝土强度倒缩期的因素，提出提高耐久性的对策；

③确认结构设计强度应当以混凝土何龄期强度为依据。

9. 该项研究的技术路线没有捷径，只有实实在在地变换条件成型统计性数量的试件，进行 20 年以上的室外自然条件下的观测，并和标准养护进行对比。同时做蒸气养护和压蒸养护试件的试验。

10. 像生物的生老病死一样，混凝土的生长、劣化和失效也是客观规律，但是既然世界上存在像古罗马万神殿那样已经屹立了 2000 多年的混凝土结构，我们就能做到尽量延长现代混凝土结构的使用寿命，以节能、降耗、减排，促进人类社会的可持续发展。即使研究结果是在一定的条件下，混凝土强度不发生倒缩，那么这个条件可能就是“取其养护四周之后强度还能显著增长的优势，高强混凝土往往以 56 天或 91 天强度进行设计”（S. Mindess，J. F. Young & D. Darwin，《Concrete》）。

11. 此项研究试验量大，持续时间长，见效晚，极少人愿意做。但是研究的结果对整个土木工程行业从混凝土材料的生产到质量检测，从结构设计到施工验收，从理论到实践，从标准、规范，一直到建设者的决策。都会有重大的冲击，其意义极其重大。

问题二　面对混凝土材料的变化和混凝土工程的发展，应当如何评价水泥的质量

1. 有人问水泥厂：“水泥除了打混凝土，还能做什么?”（砂浆是没有粗骨料的混凝土），言外之意是：水泥就是为混凝土服务的。但是现在水泥和混凝土是分离的，互相之间不了解，各行其是。水泥标准是只用于控制水泥质量的产品标准。在以往整体水平较低的情况下，混凝土对水泥的要求较单纯，水泥标准的指标较低，品种、成分也比较简单，现代混凝土拌和物因外加剂的使用而发生了具有根本性的变化，水泥和混凝土检测技术的差异，导致其间无法建立联系。水泥生产脱离用户需要，检测不符合混凝土的规律，双方相容性变差。为混凝土服务而生产的水泥应当到了需要改革和创新的时候了。

2. 回顾历史，GB 175—62 水泥标准规定两种检测水泥强度的方法，一种是混凝土法，一种是水泥胶砂法。因为当时混凝土强度最高也只是 $300kgf/cm^2$，其中“混凝土法评定水泥强度的标准，和实际应用混凝土的强度相关性很好”（中国建筑材料科学研究院水泥研究所，《水泥物理检验》，中国建筑工业出版社，1960 年）；其水泥胶砂法检测不同水泥强度所用的水灰比是不同的，是以达到相同稠度的用水量和需水特性确定用水量检测的，分为三种情况：

普通硅酸盐水泥：$W=\frac{P}{4}+2.6$　　水灰比约为 0.36

矿渣硅酸盐水泥：$W=\frac{P}{4}+1.8$　　水灰比约为 0.31

火山灰质硅酸盐水泥：$W=\frac{P}{4}+2.4$　　水灰比约为 0.38

式中　P——水泥净浆标准稠度用水量，%；

W——水泥胶砂用水量，%。

3.“在一定的施工条件下，集料的品质和配合比不变时，可塑性混凝土的强度与其他性质都是由水灰比所决定。”这一 Abrams 定律至今仍具有普适性。注意定律中的条件是“一定施工条件下的可塑性混凝土”，即拌和物性质决定混凝土最终产品质量。ACI 用了很大的力量把“混凝土配合比设计”改成“混凝土拌和物配合比设计”，其意义就在于强调拌和物的施工性能。而且，也因为搅拌站所设计和生产的确实只是拌和物。

4. 近年来美国主张在对混凝土拌和物配合比设计中，用混凝土单方用水量控制混凝土质量；日本国土交通省在 2003 年发文规定，现场每 $100m^3$ 混凝土拌和物，应检测一次单方用水量。一方面，在水灰比一定的情况下，混凝土拌和物用水量的变动表明混凝土的浆骨比的变动，这是影响混凝土耐久性的重要配合比要素；另一方面，对确定了配合比的拌和物，用水量的变化表明水灰比的变化，这是影响混凝土质量的另一配合比要素。

5. 由以上分析可见，混凝土配合比的优化都必须在相同施工性的条件下进行比较，而不是相同水灰比。这表明水泥的物理检验使用相同水灰比是不合理的，违反混凝土拌和物配合比设计的规律和原则，GB 175—62 的水泥强度检验标准考虑到不同的需水量，则比较接近于混凝土的规律。更何况那时水泥 GB 175—62 中强度检测还有“混凝土法”的标准。我国后来几次修订水泥标准，特别是“与 ISO 接轨”，水泥的生产与检验就更加脱离混凝土的应用。不知这种“接轨”的目的是什么，如果是为了国际贸易，那么出口水泥等于卖资源，是不可取的；显然也不是为我国混凝土事业的发展。外国的不一定都先进，在国外，先进的也是少数。未必能及时反映在标准上。西方国家在 10 年前就有人质疑用 0.5 作为检测水泥强度的水灰比；最近听说，ISO 近来也在研究水泥标准是否可改成用水胶比 0.35。即使这样，也还是等水灰比法，仍不能解决掺高效减水剂的现代混凝土与水泥之间的矛盾。

6. 现在的混凝土水灰比一般不会超过 0.5，水泥中的石膏与 C_3A 相比，溶解速率慢得多，水灰比越低，浆体中水越少，溶解速率很快的 C_3A 越会与石膏争夺水，则同期浆体中溶出的 SO_3 量就会越不足，这就是在水泥厂做与减水剂相容性试验合格，而在混凝土中可能不合格的原因。亦即，降低水灰比必须使用减水剂，才能更符合混凝土的规律。

7. 现代混凝土的主要特征就是广泛使用高效减水剂，按需水量或者等流动度法检测水泥的性质，必须有减水剂的参与，这就是为什么用现行标准检测的 32.5 水泥能够配制出 C60 以上泵送混凝土的原因。这才是混凝土的规律。在生产水泥时将高效减水剂和熟料及优化的石膏混磨，不仅可解决以上各种矛盾，而且可以起助磨的作用，并提高高效减水剂的效率。回顾 GB 175—62，在水泥品种中曾有过塑化硅酸盐水泥及其质量评价方法，只是当时还没有高效减水剂，使用的是纸浆废液浓缩物。20 世纪 90 年代初，前苏联、意大利和瑞典都曾分别报道过他们对掺高效减水剂生产水泥的研制和生产。

8. 我国制订标准的专家们因担心水泥生产乱掺而失控，一直禁止在水泥中掺减水剂。但是近年来我国水泥生产却出现“混合材”乱掺而失控的问题，有的水泥厂无视水泥标准中关于混合材使用的规定，为了降低成本，把一些没有经过系统试验、检验、论证和专家鉴定而纳入标准的废弃物用作混合材；为了增大掺量而不影响标称强度，就越磨越细，为此甚至要求助磨剂中添加早强的组分。近来水泥比表面积已近 $400m^2/kg$，致使混凝土强度早期高而后期增进率很低，会失去对初始不可见缺陷自愈的能力，抗冻性差，开裂敏感性大，（见图 4～图 7，图片来源于 R. W. Burrows，《The Visible and Invisible Cracking of Concrete》published by ACI，Farmington Hills borough，Michigan）。还有的水泥厂为了满足用户对“需水量要低”的要求，私自掺入高效减水剂，而又不敢公开，造成混凝土试配的困难。以上现状不仅不利于混凝土的生产，而且还会造成工程质量事故或隐患。利益的驱使是市场经济的自然规律，但是标准的正确制定和严格管理是把人的欲望关在笼子里的重要措施。美国水泥混凝土老专家说过：“改变水泥规范可提高混凝土耐久性”（引自上述 Burrows 专著）。

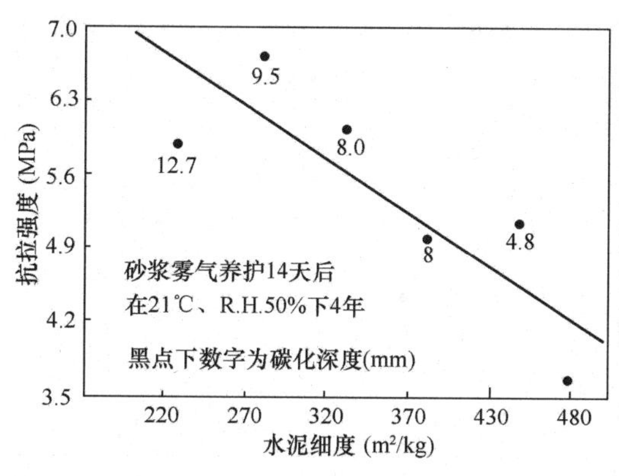

图 4　水泥细度和混凝土抗拉强度的关系

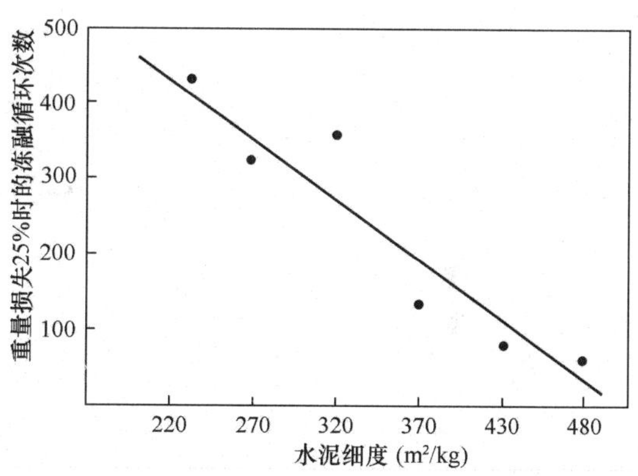

图 5　水泥细度和混凝土抗冻性的关系

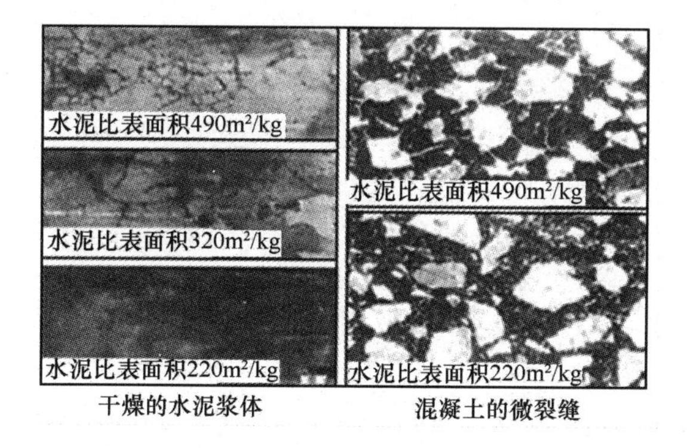

图 6　水泥细度对水泥净浆和混凝土抗裂性的影响

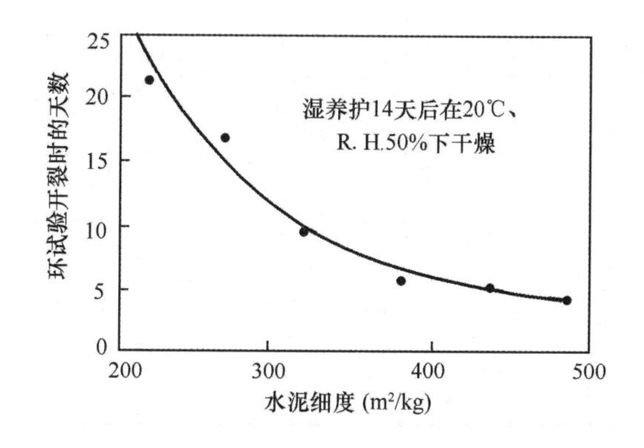

图 7　水泥细度对砂浆开裂敏感性的影响

9. 概括以上需要研究的问题：

（1）什么是水泥的质量？根据什么检测水泥的质量？怎样检测？进入普遍使用高效减水剂的现代混凝土时代，100 多年来对水泥质量标准的传统观念是否需要转变？如何转变？

（2）从我国水泥标准的变迁总结否定之否定发展规律的体现和预测。

（3）高效减水剂对检测水泥质量标准的影响是什么？

（4）怎样使水泥按混凝土需要生产并按混凝土规律检测？

（5）水泥品质及其不同检测方法与混凝土性质相关性有什么规律？

问题三　作为外行对水工大坝混凝土耐久性的疑问和思考

1. 吴中伟生前曾对我说过，大坝混凝土的耐久性主要是渗透性的问题，渗透性问题主要应为胶凝材料太少。当时我没有太理解。在工程实践中逐渐了解到，常态的大坝混凝土使用坍落度一般只有几十毫米的塑性混凝土，强度等级不高，近年发展拱坝使用混凝土最多也是C40，即被称之为高强（这也符合国际惯例：ACI 363委员会定义高强混凝土的28天圆柱体强度为7000 Psi，折合我国现行标准公制立方体强度约为50MPa。设计C40混凝土的配制强度，以标准差为5MPa计，保证率大于95%时，配制强度约为48MPa。）。因此胶凝材料用量可以很少，以便尽量降低混凝土的温升，并且大坝混凝土用量巨大，水泥用量对工程造价影响也巨大。骨料用量大、粒径很大，也是为了减小水泥用量。

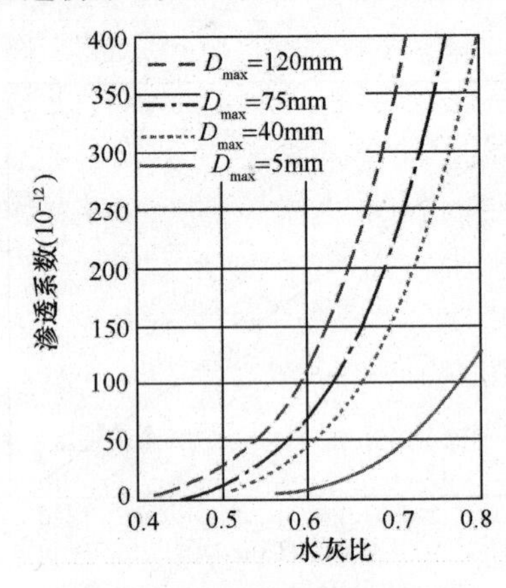

图8　骨料粒径和混凝土渗透系数的关系

2. 大坝混凝土的胶凝材料总量不足，浆骨比太小，骨料粒径又很大（最大粒径80～150mm）。在强度不高而水灰比较大的情况下，骨料和浆体的界面过渡区厚度就比较大。界面过渡区是混凝土中薄弱的区域，且存在变化的梯度——越靠近骨料表面，水灰比越大，故孔隙率越大，密实度越差，$Ca(OH)_2$越富集且取向性强。距离石子表面最近处是过渡层的最薄弱之处，为渗透性最大的通道；界面过渡区厚度随骨料粒径的增大而增大；因石子之间距离很近，相邻石子的过渡区都相交，而原设计的混凝土的砂浆本体不再存在或很少。这样的拌和物渗透性是采用湿筛去除最大颗粒石子后的较小试件所检测不出来的。骨料最大粒径越大，混凝土渗透性越大，水灰比越大，这种影响越大，如图8所示。

3. 已经有水工技术人员发现除去大骨料检测小试件与工程情况“差别很大”，他们解决的方法是使用400mm×400mm×400mm的试模。这样大的试件就得用吊车搬运，用大吨位压力机试压，这符合科技发展规律吗？不禁令人想起40多年前的计算机的体积能占据一个房间，现在连手机都能上网了，这是科技的进步。没有科技含量地越做越大，可以称作“傻大黑粗”，并不能算作技术的进步。

4. 换一种思维方法考虑设想一下：维持水泥用量不变，通过增加矿物掺和料用量（完全不必追求Ⅰ级粉煤灰）来适当增大胶凝材料总量，同时适当减小水胶比；减小骨料最大粒径，但骨料总体积不变。因水泥用量不增加，不会提高水化温升，也不会明显增加成本，可能会增加一些骨料加工费用。但是换来的是将有效减小界面过渡区厚度而提高抗渗性的效果。结构物的耐久性可大大提高，所节省的维护、加固费用、资源消耗的降低和使用寿命的增加会是可观的。这就叫做全寿命周期经济分析。

5. 本部分所选论文之廿二中引用过一个建设方要求使用寿命1000年的混凝土工程的实例。该实例引自P. K. Mehta和W. S. Langley的文章《Monolith Foundation：Built to Last a “1000 Years”》（建造维持1000年的大型整体基础）。文中介绍说，“对在20世纪建造的大多数结构物都不能企望维持100年，因为硅酸盐水泥混凝土会因为一系列彼此相关的原因而开裂和劣化。例如温度收缩、干缩，受冻融循环作用、钢筋锈蚀、碱-骨料反应以及硫酸盐侵蚀等。与此对比，2000多年以前建造的一些古罗马建筑物却仍然保持良好。因此，今后要想建造能维持长久的混凝土结构物，就要运用建造已经维持多个世纪的古代建筑物的材料和方法的基本知识。”那就是尽量减少硅酸盐水泥的用量，并且使用C_3A和C_3S含量尽量低的水泥，尽量减少用水量，精心选择骨料，其粒形和级配良好。最后，该工程混凝土所使用的硅酸盐水泥熟料矿物中含 C_3S 14%，C_2S 62%，C_3A 1%；胶凝材料中又掺入几乎 3 倍于水泥价格的粉煤灰，掺量为 60%。这样的水泥比现今的 β-C_2S 水泥中的 C_2S 含量还高。C_2S 的烧成温度比 C_3S 的烧成温度低，价格不应当高于通用硅酸盐水泥的。对于需要专供水泥的水利工程来说，是否应当能做到？

6. 按照 P. K. Mehta 的耐久性模型，即使施工良好，成型密实，没有任何可见的缺陷，混凝土硬化后也必然会存在一些不可见的缺陷。在大气干湿交替、冷热循环的作用下，那些不可见裂缝会扩展成为环境中侵蚀性介质侵入的通道。如果水泥能持续地水化，则 0.1mm 以下的微裂缝可以自愈。对于当前的通用硅酸盐水泥，因为 C_3S 含量大、粉磨过细而早期水化程度高，早期强度高，则后期自愈能力就会下降。而 C_2S 的缺点是早期强度低，强度增长缓慢，故该工程的水泥可以磨到比表面积为 $400m^2/kg$，却不会有“早长晚不长”的问题。

7. 混凝土配合比和性质如表 1 所示。细骨料使用机制的中砂，粗骨料是最大粒径为 25mm 和 10mm 两级配的碎石。该混凝土强度发展很慢，90 天强度只有 24MPa（设计要求 90 天 20MPa）。值得注意的是绝热温升只有 12℃。在配合比上，这样的混凝土温升低、可避免能引起在约束下开裂的温度收缩，水胶比为 0.40，浆骨体积比约为 0.26，且不论上述混凝土基础用于含有海水成分的地下环境中的优势，就该混凝土的本征特性来说，成型密实的混凝土会干缩很小，开裂敏感性很低。这和公元前用 1∶2 的石灰-火山灰建造的那不勒斯海港“长数百米，至今无一裂缝”的先例有些接近，是否能满足大坝建设的需要？

表 1　Kaua'i 岛用混凝土配合比与性质

参数	现场试用的混凝土拌和物配合比	参数	现场试用的混凝土拌和物配合比
硅酸盐水泥/(kg/m^3)	106	坍落度/mm	150
F 类粉煤灰/(kg/m^3)	142	含气量/%	3.5
水/(kg/m^3)	100	绝热温升/℃	12
细骨料/(kg/m^3)	944	抗压强度/MPa	
粗骨料/(kg/m^3)	1121	3d	5.5
普通减水剂/(ml/m^3)	770	7d	9.6
高效减水剂/(ml/m^3)	3480	28d	15.9
引气剂/(ml/m^3)	116	90d	24.1

8. 转变一下观念，为了人类的可持续发展，像大坝这样消耗大量资源和人力的工程，理应有 100 年以上的服役寿命，只要牺牲一点建设速度。有舍才能有得。用石灰＋火山灰为胶凝材料的混凝土建造的万神庙，就是因为混凝土“精心选择原材料，精心搅拌，精心施工”，而且硬化很缓慢，才得以在世界上屹立至今 2000 多年。

9. 据水工界人士说：“那可不行，对于水利工程，经济性最重要。”话虽如此，但如果建成后，频频修补，几十年就变成危坝，如何算经济账？

10. 我赞成 Karl R. Popper（英国，1902－1994）“猜想与反驳”的科学研究方法。我是外行，没有参与过水工结构的建造，只是从材料的角度，提出以上疑问。希望感兴趣的人反驳我的观点，参加以上问题的讨论，必要时进行试验。也许形成现状主要的原因不是技术性问题，但是历史上很多变革都不仅是技术问题，而非技术问题也不是不能解决的。关于经济性，其实也是个观念问题，其在人们心目中的接受程度也会随着时代的发展而变化。例如高效减水剂在刚开始推广时，因为每立方米混凝土会增加几十元成本而不被接受，现在已经普遍使用了，价格也随着销售量的增加而降低；其他如混凝土搅拌机的换代、钢纤维的使用，等等。任何新材料和技术一开始常常都会存在经济上的接受问题，随着社会经济的发展，对质量的要求的提高，经济性的标准也会变化。这样的过程是必然的，不影响前沿性的研究。否则何来创新？关键是是否能迈出第一步。

问题四　混凝土结构理论和设计方法的范式转换问题

1. 什么是范式?

范式（paradigm）是1962年美国著名科学哲学家托马斯·库恩（Thomas，Kuhn）提出的，发表于其著作《科学革命的结构》（The Structure of Scientific Revolutions）。库恩说："科学革命"的实质，一言以蔽之，就是"范式转换"；范式的突破导致科学革命。"按既定的用法，范式就是一种公认的模型或模式。"是"在科学实际活动中某些被公认的范例——包括定律、理论、应用以及仪器设备统统在内的范例——为某种科学研究传统的出现提供模型。"范式"是一种对本体论、认识论和方法论的基本承诺，是科学家集团所共同接受的一组假说、理论、准则和方法的总和，而形成科学家的共同信念。"

2. 传统混凝土结构理论和设计方法的范式

结构理论和设计传统的范式是依从材料力学的假设，其基本假设为：

（1）小变形的弹性体：在外力作用下，弹性体的变形应使弹性体各相邻部分既不能断开，也不能发生重叠的现象；

（2）连续性：体内充满物质，毫无空隙；

（3）匀质性：体内任何部分力学性质相同；

（4）各向同性：体内不同方向的力学性质相同。

3. 混凝土材料的特性与材料力学基本假设的对照

尽管也将材料的变形分成弹性、塑性、脆性，却只是在计算模型上有所差别，因此，材料力学和作为结构设计的基础的结构力学都不管材料种类，只要符合上述假设，都可以用材料力学的模型进行计算。大多数工程材料虽然微观上不是各向同性的，例如金属材料单个晶粒呈各向异性，但由各个晶粒形成多晶聚集体时随机取向，则整体在宏观上表现为各向同性。对于混凝土，在过去的组成相对简单，强度等级相对较低且范围不大的情况下，在构件总体的力学性质上，可认为符合材料力学假设。设计时从荷载和混凝土强度特征值两方面都取一定的安全系数，混凝土在制备中还使配制强度有95%的保证率，则可以被接受保证结构的安全性。但是，以高效减水剂的广泛使用为特征的现代混凝土，和过去相比，有了很大的变化。与材料力学的基本假设对照，以下的特点更加突出，如果说对传统混凝土来说可以不加考虑的话，对现代混凝土则需要重视了：

（1）不连续性：按材料学的分类，无论水灰比多低，混凝土都属于多孔材料，总是存在可见与不可见的孔缝等缺陷；因其抗拉强度很低，无论受外力变形还是非荷载变形，都会造成相邻部位的断开而不连续。

（2）非均质性：混凝土材料是在不同尺度上存在不同尺寸的固、液、气多相高度复杂的非均质体系。对于构件来说，其中的混凝土不同部位由于早期温度的不一致，强度是不一致的：对于使用纯硅酸盐水泥的混凝土来说，温度越高，早期强度越高，28天强度增进率越低。如图 9 中标记和曲线所示，超过 40℃，7 天以后就不再增长；超过 85℃，7 天以后就开始倒缩。掺入粉煤灰后，则不论在什么温度下。强度始终随龄期而增长。

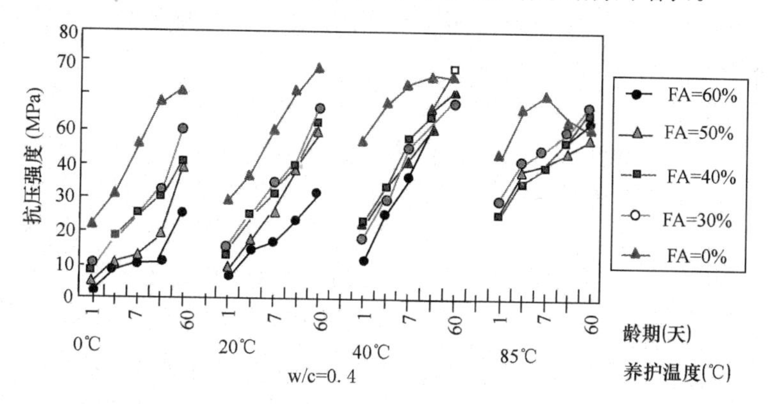

图 9　养护温度对水胶比 0.4、粉煤灰掺量不同的砂浆抗压强度发展的影响

图 9 所示试件始终处于指定温度下连续养护，实际工程中的混凝土在升温至峰值后，温度会逐渐下降直到与气温平衡。如果采取温度跟踪养护（temperature match condition）技术，则可了解到构件内混凝土强度发展的真实情况（见第四部分论文之二图 4 和论文之七的分析）。图 10 是水胶比为 0.45，在不同温度下浇注并养护到 28 天后继续在标准条件下（美国标准条件的温度是 70℉，折合为 23℃）养护，以模拟工程中纯硅酸盐水泥的试件抗压强度的发展。由图 10 可见，初始强度随温度的升高而提高，约 7 天以后则恰好相反。结构中混凝土的温度都不会低于标准养护温度，而结构设计的混凝土强度使用的是标准养护 28 天的结果，显然是不安全的。那么，结构设计使用的混凝土的强度标准值还可靠吗？

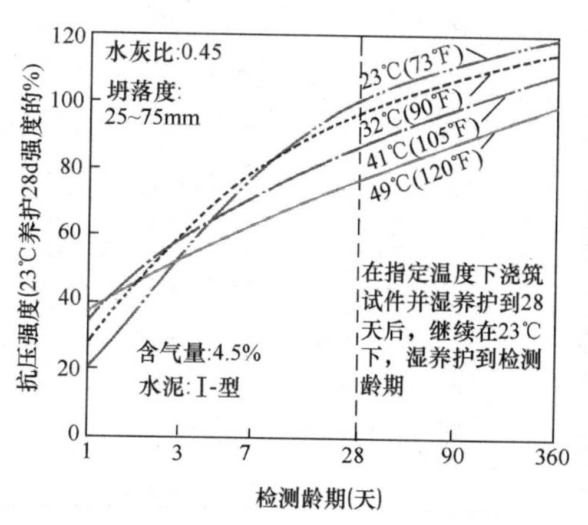

图 10　温度对纯硅酸盐水泥抗压强度发展的影响

实际上，构件表层和内部因温度不同，强度也有差别。以一根宽度超过 30cm 的矩形断面大梁为例，早期温度最高之处在水平轴线附近，温度最低之处是上下钢筋的保护层。那么，过去人们都关心的是中心强度，现在看来，结构设计应当选用哪里的强度作为标准值？对于其他构件，也与此类似。在现行生产关系下，掺入矿物掺和料后，更会增加混凝土的非均质性。与材料力学的假设差异更大。

由多种非均质材料组成混凝土拌和物，浇注成型后，在浇注方向的上、下面和侧面，其微结构和密度都必然会有差异，与材料力学“各向同性”的假定相去甚远，预拌混凝土因流动性大而尤甚。

（3）微结构和性能的不确定性：微结构和性能随时间和环境而变化。严格地说，因原材料不可提纯而成分和性质波动，造成混凝土非匀质性和复杂性，混凝土的微结构和性能还随时间和环境而变化而造成其不确知性和不确定性，属于具有蝴蝶效应的混沌体系（初始参数微小变化可能引起最终结果的巨大差异，即“差之毫厘，失之千里”）。过去的混凝土（使用

GB 175—77 以前的水泥和强度等级较低如 300＃以下）的情况下，这个问题还没有这样突出。如今水泥和混凝土都发生了很大的变化，加上现在混凝土拌和物基本上都在由不懂混凝土或只有肤浅混凝土知识的包工队去操作成型工艺，随时间的推移和因此的环境变化对混凝土微结构和性能影响的结果无法准确预测。

（4）黏弹塑性体：混凝土中水泥浆体由水泥凝胶和未水化颗粒组成，凝胶的特性和界面的影响使混凝土成为黏弹塑性体：因为凝胶体的生成而具有黏性；因为裂缝的出现及开展而在宏观上表现为塑性变形；弹性则极不明显。不同强度等级混凝土的黏弹塑性行为不同。

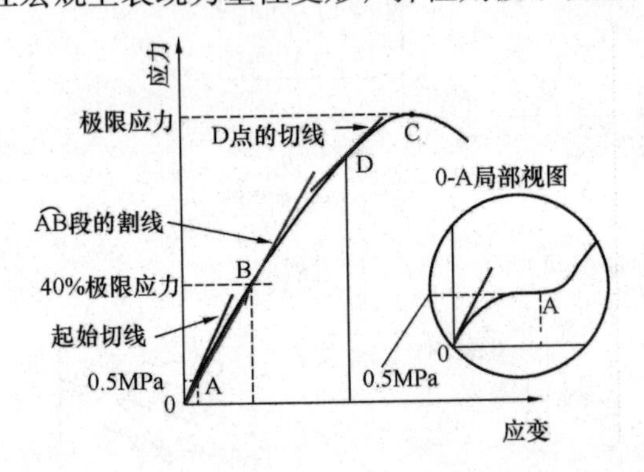

图 11　中等强度混凝土单向应力-应变关系示意

图 11 为中等强度等级混凝土应力应变关系曲线示意。混凝土的应力-应变关系不是直线，通常认为只有从 0 点到应力达 0.5MPa 时的线段为直线，而假定直线 OA 的斜率为该混凝土的起始模量。将该线段延长后可知其与其他线段相差较远。我国现行规范规定使用到达 0.4 最大应力（B 点）时卸载，再反复加载、卸载，直至应力和应变呈直线关系，用塑性变形点到最后加载点连线的斜率表示该混凝土弹性模量。日本是直接使用 OB 连线计算，二者相差不大。实际上不仅 OB 不是直线，而且 OA 也不是直线，将该段放大即清晰可见（见图 11 的 O-A 放大的视图）。所以混凝土不是真正的弹性体，也不是纯粹的脆性体或塑性体。在恒定的持续荷载下变形继续增大，具有明显的徐变性质。

4. 结构设计除了强度还考虑什么？

（1）应力集中　无限大平板在单向拉伸情况下，其中圆孔边缘的应力集中系数 $k=3$；在弯曲情况下，对于不同的圆孔半径与板厚比值，$k=1.8\sim3.0$；在扭转情况下，$k=1.6\sim4.0$；这样大的集中应力，当属于构造设计或施工要求的范畴时，在设计时预先可知，可能会考虑，但是在施工验收时因钻芯取样造成的应力集中怎么考虑？

（2）结构设计只考虑承载时的应力-应变关系　现代混凝土因水泥粉磨过细，除会增大干燥收缩之外，早期温度收缩已不可忽视，还会因低水胶比而产生自收缩等变形。这些变形在约束条件下所产生的应力，即使早期尚未造成开裂，仍然会在构件中存在并积聚。这些内部应力在外部荷载作用下会产生什么后果？

（3）徐变　过去我们对混凝土徐变认识不足，只有在预应力钢筋混凝土结构中考虑。现在发现，过大的徐变还可能引起其他的问题。A. N. Neville 曾举过两个例子说明徐变的影响：一个是“承载的柱子因徐变而缩短，则应力重分配，引起与阳台刚性连接的板变形并脱落”；另一个是“美国在太平洋一岛屿上建造的一座桥梁，由于混凝土的徐变，跨中向下挠曲，终至 1988 年垮塌。”结构设计应当如何考虑？现代混凝土徐变行为的规律是否会因不同矿物掺和料的掺入发生什么变化？

5. 混凝土性能的不可模拟性：

（1）现代混凝土的性能（performance）例如耐久性，其含义实际上是能健康服役的年限，无法在实验室模拟并建立标准的检测方法而快速检测出量化的指标；又如施工性，在实验室检测的坍落度，只是反映混凝土拌和物静态的流动度，而工程中大量采用泵送，往往在拌和物运送至工地可能出现触变性、滞后泌水、堵泵、剪胀性等传统混凝土未遇到过的故障，在实验室中也难以模拟和预测。

（2）在实验室检测出的那些试验指标，只是反映（原材料的配合比）不同混凝土力学的、物理的、化学的本征特性，并不代表结构中混凝土性质的真值。由于是由统一标准方法和条件所检测出的，就具有可比性，只用于优化时的相互比较。

6. 如何评价混凝土强度？

（1）关于破坏理论，从材料力学来说，最普遍采用的是最大拉应力（第一强度）理论；对于脆性材料，无论材料处于什么应力状态，只要微单元内的最大拉应力 σ_1 达到单向拉伸的强度极限 σ_b，就发生断裂破坏；从能量的角度，破坏始缺陷；与构件中混凝土相比，实验室强度试验小试件缺陷的概率小得多。这是一致的，但是混凝土结构设计中依据的却是混凝土的抗压强度。

（2）按照传统的定义，混凝土强度是用标准试件，在标准条件下成型并养护到 28 天的抗压强度。有的国家，如美国、日本，使用圆柱体试件；俄国、德国和我国用立方体试件。结构设计人员以为这样的得出的结果就是混凝土结构中混凝土的强度。直到近几年，一部分人才因转变了思维方法而认识到实验室检测的混凝土强度值并不等于结构中混凝土的强度；

（3）在实验室中用标准方法和标准条件检测出的混凝土强度只是实验室小试件单向受压实验的结果；抗压强度值是用所加荷载除以试件平行于荷载作用方向上的面积所得平均应力；鉴于混凝土是高度非均质的多相复杂体，该平均应力是垂直于荷载方向上无穷多个平面上无穷多个点应力的平均值，而实际上，混凝土的断裂发生于最薄弱的缺陷处，和大构件相比，显然小试件缺陷概率低得多；

（4）实践证明，在影响强度的诸因素中，抗折强度比抗压强度敏感；

（5）那么，用什么样的试件，检测什么强度更接近于真实？（见图 12）

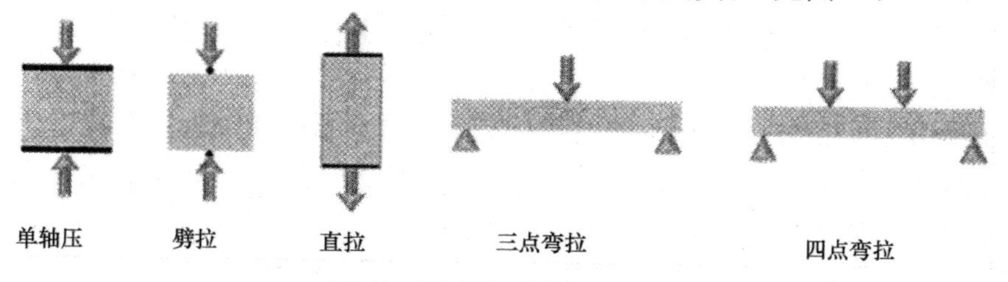

图 12　强度的不同检测方法

7. 如何进行结构理论和设计范式转换的研究？

（1）范式转换首先必须转变传统观念

①打破专业藩篱和学科壁垒　从 1901 年首次颁发以后 100 年内获得诺贝尔奖的自然科学中跨学科的成果从 36.23%，增加到 43.37%。这说明科学技术越进步，越需要相邻学科的支持和参与。人类对客观世界的研究越深入发展，就会越发现世界的复杂性，单靠个人和本学科狭窄的学科难以解决其所遇到的难题；现在很多科学技术的进展都与相邻学科的发展有关。例如，天气预报如果没有现代数学和计算机技术，很难取得现在的成绩；机器人的制造需要机械、电子、自动化、力学、数学、材料科学、生物等各学科的合作；等等。

②改变强度第一的传统思维和观念　耐久性是维系结构长久安全性的重要因素。按耐久性设计的混凝土，强度一般都会满足设计要求，而如果耐久性不足而提前劣化，则会为侵蚀

性介质提供侵入的通道，一旦钢筋锈蚀，就会威胁到安全性。

③树立可持续发展的观念。

④改变设计者不懂施工、不懂材料、混凝土生产者不懂结构和施工的现状。

⑤改变本构关系的传统研究模式，引进相邻学科知识，如不确定科学决策法、蒙特卡洛随机模拟法等，按照混沌体系特点，建立多重性、多阶段的数学方程和计算机程序。因需要多次模拟、大量反复计算来提高预计精度，故需运用大容量、高速计算机。

（2）混凝土结构设计范式转换必须懂得混凝土：不同材料的结构有不同的设计方法和施工技术，设计和施工方法必须跟随材料的变更而变更。

① 事物的发展是客观存在的，对变化了的材料不能使用不变的方法。

② 混凝土工程的质量是由各环节共同决定的。工程建设活动就是把材料做成结构物。

③结构设计人不了解混凝土，就不是一个真正完全的设计师。（A. Neville：《混凝土技术是结构设计的要素》）

（3）不懂混凝土所造成的后果

①为什么现在的结构物容易开裂？结构工程师说："混凝土构件都带缝工作，没有不裂的混凝土，有裂缝没关系。"这样说有什么问题？

对承受荷载可能没关系，对耐久性呢？按照 P. K. Mehta 的整体论分析，如果施工良好，不存在连续的可见裂缝，但存在不可见的微细孔缝，开始服役后，承受反复的或循环的甚至冲击的荷载，并在在大气冷热循环、干湿交替作用下，那些不可见的微裂缝连通并扩展，逐渐丧失不透水性，成为水和氧气侵入的通道，即使没有化学侵蚀性的环境，也会因水和氧气的侵入而引起钢筋的锈蚀，劣化开始。如果在施工阶段就出现可见裂缝，那么直接就是侵蚀性介质侵入的通道。如此的裂缝就不是没关系。结构工程师所说的"带缝工作"，不是这里所说的非荷载裂缝。有实验研究已证明非荷载裂缝比荷载裂缝对耐久性的危害更大，但是可以避免的。不了解什么是混凝土的耐久性造成结构工程师的误解，后果是会保不住设计所期望的安全性。

②为什么现在断面最小尺寸 30cm 的构件混凝土也需要控制温度？有工程师却说："构件断面最小的尺寸超过 1m，不，现在改成 80cm 就叫大体积。"

这也是结构工程师不懂得现代混凝土的变化，只是套用了日本的规定。而我国现在水泥、混凝土的变化所造成的后果是：新型干法窑的生产使水泥碱含量增大，开裂敏感性也随之增加；水泥的水化热增大，混凝土强度提高要求水灰比低，又使水泥用量增大而造成混凝土中温升大。现在工程中早期开裂的发生率约 60%即缘于此；水泥粉磨过细，改变了混凝土强度发展规律，而且使混凝土初始的可见与不可见缺陷失去后期的自愈性。

③在房屋建筑中，拌和物送到工地因工地不认识现代混凝土拌和物的触变性，往往又额外加水。农民工说："拌和物太糨，不加水不好泵"，施工工程师说："现在强度都有富余，加水没关系。"怎样看待这个问题？

这都是因为不懂混凝土。在混凝土经试配优化后，其拌和物用水量是不能随意改动的。用水量的大小有两个意义：一是代表水胶比，这是在原材料一定的情况下，决定硬化混凝土强度的主要因素；二是代表在一定水胶比下的浆骨比，这是影响硬化混凝土体积稳定性的主要因素。如果向经试配优化后生产的拌和物中再加水，则水胶比和浆骨比都会增大。更重要的是，在工地加水后不可能搅拌均匀（当前所用搅拌运输车并不具备搅拌功能），造成混凝土的匀质性下降。而匀质性是混凝土结构最重要的质量。

④不懂混凝土时，对工程中出现的异常现象（如图13所示），就会分析不出原因。往往甲方找设计，设计找施工，施工找混凝土材料。实际上如果甲方、设计、施工都懂得混凝土的特性和成型工艺的关系，这些问题就都能分析出原因。例如图中的墙，模板伊始，就发现这样规则间距的宽度很大的水平裂缝。这是因为上世纪90年代中期的混凝土拌和物一般的骨料最大粒径一律都是31.5mm，而该墙钢筋的混凝土保护层厚度只有20mm，石子就会堆积在水平钢筋上方，混凝土拌和物就会在墙体的水平筋下方沉降而留下宽度很大的水平裂缝，成为钢筋锈蚀的隐患；再如图中的桥，建设完工后只有8年，大梁下方混凝土顺筋开裂，使保护层脱落在地面，这是因为处于盐雾作用的环境下，混凝土密实度和保护层厚度不足，Cl^- 侵入导致钢筋锈蚀所致。另两个实例和这两个相同之处是：与设计、材料和施工都有关系。

图13　这些开裂都与设计不无关系

⑤在 A. Neville 的《混凝土技术是结构设计的要素》一文中举例说：对破坏情况的调查，通常是局部或部分破坏。是应力导致的开裂或剥落？还是与热性能有关？是收缩上存在差异，还是因为养护不良？一位纯材料科学家回答不了。这种问题，因为他们不了解结构的构性。结构工程师可以回答，但只有在他们懂得混凝土在所有环境条件下的表现才行。

⑥结构工程师说："没有不裂的混凝土，裂缝是混凝土因受约束而产生过大的应力，出现裂缝正好可以释放这种应力，出了裂缝没关系。"他们说的"没关系"主要针对结构承载力，即结构安全性。他们保证结构耐久性的就是混凝土的密实性。但是，混凝土结构的劣化主要体现在钢筋的锈蚀和失效，即使混凝土"很密实"，而一旦出现裂缝，就会成为一切侵蚀性介质进入的通道。导致混凝土逐渐受腐蚀而劣化，威胁到钢筋。因此，混凝土的裂缝是威胁混凝土耐久性的根源，进而威胁结构的安全性，这就是耐久性比强度更重要的原因，决不能轻视。

8. 转换结构理论和设计的范式是一项巨大的工程，有可能需要多学科和几代人的努力。但早晚会有实现的一天到来。我们现在能做的就是为这一天做准备。在上述一些问题中找到自己感兴趣的问题先做起来。例如是否能从统计学与概率的角度，在计算荷载时把早期积聚的应力和施工产生的应力集中考虑进去？能否通过微观的观测与计算和宏观行为的观测，对不同种类的损伤及其发展进行数据的积累和模拟？等等。近年来已经有少数人再开始尝试地去解决。例如有人在做掺粉煤灰后的混凝土构件尺寸效应，有人在用损伤力学建立混凝土的本构关系，这些都是开始考虑对现代混凝土本构关系研究的尝试，毕竟是个好的开端，但是首先需要转变思维方法和观念，不能再局限于传统的认识。甚至"断裂力学"、"损伤力学"，都不能从原本对金属材料的研究简单地移植过来，重要的是进一步与混凝土材料、数学、系统科学、计算机等学科专家合作，开阔思路，了解现代混凝土材料的本质，从工程的角度进行整体性研究，创造一种更符合实际的、科学的新范式。

问题五　从混凝土工程中的生产关系到整个混凝土结构工程的生产关系改革

1. 混凝土工程指的是从原材料选择、拌和物试配、生产，到运送、浇筑、振捣、收面、养护等成型工艺的全过程。现在如论文之廿三所述，原材料质量控制-混凝土试配与拌和物生产-混凝土成型工艺被割裂成各自独立的行业，正在严重地影响混凝土结构工程的质量，造成能源、资源的浪费和对环境的冲击。为了缓解当前水泥产能过剩的压力，近年来越来越多的水泥集团并购了混凝土搅拌站，但却没有解决根本的问题——生产关系问题，造成夹生饭。这实质上首先是生产关系的变革，但是变革生产关系的目的是为了推动生产力的发展，因此生产关系的创新最终还是要落实到技术的创新，才能实现生产力的发展。如果太看轻了混凝土的技术，也还会难以创新。在本部分所选有关论文中已涉及到需要进行研究的问题。重要的还是观念的转变。

2. 实际上还不只是设计和施工的从业人员充实混凝土材料知识的问题，工程的最终的目的是将预期的结构物从图纸上搬到地面上，用材料组成构件，由构件组成结构。现在不仅设计者不懂混凝土，施工者的混凝土知识欠缺，操作者的只是更只是“皮毛”；反过来，混凝土生产者既不懂设计，也不懂施工的也不少。施工者的任务是“看懂”图纸，而设计院的图纸还常常令人看不懂。为什么不能由施工工程师（建造工程师）做施工图？建筑师认为不管他们做出多么稀奇古怪的形状，结构工程师必须能算出来，施工工程师必须能做出来。其结果是造成很多隐患。为什么不能在开工前由建筑、结构、施工（包括材料）和设备（因为许多要埋在混凝土中，占据构件的体积）的各方有关工程师先进行“概念设计”，共同保证质量？这个问题的研究似应引起重视。这也是生产关系的问题。

3. 混凝土只有当拌和物经过浇筑、振捣、收面、养护等成型工艺，才能成为最终产品，拌和物如果脱离了成型工艺，其质量就毫无意义。然而当前如果说：“现在混凝土成型工艺的质量掌握在农民工手里”，并不过分，因为严格操作的成型工艺是保证结构物质量的关键。过去工地所用混凝土都是施工单位自己设计和试配、搅拌并成型，成型工艺的工长是自己的技术工人（四级以上），工人是“为自己工作的”；后来改成用农民工，由自己的工长管带；现在已经发展到把混凝土成型工艺承包给农村的包工队，工长也是人家的，也就是说“没有施工单位的什么事了”。农民工对混凝土的了解是“祖传的”陈旧的知识，认为是没有什么技术可言的“和泥”。他们不懂混凝土，更不懂现代混凝土的特性，而且当前对现代混凝土也缺少操作规程。

4. 混凝土成型工艺是混凝土工程的最后一道环节，对混凝土结构构件成品的质量有关键性的影响。以下工序看似简单，实则掌握不易，需具体问题具体分析。

浇注：结构不同部位构件的浇注的时间、浇注方式、浇注顺序和分次浇筑高度等，会影响构件的成型后的匀质性和体积稳定性；

振捣：不同构件振捣的方式（所使用工具）、振捣顺序、振点间隔、持续作用时间等，对成型后混凝土的密实性、匀质性、体积稳定性、有很大影响，不恰当的振捣会造成混凝土缺陷的产生；

收面：对板状构件（如场坪、道面等）的收面时机、收面方式如果不当，如过早或过晚，力度不够，则可使塑性阶段产生的裂缝成为后期的开裂源，甚至可能在后期形成贯穿裂缝；

养护：由于现代混凝土的特点，养护的内容已不仅是“及时浇水”那样的简单，温度控制同样重要。开始养护的时间、温控方案、保湿方式和保湿期都至关重要；

拆模：传统上根据混凝土强度确定拆模时间，现在由于混凝土强度等级水平的提高，早期强度也有较大的提高，为了周转模板以加快施工进度，往往提早拆模，甚至十几个小时就拆，造成混凝土降温过快而开裂；拆模过早而又不加保湿，常因失水而产生干燥收缩裂缝，或碳化加速。所以不仅需要转变对现代混凝土养护的概念，而且转变拆模时间的观念。

因为现代混凝土成型工艺比过去的复杂得多，另一方面混凝土成型工艺又被看成是太简单的技术，使得农民工没有学习的要求；没有章法可依，也使得监理人员监管不力。为了保证和提高混凝土工程质量，有必要进行研究，制订出适合于不同条件的混凝土成型工艺操作规程。

5. 混凝土工程中的混凝土拌和物制备环节与其上游、下游的脱离和失控，已经在严重地影响着工程的质量。从混凝土的上游来看，绝大部分混凝土工程的原材料质量不能满足混凝土的要求，无从选择，更不能控制；从下游来看，保证混凝土工程质量的关键是提高操作人员的素质和技能。但是一些属于生产关系的非技术问题更加重要，是阻碍解决技术问题的关卡，更需要进行研究和制定对策。农民工的问题不解决，我国混凝土工程质量就会得不到保障。如果报喜不报忧，所留下的隐患就很危险。这种隐患的爆发不像炸弹那样爆发得很快，而是像慢性病一样，可能挺到一、二十年，却难以实现“百年大计”的愿望。20 多年时间的建设已几乎耗尽我国天然砂资源，如果结构物短期内就破损得需要大量修补，则从资源浪费的角度来看，是不可持续发展的。不管是与上游还是下游的脱离都是阻碍我国混凝土工程技术进步的障碍，其所造成的损失可能会在不久的将来才能显现，但是将无法挽回。现在我国建设高潮尚未过去，还来得及进行研究和尽快解决。应当把这项混凝土工程生产关系的改革纳入我国总体改革的内容。

6. 研究的目的和内容

水泥、砂石等原材料和混凝土的关系问题，本集第四部分中所选论文及自评中已有较多阐述。这里单就和施工的关系提出建议。由以上所述，预拌混凝土拌和物的成型工艺转交给了施工单位，好比“自己生的孩子不能自己养”，而施工单位又分包给了农民工的包工头，这就离得更远了。因此需要做以下问题的研究：

（1）当前农民工问题的根本问题在哪里？

（2）对农民工的欠薪已经不是大问题了，混凝土的农民工日工资现在已经超过建设单位技术人员和管理人员的工资水平，为什么农民工还是不好好干？

（3）怎样能使农民工愿意接受培训？怎样培训？由谁负责培训？

（4）现有农民工由“包工头”管理并向施工单位承包的生产关系存在什么利弊？存在的问题如何解决？

（5）施工单位接受的拌和物不仅因为不是自己做的，而且自己对现代混凝土的知识也有欠缺，则把混凝土工程的关键工序又转交给了对现代混凝土更加不懂的农民包工队去做。那么，成型的质量由谁负责？

（6）预拌混凝土搅拌站和施工单位现有的关系实际上造成很大的浪费：“会生又会养”者本来能够“养好自己生的孩子”，却不得不交给不大懂得的去养，而领养者又转手交给毫无关系的第三者。在当今物欲横流的风气下，其最后的结果可想而知。预拌混凝土原有的成型工艺知识和技术完全用不上，这岂不是浪费？那么，应当是怎样的关系，才能使三方密切合作共同为混凝土工程分工合作共同负责？近年来有个别混凝土拌和物生产企业以售后服务的方式介入施工中混凝土成型工艺的管理，取得良好的效果，但是仍然是利益的问题，绝大多数难以做到。

(7) 生产关系的内容是生产资料所有关系和利润分配的制度，其实质就是利益问题。因此难度很大。但是当生产关系阻碍了生产力的发展时，就必须变革，否则损失的将是整个社会的利益。

问题六　现代混凝土拌和物流变特性的研究

1. 按照胶体化学的分析，水泥属于一种浓分散体系。浓分散体系流体流变特性主要是触变性和剪胀性。现代混凝土因减水剂的使用，拌和物浓度增加而具有不同程度的触变性；浓分散体系在很窄的浓度范围内会产生剪胀性，掺高效减水剂高强度混凝土则浓度更大，有可能在泵送中发生剪胀性。

2. 1993 年我和覃维祖在深圳天健集团混凝土搅拌站做混凝土路面混凝土试验时，建议掺用粉煤灰和引气剂以及木质素磺酸钙减水剂，以改善路面混凝土的质量。拌和物从搅拌机出料后，几乎没有坍落度，覃维祖开动振捣棒一插入堆积的拌和物，立刻出现很好的流动。这是我第一次认识到混凝土拌和物的触变性。此后在工程中逐渐体会到，触变性是使用减水剂的低水胶比混凝土拌和物的特性，这种特性可使泵送混凝土拌和物黏聚性增大而抗离析，但与泵管壁之间的粘滞性小，得以不堵泵、不沾泵。缺点是需要适当增加泵压，而且操作工易误认为“不好泵”，“不好振”而强行要求加水。

3. 2008 年天津某桥 C60 混凝土，拌和物入泵前坍落度为 215mm，坍落扩展度为 500mm；出泵后坍落度为 160mm，坍落扩展度为 220mm，流动损失严重，随后在浇筑第二车时增加了 $2kg/m^3$ 的水，结果是泵前坍落度为 255mm，坍落扩展度为 680mm，出泵后变成坍落度为 225mm，坍落扩展度为 370mm。按常规，两次浇筑所用拌和物泵前和泵后的坍落度都不小，但是坍落扩展分别都只有 220mm 和 370mm，表明坍落度已经不能代表现代混凝土拌和物的流动性了。这是我第一次遇见的剪胀性问题。当时将其称为“坍落度的动力损失”，实际上应当是流动性的动力损失，即“剪切增稠”。

4. 另外一个问题是，使用聚羧酸减水剂，当与使用条件配合不当时，可能发生滞后泌水的现象，一般发生在柱子的浇注：拌和物入泵前表现很好，出泵也正常，入模后约 3～4 小时出现严重泌水，泌水高度可达 30cm。这是对高效减水剂的选用不当造成的，不在这里讨论。

5. 侯万国在其著作《应用胶体化学》（侯万国、孙德军、张春光，1998 年，科学出版社）中说：“关于触变性产生的原因”，“比较流行的看法是：粒子靠一定方式形成网架结构，流动时结构被拆散，并在切应力作用下粒子定向，当切变速率降低或停止时，被拆散的粒子必须靠布朗运动移动到一定的几何位置，才能重新形成结构，这个过程需要时间，从而呈现出时间依赖性。”

6. 查阅《应用胶体化学》（余飙声，1948 年，中华书局），称胀性体系为“表征黏度随剪切速率的提高而增大的固液混合体”，“含有高浓度固体粉末的浆状体，在搅拌时其体积和刚性都有增大现象”，“产生胀性的原因是体系中粒子间排列很紧密，静置时粒子间液体占有空隙体积最小，而搅动时粒子发生重排，使空隙的体积增大，因而体系的总体积有所膨胀。由于空隙增大，粒子接触处的液体量减少，粒子间原有的润滑作用相应降低，因而流动阻力增大。具有胀性特点的体系必须满足两个条件：（1）分散相的浓度必须相当大，而且范围狭小；（2）粒子必须是分散的，而不是凝聚结构。”

7. 在加拿大 G. W. Govier 和 K. Aziz 的著作中对剪胀性机理的叙述为：“在剪切作用下，密集的颗粒体系变成松散的排列，因而有一些液体吸入或停留在其中。这个‘吸干’效应造成对位移阻力的增加，因而 $\tau-\dot{S}$ 曲线的斜率增加。这种效应……具有胀流性的有限浓度范围，与颗粒的形状及粒径分布有关；在更低或更高浓度中，流变特性可能是假塑性，屈服-假塑性或接近于宾汉姆体。”（G. W. Govier，K. Aziz：《复杂混合物在管道中的流动》上册，权忠舆，叶良溪译，1983）

8. 以上两著作中的描述和当前使用聚羧酸减水剂（使用萘磺酸盐减水剂较少发现）发生的“流动性动力损失”现象相符。如果能设法在实验室中预测，则可使避免混凝土泵送后因难以振捣的风险。

9. 有人做了一个实验，发现搅拌时间 60 秒时，坍落度比搅拌 30 秒时的坍落度大，再继续搅拌到 90 秒，坍落度就大为减小；在实验室试配时一般会搅拌 2～2.5min，有人发现，强度较高的混凝土，观察搅拌机中拌和物判断流动性会随搅拌时间而增大，但是全程搅拌结束后，就会发生坍落度反而很小，流动性损失很大。猜想是不是因为该拌和物本来具有剪胀性，因搅拌时间过短，拌和物不均匀，未表现出来，经过泵送的动力作用后产生剪胀性，等于继续搅拌，从而发生因受剪切力而增稠，使得坍落度或流动性严重“损失”？

10. 需要研究的问题是：

（1）现代混凝土拌和物触变性虽然可不影响泵送，但需要提高泵压，故需要进行触变性大小的试验和定量预测，建立触变性和泵压的关系；

（2）避免拌和物形成胀性流体，首先需要研究影响剪胀性的条件和因素，如水胶比、外加剂、矿物掺和料、砂率、骨料含泥量、水泥（细度、石膏品种）等；

（3）过去使用木质素磺酸盐类和萘磺酸盐类减水剂已经有几十年的经验，生产原材料单一，工艺较成熟，对其生产工艺与水泥相容性的关系比较清楚，很少发生现在遇到的剪切增稠现象而影响泵送的问题。聚羧酸减水剂的原材料就比较多样化，生产工艺也随之有变化，生产和使用至今不过五、六年，逐渐发现不少与原来所想的大不一样的问题，除从水泥和混凝土拌和物方面进行研究外，也需要从聚羧酸的生产的角度，进一步研究其对混凝土拌和物性质影响的机理，进一步提出使用的指南。

（4）试验方法：有以下设想：

①侯万国等人的《应用胶体化学》中介绍的“滞后圈法”测定触变性。即用旋转黏度计先从低到高，再从高到低，均衡地变化转速，记录相应的剪切应力 τ，作 $\tau-D$ 图。上行线 ABC 和下行线 CA 组成月牙形的 ABCA 环形曲线，称为“滞迴圈”。滞迴圈的面积可表示触变性的大小，面积越大，触变性越强。这个方法操作比较方便、可靠。

②实验室试配时搅拌与泵送的在动力作用上相似，是否可在实验室用搅拌时间的变化检测拌和物的剪胀性？首先采用预计会产生剪胀性的拌和物，记录不同搅拌时间后流动性的变化，找到发生剪胀性的临界搅拌时间；然后变化影响因素进行试验，可认为超过临界搅拌时间而仍保持流动性的拌和物不发生剪胀性。

③拌和物发生剪胀性（或剪切增稠）与剪切速率和剪切作用时间有关。上述实验先用砂浆进行变速搅拌，人工控制搅拌时间，研究用该方法预测剪胀性的可能改变上述影响剪胀性

因素，检测搅拌不同时间流动性的变化，找到临界搅拌时间。最后再用混凝土的拌和物进行验证性试验。

④拌和物适量的泌水有利于泵送和浇筑后缓解塑性收缩的开裂敏感性；当泌水量过大时，在密集的钢筋区域或弯曲、分叉部位，粗骨料在剪力作用下可能相互拥挤而成拱，则会有因拌和物离析而堵泵的风险；如果粒形不够好，则会造成出泵后的拌和物匀质性差，在结构物中留下缺陷。深圳张永秋曾用检测石子压碎指标的装置检测拌和物在压力下泌水量，确定临界泌水量值。不妨结合拌和物其他条件进行试验。

问题七　工程中出现混乱不清的问题根子在教育，首先在学校的教育

1. 现在大量设计人员和施工人员毕业于工业与民用建筑、路桥工程等专业，主业是结构，参加工作后，要想重温在校期间所学有限的混凝土知识，主要通过教材；一些工程经验较多的人，希望能够结合理论总结自己的经验，以取得更大的进步，他们接触较多的也是教材。现在的新气象是不缺少教材了，“建筑材料”的教材之多已难以统计，绝大部分内容都沿袭几十年前的老教材并互相抄袭，概念和知识陈旧，甚至有不少糊涂之处和错误之处，更谈不上具有特色。

2. 学校检查教学质量只注重教学法，严格地说也只是“讲课表达法”，很少检查教学内容。教师本身缺少工程实践，难以引导学生思维的创新。甚至有的教师害怕学生提问题。试想，学生不敢挑战课堂，将来怎么能挑战权威？

3. 不少教师说学生对建材课没有兴趣。对知识有兴趣无疑是学习的动力。首先老师要问问自己有没有兴趣？如果说学生没有兴趣是教师的责任，这应当并不为过。因为你自己没兴趣，怎么能引起学生的兴趣？老师首先要有工程经验，才会有兴趣；有兴趣，才会不断学习、思考，不断提高学术水平和加强文化功底。当然，在学习上我总告诉学生：专业是次要的，能力是主要的，但是专业是学习能力的载体，能力是通过专业学习取得的，所以学好专业也是很重要的，问题是怎么学？需要敬业的教师去引导。

4. 有人说，“本科生教学内容就只能是基础的，经典的、成熟的。”那么，什么是“最基础的”？什么是经典？什么是成熟？比如混凝土配合比只能用包罗米公式吗？混凝土强度靠的是水泥水化度吗？骨料越多混凝土强度越高吗？“水泥中 C_3S 水化生成 $C_3S_2H_3$”有什么意义吗？应该不应该与时俱进地更新观念？什么是新观念？应该不应该引导学生关心前沿的信息？这样的研究怎样进行？

5. 举两个例子来说明现在教学中的概念问题：这里所说的是概念问题，可能对配混凝土影响不大，但是概念是不能错的，否则也会有影响大的时候。

例一：现在有的教师坚信骨料越多、骨料粒径越大，混凝土强度越高，认为水胶比相同时，混凝土强度高于砂浆强度，砂浆强度高于净浆强度，还让学生做试验验证。我们都知道，从混凝土界面结构来看，对于中低强度等级的常态混凝土来说，在水胶比和浆骨比一定的情况下，净浆、不同灰砂比的砂浆和混凝土的强度是顺序递降的；1∶2 的砂浆强度高于 1∶3砂浆的强度，如图 14 所示。原因是骨料和浆体界面的影响。因骨料是亲水性材料，混凝土成型后水份会向骨料表面迁移，在骨料表面形成一层水膜，从浆体本体向骨料表面形成水灰比增大的梯度和孔隙率与 $Ca(OH)_2$ 取向度增加的梯度，成为混凝土中薄弱的过渡区；同时，对于普通混凝土，因骨料和浆体弹性模量的差异而变形不一致，界面处的浆体变形受

到骨料的约束，会在承受外部荷载之前就存在不同方向的大量微裂缝，更会弱化界面的粘结力。骨料粒径越大，单个骨料界面过渡区厚度越大，总长度越长。一个极端的例子就是活性粉末混凝土，浆骨比很大，不含粗骨料，但是抗压强度可达200MPa。因此应当是：在水胶比相同时，净浆强度高于砂浆强度，砂浆强度高于混凝土强度。这个实验很难做，学生自己是不可能做出来的。老师也应当知道其原因是不可能做出相同尺寸的试件，也就没有可比性。但是在基本概念上应当是清楚的，界面是混凝土的一个重要组成部分，这已经从历史上不少对界面问题的研究得到证实。之所以出现这个问题，说明这样一些教师对"化学成分通过结构决定物质的强度"这一材料科学的基本理论理解不足，也没有充分理解"方法比技术更重要"这一科学研究的原理。

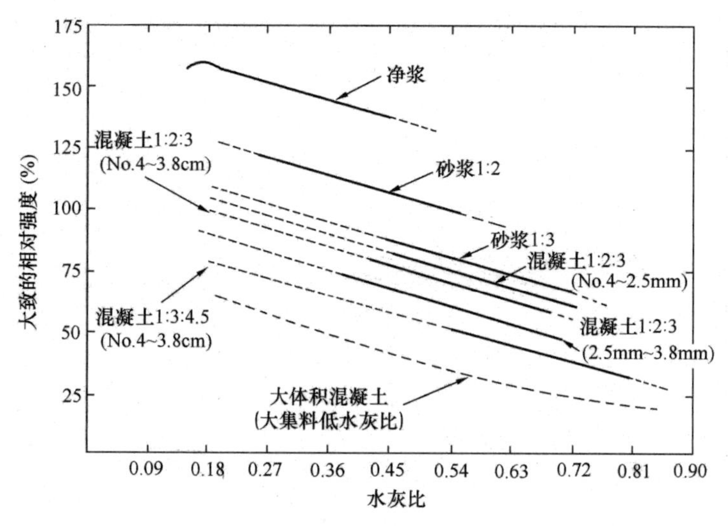

图 14　水灰比对骨料粒径与含量不同的混凝土强度的影响

（吴科如、张雄、姚武译《混凝土》，化工出版社 2005 年。
原著第二版著作者 Sameness，J.. F. Young，D. Darwin）

例二：全世界和我国水利工程都用浸水饱和面干骨料的密度计算混凝土配合比，我国很多人却不知何故，不知怎样是饱和面干。我曾举例说，砂子饱和面干含水率在数值上与砂子的吸水率相等，不懂得的人说："懂了。"但是这里面却有一笔糊涂账：按现行标准，计算砂子的吸水率和含水率都使用干基，即计算出所测水量与绝干砂子质量之的比值，在这样的规定下，二者数值是相同的，但是实际上在物理意义上，二者是不同的。吸水率检测的是不含水的干砂能吸收多少水，表征砂子吸收水分的能力，用干基计算是对的；含水率检测的是含水的砂子中含有多少水，表征砂子中含有水份的状态，用干基计算就不知所云了，物理意义不清楚。在生产混凝土上料时，为了在用水量中扣除或补充砂子中所含的水，还得把所测的干基含水率再倒成实际含水砂中的含水率，则先前用干基的计算就多此一举，岂不自找麻烦？再说，检测出砂子不同状态的含水率有不同的用途，是需要不同对待的。这也是个基本概念问题。

5. 质疑和思考对科学发展具有重要意义，在人才培养中必须给予高度关注。周光召说过："怀疑精神是科学精神的重要组成部分。"质疑与科学知识的传承是处于同样重要的地位。没有科学知识的传承，牛顿无法站在巨人肩膀上看得更远。没有质疑，世界将停留在中世纪的愚昧之中。如果我们的教师不习惯、不敢或没有能力在课堂上直面科学与工程界的问题和困惑，而是让学生认为规范标准无可质疑，老师说的肯定是对的。这样，说的小是误人子弟，说的大是误国误民。我们无法培养出勤于思考、敢于质疑的学生，是教师的失职，是教育的失败。如何在教学中引导学生思考，鼓励学生质疑是混凝土材料类课程教师需要解答的重要课题。

6. 是当前教育界对教师考核与管理的办法，误导了教师们不重视基本概念、基本方法和基本技能的浮躁行为，势必也会对学生有同样的误导。希望能促进教师们沉下心来清理一下现行教学中存在哪些基本概念不清的问题，或有哪些只“授人以鱼”而忽视同时还应“授人以渔”的问题。研究和讨论学校应教给学生什么？怎么教？如何进行工程教育？工科在校的教学如何与工程接轨？等等。各行各业都应重视教育问题。

老当益壮，宁移白首之心；穷且益坚，不坠青云之志。

——王勃

附 录

廉慧珍论文选评《求索与感悟》编者按

廉慧珍先生论文选评在出版前，曾邀请多人进行审阅，收到很多宝贵的修改意见。其中有的还有感而发撰写了感言。这些短文作者都是受过廉慧珍教诲的学生，其中有的是在清华大学攻读过研究生学位的研究生，有的是在工作和学习中受到先生的指导的技术人员。他们都勤于学习，善于思考，勇于开创，积极向上；在各自的岗位上认真负责，处事敬业。所撰感言皆发自肺腑，现随本书的出版一并刊出，与读者共享。

梁文泉　科学求真 高屋建瓴——写在廉先生《求索与感悟》出版之际

我与廉先生相识要追溯到1985年，从那时至今算起来已将近30年了。我1980年考入同济大学建材系混凝土与水泥制品专业，这一读就是5年。说来也算幸运，毕业那年正赶上当时的国家教委首次实行免试保送研究生制度，在同济大学和清华大学两个学校中选择，我毫不犹豫地选择了清华大学——我心中梦寐以求的高等学府，更加幸运的是有机会成为了廉先生的学生。我现在武汉大学水利电力学院任教，如果说工作上还取得了一点点成绩的话，这与先生多年来的悉心教诲和倾心帮助密不可分。先生的这部新作，是先生毕生从事混凝土材料教学和科研工作的总结，也是先生在不断地反思过程中逐步升华至一个全新高度的标志，更是先生启迪后人、不断探索的宝贵财富，同时，为我国混凝土学术界和工程界对混凝土技术与工程的研究和应用指明了方向。

先生的书稿我读了至少三遍，每一遍都有更深的感触，仿佛身临其境，有感而发。

在清华大学读书时，有幸参加了先生主持并负责的国家“七五”重点科技攻关专题中有关循环流化床锅炉燃煤固硫废渣建材资源化的系统研究。我的硕士论文是《循环流化床锅炉燃煤固硫渣制作水泥的研究》，在制订论文的技术路线、研究方法的工作中，倾尽了先生的心血，尤其是在萃取钙矾石并进行定量分析以证明固硫渣中石膏对水泥无害化的过程中，先生那种一丝不苟、认真求实的科研态度让我至今记忆犹新，并一直影响着我的工作和学习。

混凝土材料是迄今为止用量最大宗、用途最广泛的建筑结构材料。随着国家基本建设规模逐年增大，混凝土材料的需求量也在不断攀升，鉴于混凝土材料的复杂性、随机性和不确知性，对混凝土材料的认识相对落后于混凝土材料用量的增长速度，导致工程质量出现问题，这一切归咎于人们受到传统观念的束缚。打破既有，冲出传统观念的束缚是一件非常艰难的事情。为此，先生在她的论文中和演讲中多次提到：“思维方法和观念的转变比技术本身更重要”。先生在这部著作中就用了9篇论文来阐述这个观点，这一点我的体会较深。事

梁文泉　武汉大学副教授。1985年被上海同济大学推荐免试入清华大学，由廉慧珍先生指导获硕士学位，为廉慧珍承担的“七五”攻关专题研究的主力之一。

实上，目前在混凝土学术界和工程界仍然存在着“纯水泥的混凝土才是好的混凝土”、“矿物掺和料是水泥的廉价替代品”、“混凝土强度第一甚至唯一”等等不正确的思维和认识，产生这些错误认识的根源在于人们的认识有意或者无意地仍然停留在对传统混凝土的思维方法上，这就亟需转变观念。在对标准、规范的认识和使用方面，同样存在着不正确的思维，认为技术类的标准与规范就是法律，这是完全错误的。技术类的标准与规范只是提供了满足技术条件的最低要求，是具有可选择性的，即业主和承包人不仅可以选择不同的标准与规范，而且还可以采用不同于现行标准与规范的要求和做法。这就要求标准与规范的执行者不仅要具备专业知识和工程经验，更重要的是能够创造性地执行标准与规范，这是因为任何标准与规范都是人们对以往实践的认识和总结，其永远滞后于科学技术的发展，但不能落后于科学技术的发展，更不能成为科学技术发展的障碍。

科学技术发展的动力和本质在于质疑。廉先生在《求索与感悟》的第十八篇论文中对《回弹法检测混凝土抗压强度技术规程》(JGJ 23）提出了质疑，以回弹法检测混凝土抗压强度在工程使用过程中确实存在问题，饱受诟病。作为试块法检测混凝土抗压强度失效后的一种补充方法，作为非破损检测方法之一的回弹法确实有其一定的优点，例如操作过程简单并可反复检测。但不可回避的问题有三个：第一，该规程不能称为“回弹法检测混凝土抗压强度技术规程”，而应叫做“回弹法推定混凝土抗压强度技术规程”，因为以回弹法得到的并不是混凝土实体强度值，而是混凝土实体表面的回弹值，经过碳化层深度折减并进行回归得到的混凝土测区强度换算值；第二，这样回归的误差比较大，同时碳化层深度折减亦存在问题，与实际情况不符，而且回弹值受操作人员的人因素干扰也比较大；第三，即便以回弹法检测出来的结果是混凝土结构实体的强度，这个强度与试块法得到的强度完全是两回事，对检测实体的强度意义并不大。为什么呢？这是因为混凝土结构的设计依据不是混凝土结构实体的强度，而是抗压强度标准值，这个标准值不是采用回弹法得到的，恰恰是采用标准试块得到的。在工程实际中还存在着一种现象，无论以试块法对混凝土强度的检测是否合格，都必须进行回弹法，这的确没有任何意义。

孔子说，见贤思齐焉，见不贤而内自省也。意思是说：见到贤人，就想向他看齐；见到不贤的人，就要自我反省。反省、反思是《论语》宣扬的重要思想，也是个人修养的重要组成部分。高性能混凝土的概念自上世纪九十年代引入我国后，清华大学进行了大量的推广应用工作。为了更好地向国内同行介绍高性能混凝土，于1999年9月由中国铁道出版社出版了廉先生和吴中伟院士合编的《高性能混凝土》。高性能混凝土的引入，一方面促进了我国混凝土技术的发展，例如对混凝土耐久性的重视有所加强了，粉煤灰、矿渣等矿物掺和料的使用增多了，预拌混凝土更普遍了，混凝土拌和物的施工性能得到了改善；另一方面，由于对高性能混凝土概念的认识问题，存在着很多误区仍沿袭至今。有人认为高强的混凝土才是高性能混凝土，也有人认为大流态的混凝土才是高性能混凝土，更有人认为只要在混凝土中掺了矿物掺和料和化学外加剂的混凝土就是高性能混凝土，以至于在1995年至2000年期间，凡是涉及到混凝土的论文几乎都要冠以“高性能”这个时髦的名头，可谓无混凝土不高性能，于是采用“高性能混凝土”施工的工程开裂现象比比皆是，甚至有人怀疑是否真的存在“高性能混凝土”。1999年，我曾收到某杂志社请我审稿的一篇稿件，这论文的题目是《试析高性能混凝土的开裂及防治对策》，仅从该论文的题目上看，就已经让人啼笑皆非：既然是以耐久性为设计指标的高性能混凝土，能开裂吗？反过来，开裂的混凝土还能够称为高性能混凝土吗？针对这种乱象，廉先生坐下来就高性能混凝土在我国推广应用十几年来的成与败进行深刻的反思，在发表了《对“高性能混凝土”十年来推广应用的反思》一文后，又进行了《对“高性能混凝土”的再反思》，并从高性能混凝土的概念和界定上拨乱反正：高性能混凝土并非像高强混凝土、自密实混凝土和大体积混凝土那样是混凝土的一个品种，而是混凝土质量最终控制目标，对原材料、制作工艺和技术指标都有着严格的和具体的要求。据蒋家奋教授考证，高性能混凝土是在1986年由挪威学者首先提出研究的，这是因为挪威盛产硅灰，掺硅灰提高了混凝土的强度、抗渗性、抗氯离子扩散性，从而提高了其耐久性（但也增大了开裂风险），继而西方发达国家相继投入人力、物力、财力致力于高性能混凝土的研究和开发。当时的混凝土学者已经意识到强度并不是表征混凝土性能的唯一重要指标，而耐久性则是更重要的性能。1990年5月在美国马里兰州，由NIST（National Institute of Standards and Technology）和ACI（American Concrete Institute）主办的研讨会上，确定了高性能混凝土的定义：符合特定工程性能组合和匀质性要求的混凝土。易于浇筑、捣实而不离析；卓越的、能长期保持的力学性能；早期强度高、韧性高和体积稳定性好；在恶劣的使用条件下寿命长。吴中伟院士也指出：“高性能混凝土是一种新型高技术混凝土，它应该以耐久性作为设计的主要指标。针对不同用途要求，高性能混凝土对下列性能有重点地予以保证：耐久性、工作性、适用性、强度、体积稳定性、经济性等。为此，高性能混凝土在配制上的特点是低水胶比，选用优质原材料，并除水泥、水、集料外，必须掺加足够的矿物细掺料和高效外加剂”。

先生除了治学严谨以外，对中国混凝土与水泥制品协会的工作给予了大力支持，先后策划了“建材人生”、“混凝土中最小水泥用量讨论沙龙”等大型活动，并亲力亲为，对已经是耄耋之年但精神矍铄的老人来说，实在是难能可贵。更可贵的是，先生在帮助提携后人方面倾尽心血，在先生的朋友中，有三四十岁的青年人，也有五六十岁的中年人，且均是混凝土行业的佼佼者。先生不仅在学术上收获颇丰，而且桃李满天下，为混凝土行业培养了一大批中坚力量。

值先生新作《求索与感悟》出版之际，寥寥数言，以表祝贺之意、欣慰之情。正如先生所言：谨以此书献给热爱混凝土事业的人们！让热爱、关心和支持混凝土事业的同行们伴随着先生的大作和混凝土技术的进步一起成长吧！

陈恩义　**耕耘不懈的恩师益友**——回顾与廉先生在一起的日子

获悉廉老师最新著作《求索与感悟》就要出版了，欣喜之余，更多的是回顾与感慨！

我有10余年的时间有幸跟随廉先生在一起，这期间也正值廉先生主持国家“七五”（1986～1990年）和“八五”（1991～1995年）重点科技攻关专题中有关“循环流化床锅炉

陈恩义　新加坡昂国集团（EnGro Corporation Ltd）特种水泥部总经理。1986年重庆建工学院本科毕业后推荐免试入清华大学；1989年在廉慧珍指导下获硕士学位，同年被保留博士研究生资格留校任教三年；1995年在吴中伟（导师）和廉慧珍（副导师）指导下获得博士学位。在校十年期间，始终是廉慧珍承担的国家“七五”和“八五”攻关专题研究的主力之一。

燃煤固硫废渣建材资源化的系统研究”。这个项目不仅在建材教研室，甚至在整个土木系也小有影响。虽然不是一个阵容庞大的研发团队，但在当时的建材教研室也算是一支包含老、中、青三代教师，博士、硕士和本科生以及试验员在内的颇具典型性的高校科研队伍了。记得建材教研室除廉先生外还有郭玉顺老师，王志敏老师以及刘燕芳、梁文泉、江加标、我本人，构成这个团队的中坚。作为项目负责人，我们的大家长，廉先生事无巨细必定身先士卒，亲历亲为，尽心敬业，可谓劳心伤神！

北京的冬天很冷，路上经常结冰，下午5点不到天就黑了。在课题组加班干活是大家的习惯，从来没有人抱怨，更没有人会计较。因为，我们有一个榜样。记得我还是做研究生时，我们通常在6点下班后到附近学生食堂买饭回到教研室吃，然后工作到10点左右回宿舍休息。已经记不清楚是多少次了，都快到晚上8点了，您还在忙，我们不得不多次提醒您：“老太太，您该回去了，家里人还等着您做饭呢！”——只有在这个时候，廉先生才会从忘我的工作状态中回过神来，在照澜院超市几乎打烊的时候，急匆匆地骑车赶回家，为家人赶做已经迟到的晚餐。有时您还会突然会冒出一句：“哦！我得赶紧回去了，我们家老李明天要出差，我还得给他准备出差开会的衣服呢！”。每每到这个时候，看到先生急匆匆地骑车离开实验室，渐渐消失在黑暗中的瘦弱身影，我总不免为先生感到一丝的担忧和心疼。心疼您整日不停的辛勤操劳，担心您在回家的路上会不小心滑倒，又或是被某个冒失的学生骑车撞到。这是既是对项目组大家长的关心，更是一个孩子对于母亲的心疼和挂念。在您瘦弱的双肩上，一头扛着的是教书育人、国家攻关项目的责任和承诺，另一头扛着的则是作为母亲和妻子的职责。

尽管很辛苦，您却经常自乐其中，享受这一过程带来的喜悦、快乐与满足。每每在这个时候，我所感受的是一个坚强和厚实的家长风范，一个学者和学科带头人的朴实、严谨与乐观，它深深地感染到我们这批年轻弟子，成为我们成长中的正能量。

在清华学习和工作的10年中，有许多令人难忘的东西，而“听话出活”是廉先生给我们这批年轻弟子们提到最多的一句话，几乎成为与我们谈心交流时的口头禅，也是先生对我们团队成员的基本要求。廉先生的一言一行，也在演绎着清华这个优良传统的精髓与现实。这是先生作为一个老清华人给我作为一晚辈留下最深烙印的东西，一笔承载着百年清华传统的财富。我离开清华近20年了，但作为团队建设和人才培养，我感觉“听话出活”仍具有其现实意义，让我终生受益。“听话”代表一种积极进取的态度，既是谦逊、耐心，也是对授话者（长者/上级）的尊敬和礼貌，是我们学习、进步和成长（才）的基础和前提。“出活”表示你有能力和胆识把交办的任务完成好，是对你工作效率和质量的考量。它代表你的能力、胆识、创造性和你对工作、对团队、对公司的责任感，也是你成长、成熟、成才的一个标志，是获得信任和承担更大责任的基础。因此，我对我所负责的工作团队，其基本要求也只有简单的四个字，那就是“听话出活”；同样的，在对于我的上级时，它也同样适用与我本人。

从参加廉先生主持的国家“七五”和“八五”重点科技攻关项目研究，到给本科生和专科生开设“建筑材料”专业课，再到深圳大亚湾等地参与由先生主持的高强高性能混凝土横向合作项目，耳濡目染先生率真干练的行事作风，严谨求实、身体力行的科学研究精神，豁达亲切的人格魅力。所有这些不仅成就我事业的重要基础，更是我人生旅途中最宝贵的财富！

在工作中，与廉先生在一起，我们感受到的，既是一个严谨到把凡事做到完美成为习惯的学者、教授、课题负责人，更是一个行胜于言的榜样。在清华土木系建材教研室众多的前辈中，像廉先生这样凡事必亲历亲为的近乎“拼命三郎”的老先生不少，但已近古稀之年仍然带着年轻学生一起到大亚湾施工现场试拌高强混凝土，到深圳五建莲花北小区工地钻芯取样的“奶奶”级的拼命三郎在全国也实属罕见。——也正是这样的身体力行，重在实践的作风和传承，您让我有机会参与多项工程混凝土生产和施工的实践，真正感受到混凝土这门传统技术和辨证与发展的科学方法有机地结合在一起后的神秘和美妙，我更愿意把它称为“混凝土艺术”——一个生命和美的完美结合体，只有懂它并能驾驭它的大师，才能打造出流芳百世的作品，因为它倾注了从材料、设计、生产到施工与养护各个阶段大批艺术家的心血和爱。这些诠释，更加激起我们去探索和求真这个全世界最大宗建筑材料领域科学规律的坚持与不懈。在清华这几年的学习、实践和磨炼，您的言传身教，成就了我与水泥混凝土材料科学之缘，也是我工作学习中的乐趣所在。

廉先生开朗健谈，爱憎分明，热心帮助他人，凡事又爱深究的人格魅力让我们这批晚辈非常乐意与您相处，成为可以交心的好朋友，甚至是无话不谈的人生知己。与您的每次交流，无论是开会时的面对面还是在电话线上，与您总有说不完的话，话题一个接一个，此时只有一种感觉，那就是希望时间能够停止下来。跟着您，学到更多的是慎言、慎行、慎独。

为师品德，做人原则，重在参与，贵在坚持，享受过程，实现自我。这是我与廉先生在一起 10 年的点滴与感悟，它成为我人生旅途中的珍藏。谢谢您，廉老师！我的恩师益友。

最后，祝愿先生所孜孜以求、耕耘不懈的混凝土事业常青！

江加标　路漫漫兮其修远——读《求索与感悟》感言

廉老师的《求索与感悟》终于要正式出版了！作为廉老师的学生——我很幸运地能第一时间看到书稿，先睹为快，然后慢慢地一篇一篇去读、去领悟，直到最近才读完了一遍。该书稿收集了从上世纪七十年代到近期不同年代有代表性的论文，内容涵盖水泥基材料科学基本理论和方法的研究、工程技术的实践问题和解决方案、不同时期对材料科学理论和技术的感悟和反思再反思等广泛的领域。从理论到实践，再从实践到理论。这种理论和实践的有机结合不断形成、丰富和完善一套独特的水泥混凝土材料科学的研究方法论和认识论。

该书稿中很大一部分论文以前我在不同时期零星地读过，但每一次阅读都会有不一样的感想和收益。这次阅读整个书稿，尤其是加上论文的自评和反思，使我能得以在一个更大的时空背景中去了解廉老师的学术道路和认识历程，学习领悟到其中一些更深刻的东西，使我对过去这些年一些零零碎碎的感悟变得更加清晰。

该书稿中收集了廉老师早期研究探讨沸石岩在水泥中作用机理的系列论文，证明沸石岩的活性并非来自结晶态沸石本身的活性，结晶态的沸石在常温常压下不可能具有火山灰活性，而含结晶态沸石的凝灰岩的活性高于无定形火山灰的活性，其机理却与结晶态沸石有关——晶态

江加标　格雷斯公司主任科学家（Principal Scientist）兼亚太区产品经理。1987 年毕业于同济大学材料系，同年入学为清华大学硕士研究生，由廉慧珍教授指导于 1989 年获硕士学位，为廉教授承担的“七五”攻关专题研究的主力之一。1999 年获新加坡南洋理工大学博士学位，曾任教于新加坡国立大学。

沸石特殊结构的特殊物理作用，即含有结晶态的沸石岩具有特殊的多孔结构，通过水气交换界面作用能够促进具有火山灰活性的无定性沸石的作用，以及结晶态沸石晶体颗粒在胶凝材料体系中的微粒级配作用，而并非当时人们通常认为沸石岩晶体具有更高的火山灰反应活性，融合水泥基复合材料中心质理论，创新地提出了材料的化学组成通过物理结构发挥作用的概念，在以前很长一段时间内，一直不是很理解“化学组成通过物理结构发挥作用”这句话的真正意义。随着在实践中认识的提高和知识经验的积累，逐渐理解到该书稿中对沸石岩的研究方法，以及“材料的化学组成通过物理结构发挥作用”的原则对于通过充分挖掘利用各种活性、非活性的掺和料的物理作用具有很现实的指导意义。长期以来，人们对掺和料的研究和应用着重其化学活性即火山灰反应活性或潜在水硬性，以及如何激发掺和料的化学活性。尽管也对掺和料的物理性质和作用，如粉煤灰的形貌、微珠润滑减水作用等有所研究，但对充分挖掘利用掺和料物理作用的研究和应用不太多，尤其掺和料对水泥混凝土早期性能的贡献。但火山灰或潜在水硬性的活性掺和料常温下在早期 14 天内参与化学反应的能力几乎可以忽略不计，其化学活性主要发生在后期，对于提高混凝土后期性能发展的潜力发挥重要的作用，即使是像硅灰那样火山灰反应活性很高的掺和料也是如此，其对新拌混凝土流变性能和硬化混凝土的早期性能影响主要是通过物理作用。掺不同类型或同一类型但不同来源的掺和料的水泥混凝土早期性能有可能不一样，其差别并非来自于不同类型的掺和料的不同化学活性，而是由于不同类型的掺和料的不同程度的物理性能和作用对水泥水化（并非掺和料的火山灰反应）的不同影响以及对水泥混凝土基体颗粒级配和堆积密度的不同程度贡献。即使是常温常压下非活性的磨细石英砂在一定的掺量下对水泥浆体也有强度贡献。由于不同的活性、非活性的掺和料具有各自不同特点的物理和化学性质，其物理作用的程度不一，而且其物理作用取决于掺和料的颗粒尺寸大小和分布、表面结构和状态，体系浓度或水胶比。随水胶比降低，物理作用变得更明显。在当时由于外加剂应用不普遍和应用技术的落后，通常水泥混凝土水灰比的都较高，掺和料的物理作用不太明显。因此，在掺和料直接取代水泥的条件下，随掺量的增加，水泥混凝土早期性能大幅度降低制约着掺和料的应用。而且水灰比越高，下降得越多。在外加剂技术应用日益提高完善的今天，对材料颗粒的物理作用的认识越来越深，应用技术越来越至臻成熟。正因为如此，廉老师早期有关沸石岩在水泥中作用机理的研究尽管不尽完善，但其结果所具有的预见性和指导性说明了科学技术研究中方法论和认识论的重要性，唯有正确的方法论和认识论，才能透过现象看本质，找到问题的解决方向。另外一个启示就是创新并不是凭空而来，需要建立在深厚的理论基础以及继承前人的成果之上。

“思维方法的观念转变比技术更重要”系列论文也是科学技术研究方法论和认识论在水泥混凝土材料研究和应用中的延伸。材料科学的研究对象为材料组成、微结构、性能以及三者之间的关系，进而通过设计控制材料的组成和微结构来达到材料所设计的性能这一目标。由于水泥混凝土材料的不提纯和当地化的特点，在目前的条件下是无法做到完全控制材料的组成和微结构，因此也无法通过完全设计控制微观层次上结构来控制水泥混凝土的性能。但是，通过研究了解水泥混凝土材料组成、微结构、性能以及三者之间的关系可以为解决工程实践问题提供指导的方向，应用工程的方法在细观和宏观层次上来达到设计控制水泥混凝土的性能。但需要认识到所有具体的试验/实践结果或技术解决方案都是在一定的具体条件下取得的，其有效性与具体的客观条件密切相关。完全的经验论者因其所拥有很多的经验跟不上甚至排斥技术的进步和新技术的应用，但所拥有的经验都是在一定的条件下所产生的，随着时空条件的改变，过去所拥有的经验其有效性可能发生很大的变化。另一方面，纯粹的材料科学理论研究容易忽视水泥混凝土材料的特点，或是由于水泥混凝土材料的特点而陷入困境。从“思维方法的观念转变比技术更重要”系列论文得到的启示就是以理论为指导具体问题具体分析。脱离了具体的条件和限制，即使是很好的技术在实践中应用也很容易走向反面，如膨胀剂、微纤维等。认识到这些不容易，要做到就更不容易。

最后也是最重要的感想，《求索与感悟》论文集书稿的写法和编排很新颖，其中论文的背景介绍和自评是本论文集的一个较鲜明的特色，这在绝大多数的著作书籍中所见不到的。论文背景的介绍让读者比较容易明白了解问题的来源和每篇论文中的论点以及论文之间的相互联系和发展。此论文集并没有像绝大多数论文集一样对于论文中由于当时各观条件限制或认识上不足而存在的一些不完善的地方重新进行修改编辑，使论文看起来更好，而是一切保留当时发表时的原样，并通过对论文进行自我评论，或者反思再反思论文的方式，指出当时的条件局限甚至包括认识观点上的不足，以及未来需要更多深入细致研究的地方，这也是最难能可贵的东西。一方面反映了廉老师实事求是、一丝不苟的严谨科学态度，另一方面反映了廉老师开放包容、虚怀若谷的个人品格，可以让读者了解到廉老师在水泥混凝土材料科学技术研究和实践道路上的探索和认识发展的历程以及学术思想的形成过程。这种实事求是的科学态度，包容的品格以及终身学习不断探索创新的精神，尤其是对我们后辈科技工作者具有现实的启发示范作用。尤其令人敬佩和感动的是即使在八十岁的今天廉老师仍然如此活跃在最前沿的领域和高度，思考当前水泥混凝土材料研究和应用以及行业中的问题，探索水泥混凝土材料未来发展的方向和可持续性！

《求索与感悟》论文集涵盖的内容和启示浩瀚深隧，个人的理解和领悟只是管窥蠡测。谨以此微小感言庆贺廉老师《求索与感悟》论文集正式出版！

宋少民 **春风化雨 润物无声**——写在《求索与感悟》出版之际

获悉恩师廉慧珍教授的书即将出版，思绪万千，不能平复。其实多年前，我就一直希望廉先生写这本书，先生一直犹豫，我知道她是不重名利的人。直到有一次，我说这本书不是为您自己写的，应该是为水泥与混凝土行业的青年人写的，为他们在探索前行之路上有一些启示和帮助，先生才决定动笔。看到书名就知道这本书不是简单的论文集，它记载了一个知识分子的研究、探索之路，书中通过对论文和观点的分析、反思和否定，告诉青年人一个学者的治学态度和进步法门。

我认识廉先生十几年了，来北京建筑大学教书后，耳提面命，受到先生的教诲越来越多，我逐渐认识到廉先生对我而言不是一般的先生，她是值得我拜师的人。其实很久以来我一直觉得中国高等理工科教育存在很多问题，这些年跟随廉先生在学习过程中使我的认识越来越清晰，我们的教师大都在授之以鱼，而不是授之以渔。鱼易得而渔难求，需要长年的熏陶。我国

宋少民　北京建筑大学教授，建筑材料学科负责人，中国混凝土与水泥制品协会教育与人力资源委员会理事长。是廉慧珍接受正式拜师的弟子。多年追随和协助廉先生弘扬现代混凝土技术和人才培养的思维方法和观念。

建材行业学术成就很高的学者很多，但如果说到从思维方法和观念的层面去引领大家认识和理解现代混凝土的学者就少而又少了。这就是大家都喜欢听廉先生做报告的原因之一。

在廉先生给我的教育中，我印象最深的是养成质疑的习惯，先生那三句口头禅大家都知道，“存在的不一定都合理，外国的不一定都先进，专家和权威的不一定都正确。”学会用批判和分析的眼光去学习，去面对学术和工程问题，包括面对自己的研究和成果，是科学精神的“精髓”，是进步的阶梯。好好读读这本书你会认识到这一点。

廉先生已年过八十，但是在心理上她可不是个老人，她最爱和年轻人交朋友，最喜欢关注新观点。保林、雄利这些年轻技术人员都是廉先生的好朋友，她说从他们身上也学习到许多东西。每每听到一个好点子廉先生有时高兴得像个孩子，总是迫不及待地和我们分享。《大科技》是廉老师长年订阅的杂志，她经常给我讲其中的精彩文章和观点，我想先生活跃的思维、灵敏的触角与此大有关系。

廉先生做学问很认真，写这本书耗费了太多心血，有的章节的分析与反思真是连写再改十余遍，她总是怕对不起读者，迟迟不肯脱稿，还是我们一再催促下才肯罢笔。我和廉先生写“现代混凝土技术的哲学思考”那篇论文，关于哲学的知识和术语的理解，先生就和我讲了好几次课，指出了错误和模糊之处，生怕产生误导，当时的情景记忆犹新。前年我写教材《混凝土学》，廉老师帮我审查水泥和掺和料两章，真是字斟句酌，反复讨论，现在教材出版了，再看时其他章节的错误都比这两章多。去年我去台湾讲学，廉先生很重视，对 PPT 认真审阅，大到观点和概念，小到一个图和排版，她都指正和认真修改，手把手教。

廉先生多年来热切关注高校和行业的教育问题，中国混凝土与水泥制品协会教育与人力资源委员会（全国高等学校建筑材料学科研究会）的许多活动是先生直接策划或指导的，今年暑期在银川的全国建筑材料青年教师讲课培训活动，就是这样，廉先生不仅到会，而且亲自主持活动。她一再强调要教会青年教师改掉灌输式教学的传统方法，用问题引导学生思考和讨论，使学生更多地参与到教学活动中来。从本书的字里行间大家都会看到先生注重提出问题和思考问题的做法。

作为一个优秀的学者，必须有自己的学术观点和对工程与技术问题的独到思考，书中廉先生将自己多年的思考和感悟毫无保留地奉献给大家，先生确实从不保守，她总说正确的观念和技术让更多的人知道才会有价值、有意义，而且人应该有不断前进的自信心。

作为廉先生的弟子，我不是想吹捧先生，我是有感而发的。《求索与感悟》是一本有特点、有思考、有内容的科技著作。书中不仅向我们展示了现代混凝土的正确理念和技术进展，而且沿着作者求学和研究历程，去反思、去感悟，读者如果用心可以学会许多做学问的方法。中国是混凝土的大舞台，我们不能总是跟着外国人走，我们必须也应该引领世界混凝土科学和技术，这就需要我们学会思考、学会质疑，转变观念和思维方式。具有提出问题、分析和解决问题的能力，有了思考，就会有思路，有了思路才有可能创新。亲爱的同行和读者，希望这本书能够给你们带来改变，能够助力你们在水泥和混凝土领域勇往直前，这也是廉先生最大的愿望。

李玉琳　我向您致敬——真正的行业领军人物和值得敬佩与追随的老师

廉老师：

您好，初次认识您还是80年代初来企业对我国第一届电视大学的学生讲解建筑材料专业课，您给我们的印象是：知识丰富；平易近人；善于引导；受益匪浅。

有幸在95年由住总总承包国内第一个檐高高度超过100米的钢筋混凝土结构工程中，因当时该工程的C60高强混凝土技术路线需要国家级单位的技术支持，所以与清华大学合作，在您的直接指导下顺利完成了C60的研发与工程应用，其技术贡献主要是优化普通的萘系减水剂和配合比，科学合理利用各种材料的特性，大掺量粉煤灰，结合结构件的特点，在冬季施工中没有加入任何“防冻剂”，有效解决了冬施的一些技术难题，实时成为冬季施工的突出范例（那时整体技术与现在比还是有很大差距的）。这是我开始步入混凝土技术领域的第一次，也是我感到混凝土“魅力”的强大吸引力所在。您工作作风严谨，一丝不苟，对技术从不保守，而且善于抓住利国利民的超前的技术思路，尤其是对国家有可持续发展的技术路线。水泥是我国乃至世界使用量最大的建筑基础性材料，由于其生产过程中会产生很多对环境有破坏影响的二氧化碳，所以，科学合理地利用水泥性能，降低污染物对环境的影响，有效发挥水泥熟料的最大作用，是一个有利于行业发展的大事。利用材料叠加效应的效果，克服实时预拌混凝土生产中易出现的胶凝材料投料与计量失误的问题，提出生产高性能胶凝材料（水泥）的技术路线和工艺，这应该是最早在国内混凝土材料领域内的“高性能”材料产品。这项技术贡献是使用小于50％的熟料，发挥不同掺和料的性能优势，再结合外加剂的叠加作用，即将搅拌站的“中草药”式的胶凝材料多次分别加入，改为“中成药”加工成品的一次直接加入，充分发挥了各个材料性能的综合效应，在达到同等技术指标效果的前提下，大量减少了熟料的使用量，直接起到了对国家资源有效保护的务实作用。在多项工程的应用效果中，收到了很好的经济效益和社会效益。使得这项技术2000年被北京市科委组织的专家鉴定认可，同时获得了市级科技进步奖。虽然至今已有近二十年时间，但是在当今国家资源与能源需要更多合理使用和保护的“国策”下，应该说更具有实际意义，也更具有宣传性和提倡性，这足以证明您的技术思想先进性、实效性和超前性。所以您才是我们真正需要的行业领军人物和值得敬佩与追随的老师。

还有很多想说的收获和感慨，我会慢慢去享受这种人生真正的快乐与幸福。因为执着与兴趣是取得成就的最好的基础，是您给了我最基本的教导与教育，我会终生受益。借此在您的《我的论文选评　求索与感悟》一书出版之际，我诚挚向您致敬，祝您——我们敬佩的教师：身体健康，愉快幸福，继续引领我们在行业领域内创造更多的成就。

李玉琳　原北京中关村建设集团中宏基建筑工程公司总工。是廉慧珍在70年代企业办学受聘讲课时的学生，被廉慧珍认为人品好、责任心强、有悟性和潜力而精心培养，后在80年代又在北工大进修工民建专业。曾做过预制构件、预拌混凝土生产、试验检测及建筑施工等技术管理工作，多次协助清华大学完成建筑施工科研项目。在年届50时被破格提拔为高级工程师。退休后受聘为廊坊泳利混凝土公司总工兼总顾问。

韩小华　平实求真 厚德载物——读廉慧珍老师《求索与感悟》有感

初读廉慧珍老师的论文评选《求索与感悟》时，看到平实的语言及字里行间都体现廉老师对科学真理不懈追求的精神，像是她在跟你亲切谈论这些问题，这就是我认识的廉慧珍老师。她的文章从来不用华丽的词语，也很少用晦涩难懂或标新立异的科学名词，她提出的每一观点都会充分的论证，不是简单地把别人的观点与数据引用一下，而是把每条论据经过仔细的推敲，然后剖析清楚其中的道理。在她的思想体系中，把哲学思想（辩证法）与混凝土科学地结合到一起，深刻地体现了要做好混凝土必须要把握好度。每每读廉老师的书与文章，都不是简单地了解一个或一些观点和观念，而是感到自己的知识面在拓宽。就像廉老师的《高性能混凝土》一书是近 20 年来的经典之作，也是很多混凝土行业技术人员必备的一本参考书。

真正与廉老师相识，是我在清华读工程硕士做论文阶段，当时廉老师希望能够找一个人做一些关于水灰比测定仪的试验，我接了这项任务。当时，我并未搞明白为什么要做这个试验？怎么做？原理是什么？在完成试验过程与廉老师不断的交流，了解测试混凝土拌和物单位用水量的意义及仪器的原理，之前测试结果与实际用水量的差异的原因，还与廉老师共同写了文章。现在，廉老师还为这种检测混凝土拌和物质量的方法在呼吁和努力。从那个时候，我就从心里深深的佩服廉老师的为人。这个方法的推广和应用的好与坏对廉老师本人没有任何的利益关系，她只是出于对混凝土行业的热爱，对国家和社会的责任感，希望能有个切实有效的方法能改变工程验收结果落后于工程进度的现状。之后，在完成毕业论文过程中，感谢廉老师给予我的指导和帮助，使我深深地受到感染，认真地去做每一件事，对事物要知其然，还要知其所以然，要善于提问题等等。

最早提议出版廉慧珍老师论文集是 2010 年，在四川参加一个学术会议时。我读到廉老师的文章基本是本书中第四部分的内容。当时，我只是想简单把廉老师的文章汇总成集，便于读者阅读。时隔五年，当廉老师把她的书稿呈现在我们的面前时，我震撼了，这不仅是廉老师从事混凝土材料研究不同阶段的精选，同时在廉老师自评中，用她现在的认识角度，分析当时认识局限并提出了一些还需要更深入研究的问题。她的敢于自我批评与毫不吝惜地和我们分享她多年经验的精神，都值得我们学习。

廉老师就是这样本着对混凝土的热爱与科学求真的态度，一直在行业中呼吁大家要正视行业中存在的问题，并使用科学的方法解决这些问题。廉老师一直在协会做一个志愿者，在她的建议和倡导下，我们共同完成了九次研讨沙龙。特别关于“混凝土阴阳配合比的是与非”、“混凝土的最小水泥用量的限制”、“回弹法检测混凝土强度技术规程应用研讨沙龙”与“天然砂石资源短缺及对策研讨沙龙”等等。这些问题都是行业中存在的普遍问题，困扰着基层技术人员，制约了混凝土行业健康发展，也是大家有意回避的问题。通过这些活动以及廉老师在各种场合的宣传，现在很多地域的技术人员与工程质量管理部门已经开始逐步认识

韩小华　中国混凝土与水泥制品协会副秘书长，兼矿渣混凝土功能材料分会秘书长，2005 年考入清华大学攻读工程硕士学位，在廉慧珍和徐永模（校外导师）指导下获工程硕士学位。

和解决这些问题。

因工作关系，这些年我有很多机会聆听廉老师的报告。廉老师的很多想法都具有前瞻性，早在多年前，她就在各种场合宣传骨料对混凝土的重要性，提出了在预拌混凝土厂多级配配制混凝土的观点，尤其在《砂石质量是影响混凝土质量的关键》一文中强调骨料的重要性。近一两年因为资源短缺，使用不合格的骨料造成的工程问题频发，越来越多的大型企业投资骨料行业。廉老师在书中还写到她还想研究的 7 个问题，即“现代强度发展规律”、“面对混凝土材料与工程的发展、如何评价水泥的质量”、“对大坝耐久性的疑问与思考”、“混凝土结构设计的范式转换问题”、“混凝土工程结构生产关系的变革”、“现代混凝土拌和物流变特性的研究”等。这些问题都是存在于我们的日常工作中，当我们遇到时，常被一些固有的思维方式影响而忽略这些问题，认为别人就是这样说的，学校就是这样教的，没有办法改变。就像我刚看完廉老师《矿物掺和料有谁来掺?》，我就不能完全理解，认为现在的大规模工程建设和规模化的水泥生产模式下，廉老师提出的模式不可能实现。但是经过多次的交流和讨论，我改变了，用发展的眼光去看待这个问题，混凝土的胶凝材料体系向多元化发展，只有在这种新的生产模式下才能达到材料的匀质化及最优化，相信在不久的将来廉老师提出的观点就一定能变成现实。思想活跃的廉老师，就是这样敢于挑战传统及权威观念。作为一个年过八旬的老者，廉老师本着对混凝土科学与事业的热爱，为我们提出了问题，我希望活跃在混凝土相关各领域的技术人员共同努力研究和解决这些问题。

祝贺廉老师的新书《求索与感悟》的出版，借此向廉老师多年来给予我的教导和帮助表示由衷的感谢，希望热爱混凝土事业的人们能从廉老师书中得到更多启示。让我们为混凝土技术的进步都付出更多的努力!

师海霞　**思想和行动上的巨人**

认识廉慧珍教授是在一次讲座中，因为和主办方是朋友，所以有缘到现场聆听。廉教授讲完后，很多人就围了上去，我也挤在人群中，听着大家七嘴八舌地问问题，廉教授一直耐心微笑地做答，在主办方的催促下，大家渐渐散去，我和主办方共同陪同廉教授离开，这时我问廉教授是否可以推荐几本专业书，因为我的本科没有学过混凝土，廉教授认真地为我写下了 2 本书的名字。那是我初次认识廉教授，已经被她的学术热情所折服，渐渐在心底埋下了进一步学习的想法，后来有机会在清华大学修读材料学工程硕士，廉教授真正成为了我的老师，之后从事协会工作，我更是和廉教授密不可分了。

廉教授撰写过很多篇文章，她的思想和理论体系，我想很多人都有了解，她的敬业和热情也被很多人所喜欢。在这里，借着廉教授新书出版的机会，我想讲述两件事情来从另外的角度看教授。

廉教授退休后经常被邀请在外讲学，近年来我国的混凝土产量发展迅猛，但是从业人员

师海霞　北京东方建宇混凝土科学技术研究院副院长，中国混凝土与水泥制品协会预拌混凝土分会秘书长，2005 年考入清华大学攻读工程硕士学位，由孔祥明指导获工程硕士学位。廉慧珍曾为其授课老师。

的能力和专业素质是参差不齐的，在廉老师的报告后，就不免有些人断章取义，片面宣传，也造成了一些问题，廉教授为此非常苦恼，有一次她就和我说："我不是怕别人说我什么，而是怕他们拿着这一知半解的想法去做事，一是做不成，二是真是浪费。浪费了时间，浪费了资源，他们谁愿意做，都可以来找我，我都会指导的。"她说过，混凝土是用简单的工艺生产的最复杂的产品，任何理论和方法都不能放之四海而皆准，需要具体问题具体分析。她还说过，不要认为专家说的就都对，需要通过科学的试验和分析，才能做好的产品。

廉教授从来不会居高临下地对人，她对年轻人的喜爱和关心让人感动。一次技术交流会期间，一个企业的技术人员知道有廉教授的报告，特地带了七、八种当地的骨料。他来到廉教授所住的房间，希望向廉教授请教，当时我也在场。在交谈中得知，这位技术人员并没有读过正规的专业课程，是自己在工作中逐步自学的一些混凝土知识，当时我就心里想，这样的交流会非常吃力，我借故离开了，后来得知，那个小伙子在廉教授的房间里待到很晚，廉教授不厌其烦地为他讲解，并不断地肯定他的追求精神，鼓励他坚持做好工作。在廉教授的鼓励和影响下，很多人都在混凝土的技术和生产中孜孜以求，找到了自己人生的事业方向。

廉教授是一位平易近人的学者，她把自己的所有学识和热情都给了水泥、混凝土行业的人和事，只要是她能够做的，她都会不遗余力地、满腔热情地去做，现在一直到将来，她都会是一直奋斗在行业前列的战士，是我最尊敬的老师！

祝贺廉慧珍教授的新书顺利面世，祝愿老人家身体健康！

郭保林　我认识的廉老师

认识廉慧珍老师是在2004年5月北京举行的"International Workshop on Sustainable Development and Concrete"上，我代表导师向大会做报告，会上丁建彤博士用英文问了我两个问题，其中我一个回答错了，另外一个连问题也没听清，可想而知当时是多么的尴尬。在会议休息期间，一个老太太活跃于与会代表之间，经过询问旁边的参会代表，才知道她就是廉慧珍老师。作为一个经常翻看《高性能混凝土》的人，竟然在这里遇到了作者，心情的确有些兴奋，我怀着一颗忐忑的心向廉老师要她的电话号码，问是否可以在后续的学习中请教她，她非常痛快的给我了。当时的确没有想到，就这次简短的会面，拉开了后来10余年学术交流的序幕。

在2004年8月，我当时在做纳米材料改性水泥混凝土方面的研究，我就挑选了一个长时间困扰我的问题向廉老师请教。听完我的问题，廉老师没有直接回答我的问题，只是留给我一句话：你要把自己当作是研究人员，而不是一名学生，面对问题首先要做出假设并验证之，即使验证不了你的假设就去证实这个假设不成立。这句话成了我后来做研究的主导思想，要想证实就必须证据确凿，实在证实不了就证伪，总之得证据确凿，经得起推敲。

郭保林　山东高速青岛发展有限公司工程师，从读硕士研究生起，深受廉慧珍老师的思维方法和观点的影响，并全程参与了青岛胶州湾大桥的建设。

从廉老师的论文中，不难发现廉老师一直遵循着从最初的大胆假设、科学求证，到随着认识不断提高之后的自我否定，再到提出新的认识，并反过头来剖析当时的失误，一直处于螺旋上升的状态。2007 年，廉老师知道我在从事研究工作时，曾推荐我看“沸石岩在水泥中的作用机理探讨”一文。暂且不讨论文中讲述的内容，单就研究中使用的思维方法就非常值得我们学习，研究。廉老师自己坦言，此前也从过众，但在发现有些现象与基础理论相悖时，没有继续随声附和，而是去大胆假设，并设计合理的试验来逐一验证和排除，并用先进的实验手段证实自己的假设。有理有据，思维清晰，逻辑严密，堪称学术研究的典范。

记得廉老师的“质疑‘回弹法检测混凝土抗压强度’”，在发给出版社之前，发给我让我给看一下文章表述上是否妥当。我接到这个任务时心里很诧异，清华大学的廉慧珍教授、全国著名混凝土专家写的一篇文章，请我一个无名小辈来检查论文是否妥当，的确有些让人吃惊。我也是个实在人，不太会恭维人，就在认真阅读文章之后，从我对工程混凝土理解的角度，提出一些我认为表述过激的地方，并对文章提及的几个观点进行了质询，发邮件给廉老师了。当晚廉老师来电话，对我提出的意见进行了交流，有些接受了我的修改意见，有些详细解释了她的意图，说服了我的就保留了下来。也就是这篇文章，引发了一系列的争论，也就直接催生了与回弹法规程编制单位共同举办了一次学术沙龙，我有幸参加那次沙龙。我估计我不是唯一一个被要求对文章提意见的人，单就这件事情，我认为廉老师是在追求一种境界，一种为了真理(至少当下被认为最正确、最合理的)，早已经把世俗所看重的身价忘到脑后了。

另外，廉老师的文章还有一个非常明显的特点，那就是说常人不屑说，说常人不敢说。所谓的常人不屑说，是因为大家觉得这个事情是人都知道，短时间内难以改变的了，觉得没有必要说的，其实这恰恰是学问所在，一旦解决，那社会效益将是巨大的；所谓的常人不敢说，是因为担心说了会得罪一些领导、一些机构、会危害到自己的利益，而廉老师在学术问题面前全然不顾这些，这是需要学术胆识的，同时敢于亮剑是需要自信的。廉老师经常说的一句话就是：有些事情只有我说过了，才能对得起我自己的良心。这可能是她老人家退休 20 年来一直活跃在各大学术论坛的原因之一。

我浅薄地认为，无论是从事结构设计、试验检测还是科学研究，阅读一下本书，会拓展认识事物的维度，会体会到自我否定之后带来的认识提高，会更加激发对科研工作的热情。

杨雄利　**科学的诚实**

在混凝土技术文献中，廉慧珍教授的文章是我最喜欢的，它伴随了我的整个技术成长过程。廉老师的文章，充满哲学思考，既有高山一样的巍峨，又有森林一般的深邃，其科技价值自不必说。廉老师撰文，没有一般科技论文的刻板与说教，她选词精当，文笔优美，读来畅快淋漓，如沐春风。

科学研究是曲折地、逐步地逼近真理的过程，研究会无限接近真理，但或许我们永远得

杨雄利　天津市地下铁道集团有限公司高级工程师。酷爱读书，受廉慧珍教导而树立了科学的学习方法，并善于联系工程实际解决问题，取得了丰富的工程经验。

不到它。廉老师的这些文章是通向真理之路的一行坚实的脚印，但也必然具有时代的局限性，而非真理本身。平日的交谈中，廉老师总是提醒我要用批判的眼光看待她的文章，这是体现了她高度的责任感，是一种科学的诚实。

文集的最后呼吁大家关注“混凝土结构理论和设计方法的范式转换”。我个人认为这是廉老师提及的最重要的主张，是建筑界革命性的主张，意义极其重大。

对范式转换，我有个不成熟的观点，我认为范式转换需要更彻底。一定施工条件下获得的混凝土结构实体，其“强度”和承载能力究竟是什么关系？强度高到一定程度后是否会因弹性模量过大，徐变能力过小，温度应力、自干缩应力的增加而呈现负相关？材料的韧性对工程的坚固性是否更重要？是否应放弃对结构内有限元应力情况进行详细描述的企图，干脆从“能”而非“力”的角度来考核结构物的承载能力和持久性（实际上已经有很多力学研究避开了对应力的描述，选择了从“能”和“形变”的角度考虑问题）？茅以升先生认为，从物理学的角度讲，“能”比“力”的概念更为基本。这个观点值得深思。

混凝土是一个气、液、固三相共存的弹、黏、塑性体，钢筋的配制更增加了其复杂性。混凝土内部各点的应力随荷载的变化而不断变幻，无法准确描述。我们计算的所谓某截面的应力，实际上是该截面上各点应力的合应力。而真正导致混凝土结构破坏的，是混凝土结构缺陷处或材料性质突变处的高应力，而非与“强度”关系密切的合应力。因此，细致地纠缠混凝土强度是否真有必要？

思维方式的转变必然会带来重大科技革命，20 世纪初的量子理论和相对论就是杰出的例证。对于混凝土这种高度混沌的体系，我们是否可以尝试彻底使用整体论的思维模式来考虑问题？而不是先分解再综合。如果真能做到这样，传统的“力”、“强度”、“应力”等概念估计要被摒弃。这种大破大立的革命性创举，呼唤大师的出现，或许需要我们艰难探索 100 年，200 年……

张大康　**致广大而尽精微**

欣闻廉慧珍先生论文集即将付梓。多年来读先生文章，承先生教诲，受先生思想，受益良多。“君子尊德性而道问学，致广大而尽精微，极高明而道中庸”。此既为中国传统儒家文化之精义，亦为治学之崇高境界。先生治学，正合以致广大而尽精微概之。

书籍乃知识之载体，然知识一旦变为文字，往往容易失其血脉。是编不仅为混凝土科学技术研究成果的荟萃，更记录了作者治学的思想、方法，心路历程。编选之初，先生即向我谈及，不愿如一般论文集一样仅是编次文章，更要对每一篇论文回忆背景，反思得失。先生以耄耋之年，为文集添写数万字自评，治学之勤奋、严谨可见一斑。有了这些文字，全编方得气韵生动，方得血脉相连。读之或如当头棒喝，豁然开朗；或如醍醐灌顶，茅塞顿开。阅读妙境在于心领神会，自然科学和工程技术论文以平实、客观文字为文，读之可得知识，却难得精髓。

张大康　鑫统领建材集团有限公司副总经理兼总工程师，四川鑫统领混凝土有限公司总经理。从事水泥、混凝土及其外加剂生产 30 余年。长于水泥、混凝土性能及质量管理。

昔禅宗不立文字，或有极端之嫌，却也印证了语言对于表达思想的局限。科学研究是一过程，而科学论文却只呈现结果。廉先生的自评，述论文之背景，补论文之未有，揭研究之方法，使读者不仅了解科学研究之结论，更了解科学研究之过程，较之论文本身更加弥足珍贵。

授人以鱼，三餐之需；授人以渔，终生之用。廉先生不仅潜心学术，硕果累累，而且教书育人，春晖四方。退休之后，先生四方讲学，得益先生之教者，非仅限入室弟子。与先生相识经年，多有请教。先生诲人不倦，不仅传授知识，更传授思想与方法。后者比前者让我受益更多。亚里士多德的《工具论》，结束了古希腊哲学的混沌时代，以清晰可辨的思维方式，奠定了西方文化的思想基础。从此，西方文化基因中植入了理性与逻辑的要素。近代西方科学正诞生于此。著名的李约瑟难题，窃以为东方文化中向来缺少理性与逻辑，应为其答案之一。自青年时代为学，即有意力避中国文化的不利因素，然时感所处环境之窒息。读先生是编之自评，其思路，其方法，了然于心，如沐春风。立于巨人肩膀者，非仅知识，更兼思想、方法。

廉先生是我国最早使用显微手段研究建筑材料细观结构的学者，并编有《建筑材料物相研究基础》一书，开我国建筑材料细观研究领域之先河。先生主持完成的“水泥浆体微结构与宏观行为的关系及其定量分析”课题，将英国著名学者 H. F. W. Taylor 提出的 FBT 模型定量化，成为在细观尺度研究建筑材料，并将建筑材料之细观结构与宏观行为联系的典范。此先生之精微。受先生影响，对材料学问题，一直力求同时在细观、宏观尺度上考量，并在二者之间建立联系。近年廉先生讲学，多以宏大时空跨度来审视混凝土之科学与技术，更将哲学融入混凝土科学。混凝土耐久性之整体论观点、否定之否定规律、思维方法和观念的转变比技术更重要，皆高屋建瓴。此先生之广大。

近代国人接受西方科学之初，即将科学异化。弃科学之精神内核，而将其工具化、功利化。延续千年的道器之辨，清末的体用之争，至当代的科学技术是第一生产力之说，无不反映国人与西人在科学观念上的巨大差异。这一差异乃中国科学发展之巨大障碍。即在教育普及之今日，国人科学观念、科学精神之缺乏，仍令人胆寒。何为科学的本质，“科学就是整理事实，从中发现规律，作出结论”（达尔文）。科学的底端是事实，科学的顶端是理论。科学的主要任务乃不断探索事实，发现规律，综合众多规律构建理论大厦。先生是编，思维逻辑之严密，科学方法之精当，对规律之探究，对理论之建树，正合于科学之本质。不仅可为混凝土科学与技术之教材，亦为领悟科学思想，训练科学方法之范本。

人非圣贤，研究、写作难得完璧无瑕。先生所选论文，皆依原貌。自评中加以辨析、反思与订正。为求科学之真，不护一己之短。如此虚怀若谷，勇于自我批评，自我否定，真乃大家风范，令晚生后学，高山仰止。

文集最后，廉先生提出：我还有什么问题希望研究？旋即想起百年前的希尔伯特问题。这些题目视野开阔，立意高远，或将成为今后混凝土科学与技术发展之方向与框架。对问题一“水泥混凝土强度发展规律现状和问题”，我已于去年开始试验，并曾就试验方案、方法向先生请教。对问题二“面对混凝土材料的变化和混凝土工程的发展，应当如何评价水泥的质量”，亦不揣浅陋，撰文探究。

建筑质量，百年大计。固知此绝非易事。然前有詹天佑、茅以升诸先贤之典范，近有先生之鞭策。身处其中，虽沧海一粟，未敢懈怠，不尽绵薄之力。倘稍有所成，亦不负先生教诲。